FEATURES AND BENEFITS

Houghton Mifflin Unified Mathematics, Book 1

Comprehensive content coverage shows the interrelationship of logic, algebra, geometry, probability, and statistics with special attention to problem solving skills and concepts. (Table of Contents, pp. iii – vii)

Special end-of-chapter features include an enrichment page, Vocabulary Review, and Mixed Review Exercises. These pages accompany Computer Exercises, Chapter Review, Chapter Test, and periodic Cumulative Review pages. (pp. 124–131, 215–223)

The **Preparing for Regents Examinations** appendix helps students prepare for the exams and includes topic checklists and sample exam questions. (pp. 561–573)

Algebra Review Exercises and **Extra Practice** appendices contain additional exercises on algebra and other important topics. (pp. 546–551, 532–545)

Applications of the computer to lesson concepts can be found at the end of each chapter in **Computer Exercises** and in the new **Programming in BASIC** appendix. (pp. 27, 380, 462, 552–560)

Supplementary materials include a **Teacher's Manual with Solutions** and a **Resource Book** that contains permission-to-copy blackline master pages of tests (quizzes, chapter tests, and cumulative tests), practice exercises, enrichment activities, and diagrams.

Houghton Mifflin
Unified Mathematics
Book 1

COORDINATING AUTHOR
Gerald R. Rising

William T. Bailey
David A. Blaeuer
Robert C. Frascatore
Virginia Partridge

HOUGHTON MIFFLIN COMPANY / BOSTON
Atlanta   Dallas   Geneva, Ill.   Hopewell, N.J.   Palo Alto   Toronto

AUTHORS

Coordinating Author
Gerald R. Rising, Professor, Department of Instruction
State University of New York at Buffalo, Amherst, N.Y.

William T. Bailey, Associate Professor of Mathematics
State University College at Buffalo, Buffalo, N.Y.

David A. Blaeuer, Professor, Computer and Information Sciences,
The Ohio State University at Lima

Robert C. Frascatore, Associate Professor of Mathematics
State University College at Buffalo

Virginia Partridge, Mathematics Teacher
Suffern High School, Suffern, N.Y.

EDITORIAL ADVISERS

Andrew M. Gleason, Hollis Professor of Mathematics and Natural Philosophy
Harvard University, Cambridge, Mass.

Alice M. King, Director of Entry Level Mathematics Project,
California State Polytechnic University, Pomona, Cal.

TEACHER CONSULTANTS

John G. Balzano, Mary E. Kendall, Jean M. Waganka, Martha Brown Junior High
School, Fairport, N.Y.

James Burke, Sayville High School, Sayville, N.Y.

Janet M. Burt, Clinton Central School, Clinton, N.Y.

Copyright © 1985, 1981 by Houghton Mifflin Company

All rights reserved. No part of this work may be reproduced or transmitted in any form or by any means, electronic or mechanical, including photocopying and recording, or by any information storage or retrieval system, except as may be expressly permitted by the 1976 Copyright Act or in writing by the Publisher. Requests for permission should be addressed in writing to: Permissions, Houghton Mifflin Company, One Beacon Street, Boston, Massachusetts 02108.

Printed in U.S.A.

ISBN: 0-395-36085-4

CDEFGHIJ-D-93210/89876

Table of Contents

1 NUMBERS AND OPERATIONS

1-1	Sets	1
1-2	The Number Line	4
1-3	Addition	7
1-4	Subtraction	11
1-5	Multiplication	12
1-6	Division	15
1-7	Addition and Subtraction of Rational Numbers	17
1-8	Multiplication and Division of Rational Numbers	21
1-9	Order of Operations	23

Self-Tests 7, 17, 26   Computer Exercises 27
Chapter Review 29   Chapter Test 30
Careers — Technical Writer 31
Vocabulary Review / Mixed Review 32

2 LOGIC

2-1	Introduction to Logic	35
2-2	Negation	38
2-3	Conjunction	40
2-4	Disjunction	43
2-5	Conditional Statements	46
2-6	Constructing Truth Tables	50
2-7	Related Conditional Statements	53
2-8	Biconditional Statements	57
2-9	Valid Arguments (Optional)	60

Self-Tests 46, 60   Computer Exercises 65
Chapter Review 66   Chapter Test 67
Biographical Note — George Boole 69
Vocabulary Review / Mixed Review 70

3 SOLVING EQUATIONS

3-1	Evaluating Algebraic Expressions	73
3-2	Introduction to Equations	76
3-3	Solving Equations by Addition	80
3-4	Solving Equations by Multiplication	82
3-5	Using both Addition and Multiplication to Solve Equations	86
3-6	Translating Words into Equations	89
3-7	Solving Problems with Equations	93
3-8	Working with Formulas	97

iii

Self-Tests   79, 89, 99   Computer Exercises   100
Chapter Review   100   Chapter Test   102
Application — Balancing a Checkbook   103
Vocabulary Review / Mixed Review   104

4 SOLVING INEQUALITIES

4-1	Introduction to Inequalities	107
4-2	Equivalent Inequalities; Addition Properties	109
4-3	Inequalities Involving Multiplication	112
4-4	Compound Inequalities	116
4-5	Inequalities Involving Addition and Multiplication	119
4-6	Solving Problems with Inequalities	121

Self-Tests   116, 121, 124   Computer Exercises   124
Chapter Review   125   Chapter Test   126
Cumulative Review: 1–4   126
Careers — Epidemiologist   129
Vocabulary Review / Mixed Review   130

5 BASIC CONCEPTS OF GEOMETRY

5-1	Points, Lines, and Planes	133
5-2	Distance, Line Segments, and Rays	137
5-3	Angle Measure and Special Angles	140
5-4	Some Angle Relationships	145
5-5	Parallel Lines and Transversals	150
5-6	Angles of a Triangle	156
5-7	Polygons	160

Self-Tests   145, 164   Computer Exercises   165
Chapter Review   166   Chapter Test   167
Biographical Note — Euclid   169
Vocabulary Review / Mixed Review   170

6 CONGRUENT TRIANGLES

6-1	Constructions	173
6-2	Congruent Triangles	179
6-3	Isosceles Triangles	186
6-4	Parallelograms	188
6-5	Trapezoids and Special Parallelograms	193
6-6	Areas of Rectangles and Parallelograms	197
6-7	Areas of Triangles and Trapezoids	202
6-8	Areas of Prisms and Pyramids	207
6-9	Volumes of Prisms and Pyramids	211

Self-Tests   185, 196, 214   Computer Exercises   215
Chapter Review   216   Chapter Test   218
Application – Baseball and Mathematics   221
Vocabulary Review / Mixed Review   222

7 POLYNOMIALS AND FACTORING

7-1	Exponents and Factors	225
7-2	Introduction to Polynomials	228
7-3	Addition and Subtraction of Polynomials	232
7-4	Multiplication of Monomials	234
7-5	Multiplication of Polynomials	237
7-6	Division of Polynomials by Monomials	240
7-7	Common Monomial Factors	243
7-8	Multiplying Binomials	245
7-9	Factoring Quadratic Polynomials	248
7-10	Special Products and Factors	251
7-11	Solving Quadratic Equations	254
7-12	Solving Problems with Quadratic Equations	257

Self-Tests   237, 248, 260   Computer Exercises   261
Chapter Review   262   Chapter Test   263
Cumulative Review: 1-7   264
Biographical Note – Ada Byron Lovelace   267
Vocabulary Review / Mixed Review   268

8 RATIONAL AND IRRATIONAL NUMBERS

8-1	Writing Rational Numbers as Decimals	271
8-2	Changing from Decimal to Fractional Form	274
8-3	Rational Square Roots	276
8-4	Irrational Square Roots	279
8-5	Multiplication and Division of Radicals	283
8-6	Simplifying Radicals	285
8-7	Addition and Subtraction of Radicals	287
8-8	Binomials Containing Radicals	289
8-9	The Quadratic Formula	290

Self-Tests   276, 282, 288, 293   Computer Exercises   293
Chapter Review   294   Chapter Test   295
Careers – Automobile Mechanic   297
Vocabulary Review / Mixed Review   298

9 RATIO AND SIMILARITY

9-1	Ratios	301
9-2	Proportions	304
9-3	Percents	307
9-4	Similar Polygons	314

v

9-5	Areas and Volumes of Similar Figures	320
9-6	The Pythagorean Theorem	326
9-7	Special Right Triangles	331

Self-Tests   313, 325, 336   Computer Exercises   337
Chapter Review   338   Chapter Test   340
Application — Using a Credit Card   341
Vocabulary Review / Mixed Review   342

10 GEOMETRY OF THE CIRCLE

10-1	Basic Definitions	345
10-2	Central Angles and Arcs of Circles	349
10-3	Inscribed Angles and Intercepted Arcs	353
10-4	Other Angles and Intercepted Arcs	359
10-5	Circumference and Area of a Circle	367
10-6	Surface Areas and Volumes of Spheres	372
10-7	Surface Areas and Volumes of Cylinders and Cones	375

Self-Tests   352, 366, 379   Computer Exercises   380
Chapter Review   381   Chapter Test   382
Cumulative Review: 1–10   383
Careers — Meteorologist   387
Vocabulary Review / Mixed Review   388

11 PROBABILITY

11-1	One-Stage Experiments	391
11-2	Events and Their Probabilities	395
11-3	Many-Stage Experiments	397
11-4	The Counting Principle	404
11-5	Permutations	408
11-6	Probabilities of Many-Stage Experiments	412
11-7	General Rules about Probability	416
11-8	Drawing with and without Replacement	420
11-9	Empirical Probability	424

Self-Tests   403, 412, 428   Computer Exercises   428
Chapter Review   429   Chapter Test   431
Biographical Note — Benjamin Banneker   433
Vocabulary Review / Mixed Review   434

12 STATISTICS

12-1	Introduction to Statistics	437
12-2	Charts and Graphs	441
12-3	Measures of Central Tendency	446
12-4	Quartiles and Percentiles	449
12-5	Sampling and Simulation	455

12-6 Mathematical Expectation   458

    Self-Tests   445, 454, 461   Computer Exercises   462
    Chapter Review   464   Chapter Test   466
    Careers – Statistician   467
    Vocabulary Review / Mixed Review   468

13   LINES AND THEIR EQUATIONS

13-1   A Rectangular Coordinate System   471
13-2   Linear Equations in Two Variables   475
13-3   The Slope of a Line   480
13-4   Slope-Intercept Form   485
13-5   Graphing Linear Inequalities in Two Variables   489

    Self-Tests   479, 492   Computer Exercises   493
    Chapter Review   495   Chapter Test   496
    Application – Line of Best Fit   497
    Vocabulary Review / Mixed Review   498

14   SYSTEMS OF LINEAR EQUATIONS

14-1   Using Graphs to Solve Pairs of Linear Equations   501
14-2   Using the Substitution Method   504
14-3   Using the Addition or Subtraction Method   507
14-4   Using Systems of Equations to Solve Problems   511
14-5   Graphing Systems of Linear Inequalities   515
14-6   Linear Programming (Optional)   518

    Self-Tests   511, 517   Computer Exercises   522
    Chapter Review   523   Chapter Test   524
    Cumulative Review: 1-14   525
    Biographical Note – Emmy Noether   529
    Vocabulary Review / Mixed Review   530

EXTRA PRACTICE   532
ALGEBRA REVIEW EXERCISES   546
PROGRAMMING IN BASIC   552
PREPARING FOR REGENTS EXAMINATIONS   561
ANSWERS TO SELF-TESTS   574
INDEX   580
SELECTED ANSWERS   586
ANSWERS TO COMPUTER EXERCISES   627

SYMBOLS

Symbol	Meaning	Page	Symbol	Meaning	Page
$\{1, 2, 3\}$	the set containing the elements 1, 2, 3	p. 2	$\cong$	is congruent to	p. 137
...	and so on	p. 2	$\angle$	angle	p. 140
$\emptyset$ or $\{\ \}$	the empty set	p. 2	$m\angle HOE$	the measure of $\angle HOE$	p. 140
$\sim$	the negation of	p. 38	$35°$	35 degrees	p. 141
$\wedge$	and	p. 40	$\perp$	is perpendicular to	p. 145
$\vee$	or	p. 43	$\triangle$	triangle	p. 156
$\rightarrow$	implies	p. 46	n-gon	polygon with n sides	p. 161
$\leftrightarrow$	if and only if	p. 57	$0.4\overline{3}$	the repetend 3	p. 273
$\therefore$	therefore	p. 60	$\sqrt{\ }$	the principal square root	p. 276
$\neq$	does not equal	p. 107	$\sqrt[3]{\ }$	the cube root	p. 279
$<$	is less than	p. 107	$\approx$	is approximately equal to	p. 279
$>$	is greater than	p. 107	$\pm$	plus or minus	p. 291
$\{x : x < 2\}$	the set of all x such that x is less than 2	p. 111	$a:b$	the ratio of a to b	p. 302
			$\%$	percent	p. 308
$\leq$	is either equal to or less than	p. 116	$\sim$	is similar to	p. 314
			$\odot O$	circle O	p. 345
$\geq$	is either equal to or greater than	p. 116	$\overset{\frown}{AB}$	minor arc AB	p. 349
$\overleftrightarrow{AB}$	line AB	p. 133	$m\overset{\frown}{AB}$	the measure of $\overset{\frown}{AB}$	p. 349
$\parallel$	is parallel to	p. 135	π	pi	p. 367
$\overline{AB}$	line segment AB	p. 137	$P(A)$	the probability of A	p. 395
$\overrightarrow{AB}$	ray AB	p. 137	$!$	factorial	p. 408
AB	the length of $\overline{AB}$	p. 137	$_nP_r$	the number of permutations of n things taken r at a time	p. 409

METRIC UNITS OF MEASURE

Prefixes

kilo	hecto	deka	deci	centi	milli
1000	100	10	0.1	0.01	0.001

Length
1 centimeter (cm) = 10 millimeters (mm)
1 meter (m)   = 100 cm
1 kilometer (km) = 1000 m

Area
1 square centimeter (cm^2) = 100 square millimeters (mm^2)
1 square meter (m^2)   = 10,000 cm^2

Volume
1 cubic centimeter (cm^3) = 1000 cubic millimeters (mm^3)
1 cubic meter (m^3)   = 1,000,000 cm^3
1 liter (L)   = 1000 cm^3
1 liter   = 1000 milliliters (mL)

Mass
1 gram (g)   = 1000 milligrams (mg)
1 kilogram (kg)   = 1000 g
1 metric ton (t)   = 1000 kg

ACKNOWLEDGMENTS

Cover concept by Richard Hannus
Biographical portraits by Gary Torrisi (pages 69, 169, 267, 433, and 529)

Photographs:
page x: © Peter Menzel 1983
page 31: © Freda Leinwand / Monkmeyer Press Photo Service
page 34: © Michal Heron
page 72: © Foto Georg Gerster / Photo Researchers
page 106: © Grant White / Monkmeyer Press Photo Service
page 129: © Richard Wood 1983 / The Picture Cube
page 132: © Peter Menzel 1981
page 172: © Hazel Hankin / Stock Boston
page 224: © Rainer Binder, G + J Images / The Image Bank
page 270: © W.B. Finch / Stock Boston
page 297: © Fredrik D. Bodin / Stock Boston
page 300: © Steve Rosenthal
page 344: © Fred Ward / Black Star
page 387: ESSA Photo
page 390: © Christopher Morrow / Stock Boston
page 436: © Orville Andrews / Photo Researchers
page 467: © Eric Kroll 1979 / Taurus Photos
page 470: © George Bellerose / Stock Boston
page 500: courtesy of the Ford Motor Company

Any collection of objects can be considered to be a *set*. The players on your favorite team, the students in your class, and the whole numbers less than twenty-five are all examples of sets.

Numbers and Operations 1

What does the word "number" mean to you? You will probably think of the counting numbers (1, 2, 3, and so on). Include 0 and you have the whole numbers.

Actually, there are many kinds of numbers. In this book you will study some of the properties of numbers called the real numbers. In this chapter most of your work will be with integers and the rational numbers.

Section 1-1 SETS

Sometimes you will need to talk about a particular set of numbers, and so in this section you will learn how to:
- recognize a set
- list a set
- describe a set

In your everyday life you talk about the players on your favorite baseball team, the teachers in your school, the students in your math class, the members of the girls' volleyball team, and so on. Each of these is an example of a set. A *set* is a collection of members, or elements.

EXAMPLES OF SETS

>The 50 states of the United States
>The days of the week
>The whole numbers between 5 and 10

A set is *well-defined* if one can decide whether or not any particular object is a member of the set. In mathematics you will deal only with well-defined sets.

EXAMPLES OF WELL-DEFINED SETS

>Months of the year beginning with the letter M
>States in the United States named by two words
>The whole numbers between 1 and 10

EXAMPLES OF SETS THAT ARE *NOT* WELL-DEFINED

Good cities in which to live   Tasty food
Tall people   Lucky numbers

The two examples on the left above are not clearly described. The two on the right depend upon individual preferences.

A set can be described by listing its members between braces, { }. This method is called the *roster* method. The order in which the members are listed in a set is not important. The set $\{1, 7, 2\}$ is the same as the set $\{2, 1, 7\}$.

For convenience you may use a capital letter to represent a set.

EXAMPLES

The set A of months beginning with the letter M.
$A = \{\text{March, May}\} = \{\text{May, March}\}$

The set B of states in the United States named by two words.
$B = \{$New Hampshire, New Jersey, New Mexico, New York, North Carolina, South Carolina, North Dakota, South Dakota, Rhode Island, West Virginia$\}$

The set C of whole numbers between 1 and 10.
$C = \{2, 3, 4, 5, 6, 7, 8, 9\}$

Sometimes you will find it takes too long to list all the elements in a set. Would you like to write out the set of all 50 states in the United States? In this case you *describe* the set by a phrase or a rule that clearly defines the members of the set.

EXAMPLES

$D = \{$ states in the United States $\}$
$E = \{$ all whole numbers between 1 and 1000, including 1 and 1000 $\}$
$W = \{$ all whole numbers $\}$

The sets E and W could also be written by using three dots, which indicate that the pattern continues.

$E = \{1, 2, 3, 4, \ldots, 1000\}$
$W = \{0, 1, 2, 3, 4, \ldots\}$

Do all sets contain elements? How about the set of days of the week beginning with the letter H? The set that has no members is called the *empty set* and is written as $\emptyset$ or as { }.

ORAL EXERCISES

Tell whether or not the indicated set is well-defined.

1. { the days of the week }   2. { the seasons of the year }   3. { all intelligent people }

4. { all small cars }   5. { five-letter words }   6. { four-sided triangles }

State the members of each of the following sets.

7. { the months beginning with the letter J }
8. { the whole numbers between 5 and 10 }
9. { the states in the United States beginning with the letter J }
10. { the days of the week containing exactly six letters }
11. Tell which of the sets in Exercises 1–10 are the empty set.

WRITTEN EXERCISES

Indicate whether or not each of the following sets is well-defined.

A
1. { the provinces of Canada beginning with the letter N }
2. { the months beginning with the letter D }
3. { big fish }
4. { small animals }
5. { the best football players }
6. ∅

Write out each of the following sets using the roster method.

7. { the days of the week beginning with the letter T }
8. { the first three American presidents }
9. { the states in the United States beginning with the letter A }
10. { the last three letters of the alphabet }
11. { the whole numbers between 12 and 17 }
12. { the whole numbers between 13 and 16, including 13 and 16 }
13. { the even whole numbers less than 15 }
14. { the odd whole numbers less than 16 }
15. { the months of the year having 31 days }
16. { the days of the week containing exactly seven letters }

Describe each of the following sets by a rule.

B
17. { Saturday, Sunday }
18. { June, July }
19. { 1, 2, 3, 4, 5, ... }
20. { 3, 6, 9, 12, 15, ... }

C
21. { 2, 4, 8, 16, 32, 64, ... }
22. { 3, 9, 27, 81, 243, ... }
23. { 2, 3, 5, 7, 11, 13, ... }
24. { 4, 6, 8, 9, 10, 12, ... }

Find the next three members of each of the following patterns.

25. 5, 25, 125, 625, _?_, _?_, _?_, ...
26. 1, 1, 2, 3, 5, 8, 13, _?_, _?_, _?_, ...
27. O, T, T, F, F, S, _?_, _?_, _?_, ...
28. 4, 5, 7, 11, 19, _?_, _?_, _?_, ...

NUMBERS AND OPERATIONS

Section 1-2 THE NUMBER LINE

In this section we will construct a number line and use it to:
- graph numbers
- find the coordinates of points

You now know how to represent a set by using either a roster or a description. Another way to represent sets of numbers is by drawing a picture called a *graph*. Numbers can be represented as points on a *number line*.

To construct a number line, you draw a line and make a mark on the line which you call the *origin*. You label this mark 0 (zero).

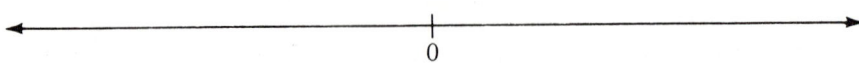

Now at some convenient distance to the right of the origin you place another mark and label it 1.

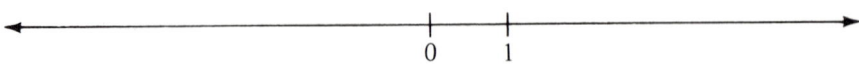

Using the distance between 0 and 1 as a unit length, mark off units to the right of 1 and label them consecutively 2, 3, 4, 5, and so on.

Using the same unit length, mark off units to the left of 0 and label them consecutively ⁻1 (negative 1), ⁻2, ⁻3, ⁻4, ⁻5, and so on. You have now constructed the number line. The number line extends indefinitely in both directions, as indicated by the arrowheads.

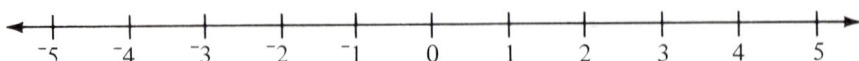

The set $\{\ldots, ^-5, ^-4, ^-3, ^-2, ^-1, 0, 1, 2, 3, 4, 5, \ldots\}$ is called the set of *integers*. The numbers 1, 2, 3, 4, ... are called the *positive integers*. The numbers ⁻1, ⁻2, ⁻3, ⁻4, ... are called the *negative integers*.

Since ⁻5 is to the left of 2 on the number line, you can write:

$$^-5 < 2 \quad \text{(read ``}^-5 \text{ is less than 2'')}$$

or

$$2 > ^-5 \quad \text{(read ``2 is greater than }^-5\text{'')}$$

Are there numbers between 0 and 1? Yes, $\frac{1}{2}$ (or 0.5) is to the right of 0 and to the left of 1. In fact, every point on the number line is paired with a number called a *real number*.

> To every point on the number line there corresponds a real number.
>
> To every real number there corresponds a point on the number line.

Each point on the number line is called the *graph* of the real number assigned to it. The number itself is called the *coordinate* of the point. Points on the number line are often labeled with capital letters.

EXAMPLE State the coordinates of the points *A*, *B*, and *C* on the number line.

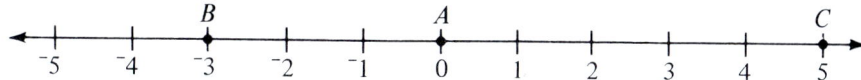

SOLUTION: The coordinates of *A*, *B*, and *C* are 0, ⁻3, and 5 respectively.

EXAMPLE On the number line show the graphs of ⁻2, 1, and $4\frac{1}{2}$. Label these points *D*, *E*, and *F* respectively.

SOLUTION:

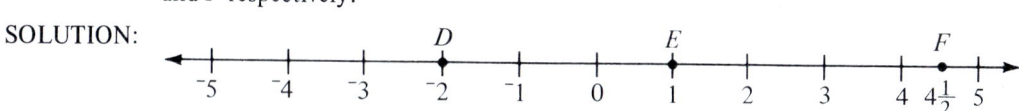

Because 4 means positive four, we sometimes write it as ⁺4. Positive numbers and negative numbers lie in opposite directions on the number line. That is, ⁺4 is four units to the right of the origin, and ⁻5 is five units to the left of the origin. For this reason, positive and negative numbers are often called *directed numbers*.

ORAL EXERCISES

Use the number line below for Exercises 1–12. Name the graph of the given number or state the coordinate of the given point.

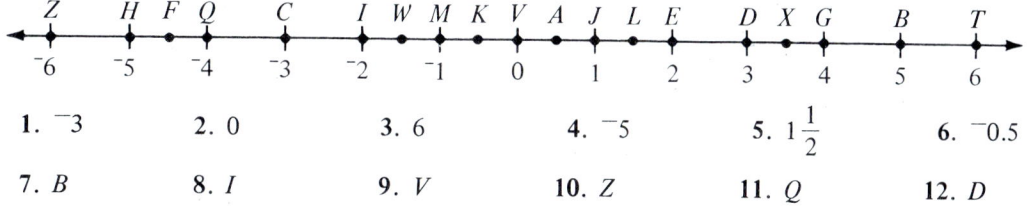

1. ⁻3
2. 0
3. 6
4. ⁻5
5. $1\frac{1}{2}$
6. ⁻0.5
7. *B*
8. *I*
9. *V*
10. *Z*
11. *Q*
12. *D*

Name each measurement as a directed number. Assume that north, east, up, above, win, over, gain, and hits are positive.

13. **a.** 25 km north   **b.** 35 km south
14. **a.** 160 m west   **b.** 210 m east
15. **a.** up by 12 points **b.** down by 8 points
16. **a.** 72 degrees above zero **b.** 12 degrees below zero
17. **a.** wins 5 games   **b.** loses 10 games
18. **a.** 10 over par   **b.** 3 under par
19. **a.** gains 3 pieces   **b.** loses 5 pieces
20. **a.** misses 7 targets **b.** hits 13 targets

NUMBERS AND OPERATIONS

WRITTEN EXERCISES

Using the number line below, name the graph of the given number or state the coordinate of the given point.

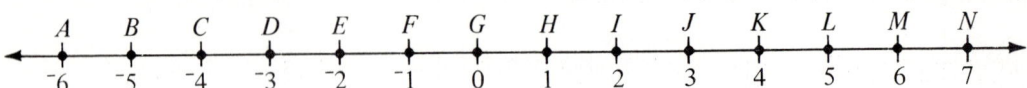

A
1. A
2. D
3. J
4. M
5. ⁻4
6. 5
7. 0
8. ⁻1
9. L
10. H
11. E
12. B
13. 2
14. ⁻3
15. 7
16. ⁻6

Draw a number line and show the graphs of the given numbers.

17. 3
18. ⁻1
19. 0
20. 7
21. ⁻5
22. 2
23. 1
24. ⁻4
25. $\frac{^-1}{2}$
26. $2\frac{1}{2}$
27. $\frac{24}{6}$
28. $\frac{100}{100}$

Graph the given set.

29. $\{^-3, 0, 3\}$
30. $\{1, 3, 5\}$
31. $\{^-6, ^-2, 3\}$
32. $\{^-4, ^-3, ^-2, ^-1\}$

EXAMPLE Graph the set of positive integers.

SOLUTION

The heavy arrowhead at the right of the number line indicates that the graph of the set continues without end in the direction indicated.

B 33. Graph the set of negative integers.
34. Graph the set of integers.
35. Graph the set of odd integers.
36. Graph the set of even integers.

Use the number line below to find the number represented by each of the following points.

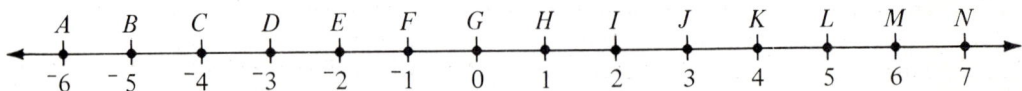

37. The point halfway between B and H.
38. The point halfway between E and M.
39. The point halfway between A and N.
40. The point halfway between D and K.

6 CHAPTER 1

EXAMPLES

$1 + (-1) = 0$ $-3 + 3 = 0$ $-12 + 12 = 0$ $0 + 0 = 0$

The order in which two numbers are added is not important. This is called the commutative property for addition.

EXAMPLES

$2 + 5 = 7$ $1 + (-6) = -5$
$5 + 2 = 7$ $-6 + 1 = -5$

You can use the number line when adding more than two numbers. You can insert parentheses to indicate which two numbers you would add first.

EXAMPLES

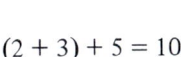
$(2 + 3) + 5 = 10$
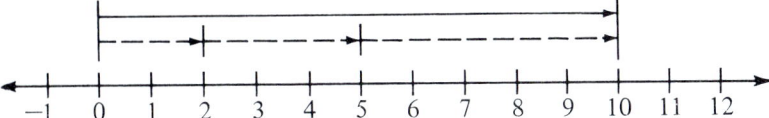

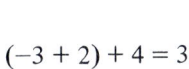
$(-3 + 2) + 4 = 3$
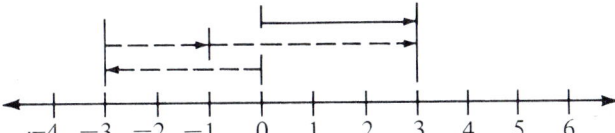

The way you pair numbers when you add more than two numbers together does not affect the sum. This is called the associative property for addition.

EXAMPLES

$(3 + 4) + 8 = 7 + 8 = 15$ $(-2 + 3) + (-5) = 1 + (-5) = -4$
$3 + (4 + 8) = 3 + 12 = 15$ $-2 + (3 + (-5)) = -2 + (-2) = -4$

ORAL EXERCISES

What number is the opposite of the given number?

1. 7   2. -25 3. 0   4. -9

Add.

5. $1 + 5$ 6. $4 + 6$ 7. $12 + 8$ 8. $21 + 9$
9. $2 + (-7)$ 10. $5 + (-9)$ 11. $-5 + 12$ 12. $-3 + 17$
13. $-3 + (-6)$ 14. $-11 + (-2)$ 15. $8 + 0$ 16. $0 + (-4)$

NUMBERS AND OPERATIONS   9

WRITTEN EXERCISES

Add.

A
1. $17 + 15$
2. $12 + 29$
3. $635 + 214$
4. $362 + 427$
5. $32 + (-18)$
6. $45 + (-19)$
7. $27 + (-53)$
8. $31 + (-47)$
9. $-23 + 56$
10. $-28 + 63$
11. $-84 + 43$
12. $-76 + 39$
13. $-13 + (-37)$
14. $-43 + (-18)$
15. $-83 + 83$
16. $125 + (-125)$
17. $(32 + 47) + 59$
18. $22 + (37 + 63)$
19. $(-16 + 39) + (-42)$
20. $(40 + (-37)) + 12$

EXAMPLE $31 + (-47) + (-26) + 14$

SOLUTION: To add several numbers, it is often convenient to group the numbers in a special way. Here we begin by grouping the negative numbers and adding them. Then we use the commutative property to help us group and add the positive numbers. (In your work you do not need to show all the steps given below.)

$$31 + ((-47) + (-26)) + 14$$
$$= 31 + (-73) + 14$$
$$= 31 + 14 + (-73)$$
$$= 45 + (-73)$$
$$= -28$$

B
21. $(-24) + (-45) + 17 + 52$
22. $45 + (-89) + 64 + (-26)$
23. $(-37) + 54 + (-23) + 43$
24. $62 + (-41) + (-53) + 12$
25. $114 + 968 + (-114) + (-968)$
26. $(-237) + (-79) + 79 + 237$
27. $45 + (-22) + 37 + (-23) + 19$
28. $-53 + 26 + 42 + (-31) + 11$

In each of the following exercises, name the property of addition that the statement illustrates.

29. $6 + (-6) = 0$
30. $(-3 + 2) + 4 = -3 + (2 + 4)$
31. $5 + 11 = 11 + 5$
32. $-4 = 0 + (-4)$
33. $(5 + 7) + 12 = 5 + (7 + 12)$
34. $3 + (-7) = -7 + 3$
35. $12 + 0 = 12$
36. $0 = (-11) + 11$

C
37. $(-8 + 8) + 10 = 0 + 10$
38. $(6 + 0) + 2 = 6 + 2$

39. By the inverse property for addition, $3 + (-3) = 0$ and $(-3) + (-(-3)) = 0$. By the commutative property for addition, the statement $3 + (-3) = 0$ can be rearranged as $(-3) + 3 = 0$. What can you conclude?

40. The statement $(2 + (-2)) + (5 + (-5)) = 0$ can be rearranged as $(2 + 5) + (-2 + (-5)) = 0$. Also, $7 + (-7) = 0$ can be rearranged as $(2 + 5) + (-(2 + 5)) = 0$. What can you conclude?

10   CHAPTER 1

Section 1-4 SUBTRACTION

You know from arithmetic that subtracting 4 from 7 is the same as finding a number whose sum with 4 is 7. Compare the addition and subtraction below.

Addition   *Subtraction*
$4 + 3 = 7$ $7 - 4 = 3$

But you know from Section 1-3 that $7 + (-4) = 3$. That is, subtracting 4 from 7 is the same as adding the opposite of 4 to 7.

> To subtract, you replace the number you are subtracting by its opposite and then add.

EXAMPLES

$$4 - 7 = 4 + (-7) = -3 \qquad -4 - 7 = -4 + (-7) = -11$$
$$-4 - (-7) = -4 + 7 = 3 \qquad 4 - (-7) = 4 + 7 = 11$$

Subtracting zero from a number does not change the number.

EXAMPLES $\qquad 9 - 0 = 9 + 0 = 9 \qquad\qquad -5 - 0 = -5 + 0 = -5$

Unlike addition, the order in which you subtract two numbers is very important.

EXAMPLES

$$5 - 3 = 5 + (-3) = 2 \qquad 11 - (-4) = 11 + 4 = 15$$
$$3 - 5 = 3 + (-5) = -2 \qquad -4 - 11 = -4 + (-11) = -15$$

When a subtraction involves more than two numbers you can insert parentheses to indicate which numbers you would subtract first. The way you pair numbers when subtracting more than two numbers is very important.

EXAMPLES

$$(2 - 3) - 4 = -1 - 4 = -5 \qquad (15 - 8) - 5 = 7 - 5 = 2$$
$$2 - (3 - 4) = 2 - (-1) = 2 + 1 = 3 \qquad 15 - (8 - 5) = 15 - 3 = 12$$

ORAL EXERCISES

Subtract.

1. $11 - 3$	**2.** $15 - 7$	**3.** $37 - 12$	**4.** $49 - 23$
5. $8 - 17$	**6.** $6 - 23$	**7.** $-9 - 15$	**8.** $-4 - 29$
9. $53 - 27$	**10.** $41 - 13$	**11.** $75 - 0$	**12.** $0 - 23$
13. $8 - (-5)$	**14.** $12 - (-7)$	**15.** $-2 - (-15)$	**16.** $-7 - (-9)$

NUMBERS AND OPERATIONS

WRITTEN EXERCISES

Subtract.

A
1. $23 - 11$
2. $47 - 25$
3. $12 - 37$
4. $8 - 59$
5. $32 - 19$
6. $55 - 28$
7. $19 - 32$
8. $28 - 55$
9. $172 - 83$
10. $234 - 97$
11. $38 - 475$
12. $53 - 294$
13. $321 - (-237)$
14. $542 - (-310)$
15. $-623 - 439$
16. $-846 - 358$
17. $-742 - (-371)$
18. $-519 - (-612)$
19. $-235 - (-829)$
20. $-637 - (-148)$

B
21. $(37 - 23) - 45$
22. $(112 - 52) - 39$
23. $37 - (23 - 45)$
24. $112 - (52 - 39)$
25. $-46 - (82 - 47)$
26. $-93 - (14 - 27)$
27. $(231 - 157) - (98 - 87)$
28. $(57 - 65) - (292 - 186)$
29. $(43 - 109) - (116 - 81)$
30. $(214 - 87) - (85 - 197)$
31. $(94 - 218) - (57 - 120)$
32. $(200 - 1000) - (750 - 925)$
33. $67 - (79 - 54) - 95$
34. $112 - (99 - 73) - 315$

35. Is there a commutative property for subtraction?
36. Is there an associative property for subtraction?

Section 1-5 MULTIPLICATION

This section discusses:
- rules for multiplying two numbers
- some properties of multiplication

You can think of a product such as $3 \times (-4)$ as the sum $(-4) + (-4) + (-4)$. Thus you can use the number line to illustrate multiplication of two numbers when at least one of them is a positive integer. The results can be generalized to the product of any two numbers.

EXAMPLE $3 \times 4 = 4 + 4 + 4 = 12$

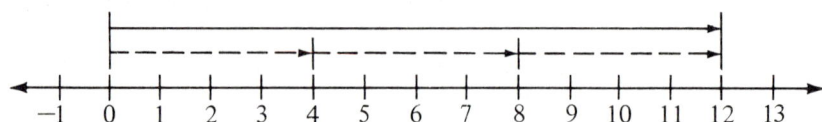

The product of two positive numbers is positive.

12   CHAPTER 1

EXAMPLE $3 \times (-4) = (-4) + (-4) + (-4) = -12$

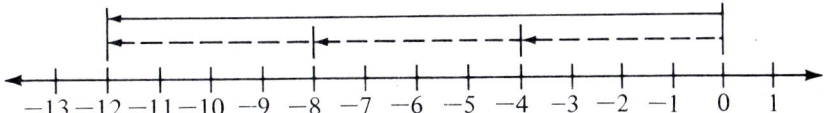

> The product of a positive number and a negative number is negative.

Notice in the examples above that changing one number in a product of two positive numbers to its opposite changes the product to its opposite. Suppose that in the multiplication $3 \times (-4) = -12$ you replace 3 by its opposite, -3. Then the product will be the opposite of -12:

$$(-3) \times (-4) = 12$$

> The product of two negative numbers is positive.

The symbols $\cdot$ and $(\ )(\ )$ are often used when two numbers are multiplied.

EXAMPLES

$5 \times 2 = 10$ $5 \cdot 2 = 10$ $(5)(2) = 10$ $5(2) = 10$

$(-3) \times 4 = -12$ $(-3) \cdot 4 = -12$ $(-3)(4) = -12$ $-3(4) = -12$

Parentheses are also used when multiplying more than two numbers to indicate which numbers you are to multiply first.

The way you pair numbers does not affect the product. This is called the associative property for multiplication.

EXAMPLES

$(2 \times 5) \times 7 = 10 \times 7 = 70$ $((-3)(5))(-6) = (-15)(-6) = 90$

$2 \times (5 \times 7) = 2 \times 35 = 70$ $(-3)(5(-6)) = (-3)(-30) = -90$

The order in which numbers are multiplied is not important. This is called the commutative property for multiplication.

EXAMPLES

$(-3)4 = -12$ $(-5)(-9) = 45$

$4(-3) = -12$ $(-9)(-5) = 45$

NUMBERS AND OPERATIONS   13

Multiplying a number by 1 does not change the number. This is called the identity property for multiplication.

EXAMPLES

$$(-17)(1) = -17 \qquad 1(-23) = -23$$

The product of a number and −1 is the opposite of the given number. This is called the multiplicative property of −1.

EXAMPLES

$$16(-1) = -16 \qquad (-1)(-4) = 4$$

The product of any number and zero is always zero. This is called the multiplicative property of 0.

EXAMPLES

$$5 \cdot 0 = 0 \qquad -7 \cdot 0 = 0 \qquad 0(-135) = 0$$

ORAL EXERCISES

Multiply.

1. 7 × 2
2. 3 × 10
3. (5)(12)
4. (13)(6)
5. 5(−4)
6. 4(−5)
7. (−2)(−15)
8. (−11)(−3)
9. 93(1)
10. 1(−72)
11. 7 · 9 · 0
12. 3 · 0 · (−11)
13. 2 · 2 · 3 · 3
14. 3 · 5 · 5 · 3
15. (−2)(−3)(−8)
16. (−3)(−5)(6)

WRITTEN EXERCISES

Multiply.

A
1. 13 × 12
2. 23 × 11
3. 37 × 45
4. 32 × 56
5. 27(−43)
6. 38(−52)
7. −18(67)
8. −29(83)
9. 17 · 19
10. 19 · 17
11. (−31)(−53)
12. (−64)(−28)
13. 325 · 579
14. 472 · 876
15. 111 · 763 · 0
16. 629 · 333 · 0
17. −423(83)
18. 625(−96)
19. 3792(1)
20. (−1)(10,793)

B
21. (13)(19)(−23)
22. (−37)(11)(29)
23. (−12)(22)(−30)
24. (15)(−18)(−16)
25. (−7)(−9)(−11)(−15)
26. (8)(12)(−14)(−24)
27. 2 · 2 · (−2)(−3)(−3) · 5 · 7
28. 2(−2)(−2) · 5 · 6 · 9 · (−11)

14   CHAPTER 1

Name the property that each statement illustrates.

29. $(12 \cdot 13) \cdot 17 = 12 \cdot (13 \cdot 17)$
30. $0 = 0(-172)$
31. $(-4)(-8) = (-8)(-4)$
32. $15 = 1 \cdot 15$
33. $(-1)6 = -6$
34. $\big((-3)(-5)\big)8 = -3\big((-5)(8)\big)$
35. $13 \cdot 1 = 13$
36. $3 \cdot 7 = 7 \cdot 3$
37. $12 \cdot 0 = 0$
38. $-9 = (-1)9$

C **39.** $(2 \cdot 4)10 = (4 \cdot 2)10$
40. $(3 \cdot 9)0 = 0$
41. $1 + (3 \cdot 7) = (3 \cdot 7) + 1$
42. $(2 \cdot 3)11 + 0 = (2 \cdot 3)11$

Section 1-6 DIVISION

Division is related to multiplication, as shown below.

Multiplication	*Division*		
$2 \cdot 6 = 12$	$12 \div 6 = 2$	and	$12 \div 2 = 6$
$-2 \cdot 6 = -12$	$-12 \div 6 = -2$	and	$-12 \div (-2) = 6$
$2 \cdot (-6) = -12$	$-12 \div (-6) = 2$	and	$-12 \div 2 = -6$
$-2 \cdot (-6) = 12$	$12 \div (-6) = -2$	and	$12 \div (-2) = -6$

> The quotient of two positive numbers is positive.
>
> The quotient of one negative and one positive number is negative.
>
> The quotient of two negative numbers is positive.

You may write the quotient of two numbers such as 12 and -4 as $12 \div (-4)$ or as $\dfrac{12}{-4}$.

EXAMPLES

$$15 \div 3 = 5 \qquad \frac{15}{3} = 5$$

$$-132 \div 11 = -12 \qquad \frac{-132}{11} = -12$$

$$72 \div (-8) = -9 \qquad \frac{72}{-8} = -9$$

$$-28 \div (-4) = 7 \qquad \frac{-28}{-4} = 7$$

If you divide a number by 1, the number is not changed.

EXAMPLES

$$5 \div 1 = 5 \qquad \frac{-17}{1} = -17$$

NUMBERS AND OPERATIONS

The order in which you divide two numbers is very important.

EXAMPLES

$$14 \div 2 = 7 \qquad \frac{25}{5} = 5$$

$$2 \div 14 = \frac{2}{14} = \frac{1}{7} \qquad \frac{5}{25} = \frac{1}{5}$$

The last two examples above show that the quotient of two integers is not necessarily an integer.

Division by 0 is not defined. To solve the division problem $5 \div 0 = \boxed{?}$, you would consider the multiplication $0 \cdot \boxed{?} = 5$. There is no solution for $0 \cdot \boxed{?} = 5$, since the product of 0 and any number is 0.

ORAL EXERCISES

Divide.

1. $8 \div 2$
2. $0 \div 5$
3. $64 \div (-4)$
4. $-64 \div 4$
5. $-81 \div 3$
6. $-81 \div (-3)$
7. $-20 \div 5$
8. $-30 \div 10$
9. $60 \div (-12)$
10. $96 \div (-8)$
11. $-36 \div (-12)$
12. $-49 \div (-7)$
13. $\frac{-84}{4}$
14. $\frac{100}{-5}$
15. $\frac{-432}{-8}$
16. $\frac{-506}{-11}$

WRITTEN EXERCISES

Divide.

A
1. $216 \div 9$
2. $315 \div 5$
3. $115 \div (-23)$
4. $224 \div (-32)$
5. $-187 \div 17$
6. $-312 \div 39$
7. $-266 \div (-14)$
8. $-132 \div (-11)$
9. $-322 \div 14$
10. $-513 \div (-19)$
11. $-495 \div (-33)$
12. $882 \div (-21)$
13. $\frac{-611}{13}$
14. $\frac{-286}{22}$
15. $\frac{255}{-15}$
16. $\frac{414}{-18}$
17. $\frac{-390}{-26}$
18. $\frac{-741}{-39}$
19. $\frac{-3185}{-49}$
20. $\frac{-3034}{-37}$

B
21. $-18{,}183 \div 209$
22. $-14{,}508 \div (-186)$
23. $16{,}046 \div (-142)$
24. $-20{,}328 \div (-154)$
25. $\frac{-134{,}534}{-137}$
26. $\frac{-161{,}357}{623}$
27. $\frac{241{,}713}{-251}$
28. $\frac{-272{,}076}{-492}$

29. Evaluate $(8 \div 4) \div 2$ and $8 \div (4 \div 2)$. What can you conclude?
30. Is there a commutative property for division?

SELF-TEST 2

Add or subtract as indicated.

1. $379 + (-542)$
2. $-128 + (-355)$ Section 1-3
3. $17{,}425 + 9756$
4. $17 + 49 + (-58) + 96$

5. $436 - 213$
6. $792 - (-145)$ Section 1-4
7. $7321 - 5117$
8. $-6124 - (-4537)$

Multiply or divide as indicated.

9. $37 \cdot 92$
10. $45(-129)$ Section 1-5
11. $(-52)(-45)(-32)$
12. $3(-5)7(-11)$

13. $1288 \div 56$
14. $-1305 \div 29$ Section 1-6
15. $\dfrac{6958}{-71}$
16. $\dfrac{-7752}{-136}$

Section 1-7 ADDITION AND SUBTRACTION OF RATIONAL NUMBERS

When you add, subtract, or multiply two integers, their sum, difference, or product is an integer. When you divide one integer by another integer, is the quotient always an integer? Is the quotient of 1 divided by 6 an integer? When the quotient of two integers is an integer, we say that the first number is *divisible* by the second number.

In this section you will review:
- rational numbers
- addition of rational numbers
- subtraction of rational numbers

A *rational number* is a number that can be expressed as the quotient of two integers. (Recall that the divisor cannot be 0.) Every integer is a rational number, since

$$0 = \frac{0}{1}, \quad 1 = \frac{1}{1}, \quad 7 = \frac{7}{1}, \quad -5 = \frac{-5}{1}, \quad \text{and so on.}$$

Mixed numbers such as $2\frac{3}{8}$ and $-4\frac{2}{5}$ are rational numbers, since each mixed number can be written as the quotient of two integers:

$$2\frac{3}{8} = \frac{19}{8} \qquad -4\frac{2}{5} = \frac{-22}{5}$$

NUMBERS AND OPERATIONS

A rational number can be written in many different ways as the quotient of two integers.

EXAMPLES

$$3 = \frac{3}{1} = \frac{-3}{-1} = \frac{6}{2} = \frac{9}{3} = \frac{-45}{-15} \qquad -\frac{2}{5} = \frac{-2}{5} = \frac{2}{-5} = \frac{-12}{30}$$

A rational number is in *lowest terms* when its numerator and denominator have no common divisor other than 1.

To *add* two rational numbers with the same denominator, you add the numerators. You usually write the sum in lowest terms.

EXAMPLES

$$\frac{2}{5} + \frac{4}{5} = \frac{6}{5} \qquad\qquad \frac{1}{6} + \frac{2}{6} = \frac{3}{6} = \frac{1}{2}$$

$$\frac{5}{11} + \left(-\frac{3}{11}\right) = \frac{5}{11} + \left(\frac{-3}{11}\right) = \frac{2}{11} \qquad -\frac{5}{8} + \frac{3}{8} = \frac{-5}{8} + \frac{3}{8} = \frac{-2}{8} = -\frac{1}{4}$$

To *subtract* two rational numbers with the same denominator, you subtract the numerators.

EXAMPLES

$$\frac{4}{5} - \frac{2}{5} = \frac{2}{5} \qquad\qquad -\frac{7}{8} - \frac{3}{8} = \frac{-7}{8} - \frac{3}{8} = \frac{-10}{8} = -\frac{5}{4}$$

$$\frac{7}{15} - \frac{4}{15} = \frac{3}{15} = \frac{1}{5} \qquad \frac{3}{13} - \left(-\frac{5}{13}\right) = \frac{3}{13} - \left(\frac{-5}{13}\right) = \frac{8}{13}$$

If you wish to add or subtract two rational numbers with different denominators, you must first find a *common denominator*. A common denominator is a multiple of each denominator. One way to find a common denominator is to multiply all the denominators together. However, the least common denominator (LCD) is often the most convenient denominator to use.

Next you express each number as an equal number whose denominator is the common denominator you have chosen. After you add or subtract, you usually rewrite your sum or difference in lowest terms.

EXAMPLES

$$\frac{1}{2} + \frac{2}{3} = \frac{3}{6} + \frac{4}{6} = \frac{7}{6} \qquad\qquad \frac{1}{2} + \left(-\frac{7}{8}\right) = \frac{4}{8} + \left(-\frac{7}{8}\right) = -\frac{3}{8}$$

$$\frac{2}{15} - \frac{3}{10} = \frac{4}{30} - \frac{9}{30} = -\frac{5}{30} = -\frac{1}{6} \qquad \frac{5}{6} - \frac{2}{3} = \frac{5}{6} - \frac{4}{6} = \frac{1}{6}$$

18   CHAPTER 1

A decimal such as 0.389 represents a rational number since it can be expressed as the quotient of two integers.

EXAMPLES

$$0.389 = \frac{389}{1000} \qquad 0.5 = \frac{5}{10} = \frac{1}{2} \qquad -2.6 = -2\frac{6}{10} = -\frac{26}{10}$$

The way you add or subtract decimals is similar to the way you add or subtract integers.

EXAMPLES

$$-0.2 + (-0.5) = -0.7 \qquad 0.90 + (-0.05) = 0.85$$
$$0.6 - 0.9 = -0.3 \qquad -2.3 - (-1.5) = -2.3 + 1.5 = -0.8$$

When you add or subtract two decimals with a different number of places after the decimal points, you may find it helpful to insert zeros to make the number of decimal places the same.

EXAMPLES

$$-2.98 + (-0.4) = -2.98 + (-0.40) = -3.38$$
$$3.6 - 2.955 = 3.600 - 2.955 = 0.645$$

ORAL EXERCISES

In Exercises 1–8, find the least common denominator (LCD) for the given rational numbers.

1. $\dfrac{1}{2}, \dfrac{2}{5}$
2. $\dfrac{1}{3}, -\dfrac{6}{7}$
3. $\dfrac{7}{8}, -\dfrac{1}{2}$
4. $-\dfrac{6}{10}, \dfrac{1}{5}$
5. $\dfrac{3}{4}, -\dfrac{1}{6}$
6. $-\dfrac{1}{4}, -\dfrac{3}{10}$
7. $\dfrac{1}{6}, \dfrac{2}{3}, \dfrac{1}{8}$
8. $\dfrac{3}{10}, -\dfrac{4}{15}, \dfrac{1}{6}$

Add or subtract. Give your answer in lowest terms.

9. $\dfrac{2}{11} + \dfrac{5}{11}$
10. $\dfrac{6}{7} + \dfrac{3}{7}$
11. $\dfrac{6}{17} - \dfrac{13}{17}$
12. $-\dfrac{3}{4} - \dfrac{6}{4}$
13. $\dfrac{7}{8} + \dfrac{5}{8}$
14. $\dfrac{1}{9} + \dfrac{2}{9}$
15. $\dfrac{7}{12} - \dfrac{4}{12}$
16. $-\dfrac{5}{6} - \dfrac{4}{6}$
17. $\dfrac{1}{2} + \dfrac{1}{6}$
18. $\dfrac{3}{4} + \dfrac{1}{8}$
19. $\dfrac{2}{3} + 0$
20. $0 + \left(-\dfrac{7}{9}\right)$

NUMBERS AND OPERATIONS

WRITTEN EXERCISES

Write each rational number in lowest terms.

A 1. $-\dfrac{2}{8}$ 2. $\dfrac{3}{9}$ 3. $-\dfrac{10}{25}$ 4. $\dfrac{16}{18}$

5. $-\dfrac{6}{16}$ 6. $-\dfrac{25}{15}$ 7. $\dfrac{18}{27}$ 8. $-\dfrac{15}{24}$

Add or subtract. Give your answer in lowest terms.

9. $\dfrac{3}{10} + \dfrac{1}{10}$ 10. $\dfrac{5}{12} - \dfrac{3}{12}$ 11. $\dfrac{5}{12} - \left(-\dfrac{3}{12}\right)$ 12. $-\dfrac{1}{4} - \left(-\dfrac{3}{4}\right)$

13. $\dfrac{1}{2} + \dfrac{2}{5}$ 14. $\dfrac{2}{3} - \dfrac{1}{7}$ 15. $-\dfrac{1}{5} - \left(-\dfrac{2}{7}\right)$ 16. $-\dfrac{1}{2} + \left(-\dfrac{2}{3}\right)$

17. $\dfrac{3}{16} + \left(-\dfrac{5}{8}\right)$ 18. $-\dfrac{1}{6} + \left(-\dfrac{1}{3}\right)$ 19. $\dfrac{1}{12} - \left(-\dfrac{1}{6}\right)$ 20. $\dfrac{7}{10} - \dfrac{2}{5}$

21. $-\dfrac{5}{12} + \dfrac{3}{8}$ 22. $-\dfrac{1}{4} - \left(-\dfrac{5}{6}\right)$ 23. $\dfrac{3}{4} - \dfrac{7}{10}$ 24. $-\dfrac{5}{6} + \left(-\dfrac{1}{8}\right)$

25. $-\dfrac{1}{12} + \left(-\dfrac{1}{16}\right)$ 26. $\dfrac{1}{18} + \left(-\dfrac{1}{12}\right)$ 27. $\dfrac{5}{9} - \left(-\dfrac{4}{15}\right)$ 28. $-\dfrac{3}{10} + \left(-\dfrac{6}{25}\right)$

Add or subtract.

29. $0.25 + 0.35$ 30. $-1.3 - (-2.5)$ 31. $0.72 - 0.81$ 32. $-2.63 + (-3.94)$

33. $-0.68 - (-0.41)$ 34. $14.58 + (-3.62)$ 35. $0.5 + (-0.25)$ 36. $-0.64 + (-0.4)$

37. $2.86 - 0.9$ 38. $5.49 + (-3.7)$ 39. $2 - 0.45$ 40. $0.86 - 4$

B 41. $2.72 + (-1.43) + (-3.57)$ 42. $-2.61 + 4.86 + (-3.29)$

43. $-4.7 + (-0.469) + (-2.32)$ 44. $8.35 + (-6.343) + (-0.8)$

45. $(0.3 - 0.03) - 0.003$ 46. $(-2.5 + 1.57) - 2.75$

Add. Give your answer in lowest terms.

47. $\dfrac{1}{6} + \dfrac{2}{3} + \dfrac{1}{8}$ 48. $\dfrac{3}{10} + \left(-\dfrac{1}{6}\right) + \left(-\dfrac{4}{15}\right)$

49. $\dfrac{5}{12} + \dfrac{7}{18} + \left(-\dfrac{4}{5}\right)$ 50. $\dfrac{3}{8} + \left(-\dfrac{3}{10}\right) + \left(-\dfrac{1}{12}\right)$

Add and give your answer in lowest terms. (Recall that the way you group the numbers does not change the sum.)

51. $\dfrac{1}{2} + \dfrac{4}{5} + \left(-\dfrac{3}{2}\right) + \left(-\dfrac{9}{5}\right)$ 52. $-\dfrac{2}{15} + \dfrac{5}{6} + \left(-\dfrac{1}{6}\right) + \dfrac{7}{15}$

53. $-\dfrac{1}{4} + \left(-\dfrac{7}{10}\right) + \dfrac{3}{4} + \dfrac{1}{5}$ 54. $\dfrac{2}{3} + \left(-\dfrac{5}{4}\right) + \left(-\dfrac{1}{6}\right) + \dfrac{3}{4}$

Section 1-8 MULTIPLICATION AND DIVISION OF RATIONAL NUMBERS

In this section you will review:
- multiplication of rational numbers
- division of rational numbers

To *multiply* two rational numbers that are expressed as the quotient of two integers, you multiply the numerators and multiply the denominators. You usually reduce the product to lowest terms.

EXAMPLES

$$\frac{1}{2} \cdot \frac{3}{5} = \frac{3}{10} \qquad \frac{2}{7}\left(-\frac{5}{3}\right) = -\frac{10}{21}$$

$$\frac{4}{7} \cdot \frac{1}{8} = \frac{4}{56} = \frac{1}{14} \qquad \left(-\frac{1}{3}\right)\left(-\frac{6}{11}\right) = \frac{6}{33} = \frac{2}{11}$$

If a numerator and a denominator have a common divisor, you can simplify before you multiply.

EXAMPLES

$$\frac{4}{7} \cdot \frac{1}{8} = \frac{4 \cdot 1}{7 \cdot 8} = \frac{1 \cdot 1}{7 \cdot 2} = \frac{1}{14} \qquad \left(-\frac{1}{3}\right)\left(-\frac{6}{11}\right) = \frac{1 \cdot 6}{3 \cdot 11} = \frac{1 \cdot 2}{1 \cdot 11} = \frac{2}{11}$$

When you multiply two decimals, the number of places to the right of the decimal point in the product is the sum of the number of places to the right of the decimal point in the two numbers being multiplied.

EXAMPLES $\quad (-0.4)(-0.2) = 0.08 \qquad 2.65(-10) = -26.50$

The *reciprocal*, or *multiplicative inverse*, of a rational number is obtained by interchanging the numerator and denominator. Every rational number except 0 has a reciprocal.

EXAMPLES

$\frac{2}{3}$ is the reciprocal of $\frac{3}{2}$. $\qquad \frac{2}{3} \cdot \frac{3}{2} = \frac{6}{6} = 1$

$\frac{1}{5}$ is the reciprocal of 5. $\qquad \frac{1}{5} \cdot \frac{5}{1} = \frac{5}{5} = 1$

$-\frac{2}{7}$ is the reciprocal of $-\frac{7}{2}$. $\qquad -\frac{2}{7} \cdot -\frac{7}{2} = \frac{2 \cdot 7}{7 \cdot 2} = 1$

The examples above suggest the inverse property for multiplication:

The product of a rational number and its reciprocal is always 1.

NUMBERS AND OPERATIONS

To divide one rational number by another, you multiply the first number by the reciprocal of the divisor.

EXAMPLES

$$\frac{1}{2} \div \frac{3}{5} = \frac{1}{2} \cdot \frac{5}{3} = \frac{5}{6} \qquad \frac{2}{7} \div \frac{1}{4} = \frac{2}{7} \cdot \frac{4}{1} = \frac{8}{7}$$

$$-\frac{3}{4} \div -\frac{2}{5} = -\frac{3}{4}\left(-\frac{5}{2}\right) = \frac{15}{8} \qquad \frac{5}{9} \div 1 = \frac{5}{9} \cdot 1 = \frac{5}{9}$$

$$-\frac{1}{2} \div 6 = -\frac{1}{2} \cdot \frac{1}{6} = -\frac{1}{12} \qquad 20 \div (-12) = \frac{20}{1}\left(-\frac{1}{12}\right) = -\frac{20}{12} = -\frac{5}{3}$$

The *powers* of 10 are 10, 100, 1000, and so on. When you divide one decimal by another, you generally begin by multiplying the dividend and divisor by the least power of 10 that makes the divisor an integer. You may find it helpful to treat the divisor and dividend as positive numbers and then insert the appropriate sign for the final quotient.

EXAMPLES

$$\begin{array}{r} 5.8 \\ 2.6\,\overline{)15.0\,8} \\ \underline{13\ 0} \\ 2\ 0\ 8 \\ \underline{2\ 0\ 8} \\ 0 \end{array} \qquad \begin{array}{r} 1\ 8000 \\ 0.0004\,\overline{)7.2000} \\ \underline{4} \\ 3\ 2 \\ \underline{3\ 2} \\ 0 \end{array}$$

$$-15.08 \div 2.6 = -5.8 \qquad -7.2 \div (-0.0004) = 18{,}000$$

ORAL EXERCISES

State the reciprocal of the given number.

1. $\frac{1}{3}$
2. -3
3. $-\frac{3}{5}$
4. $\frac{9}{5}$
5. 1
6. -1

Multiply or divide. Give your answer in lowest terms.

7. $\frac{2}{3}\left(-\frac{4}{5}\right)$
8. $\left(-\frac{3}{5}\right)\left(-\frac{2}{7}\right)$
9. $\frac{2}{3} \div -\frac{4}{5}$
10. $-\frac{3}{5} \div -\frac{2}{7}$
11. $-\frac{3}{4} \cdot \frac{2}{9}$
12. $\left(-\frac{9}{7}\right)\frac{14}{3}$
13. $-\frac{4}{9} \div -\frac{2}{3}$
14. $-\frac{5}{6} \div \frac{10}{3}$
15. $-\frac{1}{4} \div 4$
16. $\left(-\frac{1}{3}\right)3$
17. $4 \div \left(-\frac{3}{8}\right)$
18. $(-2)\left(-\frac{1}{2}\right)$
19. $(0.25)5$
20. $3(-1.7)$
21. $-0.75 \div -3$
22. $-6 \div 0.5$

22   CHAPTER 1

WRITTEN EXERCISES

Multiply or divide. Give your answer in lowest terms.

A
1. $\frac{3}{7}\left(-\frac{7}{9}\right)$
2. $-\frac{9}{11} \cdot \frac{22}{3}$
3. $-\frac{4}{15} \cdot 10$
4. $\frac{1}{9}(-6)$
5. $\left(-\frac{7}{16}\right)\left(-\frac{10}{21}\right)$
6. $\left(-\frac{14}{15}\right)\left(-\frac{20}{21}\right)$
7. $\frac{12}{7} \cdot \frac{35}{6}$
8. $-\frac{9}{13} \cdot \frac{4}{15}$
9. $(-0.15)(0.35)$
10. $(-1.4)(-5.8)$
11. $(6.4)(-0.22)$
12. $(-0.004)(-0.08)$
13. $-\frac{3}{10} \div \frac{7}{20}$
14. $-\frac{9}{11} \div -\frac{33}{35}$
15. $-\frac{28}{15} \div \frac{21}{10}$
16. $\frac{16}{15} \div -\frac{12}{35}$
17. $-0.52 \div 10$
18. $2.8 \div 4$
19. $10 \div 0.2$
20. $15 \div 0.03$

B
21. $\left(-\frac{5}{7}\right)\left(-\frac{14}{15}\right)\left(-\frac{9}{2}\right)$
22. $\left(-\frac{8}{3}\right)\left(\frac{2}{5}\right)\left(-\frac{15}{16}\right)$
23. $\left(\frac{11}{28}\right)\left(-\frac{7}{22}\right)\left(-\frac{16}{5}\right)$
24. $\left(-\frac{12}{25}\right)\left(-\frac{15}{16}\right)\left(\frac{10}{9}\right)$
25. $(2.3)(-0.4)(1.5)$
26. $(-4.8)(-5.5)(-0.2)$
27. $(-0.3)(-2.7)(0.06)$
28. $(1.25)(-5.6)(0.08)$
29. $0.656 \div 0.16$
30. $-0.02356 \div 3.8$
31. $0.0361 \div -9.5$
32. $0.02115 \div 0.47$
33. $0.04823 \div 0.265$
34. $74.46 \div -0.0255$
35. $-0.24056 \div -0.496$
36. $-3.6157 \div 0.0865$

37. How would you multiply $2\frac{5}{8}$ by $2\frac{3}{4}$?
38. How would you multiply $3\frac{2}{5}$ by 5?

Section 1-9 ORDER OF OPERATIONS

What does the expression $8 \cdot 4 - 3$ mean? If you multiply before subtracting, you get $32 - 3$, or 29. If you subtract before multiplying, you get $8 \cdot 1$, or 8. The expression $8 \cdot 4 - 3$ does not represent a unique number until an order of operations is agreed upon. This section introduces:
- an order of operations
- the distributive property

To *simplify* an expression, you do all multiplications and divisions first, working from left to right. Then do all additions and subtractions, again working from left to right.

EXAMPLES

$8 \cdot 4 - 3 = 32 - 3 = 29$ $6 \div 2 - 2 \cdot 5 = 3 - 10 = -7$

$3 - 4 \cdot 3 + 7 = 3 - 12 + 7 = -2$ $3 \cdot 4 + 7 - 5 + 12 \div 4 = 12 + 7 - 5 + 3 = 17$

NUMBERS AND OPERATIONS

You can avoid many difficulties if you use parentheses to indicate which operations you will perform first. You perform the operation within parentheses first.

EXAMPLES

$8(4 - 3) = 8 \cdot 1 = 8$ $(2 + 7)(8 - 5) = 9 \cdot 3 = 27$

$(3 \cdot 4) - (8 \div 2) = 12 - 4 = 8$ $(3 + 5) \div (2 \cdot 2) = 8 \div 4 = 2$

The example below illustrates the distributive property of multiplication over addition. The distributive property will be used frequently throughout this book.

EXAMPLE   Notice that:

$4(12 + 9) = 4 \cdot 21 = 84$ $(12 + 9)4 = 21 \cdot 4 = 84$

$4 \cdot 12 + 4 \cdot 9 = 48 + 36 = 84$ $12 \cdot 4 + 9 \cdot 4 = 48 + 36 = 84$

Thus: $4(12 + 9) = 4 \cdot 12 + 4 \cdot 9$ and $(12 + 9)4 = 12 \cdot 4 + 9 \cdot 4$

Since subtraction is the same as adding an opposite, you can also apply the distributive property to multiplication and subtraction.

EXAMPLE $7(10 - 6) = 7 \cdot 10 - 7 \cdot 6$

You may recall from your previous work in mathematics that $3^2 = 3 \cdot 3$ and $4^3 = 4 \cdot 4 \cdot 4$. When you simplify expressions involving *powers* such as these, you simplify the powers first and then proceed as before.

EXAMPLES

$2^2 + 3^2 = 4 + 9 = 13$ $50 - 5^2 = 50 - 25 = 25$

$(6^2 - 5^2) \div (10 + 1^2) = (36 - 25) \div (10 + 1) = 11 \div 11 = 1$

$(3^2 + 4^2) + (3 + 4)^2 = (3^2 + 4^2) + (7)^2 = (9 + 16) + 49 = 74$

By agreement, operations are performed in the following order:

1. Perform operations within parentheses.
2. Simplify expressions involving powers.
3. Do all multiplications and divisions, working from left to right.
4. Do all additions and subtractions, working from left to right.

ORAL EXERCISES

State the value of each of the following expressions.

1. $2(-4) - 3$
2. $7 - 5 \cdot 6$
3. $8 \cdot 2 + 3 \cdot 4$
4. $(-7)(-3) - 9 \cdot 4$

5. $18 \div 3 - 4 \div 2$
6. $10 \div 5 - 6 \div 3$
7. $(-20) \div 5 \div 2$
8. $2(-3)(-12)$
9. $(-4)^2 + 3^2$
10. $5^2 + 2^2$
11. $3^2 - 6^2$
12. $7^2 - 5^2$
13. $(2+3)(7-5)$
14. $(5+4)(6-9)$
15. $(15-8)-(12-10)$
16. $15-(8-12)-10$

WRITTEN EXERCISES

Simplify.

A
1. $(-12)(-17) - 13 \cdot 11$
2. $21(-9) - 8 \cdot 14$
3. $10 \cdot 5^2 + 3 \cdot 4^2$
4. $8 \cdot 7^2 + 5 \cdot 9^2$
5. $(10 - 4 + 18) \div (-6)$
6. $(23 - 7 + 6) \div 11$
7. $\dfrac{5(12+8)}{2}$
8. $\dfrac{7(9-12)}{3}$
9. $(2^2 + 5)(3^2 + 6)$
10. $(4 + 7^2)(11 + 3^2)$
11. $2(2+3)(3+4) - 5$
12. $5(2-6)(3-5) - 12$
13. $6 \cdot 4 \div 2 - 5 \cdot 3 + 6$
14. $-8 \div 4 \cdot 3 - 7 + 9 \cdot 2$
15. $2^2 + 3^2 - 4^2 + 5^2$
16. $2^3 - 3^3 + 4^3 + 5^3$
17. $\dfrac{(3-12)(-4)}{2}$
18. $\dfrac{(8-5)9}{-3}$
19. $\dfrac{(2 \cdot 5) + 8}{3}$
20. $\dfrac{2 - (4 \cdot 8)}{5}$

B
21. $\dfrac{7 + (3 \cdot 5) - (-8)}{6}$
22. $\dfrac{-8 + (7 \cdot 5) + 12}{13}$
23. $\dfrac{(7+3)(18-11)}{2+3}$
24. $\dfrac{(8-2)(11-7)}{3 + (-5)}$
25. $12\left(7 + (8-3) \cdot 4\right)$
26. $27 + \left(16 - 12(3+7) - 5\right)$
27. $\left(3 \cdot 4 + 7 - (-3)\right) + 64$
28. $32 + 3\left(11 + (4+3) \cdot 2\right)$
29. $5\left((2-1)^2 + (3+2)^2 - (5-1)^2\right)$
30. $3\left(2^3 + (4-2)^3 - (6-2)^3\right)$

In each of the following exercises, name the general property that the statement illustrates.

31. $13 \cdot 1 = 13$
32. $5(7+3) = 5 \cdot 7 + 5 \cdot 3$
33. $3 + (-3) = 0$
34. $-7 + 0 = -7$

In each of the following exercises, name the general property that the statement illustrates.

35. $\left(3 + \frac{12}{5}\right) + 1 = 3 + \left(\frac{12}{5} + 1\right)$
36. $5 \cdot \frac{1}{5} = 1$

37. $12 + 39 = 39 + 12$
38. $113 \cdot 27 = 27 \cdot 113$

39. $0(-1) = 0$
40. $(4 \cdot 2) \cdot 7 = 4 \cdot (2 \cdot 7)$

C 41. $8(9 + 2) = (9 + 2)8$
42. $4 \cdot 7 + 2 = 7 \cdot 4 + 2$

43. $(7 \cdot 1 + 5) + 3 = (7 + 5) + 3$
44. $0(5 + 2) + 3 = 0 + 3$

Study the following pattern. Then find the squares in Exercises 45–50 without actually multiplying the given numbers.

$$5^2 = 25$$
$$1 \cdot 1 + 1 = 1 + 1 = 2 \qquad 15^2 = 225$$
$$2 \cdot 2 + 2 = 4 + 2 = 6 \qquad 25^2 = 625$$
$$3 \cdot 3 + 3 = 9 + 3 = 12 \qquad 35^2 = 1225$$
$$4 \cdot 4 + 4 = 16 + 4 = 20 \qquad 45^2 = 2025$$
$$5 \cdot 5 + 5 = 25 + 5 = 30 \qquad 55^2 = 3025$$

45. 65^2 **46.** 75^2 **47.** 85^2

48. 105^2 **49.** 255^2 **50.** 555^2

SELF-TEST 3

Perform the indicated operations. Give your answer in lowest terms.

1. $\frac{2}{7} + \frac{1}{9}$ **2.** $\frac{3}{10} + \frac{4}{5} - \frac{2}{6}$ *Section 1-7*

3. $\frac{3}{5} - \frac{1}{4}$ **4.** $\frac{1}{12} - \frac{3}{6} + \frac{5}{8}$

5. $\frac{2}{7} \cdot \frac{5}{12}$ **6.** $\frac{3}{8}\left(-\frac{1}{6}\right)$ *Section 1-8*

7. $\frac{2}{3} \div \frac{7}{11}$ **8.** $\frac{1}{5} \div \left(-\frac{3}{2}\right)$

Simplify.

9. $4 \cdot 2 - 30 \div 6$ **10.** $12 \div 2 + 7 - 3 \cdot 5$ *Section 1-9*

11. $(5 + 2)(3 - 7)^2$ **12.** $\frac{(6 - 18)3}{-9}$

Computer Exercises

These exercises are designed for use with a computer that will accept BASIC.

BASIC uses operation symbols similar to those used in arithmetic and algebra:

+	for addition
−	for subtraction
*	for multiplication
/	for division
↑ or ^	for powers
( )	parentheses

The examples below illustrate how these symbols may be used.

Arithmetic expression	Translation into BASIC	Value
3×4	3*4	12
$32 \div 4^2$	32/4^2	2
$(3 \times 4) - (32 \div 4^2)$	(3*4) - (32/4^2)	10
$\dfrac{(3 \times 4) - (32 \div 4^2)}{5}$	((3*4) - (32/4^2))/5	2

Notice that the order of operations in BASIC is the same as the order of operations in algebra (page 24).

The following program will find the value of the last expression above:

```
10  PRINT ((3*4)-(32/4^2))/5
20  END
```

1. Type in and RUN the two-line program above.

2. Translate each of the following expressions into BASIC:
   a. $(15 - 3) \times (15 + 3)$
   b. $15 - 3 \times 15 + 3$
   c. $3 \times 4^2 - 18$
   d. $(3 \times 4)^2 - 18$
   e. $\dfrac{3 \times 8 - 4 \times 5^2}{7 + 3 \times 4}$
   f. $\dfrac{18^6 + (39 - 3) \times 2}{4}$

3. Write and RUN a program to evaluate the expressions in Exercise 2. (You can use the two-line program above, changing only line 10.)

A *loop* can be used to print out a set of numbers. The FOR and NEXT statements in the program on the next page define a loop.

4. Type in and RUN the following program. Use several whole-number values for A and B, with A less than B.

```
10   PRINT "TO PRINT OUT THE MEMBERS OF THE SET OF WHOLE NUMBERS"
20   PRINT "     FROM A TO B, INCLUDING A AND B:"
30   PRINT
40   PRINT "A =";
50   INPUT A
60   PRINT "B =";
70   INPUT B
80   PRINT
90   FOR I=A TO B
100  PRINT I;  ;
110  NEXT I
120  END
```

5. Change lines 20 and 90 in the preceding program as follows and then RUN the revised program for several whole-number values of A and B.

```
20   PRINT "     BETWEEN A AND B:"
90   FOR I=A+1 TO B-1
```

6. Type 10 with nothing after it in order to delete that line from the program of Exercise 5. Do the same for lines 20 and 30. Then change line 90 to:

```
90   FOR I=A TO B STEP 2
```

RUN the resulting program, making A an even number. What numbers are printed out? What happens when B is even? When B is odd?

7. RUN the program of Exercise 6, making A an odd number. What numbers are printed out? What happens when B is even? When B is odd?

8. In the program of Exercises 6 and 7, change line 90 to:

```
90   FOR I=A+1 TO B STEP 2
```

Try odd and even values for A. Describe what happens.

9. Change line 10 of the program of Exercise 5 to:

```
10   PRINT "TO PRINT OUT THE MEMBERS OF THE SET OF INTEGERS"
```

RUN the program with $A = -10$ and $B = 10$. Try other integer values.

10. Change line 90 of the program of Exercise 9 to:

```
90   FOR I=A-1 TO B+1 STEP -1
```

Try several integer values for A and B, with A greater than B.

28   CHAPTER 1

CHAPTER REVIEW

1. Write out using the roster method: {the first five letters of the alphabet}.   Section 1-1
2. Describe by a rule: {5, 6, 7}.
3. Which ones, if any, of the following describe or represent the set that contains no members?
   a. {the whole numbers less than 0}   b. ∅   c. { }

Use the number line below in Exercises 4 and 5.

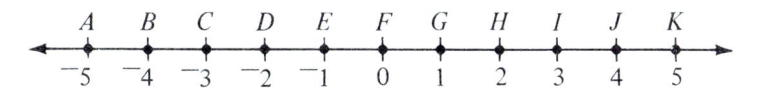

```
      A   B   C   D   E   F   G   H   I   J   K
   ───●───●───●───●───●───●───●───●───●───●───●───▶
      ⁻5  ⁻4  ⁻3  ⁻2  ⁻1   0   1   2   3   4   5
```

4. Name the coordinate of: a. D   b. F   Section 1-2
5. Name the graph of: a. 2   b. ⁻3
6. Graph {⁻3, ⁻2, 0, 1}.

Simplify.

7. $-68 + (-93)$
8. $(-81 + 94) + (-656)$ Section 1-3
9. $29 + (-57) + (-29) + 57$
10. $35 + (-15) + (-52) + 24$
11. $580 - (-141)$
12. $733 - 959$ Section 1-4
13. $-48 - 244$
14. $-1000 - (-850)$
15. $25(-15)$
16. $(-62)(-70)$ Section 1-5
17. $38 \cdot 0 \cdot 105$
18. $(-3)(-16)(-10)$
19. $-437 \div -19$
20. $-2166 \div 57$ Section 1-6
21. $\dfrac{3710}{106}$
22. $\dfrac{3784}{-86}$
23. $0.38 + (-0.29)$
24. $-0.06 - (-3)$ Section 1-7

Simplify. Give your answers in lowest terms.

25. $\dfrac{7}{9} - \dfrac{5}{18}$
26. $-\dfrac{1}{6} + \dfrac{1}{2} + \left(-\dfrac{3}{8}\right)$ Section 1-8
27. $\dfrac{15}{4} \cdot \dfrac{2}{5}$
28. $-\dfrac{3}{10} \cdot 4$ Section 1-9
29. $-\dfrac{1}{8} \div -\dfrac{3}{4}$
30. $\dfrac{12}{7} \div -\dfrac{6}{35}$

NUMBERS AND OPERATIONS

CHAPTER TEST

1. Write out using the roster method: {the even whole numbers between 1 and 5}.
2. Describe by a rule: {March, May}.
3. Graph $\{-5, -\frac{4}{2}, 0, \frac{21}{7}\}$.
4. Graph { the negative integers between −4 and 1 }.
5. What number is the opposite of $-\frac{1}{2}$?
6. What number is the reciprocal of $-\frac{1}{2}$?

Simplify.

7. $(-48) + (-85) + 101 + 32$
8. $72 + (-76) + (-28) + 30$
9. $-102 - (-201)$
10. $319 - 527$
11. $-860 - 445$
12. $(92 - 71) - (-10)$
13. $(24)(-24)$
14. $(-1)(4278)(-1)$
15. $-1332 \div -18$
16. $\frac{9430}{-205}$
17. $3.9 - 0.48$
18. $1.77 + (-2.51)$
19. $(1.6)(-0.03)$
20. $2.28 \div 9.5$

Simplify. Give each answer in lowest terms.

21. $\frac{5}{8} + \frac{1}{6}$
22. $\frac{3}{5} - \frac{14}{15}$
23. $\left(-\frac{2}{9}\right)\left(-\frac{15}{2}\right)$
24. $-\frac{16}{15} \div \frac{4}{3}$

Name the property that each statement illustrates.

25. $-4 + (-6) = -6 + (-4)$
26. $4(-5 + 3) = 4(-5) + 4(3)$
27. $\left(-\frac{7}{8}\right)\left(-\frac{8}{7}\right) = 1$
28. $1025 + 0 = 1025$

Careers — Technical Writer

Every time you take apart an electrical appliance or analyze the structure of a novel you use skills a technical writer needs. Technical writers write instructions for installing and using many types of instruments and machines, ranging from radios to electron microscopes.

The computer revolution added a new dimension to technical writing. What are called hardware technical writers describe the workings of the computers themselves. Software technical writers explain the programs that run the computers. Because so many different types of people use programs, the articles and manuals software technical writers create have to be geared for specific audiences. Some of them are written for computer users, others for marketing purposes, still others for programmers.

Technical writers take very complicated data and break it down into small, clearly explained units. They spend a lot of time asking programmers and engineers how things work. Then they have to translate the answers into language anyone could understand.

If you want to become a technical writer, you should concentrate on learning how to write and on gaining an understanding of the field of your choice. Some industries prefer writers with a degree in science or engineering. Software technical writers, however, often have liberal arts degrees. Courses in technical writing are available at many colleges and universities.

VOCABULARY REVIEW

Be sure that you understand the meaning of these terms:

set, p. 1
well-defined set, p. 1
empty set, p. 2
number line, p. 4
origin, p. 4
integer, p. 4
real number, p. 4
graph of a real number, p. 4
coordinate of a point, p. 5

opposite (additive inverse) of a number, p. 8
divisible, p. 17
rational number, p. 17
lowest terms, p. 18
least common denominator, p. 18
reciprocal (multiplicative inverse) of a rational number, p. 21
simplify an expression, p. 23

MIXED REVIEW

Add or subtract.

1. $(0.2 - 0.35) + 0.73$
2. $-128 + (-54)$
3. $(12 - 73) - (-40)$
4. $-\frac{5}{6} + \frac{3}{10}$
5. $34 - 412$
6. $-0.64 - 0.32$

Multiply or divide.

7. $(-2)(-3)(4)(-5)$
8. $3551 \div (-53)$
9. $-180 \div (-0.24)$
10. $\left(\frac{8}{15}\right)\left(-\frac{9}{14}\right)\left(-\frac{35}{36}\right)$
11. $\frac{6384}{-112}$
12. $(-11)(30)(-14)$

13. Describe the set $\{-1, -2, -3, -4, \ldots\}$ by a rule.
14. Graph the set of negative odd integers.
15. Write $-\frac{30}{84}$ in lowest terms.
16. Does 0 have an opposite? a reciprocal?
17. Write out the set {colors with names beginning with the letter X} using the roster method.

Exercises 18–20 refer to the number line.

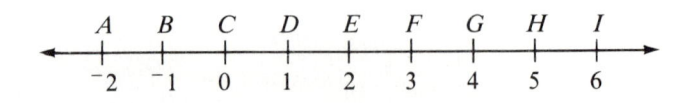

```
      A   B   C   D   E   F   G   H   I
    ──┼───┼───┼───┼───┼───┼───┼───┼───┼──
     -2  -1   0   1   2   3   4   5   6
```

18. The origin is point __?__.
19. The coordinate of B is __?__.
20. The graph of 5 is __?__.

32   CHAPTER 1

Simplify.

21. $(2^2 + 2)(3^3 \div 9)$

22. $\dfrac{-15 + 16 \div 2}{-7 \cdot 2}$

23. $(5^2 - 1) \div (-3) \cdot \dfrac{3}{4}$

24. $-1(2 - (-9 + 4 \cdot 2))$

Name the property that each statement illustrates.

25. $\left(-\dfrac{5}{4}\right)\left(-\dfrac{4}{5}\right) = 1$

26. $0(-9) = 0$

27. $(2 + (-3)) + 7 = 2 + (-3 + 7)$

28. $1.6 + 0 = 1.6$

29. $5(-7 + 21) = 5(-7) + 5(21)$

30. $(-27)(-15) = (-15)(-27)$

Indicate whether or not the set is well-defined.

31. {tall people}

32. {people 6 ft tall or taller}

33. {people 12 ft tall or taller}

34. {integers between 5 and 7}

NUMBERS AND OPERATIONS

Whether you are running for office or just presenting an idea to your friends, a well-constructed argument can help persuade your audience. As a listener, you can evaluate an argument more effectively if you can analyze its logical structure. This chapter will help you understand logical reasoning.

Logic   2

What does it mean to reason accurately? Does a television commercial persuade a viewer by logical arguments to buy an advertised product? Does a politician use valid reasoning to win votes?

An understanding of the principles of valid reasoning will help you draw accurate conclusions, both in mathematics and in everyday life. In this chapter you will study concepts that are related to logical thinking.

Section 2-1 INTRODUCTION TO LOGIC

This section will introduce some basic terms in *logic*, which is the study of valid reasoning:
- truth value
- statement
- variable
- replacement set, or domain
- open sentence
- solution set

In logic, we are interested in the *truth value* of a given sentence; that is, in whether the sentence is true or false. If a sentence can be meaningfully assigned a truth value, it is called a *statement*. The following are statements.

EXAMPLES

5 is between 2 and 7.	T
Abraham Lincoln was a president.	T
$2 + 3 = 3 + 2$	T
4 is greater than 9.	F
The United States has 60 states.	F
$3 + 1 \neq 4$	F

Note that the truth values of the first three statements are true (T) and of the last three are false (F).

The truth values of some sentences cannot be determined without additional information.

EXAMPLES

$x + 2 = 5$ *It* is an ocean.

When a replacement is made for x or for *it*, the sentence becomes a statement; that is, the sentence becomes true or false. If $x = 7$, then the statement $7 + 2 = 5$ is false.

The x and the *it* are called *variables*. Variables are symbols used to represent any element from a specified set called the *replacement set* or *domain* of the variable. A sentence that contains a variable is called an *open sentence*.

EXAMPLE   Open sentence: $x + 2 = 5$ Replacement set: $\{1, 2, 3, \ldots\}$

If $x = 3$, the sentence is true.

If $x = 1$, the sentence is false.

EXAMPLE   Open sentence: It is an ocean.   Replacement set: $\{$bodies of water$\}$

If *it* is replaced by "the Atlantic Ocean," the sentence is true.

If *it* is replaced by "Lake Erie," the sentence is false.

The *solution set* for an open sentence with one variable is the set of all replacements for the variable that make the sentence true. If there is no replacement that makes the sentence true, the solution set is the empty set, denoted by $\emptyset$ or $\{\ \}$.

EXAMPLES

Open sentence: x is an integer between 4 and 8.
Solution set: $\{5, 6, 7\}$
Open sentence: It is a day of the week beginning with S.
Solution set: $\{$Saturday, Sunday$\}$

Some sentences are neither statements nor open sentences, as shown below.

EXAMPLES

Shut the door.   Math is fun.

Please listen carefully.   Squares are nice.

It is not meaningful to assign a truth value to either of the two sentences on the left. Two people might disagree about the truth value of each of the two on the right.

CHAPTER 2

ORAL EXERCISES

Determine whether each of the following is a statement, an open sentence, or neither. Assign a truth value to each statement.

EXAMPLE 1 $3 + 4 = 6$ SOLUTION Statement, F

EXAMPLE 2 $x + 2 = 3$ SOLUTION Open sentence

EXAMPLE 3 Mona Lisa was pretty. SOLUTION Neither

1. $2 + 3 = 5$
2. The weather is cold.
3. $x + 1 = 9$
4. Ferris wheels are fun.
5. The day after Sunday is Monday.
6. $2 + 3 = 3 + 2$
7. What time is it?
8. New York is a state.
9. $x + 4 > 5$
10. $7 + 3 > 12$
11. February has 30 days.
12. She wears glasses.
13. 24 is an even integer.
14. x is an even integer.
15. Wait for me.

WRITTEN EXERCISES

Identify the variable. Then find three values for the variable that make the open sentence true and three values that make it false.

A
1. $x + 2 > 8$
2. It is green.
3. y is a day of the week.
4. z is less than 24.
5. x is divisible by 2.
6. It is made out of wood.
7. It is longer than 12 m.
8. Two times a number is greater than 100.
9. It is a prime number.
10. x is a number that is a multiple of 24.

Classify each of the following as always true, always false, or sometimes true and sometimes false. Assume that the domain of x and of y is the set of positive integers.

11. x is both even and odd.
12. $x + 2 = 7$
13. $x + 1 < x$
14. $x + 4 = x + 5$
15. He has brown eyes.
16. She is less than five feet tall.
17. x is greater than 1.
18. $x + y = y + x$.

Given the replacement set consisting of the counting numbers, $\{1, 2, 3, 4, \ldots\}$, find the solution set for the open sentence.

B
19. $x < 7$
20. $x + 3 = 6$
21. x is between 3 and 9.
22. x is an even number less than 10.
23. x is odd.
24. x is greater than 5 and less than 4.
25. $3 + x = 11$
26. x is divisible by 4.
27. $(5 \cdot 3) + 2 = x$
28. $5(3 + 2) = x$
29. $5 + x = 5$
30. $5 + x = x + 5$

LOGIC   37

Find the truth value of the statement.

31. $(2 \cdot 4) + (6 \cdot 3) = (5 \cdot 4) + (2 \cdot 1)$
32. $(7 + 3)6 = (7 \cdot 6) + (3 \cdot 6)$
33. $4(3 + 5) = (4 \cdot 3) + 5$
34. $(5 \div 5)4 = 4$
35. $12 - 5 = 5 - 12$
36. $(12 \div 4) \div 2 = 12 \div (4 \div 2)$

C 37. Write a sentence that is true for all real numbers.

38. Write a sentence that is false for all real numbers.

39. Is the following sentence true or false? "This sentence is false."

Section 2-2 NEGATION

This section is concerned with:
- notation used in logic
- negation of statements
- the truth table for a negation

You learned in Section 2-1 that the study of logic begins with statements that are either true or false. Each of these statements can be represented by a letter such as $p, q,$ or r.

EXAMPLES

$p:$ Mars has two moons. $r:$ $-1 < 4$
$q:$ $2 + 3 = 5$ $s:$ Today is Monday.

To represent the statement "Mars has two moons," you can simply write p. If you write q, you mean the statement "$2 + 3 = 5$."

Once statements are introduced, new statements can be formed from the original ones. The *negation* of p, formed from the original statement p, is the statement "It is not true that p." This negation of p is written $\sim p$ and is read "not p."

The use of the form "It is not true that p" sometimes results in an awkward sentence. A negation is generally written in a simpler way.

EXAMPLES

$p:$ Paris is the capital of France.
$\sim p:$ It is not true that Paris is the capital of France.
or $\sim p:$ Paris is not the capital of France.

$q:$ Wednesday is the day after Monday.
$\sim q:$ It is not true that Wednesday is the day after Monday.
or $\sim q:$ Wednesday is not the day after Monday.

$r:$ $2 + 3 = 7$
$\sim r:$ It is not true that $2 + 3 = 7$.
or $\sim r:$ $2 + 3 \neq 7$

Notice that the negation of a true statement is false and the negation of a false statement is true. This can be stated concisely in a table called a *truth table*, as shown at the right.

The first column lists the possible truth values, T or F, for any statement *p*. The second column gives the corresponding truth values of ~*p*.

TRUTH TABLE FOR NEGATION

p	~p
T	F
F	T

ORAL EXERCISES

Give the negation.

1. $1 = 2$
2. $5 + 4 = 7$
3. Tom has brown hair.
4. My brother is younger than I am.
5. I don't like homework.
6. A pentagon has four sides.
7. April comes after May.
8. 0 is less than 7.
9. $2 \cdot 5$ is an even integer.
10. 5 is not an integer.

WRITTEN EXERCISES

Write the negation.

A 1. Friday is the last day of the week.
2. Kim has brown hair.
3. There's a pot of gold at the end of the rainbow.
4. The sun is shining.
5. I am not older than Bob.
6. 13 is an unlucky number.

Give the negation. Then find the truth values of the statement and of its negation.

7. $2 + 3 = 5$
8. $3 \cdot 7 = 27$
9. 31 is between 5 and 40.
10. $0 \neq 27$
11. Monkeys do not live in trees.
12. February is a short month.
13. 5 is not less than 2.
14. Every triangle has 3 sides.

Let *p* be the statement "It is raining."

B 15. Write the negation of *p*.
16. Write the negation of ~*p*.
17. In Exercise 16, is your answer the same as the original statement *p*?
18. Do you think it is true that the negation of a negation is the same as the original statement? Is ~(~*p*) the same as *p*?
19. Copy and complete the table. Then compare the first and third columns. What have you shown?

p	~p	~(~p)
T	?	?
F	?	?

LOGIC 39

Give the negation. Then find the truth values of the statement and of its negation.

EXAMPLE   2 is greater than 3.

SOLUTION   2 is not greater than 3.
The original statement is false and the negation is true.

20. 3 is not equal to 4.
21. 6 is less than 9.
22. 5 is not less than 10.
23. 4 is not greater than 7.
24. 20 is not less than 3.
25. 16 is not greater than 10.

Negate each of the following statements in the two ways described. First put "It is not true that" in front of the statement and then reword the statement.

C
26. All apples are red.
27. All people have blond hair.
28. Some winter days are warm.
29. Some students do not work hard.
30. No man is immortal.
31. No other animal is taller than a giraffe.

If p is a statement:

32. What do you think is the negation of "all p"?
33. What do you think is the negation of "some p"?
34. What do you think is the negation of "no p"?

Section 2-3   CONJUNCTION

In Section 2-2 you obtained a new statement from a given one by forming its negation. In the next three sections you will form new statements by *combining* two given ones. This section discusses:
- conjunction
- compound statements
- the truth table for conjunction

The *conjunction* of the statements p and q is the statement formed by inserting the word *and* between the two statements. The symbol $\land$ is used for the word *and*. Thus the conjunction can be written as $p \land q$ and read as "p and q." When two or more statements are combined, as in $p \land q$, the new statement is a *compound statement*.

EXAMPLES

r: $1 + 1 = 2$
s: $3 < 5$
t: A square has three sides.
u: All apples are red.

$r \wedge s$: $1 + 1 = 2$ and $3 < 5$.
$r \wedge t$: $1 + 1 = 2$ and a square has 3 sides.
$u \wedge s$:   All apples are red and $3 < 5$.
$t \wedge u$:   A square has three sides and all apples are red.

What truth value would you assign to each of the four conjunctions above? Notice that statements r and s are true, and that t and u are false.

If you said that only the first conjunction is true, then you agree with the truth values we assign to each combination of truth values. In general,

$p \wedge q$ is true only when p and q are both true.

Otherwise, $p \wedge q$ is false. This is stated in the following truth table.

TRUTH TABLE FOR CONJUNCTION

p	q	$p \wedge q$
T	T	T
T	F	F
F	T	F
F	F	F

The first two columns list all combinations of truth values for p and q, and the third column gives the corresponding truth value for $p \wedge q$. For example, the third row indicates that if p is false and q is true, then $p \wedge q$ is false.

EXAMPLE   Is the statement "4 is an even integer and 10 is an odd integer" true or false?

SOLUTION   Let p be "4 is an even integer" and q be "10 is an odd integer." The given statement in symbolic form is $p \wedge q$. Since p is true and q is false, the second row of the table indicates that $p \wedge q$ is false.

ORAL EXERCISES

Given p, q, r, and s as follows, read each compound statement in words and give its truth value.

     p: 5 is an odd integer. r: All roses are red.
     q: $2 + 4 = 6$ s: 24 is divisible by 3.

1. $p \wedge q$
2. $p \wedge r$
3. $s \wedge q$
4. $s \wedge r$
5. $r \wedge s$
6. $p \wedge s$
7. $q \wedge r$
8. $(p \wedge q) \wedge r$
9. $(s \wedge r) \wedge q$
10. $q \wedge (p \wedge r)$
11. $r \wedge (s \wedge p)$
12. $(r \wedge s) \wedge p$

LOGIC

WRITTEN EXERCISES

For Exercises 1–16 take p, q, r, and s as follows.

p: April has 30 days. r: $4 + 3 = 7$
q: $3 \cdot 4 = 12$ s: "Mathematics" has 10 letters.

Write each compound statement in words and give its truth value.

A
1. $p \wedge q$
2. $p \wedge r$
3. $r \wedge q$
4. $s \wedge q$
5. $p \wedge s$
6. $q \wedge r$
7. $(p \wedge q) \wedge r$
8. $(q \wedge s) \wedge p$
9. $r \wedge (q \wedge p)$
10. $s \wedge (r \wedge q)$

Represent each compound statement in symbolic form and determine its truth value.

EXAMPLE $4 + 3 = 7$ and April has 30 days. **SOLUTION** $r \wedge p$ True

11. $4 + 3 = 7$ and "mathematics" has 10 letters.
12. $3 \cdot 4 = 12$ and April has 30 days.
13. $3 \cdot 4 = 12$ and $4 + 3 = 7$.
14. April has 30 days and "mathematics" has 10 letters.
15. $3 \cdot 4 = 12$ and $4 + 3 = 7$ and April has 30 days.
16. "Mathematics" has 10 letters and $4 + 3 = 7$ and $3 \cdot 4 = 12$.

Determine the truth value.

17. 7 is an odd integer and $5 < 10$.
18. 12 is a multiple of 8 and $25 > 2$.
19. Abraham Lincoln was a president of the United States and Monday is the last day of the week.
20. $4 + 7 < 11$ and $8 \cdot 7 = 65$.

Given p, q, r, and s as follows, write each compound statement in words and give its truth value.

p: The moon is always full. r: Motorcycles have four wheels.
q: The Yankees are a baseball team. s: $2 + 2 = 2 \cdot 2$

EXAMPLE $\sim p \wedge \sim q$

SOLUTION The moon is not always full and the Yankees are not a baseball team. False (Since p is false, q is true, $\sim p$ is true, and $\sim q$ is false.)

B
21. $p \wedge \sim q$
22. $\sim r \wedge s$
23. $(p \wedge s) \wedge \sim r$
24. $(q \wedge \sim p) \wedge r$
25. $\sim q \wedge \sim p$
26. $\sim s \wedge \sim r$

Given that p is true, q is false, and r is true, determine the truth value.

27. $p \wedge q$
28. $q \wedge r$
29. $\sim(p \wedge r)$
30. $[\sim(p \wedge q)] \wedge r$
31. $r \wedge (q \wedge p)$
32. $\sim p \wedge q$
33. $r \wedge \sim q$
34. $(\sim r \wedge q) \wedge p$

C 35. Suppose $p \wedge \sim q$ is true. What can you say about the truth values of p and q?

36. Suppose $(\sim p \wedge q) \wedge \sim r$ is true. What can you say about the truth values of p, q, and r?

37. Suppose $\sim(p \wedge q)$ is false. What can you say about the truth values of p and q?

Section 2-4 DISJUNCTION

This section introduces:
- disjunction
- the truth table for disjunction

The *disjunction* of the statements p and q is the compound statement formed by inserting the word *or* between the two statements. The symbol $\vee$ is used for the word *or*. Thus the disjunction can be written as $p \vee q$ and read as "p or q."

EXAMPLES

r: $1 + 1 = 2$
s: $3 < 5$
t: A square has 3 sides.
u: All apples are red.

$r \vee s$: $1 + 1 = 2$ or $3 < 5$.
$r \vee t$: $1 + 1 = 2$ or a square has 3 sides.
$u \vee s$:   All apples are red or $3 < 5$.
$t \vee u$:   A square has 3 sides or all apples are red.

In the examples above, r and s are true, but t and u are false. It seems reasonable to accept the first three disjunctions as true and the last one as false. In general,

$p \vee q$ is true except when p and q are both false.

TRUTH TABLE FOR DISJUNCTION

p	q	$p \vee q$
T	T	T
T	F	T
F	T	T
F	F	F

LOGIC   43

EXAMPLES

p: 9 is divisible by 3. (T)
q: 12 is an even integer. (T)
r: $7 + 5 = 11$ (F)
s: 8 is between 11 and 17. (F)

$p \lor q$ is true. (Both p and q are true.)
$q \lor r$ is true. (q is true.)
$s \lor p$ is true. (p is true.)
$r \lor s$ is false. (Both r and s are false.)

EXAMPLE Is the statement "0 is a whole number or 0 is a counting number" true or false?

SOLUTION Let p be "0 is a whole number" and q be "0 is a counting number." The given statement in symbolic form is $p \lor q$. Since p is true and q is false, the second line of the truth table indicates that the original statement $p \lor q$ is true.

ORAL EXERCISES

Given p, q, r, and s as follows, read each compound statement in words and give its truth value.

p: Today is Saturday. r: Lizards are insects.
q: $8 + 5 = 13$ s: Bicycles have five wheels.

1. $p \lor q$
2. $p \lor r$
3. $r \lor q$
4. $s \lor q$
5. $s \lor r$
6. $r \lor s$
7. $p \lor s$
8. $(p \lor q) \lor r$
9. $(s \lor r) \lor q$
10. $q \lor (p \lor r)$
11. $r \lor (s \lor p)$
12. $(r \lor s) \lor p$

WRITTEN EXERCISES

For Exercises 1–16 take p, q, r, and s as follows.

p: $12 < 9$ r: 7 is between 2 and 5.
q: A triangle has three sides. s: 32 is an odd integer.

Write each compound statement in words and give its truth value.

A
1. $p \lor q$
2. $p \lor r$
3. $r \lor q$
4. $s \lor q$
5. $p \lor s$
6. $q \lor r$
7. $(p \lor q) \lor r$
8. $(q \lor s) \lor p$
9. $r \lor (q \lor p)$
10. $s \lor (q \lor r)$

Represent each compound statement in symbolic form and determine its truth value.

11. $12 < 9$ or 7 is between 2 and 5.
12. A triangle has 3 sides or 32 is an odd integer.

13. 7 is between 2 and 5 or 32 is an odd integer.
14. $12 < 9$ or a triangle has 3 sides.
15. 32 is an odd integer or $12 < 9$.
16. A triangle has 3 sides or 7 is between 2 and 5.

Determine the truth value.

17. 20 is divisible by 3 or February is the second month of the year.
18. 11 is an integer or 11 is a prime number.
19. One meter is 100 cm or a meter is longer than a kilometer.
20. $24 > 12 + 8$ or 24 is a multiple of 6.

Given p, q, r, and s as follows, write each compound statement in words and give its truth value.

p: $2 - 5 = 3$ r: Earth is the planet closest to the sun.
q: Fires are hot. s: Potholes are good for your car.

EXAMPLE $\sim p \vee q$
SOLUTION $2 - 5 \neq 3$ or fires are hot.
           True (Both $\sim p$ and q are true.)

B 21. $p \vee \sim q$ 22. $\sim r \vee s$ 23. $(p \vee s) \vee \sim r$
24. $(q \vee \sim p) \vee r$ 25. $\sim q \vee \sim p$ 26. $\sim s \vee \sim r$

Given that p is true, q is false, and r is true, determine the truth value of the following.

27. $p \vee q$ 28. $q \vee r$ 29. $\sim(p \vee r)$ 30. $r \vee (q \vee p)$
31. $\sim p \vee r$ 32. $(r \vee \sim q) \vee p$ 33. $(r \vee q) \vee \sim p$ 34. $(\sim r \vee q) \vee p$

C 35. Suppose $p \vee \sim q$ is false. What can you say about the truth values of p and q?
36. Suppose $(\sim p \vee q) \vee r$ is false. What can you say about the truth values of p, q, and r?
37. Suppose $(p \vee q) \wedge r$ is true. What can you say about the truth values of p, q, and r?
38. Negate the statement "Roses are red or violets are blue" without saying, "It is not true that. . . ."
39. Negate $p \vee q$ without using parentheses. That is, find an expression in terms of p and q that is the same as $\sim(p \vee q)$. Try some examples.
40. Negate the statement "Roses are red and violets are blue" without saying, "It is not true that. . . ."
41. Negate $p \wedge q$ without using parentheses. That is, find an expression in terms of p and q that is the same as $\sim(p \wedge q)$. Try some examples.

LOGIC

SELF-TEST 1

1. If the domain of x is {whole numbers}, find two values for x that make the statement $x > 11$ true and two values that make it false. *Section 2-1*

2. Taking the replacement set to be {counting numbers}, find the solution set for "24 is divisible by x."

3. Write the negation of the statement "Wednesday is the day in the middle of the week." *Section 2-2*

4. Write the negation of the statement "Elephants are not small."

For Exercises 5 and 7 take p to be "$5 > 7$" and q to be "$5 + 4 = 9$."

5. Write $p \wedge q$ in words and give its truth value. *Section 2-3*

6. Let r represent "12 is an integer" and s represent "A zip code has 5 digits." Write "12 is an integer and a zip code has 5 digits" in symbolic form and give its truth value.

7. Write $p \vee q$ in words and give its truth value. *Section 2-4*

8. Let r represent "A triangle has 5 sides" and s represent "A potato is a fruit." Write "A triangle has 5 sides or a potato is a fruit" in symbolic form and give its truth value.

Section 2-5 CONDITIONAL STATEMENTS

Concepts to be considered in this section are:
- conditional statements (implications)
- the truth table for the conditional

A *conditional* is a compound statement of the form "if p, then q," where p and q are given statements. It is written symbolically as $p \rightarrow q$ and is read "if p, then q" or "p implies q." Conditionals are also called *implications*. The first statement, p, is called the *hypothesis* or *antecedent*. The second statement, q, is called the *conclusion* or *consequent*.

EXAMPLE p: Today is Monday. q: Tomorrow is Tuesday.
$p \rightarrow q$: If today is Monday, then tomorrow is Tuesday.
$q \rightarrow p$: If tomorrow is Tuesday, then today is Monday.

It is important to note that the *if* goes in front of the statement before the arrow and the *then* goes before the statement that follows the arrow. However, when an implication is written in words, the hypothesis does not have to be written before the conclusion. Other ways to word $p \rightarrow q$ are "p only if q" and "q if p." In the example above, the conditional $p \rightarrow q$ could be written as

"Today is Monday only if tomorrow is Tuesday."

or as

"Tomorrow is Tuesday if today is Monday."

To determine the truth values of a conditional, let us consider an example.

EXAMPLE Suppose Rita promises you that if she gets her homework done, then she'll go with you to the football game. Taking p as "Rita gets her homework done" and q as "Rita goes with you to the football game," the conditional statement is written $p \rightarrow q$. Think of $p \rightarrow q$ as a promise and consider the four cases when p and q are true and false.

Case 1: p and q are both true. Rita gets her homework done and goes with you to the football game. She has kept her promise, and her statement is true.

Case 2: p is true and q is false. Rita gets her homework done, but doesn't go with you to the game. Here her statement is false.

Case 3: p is false and q is true. Rita doesn't get her homework done, but she still goes to the game with you. Her statement was not false, since her only commitment concerned what she would do *if* she got her homework done. She did not break her promise, and the conditional statement is true.

Case 4: p is false and q is false. Rita doesn't get her homework done and doesn't go to the game with you. As in Case 3, she has not broken her promise to you, and the conditional statement is again true.

Therefore, the only time a conditional statement is false is when the hypothesis is true and the conclusion is false. This is represented in the following truth table.

TRUTH TABLE FOR THE CONDITIONAL

p	q	$p \rightarrow q$
T	T	T
T	F	F
F	T	T
F	F	T

EXAMPLE
p: $4 + 5 > 7$ (T)
q: 21 is divisible by 7.   (T)
r: A circle is a square.   (F)
s: A pentagon has 6 sides.   (F)

$p \rightarrow q$: If $4 + 5 > 7$, then 21 is divisible by 7.
(True, since p and q are both true.)

$q \rightarrow s$: If 21 is divisible by 7, then a pentagon has 6 sides.
(False, since q is true and s is false.)

$r \rightarrow p$: If a circle is a square, then $4 + 5 > 7$.
(True, since r is false and p is true.)

$s \rightarrow r$: If a pentagon has 6 sides, then a circle is a square.
(True, since s and r are both false.)

Open sentences become statements when their variables are replaced by specific values from their domains. Since these sentences are true or false for specific replacements, it is possible to extend our discussion of truth values to conditionals involving open sentences. Except where otherwise indicated, we will assume that the domain of each variable is the set of real numbers.

EXAMPLE r: $x + 3 = 12$ Domain: $\{0, 1, \ldots, 9, 10\}$
   s: $x = 9$
   $r \rightarrow s$:   If $x + 3 = 12$, then $x = 9$.
   $s \rightarrow r$:   If $x = 9$, then $x + 3 = 12$.

r and s are both true when x is replaced by 9 and are both false for any other replacement from the domain of x. The two conditionals $r \rightarrow s$ and $s \rightarrow r$ are always true.

ORAL EXERCISES

Taking p, q, r, and s as follows, read the conditional in words, give its truth value, and identify the hypothesis and conclusion.

   p: 7 is less than 11. r: "Friday" has 5 letters.
   q: All circles are round. s: Earth has 2 moons.

1. $p \rightarrow q$ 2. $p \rightarrow r$ 3. $p \rightarrow s$ 4. $q \rightarrow r$
5. $q \rightarrow s$ 6. $r \rightarrow p$ 7. $r \rightarrow s$ 8. $q \rightarrow p$

WRITTEN EXERCISES

State the hypothesis and the conclusion of each statement.

A 1. If k is an even integer, then $k + 1$ is an odd integer.
2. If the sky is blue, then the grass is green.
3. If $z = -5$, then $2z = -10$.
4. If j is a negative number, then $-j$ is a positive number.
5. Jack is Karen's uncle if Karen is Jack's niece.
6. Six tickets cost $18 if one ticket costs $3.

If p, q, r, and s represent the statements indicated, translate the given conditional into words.

7. p: $3y - 2 = 7$ q: $3y = 9$
   $p \rightarrow q$

8. p: The figure is a rectangle. q: The figure is a square.
   $q \rightarrow p$

9. r: The moon is made of green cheese. s: Mice will nibble.
   $r \rightarrow s$

10. r: 2 is greater than 1. s: 2 is greater than 0.
     $s \to r$

For Exercises 11-24 take $p, q, r,$ and s as follows.

     p: A rectangle has 4 sides. r: 26 is between 20 and 24.
     q: $2 + 3 = 3 + 2$ s: 10 is not an integer.

Write each compound statement in words and give its truth value.

11. $p \to q$ **12.** $q \to p$ **13.** $q \to s$ **14.** $s \to q$

15. $r \to s$ **16.** $r \to q$ **17.** $s \to p$ **18.** $q \to r$

Write in symbolic form and determine the truth value.

19. If 10 is not an integer, then $2 + 3 = 3 + 2$.

20. If $2 + 3 = 3 + 2$, then a rectangle has 4 sides.

21. 26 is between 20 and 24 only if $2 + 3 = 3 + 2$.

22. $2 + 3 = 3 + 2$ only if a rectangle has 4 sides.

23. 10 is not an integer if 26 is between 20 and 24.

24. 26 is between 20 and 24 if a rectangle has 4 sides.

If $p, q, r,$ and s represent the statements indicated, translate the given conditional into symbolic form.

25. p: -5 is less than 0. q: 0 is greater than -5.
    If -5 is less than 0, then 0 is greater than -5.

26. p: Barbara is Jack's niece. q: Jack is Barbara's uncle.
    If Jack is Barbara's uncle, then Barbara is Jack's niece.

27. r: $-2x = 4$ s: $x = -2$
    $-2x = 4$ only if $x = -2$.

28. r: Eric works hard. s: Eric gets a raise.
    Eric gets a raise only if Eric works hard.

Given $p, q, r,$ and s as follows, write each conditional in words and give its truth value.

p: South America is a continent. r: Cows can fly.
q: $3(6 - 1) = (3 \cdot 6) - (3 \cdot 1)$ s: -3 is not less than 0.

B   **29.** $p \to \sim q$ **30.** $r \to \sim s$ **31.** $\sim r \to \sim q$ **32.** $\sim q \to \sim p$

33. $(p \land q) \to r$ **34.** $(r \lor q) \to s$ **35.** $(q \lor r) \to p$ **36.** $(q \land r) \to p$

For what values of the domain is the conditional true?

37. If $x = 9$, then $x + 2 = 11$.   Domain: $\{0, 9\}$

38. If $y = 5$, then $y - 8 = -3$.   Domain: $\{-5, 5\}$

39. If $x = 7$, then $x - 7 = -7$.   Domain: $\{-7, 0, 7\}$

For what values of the domain is the conditional true?

40. If $y = -3$, then $-3y = 9$. Domain: $\{-3, 0, 3\}$
41. If $x = -5$, then $-5x = -25$. Domain: $\{-5, 0, 5\}$
42. If $y = 0$, then $2y = 2$. Domain: $\{0, 1, 2\}$

Write each statement in if-then form.

EXAMPLE All triangles are polygons.

SOLUTION If a figure is a triangle, then it is a polygon.

43. All students like algebra.

44. All squares have 4 sides.

45. Numbers equal to the same number are equal to each other.

46. Too many cooks spoil the broth.

Rewrite each conditional as a disjunction.

EXAMPLE If you do not study, then you will not pass.

SOLUTION Either you study or you will not pass.

C 47. If a rhombus is not a rectangle, then it is not a square.

48. If it's Friday, then there's no school tomorrow.

49. If these shoes are too small, then my feet will hurt.

50. If Peter Piper picked a peck of pickled peppers, then Terry and Tommy twisted their tongues.

51. Write $p \to q$ as a disjunction.

52. Use your answer to Exercise 51 above and to Exercise 39 of Section 2-4 to write $\sim(p \to q)$ as a conjunction.

Section 2-6 CONSTRUCTING TRUTH TABLES

To determine the truth value of a complicated statement such as $(\sim p \vee q) \to (p \wedge q)$, an organized procedure would be helpful. This section shows you how to construct a truth table for any compound statement.

The basic truth tables you have already studied are shown below.

p	$\sim p$
T	F
F	T

p	q	$p \wedge q$	$p \vee q$	$p \to q$
T	T	T	T	T
T	F	F	T	F
F	T	F	T	T
F	F	F	F	T

Notice that we have listed conjunction, disjunction, and the conditional in one truth table. For example, to find the truth values for disjunction, refer to the fourth column of the truth table on the right.

The first step in constructing a truth table is naming the columns. Suppose, for example, that you want a truth table for $p \wedge \sim q$. There are two different letters, p and q, so there will be four lines of T's and F's. Before you can get a column for $p \wedge \sim q$, you need one for p, one for q, and one for $\sim q$. The headings are as follows:

p	q	$\sim q$	$p \wedge \sim q$

Next you list the four possible combinations of truth values for p and q.

p	q	$\sim q$	$p \wedge \sim q$
T	T		
T	F		
F	T		
F	F		

To determine the truth values for $\sim q$, remember that $\sim q$ is false when q is true and $\sim q$ is true when q is false. List the truth values for $\sim q$ in the third column.

p	q	$\sim q$	$p \wedge \sim q$
T	T	F	
T	F	T	
F	T	F	
F	F	T	

To find the truth values for the last column, take the conjunction of the first and third columns. Here you are connecting using $\wedge$, so refer to that table if necessary. You may recall that the only time *and* is true is when both statements are true.

p	q	$\sim q$	$p \wedge \sim q$
T	T	F	F
T	F	T	T
F	T	F	F
F	F	T	F

You now have the truth values of $p \wedge \sim q$ for the various truth values of p and q. For example, the third line states that if p is false and q is true, then $p \wedge \sim q$ is false.

LOGIC   51

EXAMPLE Construct a truth table for $\sim(p \to q)$.

SOLUTION First set up columns for p, q, $p \to q$, and $\sim(p \to q)$.

p	q	$p \to q$	$\sim(p \to q)$

Now put in the truth values, referring to the basic tables if necessary.

p	q	$p \to q$	$\sim(p \to q)$
T	T	T	F
T	F	F	T
F	T	T	F
F	F	T	F

If, for example, p is true and q is false, the second line tells you that $\sim(p \to q)$ is true.

EXAMPLE Construct a truth table for $(\sim p \vee q) \to (p \wedge q)$.

SOLUTION Naming the columns is like working out an expression with numbers—always perform the operations inside parentheses first. The column headings will be:

p	q	$\sim p$	$\sim p \vee q$	$p \wedge q$	$(\sim p \vee q) \to (p \wedge q)$

To substitute truth values, notice the following:
The first two columns list all possible combinations of truth values for p and q.
The third column is the negation of the first.
The fourth column is the disjunction of the third and second columns.
The fifth column is the conjunction of the first and second columns.
The sixth column is the implication formed by the fourth and fifth columns.
The final table is:

p	q	$\sim p$	$\sim p \vee q$	$p \wedge q$	$(\sim p \vee q) \to (p \wedge q)$
T	T	F	T	T	T
T	F	F	F	F	T
F	T	T	T	F	F
F	F	T	T	F	F

Notice in the example above that p and $(\sim p \vee q) \to (p \wedge q)$ have the same truth values.

ORAL EXERCISES

If p and q are each true, and r and s are each false, give the truth value of the following.

1. $p \land q$
2. $p \land r$
3. $q \land p$
4. $s \lor r$
5. $q \lor s$
6. $p \rightarrow r$
7. $q \rightarrow r$
8. $r \rightarrow p$
9. $r \land s$
10. $s \rightarrow q$
11. $r \land p$
12. $q \lor p$
13. $q \rightarrow p$
14. $s \rightarrow r$
15. $r \lor q$
16. $p \rightarrow s$

WRITTEN EXERCISES

Construct a truth table for each of the following.

A
1. $\sim(\sim p)$
2. $p \lor \sim p$
3. $\sim p \land p$
4. $q \rightarrow p$
5. $\sim p \land q$
6. $p \lor \sim q$
7. $\sim p \land \sim q$
8. $\sim(p \land q)$
9. $\sim q \rightarrow p$
10. $\sim q \lor p$
11. $\sim(\sim p \lor q)$
12. $\sim(p \rightarrow \sim q)$
13. $\sim p \lor \sim q$
14. $q \lor \sim(\sim p)$

B
15. $(p \land q) \rightarrow q$
16. $p \land (q \lor p)$
17. $(p \lor \sim q) \rightarrow \sim p$
18. $p \rightarrow (q \lor \sim p)$
19. $(p \lor q) \rightarrow (p \lor q)$
20. $\sim(p \lor q) \rightarrow \sim q$
21. $(p \rightarrow q) \rightarrow (\sim p \lor q)$
22. $(q \lor p) \rightarrow (\sim q \land \sim p)$
23. $(\sim p \lor \sim q) \rightarrow \sim(p \land q)$
24. $(\sim q \rightarrow \sim p) \rightarrow (p \rightarrow q)$

C
25. $[(p \rightarrow q) \land p] \rightarrow q$
26. $[(p \rightarrow q) \land \sim q] \rightarrow \sim p$

In Exercises 27-30, each truth table should have 8 lines.

27. $p \land (q \lor r)$
28. $(p \land q) \lor r$
29. $(p \lor q) \rightarrow r$
30. $(p \rightarrow q) \land \sim r$

Section 2-7 RELATED CONDITIONAL STATEMENTS

Three other conditionals that are closely associated with a given conditional statement will be considered in this section. They are:
- converse
- inverse
- contrapositive

We will also study these three topics:
- tautology
- contradiction
- logical equivalence

LOGIC 53

Suppose you are given a conditional statement $p \to q$. The *converse* of this statement is the conditional $q \to p$. The *inverse* of the statement is $\sim p \to \sim q$, and the *contrapositive* is $\sim q \to \sim p$.

EXAMPLE

p: It is snowing.
q: It is cold.

Statement: $p \to q$: If it is snowing, then it is cold.
Converse: $q \to p$: If it is cold, then it is snowing.
Inverse: $\sim p \to \sim q$: If it is not snowing, then it is not cold.
Contrapositive: $\sim q \to \sim p$: If it is not cold, then it is not snowing.

EXAMPLE

r: $x + 3 = 12$
s: $x = 9$

Statement: $r \to s$: If $x + 3 = 12$, then $x = 9$.
Converse: $s \to r$: If $x = 9$, then $x + 3 = 12$.
Inverse: $\sim r \to \sim s$: If $x + 3 \neq 12$, then $x \neq 9$.
Contrapositive: $\sim s \to \sim r$: If $x \neq 9$, then $x + 3 \neq 12$.

The following truth table displays the truth values for a conditional statement and its related conditionals.

p	q	$\sim p$	$\sim q$	Statement $p \to q$	Converse $q \to p$	Inverse $\sim p \to \sim q$	Contrapositive $\sim q \to \sim p$
T	T	F	F	T	T	T	T
T	F	F	T	F	T	T	F
F	T	T	F	T	F	F	T
F	F	T	T	T	T	T	T

Two statements are *logically equivalent* if they have the same truth values. This means that you could replace one by the other without changing the meaning. Notice from the table above that a conditional statement and its contrapositive are logically equivalent. Also, the converse and the inverse are logically equivalent.

EXAMPLE Form a statement that is logically equivalent to "If it is sunny, then I will go swimming."

SOLUTION Take p as "it is sunny" and q as "I will go swimming."
The given statement is written as $p \to q$. Since the contrapositive $\sim q \to \sim p$ is logically equivalent to $p \to q$, the answer is "If I don't go swimming, then it's not sunny."

EXAMPLE Use a truth table to show that for any statement p, the two statements p and $\sim(\sim p)$ are logically equivalent.

SOLUTION

p	$\sim p$	$\sim(\sim p)$
T	F	T
F	T	F

Since p and $\sim(\sim p)$ are logically equivalent, the conditional $p \rightarrow \sim(\sim p)$ is always true. A compound statement such as $p \rightarrow \sim(\sim p)$ that is true no matter what truth value is assigned to p is called a *tautology*. If a compound statement is always false, it is called a *contradiction*.

EXAMPLES

$p \vee \sim p$ is a tautology since its truth table is:

p	$\sim p$	$p \vee \sim p$
T	F	T
F	T	T

$p \wedge \sim p$ is a contradiction since its truth table is:

p	$\sim p$	$p \wedge \sim p$
T	F	F
F	T	F

You might expect the two results in the examples above without using a truth table. The first states that either a statement or its negation is true. The second states that a statement *and* its negation are never both true.

EXAMPLE Construct a truth table and use it to determine whether $(p \wedge q) \rightarrow (p \vee q)$ is a tautology, a contradiction, or neither.

SOLUTION

p	q	$p \wedge q$	$p \vee q$	$(p \wedge q) \rightarrow (p \vee q)$
T	T	T	T	T
T	F	F	T	T
F	T	F	T	T
F	F	F	F	T

The conditional $(p \wedge q) \rightarrow (p \vee q)$ is a tautology.

LOGIC

ORAL EXERCISES

State in the if-then form the converse, inverse, and contrapositive of each of the following conditionals.

1. $p \rightarrow q$
2. $s \rightarrow r$
3. $q \rightarrow w$
4. $t \rightarrow p$
5. $\sim p \rightarrow q$
6. $r \rightarrow \sim s$
7. $\sim t \rightarrow u$
8. $\sim u \rightarrow \sim v$

WRITTEN EXERCISES

Write in symbols the converse, inverse, and contrapositive of the following.

A 1. $r \rightarrow s$
2. $q \rightarrow \sim p$
3. $\sim p \rightarrow r$
4. $\sim t \rightarrow \sim u$

For each of the following conditionals, write the converse, inverse, and contrapositive.

5. If $x = 1$, then $x + 2 = 3$.
6. If $x = 0$, then $x < 10$.
7. If we can score a run, then we will win the game.
8. If Jane is tall, then she can play basketball.
9. If $x > 2$, then $x > 4$.
10. If you use Brand X soap, then you will have more suds.

Write a statement that is logically equivalent to each of the following.

B 11. If a figure has 3 sides, then it's a triangle.
12. If today is Saturday, then there is no school.
13. If Bill passed his test, then he is happy.
14. If it's warm out, then I am going to the beach.

Construct a truth table for each of the following and determine whether the statement is a tautology, a contradiction, or neither.

15. $\sim(\sim p) \rightarrow p$
16. $\sim s \wedge s$
17. $(p \wedge q) \rightarrow p$
18. $(p \vee q) \rightarrow (p \wedge q)$
19. $(p \vee q) \rightarrow (\sim p \wedge \sim q)$
20. $\sim(p \wedge q) \rightarrow (\sim p \vee \sim q)$

Use a truth table to show that the two statements are logically equivalent.

21. $p \rightarrow q$ and $\sim p \vee q$
22. $\sim(\sim p \vee q)$ and $p \wedge \sim q$

Give an example of a true conditional statement whose:

23. converse is false.
24. converse is true.
25. inverse is false.
26. inverse is true.

C 27. What is the converse of the inverse of a conditional statement?
28. What is the contrapositive of the converse of a conditional statement?
29. Show that $[(p \vee q) \wedge \sim p] \rightarrow q$ is a tautology.
30. Show that $[(p \rightarrow q) \wedge (q \rightarrow r)] \rightarrow (p \rightarrow r)$ is a tautology.

Section 2-8 BICONDITIONAL STATEMENTS

The *biconditional* of two statements p and q is the statement "p if and only if q," and is written $p \leftrightarrow q$. The biconditional is a combination of the two conditionals "if p, then q" and "if q, then p." That is, $p \leftrightarrow q$ is the same as $(p \rightarrow q) \wedge (q \rightarrow p)$. Notice that $q \leftrightarrow p$ and $p \leftrightarrow q$ are exactly the same.

EXAMPLE p: Yesterday was Thursday. q: Today is Friday.

$p \leftrightarrow q$: Yesterday was Thursday if and only if today is Friday.

EXAMPLE r: $x + 5 = 12$ s: $x = 7$

$r \leftrightarrow s$: $x + 5 = 12$ if and only if $x = 7$.

Since $p \leftrightarrow q$ is the same as $(p \rightarrow q) \wedge (q \rightarrow p)$, the two statements must be logically equivalent and hence have the same truth values. To determine the truth values of $p \leftrightarrow q$, look at the truth table for $(p \rightarrow q) \wedge (q \rightarrow p)$.

p	q	$p \rightarrow q$	$q \rightarrow p$	$(p \rightarrow q) \wedge (q \rightarrow p)$
T	T	T	T	T
T	F	F	T	F
F	T	T	F	F
F	F	T	T	T

The last column gives us the truth values for $p \leftrightarrow q$. The biconditional is true when p and q are both true or both false. Otherwise it is false. The truth table is shown below.

p	q	$p \leftrightarrow q$
T	T	T
T	F	F
F	T	F
F	F	T

EXAMPLE p: "Friday" has 5 letters.   (F)
q: A rectangle has 4 sides.   (T)
r: All apples are red.   (F)
s: Some cars are red.   (T)

$p \leftrightarrow q$ is false.   (p is false and q is true.)
$q \leftrightarrow r$ is false.   (q is true and r is false.)
$p \leftrightarrow r$ is true.   (p and r are both false.)
$q \leftrightarrow s$ is true.   (q and s are both true.)

LOGIC

We know that two statements are logically equivalent if and only if they have the same truth values, that is, when they are both true or both false at the same time. If we connect two logically equivalent statements with $\leftrightarrow$, the new statement is a tautology, since the statements on either side are either both true or both false. This gives you another method for determining whether two statements are logically equivalent. Connect the two statements with $\leftrightarrow$ and see if the new statement is always true.

EXAMPLE Show that $p \rightarrow q$ and $\sim p \vee q$ are logically equivalent.

SOLUTION Construct a truth table for the biconditional $(p \rightarrow q) \leftrightarrow (\sim p \vee q)$.

p	q	$p \rightarrow q$	$\sim p$	$\sim p \vee q$	$(p \rightarrow q) \leftrightarrow (\sim p \vee q)$
T	T	T	F	T	T
T	F	F	F	F	T
F	T	T	T	T	T
F	F	T	T	T	T

Since the biconditional is always true, the statements $p \rightarrow q$ and $\sim p \vee q$ are logically equivalent.

ORAL EXERCISES

Given p, q, r, and s as follows, read each statement in words and give its truth value.

p: a is a vowel.
q: Some fish swim in the ocean.
r: $(8)(7) = 48$
s: 36 is not a positive integer.

1. $p \leftrightarrow q$
2. $p \leftrightarrow r$
3. $p \leftrightarrow s$
4. $q \leftrightarrow r$
5. $q \leftrightarrow s$
6. $r \leftrightarrow s$

WRITTEN EXERCISES

If p, q, r, and s represent the statements indicated, translate the given biconditional into symbolic form.

A 1. p: Dave is Debby's brother. q: Debby is Dave's sister.
Dave is Debby's brother if and only if Debby is Dave's sister.

2. p: 12 is a multiple of x. q: 12 is divisible by x.
12 is a multiple of x if and only if 12 is divisible by x.

3. r: $x = 10$ s: $x - 3 = 7$
$x = 10$ if and only if $x - 3 = 7$.

4. r: n is an even integer. s: $n + 2$ is an even integer.
n is an even integer if and only if $n + 2$ is an even integer.

In Exercises 5-16 use *p, q, r,* and *s* as follows:

p: 8 is between 5 and 10. *r*: 12 > 7
q: A square has 5 sides. *s*: A car has 6 wheels.

Write the statement in words and give its truth value.

5. $p \leftrightarrow q$
6. $p \leftrightarrow r$
7. $p \leftrightarrow s$
8. $q \leftrightarrow r$
9. $q \leftrightarrow s$
10. $r \leftrightarrow s$

Write in symbolic form and give the truth value.

11. A square has 5 sides if and only if 12 > 7.
12. A car has 6 wheels if and only if 8 is between 5 and 10.
13. 12 > 7 if and only if 8 is between 5 and 10.
14. 8 is between 5 and 10 if and only if 12 > 7.
15. 12 > 7 if and only if a square has 5 sides.
16. 8 is between 5 and 10 if and only if a square has 5 sides.

In Exercises 17-22 use *p, q, r,* and *s* as follows. Write the statement in words and give its truth value.

p: London is the capital of England. *r*: $\frac{2}{3}$ is a rational number.
q: Basketballs are not bigger than baseballs. *s*: $3(6+2) = (3 \cdot 6) + 2$

B 17. $\sim p \leftrightarrow q$
18. $\sim q \leftrightarrow s$
19. $r \leftrightarrow \sim s$
20. $\sim s \leftrightarrow \sim p$
21. $q \leftrightarrow \sim r$
22. $s \leftrightarrow \sim p$

Using a truth table, determine if the following are logically equivalent.

23. $p \lor q$ and $\sim p \rightarrow q$
24. $\sim(q \lor p)$ and $\sim p \lor \sim q$
25. $\sim p \rightarrow q$ and $\sim q \rightarrow p$
26. $\sim(p \land \sim q)$ and $q \lor \sim p$
27. $q \rightarrow \sim p$ and $\sim q \lor p$
28. $p \land q$ and $\sim(\sim p \lor \sim q)$

Translate the statement into symbolic form using the letters *p, q, r,* and *s*. (Begin by telling what statement each letter represents.)

EXAMPLE $|x| = 2$ if and only if $x = 2$ or $x = -2$.

SOLUTION Let *p* be $|x| = 2$, *q* be $x = 2$, and *r* be $x = -2$.
The symbolic form is $p \leftrightarrow (q \lor r)$.

C 29. 4 > 3 and $x + 2 = 7$ if and only if today is Tuesday and it's hot.

30. If I either do my homework or study my book, then I'll pass the test.
31. It's raining or it's snowing if and only if I don't go.
32. If I'm late, then I'll get into trouble, and if I'm not late, then I won't get into trouble.

LOGIC 59

SELF-TEST 2

1. Let p represent "12 is even" and q represent "36 is odd." Write $p \rightarrow q$ in words and give its truth value.   Section 2-5

2. Let r represent "$7 < 3$" and s represent "$5 = 5$." Write in symbolic form and give the truth value: "If $7 < 3$, then $5 = 5$."

3. Complete: If r is false, then $\sim r$ is __?__ and $r \vee \sim r$ is __?__.   Section 2-6

4. Construct a truth table for $(p \wedge q) \rightarrow \sim p$.

5. Write the converse, inverse, and contrapositive of "If June is warm, then January is cold."   Section 2-7

6. Write an implication that is logically equivalent to $r \rightarrow \sim w$.

7. Let p represent "5 is between 7 and 8" and q represent "A square has 6 sides." Write in symbolic form and give the truth value: "5 is between 7 and 8 if and only if a square has 6 sides."   Section 2-8

8. If $\sim(p \vee q) \leftrightarrow (\sim p \wedge \sim q)$ is a tautology, what is true about $\sim(p \vee q)$ and $\sim p \wedge \sim q$?

Section 2-9 VALID ARGUMENTS (Optional)

Logic is the study of the principles of *valid* reasoning. In a valid argument, you use statements called *premises* to deduce correctly other statements called *conclusions*. If there is an error in the reasoning, the argument is said to be *invalid*. In this section you will study several examples of valid and invalid reasoning.

Suppose you begin with the following two premises.

    P_1:   If it is snowing, then I will slip and fall.
    P_2:   It is snowing.

You expect the conclusion to be:

    C: I will slip and fall.

These statements can be put into an argument pattern as follows:

    p: It is snowing.
    q: I will slip and fall.

    Premise P_1: $p \rightarrow q$
    Premise P_2: p
    ―――――――――――
    Conclusion C: $\therefore q$ ($\therefore$ means therefore)

One way to show that the reasoning is valid is to demonstrate that the argument expressed as an implication is always true. The following truth table shows that $[(p \rightarrow q) \wedge p] \rightarrow q$ is a tautology.

TRUTH TABLE FOR THE LAW OF DETACHMENT

p	q	$p \to q$	$(p \to q) \wedge p$	$[(p \to q) \wedge p] \to q$
T	T	T	T	T
T	F	F	F	T
F	T	T	F	T
F	F	T	F	T

The truth table indicates that the conclusion always follows from the statements $p \to q$ and p. It does *not* mean that the conclusion is true or false, only that given the two premises, we may logically deduce that conclusion. This pattern is called the *Law of Detachment*.

In general, any argument will fit the following pattern:

$$\begin{array}{l} \text{Premise } P_1 \\ \text{Premise } P_2 \\ \quad \vdots \\ \text{Premise } P_n \\ \hline \text{Conclusion } C \end{array}$$

An argument is valid if the implication $(P_1 \wedge P_2 \wedge \ldots \wedge P_n) \to C$ is a tautology.

Now we will examine the following argument to determine if the reasoning is valid.

If Charles is happy, then he is smiling.
Charles is not smiling.
Therefore, Charles is not happy.

To express the argument in symbolic form, we let p be "Charles is happy" and q be "Charles is smiling."

$$\begin{array}{ll} P_1: & p \to q \\ P_2: & \sim q \\ \hline C: & \therefore \sim p \end{array}$$

The argument is valid if the implication $[(p \to q) \wedge \sim q] \to \sim p$ is always true.

The truth table for the implication $[(p \to q) \wedge \sim q] \to \sim p$ is shown at the top of the next page.

LOGIC

TRUTH TABLE FOR THE LAW OF CONTRAPOSITIVE INFERENCE

p	q	$p \to q$	$\sim q$	$(p \to q) \land \sim q$	$\sim p$	$[(p \to q) \land \sim q] \to \sim p$
T	T	T	F	F	F	T
T	F	F	T	F	F	T
F	T	T	F	F	T	T
F	F	T	T	T	T	T

Since the implication is a tautology, the conclusion that Charles is not happy follows logically from the given premises. This pattern is called the *Law of Contrapositive Inference*.

Many advertisements on television sound like the following: "If you use Brand X shampoo, then your hair will smell nice." The advertiser would like you to conclude that if you do not use Brand X shampoo, then your hair won't smell nice. To determine whether this reasoning is valid, we will put the argument into symbolic form.

p: You use Brand X shampoo.
q: Your hair smells nice.

P_1: $p \to q$
P_2: $\sim p$
C: $\therefore \sim q$

The argument can be written as the implication $[(p \to q) \land \sim p] \to \sim q$. The truth table for the implication follows.

p	q	$p \to q$	$\sim p$	$(p \to q) \land \sim p$	$\sim q$	$[(p \to q) \land \sim p] \to \sim q$
T	T	T	F	F	F	T
T	F	F	F	F	T	T
F	T	T	T	T	F	F
F	F	T	T	T	T	T

Notice that the implication is not a tautology. Thus, the conclusion does not follow logically from the given premises.

Suppose you start with the following statements.

If you study, then you will pass.
If you pass, then you will get a surprise.

Can you conclude that if you study, then you'll get a surprise?

The truth table for the implication is shown at the top of the next page.

p: You study.
q: You pass.
r: You get a surprise.

P_1: $p \to q$
P_2: $q \to r$
C: $\therefore p \to r$

The implication is $[(p \to q) \land (q \to r)] \to (p \to r)$, and its truth table follows. Notice that there are 8 lines for the statements p, q, and r.

TRUTH TABLE FOR THE LAW OF THE SYLLOGISM

p	q	r	$p \to q$	$q \to r$	$(p \to q) \land (q \to r)$	$p \to r$	$[(p \to q) \land (q \to r)] \to (p \to r)$
T	T	T	T	T	T	T	T
T	T	F	T	F	F	F	T
T	F	T	F	T	F	T	T
T	F	F	F	T	F	F	T
F	T	T	T	T	T	T	T
F	T	F	T	F	F	T	T
F	F	T	T	T	T	T	T
F	F	F	T	T	T	T	T

This argument pattern is valid and is called the *Law of the Syllogism*.

If an argument fits one of the three patterns of the Law of Detachment, the Law of the Contrapositive, or the Law of the Syllogism, you may simply say that the reasoning is valid. The truth tables have already been given and there is no need to duplicate them. There are other valid argument patterns that will not be studied at this time.

WRITTEN EXERCISES

Put the argument into symbolic form. Tell if it is valid. If it is, specify which argument pattern is used.

A 1. If $x + 1 = 10$, then $x = 9$.
   $x + 1 = 10$.
   Therefore $x = 9$.

2. If he's a glob, then he's a nob.
   He's not a nob.
   Therefore he's not a glob.

3. If he's a glob, then he's a nob.
   He's not a glob.
   Therefore he's not a nob.

LOGIC 63

4. If the butler committed the crime, then the maid didn't do it.
   The maid did it.
   Therefore the butler didn't do it.

5. If the butler committed the crime, then the maid didn't do it.
   The butler didn't do it.
   Therefore the maid did it.

6. If the butler did it, then the maid didn't.
   The maid didn't do it.
   Therefore the butler did it.

7. If the butler did it, then the maid didn't.
   The butler did it.
   Therefore the maid didn't do it.

8. If the butler did it, then the maid didn't.
   If the maid didn't do it, then the cook didn't do it.
   Therefore if the butler did it, then the cook didn't do it.

9. If it's a hob, then it's a nob.
   If it's a nob, then it's a bob.
   Therefore if it's a hob, then it's a bob.

10. If it rains, then the grass will grow.
    The grass is not growing.
    Therefore it doesn't rain.

Determine if the argument is valid by expressing it as an implication and then using a truth table.

B 11. $p \vee q$
$\underline{p}$
$\therefore q$

12. $p \rightarrow q$
$\underline{q}$
$\therefore p$

13. $\sim p \vee q$
$\underline{p}$
$\therefore q$

14. $p \rightarrow q$
$\underline{q \rightarrow p}$
$\therefore p \wedge q$

15. $\sim p \rightarrow \sim q$
$\underline{q}$
$\therefore p$

16. $p \vee q$
$\underline{\sim p}$
$\therefore q$

Put into symbolic form and draw a valid conclusion if possible.

C 17. If x is a whole number, then x is an integer.
If x is not a real number, then it's not an integer.
x is a whole number.
(*Hint:* An implication can be replaced by its contrapositive.)

18. If Mary had a little lamb, then Jack and Jill went up the hill.
If Humpty sat on the wall, then Jack and Jill didn't go up the hill.
Humpty sat on the wall.

19. All babies are small.
If you're small, then you're not tall.
You're tall.

64   CHAPTER 2

Computer Exercises

1. **a.** Copy and RUN the following program on your computer.

   ```
   10   PRINT "P","Q","P ? Q"
   20   PRINT
   30   FOR P=1 TO 0 STEP -1
   40   IF P=1 THEN 70
   50   LET P$="F"
   60   GOTO 80
   70   LET P$="T"
   80   FOR Q=1 TO 0 STEP -1
   90   IF Q=1 THEN 120
   100  LET Q$="F"
   110  GOTO 130
   120  LET Q$="T"
   130  IF P=0 THEN 170
   140  IF Q=0 THEN 170
   150  LET C$="T"
   160  GOTO 180
   170  LET C$="F"
   180  PRINT P$,Q$,C$
   190  NEXT Q
   200  NEXT P
   210  END
   ```

   b. The truth table of what logical connective ("and" or "or") is printed by this program?

2. **a.** Substitute the new lines listed below into the preceding program and RUN it.

   ```
   130  IF P=1 THEN 170
   140  IF Q=1 THEN 170
   150  LET C$="F"
   170  LET C$="T"
   ```

   b. The truth table of what logical connective is printed by this program?

3. (Optional) The program in Exercise 2 can be made to print the truth table for the biconditional if certain changes are made in lines 130 and 140. Type 140 with nothing after it in order to delete that line from the program. Change line 130 by substituting something appropriate for "P = 1." RUN the new program.

LOGIC   65

4. (Optional) If the version of BASIC that you are using includes the logical operators "and" and "or," copy and RUN the following program. It will print 1 for true and 0 for false.

```
10  PRINT " P"," Q","P AND Q","P OR Q"
20  PRINT
30  FOR P=1 TO 0 STEP -1
40  FOR Q=1 TO 0 STEP -1
50  PRINT P,Q,P AND Q,P OR Q
60  NEXT Q
70  NEXT P
80  END
```

CHAPTER REVIEW

Determine whether each of the following is a statement, an open sentence, or neither. Assign a truth value to each statement.

1. $2(8 + 3) = 2(8) + 2(3)$ 2. It is a planet.   Section 2-1

3. If the domain of x is $\{$ the counting numbers $\}$, find the solution set for "x is an odd integer less than 4."

Give the negation. Then find the truth values of the statement and of its negation.

4. 1 is greater than 2.   5. Gold is a mineral.   Section 2-2

6. Let p represent "Division by zero is not defined."
    a. Write the negation of p in words and in symbolic form.
    b. Give the truth value of the negation.

In Exercises 7–14, take $p, q, r,$ and s as follows.

p: $38 \div 2 \neq 19$ F q: 7 is an even integer. F
r: $5 + 2 = 2 + 5$ T s: Peas grow in pods. T

Give the truth value of each statement.

7. $p \wedge q$ 8. $q \wedge r$ 9. $\sim p \wedge s$ Section 2-3

10. Represent "Peas grow in pods and $5 + 2 = 2 + 5$" in symbolic form and determine its truth value.

Give the truth value of each statement.

11. $r \vee s$ 12. $q \vee p$ 13. $p \vee (q \vee r)$ Section 2-4

14. Represent "$5 + 2 = 2 + 5$ or $38 \div 2 \neq 19$" in symbolic form and determine its truth value.

15. State the hypothesis and the conclusion of "If $2z = -8$, then $z = -4$."   Section 2-5

Let p represent "Trenton is the capital of New Jersey" and q represent "$\frac{1}{2} = \frac{2}{5}$."

16. Write $p \to q$ in words and give its truth value.
17. Write in symbolic form and give the truth value: "Trenton is the capital of New Jersey only if $\frac{1}{2} = \frac{2}{5}$."

Assume that r and s are each true. Complete each statement.

18. $\sim r$ is __?__, and $\sim r \to s$ is __?__. Section 2-6
19. $\sim s$ is __?__, $r \wedge \sim s$ is __?__, and $\sim(r \wedge \sim s)$ is __?__.

20. Construct a truth table for $p \to (p \vee q)$.

Complete.

21. A conditional and its __?__ are logically equivalent. Section 2-7
22. The inverse and the __?__ of a conditional statement are logically equivalent.
23. Give the inverse of "If $x = -1$, then $x + 5 = 4$."
24. Are the statements $p \to \sim q$ and $q \to \sim p$ logically equivalent? Explain.

25. Let r represent "$5 \cdot 5 = 5$" and s represent "$5\left(\frac{4}{5} + \frac{1}{5}\right) = 5$." Write $r \leftrightarrow s$ in words and give its truth value. Section 2-8
26. Write the truth table for the biconditional $p \leftrightarrow q$.
27. If two statements r and s are logically equivalent, what can you say about $r \leftrightarrow s$?

Put the argument into symbolic form. Tell if it is valid. If it is, specify which argument pattern is used.

28. If I conserve energy, then I'll save money. Section 2-9
    I save money.
    Therefore I conserve energy.
29. If $x = 2$, then $x(x - 2) = 0$.
    $x(x - 2) \neq 0$
    Therefore $x \neq 2$.

CHAPTER TEST

1. If the domain of x is {the positive integers}, find two values of x that make the statement "x is a multiple of 4" true. Find two values of x that make the statement false.
2. If the domain of y is {−1, 0, 1}, is the statement $y = -y$ always true, always false, or sometimes true and sometimes false?

LOGIC

Let p represent the statement "Five is greater than two."

3. Write the negation of p in symbolic form and in words.
4. Give the truth values of p and of its negation.

For Exercises 5-16, take p, q, r, and s as follows:

p: $3 \neq -3$ $\qquad$ q: Rhode Island is larger than Texas.
r: $-1 + 1 = 0$ $\qquad$ s: February has 30 days.

Write each compound statement in words and give its truth value.

5. $s \lor q$ $\qquad$ 6. $r \to q$ $\qquad$ 7. $r \land s$ $\qquad$ 8. $s \leftrightarrow q$
9. $\sim q \land p$ $\qquad$ 10. $s \leftrightarrow r$ $\qquad$ 11. $(q \lor r) \lor s$ $\qquad$ 12. $p \to s$

Represent each compound statement in symbolic form and give its truth value.

13. If $-1 + 1 = 0$, then $3 \neq -3$.
14. $3 \neq -3$ and February has 30 days.
15. Rhode Island is larger than Texas if and only if $-1 + 1 = 0$.
16. $-1 + 1 = 0$ or $3 \neq -3$.

Construct a truth table for each statement. Determine whether the statement is a tautology, a contradiction, or neither.

17. $\sim(p \land \sim p)$ $\qquad\qquad$ 18. $(p \land \sim q) \to q$

19. Write the converse, the inverse, and the contrapositive of the following statement: "If you enjoy dancing, then we have a common interest."
20. Write, in symbols, a statement that is logically equivalent to $\sim p \to \sim q$.
21. Show that $p \to q$ and $\sim(p \land \sim q)$ are logically equivalent.

(Optional) Put each argument into symbolic form. Tell if it is valid. If it is, specify which argument pattern is used.

22. If $x = 2$, then $3x = 6$.
    If $3x = 6$, then $3x + 1 = 7$.
    Therefore, if $x = 2$, then $3x + 1 = 7$.

23. If it rains, then it pours.
    It doesn't rain.
    Therefore it doesn't pour.

Biographical Note — George Boole

Have you ever tried to describe in words the process by which you reasoned from given facts to a conclusion? If so, you probably found that it wasn't easy. For a long time the study of logic was difficult because logicians had to use words in their analysis of the reasoning process, and the meaning of words is often vague and imprecise.

The English mathematician George Boole (1815-1864) was the first to see the close resemblance between the propositions of logic and mathematical equations. In a work commonly referred to as *Laws of Thought*, Boole set out "to investigate the fundamental laws of those operations of the mind by which reasoning is performed." The innovative element in his work was not the investigation itself but his attempt to express the operations of logical reasoning "in the language of a Calculus" — that is, by means of algebraic symbols.

First Boole showed that any class of objects — for example, hamburgers or all green things — can be represented by a symbol such as x. Then he showed that such symbols can be combined by the same rules as those that govern the operations of algebra.

Boole was the son of a shoemaker who could not afford to send him to a university. He was largely self-taught, and because of his lack of formal education some of the professions were closed to him. However, his ability brought him into the teaching profession. Disturbed by the poor quality of the mathematics textbooks then available, Boole felt that he was challenged to produce something better. He turned to further study, and after a few years published a paper on calculus that won the Royal Society's gold medal. Next he published a booklet called *The Mathematical Analysis of Logic*. The originality of these works brought him recognition at last, and in 1849 he received an appointment as professor of mathematics at Queen's College in Ireland.

Boolean algebra, as the algebra of classes has come to be called, has found many applications. It is the basis of switching theory, computer design, and information theory.

VOCABULARY REVIEW

Be sure that you understand the meaning of these terms:

variable, p. 36
domain (replacement set) of
  a variable, p. 36
open sentence, p. 36
solution set, p. 36
negation, p. 38
truth table, p. 39
conjunction, p. 40
compound statement, p. 40

disjunction, p. 43
conditional, p. 46
converse, p. 54
inverse, p. 54
contrapositive, p. 54
logically equivalent statements, p. 54
tautology, p. 55
contradiction, p. 55
biconditional, p. 57

MIXED REVIEW

1. **a.** Construct a truth table for the statement $(p \lor q) \rightarrow q$.
   b. Is the statement a tautology, a contradiction, or neither?

2. Let r represent "-7 is an integer" and let s represent "$-\frac{1}{2}$ is the reciprocal of $\frac{1}{2}$." Write $r \lor s$ in words and give its truth value.

Add or subtract.

3. $\frac{3}{8} + \frac{7}{12} + \frac{1}{6}$

4. $-9 + 24 + (-77)$

5. $-15 - (-2)$

6. $-71 + 53 + 71$

7. $\frac{12}{5} - \frac{3}{5}$

8. $0.75 - 3$

9. Describe the set $\{B, C, D, F, G, H, J, K, L, M, N, P, Q, R, S, T, V, W, X, Y, Z\}$.

In Exercises 10–12, take p to be "$x > 0$" and q to be "$x < 0$".

10. The statement $p \leftrightarrow q$ is a __?__ statement and its truth value is __?__.

11. Write the negation of p.

12. Write $q \rightarrow \sim p$ in words and give its truth value.

13. Graph: {whole numbers}

14. Give the opposite of: **a.** $-\frac{5}{8}$ **b.** 0

15. Taking the replacement set to be {counting numbers}, find the solution set for $x \leqslant 4$.

Multiply or divide.

16. $-30(21)$

17. $(-5.2)(-0.002)$

18. $\frac{2}{3}\left(\frac{3}{8}\right)\left(-\frac{7}{12}\right)$

70   CHAPTER 2

19. $-\dfrac{15}{16} \div \left(-\dfrac{19}{20}\right)$ 20. $1.74 \div (-60)$ 21. $0.245 \div (-0.35)$

22. Write the inverse of the statement "If you attend high school, then you are a student."

23. Are a statement and its converse logically equivalent?

24. Name the property illustrated by the statement $2(5 \cdot 7) = (2 \cdot 5)7$.

25. The statement $p \wedge \sim q$ is true only when p is __?__ and q is __?__.

Simplify.

26. $-1 \cdot 0 \cdot 7$ 27. $\dfrac{3}{4} - 0 + \left(-\dfrac{5}{4}\right)$

28. $-18 + (5^2 - 3^3)(-1)$ 29. $2^2 + 10 \cdot 3 \div 6$

30. The number associated with a point on the number line is called the __?__ of the point.

31. If the domain of x is $\{1, 2, 3, \ldots\}$, is the statement $\dfrac{1}{2}x = 8$ always true, always false, or sometimes true and sometimes false?

32. (Optional) Put the argument into symbolic form. Tell if it is valid. If it is, specify which argument pattern is used.

    If I don't get enough sleep, then I'll be too tired to play tennis.
    I'm not too tired to play tennis.
    Therefore I got enough sleep.

Given that p is true, q is true, and r is false, determine the truth value of each statement.

33. $\sim p \vee r$ 34. $\sim(p \wedge r)$ 35. $p \wedge (q \vee r)$

36. Using the number line below, name the graph of the given number or state the coordinate of the given point.

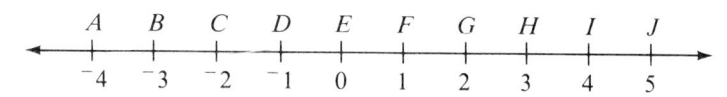

```
      A   B   C   D   E   F   G   H   I   J
   ───┼───┼───┼───┼───┼───┼───┼───┼───┼───┼──→
      -4  -3  -2  -1   0   1   2   3   4   5
```

a. C b. -2 c. I d. 0
e. The point halfway between B and J.

LOGIC 71

Equations can be used to solve a variety of geometric problems. For example, the formula for the perimeter of a rectangle can be used to solve problems involving rectangular fields such as those shown above.

Solving Equations 3

Algebra is a generalization of arithmetic in which variables play an important part. In this chapter you will learn to solve simple equations involving variables. You will also learn how to solve various kinds of problems in a systematic, logical way by translating them into equations and then solving the equations.

Section 3-1 EVALUATING ALGEBRAIC EXPRESSIONS

How old are you today? How old will you be 6 years from today? The answer to the second question depends on your answer to the first question. In this section you will study some words and phrases that will help you describe problems mathematically:

- algebraic expression
- evaluating the expression

Suppose you let the variable a represent your age today. The *domain*, or replacement set, for a might be $\{13, 14, 15, 16, 17\}$. Each number in the domain is called a *value* of the variable.

You could represent your age 6 years from now by $a + 6$. When you *substitute* a value for a, you get a value for $a + 6$. You *evaluate* $a + 6$ to find your age 6 years from now.

Both a and $a + 6$ are examples of algebraic expressions. An *algebraic expression* is an expression containing variables and operations such as multiplication and addition.

When a product contains a variable, the multiplication sign is usually omitted. That is, "$10 \times t$" is written "$10t$" and "$a \times b$" is written "ab."

EXAMPLES OF ALGEBRAIC EXPRESSIONS

$2a + 3b + 5$ $\qquad$ $\dfrac{rs}{2t}$

$ab - 7$ $\qquad$ y

73

An algebraic expression takes on a specific value when a substitution is made for the variable or variables. This process is called *evaluating the expression* or *finding the value of the expression.*

EXAMPLE   Find the value of the expression $2a + 3b + 5$ if $a = 4$ and $b = 3$.

SOLUTION   1. Replace the letter a by 4 and the letter b by 3.
2. Simplify the expression.

$$2a + 3b + 5$$
$$= 2(4) + 3(3) + 5$$
$$= 8 + 9 + 5$$
$$= 22$$

EXAMPLE   Find the value of $a + \dfrac{3}{b}$ if $a = 5$ and $b = -2$.

SOLUTION   1. Replace a by 5 and b by -2.

2. Simplify the expression.

$$a + \frac{3}{b} = 5 + \left(-\frac{3}{2}\right)$$
$$= 5 - \frac{3}{2}$$
$$= \frac{10}{2} - \frac{3}{2}$$
$$= \frac{7}{2}$$

EXAMPLE   Find the value of $5x - 3x$ if $x = -\dfrac{1}{2}$.

SOLUTION   Replace x by $-\dfrac{1}{2}$ and simplify.

$$5x - 3x = 5\left(-\frac{1}{2}\right) - 3\left(-\frac{1}{2}\right)$$
$$= -\frac{5}{2} + \frac{3}{2}$$
$$= -\frac{2}{2}$$
$$= -1$$

In Section 1-9 you saw an example of the *distributive property.* Another example would be:

$$5 \cdot 2 - 3 \cdot 2 = (5 - 3)2$$

The distributive property can also be used to combine expressions involving the same variable. For example:

$$5x - 3x = (5 - 3)x$$
$$= 2x$$

The distributive property guarantees that $5x - 3x$ and $2x$ have the same value no matter what replacement is made for x. Thus, we could evaluate the expression $5x - 3x$ by *first* using the distributive property and *then* substituting $-\dfrac{1}{2}$ for x.

$$5x - 3x = (5 - 3)x$$
$$= 2x$$
$$= 2\left(-\frac{1}{2}\right)$$
$$= -1$$

Some different forms of the distributive property are:

$$a(b + c) = ab + ac \qquad a(b - c) = ab - ac$$
$$(b + c)a = ba + ca \qquad (b - c)a = ba - ca$$

ORAL EXERCISES

In Exercises 1-8 the expression indicates one or more operations (addition, subtraction, multiplication, division) to be performed. Describe the operations for each expression.

EXAMPLE $ac - 2$

SOLUTION Multiply a and c and then subtract 2 from that product.

1. $5a + 7$
2. $3c - 1$
3. $2a + 5b$
4. $6 + ab$
5. $9 - 4a$
6. $\dfrac{b}{3}$
7. $\dfrac{a}{b} + 3$
8. $\dfrac{10}{ab}$

9-16. What is the value of each expression in Exercises 1-8 if $a = 2, b = 3$, and $c = 1$?

WRITTEN EXERCISES

Evaluate the algebraic expression if $m = 1, n = 3, r = -1, s = 7$, and $x = \dfrac{1}{2}$.

A
1. $3m$
2. $5n$
3. $-4s$
4. $-8n$
5. mn
6. ns
7. $n + r - 2$
8. $s + m - 3$
9. $n - s$
10. $r - n$
11. $2s + 4$
12. $3n + x$
13. $s + 4r$
14. $5m + 2r$
15. $4x - n$
16. $8x - 4m$
17. $\dfrac{s}{r}$
18. $\dfrac{m}{n}$
19. $18x$
20. $\dfrac{18}{x}$
21. $3n + 4n$
22. $6r + 2r$
23. $3s - 5s$
24. $4r - 3r$

B
25. $4r - 3r + 2r$
26. $2n + 5n - 2n$
27. $2s + 5s - 2s$
28. $4x - 3x + 2x$
29. $3m + ns$
30. $rs - mn$
31. $(rs)(mn)$
32. $x(s + r)$
33. $\dfrac{rs}{mn}$
34. $\dfrac{s + m}{n}$
35. $\dfrac{m + n}{r + s}$
36. $\dfrac{m - n}{r - s}$

SOLVING EQUATIONS

Write without parentheses an expression that always has the same value as the given one.

EXAMPLE $5(x + 2)$ SOLUTION $5x + 10$

37. $3(y + 6)$
38. $4(b - 3)$
39. $-2(s + 4)$
40. $-3(2 + m)$
41. $-2(c - 2)$
42. $-5(4 - d)$
43. $5(a + b)$
44. $7(s - t)$
45. $x(y - z)$
46. $(c - d)b$
47. $2(3a + 9)$
48. $4(4r - 5)$

Translate the word phrase into an algebraic expression.

EXAMPLE Five times x plus the product of a and b
SOLUTION $5x + ab$

C 49. The sum of eleven and twice b
50. Twenty less than the product of six and m
51. Eight divided by the product of x and y
52. Seven more than the quotient of r divided by eight

Give a real-life interpretation suggested by the algebraic expression.

EXAMPLE $24d$
SOLUTION The number of hours in d days

53. $7w$
54. $12m$
55. $60 + a$
56. $10t + u$
57. $15 + 13x$
58. $p + 0.07p$

Evaluate the expression if $x = 3$, $y = 4$, $z = \frac{1}{2}$, and $a = 0$.

59. $\dfrac{(x + 21)(y - ax)}{3x + 14z}$
60. $\dfrac{a(7x + 3y - 8)}{2z + 41}$

Section 3-2 INTRODUCTION TO EQUATIONS

In the rest of this chapter you will solve equations by changing them to simpler equivalent equations whose solution is obvious. In this section you will learn about:
- equations
- the solution of an equation
- equivalent equations

An *equation* is a sentence containing an equality symbol (=) between two algebraic expressions or expressions involving numbers. The expression to the left of the equality symbol is the *left side of the equation*. The other is the *right side of the equation*.

EXAMPLES OF EQUATIONS

$$3 + 4 = 7 \qquad a + 6 = 20 \qquad 2x + 3y = 8$$

Recall from Chapter 2 that $3 + 4 = 7$ is an example of a statement. The open sentences $a + 6 = 20$ and $2x + 3y = 8$ are equations in one and two variables respectively. In this chapter you will work only with equations in one variable. In Chapters 13 and 14 you will work with equations in two variables.

You *solve* an equation containing a variable when you find all values from the domain of the variable that make the equation true. Each of these values is called a *solution* of the equation. The set of all solutions from a given domain is called the *solution set* of the equation over the given domain.

EXAMPLE If x is an integer, what is the solution set of $x + 2 = 2$?

SOLUTION Since $0 + 2 = 2$ and no other value of x makes the equation true, the solution set of $x + 2 = 2$ is $\{0\}$.

EXAMPLE If the domain of x is $\{1, 2, 3, 4\}$, what is the solution set of $x + 2 = 2$?

SOLUTION Substitute the members of the domain for x:

$$1 + 2 = 3 \qquad 2 + 2 = 4 \qquad 3 + 2 = 5 \qquad 4 + 2 = 6$$

There is no member of the domain for which $x + 2 = 2$. Thus, the solution set is $\emptyset$, the empty set.

When the solution set of an equation contains only one member, we sometimes talk simply about the solution of the equation.

EXAMPLE Solve $a + 6 = 20$ if the domain of a is $\{13, 14, 15, 16, 17\}$.

SOLUTION Try replacing a with different values from the domain. The solution is 14.

Unless otherwise indicated, in this book the domain of each variable will be assumed to be the set of real numbers.

EXAMPLE Find the solution set of $x + 5 = 16$.

SOLUTION If x is replaced by 11, then the equation is true. Since no other replacement for x makes the equation true, the solution set is $\{11\}$.

EXAMPLE Solve $2x - 18 = 4$.

SOLUTION You could simply try replacing x with different values until you find that 11 is a solution and there appear to be no other solutions.

Two equations are called *equivalent* if they have the same solution set. The equations $x + 5 = 16$ and $2x - 18 = 4$ are equivalent equations because the solution set of each is $\{11\}$.

SOLVING EQUATIONS

In general, to solve an equation you will write a series of equivalent equations aimed at finding the solution. When solving $2x - 18 = 4$, for example, your final equation would be $x = 11$. In the next three sections you will learn this method of solving equations.

ORAL EXERCISES

Tell whether the given number is a solution of the given open sentence.

1. $x + 2 = 4$; 2
2. $3y = -6$; -2
3. $2z - 4 = 2$; 6
4. $4 - t = 3$; 7
5. $4a = 2$; $\frac{1}{2}$
6. $2c + 3 = 3c - 2$; 5

If the domain of b is $\{0, 1, 2, 3\}$, what is the solution set of the open sentence?

7. $b + 3 = 4$
8. $b - 2 = 1$
9. $2b = 4$
10. $3b = 0$
11. $3b + 2 = 8$
12. $4b - 5 = 7$

Are the two equations equivalent? If so, what is their solution set?

13. $x = 5$ and $x + 4 = 9$
14. $y = -2$ and $y + 4 = 6$
15. $5a = 10$ and $a = 5$
16. $2b = 15$ and $b = \frac{15}{2}$
17. $c + 4 = 7$ and $c - 5 = 0$
18. $d + 2 = 6$ and $3d = 12$

WRITTEN EXERCISES

If the domain of x is $\{-1, 0, 1, 2, 3\}$, write the solution set of the open sentence.

A
1. $x - 6 = -5$
2. $x + 3 = 2$
3. $5x = 15$
4. $-3x = 3$
5. $6x - 2 = 10$
6. $2x - 5 = -1$

Are the two equations equivalent? If so, what is their solution set?

7. $a = -1$ and $a + 1 = 2$
8. $b = 3$ and $5b = 15$
9. $3x + 7 = 10$ and $x = 1$
10. $2y - 5 = 9$ and $y = 7$
11. $4 - w = 6$ and $2 = w$
12. $8 - 3z = 2$ and $-2 = z$

Solve if the given set is the domain.

13. $3x + 7 = 4$; $\{-3, -2, -1\}$
14. $10y - 15 = 65$; $\{6, 8, 10\}$
15. $8 - x = 5$; $\{2, 3, 4\}$
16. $9 - y = -3$; $\{6, 9, 12\}$
17. $4 = 13 - 3y$; $\{1, 2, 3\}$
18. $2 = 14 - 4x$; $\{3, 4, 5\}$

B
19. $a - \frac{1}{2} = 1$; $\left\{-\frac{1}{2}, \frac{1}{2}, \frac{3}{2}\right\}$
20. $b + \frac{1}{2} = -1$; $\left\{-\frac{3}{2}, -\frac{1}{2}, \frac{1}{2}\right\}$
21. $\frac{1}{4}a = 8$; $\{2, 4, 32\}$
22. $-\frac{1}{5}b = -20$; $\{-100, -4, 4, 100\}$

23. $m + 2.5 = 7$; $\{4.5, 5, 5.5\}$
24. $n - 2.2 = 3.9$; $\{1.7, 5.7, 6.1\}$
25. $0.5m = 3.5$; $\{1.75, 7, 70\}$
26. $-2.5n = 5$; $\{-12.5, -2, 2, 12.5\}$
27. $3x + 2 = 4$; $\left\{\frac{1}{3}, \frac{2}{3}, 1\right\}$
28. $2y + 5 = 4$; $\left\{-\frac{4}{5}, -\frac{1}{2}, \frac{9}{2}\right\}$
29. $5x - 8 = -2$; $\{-2, -1.6, 1.2\}$
30. $4y + 7 = 6$; $\{-0.75, -0.5, -0.25\}$

Write an equation that has the given solution set.

31. $\{3\}$
32. $\{-2\}$
33. $\{0\}$
34. $\left\{\frac{1}{3}\right\}$

C 35. What is the solution set of $x = x$?
36. What is the solution set of $x = x + 1$?

The solution set of a conjunction of two open sentences contains all values of the variable that make *both* open sentences true. The solution set of a disjunction of two open sentences contains all values of the variable that make *either* open sentence true.

Find the solution set of the compound sentence.

EXAMPLE: $(2x = 6) \wedge (x + 2 = 5)$ SOLUTION: $\{3\}$

EXAMPLE: $(y + 1 = 7) \vee (3y = -9)$ SOLUTION: $\{-3, 6\}$

37. $x = 10$ or $x = 0$
38. $y = -2$ and $y = 4$
39. $(x = 3) \wedge (x - 2 = 1)$
40. $(x = 3) \vee (x = -5)$
41. $(2x = 5) \vee (x + 2 = 5)$
42. $(5y = 15) \wedge (y - 5 = -2)$

SELF-TEST 1

1. Evaluate the algebraic expression $x + r - 2s$ if $x = 1, r = -2$, and $s = 4$. Section 3-1

2. Evaluate $xt - y$ if $x = 4, y = -3$, and $t = \frac{1}{2}$.

3. Write without parentheses an expression that always has the same value as $5(w - 2)$.

Are the two equations equivalent? If so, what is their solution set?

4. $3y = -9$ and $y = 3$
5. $x = \frac{5}{2}$ and $4x - 2 = 8$ Section 3-2

6. If the domain of z is $\{-2, -1, 0, 1, 2\}$, solve $2z - 3 = -1$.

Section 3-3 SOLVING EQUATIONS BY ADDITION

Can you write an equation that is equivalent to the equation $x = 5$? There are many equations that have the same solution set as $x = 5$.

EXAMPLES OF EQUATIONS EQUIVALENT TO $x = 5$

$$x + 1 = 6 \qquad x + 100 = 105$$
$$x + 2 = 7 \qquad x + (-5) = 0$$

These examples illustrate a property of equivalent equations that will be very helpful in solving certain kinds of equations.

ADDITION PROPERTY OF EQUIVALENT EQUATIONS

If the same number is added to both sides of an equation, the new equation and the original equation are equivalent equations.

The following example shows how you can solve an equation by writing a series of equivalent equations. The final equation is one whose solution set is obvious.

EXAMPLE Find the solution set of $x - 7 = 2$.

SOLUTION Add 7 to both sides of the equation.
$$(x - 7) + 7 = 2 + 7$$
$$x + (-7 + 7) = 2 + 7$$
$$x + 0 = 9$$
$$x = 9$$

The solution set is $\{9\}$.

In the example above, we chose 7 as the number to add to each side of the equation because 7 is the opposite, or additive inverse, of -7.

In general, to solve an equation of the form $x + c = d$, where c and d both stand for numbers, you add the opposite of c to both sides of the equation.

EXAMPLE Solve $x + 4 = 10$.

SOLUTION Add -4, the opposite of 4, to both sides of the equation.
$$(x + 4) + (-4) = 10 + (-4)$$
$$x + (4 - 4) = 10 - 4$$
$$x + 0 = 6$$
$$x = 6$$

Once you understand the process of solving this kind of equation, you may shorten your work, as follows. Recall that subtracting 8 is the same as adding the opposite of 8.

EXAMPLE Solve $y + 8 = 11$.
SOLUTION $(y + 8) - 8 = 11 - 8$
$y = 3$

When you solve an equation you can check your solution by substituting it for the variable in the original equation. Checking your solution is especially important when the original equation is somewhat complicated.

EXAMPLE Solve $4x - 3x + 5 = 19$.
SOLUTION $(4 - 3)x + 5 - 5 = 19 - 5$
$1 \cdot x = 14$
$x = 14$
Check $4(14) - 3(14) + 5 = 56 - 42 + 5 = 19$
Therefore the solution is 14.

Notice that in solving the example above, we used the distributive property to replace $4x - 3x$ by $(4 - 3)x$.

ORAL EXERCISES

Name the opposite of each number.

1. 6
2. -10
3. $\frac{1}{2}$
4. $-\frac{1}{4}$
5. $\frac{2}{3}$
6. $-\frac{3}{4}$

Tell what you would do to solve the equation. Then solve it.

7. $x + 4 = 7$
8. $x - 3 = 8$
9. $7 = y + 12$
10. $10 = y - 7$
11. $x + 4 = 0$
12. $y - 5 = -10$

WRITTEN EXERCISES

Solve.

A
1. $x + 8 = 13$
2. $x + 1 = 4$
3. $y - 5 = 0$
4. $y - 2 = 7$
5. $a - 3 = -4$
6. $a - 2 = -7$
7. $b - 13 = -4$
8. $b - 17 = -12$
9. $d + 14 = -11$
10. $d + 27 = -18$
11. $z + 20 = 20$
12. $z - 5 = -5$
13. $4 + m = -8$
14. $20 + m = 4$
15. $18 = x - 15$
16. $-22 = x + 9$
17. $-19 = y - 4$
18. $-10 = y + 17$
19. $a + \frac{1}{2} = \frac{5}{2}$
20. $a - \frac{3}{2} = \frac{9}{2}$
21. $b - 4 = \frac{3}{5}$
22. $b + 7 = \frac{2}{3}$
23. $c + \frac{1}{2} = 3$
24. $c + \frac{2}{7} = 5$

SOLVING EQUATIONS

Solve.

B 25. $x + 0.35 = 2.67$ 26. $y - 1.69 = 3.42$ 27. $z - 1.25 = -4.8$

28. $w + 0.6 = 1.79$ 29. $-2.93 + a = 4.52$ 30. $1.87 = 3.68 + b$

31. $3y - 2y + 5 = -16$ 32. $3y - 2y + 2 = 10$ 33. $8z - 7z + 5 = -16$

34. $6a - 5a - 2 = 12$ 35. $6 = 10x - 9x - 6$ 36. $-4 = 4w - 3w - 2$

C 37. $4(x - 2) - 3x = 4$ 38. $2(y + 7) - y = 20$

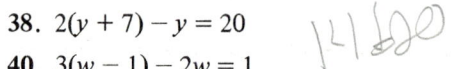

39. $6(z + 2) - 5z = 8$ 40. $3(w - 1) - 2w = 1$

The addition property of equality can be stated as:
$$\text{If } a = b, \text{ then } a + c = b + c.$$
This conditional statement can be written as $(a = b) \rightarrow (a + c = b + c)$.

41. What equation do you get if you use the addition property of equality to add $-c$ to both sides of $a + c = b + c$?

42. Write your result from Exercise 41 as a conditional statement.

43. Is your answer to Exercise 42 the inverse, converse, or contrapositive (or none of these) of the addition property of equality?

44. Combine the statement of the addition property of equality with your answer to Exercise 42 to write a biconditional statement.

45. Explain how your answer to Exercise 44 is related to the addition property of equivalent equations.

Section 3-4 SOLVING EQUATIONS BY MULTIPLICATION

Do you think you would get an equation equivalent to $x = 5$ if you multiplied each side of the equation by the same number? Try multiplying both sides of the equation by 3, by $\frac{1}{5}$, and by 0.

$x = 5$ and $3x = 15$ have the same solution set, $\{5\}$.

$x = 5$ and $\frac{1}{5}x = 1$ have the same solution set, $\{5\}$.

Do $x = 5$ and $0 \cdot x = 0 \cdot 5$ have the same solution set?

In order to answer the last question, you will have to find the solution set for $0 \cdot x = 0 \cdot 5$. If you substitute 5 for x in this equation, you will find that 5 is a solution of the equation. But so is every other real number! For instance, $0 \cdot 1 = 0 \cdot 5$ and $0(-2) = 0 \cdot 5$. Thus, the equations $x = 5$ and $0 \cdot x = 0 \cdot 5$ do not have the same solution set. That is why the word "nonzero" appears in the property below.

The following property of equivalent equations is helpful in solving certain kinds of equations that can't be solved by addition alone.

> **MULTIPLICATION PROPERTY OF EQUIVALENT EQUATIONS**
> If each side of an equation is multiplied by the same nonzero number, the new equation and the original equation are equivalent equations.

EXAMPLE Solve $3x = 12$.

SOLUTION Multiply both sides of the equation by $\frac{1}{3}$.

$$\frac{1}{3}(3x) = \frac{1}{3}(12)$$
$$\left(\frac{1}{3} \cdot 3\right)x = 4$$
$$1 \cdot x = 4$$
$$x = 4$$

In the example above, we chose $\frac{1}{3}$ as the number to multiply both sides by because it is the reciprocal, or multiplicative inverse, of 3.

In general, to solve an equation of the form $cx = d$, where c and d stand for numbers and $c \neq 0$, you multiply both sides of the equation by $\frac{1}{c}$, the reciprocal of c.

EXAMPLE Solve $7y = -10$.

SOLUTION Multiply both sides of the equation by $\frac{1}{7}$, the reciprocal of 7.

$$\frac{1}{7}(7y) = \frac{1}{7}(-10)$$
$$\left(\frac{1}{7} \cdot 7\right)y = -\frac{10}{7}$$
$$1 \cdot y = -\frac{10}{7}$$
$$y = -\frac{10}{7}$$

Once you understand the process of solving this kind of equation, you may shorten your work, as follows.

$$7y = -10$$
$$\frac{1}{7}(7y) = \frac{1}{7}(-10)$$
$$y = -\frac{10}{7}$$

SOLVING EQUATIONS

Recall that dividing by an integer is the same as multiplying by its reciprocal. Thus, you may solve many kinds of equations by division instead of multiplying by a reciprocal.

EXAMPLE   Solve $30z = -15$.

SOLUTION
$$\frac{30z}{30} = \frac{-15}{30}$$
$$z = -\frac{1}{2}$$

EXAMPLE $0.05x = 1.25$

SOLUTION
$$\frac{0.05x}{0.05} = \frac{1.25}{0.05}$$
$$x = 25$$

However, there are some equations that can't be solved easily by division. Two examples are given below.

EXAMPLE   Solve $\frac{3}{4}w = 15$.

SOLUTION   Multiply both sides of the equation by $\frac{4}{3}$, the reciprocal of $\frac{3}{4}$.
$$\frac{4}{3}\left(\frac{3}{4}w\right) = \frac{4}{3}(15)$$
$$w = 20$$

EXAMPLE   Solve $-\frac{a}{4} = -10$.

SOLUTION
$$-4\left(-\frac{a}{4}\right) = -4(-10)$$
$$a = 40$$

ORAL EXERCISES

Name the reciprocal of each number.

1. 2
2. -4
3. $-\frac{1}{2}$
4. $\frac{1}{3}$
5. $-\frac{2}{5}$
6. $-\frac{8}{3}$
7. $\frac{2}{7}$
8. $-\frac{3}{4}$

Tell what you would do to solve the equation. Then solve it.

9. $2y = 6$
10. $3x = -15$
11. $4x = 9$
12. $-2x = 6$
13. $-3y = -15$
14. $\frac{1}{4}x = 8$
15. $\frac{2}{3}y = 12$
16. $\frac{x}{2} = -6$

WRITTEN EXERCISES

Solve.

A
1. $6x = 12$
2. $2x = -8$
3. $5y = -7$
4. $3y = 2$
5. $-2a = 10$
6. $-4a = 12$
7. $-5b = -45$
8. $-3b = -21$
9. $7c = -35$
10. $-8c = -48$
11. $90 = -10d$
12. $-108 = 9d$
13. $105 = 15w$
14. $-96 = 12w$
15. $-8z = 120$
16. $-14z = -196$
17. $\frac{1}{2}c = 10$
18. $\frac{1}{3}c = 4$
19. $\frac{1}{7}d = 5$
20. $\frac{1}{10}d = -1$
21. $-\frac{1}{4}a = -6$
22. $-\frac{1}{5}a = -10$
23. $-\frac{1}{3}b = 7$
24. $-\frac{1}{9}b = 2$
25. $\frac{x}{4} = 8$
26. $\frac{z}{2} = 10$
27. $\frac{a}{3} = -3$
28. $\frac{b}{4} = -20$
29. $-\frac{c}{2} = 9$
30. $-\frac{d}{3} = 5$
31. $-\frac{r}{8} = -2$
32. $-\frac{s}{5} = -5$

B
33. $0.2y = 2.7$
34. $0.7z = -3.5$
35. $0.03w = 1.95$
36. $0.07x = 4.83$
37. $-0.05a = 4.65$
38. $-0.25b = -4.75$
39. $-1.5c = -2.55$
40. $1.8d = 0.99$
41. $\frac{3}{4}x = 12$
42. $\frac{2}{3}x = -6$
43. $\frac{4}{9}y = -4$
44. $\frac{3}{5}y = 6$
45. $\frac{1}{2}w = \frac{3}{4}$
46. $\frac{1}{3}w = \frac{2}{5}$
47. $-\frac{1}{5}z = -\frac{4}{3}$
48. $-\frac{1}{4}z = -\frac{2}{9}$
49. $\frac{5}{6}a = -\frac{2}{5}$
50. $\frac{3}{5}a = -\frac{2}{7}$
51. $-\frac{2}{3}b = \frac{4}{9}$
52. $-\frac{5}{9}b = -\frac{25}{18}$
53. $\frac{5}{7}c = -\frac{15}{28}$
54. $\frac{3}{11}c = \frac{27}{22}$
55. $-\frac{18}{7}d = -\frac{9}{35}$
56. $-\frac{16}{5}d = \frac{8}{15}$

The multiplication property of equality can be stated as:
$$\text{If } a = b, \text{ then } ca = cb.$$
This conditional statement can be written as $(a = b) \rightarrow (ca = cb)$.

C 57. What equation do you get if you use the multiplication property of equality to multiply both sides of $ca = cb$ by the reciprocal of c when $c \neq 0$?

58. Write $ca = cb$ and $c \neq 0$ as a conjunction using logical notation. Then write your result from Exercise 57 as a conditional statement.

59. Is the conditional statement you wrote in Exercise 58 the inverse, converse, or contrapositive (or none of these) of the multiplication property of equality?

60. Is $(a = b) \leftrightarrow (ca = cb)$ true for all values of a, b, and c? Explain.

61. Is $(c \neq 0) \leftrightarrow [(a = b) \leftrightarrow (ca = cb)]$ true for all values of a, b, and c? Explain.

SOLVING EQUATIONS

Section 3-5 USING BOTH ADDITION AND MULTIPLICATION TO SOLVE EQUATIONS

You can combine the methods of the last two sections to solve more complicated equations.

EXAMPLE Solve $2x + 5 = 17$.

SOLUTION Add -5 to both sides.

$$(2x + 5) + (-5) = 17 + (-5)$$
$$2x = 12$$

Now multiply both sides by $\frac{1}{2}$.

$$\frac{1}{2}(2x) = \frac{1}{2}(12)$$
$$x = 6$$

We solved the example above in two steps. First we added an opposite and then we multiplied by a reciprocal. We could have multiplied first and then added, as the following example shows.

EXAMPLE Solve $2x + 5 = 17$.

SOLUTION First multiply by $\frac{1}{2}$.

$$\frac{1}{2}(2x + 5) = \frac{1}{2}(17)$$
$$\frac{1}{2}(2x) + \frac{1}{2}(5) = \frac{17}{2}$$
$$x + \frac{5}{2} = \frac{17}{2}$$

Now add $-\frac{5}{2}$ to both sides.

$$x + \frac{5}{2} + \left(-\frac{5}{2}\right) = \frac{17}{2} + \left(-\frac{5}{2}\right)$$
$$x + \left(\frac{5}{2} - \frac{5}{2}\right) = \frac{17}{2} - \frac{5}{2}$$
$$x = \frac{12}{2}$$
$$x = 6$$

It is usually easier to do the addition first and then the multiplication, as the two examples above show.

EXAMPLE Solve $\frac{x}{5} + 7 = 9$.

SOLUTION

$$\left(\frac{x}{5} + 7\right) - 7 = 9 - 7$$
$$\frac{x}{5} = 2$$
$$5\left(\frac{x}{5}\right) = 5 \cdot 2$$
$$x = 10$$

EXAMPLE Solve $2x - 4(x - 3) = 0$.

SOLUTION Use the distributive property.
$$2x - 4x - 4(-3) = 0$$
$$2x - 4x + 12 = 0$$

Use the distributive property again.
$$(2 - 4)x + 12 = 0$$
$$-2x + 12 = 0$$
$$-2x = -12$$
$$\frac{-2x}{-2} = \frac{-12}{-2}$$
$$x = 6$$

Since variables represent numbers, you get an equivalent equation when you add an expression containing a variable to each side of the original equation.

EXAMPLE Solve $5x - 6 = 3x + 2$.

SOLUTION
$$(5x - 6) + 6 = 3x + 2 + 6$$
$$5x = 3x + 8$$
$$-3x + 5x = -3x + (3x + 8)$$ (Note that we added $-3x$ to each side.)
$$2x = 8$$
$$x = 4$$

EXAMPLE Solve $3(2x + 8) = 4x + 20$.

SOLUTION
$$6x + 24 = 4x + 20$$
$$(6x + 24) - 24 = (4x + 20) - 24$$
$$6x = 4x - 4$$
$$-4x + 6x = -4x + (4x - 4)$$
$$2x = -4$$
$$x = -2$$

The kind of equation you have been studying in this chapter does not always have exactly one solution. For instance, the equation $y = y + 4$ is false for all values of y and hence has no solution. Its solution set is ∅. On the other hand, the equation $y + 4 = 4 + y$ is true for all values of y. Its solution set is the set of real numbers.

ORAL EXERCISES

Describe how you would solve each equation.

1. $3x + 2 = 8$
2. $3x - 5 = 10$
3. $4x - 1 = 3$
4. $-2x + 1 = 3$
5. $-3x + 1 = -5$
6. $\frac{5}{2}x + \frac{1}{2} = \frac{3}{2}$

Find the solution set.

7. $3(x + 5) = 3x + 15$
8. $2(x - 1) = 2x + 7$

SOLVING EQUATIONS

WRITTEN EXERCISES

Solve.

A
1. $3x + 7 = 10$
2. $3y + 2 = 8$
3. $4a - 20 = 20$
4. $6b - 16 = 2$
5. $2c - 9 = -13$
6. $5d + 12 = -3$
7. $14w + 6 = -8$
8. $12z - 3 = -27$
9. $6y - 13 = 20$
10. $4x - 8 = 15$
11. $\dfrac{b}{2} + 10 = 12$
12. $\dfrac{a}{3} - 4 = 5$
13. $4z - \dfrac{1}{2} = -\dfrac{3}{2}$
14. $2w + \dfrac{1}{2} = \dfrac{3}{2}$
15. $3d + 1 = 17$
16. $4c + 1 = 10$
17. $13 - 8x = -11$
18. $20 - 12x = -4$
19. $15 - 4b = 7$
20. $3 - 5a = 8$
21. $17 = 2x - 5$
22. $20 = 11x - 13$
23. $6 = 9x + 9$
24. $10 = 3x + 15$
25. $y + 6 = 6 + y$
26. $w + 3 = w$
27. $\dfrac{s}{5} + 8 = 7$
28. $\dfrac{r}{3} - 9 = -12$
29. $8 + \dfrac{z}{2} = \dfrac{z}{2}$
30. $3 - \dfrac{k}{6} = -\dfrac{k}{6} + 2$

B
31. $4(t + 1) = 8$
32. $3(2s - 1) = 9$
33. $\dfrac{1}{2} + 2y = \dfrac{7}{2}$
34. $\dfrac{3}{2} - 3x = \dfrac{7}{2}$
35. $\dfrac{2}{3}t - 5 = -17$
36. $\dfrac{3}{5}z + 2 = 8$
37. $\dfrac{6}{7}r - 2 = -5$
38. $\dfrac{5}{4}v + 4 = -11$
39. $a + (a + 1) = 37$
40. $b - (b + 1) = 10$
41. $2x + 10 = x$
42. $5y - 9 = 4y - 9$
43. $9z + 5 = 5z - 2$
44. $3w - 1 = 6w - 5$
45. $8j - 8 = 5(j - 2)$
46. $4(k + 2) = 4k + 8$
47. $-4(3 - 3y) = 11(y + 2)$
48. $2(5z - 4) = 3(3z + 6)$
49. $\dfrac{1}{3}(6a + 9) = \dfrac{1}{2}(8a - 4)$
50. $\dfrac{1}{5}(10w - 15) = 4(2w - 1)$
51. $0.3(k + 30) = 9.45$
52. $0.2j = 0.6(8 - j)$
53. $1.5(9 - 6r) = -4.5r$
54. $2.4s = -0.8(2s - 1.2)$

C
55. $x + (3x - 20) = 32$
56. $(2 - 7y) + 4y = 0$
57. $(5t + 11) + (3t - 7) = -4$
58. $(2w + 19) + (5 - w) = 6$
59. $(180 - m) + 5(90 - m) = 60$
60. $2(45 - n) + 5(15 - 2n) = -15$
61. $0.05(x + 6) = 0.13(3 - x)$
62. $5(y - 2.36) = 0.25(10 - 2y)$
63. $2w + 3\left(w - \dfrac{1}{2}\right) = \dfrac{1}{6}(2w + 33)$
64. $\dfrac{1}{9}(3z - 1) = \dfrac{1}{15}(5z + 3)$
65. $\dfrac{1}{5}(15s - 10) + 4 = \dfrac{1}{3}(9s + 6)$
66. $\dfrac{1}{3}(7t + 1) = \dfrac{1}{5}(14t - 25) + \dfrac{1}{7}t$

SELF-TEST 2

Solve.

1. $y + 2 = 17$	2. $a - 5 = 6$	Section 3-3
3. $b - 3 = -7$	4. $-4 = t + 10$	
5. $2c = 16$	6. $3z = -21$	Section 3-4
7. $-4x = 7$	8. $-\frac{1}{4}k = -2$	
9. $2z + 11 = 3$	10. $5r - 1 = 6$	Section 3-5
11. $-14 = 3s - 5$	12. $\frac{x}{5} + 12 = 8$	

Section 3-6 TRANSLATING WORDS INTO EQUATIONS

An important part of studying mathematics is learning to solve problems. In this section, you will practice translating:
- word phrases into algebraic expressions
- word sentences into equations

You can use algebra to solve many real-life problems. To solve a word problem, you translate the words into an equation and then solve the equation. Some frequently used words and their translations are given below.

Words	Translations	
sum, increased by	$+$	(add)
difference, decreased by	$-$	(subtract)
product	$\cdot$	(multiply)
quotient	$\div$	(divide)
3 more than	$+\,3$	(add 3)
3 less than	$-\,3$	(subtract 3)
3 times as much	$\cdot\,3$	(multiply by 3)
$\frac{1}{3}$ of	$\cdot\,\frac{1}{3}$	(multiply by $\frac{1}{3}$)
is, are	$=$	(equals)

Before we translate complete sentences into equations, we will translate some word phrases into algebraic expressions. Notice that when you translate "add" or "more than," the order in which the numbers are added is not important. However, order is very important when you translate "subtract" or "less than."

SOLVING EQUATIONS

EXAMPLES

Word Phrase	Algebraic Expression
a number five more than x	$x + 5$
a number twelve less than x	$x - 12$
a number twice as great as x	$2x$
a number five times x	$5x$
a number seven more than twice x	$2x + 7$
a number one-fifth of x	$\frac{1}{5}x \quad \left(\text{or } \frac{x}{5}\right)$
a number four less than half of x	$\frac{1}{2}x - 4 \quad \left(\text{or } \frac{x}{2} - 4\right)$
twice the sum of x and five	$2(x + 5)$
four times the difference x minus 7	$4(x - 7)$
the sum of x and twice x	$x + 2x$

EXAMPLE Translate into an equation: The sum of the number n and five times n is 18.

SOLUTION $n + 5n = 18$

Before you translate a word problem into an equation, you will usually have to choose a variable to represent a number.

EXAMPLE Translate into an equation: The product of a number and 5 is 12.

SOLUTION Let the number be n.
$5n = 12$

In the example above, we could have chosen a different variable to represent the number. That is, the equation could also be written as $5x = 12$ or as $5a = 12$.

EXAMPLE Translate into an equation: The sum of two consecutive integers is 37.

SOLUTION Let x be the first integer.
Then $x + 1$ represents the next consecutive integer.
$x + (x + 1) = 37$

Many geometric problems can be solved using equations. Recall from your previous work in mathematics that the perimeter of a rectangle is the sum of the lengths of the sides.

EXAMPLE Translate into an equation: The length of a rectangle is twice the width, and the perimeter is 24.

SOLUTION Let $w =$ the width.
Then $2w =$ the length.
$w + 2w + w + 2w = 24$

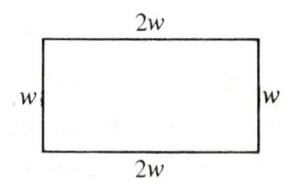

ORAL EXERCISES

Translate the word phrase into an algebraic expression.

1. Eight less than x
2. Five more than y
3. The product of seven and z
4. x increased by ten
5. Twenty decreased by n
6. The sum of w and half of w

Translate the sentence into an equation.

7. One less than the product of five and z is eight.
8. The difference when w is subtracted from ten is five.
9. The quotient when x is divided by nine is negative six.
10. The sum of negative seven and five times y is two.

Complete.

11. Larry is 6 years older than Maria.
    If m = Maria's age, then ___?___ = Larry's age.
12. The temperature today is 7° lower than yesterday.
    If t = yesterday's temperature, then ___?___ = today's temperature.
13. The length of a rectangle is 5 more than twice the width.
    If w = the width, then ___?___ = the length.
14. Kim's height is 1m more than half Jerry's height.
    If j = Jerry's height in meters, then ___?___ = Kim's height in meters.
15. The sum of two numbers is 24.
    If ___?___ = one number, then ___?___ = the other number.
16. Andy is four years older than Barbara. Carlos is twice as old as Andy.
    If Barbara is ___?___ years old, then Andy is ___?___ years old and Carlos is ___?___ years old.

WRITTEN EXERCISES

Write as an algebraic expression. Use n to represent the given number.

A
1. Six more than a number
2. The quotient of a number divided by ten
3. One less than twice a number
4. Five decreased by a number
5. Three times the difference of a number minus five
6. The difference of three times a number minus five

Write as an algebraic expression.

EXAMPLE The value of n nickels (a) in cents; (b) in dollars

SOLUTION (a) $5n$ (b) $0.05n$

7. The value in cents of q quarters
8. The value in cents of d dimes

SOLVING EQUATIONS

9. The value in dollars of d dimes
10. The value in dollars of q quarters
11. The value in dollars of p pennies
12. The value in cents of x dollars
13. The value in cents of $3x$ dimes
14. The value in cents of $2y + 3$ nickels
15. The cost in dollars of r record albums at $5.98 each
16. The cost in dollars of s stamps at 15¢ each
17. The cost in dollars of c cars at $5200 each
18. The cost in dollars of t football tickets at $2 each

Complete.

19. If f is Faith's age now, then ___?___ is Faith's age six years from now.
20. If s is Sam's age now, then ___?___ was Sam's age three years ago.
21. If x is an integer, then the next two consecutive integers are ___?___ and ___?___.
22. If x is an even integer, then ___?___ is the next consecutive even integer.
23. If x is an odd integer, then ___?___ and ___?___ are the next two consecutive odd integers.
24. If x is an integer, then the two integers that differ by one from x are $x + 1$ and ___?___.

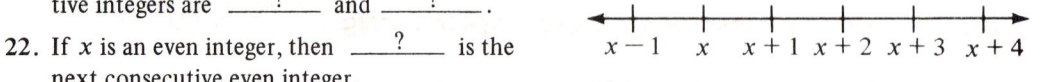

Write the sentence as an equation. Choose a variable to use.

25. Five more than twice a number is 25.
26. Six less than five times a number is 29.
27. The quotient when three times a number is divided by seven is six.
28. The sum of two consecutive integers is 45.
29. When five times a number is decreased by eight, the result is 17.
30. When six times a number is increased by three, the result is 21.
31. Three more than twice a number is sixteen less than three times the number.
32. Twice the sum of a number and four is four more than the number.

B 33. The width of a rectangle is half the length, and the perimeter is 32.
34. The length of a rectangle is five more than four times the width, and the perimeter is 40.
35. One number is six less than another one, and their sum is 16.
36. One number is half another one, and their difference is 10.
37. Luis is three years older than Doris, and the sum of their ages is 29.
38. Carol is four years older than one-third her father's age, and the difference in their ages is 22.

39. Ten years ago Joan was $\frac{1}{3}$ her present age.

40. In thirty years Leo will be three times his present age.

41. The sum of three consecutive integers is -18.

42. The sum of three consecutive odd integers is 33.

43. A number is six more than half its opposite.

44. When three times the opposite of a number is subtracted from the number itself, the difference is six.

45. You have five more dimes than quarters, and the total value is $2.95.

46. You have twice as many quarters as nickels, and the total value is $9.35.

Section 3-7 SOLVING PROBLEMS WITH EQUATIONS

To solve a problem, translate it into an equation and solve the equation. Then you must interpret the solution to the equation in terms of the statement of the problem. Finally, you check that your solution makes sense.

If drawing a picture or sketch helps you to translate the problem, you should draw one. People frequently solve the same problem in different ways. Any technique that you find helps you solve problems is worth using.

We will start with very easy problems, which you may be tempted to solve in your head. But you need to develop a method that you can also use to solve much harder problems.

EXAMPLE Suppose that you want to fence in a square portion of your back yard to have a safe place for your new puppy to play. You have 36 m of fencing in the garage, and you wish to use all of it. How long should each side of the square region be?

DISCUSSION The basic facts are:
(1) The fenced region must be square.
(2) You want to use 36 m of fencing.
The information that isn't known is the length of a side of the square (in meters).

SOLUTION Let the length of a side be s. Since the perimeter is 36 m, we can write:
$$s + s + s + s = 36$$
$$4s = 36$$
$$\frac{1}{4}(4s) = \frac{1}{4}(36)$$
$$s = 9$$

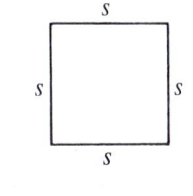

To check this answer, we need to show that if each side of a square is 9 m long, the perimeter is 36 m.
$$9 + 9 + 9 + 9 = 36$$
Each side of the square region is 9 m long.

SOLVING EQUATIONS

When you want to solve a word problem, you should:

1. Read the statement of the problem carefully to determine what facts are known and what is asked for.
2. Choose a variable. Use it to represent the number or numbers asked for.
3. Translate the words that give the facts into an equation.
4. Solve the equation you have written.
5. Check your solution to make sure it fits the statement of the problem.

EXAMPLE George wants to fence in his rectangular vegetable garden at the back of his house to keep out the neighborhood rabbits. He needs to fence only three sides because the back of the house will serve as one side. Also, he would like the side along the house to be twice the length of the two adjoining sides. If George uses 12 m of fencing, what will be the dimensions of the enclosure?

SOLUTION Let w = width (in meters).
Then $2w$ = length (in meters). 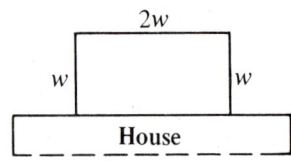
$$w + 2w + w = 12$$
$$4w = 12$$
$$w = 3$$

Check $3 + 6 + 3 = 12$ Therefore the dimensions are 3 m and 6 m.

EXAMPLE The sum of three consecutive integers is 57. What are the integers?

SOLUTION Let x = first integer.
Then $x + 1$ = second integer, and $x + 2$ = third integer.
$$x + (x + 1) + (x + 2) = 57$$
$$3x + 3 = 57$$
$$3x = 54$$
$$x = 18$$
$$x + 1 = 19 \text{ and } x + 2 = 20$$

Check 18, 19, and 20 are consecutive integers whose sum is 57. The solution is correct.

The next example shows one reason why it is important to check your solution to make sure it fits the statement of the problem. Can you think of another important reason why it is not enough to check your solution in the equation that you wrote?

EXAMPLE Find two consecutive odd integers whose sum is 50.

SOLUTION Let x = first odd integer.
Then $x + 2$ = second odd integer. 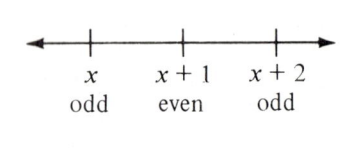
$$x + (x + 2) = 50$$
$$2x + 2 = 50$$
$$2x = 48$$
$$x = 24$$
$$x + 2 = 26$$

Check   24 is a solution of the equation $x + (x + 2) = 50$, *but* 24 and 26 are not odd integers. There are no two consecutive odd integers whose sum is 50.

ORAL EXERCISES

1. Why is it important that you check your answer to a problem with the facts of the original problem?

Tell what you would choose to let the variable x represent.

2. The sum of three consecutive integers is twice the first integer. What are the three integers?
3. The sum of two consecutive even integers is 86. What are the two integers?
4. A board is 5 m long. You want to cut it into two pieces so that one piece is 1 m less than twice the length of the other piece. Find the lengths of the two pieces.
5. Ann is twice as old as Jim was two years ago. The difference in their present ages is 7. Find both ages now.

Make up a problem dealing with the given topic that you feel can be solved with an equation.

6. a rectangle and its perimeter
7. money or coins
8. the ages of two people

WRITTEN EXERCISES

Solve the problem by writing an equation and solving it. Check your answer.

A 1. The sum of two numbers is 24. One number is three times the other. Find both numbers.
2. The length of a rectangle is 6 m more than the width. The perimeter of the rectangle is 20 m. Find the length and width.
3. David rented a car for a flat rate of $21 a day. If the total charge was $147, how many days did he have the car?
4. Twelve packages of weatherstripping cost $13.80. What is the price of each package?
5. Ellen traveled six times as far as Beth. If Ellen traveled 150 km, how far did Beth travel?
6. The school cafeteria tables are all the same length. When 14 tables are placed end to end, they reach the length of the room. If the length of the room is 31.5 m, find the length of each table.
7. Judy is five years older than Punch. If the sum of their ages is 47, how old is each?

SOLVING EQUATIONS

8. The difference in the ages of two people is 8 years. The older person is 3 times the age of the younger. How old is each?

9. A dress on sale was reduced to $15. If that was one-sixth of the original price, find the original price.

10. Ten more than twice a number is equal to four times the number. Find the number.

11. Sixteen more than a number is four less than five times the number. Find the number.

12. The 200 employees of the Ace Paper Company are 7 less than three times the number of workers at Uniclad Electroplating. How many people work at Uniclad?

13. Suppose you enter a number on a hand calculator, double it, and add 27. The number now on the calculator is 49. What was the number you originally entered?

14. Carol rented a car for a day. The charge was $16 plus 5¢ a kilometer. If the total charge was $27.90, how many kilometers did she drive?

15. Five dimes plus some nickels are worth $1.30. How many nickels are there?

16. The total value of some dimes and quarters is $4.95. If there are three times as many dimes as quarters, how many of each coin are there?

17. The sum of two consecutive odd integers is 40. What are the integers?

18. The sum of two consecutive even integers is 86. What are the two integers?

19. The sum of four consecutive integers is 110. Find each integer.

20. The sum of three consecutive odd integers is −75. Find each integer.

B 21. The sum of three consecutive integers is twice the first integer. What are the three integers?

22. Ten decreased by three times a number is equal to the original number increased by 2. Find the original number.

23. A board is 5 m long. You want to cut it into two pieces so that one piece is 1 m less than twice the length of the other piece. Find the lengths of the two pieces.

24. A club decided to have a picnic. Six less than $\frac{4}{5}$ of the members attended. If 74 members were at the picnic, how many people belong to the club?

25. A runner entered a marathon race. If the runner's total time was 2 h 38 min, what was the time for each quarter of the race? (Assume that each quarter was run in the same time.)

26. Commercials make up $\frac{1}{5}$ of the programming time on many television shows. After how many hours of watching these shows would you have seen a total of 144 min of commercials?

27. The quotient of a number divided by 4 is the same as the number decreased by 33. Find the number.

28. When a number is increased by 15, the new number is equal to $\frac{5}{2}$ the original number. Find the original number.

29. Assume you have exactly 20 coins, consisting of nickels, dimes, and quarters. There are 10 nickels. If the total value of all 20 coins is $1.95, how many dimes and how many quarters are there?

30. Ann is twice as old as Jim was two years ago. The difference in their present ages is 7. Find both ages now.

31. Jake is paid 1.5 times his usual hourly rate for overtime work. One week he earned $227.50 for working 42 h, of which 7 h were overtime. Find his usual hourly rate and his overtime rate.

32. A red maple tree was reported to be 72 m shorter than a certain redwood tree. If the height of the redwood was 4 m less than 3 times the height of the red maple, what was the height of the redwood?

Solve each problem if possible. If there is no solution, tell why.

33. Find three consecutive integers whose sum is −60.

34. Find two consecutive integers whose sum is 22.

35. Find two consecutive integers whose difference is 2.

36. The length of a rectangle is 2 m less than twice the width. If the perimeter is 11 m, find the length and width.

37. The length of a rectangle is 2 more than half the width. If the perimeter is 2, find the length and width.

38. The length of a rectangle is 3 less than half the width. If the perimeter is 6, find the length and width.

39. Lee's age is 15 years less than half Bob's age. If the sum of their ages is 9, find the ages.

40. Find two consecutive odd integers such that five times the first integer is 25 less than twice the greater integer.

Section 3-8 WORKING WITH FORMULAS

In the last two sections you used the fact that the perimeter of a rectangle is the sum of the lengths of the sides. If *P, l,* and *w* represent the perimeter, length, and width of the rectangle, another way of stating this fact is to write:

$$P = 2l + 2w$$

This equation is an example of a formula. A *formula* is an equation that shows the relationship between two or more quantities represented by variables. Every formula contains at least two variables.

SOLVING EQUATIONS

When working with an equation in more than one variable, you will sometimes need to express a particular variable in terms of the other variables.

EXAMPLE Solve $P = 2l + 2w$ for l.

SOLUTION
$$P = 2l + 2w$$
$$P - 2w = (2l + 2w) - 2w$$
$$P - 2w = 2l$$
$$\frac{P - 2w}{2} = \frac{2l}{2}$$
$$\frac{P - 2w}{2} = l$$

Check
$$P = 2l + 2w$$
$$P = 2\left(\frac{P - 2w}{2}\right) + 2w$$
$$P = (P - 2w) + 2w$$
$$P = P$$

EXAMPLE Solve $a = 2s - rs$ for s.

SOLUTION
$$a = 2s - rs$$
$$a = (2 - r)s$$
$$\frac{a}{2 - r} = \frac{(2 - r)s}{2 - r}$$
$$\frac{a}{2 - r} = s$$

Is this solution valid for all values of a and r? Since division by zero is not possible, $2 - r \neq 0$. But $2 - r = 0$ is equivalent to $2 = r$. Thus $r \neq 2$.

Check
$$a = 2s - rs$$
$$a = 2\left(\frac{a}{2 - r}\right) - r\left(\frac{a}{2 - r}\right)$$
$$a = \frac{2a}{2 - r} - \frac{ra}{2 - r}$$
$$a = \frac{2a - ra}{2 - r}$$
$$a = \frac{(2 - r)a}{2 - r}$$
$$a = a$$

ORAL EXERCISES

For what value of x is the expression not defined?

1. $\dfrac{1}{x}$

2. $-\dfrac{x}{4x}$

3. $\dfrac{a}{x + 2}$

4. $\dfrac{2b}{3 - x}$

5. $\dfrac{cd}{2x + 10}$

6. $\dfrac{1}{9 - 6x}$

CHAPTER 3

Solve for x or y. State any restrictions on possible values of the variables.

7. $x + a = b$
8. $y - 3 = -d$
9. $ay = b$
10. $-\frac{1}{3}x = d$
11. $4a - 3x = 12$
12. $c = \frac{ay}{2}$

WRITTEN EXERCISES

Solve for the variable indicated. State any restrictions on possible values of the variables.

A
1. $A = bh$; b
2. $D = rt$; r
3. $V = lwh$; h
4. $V = lwh$; l
5. $3x = 2a$; x
6. $3x = 2a$; a
7. $r = s - t$; s
8. $r = s - t$; t
9. $3x - 2y = 1$; x
10. $3x - 2y = 1$; y
11. $2x - b = 3$; x
12. $2x - b = 3$; b
13. $S = \frac{n}{2}(f + l)$; f
14. $A = \frac{h}{2}(b + c)$; c
15. $A = \frac{bh}{2}$; b
16. $A = P + Prt$; r
17. $x - y = 3x - y$; x
18. $2r + st = 2q$; r

B
19. $A = P + Prt$; P
20. $a = b - bc$; b
21. $1 - x = ax$; x
22. $t = \frac{V - k}{g}$; g
23. $t = \frac{V - k}{g}$; V
24. $t = \frac{V - k}{g}$; k

SELF-TEST 3

Write as an algebraic expression.

1. Six less than three times a number x Section 3-6
2. The value in cents of $3n$ quarters
3. Frank's age eight years ago if his present age is a
4. Write as an equation: Eight more than twice a number is 16.

Solve by writing an equation. Check your answer.

5. The sum of three consecutive even integers is 54. Find the integers.   Section 3-7
6. The length of a rectangle is four more than three times the width. The perimeter is 64. Find the length and the width.
7. Six years from now Beth will be twice as old as she was two years ago. Find her age now.
8. One-third of a number is -5. Find the number.

Solve for the variable indicated.

9. $a = 2b - c$; b
10. $a = 2b - c$; c
11. $A = \frac{bh}{2}$; h Section 3-8

SOLVING EQUATIONS

Computer Exercises

1. a. Copy and RUN the following program.

   ```
   10  PRINT "TO EVALUATE 14A - 12B + 17C,"
   20  PRINT "    INPUT A, B, C."
   30  PRINT "TYPE ALL THREE VALUES ON ONE LINE,";
   40  PRINT "    WITH COMMAS AFTER A AND B."
   50  INPUT A,B,C
   60  PRINT
   70  PRINT "14A - 12B + 17C =";
   80  PRINT 14*A-12*B+17*C
   90  END
   ```

   b. Make up another expression involving A, B, and C. Change lines 10, 70, and 80 to match your expression. Notice the differences between lines 70 and 80. RUN your new program.

2. RUN the following program to solve the given equations. Observe that you must first express the equation in the form $Ax + B = Cx + D$. For example, you must rewrite the equation $5x - 6 = 2 - 3x$ as $5x + -6 = -3x + 2$, so that $A = 5, B = -6, C = -3$, and $D = 2$.

   ```
   10  PRINT "TO SOLVE AN EQUATION OF THE FORM"
   20  PRINT "    AX + B = CX + D (A<>C)"
   30  PRINT "    INPUT THE VALUES A,B,C,D"
   40  INPUT A,B,C,D
   50  LET X=(D-B)/(A-C)
   60  PRINT "X = ";X
   70  END
   ```

   (*Note*: "<>" means "≠" in BASIC.)

   a. $5x - 6 = 2 - 3x$

   b. $-2x + 7 = x - 5$

   c. $3x + 8 = 5$

3. The program in Exercise 2 cannot be used to solve equations such as $2x + 3 = 2x + 5$ or $3x - 8 = 3x - 8$. Explain. (*Hint*: Examine line 50.)

CHAPTER REVIEW

Evaluate the algebraic expression if $a = 3$, $b = -1$, $c = \dfrac{1}{3}$, and $d = -5$.

1. $5a - 2b$ 2. $\dfrac{a + d}{b}$ 3. $c(a + b + d)$ Section 3-1

4. Write without parentheses an expression that always has the same value as $7(k - 2)$.

If the domain of x is $\{-2, -1, 0, 1, 2\}$, write the solution set of the open sentence.

5. $2 - 5x = 7$
6. $2x + 6 = 10$
7. $-5 = 1 + 3x$

Section 3-2

Are the two equations equivalent? If so, what is their solution set?

8. $t = 4$ and $-10 - 3t = 2$
9. $-3 = k$ and $1 - k = 4$

Solve.

10. $-12 + c = 19$
11. $4 + n = -1.5$
12. $3y - 2y + 6 = -1$ Section 3-3

13. $-7w = 175$
14. $0.9m = -68.4$
15. $-\frac{5}{6}a = -35$ Section 3-4

16. $22 - 7d = -6$
17. $\frac{s}{5} + 3 = -7$
18. $\frac{3}{8} - x = \frac{5}{8}$ Section 3-5

19. $k - 5 = k + 5$
20. $4r - 9 = r - 3$
21. $-2(2y - 3) = 6 - 4y$

Write as an algebraic expression.

22. The sum of a number, x, and $\frac{2}{3}$ of x Section 3-6

23. The value in cents of p pennies and n nickels

24. The perimeter of a rectangle whose width is 4 less than its length, l

25. Write as an equation: Three times the sum of a number, n, and 4 is 10 more than the number.

Solve the problem by writing an equation and solving it. Check your answer.

26. Eight more than the product of 5 and a number is equal to the number. Find the number. Section 3-7

27. Manuel had $60 in $5 bills and $10 bills. If he had 4 times as many $5 bills as $10 bills, how many of each did he have?

28. Becky's score on her first English test was 12 points less than her score on the second test. If half the sum of the two scores was 79, what were her scores?

Solve $A = \frac{1}{2}h(x - y)$ for the variable indicated. State any restrictions on possible values of the variables.

29. h
30. x
31. y Section 3-8

SOLVING EQUATIONS 101

CHAPTER TEST

Evaluate the algebraic expression if $w = 2$, $x = 1$, $y = -4$, and $z = -\frac{1}{2}$.

1. $x + 2z$
2. $\frac{w - y}{w + x}$
3. $z(w + y)$
4. $3z + 4z - 5z$

Solve if $\{-4, -2, 0, 2, 4\}$ is the domain of n.

5. $5n - 4 = -14$
6. $3 - 2n = 3$
7. $-\frac{1}{2}n = 2$

8. Are the two equations equivalent? If so, what is their solution set?
$$15 - x = 7 \quad \text{and} \quad \frac{x}{4} = 2$$

Solve.

9. $\frac{1}{3}m = -5$
10. $2y - y - 1 = -2$
11. $5.02 + a = 8.51$
12. $3b - 4 = 14$
13. $-7c = -91$
14. $j + 5 - j = 5$
15. $-\frac{x}{5} = 15$
16. $d - \frac{5}{6} = -\frac{17}{6}$
17. $\frac{3}{4}f + 5 = 2$
18. $12q - 1 = 6q + 2$
19. $\frac{1}{2}(6 - 4h) = -3 + h$
20. $\frac{z}{4} - 2 = -1$

Write as an algebraic expression.

21. Five times a number, y, increased by half the number
22. The value in cents of n nickels and $2n + 1$ dimes

Solve the problem by writing an equation and solving it. Check your answer.

23. Rafael is 5 years younger than Seth, and the sum of their ages is 19. What are their ages?
24. The opposite of a number is 21 less than twice the number. Find the number.
25. The sum of four consecutive odd integers is -8. What are the integers?
26. The length of a rectangle is 3 more than the width. The perimeter of the rectangle is 44. Find the length and the width.

Solve for the indicated variable. State any restrictions on possible values of the variables.

27. $3r - st = 2$; s
28. $V = \frac{1}{2}abh$; b

Application — Balancing a Checkbook

To balance a checkbook you must first have kept an accurate record of all withdrawals and deposits for your account. You can calculate your account's running balance in your checkbook register by subtracting each withdrawal and adding each deposit to the previous balance. This running balance is checked every time you receive a monthly statement, along with your canceled checks, from your bank. Elizabeth's bank statement looks something like this:

Super Now Account 14881233214   Service Charges $0.80   Items 2

Date	Description	Amount	Balance
02/10/8-	Opening Balance		$802.59
02/13/8-	Check Paid #129	18.00—	784.59
02/14/8-	Check Withdrawal #134	25.00—	759.59
02/15/8-	Deposit—Interest	3.12	762.71
02/16/8-	Deposit	20.00	782.71
02/17/8-	DP—24 HR Teller 129 Trem.	50.00	832.71
02/18/8-	WD—24 HR Teller 129 Trem.	46.31—	786.40
02/19/8-	Withdrawal—General Insur.	100.00—	686.40
02/20/8-	Deposit—Little, Green	250.00	936.40
03/09/8-	Fee—Service Charge	0.80—	935.60
03/09/8-	Fee—Maintenance Charge	1.00—	934.60

At the top of the statement is the account number and the service charges and number of checks for that month. On 2/13 the bank paid check #129; on 2/14, Elizabeth cashed check #134 on her account; on 2/16, she made a deposit with a teller at the bank; on 2/17, she deposited $50 through an automatic teller; on 2/18, she withdrew $46.31 through an automatic teller; on 2/19, $100 was automatically withdrawn from her account and paid to General Insurance; on 2/20, her pay check from Little, Green was automatically deposited in her account.

These are the steps for balancing a checkbook record with a statement:

1. Enter your checkbook balance. $ _____
2. Subtract all bank charges, such as service fees.   − _____
3. Add interest and any other credits.   + _____
   CHECKBOOK TOTAL $ _____

(Record items 2 and 3 and new running balance in checkbook.)

4. Enter statement closing balance.   _____
5. Add any deposits not shown on the statement.   + _____
6. Subtract all checks not yet paid by the bank.   − _____
   STATEMENT TOTAL, which should agree with
   checkbook total $ _____

SOLVING EQUATIONS

VOCABULARY REVIEW

Be sure that you understand the meaning of these terms:

value of a variable, p. 73
algebraic expression, p. 73
evaluating an expression, p. 74
equation, p. 76

solution set of an equation, p. 77
equivalent equations, p. 77
formula, p. 97

MIXED REVIEW

Given that p is false, q is false, and r is true, determine the truth value.

1. $\sim(p \vee q)$
2. $(q \rightarrow r) \leftrightarrow p$
3. $\sim q \wedge r$

Simplify.

4. $0.03(-4.12)$
5. $\frac{3}{4} + \frac{7}{12} + \frac{1}{3}$
6. $\frac{-341.04}{812}$

7. $-\frac{1}{2} - \frac{7}{8}$
8. $-0.2 \div 1.28$
9. $\left(-\frac{18}{35}\right)\left(-\frac{77}{39}\right)$

10. a. Write the contrapositive of "If $x^2 \neq 1$, then $x \neq 1$".
    b. Are the given statement and its contrapositive logically equivalent?

11. Solve $y = mx + b$ for x.

12. The lengths of the sides of a triangle are consecutive integers. The perimeter of the triangle is 39 cm. Write and solve an equation to find the length of each side.

13. a. Construct a truth table for $(p \wedge \sim q) \rightarrow \sim p$.
    b. Is the statement a tautology, a contradiction, or neither?

14. Simplify: $1 + (2^2 - 1)^2 \cdot 3$

15. Evaluate $\frac{6x - y}{z}$ if $x = \frac{2}{3}$, $y = 1$, and $z = -6$.

16. Write out {the additive inverses of the integers between 1 and 5} using the roster method.

17. Solve $-\frac{2}{3}y = \frac{8}{9}$ if the domain of y is $\left\{\frac{16}{27}, \frac{4}{3}, -\frac{16}{27}, -\frac{4}{3}\right\}$.

18. Graph: {integers between -2 and 3}

19. Classify $x \cdot y = y \cdot x$ as always true, always false, or sometimes true and sometimes false. Assume that x and y are positive integers.

Solve.

20. $-\frac{1}{8}a = 2$
21. $t - 0.18 = 7.1$

22. $m + \dfrac{4}{5} = -\dfrac{6}{5}$
23. $4z - 2(1 + 2z) = 0$
24. $\dfrac{k}{4} + 1 = 6$
25. $12 - 3j = -3(j - 4)$

In Exercises 26 and 27, let p represent "$2(3 + 5) = (2 + 3)(2 + 5)$" and let q represent "$6(9 - 4) = (9 - 6)(6 - 4)$."

26. Write $q \rightarrow \sim p$ in words and give its truth value.

27. Write $\sim q \vee \sim p$ in words and give its truth value.

28. a. Write the converse, inverse, and contrapositive of "If $x^2 > y^2$, then $x > y$."
    b. Give the truth value of each statement if x and y are negative.

29. Use a truth table to show that $\sim p \wedge q$ and $\sim(p \vee \sim q)$ are logically equivalent.

Name the property that each statement illustrates.

30. $-\dfrac{1}{3}\left(\dfrac{3}{5} + \dfrac{2}{3}\right) = -\dfrac{1}{3}\left(\dfrac{3}{5}\right) + \left(-\dfrac{1}{3}\right)\left(\dfrac{2}{3}\right)$
31. $-\dfrac{1}{3}\left(\dfrac{3}{5} + \dfrac{2}{3}\right) = \left(\dfrac{3}{5} + \dfrac{2}{3}\right)\left(-\dfrac{1}{3}\right)$
32. $0(0.8 + 4.3) = 0$
33. $-\dfrac{3}{7}\left(-\dfrac{7}{3}\right) = 1$

34. Write as an algebraic expression: the wages paid for h hours of work if the pay rate is \$12 per hour

35. Given {integers} as the replacement set for x, find three values of the variable that make $x^2 > x^3$ true and three values that make it false.

36. Write the negation of "3 is not greater than 4."

37. a. Write as an equation: Five more than half of a number is one less than the number.
    b. Find the number.

38. (Optional) Put the following argument into symbolic form. Tell if it is valid. If it is, specify which argument pattern is used.
    If this is true, then that is false.
    That is false.
    Therefore this is true.

Some everyday problems involve inequalities. For example, as discussed on the facing page, the driver of a truck approaching a bridge must consider whether the weight of the loaded truck is greater than the bridge's load limit.

4
Solving Inequalities

A truck weighing 6 metric tons is carrying a load of TV sets weighing 3 metric tons. It approaches a bridge which has a load limit of 10 metric tons. Is it safe for the truck to cross the bridge? Could it cross safely carrying a 5-ton load? What loads could it carry safely across the bridge?

To answer problems such as this, we use inequalities rather than equations. In this chapter you will learn how to handle inequalities and how to use them in solving problems.

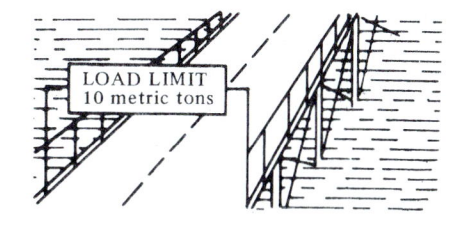

Section 4-1 INTRODUCTION TO INEQUALITIES

You are already familiar with these inequality symbols:

$\neq$ means "does not equal," or "is not equal to"
$<$ means "is less than"
$>$ means "is greater than"

Any sentence containing one of these signs is called an *inequality*.

EXAMPLES

$2 \neq 7$ Two does not equal 7.
$2 < 7$ Two is less than 7.
$-5 < 4$ Negative five is less than four.
$8 > 2$ Eight is greater than 2.

You can use a number line as an aid in determining whether an inequality is true or false. Recall from Section 1-2 that since -5 is to the left of 2 on the number line, the following inequalities are true:

$-5 < 2$ $2 > -5$

In general terms,

1. The number a is less than the number b if a is to the left of b on the number line.
2. The number a is greater than the number b if a is to the right of b on the number line.

When you first studied equations you found that easy ones could be solved at sight or by trial and error. The same is true for inequalities.

EXAMPLE Solve the inequality $x - 5 < 12$, where the domain of x is {positive integers}.

SOLUTION A little thinking should help you to see that several different values of x would make the inequality true. For example, substituting 1 for x produces $1 - 5 < 12$, which is true because $1 - 5$ is -4, and certainly -4 is less than 12. Of course, other solutions for x exist, such as $x = 10$, since $10 - 5$ is 5, and $5 < 12$. Trying values for x of $1, 2, 3, \ldots, 17$ leads to the discovery that substituting any positive integer less than 17 for x produces a true sentence.

The solution set of $x - 5 < 12$ where the domain of x is {positive integers} is $\{1, 2, 3, \ldots, 16\}$.

In the next three sections, you will study some systematic methods for solving inequalities.

ORAL EXERCISES

Tell whether the inequality is true or false.

1. $2 < 4$
2. $6 > 10$
3. $-5 > 1$
4. $-3 < -2$
5. $2 < 2$
6. $0 > 3$
7. $3 \neq \frac{9}{3}$
8. $10 > 0$
9. $-6 < 6$
10. $7 \neq 7$

Tell whether the given solution set is correct. Assume that the domain of the variable is $\{-2, -1, 0, 1, 2\}$.

11. $x - 2 < 5$ $\{0, 1, 2\}$
12. $y + 1 > 3$ $\{0\}$
13. $x + 3 > -10$ $\{-2, -1, 0, 1, 2\}$
14. $n - 5 < 0$ $\{-2, -1, 0, 1, 2\}$
15. $y - 4 < 2$ $\{0, 1, 2\}$
16. $t - 2 > -6$ $\emptyset$

WRITTEN EXERCISES

Express with an inequality symbol.

A 1. y is less than 10.
2. x is not equal to 18.
3. n is greater than 3.
4. y is greater than 0.
5. x is less than -7.
6. y is less than 6 subtracted from 14.
7. n is not equal to the product of -4 and 9.
8. x is greater than the sum of 8 and -5.

Translate into a word sentence.

9. $x < 4$ 10. $y < -5$ 11. $n > -3$ 12. $x > 0$

Solve. Assume that the domain of the variable is {positive integers}.

B 13. $y - 8 < 12$ 14. $x + 5 < 20$ 15. $d + 6 > 0$
   16. $x + 7 > -3$ 17. $n - 4 > 0$ 18. $y + 12 < 30$
   19. $t - 9 < -5$ 20. $x + 4 > 4$ 21. $m + \frac{1}{2} > 8\frac{1}{2}$
   22. $y + 2\frac{1}{4} < \frac{1}{4}$ 23. $n + 2 < 0$ 24. $t - \frac{1}{3} > 8\frac{2}{3}$

For Exercises 25-27, express each fact using an inequality.

C 25. The temperature of the Health Center swimming pool is always greater than 18°C.

26. A certain car gets less than 12 kilometers per liter of gasohol.

27. In a certain mathematics class all the students are less than 17 years old.

28. The Comparison Property (also called the Trichotomy Property) states that if a and b are real numbers, then one and only one of the following statements is true: $a < b$, $a = b$, $a > b$. Explain in your own words what this property means.

29. The Transitive Property of Order states that if a, b, and c are real numbers, and if $a < b$ and $b < c$, then $a < c$. Illustrate this property with some real numbers.

Section 4-2 EQUIVALENT INEQUALITIES; ADDITION PROPERTIES

When you learned to solve equations you found that many which look different are actually equivalent. You used equivalent equations to arrive at the solution of a given equation. In the solution of inequalities likewise we make use of equivalent inequalities. In this section you will learn how to:
- draw graphs of inequalities
- write equivalent inequalities
- use addition properties of inequalities

Remember that, as stated in Chapter 3, in this book we assume that the domain of each variable is the set of real numbers unless otherwise indicated.

EXAMPLE   Graph $x < 6$.

SOLUTION

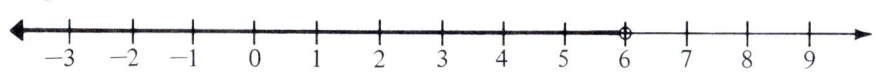

```
◄——+——+——+——+——+——+——+——+——ο——+——+——+——►
   -3  -2  -1   0   1   2   3   4   5   6   7   8   9
```

To show all the numbers less than 6, we start at 6 and draw a heavy line to the left, as shown. The open dot at 6 means that 6 is not included. But all the numbers to the left of 6 are included.

EXAMPLE Graph the solution set of $y - 2 > 6$.

SOLUTION Try a few values for y.

$y = -1$	$-1 - 2 > 6?$	False
$y = 0$	$0 - 2 > 6?$	False
$y = 2$	$2 - 2 > 6?$	False
$y = 7$	$7 - 2 > 6?$	False
$y = 8$	$8 - 2 > 6?$	False
$y = 8\frac{1}{2}$	$8\frac{1}{2} - 2 > 6?$	True
$y = 10$	$10 - 2 > 6?$	True

It appears that every value of y greater than 8 will make $y - 2 > 6$ a true statement. Hence we say that $y > 8$ and $y - 2 > 6$ have the same solution set.

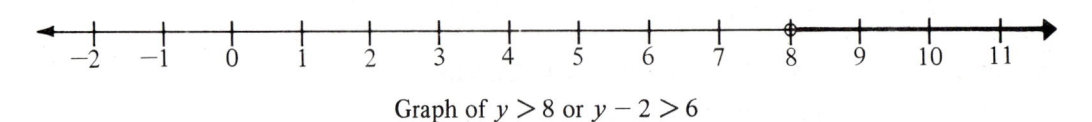

Graph of $y > 8$ or $y - 2 > 6$

Inequalities having the same solution set are said to be *equivalent inequalities*. $y - 2 > 6$ and $y > 8$ are equivalent inequalities.

Of course we need rules for solving inequalities. Perhaps you have guessed that $y - 2 > 6$ can be solved by adding 2 to each side of the inequality.

$$y - 2 > 6$$
$$y - 2 + 2 > 6 + 2$$
$$y - 0 > 8$$
$$y > 8$$

This is an illustration of the following property.

ADDITION PROPERTY FOR $a > b$

If a, b, and c are real numbers and if $a > b$, then $a + c > b + c$.

Again, take the example $x - 1 < 6$. We may write

$$x - 1 + 1 < 6 + 1$$
$$x < 7$$

We have used the second addition property:

ADDITION PROPERTY FOR $a < b$

If a, b, and c are real numbers and if $a < b$, then $a + c < b + c$.

EXAMPLE Solve $x - 5 > 2$ and graph the solution set.

SOLUTION Add the opposite of -5 (that is, 5) to both sides.
$$(x - 5) + 5 > 2 + 5$$
$$x + (-5 + 5) > 2 + 5$$
$$x + 0 > 7$$
$$x > 7$$

The solution set is {all real numbers greater than 7}.

A shorthand way to write {all real numbers greater than 7} is $\{x: x > 7\}$. We read this as "the set of all x such that x is greater than 7."

EXAMPLE Solve $y + 4 < 10$ and graph the solution set.

SOLUTION Add the opposite of 4 (that is, -4) to both sides.
$$y + 4 + (-4) < 10 + (-4)$$
$$y + (4 - 4) < 10 - 4$$
$$y + 0 < 6$$
$$y < 6$$

The solution set is $\{y: y < 6\}$.

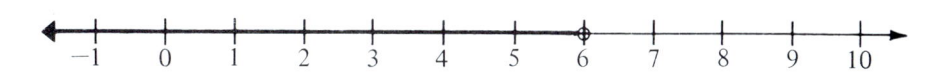

These two examples illustrate the fact that, in general, to solve an inequality of the form $x + a > b$ or $x + a < b$, you add the opposite of a to both sides of the inequality.

ORAL EXERCISES

Tell whether or not the two inequalities are equivalent.

1. $x + 1 > 4$, $x > 3$
2. $y - 6 < 2$, $y < 4$
3. $x - 5 < 0$, $x < 5$
4. $n + 7 > -1$, $n > -8$
5. $y + 10 > 10$, $y > 20$
6. $4 > n$, $n < 4$

Tell what number should be added to both sides of the inequality to solve it.

7. $t + 4 < 10$
8. $x - 3 > 8$
9. $n - 6 > -3$
10. $y + 3 > 0$
11. $x + 20 < -8$
12. $y - 4 < 0$

SOLVING INEQUALITIES

WRITTEN EXERCISES

Write an inequality for the graph. Use x for the variable.

A 1. [number line with open circle at -2, arrow extending right from -3 to 4]

2. [number line with open circle at 5, from 0 to 7]

3. [number line with open circle at 6, from 0 to 7]

4. [number line with open circle at -18, from -20 to -14]

Graph the inequality.

5. $x < 6$ 6. $y > -3$ 7. $n < 0$ 8. $x < -5$

Solve by adding an opposite.

9. $x + 8 < 13$
10. $x + 1 < 4$
11. $x - 5 > 0$
12. $x - 2 > 7$
13. $y - 3 < -4$
14. $y - 2 < -7$
15. $y - 13 > -4$
16. $y + 7 > -12$
17. $x + 20 < 20$
18. $y - 5 > -5$
19. $4 + y > 8$
20. $20 + x < 4$

B 21. $x - 20 > \dfrac{3}{4}$
22. $x - 4 < \dfrac{1}{2}$
23. $y - 6 > -\dfrac{1}{2}$
24. $y + \dfrac{1}{2} < 3$
25. $y - \dfrac{2}{3} < \dfrac{1}{2}$
26. $x + \dfrac{5}{2} < \dfrac{5}{4}$

EXAMPLE Solve the inequality $x + a < b$ for x.

SOLUTION Add the opposite of a (that is, $-a$) to both sides.
$$(x + a) + (-a) < b + (-a)$$
$$x + [a + (-a)] < b - a$$
$$x + 0 < b - a$$
$$x < b - a$$
The solution set is $\{x : x < b - a\}$.

Solve for x.

C 27. $x + b > a$ 28. $x - c < -d$
29. $a < b + x$ 30. $c > x + d$

Section 4-3 INEQUALITIES INVOLVING MULTIPLICATION

When an inequality involves multiplication we must pay attention to the inequality signs as we solve. The following examples will show why.

EXAMPLES

$3 < 5$	$3 \cdot 2 < 5 \cdot 2$	True
$6 > 2$	$6 \cdot 2 > 2 \cdot 2$	True
$3 < 5$	$3(-2) < 5(-2)$	False
$10 > 2$	$10(-3) > 2(-3)$	False

To make the results in the last two examples true statements, we must reverse the inequality signs.

$$3(-2) > 5(-2) \quad \text{True}$$
$$10(-3) < 2(-3) \quad \text{True}$$

In other words, multiplying both sides of an inequality by a negative number reverses the direction of the inequality.

We use the following properties:

MULTIPLICATION PROPERTIES OF INEQUALITIES

Given the real numbers a, b, and c:
If c is a positive number and if $a < b$, then $c \cdot a < c \cdot b$
If c is a positive number and if $a > b$, then $c \cdot a > c \cdot b$
If c is a negative number and if $a < b$, then $c \cdot a > c \cdot b$
If c is a negative number and if $a > b$, then $c \cdot a < c \cdot b$

Notice that these properties involve only nonzero multipliers. Multiplying both sides of an inequality by zero always produces the equation $0 = 0$.

EXAMPLE Solve $3x < 12$ and graph the solution set.

SOLUTION Multiply both sides by $\frac{1}{3}$, the reciprocal of 3.

$$\frac{1}{3}(3x) < \frac{1}{3}(12)$$
$$\left(\frac{1}{3} \cdot 3\right)x < 4$$
$$1 \cdot x < 4$$
$$x < 4$$

The solution set is $\{x: x < 4\}$.

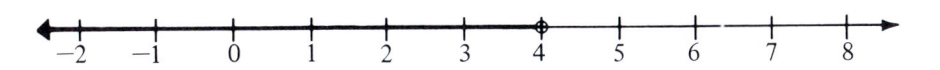

SOLVING INEQUALITIES

EXAMPLE Solve $\frac{1}{5}y > 2$ and graph the solution set.

SOLUTION Multiply both sides by 5, the reciprocal of $\frac{1}{5}$.

$$5\left(\frac{1}{5}y\right) > 5(2)$$
$$\left(5 \cdot \frac{1}{5}\right)y > 10$$
$$1 \cdot y > 10$$
$$y > 10$$

The solution set is $\{y: y > 10\}$.

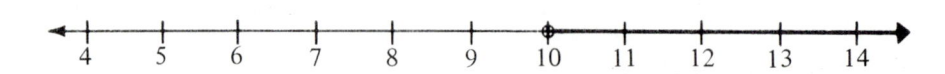

EXAMPLE Solve $-6x > 24$.

SOLUTION Multiply both sides by $-\frac{1}{6}$, the reciprocal of -6. Since the multiplier is negative, we reverse the direction of the inequality sign.

$$-6x > 24$$
$$-\frac{1}{6}(-6x) < -\frac{1}{6}(24)$$
$$\left[-\frac{1}{6}(-6)\right]x < -4$$
$$1 \cdot x < -4$$
$$x < -4$$

The solution set is $\{x: x < -4\}$.

EXAMPLE Solve $-20 < -4y$.

SOLUTION Multiply both sides by $-\frac{1}{4}$, the reciprocal of -4, and reverse the direction of the inequality sign.

$$-\frac{1}{4}(-20) > -\frac{1}{4}(-4y)$$
$$5 > \left[-\frac{1}{4}(-4)\right]y$$
$$5 > 1 \cdot y$$
$$5 > y$$

The solution set is $\{y: 5 > y\}$.

CHAPTER 4

ORAL EXERCISES

By what number would you multiply both sides of the inequality in order to solve it? Would the direction of the inequality sign be reversed?

1. $2x < 6$
2. $3n < 15$
3. $-\frac{1}{4}y < 6$
4. $7n < 7$
5. $-2x < 6$
6. $-3n > 15$
7. $\frac{1}{3}x > 8$
8. $\frac{3}{4}y < 9$
9. $-\frac{3}{5}t < 4$
10. $-4t > 9$
11. $9x > 2$
12. $8y > 3$

WRITTEN EXERCISES

A 1–12. Solve Oral Exercises 1–12 and graph the solution sets.

Solve.

13. $6x < 12$
14. $3y < 8$
15. $7 > 5t$
16. $3n < 2$
17. $8 < -2x$
18. $-4x > 12$
19. $\frac{1}{2}x > 10$
20. $4 < \frac{1}{3}d$
21. $\frac{3}{4}y < -9$
22. $5x < -5$
23. $-44 > 11h$
24. $-3m > -12$

B 25. $8x < 14$
26. $-10 > -\frac{1}{5}x$
27. $-2x > \frac{1}{3}$
28. $8 > -21y$
29. $-17 < -\frac{3}{2}x$
30. $-4x > -\frac{6}{5}$

EXAMPLE Solve $cx < b$ for x if $c \neq 0$.

SOLUTION *Case 1.* $c > 0$. $cx < b$
$\frac{1}{c}(cx) < \frac{1}{c} \cdot b$
$\left(\frac{1}{c} \cdot c\right)x < \frac{b}{c}$
$1 \cdot x < \frac{b}{c}$
$x < \frac{b}{c}$

Case 2. $c < 0$. $cx < b$
$\frac{1}{c}(cx) > \frac{1}{c} \cdot b$
$\left(\frac{1}{c} \cdot c\right)x > \frac{b}{c}$
$1 \cdot x > \frac{b}{c}$
$x > \frac{b}{c}$

Thus, $x < \frac{b}{c}$ if $c > 0$; $x > \frac{b}{c}$ if $c < 0$.

Solve for y if $b \neq 0$ and $c \neq 0$.

C 31. $by > a$
32. $-cy < d$
33. $\frac{1}{b} \cdot y < a$
34. $-\frac{1}{c} \cdot y > d$

SOLVING INEQUALITIES

SELF-TEST 1

Express with an inequality symbol.

1. n is greater than 7.
2. y is less than 0.
3. x is not equal to 36.
4. t is greater than -14.

Section 4-1

Graph the inequality.

5. $x < 3$
6. $y > -4$

Section 4-2

Solve.

7. $t - 5 < 7$
8. $n + 3 > -4$
9. $4x < 12$
10. $\frac{1}{5}y > 2$

Section 4-3

11. $-6x > 18$
12. $\frac{1}{3}t < -7$

Section 4-4 COMPOUND INEQUALITIES

In the chapter on logic you studied compound sentences. Some very useful examples of compound sentences are compound inequalities. In this section you will learn how to solve:
- compound inequalities which are disjunctions
- compound inequalities which are conjunctions

The following examples show two compound inequalities which are disjunctions and their graphs. Notice that in each graph a solid black dot is used to indicate that the number associated with the point is included.

EXAMPLE $x \leqslant 4$ This means that x is either equal to 4 or less than 4. Thus, $x \leqslant 4$ is a short form for the disjunction $(x = 4) \lor (x < 4)$.

EXAMPLE $t \geqslant -3$ This means that t is either equal to -3 or greater than -3. That is, $(t = -3) \lor (t > -3)$.

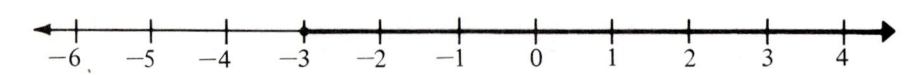

116   CHAPTER 4

Since a compound inequality such as $t \geq -3$ is a disjunction, it is true if either $t = -3$ or $t > -3$, and false if both these sentences are false. The truth table at the right shows the truth value of $t \geq -3$ for some replacements for t.

t	$t = -3$	$t > -3$	$t \geq -3$
-4	F	F	F
-3	T	F	T
0	F	T	T

You may think of solving a compound inequality involving the symbol $\geq$ as solving two sentences, an equation and an inequality. You may also solve directly using the same properties as for simple inequalities.

EXAMPLE Solve $x - 3 \leq 2$ and graph the solution set.

SOLUTION 1 $(x - 3 = 2) \vee (x - 3 < 2)$

$$
\begin{array}{l|l}
x - 3 = 2 & x - 3 < 2 \\
x - 3 + 3 = 2 + 3 & x - 3 + 3 < 2 + 3 \\
x + 0 = 5 & x + 0 < 5 \\
x = 5 & x < 5
\end{array}
$$

The solution set is $\{x : (x = 5) \vee (x < 5)\}$, that is, $\{x : x \leq 5\}$.

SOLUTION 2
$$x - 3 \leq 2$$
$$x - 3 + 3 \leq 2 + 3$$
$$x + 0 \leq 5$$
$$x \leq 5$$

The solution set is $\{x : x \leq 5\}$.

We can also form compound inequalities that are conjunctions.

EXAMPLE $7 < x < 11$ This means that 7 is less than x and x is less than 11. Thus, $7 < x < 11$ is a short form for the conjunction $(7 < x) \wedge (x < 11)$. In other words, x is *between* 7 and 11.

EXAMPLE $3 \geq x > -2$ This means that 3 is equal to or greater than x and x is greater than -2. That is, $(3 \geq x) \wedge (x > -2)$.

SOLVING INEQUALITIES   117

You may solve a compound inequality that is a conjunction in much the same way you solve one that is a disjunction.

EXAMPLE Solve $5 < y - 3 \leq 8$ and graph the solution set.

SOLUTION 1
$$(5 < y - 3) \wedge (y - 3 \leq 8)$$

$$\begin{array}{l|l} 5 < y - 3 & y - 3 \leq 8 \\ 5 + 3 < y - 3 + 3 & y - 3 + 3 \leq 8 + 3 \\ 8 < y + 0 & y + 0 \leq 11 \\ 8 < y & y \leq 11 \end{array}$$

The solution set is $\{y: (8 < y) \wedge (y \leq 11)\}$, that is, $\{y: 8 < y \leq 11\}$.

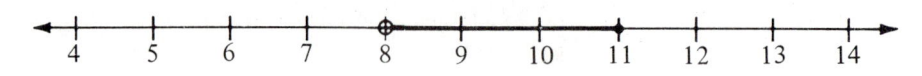

SOLUTION 2
$$5 < y - 3 \quad \leq 8$$
$$5 + 3 < y - 3 + 3 \leq 8 + 3$$
$$8 < y + 0 \quad \leq 11$$
$$8 < y \leq 11$$

The solution set is $\{y: 8 < y \leq 11\}$.

ORAL EXERCISES

Read the compound inequality as a disjunction or a conjunction.

1. $x \geq 1$
2. $5 < x < 15$
3. $y \leq -6$
4. $-7 > t > -9$

Tell whether the inequality is true or false.

5. $7 \leq 0$
6. $9 \geq 8 > 5$
7. $-3 \leq 3$
8. $5 \geq 5$
9. $-6 < 0 \leq 1$
10. $13 \leq 13 < 15$
11. $-2 \leq 0 < -1$
12. $16 \leq 20 \leq 24$

Tell whether the given solution set is correct.

13. $x + 2 \geq 3$ $\quad \{x: x \geq 5\}$
14. $4y \leq 8$ $\quad \{y: y \leq 2\}$
15. $n + 7 \leq -3$ $\quad \{n: n \leq -10\}$
16. $\frac{1}{3}t \geq 5$ $\quad \{t: t \geq \frac{5}{3}\}$

If x is an integer, what is the solution set of the inequality?

17. $5 < x < 11$
18. $8 \geq x > 0$
19. $-7 < x \leq 0$
20. $4 > x \geq -4$

WRITTEN EXERCISES

A Express with an inequality symbol or symbols.

1. d is equal to or less than 10.
2. y is equal to or greater than -6.
3. 3 is less than x and x is less than 9.
4. x is between 20 and 30.

5. 12 is greater than t and t is equal to or greater than 7.
6. -9 is equal to or less than n and n is less than -8.
7. 0 is greater than x and x is equal to or greater than -11.
8. d is between 5 and -5.

Solve, and graph the solution set.

B 9. $y - 8 \leq 1$ 10. $t + 3 \geq 0$ 11. $12 \leq 6x$ 12. $y - 12 \leq -2$
13. $-2n \leq 8$ 14. $x + 5 \geq -10$ 15. $3 \geq y - 4$ 16. $3d \leq -12$
17. $x - 6 \geq -3$ 18. $-7n \geq 7$ 19. $-\frac{1}{5}t \leq \frac{2}{5}$ 20. $-24 \leq 8y$
21. $12 < 3x < 18$ 22. $16 \geq x - 4 > 3$ 23. $5 < y + 3 \leq 9$ 24. $7 > t - 8 \geq 0$

Section 4-5 INEQUALITIES INVOLVING ADDITION AND MULTIPLICATION

You can now combine the methods you have studied in the preceding sections and learn to solve more involved inequalities.

EXAMPLE Solve $-2x + 5 > 17$ and graph the solution set.

SOLUTION Add -5 to both sides.
$$(-2x + 5) + (-5) > 17 + (-5)$$
$$-2x + (5 - 5) > 12$$
$$-2x + 0 > 12$$
$$-2x > 12$$

Now multiply both sides by $-\frac{1}{2}$.
$$-\frac{1}{2}(-2x) < -\frac{1}{2}(12)$$
$$\left(-\frac{1}{2} \cdot -2\right)x < -6$$
$$x < -6$$

The solution set is $\{x: x < -6\}$.

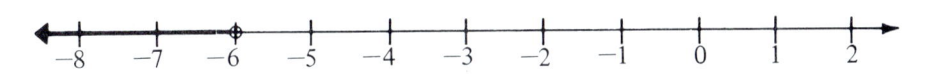

In this example an inequality of the form $ax + b > c$ was solved in two steps. First we added the opposite of b to both sides and then we multiplied both sides by the reciprocal of a. Since a is negative in this example, we reversed the direction of the inequality when we multiplied by $\frac{1}{a}$, the reciprocal of a.

When a variable appears on both sides of the inequality sign, the first step in solving the inequality is to eliminate the variable from one side.

SOLVING INEQUALITIES

EXAMPLE Find the solution set of $3x + 5 > 5x + 17$.

SOLUTION Add $-5x$ to both sides.
$$(-5x) + (3x + 5) > (-5x) + (5x + 17)$$
$$(-5x + 3x) + 5 > (-5x + 5x) + 17$$
$$-2x + 5 > 0 + 17$$
$$-2x + 5 > 17$$

Proceeding as in the previous example, we find that the solution set is $\{x : x < -6\}$.

ORAL EXERCISES

Describe the steps you would take to solve the inequality.

1. $3x + 2 < 8$
2. $3y - 4 \leq 10$
3. $4n - 1 > 3$
4. $-2n + 1 \geq 3$
5. $-3t + 1 \leq -5$
6. $2x + \frac{1}{2} < 3$
7. $3y + 7 \leq 2y + 2$
8. $6 \leq 4m + 2$

WRITTEN EXERCISES

A 1–8. Solve Oral Exercises 1–8 and graph the solution sets.

Solve.

9. $3x + 7 < 10$
10. $2 > 8 - 3y$
11. $-4n - 20 < 20$
12. $6y - 16 \geq 20$
13. $-\frac{1}{2}t + 10 > 12$
14. $2r + \frac{1}{2} \leq \frac{3}{2}$
15. $14 \geq 10 + 4d$
16. $5x - 3 < -3$
17. $4n + 2 < 10$
18. $8x - 13 > 11$
19. $-12x + 20 \leq -4$
20. $0 \leq \frac{1}{5}y$

B 21. $\frac{1}{4}t + 5 < 10$
22. $4(m + 1) < 8$
23. $-3(2x - 1) > 9$
24. $\frac{1}{2} - 2y > -\frac{7}{2}$

25. $4s + 10 > 3s$
26. $8k + 4 < 5k - 11$
27. $3t + 6 \geq 7t + 2$
28. $-6(r - 5) \leq 9r$
29. $8d + 8 \leq 2(d + 7)$
30. $4 \leq 2x + 4 < 6$
31. $-5 < 3y + 1 \leq 10$
32. $12 > \frac{1}{2}x + 4 \geq 0$

C 33. $-3x - \frac{10}{3} \leq 4(x - 5)$
34. $11t - \frac{1}{3}(t + 1) \geq \frac{1}{2}t + 3$

35. $3 \geq -\frac{1}{3}x + 2 \geq -3$
36. $0 < 4y - \frac{1}{2} < 2$

37. $-\frac{1}{3}(x + 7) \leq \frac{1}{2}(3x - 1)$
38. $\frac{4}{5}(6z - 5) + \frac{9}{10}(-z + 8) \leq 2z - \frac{3}{5}$

CHAPTER 4

SELF-TEST 2

Express with an inequality symbol or symbols.

1. x is equal to or less than -3.

2. y is between 8 and 3.

Section 4-4

Solve, and graph the solution set.

3. $t + 2 \geqslant 8$

4. $n - 7 \leqslant 3$

5. $2x + 3 \leqslant 5$

6. $-\frac{1}{3}y - 6 \geqslant \frac{2}{3}$

Section 4-5

7. $5 \leqslant 2x - 7 < 9$

8. $3 \leqslant 5x + 3 < 13$

Section 4-6 SOLVING PROBLEMS WITH INEQUALITIES

Now you will have the opportunity to solve some problems which involve inequalities. The same basic methods you used in solving problems with equations will also apply here with just a few changes. In each of the following examples you will see that we find an inequality to fit the problem situation, solve the inequality, and then check the solution to the inequality with the original problem situation to make sure the solution is suitable.

EXAMPLE   There are 3 more goldfish in a large aquarium than in a small one. There are fewer than 25 goldfish in all. How many goldfish might there be in each aquarium?

SOLUTION   Let f = number of goldfish in small aquarium.
Then $f + 3$ = number of goldfish in large aquarium.

$$f + (f + 3) < 25$$
$$2f + 3 < 25$$
$$2f < 22$$
$$f < 11$$
$$f + 3 < 14$$

The following table shows the pairs of numbers that fit the conditions of the problem.

f	0	1	2	3	4	5	6	7	8	9	10
$f + 3$	3	4	5	6	7	8	9	10	11	12	13

Check   The first pair of numbers indicates that there are no fish in the small aquarium and 3 fish in the large aquarium. Since 3 is 3 greater than 0, and the sum of 0 and 3 is less than 25, the number pair 0 and 3 is a solution to the problem. Each of the other number pairs should be checked in the same way. You might also like to check the numbers 11 and 14 to confirm that they do *not* fit the conditions of the problem.

SOLVING INEQUALITIES

EXAMPLE The sum of the length and width of a rectangle is equal to or less than 60 cm and the length is twice the width. Find the greatest possible dimensions of the rectangle.

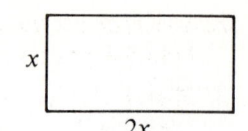

SOLUTION Let x = width (in centimeters) of the rectangle.
Then $2x$ = length (in centimeters).

$$x + 2x \leqslant 60$$
$$3x \leqslant 60$$
$$\tfrac{1}{3}(3x) \leqslant \tfrac{1}{3}(60)$$
$$x \leqslant 20$$
$$2x \leqslant 40$$

Thus the greatest possible width of the rectangle is 20 cm and the greatest possible length is 40 cm.

Check Is the sum of the length and the width of the rectangle equal to or less than 60? Yes, since $40 + 20 \leqslant 60$.
Is the length twice the width? Yes, since $40 = 2(20)$.

There are some special phrases or clue words to look for in setting up an inequality. Here are a few of the more common ones, and their translations.

Clue words	Translations
is less than; is smaller than	<
is more than; is greater than	>
is at least	⩾
is at most	⩽
is not more than	⩽
is not less than	⩾

ORAL EXERCISES

Do the two phrases have the same meaning?

1. "is equal to or greater than" and "is at least"
2. "is not less than" and "is at most"
3. "is equal to or greater than" and "is not less than"
4. "is at least" and "is not greater than"

WRITTEN EXERCISES

Solve the problem by using an inequality. Check your answer.

A 1. The sum of two numbers is at least 30. One number is five times the other number. Find the least possible values for these numbers.

2. The sum of two numbers is at most 10. One number is four times the other. Find the greatest possible values for the two numbers.

3. Ellen is 7 years younger than Carol. The sum of their ages is at least 39. At the very least, how old are Ellen and Carol?

4. The dimensions of a rectangle are whole numbers. Its length is 3 times its width. Its perimeter is less than 72 cm. Find the greatest possible dimensions of the rectangle.

5. There are 38 classrooms in a school with at least 15 students and at most 34 students in each classroom. What can you say about the total number of students in the school?

6. If the combined salaries of two workers are not less than $38,000 and one person's salary is $3000 less than the other's, what is the least possible salary of each worker?

7. If a liter of gasoline costs 35 cents, at most how many full liters of gasoline can be bought for $10?

8. Five years from now Rosa will be more than twice as old as she is now. What can you say about her age today?

B 9. How long must Gerald drive at 84 km/h in order to travel at least 8 km more than Reiko, who drove at 73 km/h for 2 h?

10. You earn $45 per week and would like to save enough each week so that you will have at least $1200 at the end of the year not including bank interest. At the very least, how much should you plan to save each week? (Assume that you save the same amount each week.)

11. The sum of two consecutive odd integers and their doubles is not more than 132. What are the greatest values these numbers might have?

12. You plan to save a certain amount of money one week and then to keep doubling your savings for each of the next three weeks. If you wish to save a total of at least $12, what is the least amount you should save the first weck?

C 13. Pat Johnson is going to fence a rectangular plot using fifty boards each 30 cm wide. The desired length of the plot must be three times the width, and the boards are not to be cut. What is the length and the width in meters of the largest plot that can be fenced? (The plot will be fenced in by placing the boards vertically.)

14. The perimeter of a rectangle, whose length is 8 m more than its width, must be at least 20 m and at most 32 m. Find all possible whole-number solutions.

15. The lengths of the sides of a triangle are whole numbers. Two of the sides are equal in length. The third side is 6 cm less than one half the length of one of the other sides. If the perimeter is less than 78 cm, find all the possible values for the lengths of the sides of the triangle.

16. A certain job can be finished by Allan in t hours, while it takes David $\frac{1}{2}t + 3$ hours to complete the same job. You know that t is less than $\frac{1}{2}t + 3$. You also know that the sum of the two individuals' times is at least 9 hours. Could either Allan or David start and complete the job between the hours of 9 A.M. and 3 P.M. of the same day? If both Allan and David begin their jobs at 9 A.M., what is the earliest possible time each worker may complete his job?

17. There are 5 exams in a marking period. Gerry received grades of 70, 78, 90, and 95 on the first 4 exams. What grade must she get on the last exam so that her average for the marking period will be at least 85? (Remember, to find the average of 5 scores you add the scores and divide their sum by 5.)

SELF-TEST 3

Solve and check.

Section 4-6

1. Two numbers, one 5 greater than the other, have a sum of at least 13. What are the least possible values for these numbers?
2. Doris is 3 years younger than twice Diana's age. The sum of their ages is at most 42. What are their oldest possible ages?
3. The difference of two numbers is at most 27. The larger of the numbers is equal to 5 more than twice the smaller. Find the greatest possible values for the numbers.

Computer Exercises

The following program solves inequalities of the form $Ax + B < C$ and shows the steps in the solution process.

```
10   PRINT "THIS PROGRAM SOLVES INEQUALITIES OF THE FORM:"
20   PRINT "    AX + B < C, A<>0"
30   PRINT "WHAT ARE THE VALUES: A, B, C"
40   PRINT "    (TYPE COMMAS AFTER A AND B)";
50   INPUT A,B,C
60   PRINT
70   PRINT A;"X +";B;" <";C
80   PRINT A;"X <";C-B
90   IF A<0 THEN 120
100  PRINT "X <";(C-B)/A
110  STOP
120  PRINT "X >";(C-B)/A
130  END
```

124   CHAPTER 4

1. Why are there two different PRINT statements after line 90?
2. RUN the program to solve:
   a. $3x + 8 < 2$
   b. $-8x + 17 < -7$
   c. $2x - 8 < 13$
3. Try some other examples.
4. Revise lines 20, 70, 80, and 90 so that the program will solve inequalities of the form $Ax + B > C$.

CHAPTER REVIEW

1. Express in symbols: t is greater than -8.   Section 4-1
2. Translate into a word sentence: $r < 5$

Solve. The domain of x is {positive integers}.
3. $x - 1 > 3$
4. $7 > x + 5$

Graph the inequality.
5. $m < -2$
6. $r > 0$ Section 4-2

Solve. The domain of each variable is {real numbers}.
7. $y + 6 > 10$
8. $-2 + z < -3$
9. $7a < 42$
10. $-9m < 27$ Section 4-3
11. $4x > -\frac{1}{2}$
12. $-8 > -\frac{2}{3}b$

13. Write as a conjunction or a disjunction: $1 > z > -1$ Section 4-4
14. Express with an inequality symbol or symbols:
    p is equal to or less than 14

Solve, and graph the solution set.
15. $c - 7 \geqslant 0$
16. $-5w \leqslant 10$
17. $2n - 7 \leqslant -13$
18. $2 - \frac{1}{3}k > 2$ Section 4-5
19. $3w - 4 > 4w + 1$
20. $-7 < 4x - 3 \leqslant -3$

Solve and check.
21. One number is 4 times another. The sum of the numbers is at least -15. Find the least possible values of the numbers.   Section 4-6
22. The length of a rectangle is 15 cm less than twice the width. If the perimeter of the rectangle is at most 300 cm, find the greatest possible length of the rectangle.

SOLVING INEQUALITIES   125

CHAPTER TEST

Write each inequality in symbols.

1. a is greater than 17.
2. -4 is greater than b and b is equal to or greater than -8.
3. 1 is equal to or less than c.
4. d is between -2 and 2.

Write each inequality as a word sentence.

5. $-2 < w$
6. $x \geqslant 8$
7. $0 < y \leqslant 5$

Solve. The domain of t is {positive integers}.

8. $9 + t > 18$
9. $3 \geqslant t + 4$
10. $t - \dfrac{1}{5} < \dfrac{14}{5}$

Solve. The domain of each variable is {real numbers}.

11. $8q \leqslant 40$
12. $6s + 5 \geqslant 3s - 1$
13. $-5 > t - 2 > -10$

Solve, and graph the solution set.

14. $x - 4 \geqslant -2$
15. $4 > \dfrac{2}{7} y$
16. $-8z + 3 \leqslant 11$

Solve and check.

17. The sum of two consecutive odd integers is not less than -12. Find the least possible values for the integers.
18. Karen would like to make a long-distance call from a pay telephone. The charge is $1.10 for the first three minutes and $.15 for each additional minute (or fraction of a minute). If Karen has only enough change to pay for any charge up to $3.00, what is the greatest number of minutes that she may talk?
19. The sum of a number and twice its opposite is less than twelve. What can you say about the number?

CUMULATIVE REVIEW: CHAPTERS 1-4

Indicate the best answer by writing the appropriate letter.

1. Simplify: $-38 + (-80 + 38)$
   a. 0
   b. 80
   c. -80
   d. -42
2. The converse of $q \to p$ is
   a. $\sim q \to \sim p$
   b. $p \to q$
   c. $q \leftrightarrow p$
   d. $\sim p \to \sim q$
3. Simplify: $-128 - (-135)$
   a. -263
   b. -7
   c. 263
   d. 7

126 CHAPTER 4

4. Solve: $-\dfrac{x}{8} = 4$
   a. 2
   b. $-\dfrac{1}{2}$
   c. 32
   d. -32

5. Simplify: $-221 \div (-13)$
   a. 17
   b. -17
   c. -2873
   d. $-\dfrac{1}{17}$

6. Solve: $-5d < 15$
   a. $\{d:d > -3\}$
   b. $\{d:d < -3\}$
   c. $\{d:d > 3\}$
   d. $\{d:d < 3\}$

7. On the number line shown, what is the coordinate of the point halfway between H and B?

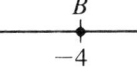

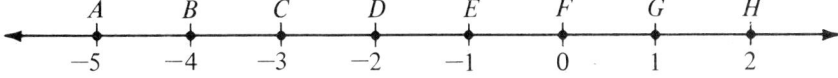

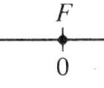

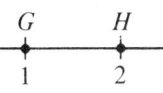

A	B	C	D	E	F	G	H
-5	-4	-3	-2	-1	0	1	2

   a. -1
   b. F
   c. E
   d. 0

8. $p \wedge q$ is true when p is __?__ and q is __?__.
   a. T; T
   b. T; F
   c. F; T
   d. F; F

9. Find the value of $(a + b) \div (-c)$ when $a = -3$, $b = -7$, and $c = 2$.
   a. 5
   b. -5
   c. -6
   d. 4

10. If the domain of y is {positive integers}, find the solution set of $y - 3 < 1$.
    a. $\{0, 1, 2, 3\}$
    b. $\{1, 2, 3, 4\}$
    c. $\{0, 1, 2, 3, 4\}$
    d. $\{1, 2, 3\}$

11. Solve: $\dfrac{1}{3} - r = \dfrac{1}{3}$
    a. $\dfrac{2}{3}$
    b. $\emptyset$
    c. 0
    d. $\dfrac{1}{9}$

12. Let p be "I volunteer at the hospital" and q be "I meet interesting people." Then "I meet interesting people if I volunteer at the hospital" can be written symbolically as
    a. $p \rightarrow q$
    b. $q \rightarrow p$
    c. $p \leftrightarrow q$
    d. $q \rightarrow \sim p$

13. Simplify: $-\dfrac{24}{25} \div \dfrac{9}{10}$
    a. $\dfrac{15}{16}$
    b. $-\dfrac{108}{125}$
    c. $-\dfrac{16}{15}$
    d. $-\dfrac{15}{16}$

14. $\sim(p \vee \sim q)$ is true when p is __?__ and q is __?__.
    a. T; T
    b. T; F
    c. F; T
    d. F; F

15. The graph at the right is the graph of the solution set of which inequality?
    a. $x + 2 < 4$
    b. $-4 > x - 2$
    c. $2 < x + 4$
    d. $2 - x < 4$

16. Find the solution set of $7 - 3n = 10$.
    a. $\emptyset$
    b. $\{1\}$
    c. $\{-1, 1\}$
    d. $\{-1\}$

SOLVING INEQUALITIES

17. If p and q are both false, which of the following is false?
    a. the converse of $p \rightarrow q$
    b. the inverse of $p \rightarrow q$
    c. the contrapositive of $p \rightarrow q$
    d. none of these

18. The length of a rectangle is 4 more than the width, w. Which expression represents the perimeter of the rectangle?
    a. $w + (w + 4)$
    b. $2(w + 4)$
    c. $2(w + (w + 4))$
    d. $2(w + (w - 4))$

19. The inequality $-1 < y - 2 \leqslant 4$ is equivalent to:
    a. $-3 < y \leqslant 2$
    b. $1 < y \leqslant 6$
    c. $-2 < y \leqslant 3$
    d. $-6 \leqslant y < -1$

20. Which of the following does *not* describe or represent the empty set?
    a. $\emptyset$
    b. $\{$ an integer between 1 and 2 $\}$
    c. $\{x : x = x + 3\}$
    d. $\{$ the opposite of 0 $\}$

21. $(p \rightarrow q) \wedge (q \rightarrow p)$ is logically equivalent to
    a. $p \wedge \sim q$
    b. $p \leftrightarrow q$
    c. $\sim(p \vee q)$
    d. $(p \rightarrow q) \vee (q \rightarrow p)$

22. Which graph shows the solution to $(x < -2) \vee (x \geqslant 0)$?

    a. [number line from -3 to 1]
    b. [number line from -3 to 1]
    c. [number line from -3 to 1]
    d. [number line from -3 to 1]

23. Solve: $-2(m - 3) - m = 9$
    a. -1
    b. -5
    c. 5
    d. 15

24. Find the solution set of $3 \leqslant 2s + 1 < 7$ if s is an integer.
    a. $\{1, 2\}$
    b. $\{1, 2, 3\}$
    c. $\{2, 3\}$
    d. $\{2, 3, 4, 5\}$

25. Solve $ba = bc - 2$ for b.
    a. $b = \dfrac{c - 2}{a}, a \neq 0$
    b. $b = -\dfrac{2}{a - c}, a \neq c$
    c. $b = \dfrac{2}{a - c}, a \neq c$
    d. $b = \dfrac{c - a}{2}$

26. What is the solution set of $2k + 3 = 2(k - 4) + 1$?
    a. $\emptyset$
    b. $\{3\}$
    c. $\{-7\}$
    d. $\{10\}$

27. The dimensions of a rectangle are consecutive odd integers. Its perimeter is greater than 56 cm. What are the least possible dimensions of the rectangle?
    a. 27 cm by 29 cm
    b. 15 cm by 13 cm
    c. 16 cm by 14 cm
    d. 17 cm by 15 cm

28. Given the true statement: "If a and b are both integers, then $a + b$ is an integer," which of the following *must* be true?
    a. If $a + b$ is an integer, then a and b are both integers.
    b. If $a + b$ is not an integer, then a and b are not both integers.
    c. If a and b are not both integers, then $a + b$ is not an integer.
    d. If $a + b$ is not an integer, then a and b are both integers.

CHAPTER 4

Careers — Epidemiologist

The term *epidemiology* means the study of epidemics. Twenty or thirty years ago epidemiologists studied large outbreaks of infectious diseases and tried to control epidemics with such programs as vaccination and quarantine. As the Western world experienced fewer epidemics, these scientists found that their skills were applicable to the modern problems of chronic disease. Most epidemiologists now study the causes of such public health problems as cancer, heart disease, and lung failure.

Their work can be broken down into three types: research directed at finding disease causes, teaching, and service activities such as consulting with industry and city and state health departments. Epidemiologists do not do research in laboratories. Their test groups are human populations that are studied by community surveys.

One example of a recent epidemiologic study is a heart study that was carried out in Framingham, Massachusetts. Most of the population of the town was monitored on a regular basis. The scientists were looking for possible causes of cardiovascular disease, such as poor diets and high salt intake. Another group of epidemiologists followed the population of Harrisburg, Pennsylvania, for five years following the accident at Three Mile Island Nuclear Power Plant. They were looking for illness that could be associated with exposure to radiation.

Schools of public health, state departments of public health, and industries involved with toxic chemicals or radiation all employ epidemiologists. Other large employers are the National Institute for Occupational Safety and Health (NIOSH), and the World Health Organization (WHO).

Thirty years ago, all epidemiologists were physicians. Now only 50 percent have M.D.'s; the rest have a general science background and a Ph.D. in epidemiology from a school of public health. Good background for the field includes courses in mathematics, biologic sciences, chemistry, computer science, and biostatistics.

SOLVING INEQUALITIES

VOCABULARY REVIEW

Be sure that you understand the meaning of these terms:

inequality, p. 107
equivalent inequalities, p. 110

MIXED REVIEW

1. Give the negation of "New York City is the capital of New York" and the truth value of the negation.

Graph the solution set.

2. $x < -2$

3. $y - 7 \geqslant -3$

4. a. Name the opposite of $-\frac{4}{5}$. b. Name the reciprocal of $-\frac{4}{5}$.

5. Use a truth table to show that $\sim(\sim p \vee q)$ and $p \wedge \sim q$ are logically equivalent.

Evaluate if $a = -4$, $b = 2$, $c = -1$, and $d = \frac{1}{4}$.

6. $(a - 2c - b)d$

7. $\frac{b-a}{cd}$

Is the open sentence always true, sometimes true, or never true? Assume that x and y are counting numbers.

8. $\left(x + \frac{2}{3}\right) + y = \left(\frac{2}{3} + y\right) + x$

9. $xy \leqslant 0$

10. $x - y < x + y$

11. $\frac{x}{y}$ is a counting number.

Simplify.

12. $(7 + (-3))^2 - 4 \cdot 5$

13. $0.16 \div (-80)$

14. $\left(\frac{8}{3} - \frac{4}{9}\right) \div \frac{5}{27}$

Solve.

15. $-0.2x = -4.24$

16. $9 + 4z = 2(2z - 3)$

17. $3s - 4 \geqslant 14$

18. $1 \geqslant 11 - j > -5$

19. $-\frac{11}{6} = \frac{13}{15} + b$

20. $-\frac{3}{4}d \leqslant 81$

21. $2p - 3 = 10p + 1$

22. $0.08 - 3r < r$

23. $n - 23 > -14$

Express as an equation or an inequality. Choose a variable to use, if necessary.

24. x is greater than 12.

25. The product of 2 consecutive even integers is 168.

Solve if the domain of x is {positive integers}.

26. $7x - 6 - 4x = 3(x - 2)$
27. $5 - 3x < 1$

28. Solve $A = \frac{1}{2}st$ for s.

Solve and check.

29. Hua jogs at the rate of 8 km/h. One day he jogged more than 25 km. What is the least possible number of minutes he spent jogging?

30. The sum of two consecutive integers is at most -25. Find the greatest possible values for the numbers.

31. In 30 years, Martina will be three times as old as she is now. How old is she now?

32. A bank contains 90 coins, consisting of nickels, dimes, and quarters. There are 42 dimes and 3 times as many nickels as quarters. If the total value of the coins is $9, how many quarters are there?

In Exercises 33-35, let p represent "$-\frac{1}{2} > -\frac{2}{3}$," q represent "$x \cdot (-1) = -x$," and r represent "$4^2 = 32$." Write each statement in words and give its truth value.

33. $p \wedge q$
34. $\sim r \rightarrow \sim p$
35. $p \leftrightarrow q$

36. Graph: {the whole numbers less than 5}

Construct a truth table for each statement and determine whether it is a tautology, a contradiction, or neither.

37. $\sim(p \wedge q) \rightarrow \sim p \vee q$
38. $\sim(\sim p \vee q) \rightarrow q$

39. Write using the roster method: {states of the United States with names beginning with N}

Given that p and q are true and r is false, determine the truth value of each statement.

40. $p \vee (\sim q \wedge r)$
41. $(\sim p \wedge q) \vee r$

Points, line segments, and planes — basic geometric figures that you will study in this chapter — can be seen in this photograph of a pumping station for an aqueduct in California.

Basic Concepts of Geometry 5

More than 2000 years ago, Euclid and other Greek mathematicians recognized that the concepts of *point*, *line*, and *plane* are the basic "building blocks" necessary for the study of geometry. Euclid's *Elements* presented the principles of geometry in a systematic manner that is remarkably similar to the way geometry is studied in many high schools today.

Section 5-1 POINTS, LINES, AND PLANES

In this section you will learn about:
- points
- lines
- planes
- some properties that relate points, lines, and planes

A *point* is the simplest figure you can study in geometry. All other geometric figures consist of sets of points. A point is usually pictured as a dot and is named by a capital letter such as *A* or *B*. Although a dot must have some size, the point it represents has no size.

A *line* extends without end in opposite directions, as the arrowheads in the picture suggest. Any two points that lie on a line can be used to name a line. For example, you could refer to the line shown as $\overleftrightarrow{CD}$ (read "line CD"), or as $\overleftrightarrow{DE}$, $\overleftrightarrow{CE}$, $\overleftrightarrow{DC}$, $\overleftrightarrow{ED}$, or $\overleftrightarrow{EC}$. A line can also be named by a lower-case letter such as *l*.

A *plane* is a flat surface that continues without end in all directions. If you can imagine the surface of the floor of a room extending indefinitely in all directions, you will have a good mental picture of a plane.

Since it is impossible to draw a picture that extends indefinitely, it is customary to use a figure such as that shown to represent a plane.

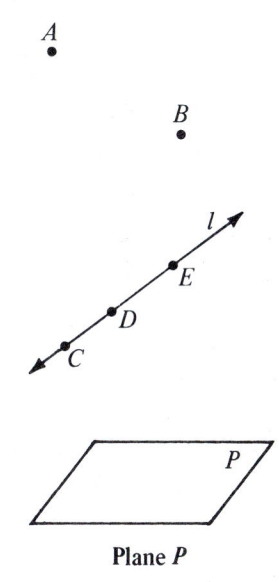

Plane *P*

The *intersection* of two geometric figures is the set of points that the figures have in common. In the figure, the intersection of plane Q and line k is point F. (You might think of a wire piercing a sheet of paper.) If two figures contain at least one common point, the figures *intersect*.

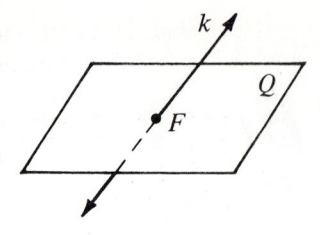

The dashes indicate the part of the figure that is hidden from view.

Collinear points are points that lie on the same line.

Coplanar points are points that lie in the same plane.

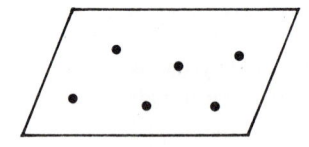

Some properties of points, lines, and planes are stated below.

Through any two points there is exactly one line.
(Two points determine a line.)

Through any three noncollinear points there is exactly one plane.
(Three noncollinear points determine a plane.)

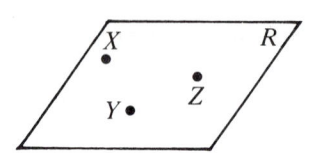

If two planes intersect, then their intersection is a line.

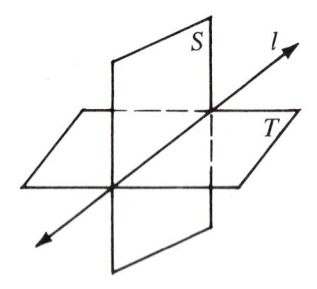

If two lines intersect, then their intersection is a point.

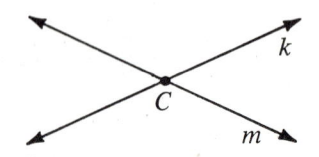

134   CHAPTER 5

If two lines lie in the same plane and do not intersect, they are *parallel*. The symbol ∥ is used for "is parallel to." In the figure on the left below, *n* ∥ *m*.

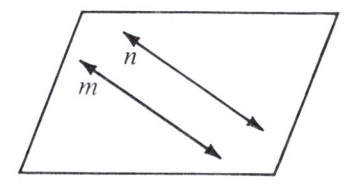

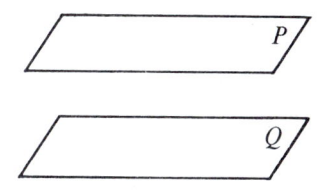

Parallel planes are planes that do not intersect. In the figure on the right above, *P* ∥ *Q*.

ORAL EXERCISES

In the figure shown, name:

1. five points
2. two lines
3. a plane
4. three noncollinear points
5. four noncoplanar points

In Exercises 6–12 consider *all* points, lines, or planes that satisfy the given condition, not just those shown in the figure.

6. How many points lie on $\overleftrightarrow{RU}$? in plane *P*?
7. How many lines does plane *P* contain?
8. How many lines contain *R* and *S*?
9. How many lines contain *R* and *T*?
10. How many lines contain *R*, *U*, and *W*?
11. How many planes contain *S*, *U*, and *W*?
12. How many planes contain *R*, *S*, and *U*?

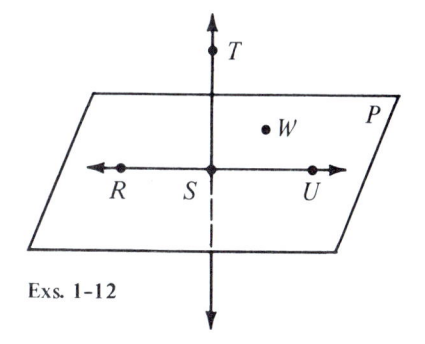

Exs. 1–12

13. Find an example in your classroom of two parallel lines.
14. Find an example in your classroom of two lines that are neither parallel nor intersecting.
15. Find an example in your classroom of two parallel planes.

BASIC CONCEPTS OF GEOMETRY

WRITTEN EXERCISES

Classify each statement as true or false.

A 1. $\overleftrightarrow{BC}$ and $\overleftrightarrow{AB}$ are the same line.
2. Lines *l* and *m* contain point *B*.
3. Line *m* lies in plane *Q*.
4. $\overleftrightarrow{BD}$ and $\overleftrightarrow{AC}$ interesect in point *B*.
5. Points *B*, *C*, and *D* are collinear.
6. Points *E*, *A*, and *C* are coplanar.
7. Points *A*, *D*, and *E* determine a plane.
8. There are points that are on line *l* but are not in plane *Q*.
9. Plane *Q* and plane *P* (not shown) both contain points *B* and *C*. What is their intersection?
10. What is the intersection of $\overleftrightarrow{AC}$ and $\overleftrightarrow{BE}$?
11. How many lines contain points *D* and *C*?
12. How many planes contain point *E*?
13. How many planes contain points *C* and *E*?
14. How many planes contain line *l*?

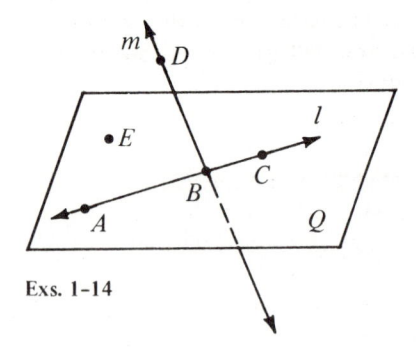

Exs. 1–14

B 15. How many lines contain a given point?
16. How many planes contain three given collinear points?
17. If a line and a plane intersect, can their intersection be a point? a line? a plane?
18. If two planes intersect, can their intersection be a point?
19. If a point lies outside a line, how many planes contain both the point and the line?
20. If two lines intersect, how many planes contain both lines?
21. If two lines are parallel, how many planes contain both lines?
22. If a line intersects one of two parallel lines, must it intersect the other?

Sketch a figure to represent each description.

23. Points *W*, *X*, *Y*, and *Z* are collinear.
24. $\overleftrightarrow{GH}$ and point *J* are coplanar.
25. $\overleftrightarrow{AC}$ and $\overleftrightarrow{CD}$ are different lines.
26. Planes *P* and *Q* intersect in $\overleftrightarrow{RS}$.
27. $\overleftrightarrow{AB}$, $\overleftrightarrow{CD}$, and $\overleftrightarrow{EF}$ are coplanar and do not intersect.
28. Planes *R*, *S*, and *T* are parallel.

136   CHAPTER 5

Section 5-2 DISTANCE, LINE SEGMENTS, AND RAYS

In the remainder of this book you may assume that all figures lie in a plane, unless otherwise specified. In this section you will learn about some special parts of lines that you will use throughout your study of geometry. The topics of this section include:
- line segments
- rays
- the length of a segment
- congruent segments
- the midpoint of a segment

A *line segment* consists of two points on a line and all the points between them. You name a line segment by its *endpoints*. Since the endpoints of the segment shown are A and B, you could refer to it as $\overline{AB}$ (read "line segment AB") or as $\overline{BA}$.

A *ray* is the part of a line that starts at a point on the line and extends indefinitely in one direction. The *endpoint* of the ray is the first point listed in naming a ray. As shown on the right, $\overrightarrow{AB}$ (read "ray AB") and $\overrightarrow{BA}$ are different rays.

The *length* of a line segment is the distance between the endpoints of the segment. The symbol AB is used to represent the length of $\overline{AB}$.

Two line segments are *congruent* if they have the same length. To indicate that $\overline{AB}$ and $\overline{CD}$ have the same length, we could write either

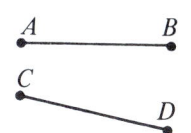

$AB = CD$ or $\overline{AB} \cong \overline{CD}$ (read "$\overline{AB}$ is congruent to $\overline{CD}$").

EXAMPLE Find the lengths of $\overline{RS}$ and $\overline{ST}$. Are $\overline{RS}$ and $\overline{ST}$ congruent?

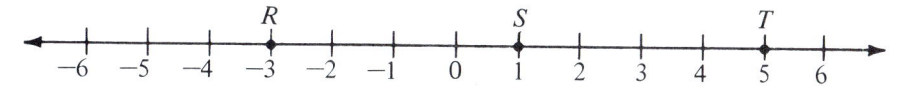

SOLUTION $\overline{RS}$ is the set of points that is the graph of $-3 \leqslant x \leqslant 1$.
$\overline{ST}$ is the set of points that is the graph of $1 \leqslant x \leqslant 5$.
We can find the length of $\overline{RS}$, for example, by counting the number of units between R and S or by subtracting the coordinate of R from the coordinate of S. (Notice that if we subtract the coordinate of S from that of R, we get a negative number, and a length can't be negative.)

$$RS = 1 - (-3) \qquad ST = 5 - 1$$
$$= 4 \qquad\qquad\quad = 4$$

Since $RS = ST$, $\overline{RS} \cong \overline{ST}$.

The *midpoint* of a line segment is the point that separates it into two congruent line segments. In the example above, point S is the midpoint of $\overline{RT}$.

BASIC CONCEPTS OF GEOMETRY

ORAL EXERCISES

1. Is $\overleftrightarrow{AD}$ the same as $\overleftrightarrow{DA}$?
2. Is $\overline{CD}$ the same as $\overline{DC}$?
3. Is $\overrightarrow{CD}$ the same as $\overrightarrow{DC}$?
4. Is BC the same as CB?
5. How many endpoints does each figure have? $\overline{AD}, \overrightarrow{AD}, \overleftrightarrow{AD}$
6. Give two other names for $\overrightarrow{AB}$.
7. Give as many names as you can for $\overleftrightarrow{AB}$.
8. Name six different line segments shown.
9. Name six different rays shown.

Exs. 1–9

10. What is the length of $\overline{HJ}$? of $\overline{GI}$? of $\overline{GE}$?
11. What is the midpoint of $\overline{FH}$? of $\overline{IE}$?
12. What is the coordinate of the midpoint of $\overline{FJ}$? of $\overline{JH}$?
13. Name a line segment that is congruent to $\overline{JF}$.

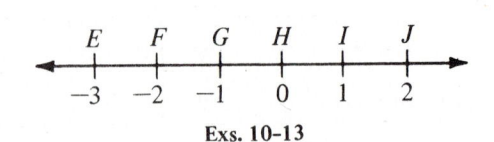

Exs. 10–13

WRITTEN EXERCISES

Name each of the following.

A 1. $K \quad L$ 2. $K \quad L$ 3. $K \quad L$ 4. $K \quad L$

Use the figure below for Exercises 5–16.

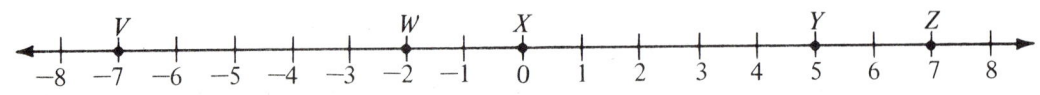

Find each length.

5. VW
6. WX
7. VX
8. WY
9. YZ
10. WZ
11. YX
12. ZX

13. Name three pairs of congruent line segments.
14. Name the midpoint of $\overline{VZ}$.
15. Name the coordinate of the midpoint of $\overline{WX}$.
16. Name the coordinate of the midpoint of $\overline{XZ}$.

In the figure, $MO = OQ$ and P is the midpoint of $\overline{OQ}$.

17. Does $\overleftrightarrow{MQ}$ have a midpoint?
18. What is the length of $\overline{PQ}$?
19. What is the length of $\overline{MQ}$?
20. What is the length of $\overline{MN}$?

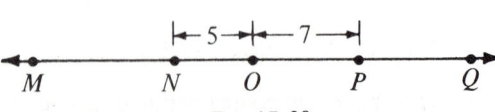

Exs. 17–20

138   CHAPTER 5

Do the two figures named intersect? If so, what is the intersection?

21. $\overleftrightarrow{AB}$ and $\overleftrightarrow{GF}$
22. $\overrightarrow{AB}$ and $\overrightarrow{ED}$
23. $\overrightarrow{BA}$ and $\overrightarrow{FC}$
24. $\overline{BC}$ and $\overrightarrow{GF}$
25. $\overrightarrow{AB}$ and $\overrightarrow{CB}$
26. $\overrightarrow{BE}$ and $\overrightarrow{BD}$

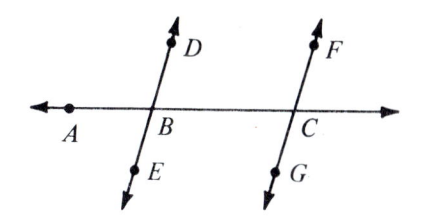

Given that M is the midpoint of $\overline{LN}$, draw a diagram and then copy and complete the table.

B

	27.	28.	29.	30.	31.	32.	33.	34.
coordinate of L	8	10	2	?	−4	?	a	?
coordinate of N	−4	5	?	−3	?	−1.5	$a+4$	$b-3$
length of $\overline{LN}$	?	?	?	?	?	?	?	?
coordinate of M	?	?	5	−10	3	−6.5	?	b

Graph the equation or inequality. Is the graph a point, a line, a line segment, a ray, or none of these?

EXAMPLE $x \geqslant 3$

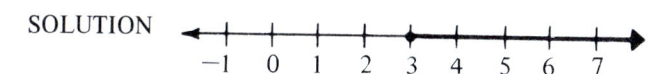

SOLUTION The graph is a ray.

35. $x + 1 = 1 - x$
36. $x + 3 \geqslant -1$
37. $-2 \leqslant x \leqslant 2$
38. $x + 2 \neq 4$
39. $x - 1 = -(1 - x)$
40. $3x + 1 = 5x - 7$
41. $x \leqslant 0$ or $x \geqslant 0$
42. $x \geqslant 3$ and $x \leqslant 4$
43. $x \leqslant -1$ or $x < 2$

Graph A, B, and C on a number line so that the given conditions are satisfied.

44. $AB = BC$ and $AC = 6$
45. $AB + BC = AC$, $AB = 2$, and $BC = 1$
46. $AB = 5$, $BC = 2$, and $AB + BC \neq AC$
47. C is the midpoint of $\overline{AB}$ and $AC = 2$

48. How many points are there on $\overleftrightarrow{DE}$ whose distance from D is 5?
49. How many points are there on $\overrightarrow{DE}$ whose distance from D is 5?

C 50. How many points are there on $\overrightarrow{DE}$ whose distance from E is 5?

51. If the coordinates of the endpoints of a line segment are a and b, express the coordinate of the midpoint of the segment in terms of a and b.

BASIC CONCEPTS OF GEOMETRY

Section 5-3 ANGLE MEASURE AND SPECIAL ANGLES

You know that every line segment has a length that is a positive number. An angle also has a measurement associated with it. This section discusses the measures of angles in general and some special angles in particular:
- congruent angles
- right angle
- acute angle
- obtuse angle
- straight angle

An *angle* is formed by two rays with a common endpoint. The endpoint is the *vertex* of the angle and the rays are the *sides*.

You could name the angle shown as ∠BAC ("angle BAC"), ∠CAB, ∠1, or ∠A. Notice that when three letters are used to name an angle, the letter that names the vertex is placed between the other two letters.

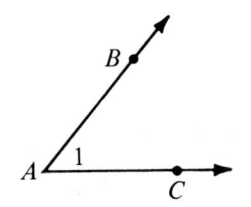

The size of an angle is often stated in terms of a unit called a *degree*. The outer scale of the protractor below shows that the *measure in degrees* of ∠HOE is 40. In symbols, this is written $m\angle HOE = 40$. The inner scale of the protractor shows that $m\angle FOG = 60$.

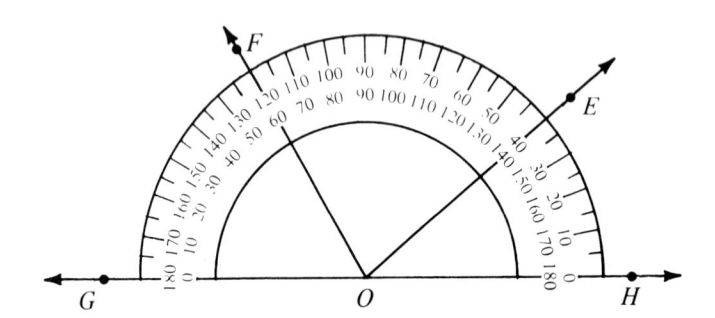

The figure shows that $m\angle HOE + m\angle EOF + m\angle FOG = 180$. Without moving the protractor, you can determine the measure of ∠EOF in either of two ways, as shown below.

$m\angle EOF + m\angle HOE = m\angle HOF$ or $m\angle EOF + m\angle FOG = m\angle EOG$
$m\angle EOF + 40 = 120$ $m\angle EOF + 60 = 140$
$m\angle EOF = 120 - 40$ $m\angle EOF = 140 - 60$
$m\angle EOF = 80$ $m\angle EOF = 80$

Notice that as a short cut for the above calculations you could use the outer scale of the protractor to get $m\angle EOF = 120 - 40$ or the inner scale to get $m\angle EOF = 140 - 60$.

140   CHAPTER 5

Congruent angles are angles whose measures are equal. Since ∠LKM and ∠MKN each have measure 30, we could write either

$$m\angle LKM = m\angle MKN \quad \text{or} \quad \angle LKM \cong \angle MKN.$$

A ray that divides the angle into two congruent angles is called a *bisector* of an angle. In the figure, $\overrightarrow{KM}$ bisects ∠LKN.

The measure of an angle is greater than 0 and less than or equal to 180. Angles can be classified in terms of their measures:

>*Acute angle:* Measure is between 0 and 90.
>*Right angle:* Measure is 90.
>*Obtuse angle:* Measure is between 90 and 180.
>*Straight angle:* Measure is 180.

EXAMPLE ∠AOB is a right angle.
∠BOC and ∠COD are acute.
∠AOD is obtuse.
∠AOC is obtuse, since $m\angle AOC = 90 + 50$, or 140.
∠BOD is obtuse, since $m\angle BOD = 50 + 60$, or 110.

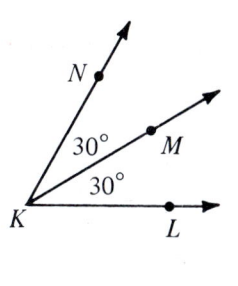

ORAL EXERCISES

Complete.

1. If two angles are congruent, then their measures are __?__.
2. If ∠A ≅ ∠B and $m\angle A = 21$, then $m\angle B =$ __?__.
3. If two angles have the same measure, then the angles are __?__.
4. If $m\angle C = 99$ and $m\angle D = 99$, then ∠C and ∠D are __?__.
5. If $m\angle E = 80$, the bisector of ∠E divides ∠E into two congruent angles, each with measure __?__.

Exercises 6–16 refer to the figure shown.

6. $m\angle KJL =$ __?__
7. $m\angle NJM =$ __?__
8. $m\angle MJL =$ __?__
9. $m\angle KJM =$ __?__

10. Name two acute angles shown.
11. Name a right angle shown.
12. Name two obtuse angles shown.
13. Which angle shown has the greatest measure?
14. How many angles shown have J as a vertex?
15. Why would it be confusing to talk about ∠J in the figure?
16. Name three angles that have $\overrightarrow{JL}$ as a side.

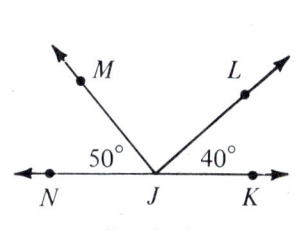

BASIC CONCEPTS OF GEOMETRY

WRITTEN EXERCISES

State the measure of each angle.

A
1. ∠AOB
2. ∠EOF
3. ∠BOD
4. ∠COE
5. ∠EOB
6. ∠BOF
7. ∠EOA
8. ∠DOE
9. ∠AOD
10. ∠DOF
11. ∠BOC
12. ∠AOC

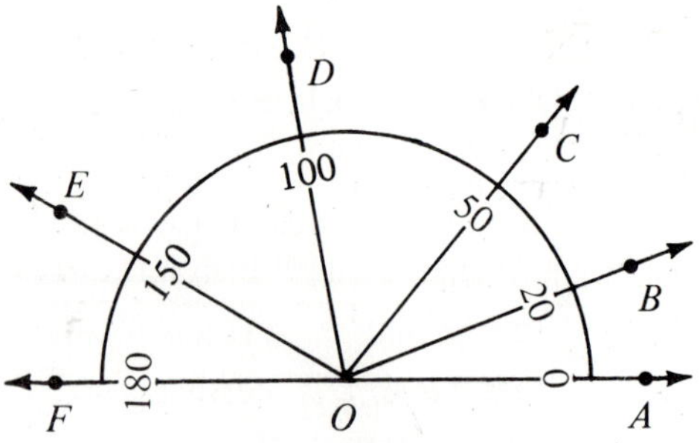

13. How many right angles are shown?
14. How many straight angles are shown?
15. Name an angle whose bisector is shown. Then name the bisector.

Name the angles that have the given measure.
16. 30
17. 80
18. 100
19. 130
20. 50 (three angles)

21. Name the acute angles shown.
22. Name the obtuse angles shown.

Exs. 1–22

Use a protractor to measure each angle.
23. 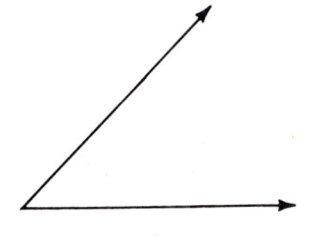
24. 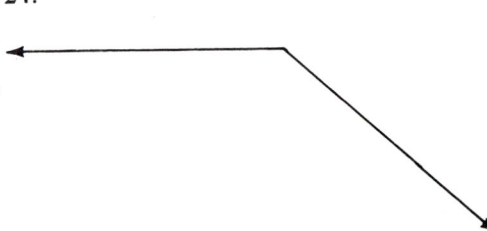
25. 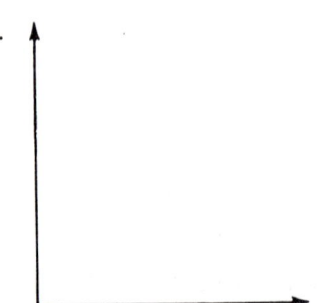
26.

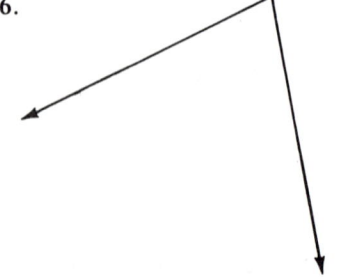

142 CHAPTER 5

Use a protractor to draw ∠ABC to fit the description.

27. m∠ABC = 35
28. m∠ABC = 82
29. m∠ABC = 120
30. m∠ABC = 90
31. m∠ABC = 59
32. m∠ABC = 155
33. ∠ABC is acute.
34. ∠ABC is obtuse.
35. ∠ABC is a right angle.
36. ∠ABC is a straight angle.

Complete.

B 37. m∠AEB + m∠BEC = m∠ ?
38. m∠AED = m∠AEB + m∠ ? + m∠CED
39. If m∠AEB = m∠DEC, then m∠AEC = m∠ ? .
40. If ∠AEB ≅ ∠DEC, then ∠AEC ≅ ∠ ? .
41. If m∠AEC = m∠DEB, then m∠AEB = m∠ ? .
42. If ∠AEC ≅ ∠DEB, then ∠ ? ≅ ∠DEC.
43. If ∠BEC ≅ ∠CED, then ? bisects ∠ ? .
44. If ∠AEB ≅ ∠BEC and m∠BEC = 55, then m∠AEC = ? .
45. If ∠AEB ≅ ∠BEC, then m∠AEB = $\frac{1}{2}$m∠ ? .
46. If ∠AEB ≅ ∠BEC ≅ ∠CED, then m∠AED = ? m∠AEB.

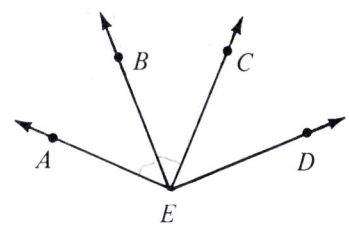

Exs. 37–46

$\overrightarrow{OY}$ bisects ∠XOZ. Use the given information to find m∠1.

47. m∠XOZ = 50
48. m∠XOZ = 68

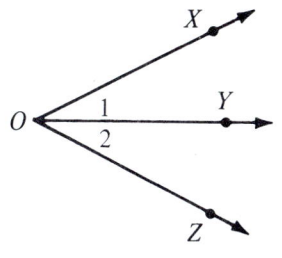

Exs. 47–54

If ∠1 ≅ ∠2, find the measure of ∠1.

EXAMPLE m∠1 = 4x; m∠2 = x + 27
SOLUTION m∠1 = m∠2
 4x = x + 27
 3x = 27
 x = 9
 m∠1 = 4(9)
 m∠1 = 36

49. m∠1 = 2z; m∠2 = 3z − 16
50. m∠1 = 9y; m∠2 = 7y + 18
51. m∠1 = 2x + 7; m∠2 = 5x − 8
52. m∠1 = −3k + 60; m∠2 = 5k − 4
53. m∠1 = 1.2y − 9; m∠2 = 2.6y − 72
54. m∠1 = $\frac{1}{2}$w + 16; m∠2 = 3w − 9

BASIC CONCEPTS OF GEOMETRY

Find the measures of ∠1, ∠2, and ∠3.

EXAMPLE $m\angle 1 = x;\ m\angle 2 = 2x - 12;\ m\angle 3 = 60$

SOLUTION $m\angle 1 + m\angle 2 = m\angle 3$
$$x + (2x - 12) = 60$$
$$3x - 12 = 60$$
$$3x = 72$$
$$x = 24$$
$m\angle 1 = 24,\ m\angle 2 = 36,\ m\angle 3 = 60$

	55.	56.	57.	58.	59.	60.
m∠1	y − 6	w	4z	6a	3b	2c
m∠2	y	2w	5z	3a + 12	66	45
m∠3	78	87	81	39	9b	c + 60

61. The sum of the measures of two angles is 127. The measure of one angle is 19 more than three times the measure of the other angle. Find the measure of each angle.

62. The sum of the measures of two angles is 105. The measure of one angle is 15 less than twice that of the other angle. Find the measure of each angle.

63. If the sum of the measures of two angles is 150, does one of the angles have to be an obtuse angle? Can it be obtuse?

Find the measures of ∠1, ∠2, and ∠3.

C

	64.	65.	66.	67.
m∠1	2w	x	y + 15	5z − 8
m∠2	3w	5x − 72	2y − 8	5z + 16
m∠3	7w − 18	2x	4y	20z − 34

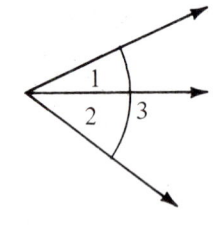

68. If ∠ABC, ∠CBD, and ∠ABD are angles such that m∠ABC = 60 and m∠CBD = 30, find all possible values of m∠ABD.

69. If m∠A + m∠B = 138 and ∠A is known to be an acute angle, what are the possible values for the measure of ∠B?

70. If m∠C + m∠D = 175 and ∠C is known to be an obtuse angle, what are the possible values for the measure of ∠D?

144   CHAPTER 5

SELF-TEST 1

1. Are points *A*, *B*, and *C* collinear? Are they coplanar?

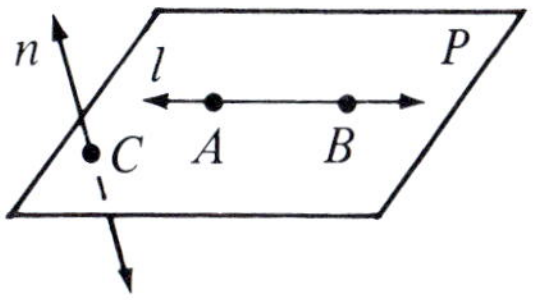

Section 5-1

Complete.

2. Line *n* intersects plane *P* in __?__.
3. Two names for the same line are *l* and __?__.
4. The number of lines through points *B* and *C* is __?__.

For Exercises 5-7 use the number line shown.

Section 5-2

5. Name a point that is on $\overrightarrow{GF}$, but not on $\overline{GF}$.
6. Name a pair of congruent line segments.
7. Name the coordinate of the midpoint of $\overline{FH}$.

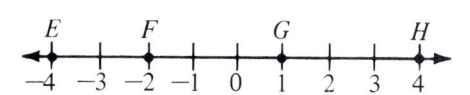

8. Name:

Section 5-3

   (a) an acute angle
   (b) a right angle
   (c) a straight angle
   (d) an obtuse angle
9. Find the measure of $\angle AOC$.

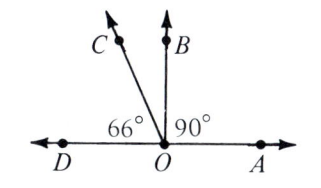

10. Use a protractor to draw an angle *XYZ* that has measure 68 and its bisector, $\overrightarrow{YV}$. Name two congruent angles shown in your diagram.

Section 5-4 SOME ANGLE RELATIONSHIPS

This section considers perpendicular lines and some special pairs of angles:
- complementary angles
- supplementary angles
- vertical angles

Two lines are *perpendicular* if they intersect to form a right angle. The symbol ⊥ is used for "is perpendicular to." In the figure, $\overleftrightarrow{AB} \perp \overleftrightarrow{CD}$ since $m\angle BOC = 90$.

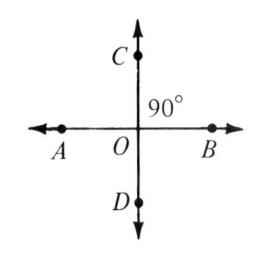

Can you guess what the measure of $\angle AOC$ is?

$$m\angle AOC + m\angle BOC = 180$$
$$m\angle AOC + 90 = 180$$
$$m\angle AOC = 90$$

Similarly, you could show that $\angle BOD$ and $\angle AOD$ are also right angles. Thus, if two lines intersect to form one right angle, they actually form four right angles.

BASIC CONCEPTS OF GEOMETRY   145

We also say that two line segments or rays are perpendicular if they intersect to form a right angle. In the figure, $\overline{NK} \perp \overline{KL}$, $\overline{NK} \perp \overline{NM}$, $\overline{NM} \perp \overline{ML}$, and $\overline{ML} \perp \overline{KL}$. Notice that a small square can be used to indicate that an angle is a right angle.

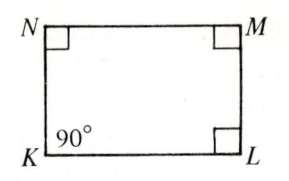

Two angles are *adjacent* if they share a vertex and a common side but have no common interior points. Notice that $\angle XWY$ and $\angle YWZ$ are adjacent angles, but that $\angle XWY$ and $\angle XWZ$ are not.

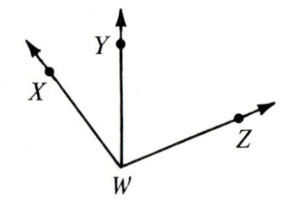

If the sum of the measures of two angles is 90, the angles are *complementary*. Each angle is called a *complement* of the other.

EXAMPLES

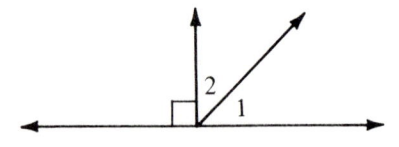

$\angle 1$ and $\angle 2$ are adjacent complementary angles.

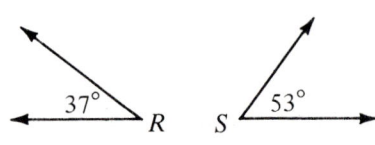

$\angle R$ is a complement of $\angle S$, since $37 + 53 = 90$.

If the sum of the measures of two angles is 180, the angles are *supplementary*. Each angle is called a *supplement* of the other.

EXAMPLES

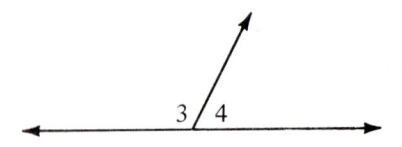

$\angle 3$ and $\angle 4$ are adjacent supplementary angles.

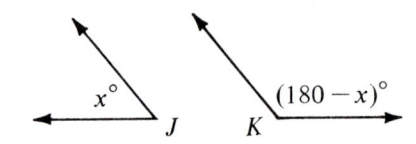

$\angle J$ is a supplement of $\angle K$, since $x + (180 - x) = 180$.

EXAMPLE Name: (a) a supplement of $\angle ABD$
(b) a complement of $\angle A$
(c) a complement of $\angle C$

SOLUTION (a) $\angle CBD$ is a supplement of $\angle ABD$, since $110 + 70 = 180$.
(b) $\angle BDC$ is a complement of $\angle A$, since $40 + 50 = 90$.
(c) $\angle ADB$ is a complement of $\angle C$, since $60 + 30 = 90$.

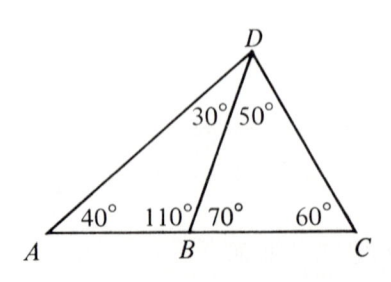

146   CHAPTER 5

EXAMPLE The measure of the supplement of an angle is four times the measure of the complement of the angle. Find the measures of all three angles.

SOLUTION Let x be the measure of the angle. Then $90 - x$ is the measure of its complement, and $180 - x$ is the measure of its supplement.

$180 - x = 4(90 - x)$
$180 - x = 360 - 4x$

$\quad 3x = 180 \qquad\qquad 90 - x = 90 - 60 \qquad\qquad 180 - x = 180 - 60$
$\quad\; x = 60 \qquad\qquad\qquad\quad\; = 30 \qquad\qquad\qquad\qquad\quad\; = 120$

The measures of the angle, its complement, and its supplement are 60, 30, and 120, respectively.

When two lines intersect, they form four angles. Two nonadjacent angles formed by two intersecting lines are called *vertical* angles. In the figure below, $\angle 1$ and $\angle 3$ are vertical angles, as are $\angle 2$ and $\angle 4$. Notice that:

$m\angle 1 + m\angle 2 = 180$
and $m\angle 3 + m\angle 2 = 180$
$\quad m\angle 1 + m\angle 2 = m\angle 3 + m\angle 2$ (By substitution.)
$\quad\qquad m\angle 1 = m\angle 3$ (By subtraction.)

Similarly, you could show that $m\angle 2 = m\angle 4$.

Vertical angles have equal measures, and thus are congruent.

ORAL EXERCISES

In the diagram, $\overleftrightarrow{CF} \perp \overleftrightarrow{DA}$ and $m\angle 1 = 30$. Classify each statement as true or false.

1. $\overline{OC} \perp \overline{OA}$
2. $m\angle 3 = 90$
3. $\angle 1 \cong \angle 5$
4. $m\angle 2 = 60$
5. $m\angle 5 = 60$
6. $\overleftrightarrow{OF} \perp \overleftrightarrow{BE}$
7. $\angle 1$ and $\angle 2$ are supplementary.
8. $\angle 1$ and $\angle 4$ are vertical angles.
9. $m\angle 4 + m\angle 5 = 90$
10. $\angle AOB$ and $\angle BOD$ are supplementary.
11. $\angle AOC$ and $\angle BOC$ are adjacent angles.
12. $\angle 5$ is a complement of $\angle 1$.
13. $\angle 3$ and $\angle AOF$ are vertical supplementary angles.
14. $\angle 2$ and $\angle 4$ are complementary angles.
15. $\angle BOF \cong \angle COE$
16. Both $\angle 1$ and $\angle 4$ are supplements of $\angle BOD$.

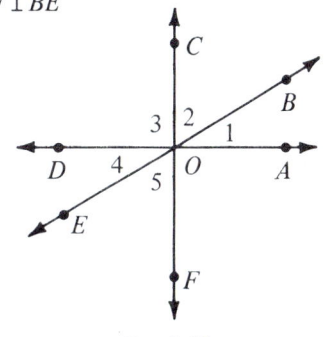

Exs. 1–16

BASIC CONCEPTS OF GEOMETRY

WRITTEN EXERCISES

A
1. Name two right angles.
2. Name two segments that are perpendicular to $\overline{JM}$.
3. Name two adjacent complementary angles.
4. Name two nonadjacent complementary angles.
5. Name two pairs of congruent angles.
6. Name two pairs of supplementary angles.

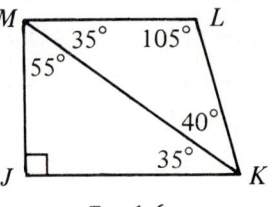

Exs. 1–6

Find the measure of a complement of an angle with the given measure.

7. 53
8. 74
9. j
10. $3k$

Find the measure of a supplement of an angle with the given measure.

11. 12
12. 107
13. y
14. $4z$

Use the measure given for one angle to find the measures of the other numbered angles.

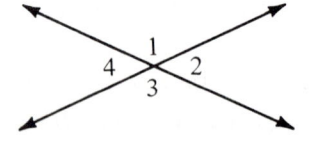

15. $m\angle 1 = 44$
16. $m\angle 2 = 129$
17. $m\angle 3 = 48$
18. $m\angle 4 = 132$

In Exercises 19–24, draw a diagram that illustrates the angles described. If no diagram is possible, write "not possible."

B
19. Two adjacent acute angles
20. Two adjacent vertical angles
21. Two congruent supplementary angles
22. Two complementary obtuse angles
23. Two complementary vertical angles
24. Two congruent adjacent angles formed by intersecting lines

Find the value of x.

25. $\left(\dfrac{x}{3}\right)^\circ$, $(x - 52)^\circ$

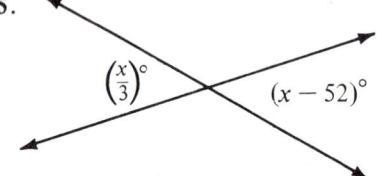

26. $(2x - 33)^\circ$, $(3x - 163)^\circ$

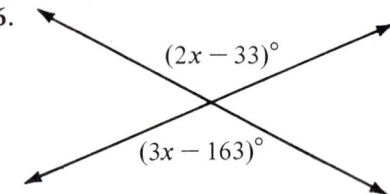

27. x°, $(3x - 42)^\circ$

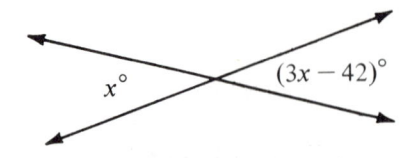

28. $(6x - 245)^\circ$, $(210 - x)^\circ$

148 CHAPTER 5

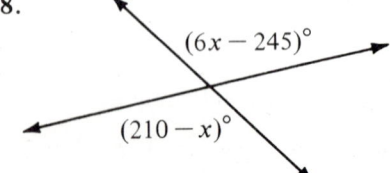

Find the measures of the two angles described.

29. The measure of an angle is twice the measure of its complement.
30. The measure of an angle is one-fourth the measure of its supplement.
31. The measure of an angle is 38 less than the measure of its supplement.
32. The measure of an angle is 22 more than the measure of its complement.
33. The measures of two vertical angles are $3z + 18$ and $8z + 3$.
34. The measures of two vertical angles are $\frac{1}{2}y$ and $y - 39$.
35. The measures of two complementary angles are $2a - 10$ and $4a - 2$.
36. The measures of two supplementary angles are $\frac{3}{2}b + 7$ and $\frac{1}{2}b + 9$.
37. The measures of two supplementary angles are $\frac{2}{3}c - 12$ and $\frac{5}{3}c + 17$.

In Exercises 38–40 find the measures of the angle, its complement, and its supplement.

38. The sum of the measures of a complement and a supplement of an angle is 214.
39. The measure of a supplement of an angle is 18 more than 4 times the measure of a complement of the angle.
40. The sum of the measure of a complement of an angle and twice the measure of a supplement of the angle is 240.

41. The sum of the measures of an angle, a complement of the angle, and a supplement of the angle is less than 4 times the measure of the angle. What can you determine about the measure of the angle?
42. If the measure of an angle is less than 30, what can you determine about the measures of its complement and its supplement?
43. If the measure of an angle is greater than 45, what can you determine about the measures of its complement and its supplement?

44. a. In the figure, $m\angle 1 = 95$, $m\angle 2 = 105$, and $m\angle 3 = 160$. Does $m\angle AOB + m\angle BOC = m\angle COA$?
    b. Suppose that $\angle AOB$ and $\angle BOC$ are adjacent angles such that
    $$m\angle AOB + m\angle BOC = m\angle COA.$$
    What can you determine about $m\angle AOB + m\angle BOC$?

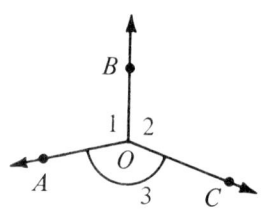

45. If both $\angle A$ and $\angle B$ are complements of $\angle C$, what can you determine about the measures of $\angle A$ and $\angle B$?
46. If $\angle 1$ and $\angle 3$ are supplements of $\angle 2$ and $\angle 4$ respectively, and $m\angle 2 = m\angle 4$, what can you determine about the measures of $\angle 1$ and $\angle 3$?

BASIC CONCEPTS OF GEOMETRY

C 47. The measure of a supplement of an angle is more than three times the measure of a complement of the angle and is less than four times the measure of the complement. What can you determine about the measure of the angle?

EXAMPLE In the figure, $m\angle A + m\angle B + m\angle C + m\angle D = 360$. If $m\angle A = m\angle C$ and $m\angle B = m\angle D$, find $m\angle A + m\angle B$.

SOLUTION
$$m\angle A + m\angle B + m\angle C + m\angle D = 360$$
$$m\angle A + m\angle B + m\angle A + m\angle B = 360$$
$$2m\angle A + 2m\angle B = 360$$
$$2(m\angle A + m\angle B) = 360$$
$$m\angle A + m\angle B = 180$$

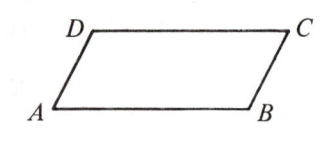

48. In the figure, $m\angle D + m\angle E + m\angle F + m\angle G = 360$. If $m\angle E = m\angle G$ and $m\angle D = m\angle F$, show that $m\angle D + m\angle E = 180$.

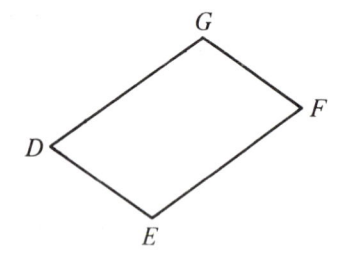

Can you conclude from the given information that $l \perp n$? Explain.

49. $m\angle 1 = 43$ and $m\angle 4 = 47$
50. $m\angle 1 = m\angle 2$ and $m\angle 3 = m\angle 4$
51. $m\angle 1 = m\angle 3$ and $m\angle 2 = m\angle 4$
52. $m\angle 1 = m\angle 4$ and $m\angle 2 = m\angle 3$

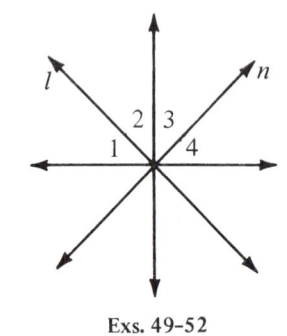

Exs. 49–52

Section 5-5 PARALLEL LINES AND TRANSVERSALS

You know that parallel lines are lines that lie in the same plane and do not intersect. If two coplanar lines are not parallel, then they must intersect, and their intersection is a point. In this section you will study some ideas associated with intersecting and parallel lines:
- transversal
- alternate interior angles
- corresponding angles
- properties relating parallel lines and certain angles

150   CHAPTER 5

A line that intersects, or cuts, two or more coplanar lines in different points is called a *transversal*. In the figure, line *t* is a transversal of lines *p* and *q*. Angles 3, 4, 5, and 6 are called *interior* angles. Angles 1, 2, 7, and 8 are *exterior* angles. Special names are given to certain pairs of angles shown in the figure.

Alternate interior angles	*Corresponding angles*	
∠3 and ∠5	∠1 and ∠5	∠2 and ∠6
∠4 and ∠6	∠3 and ∠7	∠4 and ∠8

Lines *c* and *d* in the figure are parallel. What appears to be true about the measures of the corresponding angles? Use a protractor to see if your guess is correct.

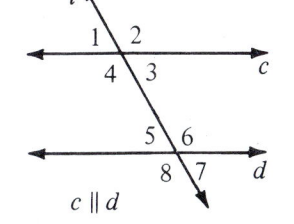

Take a sheet of ruled paper and draw any transversal intersecting two of the lines. Measure corresponding angles with a protractor. What do you find to be true?

c ∥ *d*

The following geometric property is true for *any* two parallel lines cut by a transversal.

If two parallel lines are cut by a transversal, then corresponding angles have equal measures.

EXAMPLE Suppose that $p \parallel q$ and $r \parallel s$.
Find the value of x.

SOLUTION Since $p \parallel q$, $m\angle 1 = 60$.
Since $r \parallel s$, $m\angle 1 = x + 10$.
$$x + 10 = 60$$
$$x = 50$$

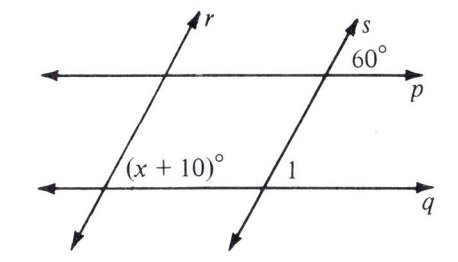

Although the converse of a true statement is not necessarily true, the converse of the property stated above is true.

If two lines are cut by a transversal so that corresponding angles have equal measures, the lines are parallel.

These properties can be written as the following biconditional.

Two lines cut by a transversal are parallel if and only if corresponding angles have equal measures.

BASIC CONCEPTS OF GEOMETRY

EXAMPLE   Are lines p and q in the figure parallel?

SOLUTION $m\angle 1 + x = 180$
$m\angle 1 = 180 - x$

$p \parallel q$, since we have shown that corresponding angles have equal measures.

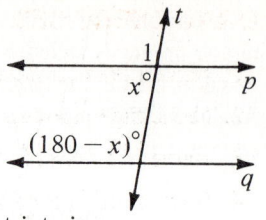

The example shows that if two lines are cut by a transversal so that interior angles on the same side of a transversal are supplementary, then the lines are parallel.

Just as you measured corresponding angles, you could measure $\angle 3$ and $\angle 5$ to determine that these particular alternate interior angles have equal measures. However, once you accept as true the property for corresponding angles formed by a transversal cutting parallel lines, you can show that a similar property is true for alternate interior angles.

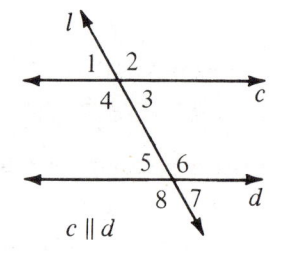

$m\angle 1 = m\angle 5$ (If two parallel lines are cut by a transversal, then corresponding angles have equal measures.)

$m\angle 1 = m\angle 3$ (Vertical angles have equal measures.)

$m\angle 3 = m\angle 5$ (By substitution.)

If two parallel lines are cut by a transversal, then alternate interior angles have equal measures.

The converse of this property is also true.

If two lines are cut by a transversal so that alternate interior angles have equal measures, the lines are parallel.

These properties can be combined as a biconditional.

Two lines cut by a transversal are parallel if and only if alternate interior angles have equal measures.

EXAMPLE   Are lines a and b in the figure parallel?

SOLUTION   Since vertical angles have equal measures, $m\angle 4 = 48$ and $m\angle 6 = 48$.
Thus $m\angle 4 = m\angle 6$.
Since $\angle 4$ and $\angle 6$ are alternate interior angles, $a \parallel b$.

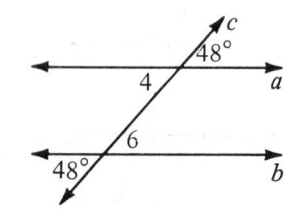

ORAL EXERCISES

1. Name the interior angles shown.
2. Name the exterior angles shown.
3. Name two pairs of alternate interior angles.
4. Name four pairs of corresponding angles.
5. Name two pairs of angles that you would describe as alternate exterior angles.

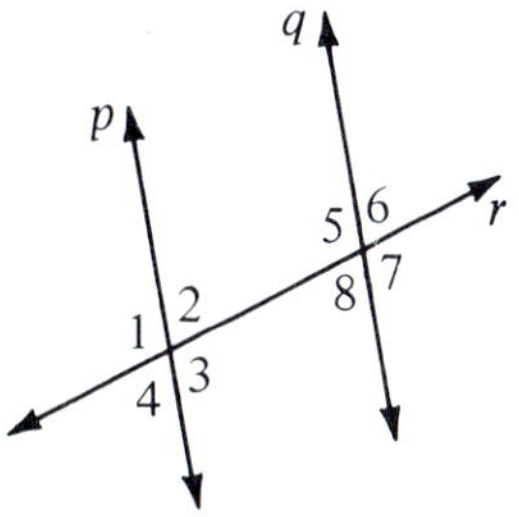

Exs. 1–16

If $p \parallel q$, classify each statement as true or false.

6. $m\angle 1 = m\angle 5$
7. $\angle 2 \cong \angle 8$
8. $m\angle 4 = m\angle 7$
9. $\angle 3 \cong \angle 7$
10. $m\angle 2 + m\angle 5 = 180$
11. $\angle 7 \cong \angle 1$

State the property that justifies each statement.

12. If $m\angle 4 = m\angle 8$, then $p \parallel q$.
13. If $p \parallel q$, then $m\angle 2 = m\angle 8$.
14. If $m\angle 3 = m\angle 5$, then $p \parallel q$.
15. If $p \parallel q$, then $m\angle 3 = m\angle 7$.

16. If $m\angle 1 = 110$, what are the measures of the other angles shown?

WRITTEN EXERCISES

In Exercises 1–6, can you conclude from the given information that $l \parallel n$?

A 1.

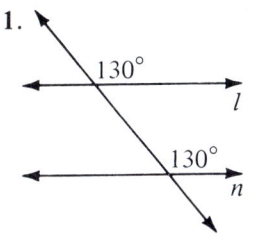

130°
130°

2.

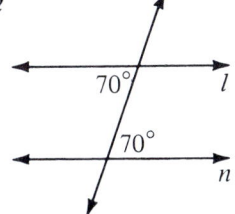

70°
70°

3.

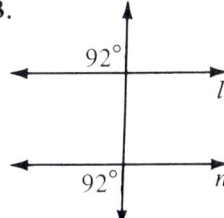

92°
92°

4.

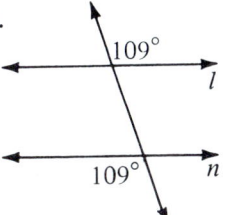

109°
109°

5.

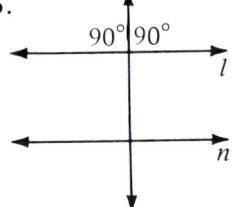

90° 90°

6.

137°
43°

BASIC CONCEPTS OF GEOMETRY   153

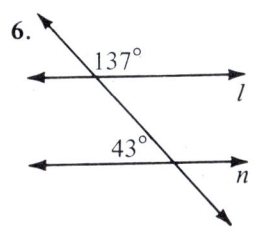

In Exercises 7-9, $r \parallel s$ and the measure of one angle is given. Find the measures of the numbered angles.

7.

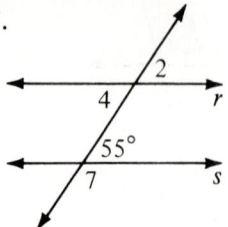

8.

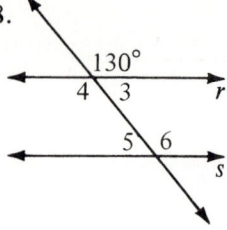

9.

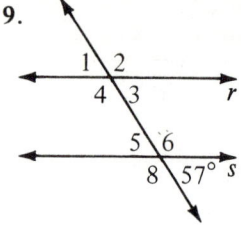

For Exercises 10-18, assume that $a \parallel b$ and $c \parallel d$.

10. List all the angles that are congruent to $\angle 1$.
11. List all the angles that are supplementary to $\angle 1$.
12. List all the angles that have the same measure as $\angle 14$.
13. If $m\angle 1 = 108$, then $m\angle 11 = \underline{\ ?\ }$.
14. If $m\angle 2 = 82$, then $m\angle 13 = \underline{\ ?\ }$.
15. If $m\angle 11 = 120$, then $m\angle 7 = \underline{\ ?\ }$.
16. If $m\angle 4 + m\angle 6 = 126$, then $m\angle 10 = \underline{\ ?\ }$.
17. If $m\angle 2 + m\angle 4 + m\angle 10 = 144$, then $m\angle 12 = \underline{\ ?\ }$.
18. If $m\angle 6 + m\angle 10 = 114$, then $m\angle 11 = \underline{\ ?\ }$.

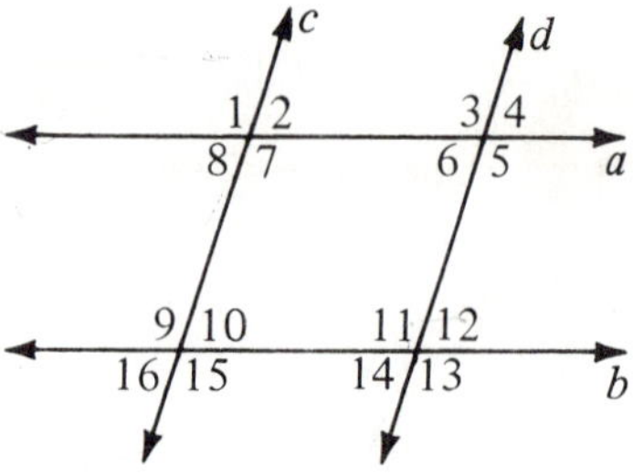

Exs. 10-34

19. If $a \parallel b$, list the angles that are congruent to $\angle 2$.
20. If $c \parallel d$, list the angles that are congruent to $\angle 2$.
21. If $a \parallel b$, list the angles that are supplementary to $\angle 8$.
22. If $c \parallel d$, list the angles that are supplementary to $\angle 8$.

In Exercises 23-28, can you conclude from the given information that $a \parallel b$?

23. $m\angle 2 = m\angle 10$
24. $m\angle 6 = m\angle 16$
25. $m\angle 7 = m\angle 11$
26. $m\angle 6 = m\angle 12$
27. $m\angle 9 = m\angle 11$
28. $m\angle 5 + m\angle 12 = 180$

In Exercises 29-34, can you conclude from the given information that $c \parallel d$?

29. $m\angle 10 = m\angle 14$
30. $m\angle 9 = m\angle 1$
31. $m\angle 3 = m\angle 11$
32. $m\angle 4 = m\angle 8$
33. $m\angle 12 + m\angle 9 = 180$
34. $m\angle 9 + m\angle 11 = 180$

In Exercises 35-43, find the value of x for which $r \parallel s$.

B 35.

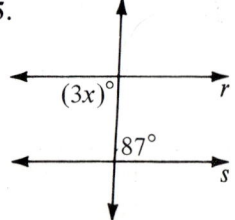

36.

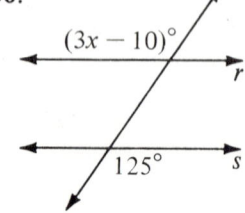

37.

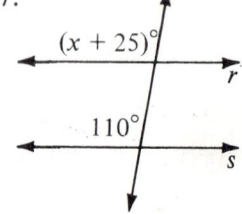

154   CHAPTER 5

38. 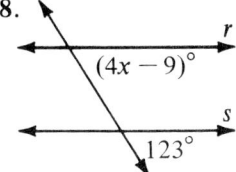 $(4x - 9)°$, $123°$

39. 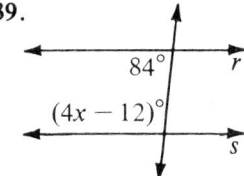 $84°$, $(4x - 12)°$

40. 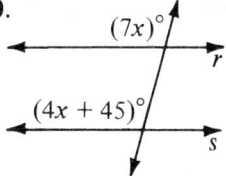 $(7x)°$, $(4x + 45)°$

41. 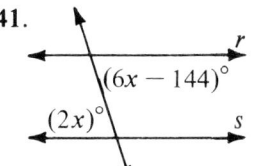 $(6x - 144)°$, $(2x)°$

42. 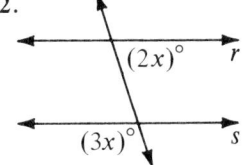 $(2x)°$, $(3x)°$

43. 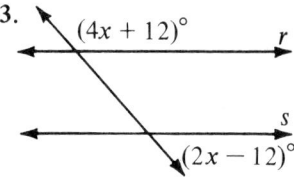 $(4x + 12)°$, $(2x - 12)°$

44. Find the measures of ∠5, ∠6, and ∠7 if $a \parallel b$ in the figure.

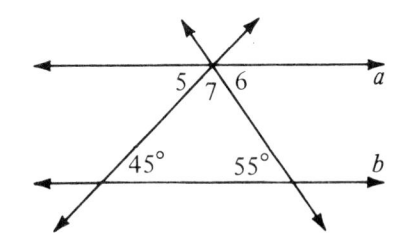

In Exercises 45-48, what can you conclude from the given information?

45. $\overleftrightarrow{AB} \parallel \overleftrightarrow{DC}$

46. $\overleftrightarrow{AD} \parallel \overleftrightarrow{BC}$

47. $m\angle 1 = m\angle 3$

48. $m\angle 2 = m\angle 4$

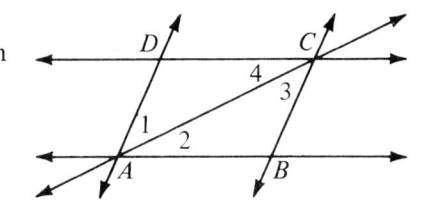

C 49. If $p \parallel q$ and $q \parallel r$, what can you conclude about p and r? Explain in terms of the measures of ∠5, ∠6, and ∠7 in the figure.

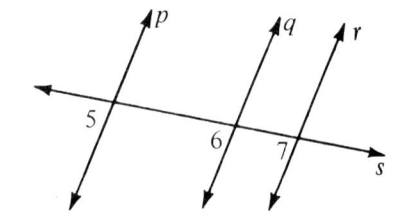

50. If $r \parallel s$ and $t \perp r$, what can you conclude about t and s? Explain in terms of ∠1 and ∠2 in the figure.

51. If $t \perp r$ and $t \perp s$, what can you conclude about r and s? Explain in terms of ∠1 and ∠2 in the figure.

BASIC CONCEPTS OF GEOMETRY

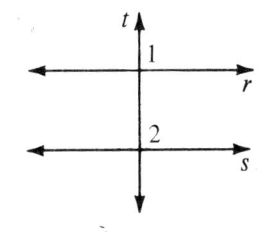

Section 5-6 ANGLES OF A TRIANGLE

If A, B, and C are noncollinear points, then the figure formed by $\overline{AB}$, $\overline{BC}$, and $\overline{CA}$ is *triangle ABC*, written $\triangle ABC$. $\angle A$, $\angle B$, and $\angle C$ are the *angles* of the triangle.

What is the sum of the measures of the angles of $\triangle ABC$? Would the sum be different for a different triangle? In this section you will investigate these questions. Then a property of parallel lines will be used to show that the sum of the measures of the angles is the same for all triangles.

Use a protractor to measure the angles of $\triangle ABC$. What is the sum of the three measures? Then measure the angles of $\triangle DEF$ below. Is the sum of the measures the same as for $\triangle ABC$?

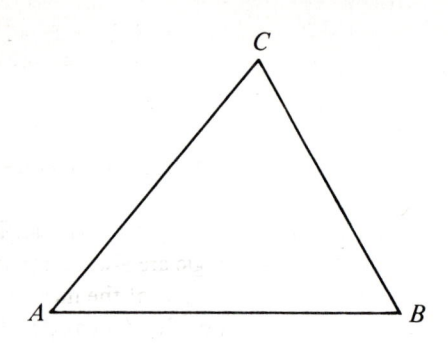

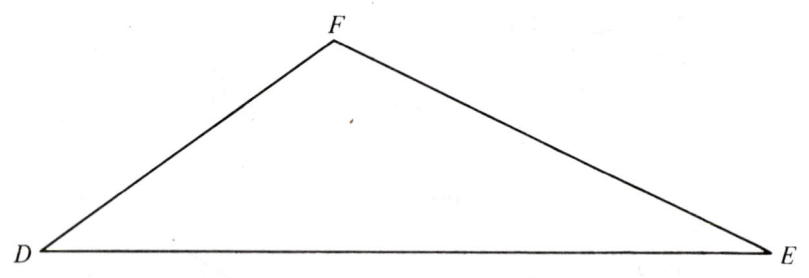

Suppose you cut a triangle out of paper and tear it into three parts, as shown below. Do you think that the angles of the triangle can be rearranged as shown? Try it!

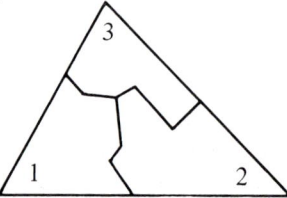

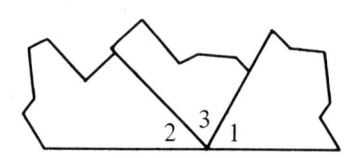

In the figure, line l is parallel to the line determined by $\overline{AB}$. Think of $\overline{AC}$ and $\overline{BC}$ as transversals of line l and $\overline{AB}$. Then $\angle A$ and $\angle 1$ are alternate interior angles, as are $\angle B$ and $\angle 2$.

$m\angle A = m\angle 1$ (If two parallel lines are cut by a transversal, alternate interior angles have equal measures.)

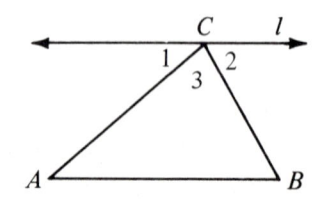

$m\angle B = m\angle 2$ (Why?)

$m\angle 1 + m\angle 3 + m\angle 2 = 180$ (The measure of a straight angle is 180.)

$m\angle A + m\angle ACB + m\angle B = 180$ (By substitution.)

The sum of the measures of the angles of a triangle is 180.

EXAMPLE Find the measures of $\angle R$, $\angle S$, and $\angle T$.

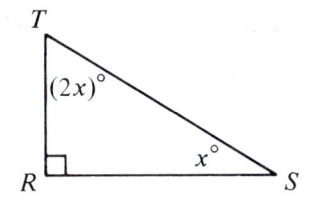

SOLUTION
$$m\angle R + m\angle S + m\angle T = 180$$
$$90 + 2x + x = 180$$
$$3x = 90$$
$$x = 30$$
$$2x = 60$$

The measures of the angles are 30, 60, and 90.

A triangle that contains a right angle is called a *right triangle*. The two acute angles of a right triangle are always complementary.

Knowing that the sum of the measures of the angles of a triangle is 180 can help you find the measures of certain angles that are *not* angles of a triangle.

EXAMPLE Find the measure of $\angle 1$.

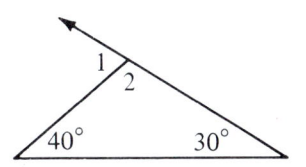

SOLUTION
$$m\angle 1 + m\angle 2 = 180$$
Also: $m\angle 2 + 30 + 40 = 180$
Substituting: $m\angle 1 + m\angle 2 = m\angle 2 + 30 + 40$
Subtracting: $m\angle 1 = 70$

EXAMPLE Find the measure of $\angle 6$ in terms of the measures of $\angle 3$ and $\angle 4$.

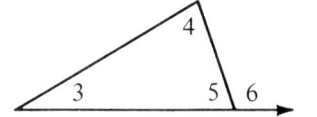

SOLUTION
$$m\angle 5 + m\angle 6 = 180$$
Also: $m\angle 5 + m\angle 3 + m\angle 4 = 180$
Substituting: $m\angle 5 + m\angle 6 = m\angle 5 + m\angle 3 + m\angle 4$
Subtracting: $m\angle 6 = m\angle 3 + m\angle 4$

Notice that $\angle 5$ and $\angle 6$ in the example above are both adjacent and supplementary. An angle such as $\angle 6$ is called an *exterior angle* of the triangle. $\angle 3$ and $\angle 4$ are *remote interior angles* with respect to $\angle 6$. The example shows that the measure of an exterior angle of a triangle is equal to the sum of the measures of the two remote interior angles.

ORAL EXERCISES

Classify each statement as true or false.

1. A triangle can have two right angles.
2. A triangle can have at most one obtuse angle.
3. A triangle can have at most one acute angle.
4. A right triangle can have an obtuse angle.
5. If the three angles of a triangle are congruent, the measure of each is 60.
6. The measures of the complementary angles of a right triangle must be 30 and 60.

BASIC CONCEPTS OF GEOMETRY

Complete.

7. If $m\angle 1 = 107$, then $m\angle 2 = \underline{\ ?\ }$.
8. If $m\angle 1 = 107$, then $m\angle 5 + m\angle 3 = \underline{\ ?\ }$.
9. If $m\angle 4 = 143$, then $m\angle 2 + m\angle 5 = \underline{\ ?\ }$.
10. If $m\angle 1 = 110$ and $m\angle 3 = 40$, then $m\angle 5 = \underline{\ ?\ }$.

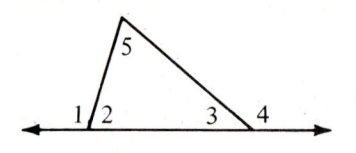

Exs. 7–10

WRITTEN EXERCISES

Find the measure of each numbered angle.

A 1.

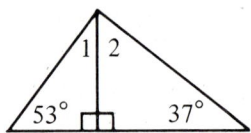

2.

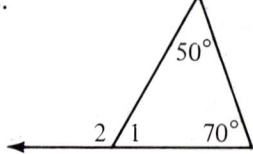

3.

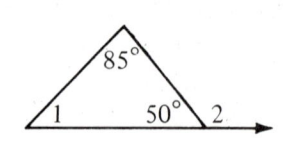

4.

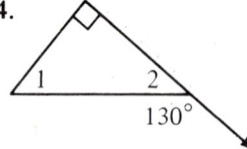

5.

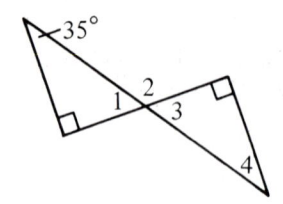

6.

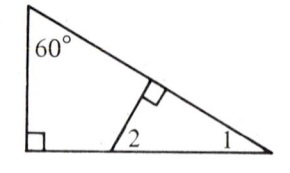

In Exercises 7–12, the measures of $\angle A$ and $\angle B$ of $\triangle ABC$ are given. Find the numerical value of the measure of $\angle C$.

	7.	8.	9.	10.	11.	12.
$m\angle A$	88	34	x	$45 - z$	$61 + k$	$3n + 56$
$m\angle B$	42	78	$90 - x$	z	$61 - k$	$59 - 3n$

In Exercises 13–18, find the value of x.

13.

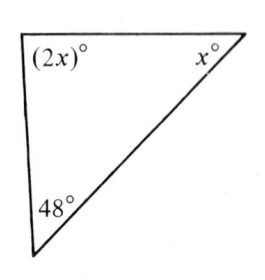

14.

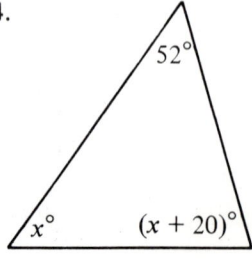

15.

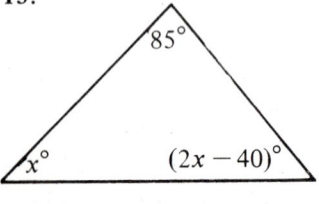

158   CHAPTER 5

16.

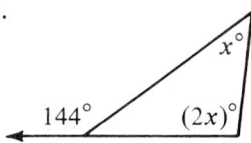

17.

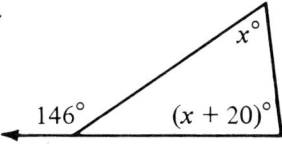

18.

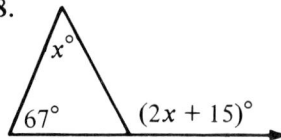

(144°, (2x)°, x°) (146°, (x + 20)°, x°) (67°, (2x + 15)°, x°)

In Exercises 19-24 the measures of the three angles of △DEF are given in terms of a variable. Find the numerical value of the measure of each angle.

B

	19.	20.	21.	22.	23.	24.
m∠D	x	3z	k	n − 3	6f + 18	2w − 16
m∠E	2x	8z	k	2n + 14	100 − 5f	3w
m∠F	3x	4z	k + 18	3n + 7	9f − 8	3w + 4

In Exercises 25-28, the measures of the three angles of △ABC are described. Find all three measures.

25. The measure of ∠B is three times that of ∠A, and the measure of ∠C is five times that of ∠A.

26. The measure of ∠A is twice that of ∠B, and the measure of ∠C is 40 more than that of ∠B.

27. The measure of ∠B is three less than the measure of ∠A, and the measure of ∠C is nine less than the measure of ∠A.

C 28. The measure of ∠A is three times that of ∠C, and the measure of ∠B is twice the sum of the measures of ∠A and ∠C.

29. If you know that DEF is a triangle such that m∠D = 60 and m∠E > 60, what can you conclude about the measure of ∠F?

30. If you know that RST is a triangle such that m∠R < 42 and m∠S < 58, what can you conclude about the measure of ∠T?

31. If m∠A = m∠D and m∠B = m∠E, show that m∠C = m∠F.

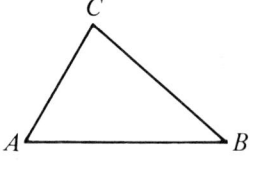

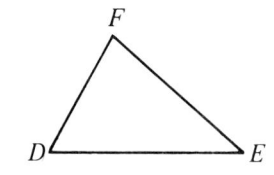

32. Show that m∠1 = m∠2.

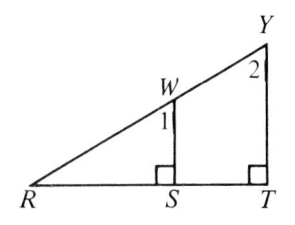

33. Show that m∠3 = m∠4.

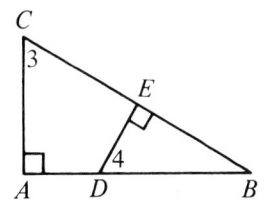

BASIC CONCEPTS OF GEOMETRY

Section 5-7 POLYGONS

In this section you will study:
- polygons
- perimeters of polygons

Imagine a circle made of string. Any figure that can be formed by moving or twisting the string is called a *closed curve*.

EXAMPLES OF CLOSED CURVES

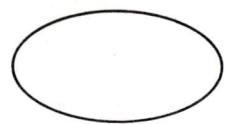

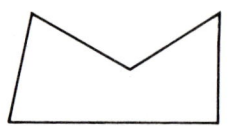

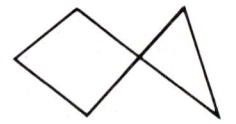

The two figures on the left are *simple closed curves*, since in each case the curve does not intersect itself. The two figures on the right are not simple closed curves.

A *polygon* is a simple closed curve formed by line segments. Suppose we take a long piece of string and wrap it tightly around the polygons shown below.

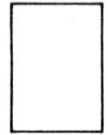

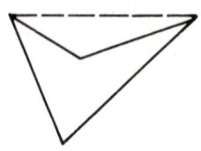

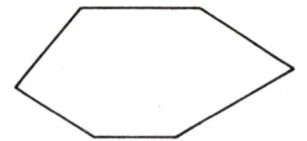

The string would fit snugly around the polygons on the left and right, but it would leave a gap around the middle polygon. The polygons on the left and right are called *convex polygons*. When we refer to a polygon in this book, we will mean a convex polygon.

The line segments that form a polygon are called the *sides* of the polygon, and the endpoints of the segments are the *vertices* of the polygon. $\overline{AB}$ is one side of the polygon shown, and point C is one vertex.

A polygon is named by listing *consecutive* vertices. One name for the polygon shown is *ABCD*. Can you give seven other names for the polygon?

The *angles* of a polygon are the angles determined by the sides of the polygon. $\angle DAB$, $\angle ABC$, $\angle BCD$, and $\angle CDA$ are the angles of polygon *ABCD*. We say that $\angle DAB$ is *included* by $\overline{DA}$ and $\overline{AB}$. Side $\overline{AB}$ is *included* by $\angle DAB$ and $\angle ABC$.

$\overline{AC}$ and $\overline{BD}$ are *diagonals* of *ABCD*. A diagonal of a polygon is a line segment that joins nonconsecutive vertices.

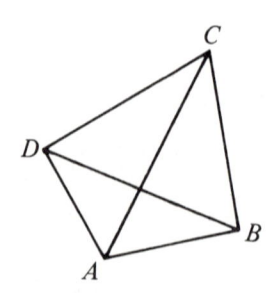

Polygons are usually classified by the number of sides they have. Names for the most common polygons are given below.

3 sides	*triangle*	6 sides	*hexagon*
4 sides	*quadrilateral*	8 sides	*octagon*
5 sides	*pentagon*	10 sides	*decagon*

A polygon with *n* sides is called an *n*-gon.

If all the sides of a polygon are congruent, the polygon is *equilateral*. If all the angles are congruent, the polygon is *equiangular*. If all the sides *and* all the angles are congruent, the polygon is a *regular polygon*.

EXAMPLES

neither equilateral nor equiangular

equilateral

equiangular

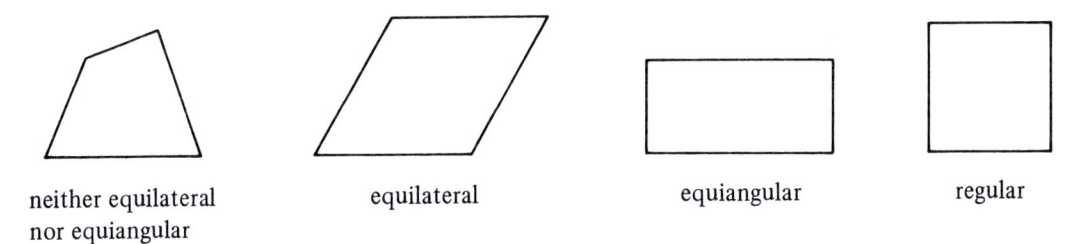

regular

The *perimeter* of a polygon is the sum of the lengths of its sides.

EXAMPLES

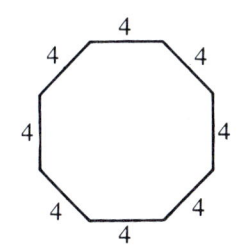

Perimeter = 4 + 4 + 4 + 4 + 4 + 4 + 4 + 4
         = 8 · 4
         = 32

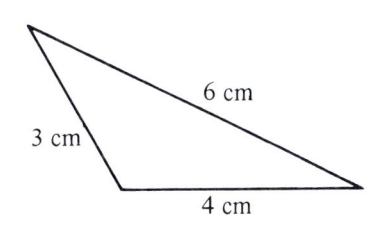

Perimeter = 3 + 4 + 6
         = 13 (cm)

ORAL EXERCISES

Is the figure a simple closed curve? a polygon? a convex polygon?

1.    2.    3. 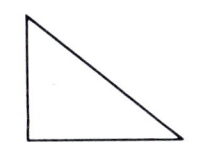   4.

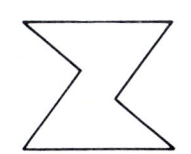

BASIC CONCEPTS OF GEOMETRY

Is the figure a simple closed curve? a polygon? a convex polygon?

5.   6. 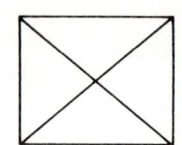   7. 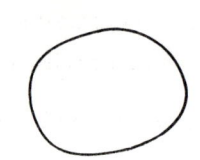   8.

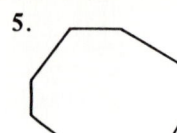

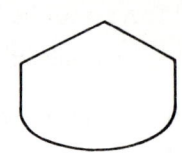

9. Give six names for the triangle shown.
10. Name the angle included by $\overline{XZ}$ and $\overline{ZY}$.
11. Name the side included by $\angle Z$ and $\angle Y$.
12. Give the other seven correct names for polygon $ABCD$ on page 160.

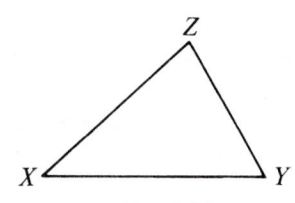

Exs. 9–11

Tell how many sides each polygon has.

13. quadrilateral   14. decagon   15. hexagon   16. octagon

17. What is the perimeter of a triangle whose sides have lengths 10, 14, and 20?
18. What is the perimeter of a regular pentagon if the length of each side is 10 cm?

WRITTEN EXERCISES

Name the polygon. Is it equilateral, equiangular, both, or neither?

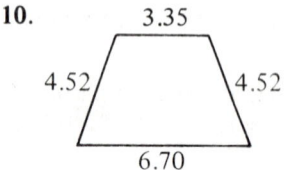

A 1.   2. 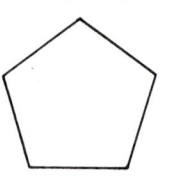   3.    4.

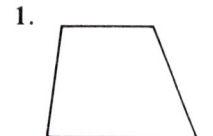

5. 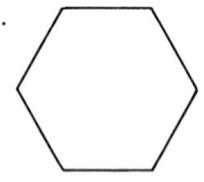   6. 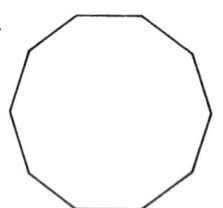   7. 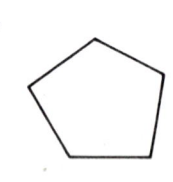   8.

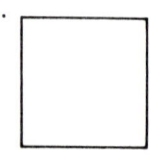

Find the perimeter.

9. (pentagon with sides 5.5, 5.5, 4.4, 4.4, 4.4)
10. (trapezoid with sides 3.35, 4.52, 4.52, 6.70)
11. (triangle with sides $\frac{1}{2}$, $\frac{1}{4}$, $\frac{1}{3}$)

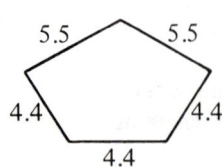

162   CHAPTER 5

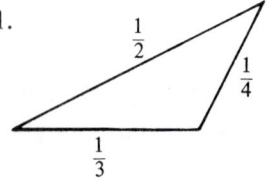

Find the perimeter of the polygon described.

12. A regular hexagon whose sides are 4.9 cm long
13. An equilateral pentagon whose sides are 7.37 cm long
14. A regular decagon whose sides are $3\frac{1}{5}$ units long

Find the length of each side of the polygon described.

EXAMPLE An equilateral pentagon with perimeter 4.5 m
SOLUTION $4.5 \div 5 = 0.9$ (m)

15. An equilateral triangle with perimeter 16.5 cm
16. An equilateral octagon with perimeter 2.4 m
17. An equilateral hexagon with perimeter 2.4 m
18. An equilateral quadrilateral with perimeter $\frac{8}{9}$

Sketch the figure described.

B 19. A simple closed curve that is not a polygon
20. A pentagon that is not convex
21. A regular triangle
22. A hexagon and all its diagonals
23. Sketch an equiangular quadrilateral that is not equilateral. This figure has a special name, which you may recall from your previous work in mathematics. What is it?
24. Sketch a regular quadrilateral. What is the special name for this figure?

25. If the length of one side of an equilateral triangle is represented by $2x + 1$, find the perimeter of the triangle in terms of x.
26. If the length of one side of a regular decagon is represented by $5y - 1$, find the perimeter of the decagon in terms of y.
27. If the perimeter of an equilateral octagon is represented by $8z - 16$, find the length of one side in terms of z.
28. If the perimeter of a regular quadrilateral is represented by $12w + 8$, find the length of one side in terms of w.
29. $\triangle ABC$ and $\triangle DEF$ are equilateral, and the length of $\overline{DE}$ is 4 cm more than twice the length of $\overline{AB}$. If the perimeter of $\triangle DEF$ is 27 cm, find the length of $\overline{AB}$.

BASIC CONCEPTS OF GEOMETRY

EXAMPLE Find the sum of the measures of the angles of a quadrilateral.

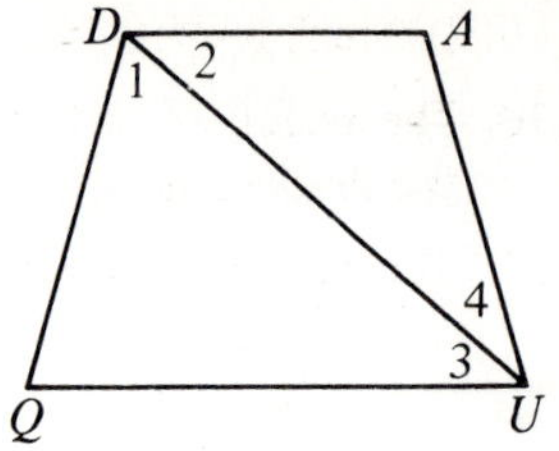

SOLUTION Draw a diagonal of the quadrilateral. The diagonal divides the quadrilateral into two triangles.

$m\angle Q + m\angle 3 + m\angle 1 = 180$

$m\angle 4 + m\angle A + m\angle 2 = 180$

$m\angle Q + m\angle 3 + m\angle 1 + m\angle 4 + m\angle A + m\angle 2 = 360$

$m\angle ADQ = m\angle 1 + m\angle 2$

$m\angle QUA = m\angle 3 + m\angle 4$

Thus, $m\angle Q + m\angle QUA + m\angle A + m\angle ADQ = 360$.

Find the sum of the measures of the angles of a polygon with the given number of sides. (*Hint*: Pick one vertex and draw all the diagonals from that vertex.)

C 30. 5   31. 6   32. 7   33. 8

34. Write a formula for the sum of the measures of the angles of a polygon with *n* sides.

35. What is the sum of the measures of the angles of a polygon with 20 sides?

SELF-TEST 2

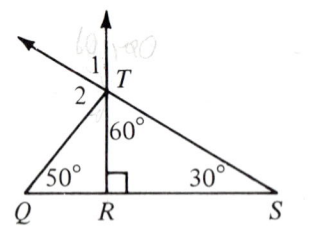

1. Name two perpendicular line segments.
2. Name a complement of $\angle RTS$.
3. Name two pairs of supplementary angles.
4. If $m\angle 1 = 3x - 9$, what is the value of x?

Section 5-4

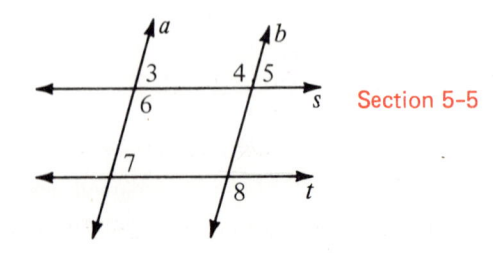

Complete.

5. If $a \parallel b$, then $m\angle 3 = m\angle$ __?__.
6. If $s \parallel t$ and $m\angle 5 = 78$, then $m\angle 8 = $ __?__.
7. If $m\angle 6 = m\angle 4$, then __?__ $\parallel$ __?__.
8. If $m\angle 6 + m\angle 7 = 180$, then __?__ $\parallel$ __?__.

Section 5-5

164   CHAPTER 5

9. In the figure for Exercises 1-4, find the measure of ∠2.   Section 5-6

10. The measures of two angles of a triangle are 49 and 73. Find the measure of the third angle.

11. The measures of the two acute angles of a right triangle are represented by x and $3x + 2$. Find the measures of the angles of the triangle.

12. Is the figure at the right a simple closed curve? Is it a polygon?   Section 5-7

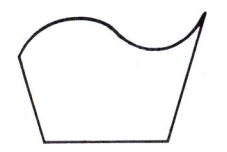

13. The sum of the measures of the angles of an equiangular octagon is 1080. What is the measure of each angle?

14. If each side of a regular hexagon is 1.3 cm long, find its perimeter.

Computer Exercises

1. **a.** This program provides practice in finding complements and supplements of angles. Copy and RUN the program.

```
10   READ D
15   PRINT
20   PRINT "MEASURE OF ANGLE:";D
30   IF D >= 90 THEN 100
40   PRINT "MEASURE OF COMPLEMENT";
50   INPUT C
60   IF C=90-D THEN 90
70   PRINT "SORRY, MEASURE OF COMPLEMENT IS";90-D
80   GOTO 100
90   PRINT "RIGHT!"
100  PRINT "MEASURE OF SUPPLEMENT";
110  INPUT S
120  IF S=180-D THEN 150
130  PRINT "SORRY, MEASURE OF SUPPLEMENT IS";180-D
140  GOTO 10
150  PRINT "RIGHT!"
160  GOTO 10
170  DATA 28,83,45,1,30,90,66,124,56,131
180  END
```

b. For more practice, revise line 170 and supply your own data.

2. Copy and RUN the program below to find the sum of the measures of the angles of:

   a. A triangle
   b. A rectangle
   c. A pentagon
   d. A decagon
   e. A polygon with 20 sides
   f. A polygon with 100 sides

```
10  PRINT "THIS PROGRAM WILL FIND THE SUM OF THE MEASURES"
20  PRINT "   OF THE ANGLES OF A POLYGON OF N SIDES."
25  PRINT
30  PRINT "HOW MANY SIDES";
40  INPUT N
50  LET S=(N-2)*180
60  PRINT "THE SUM OF THE MEASURES OF THE ANGLES OF A"
70  PRINT "   POLYGON OF";N;" SIDES IS";S;" DEGREES."
80  END
```

CHAPTER REVIEW

Exercises 1-4 refer to the diagram.

1. Give another name for line m.

2. How many planes contain points A, B, C, and E?

3. Name three collinear points and four non-coplanar points.

4. Name a ray that is contained in line l.

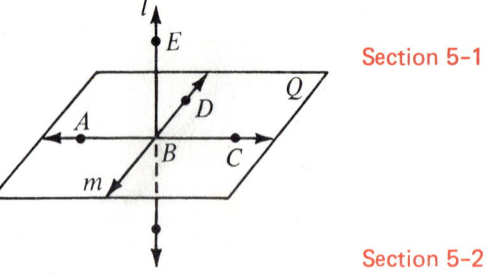

Section 5-1

Section 5-2

5. If $BA = BC$ and $BC = BD$, name three line segments that are congruent.

6. B is the midpoint of $\overline{AC}$, $AB = 8 - x$ and $BC = 2x + 2$. Find AC.

Exercises 7-12 refer to the diagram. Find the measure of the given angle. Classify it as acute, obtuse, right or straight.

7. $\angle VOY$
8. $\angle WOZ$
9. $\angle WOY$

Section 5-3

10. Name two pairs of congruent acute angles shown.

11. Name three pairs of perpendicular rays shown.

Section 5-4

12. Name two complements and a supplement of $\angle VOX$.

13. If $\angle 1$ and $\angle 2$ are vertical angles and $\angle 1$ is an obtuse angle, what can you say about $m\angle 2$?

In the diagram, $a \parallel b$.

14. State the biconditional that justifies the statement "$m\angle 2 = m\angle 6$."

15. If $m\angle 3 = 138$, find the measures of the other numbered angles.

16. If $m\angle 2 = y$ and $m\angle 3 = 3y$, what is the value of y?

Section 5-5

17. Find $m\angle 1$, $m\angle 2$, and $m\angle 3$ in the diagram.

18. If an acute angle of a right triangle has measure 24, find the measures of the other two angles of the right triangle.

19. If the measures of the three angles of a triangle are consecutive integers, find the measures of the angles.

Section 5-6

Ex. 17

20. Name the polygon shown.

21. Does the polygon appear to be equilateral, equiangular, regular, or none of these?

22. If the perimeter of the polygon shown is 30, find the value of x.

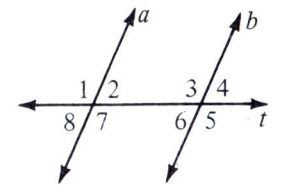

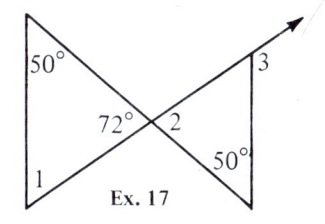

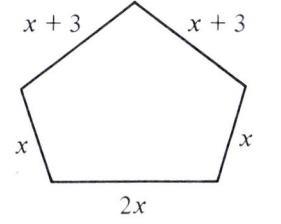

Section 5-7

CHAPTER TEST

Exercises 1-15 refer to the figure at the right.

1. Give another name for $\overrightarrow{BE}$.
2. How many lines contain all three points A, B, and C?
3. Find $m\angle DAB$.
4. Complete: Polygon $ABCD$ is a __?__.
5. Name three line segments contained in $\overleftrightarrow{BD}$.
6. Name three collinear points.
7. Name two congruent adjacent angles.
8. Name the intersection of $\overrightarrow{BD}$ and $\overrightarrow{EC}$.
9. Name a complement of $\angle BCE$.
10. Name the diagonals of polygon $ABCD$.
11. If $ABCD$ is equilateral and its perimeter is 22 cm, find AD.
12. If E is the midpoint of $\overline{BD}$, and $\overline{BE}$ has length $2x - 3$, find the length of $\overline{BD}$ in terms of x.

Classify the angle as acute, right, obtuse, or straight.

13. $\angle EBC$
14. $\angle CED$
15. $\angle DEB$

BASIC CONCEPTS OF GEOMETRY   167

16. If $r \parallel s$, name three angles congruent to $\angle 2$.

17. State the biconditional that justifies the statement "If $m\angle 1 = m\angle 6$, then $r \parallel s$."

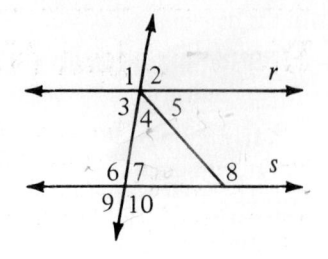

18. If $r \parallel s$, $m\angle 2 = 70$, and $m\angle 5 = 55$, find $m\angle 4$, $m\angle 7$, and $m\angle 8$.

In the figure, $\overrightarrow{PY}$ bisects $\angle XPZ$.

19. Complete: $m\angle XPY = \frac{1}{2} m\angle \underline{\quad ? \quad}$.

20. If $m\angle 1 = 8x + 2$ and $m\angle 2 = 3x + 47$, find the numerical value of the measure of $\angle XPZ$.

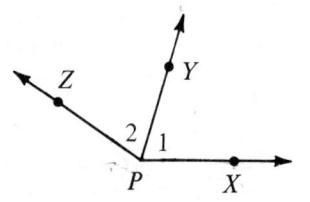

21. The measure of the supplement of an angle is 80 more than the sum of the measures of the angle and its complement. Find the measures of the three angles.

22. The measures of the angles of a triangle are given by z, $4z$, and $3z - 12$. Find the numerical value of the measure of each angle.

Biographical Note — Euclid

Although historians believe that the Greek mathematician Euclid was active, and perhaps teaching in Alexandria, about 300 B.C., we have no established facts about his life — just a few anecdotes whose truth is questionable. Euclid's influence, which extends to our own time, rests on his published work, much of which has been preserved. His *Elements*, in 13 books, is the oldest surviving work of Western mathematics.

Elements contains nearly all that was known about mathematics in Euclid's day. Because it made earlier works unnecessary, they were not preserved, and so we don't know how much of it was original with Euclid and how much was collected from the work of others. It is clear, however, that the organization and the eloquent style of *Elements* are Euclid's own contribution.

In *Elements*, Euclid discusses plane geometry, the theory of proportions, the theory of numbers, the theory of incommensurables, and solid geometry. Each book begins with a list of definitions and moves on to theorems or propositions that use these definitions. The first two definitions in Book 1 are "A point is that which has no part," and "A line is breadthless length." The first two propositions deal with constructions, a topic of particular interest to the Greeks of that time. Although textbooks today use a different approach, most plane geometry courses are based on Book 1 of *Elements*.

Euclid's method of proof, which rests on his lists of definitions and a small number of statements that are assumed to be true, has formed the basis of Western scientific thinking. A proof starts with a statement of the proposition to be proved, the given data, and the use to be made of the data. It then builds up a step-by-step argument to show that the proposition is a logical consequence of statements previously proved or accepted as true.

The first printed edition of *Elements* appeared in Europe at the end of the 15th century. Each new wave of interest in science and mathematics since that time has brought renewed appreciation of Euclid's great work.

BASIC CONCEPTS OF GEOMETRY

VOCABULARY REVIEW

Be sure that you understand the meaning of these terms:

point, p. 133
line, p. 133
plane, p. 133
collinear points, p. 134
coplanar points, p. 134
parallel lines, p. 135
line segment, p. 137
congruent line segments, p. 137
ray, p. 137
midpoint of a line segment, p. 138

angle, p. 140
congruent angles, p. 141
bisector of an angle, p. 141
perpendicular lines, p. 145
complementary angles, p. 146
supplementary angles, p. 146
vertical angles, p. 147
right triangle, p. 151
convex polygon, p. 160
regular polygon, p. 161

MIXED REVIEW

Solve.

1. $-9 + a = 19$
2. $5 - 3x \leq -10$
3. $3(b - 2) = b - 6$
4. $2 < \frac{1}{2}x + 2 \leq 3$
5. $-0.08b = -5$
6. $-12 < \frac{4}{3}z$

If $p, q,$ and r are false statements, give the truth value of each of the following.

7. $(\sim p \wedge \sim q) \rightarrow r$
8. $\sim p \leftrightarrow \sim r$
9. $(p \vee q) \rightarrow \sim r$

10. If the measures of the three angles of a triangle are consecutive even integers, find the measure of each angle.

11. Points L, M, and N have coordinates -7, 2, and k, respectively, on a number line. If M is the midpoint of $\overline{LN}$, then $k = $ __?__ .

12. Is the open sentence $-1 \cdot x = x$ always, sometimes, or never true?

Simplify.

13. $-\frac{11}{12} - \frac{3}{4}$
14. $-\frac{11}{12} \div \left(-\frac{3}{4}\right)$
15. $2^4 \div 4 - 5$

16. How many planes contain two given points?

17. Find the measure of an angle if the measure of a complement of the angle is one-fourth the measure of a supplement of the angle.

18. Use truth tables to show that the conditional $p \rightarrow q$ is logically equivalent to its contrapositive.

19. Graph the solution set of $x + 8 > 11$.

20. If $j = 4$ and $k = -3$, evaluate $2j - 5k$.

21. Two perpendicular lines intersect to form adjacent angles, $\angle 1$ and $\angle 2$. Are the angles congruent? complementary? supplementary? vertical?

22. Solve $r = 2(s - t)$ for t.

Exercises 23 and 24 refer to the diagram.

23. If $p \parallel q$ and $m\angle 10 = 95$, give the measures of as many angles as possible.

24. If $r \parallel s$, $m\angle 3 = 5t + 5$, and $m\angle 13 = 6t - 10$, find the value of t.

25. Use a protractor to draw an $\angle XYZ$ with measure 134 and $\overrightarrow{YP}$, the bisector of $\angle XYZ$.

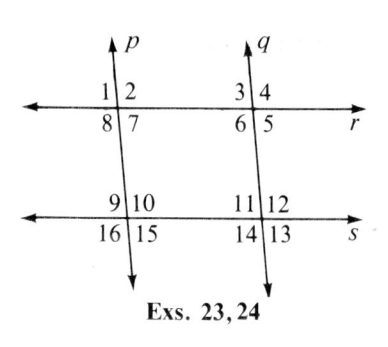

Exs. 23, 24

26. Explain why a triangle cannot have both a right angle and an obtuse angle.

27. **a.** Draw an equilateral octagon that is not regular.
    b. If each side has length x, find the perimeter in terms of x.

Solve and check.

28. The sum of three consecutive odd integers is greater than 51. Find the least possible values for the numbers.

29. The product of an integer, n, and 4 is less than -10. Find the greatest possible value for n.

30. Jack is twice as old as Jill was three years ago. He is five years older than she is. Find both ages now.

In Exercises 31–33, use p, q, r, and s as follows:

    p: $5 > 3$ r: A week has 7 days.
    q: 3 is an even number. s: The sun rises in the east.

Write in symbolic form and give the truth value.

31. The sun does not rise in the east only if $5 \leqslant 3$.

32. 3 is an even number and a week does not have 7 days.

33. $5 > 3$ if and only if a week has 7 days.

BASIC CONCEPTS OF GEOMETRY

Congruent triangles can be seen in this view of the roof of the East Wing of the National Gallery of Art in Washington, D.C. Notice, also, the squares, rectangles, and trapezoids.

Congruent Triangles

6

The two triangles shown are related in a special way. Can you guess what that way is?

The triangles are *congruent* because they have the same size and shape. (Think of turning one triangle 180° and sliding it on top of the other.) In this chapter you will learn some ways to determine whether two triangles that appear to be congruent actually are congruent.

If you could slide the two triangles together, you would get a figure congruent to the quadrilateral shown. Congruent triangles are used in this chapter to investigate certain properties of quadrilaterals such as the one shown.

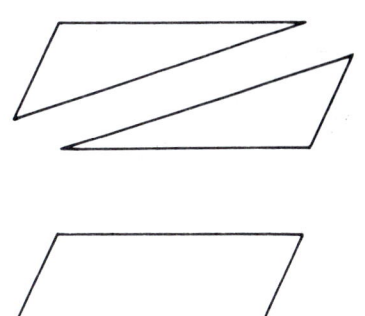

Section 6-1 CONSTRUCTIONS

You know that two line segments are congruent if their lengths are equal and that two angles are congruent if their measures are equal. If you were given a line segment or angle, could you draw a line segment or angle congruent to the given one?

You could use a ruler to measure a given segment and then draw a segment congruent to it. Similarly, you could use a protractor to copy a given angle.

The ancient Greeks described how to make exact copies of angles and line segments without actually measuring them. They used only a *compass* and a *straightedge* (an unmarked ruler) for their *constructions*. In this section you will learn how to *construct*:

- a line segment congruent to a given line segment
- an angle congruent to a given angle
- triangles with certain sides and angles congruent to given line segments and angles

A compass is a convenient tool for drawing a *circle*. The point of the compass is at the *center O* of the circle, and the distance *OM* between the point and the tip of the pencil is called the *radius* of the circle. In a construction, you often use the compass to draw only part of a circle, called an *arc* of the circle.

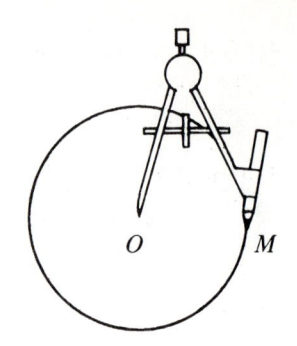

CONSTRUCTION 1   Given a line segment, construct a line segment congruent to it.

Given: $\overline{AB}$

1. Use a straightedge to draw a ray with endpoint *C*.

2. Put the point of a compass at *A* and adjust the pencil tip so that it lies on point *B*. Every circle and arc drawn with this setting will have radius *AB*.

3. With *C* as center and radius *AB*, draw an arc intersecting the ray. Label the point of intersection *D*.

$$\overline{AB} \cong \overline{CD}$$

CONSTRUCTION 2   Given an angle, construct an angle congruent to it.

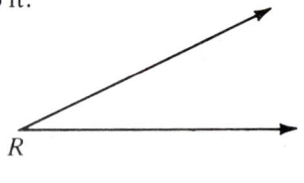

Given: ∠*R*

1. Draw a ray with endpoint *X*.

2. With *R* as center and *any* radius, draw an arc that intersects the sides of ∠*R*. Label the points of intersection *S* and *T*.

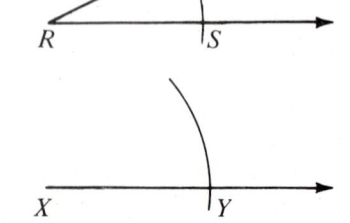

3. With *X* as center and the same radius as in Step 2, draw an arc that intersects the ray at a point *Y*.

174   CHAPTER 6

4. With *Y* as center and *ST* as radius, draw an arc that intersects the first arc at *Z*. (If necessary, go back and extend the arc you drew in Step 3.)

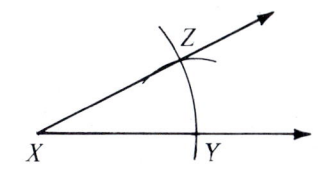

5. Draw $\overrightarrow{XZ}$.

$$\angle X \cong \angle R$$

CONSTRUCTION 3 Given three line segments, construct a triangle whose sides are congruent to the line segments.

Given: $\overline{AB}$, $\overline{CD}$, and $\overline{EF}$

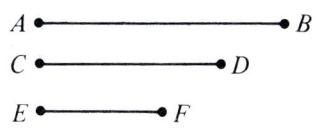

1. Use Construction 1 to construct $\overline{TR}$ so that $\overline{TR} \cong \overline{AB}$.

2. With *T* as center and *CD* as radius, draw an arc.

3. With *R* as center and *EF* as radius, draw an arc that intersects the first arc at point *I*. (If necessary, go back and extend the first arc.)

4. Connect points *T* and *I*. Connect points *R* and *I*.

△*TRI* is a triangle such that $\overline{TR} \cong \overline{AB}$, $\overline{TI} \cong \overline{CD}$, and $\overline{RI} \cong \overline{EF}$.

CONSTRUCTION 4 Given two line segments and an angle, construct a triangle with two sides and an included angle congruent to the given line segments and angle.

Given: $\overline{JK}$, $\overline{MN}$, and $\angle O$

1. Use Construction 2 to construct $\angle X$ congruent to $\angle O$.

2. On one side of $\angle X$, construct $\overline{XY}$ so that $\overline{XY} \cong \overline{JK}$.

3. On the other side of $\angle X$, construct $\overline{XZ}$ so that $\overline{XZ} \cong \overline{MN}$.

4. Connect points *Y* and *Z*.

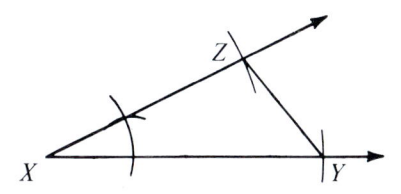

△*XYZ* is a triangle such that $\overline{XY} \cong \overline{JK}$, $\overline{XZ} \cong \overline{MN}$, and $\angle X \cong \angle O$.

CONGRUENT TRIANGLES

CONSTRUCTION 5 Given two angles and a line segment, construct a triangle with two angles and an included side congruent to the given angles and line segment.

Given: $\overline{DE}$, $\angle F$, and $\angle G$

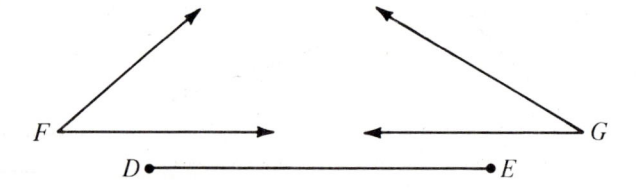

1. Use Construction 1 to construct $\overline{AB}$ so that $\overline{AB} \cong \overline{DE}$.

2. With A as vertex and $\overrightarrow{AB}$ as one side, construct an angle congruent to $\angle F$.

3. With B as vertex and $\overrightarrow{BA}$ as one side, construct an angle congruent to $\angle G$. Let C be the point where the sides of $\angle A$ and $\angle B$ intersect.

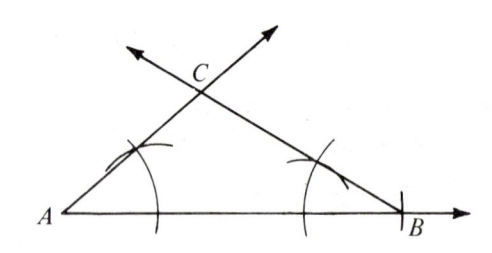

$\triangle ABC$ is a triangle such that $\angle A \cong \angle F$, $\angle B \cong \angle G$, and $\overline{AB} \cong \overline{DE}$.

WRITTEN EXERCISES

A 1. Draw an obtuse angle. Then construct an angle congruent to it.

2. Draw an acute angle. Then construct an angle congruent to it.

In Exercises 3–8 begin by drawing line segments and angles like those shown.

In Exercises 3 and 4 construct a triangle whose sides are congruent to the line segments you have drawn.

3.

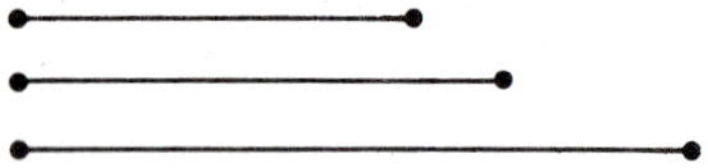

4.

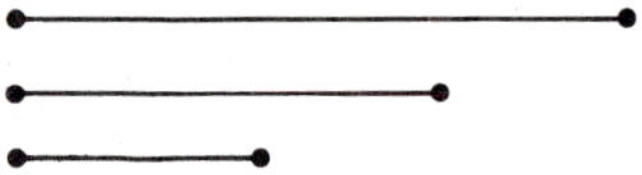

In Exercises 5 and 6 construct a triangle with two sides and an included angle congruent to the line segments and angle you have drawn.

5.

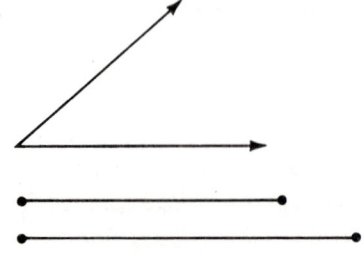

6.

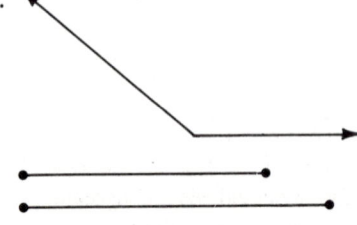

176   CHAPTER 6

In Exercises 7 and 8 construct a triangle with two angles and an included side congruent to the angles and line segment you have drawn.

7.

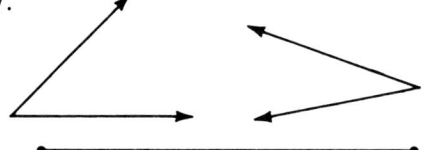

8.

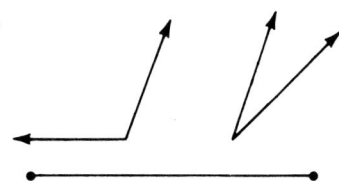

In Exercises 9-11 begin by drawing a triangle like the one shown.

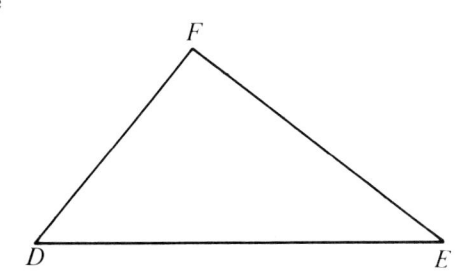

B 9. Construct △ABC so that:
$\overline{AB} \cong \overline{DE}$, $\overline{BC} \cong \overline{EF}$, $\overline{AC} \cong \overline{DF}$

10. Construct △ABC so that:
$\overline{AB} \cong \overline{DE}$, $\overline{AC} \cong \overline{DF}$, $\angle A \cong \angle D$

11. Construct △ABC so that:
$\angle A \cong \angle D$, $\angle B \cong \angle E$, $\overline{AB} \cong \overline{DE}$

12. a. Use a protractor to measure the angles of the △DEF you drew and the △ABC you constructed in Exercise 9.
    b. List three pairs of congruent angles in △DEF and △ABC in Exercise 9.
    c. Compare the three triangles you constructed in Exercises 9-11. What appears to be true?

In Exercises 13 and 14 draw a triangle like the one shown. Then use the construction of your choice to construct △ABC with the same size and shape as △DEF.

13.

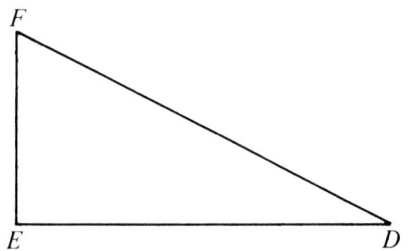

14.

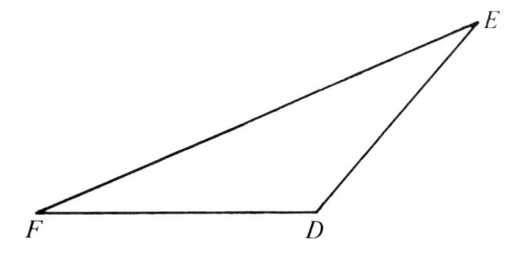

15. Can you construct a triangle whose sides are congruent to the line segments shown? Explain.

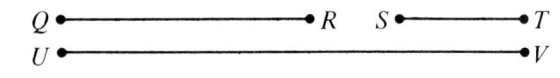

16. Can you construct a triangle with two angles congruent to the angles shown and with the included side congruent to the line segment shown? Explain.

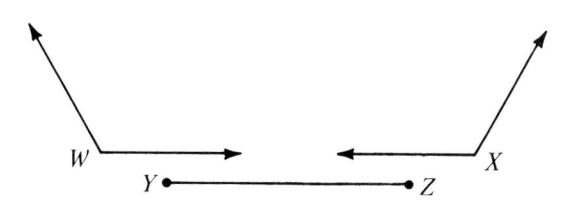

CONGRUENT TRIANGLES   177

In Exercises 17 and 18 use a protractor and a ruler to draw the triangle described.

17. The measures of two angles are 40 and 70, and the included side is 5 cm long.

18. Two sides are 4 cm and 6 cm long, and the measure of the included angle is 60.

19. Use a ruler to draw a line segment 4 cm long. Then *construct* an equilateral triangle whose sides are all congruent to the line segment.

20. Use a ruler to draw a line segment 3 cm long. Call it $\overline{AB}$. Then *construct* $\overline{CD}$ so that $CD = 2 \cdot AB$.

C 21. Draw a triangle roughly like the one shown. Then construct $\triangle DEF$ such that $DE = 2 \cdot AB$, $EF = 2 \cdot BC$, and $DF = 2 \cdot AC$.

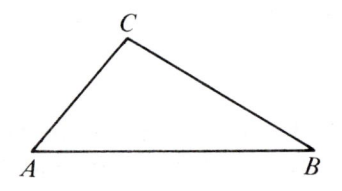

EXAMPLE Given $\angle 1$, construct an angle with measure $2 \cdot m\angle 1$.

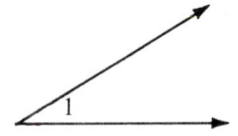

SOLUTION Use Construction 2 to construct $\angle XYZ$ such that $m\angle XYZ = m\angle 1$. Then construct $\angle WYX$ so that $m\angle WYX = m\angle 1$.

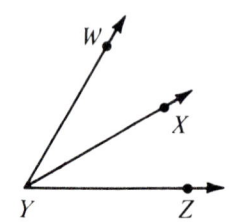

$$m\angle WYZ = 2 \cdot m\angle 1$$

22. Draw angles roughly like those shown. Then construct an $\angle 6$ so that $m\angle 6 = m\angle 4 + m\angle 5$.

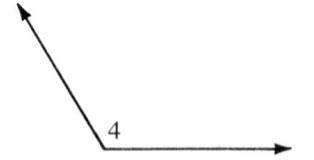

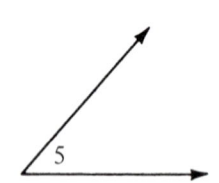

23. Draw angles roughly like those shown. Then construct $\angle 3$ so that $m\angle 1 + m\angle 2 + m\angle 3 = 180$.

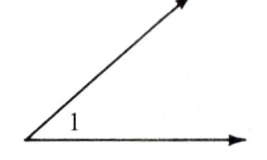

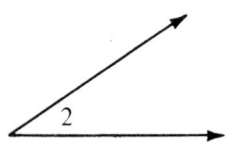

Section 6-2 CONGRUENT TRIANGLES

If two figures have the same size and shape, they are called *congruent*. However, two figures that *appear* to be congruent may not actually *be* congruent. In this section you will learn how to show that two triangles are congruent by showing that the following parts of one triangle are congruent to the *corresponding parts* of the second triangle:
- all three sides and all three angles
- all three sides (SSS)
- two sides and the included angle (SAS)
- two angles and the included side (ASA)

Line segments and angles marked alike in the figure are congruent. $\triangle ABC$ and $\triangle DEF$ have six pairs of congruent parts.

Corresponding Angles	Corresponding Sides
$\angle A \cong \angle D$	$\overline{AB} \cong \overline{DE}$
$\angle B \cong \angle E$	$\overline{BC} \cong \overline{EF}$
$\angle C \cong \angle F$	$\overline{AC} \cong \overline{DF}$

Since all six pairs of corresponding parts are congruent, the triangles are *congruent*:

$$\triangle ABC \cong \triangle DEF$$

You can write the congruence in other ways, as long as you write corresponding vertices in the same order. Three other correct ways of writing the congruence are given below. Can you name two other ways?

$\triangle BCA \cong \triangle EFD$ $\triangle CAB \cong \triangle FDE$ $\triangle CBA \cong \triangle FED$

Two triangles are congruent if and only if the three angles and three sides of one are congruent to the corresponding parts of the second. However, to show that two triangles are congruent, you do not need to show that all pairs of corresponding parts are congruent. In Section 6-1 you saw that when you construct a triangle, its size and shape are completely determined by any one of the following:

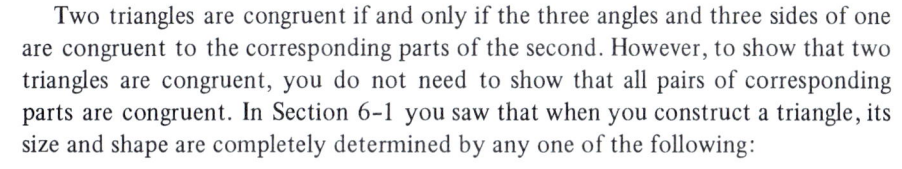

1. The lengths of the three sides
2. The lengths of two sides and the measure of the included angle
3. The measures of two angles and the length of the included side

> If three sides of one triangle are congruent to three sides of another triangle, then the triangles are congruent.   (SSS)

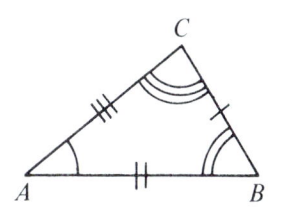

CONGRUENT TRIANGLES

EXAMPLES

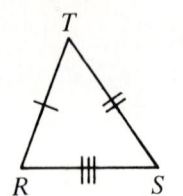

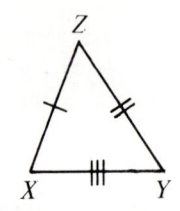

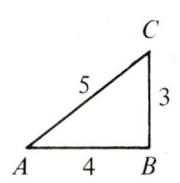

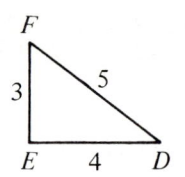

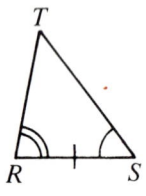

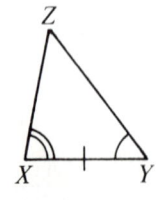

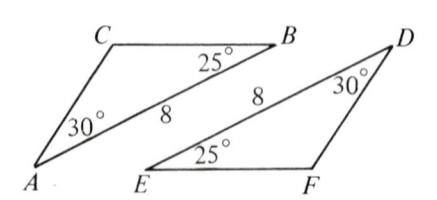

By SSS, $\triangle RST \cong \triangle XYZ$ and $\triangle ABC \cong \triangle DEF$.

> If two sides and the included angle of one triangle are congruent to two sides and the included angle of another triangle, then the triangles are congruent. (SAS)

EXAMPLES

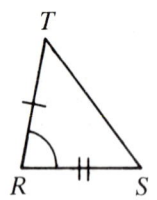

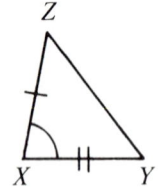

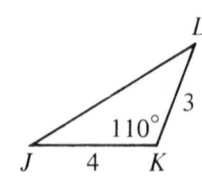

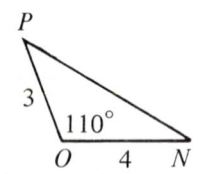

By SAS, $\triangle RST \cong \triangle XYZ$ and $\triangle JKL \cong \triangle NOP$.

> If two angles and the included side of one triangle are congruent to two angles and the included side of another triangle, then the triangles are congruent. (ASA)

EXAMPLES

By ASA, $\triangle RST \cong \triangle XYZ$ and $\triangle ABC \cong \triangle DEF$.

Whenever a line segment is a side of each of two triangles, you can use the fact that a line segment is congruent to itself.

180   CHAPTER 6

EXAMPLE Is △JKL congruent to △LMJ? If so, give a reason.

SOLUTION ∠MJL ≅ ∠KLJ, ∠MLJ ≅ ∠KJL, $\overline{JL} \cong \overline{JL}$.
By ASA, △JKL ≅ △LMJ.

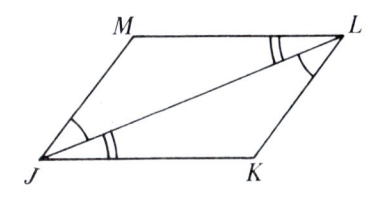

EXAMPLE Is △WAH congruent to △EAL? If so, give a reason.

SOLUTION ∠WAH ≅ ∠EAL, since they are vertical angles.
Also, $\overline{WA} \cong \overline{EA}$ and $\overline{HA} \cong \overline{LA}$.
By SAS, △WAH ≅ △EAL.

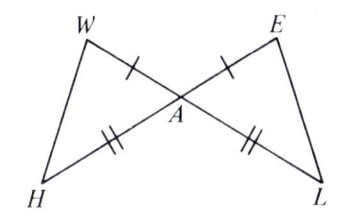

ORAL EXERCISES

If △RST ≅ △XYZ, complete the statements.

1. ∠R ≅ __?__
2. $\overline{RS} \cong$ __?__
3. $\overline{YZ} \cong$ __?__
4. ∠Y ≅ __?__
5. ∠Z ≅ __?__
6. $\overline{RT} \cong$ __?__
7. △STR ≅ __?__
8. △ZYX ≅ __?__

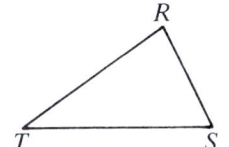

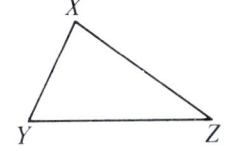

Can you tell from the given information that the two triangles are congruent? If so, give a reason (SSS, SAS, or ASA).

9. 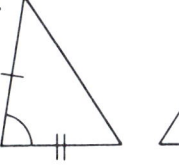 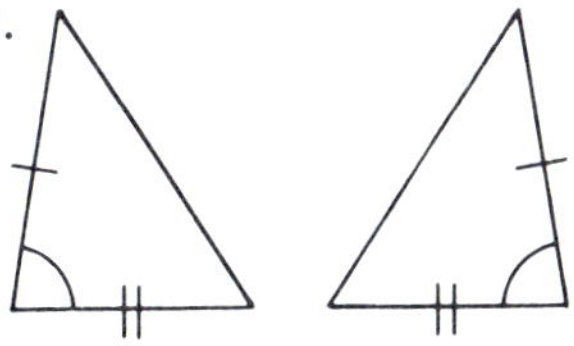 10. 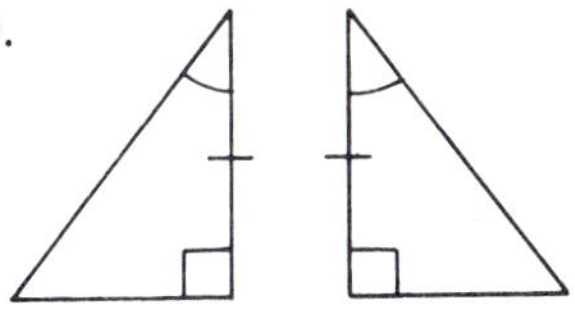 11.

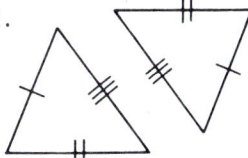

12. 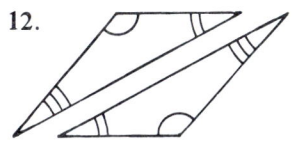 13. 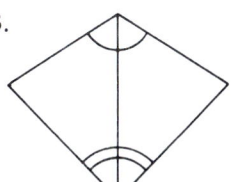 14.

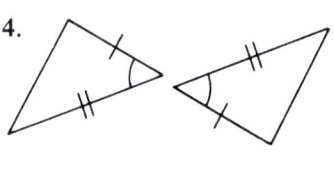

WRITTEN EXERCISES

A 1. If △ROY ≅ △JEN, list the pairs of congruent corresponding parts.

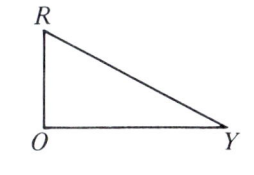

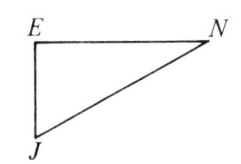

CONGRUENT TRIANGLES 181

2. If △CAT ≅ △KIT, list the pairs of congruent corresponding parts.

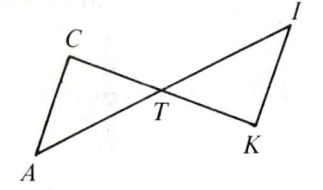

Can you tell from the given information that the two triangles are congruent? If so, give a reason (SSS, SAS, or ASA).

3. 4. 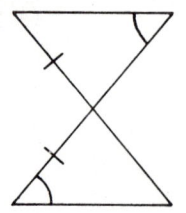 5.

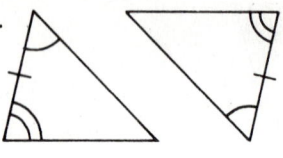

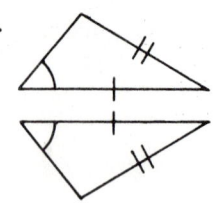

6. 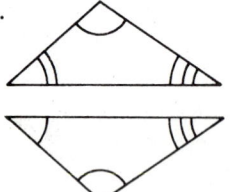 7. 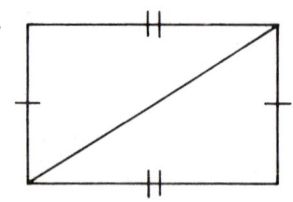 8.

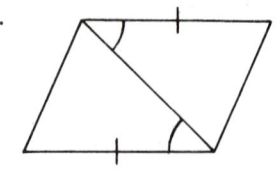

B 9. If △ABC ≅ △DCB, list the pairs of congruent corresponding parts.

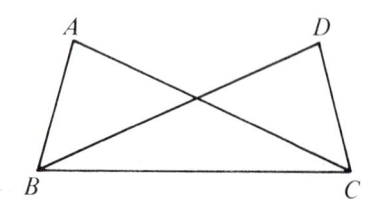

10. If △MOL ≅ △KOJ, list the pairs of congruent corresponding parts.

11. If △JML ≅ △LKJ, list the pairs of congruent corresponding parts.

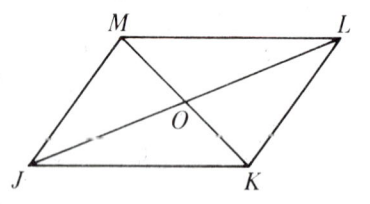

In Exercises 12-17 some congruent line segments and angles are listed. Tell which triangles shown must be congruent and give a reason.

EXAMPLE Given: $\overline{RB} \cong \overline{IB}, \overline{OB} \cong \overline{NB}$
SOLUTION ∠RBO ≅ ∠IBN, since they are vertical angles.
△RBO ≅ △IBN by SAS.

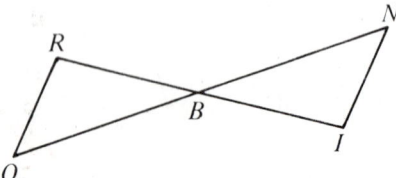

182 CHAPTER 6

12. Given: $\overline{BC} \cong \overline{DC}, \angle 1 \cong \angle 2$

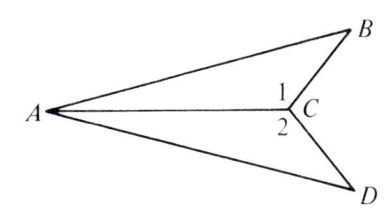

13. Given: $\angle 3 \cong \angle 4, \angle 5 \cong \angle 6$

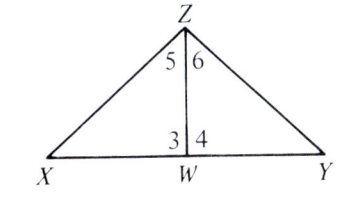

14. Given: $\overline{JK} \cong \overline{ML}, \overline{JL} \cong \overline{MK}$

15. Given: $\angle J \cong \angle M, \overline{JN} \cong \overline{MN}$

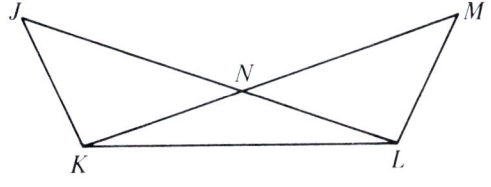

16. Given: $\overline{AB} \cong \overline{CD}, \overline{BC} \cong \overline{DA}$
    (*Hint*: There are two pairs of congruent triangles.)

17. Given: $\overline{DE} \cong \overline{BE}, \overline{AE} \cong \overline{CE}$
    (*Hint*: There are two pairs of congruent triangles.)

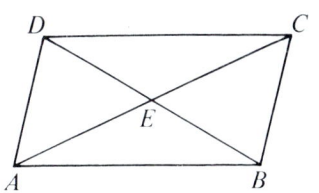

In Exercises 18-21 some congruent parts are marked. What additional information would you need in order to use the given reason for the triangles being congruent?

EXAMPLE (a) ASA
(b) SAS

SOLUTION (a) $\angle B \cong \angle E$
(b) $\overline{AC} \cong \overline{DF}$

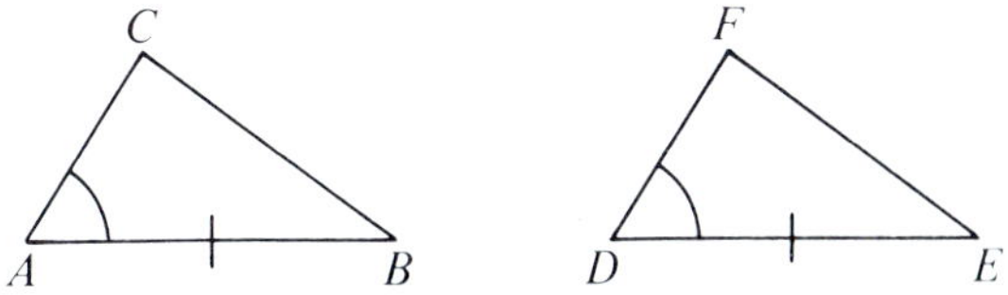

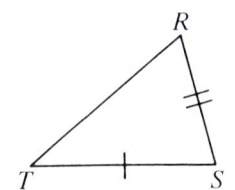

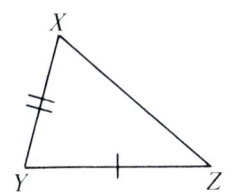

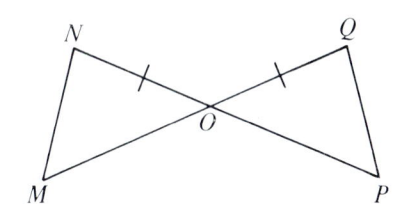

18. SSS

19. SAS

20. SAS

21. ASA

CONGRUENT TRIANGLES 183

In Exercises 22-25 some congruent parts are marked. What additional information would you need in order to use the given reason for the triangles being congruent?

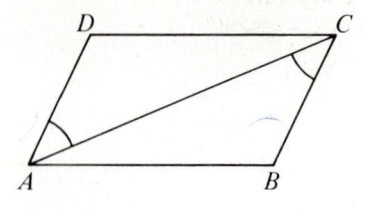

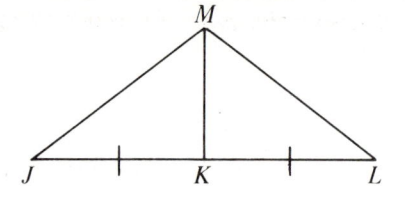

22. ASA   23. SAS   24. SAS   25. SSS

In Exercises 26-28 you will investigate the possibility of having AAS, AAA, and SSA properties for showing that two triangles are congruent.

C 26. Given: $\angle A \cong \angle D$, $\angle B \cong \angle E$, $\overline{AC} \cong \overline{DF}$
   a. Express the measure of $\angle C$ in terms of the measures of $\angle A$ and $\angle B$.
   b. Express the measure of $\angle F$ in terms of the measures of $\angle D$ and $\angle E$.
   c. Is $\angle C$ congruent to $\angle F$? Why?
   d. Is $\triangle ABC$ congruent to $\triangle DEF$? Why?
   e. If two angles and a non-included side of one triangle are congruent to the corresponding parts of a second triangle, must the triangles be congruent?

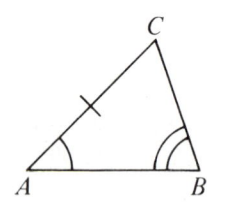

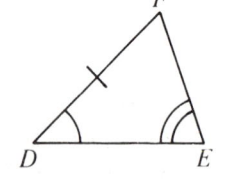

27. Given: $\overleftrightarrow{RS} \parallel \overleftrightarrow{XY}$, $RS = 11$, $XY = 18$
   a. What angle shown is congruent to $\angle 1$? Why?
   b. What angle shown is congruent to $\angle 2$?
   c. What angle shown is congruent to $\angle 3$? Why?
   d. Are $\triangle RST$ and $\triangle YXT$ congruent? Explain.
   e. If three angles of one triangle are congruent to three angles of a second triangle, must the two triangles be congruent?

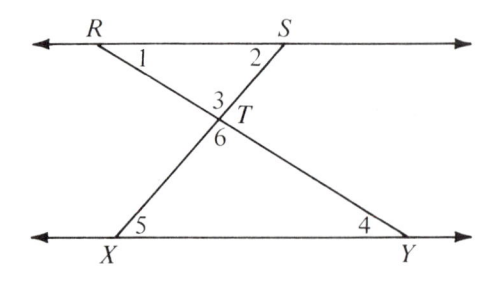

28. On your paper draw a $\triangle DEF$ like the one shown.
   a. Construct $\overline{XR}$ so that $\overline{XR}$ is congruent to $\overline{DE}$.
   b. Construct $\angle RXY$ congruent to $\angle D$.
   c. Open your compass to radius EF. With R as center, draw an arc of radius EF intersecting $\overrightarrow{XY}$ in two points, S and T. Draw $\overline{RS}$ and $\overline{RT}$.
   d. List the congruent corresponding parts of $\triangle DEF$ and $\triangle XRS$.

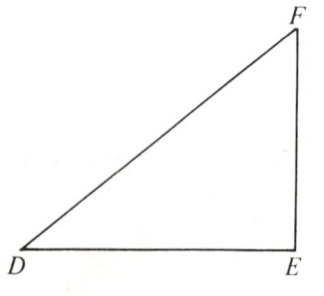

e. List the congruent corresponding parts of △DEF and △XRT.
f. Is △DEF congruent to both △XRS and △XRT?
g. If two sides and a non-included angle of one triangle are congruent to the corresponding parts of a second triangle, must the triangles be congruent?

SELF-TEST 1

In Exercises 1 and 2 begin by drawing line segments and angles like those shown below.

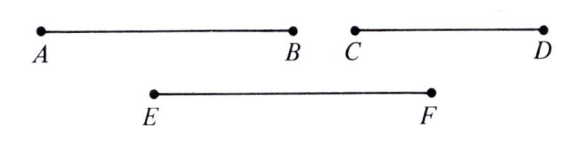

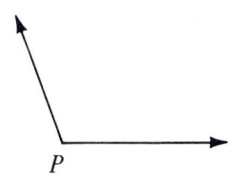

1. Construct a triangle whose sides are congruent to $\overline{AB}$, $\overline{CD}$, and $\overline{EF}$.   Section 6-1
2. Construct a triangle with two sides congruent to $\overline{CD}$ and $\overline{EF}$ and the included angle congruent to $\angle P$.

Can you tell from the given information that the two triangles are congruent? If so, give a reason (SSS, SAS, or ASA).

3. 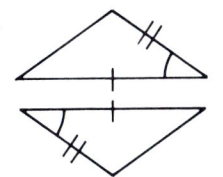   4. 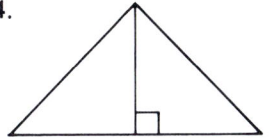   Section 6-2

5. 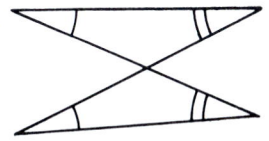   6.

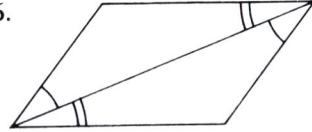

7. If △ABD ≅ △CDB, list the pairs of congruent corresponding parts.

CONGRUENT TRIANGLES   185

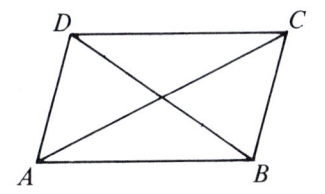

Section 6-3 ISOSCELES TRIANGLES

A triangle with at least two congruent sides is called an *isosceles triangle*. The angle included between the congruent sides is the *vertex angle* of the triangle. The other angles are the *base angles*. If $\overline{AB}$ and $\overline{AC}$ are congruent sides of $\triangle ABC$, then $\angle A$ is the vertex angle. The base angles are $\angle B$ and $\angle C$.

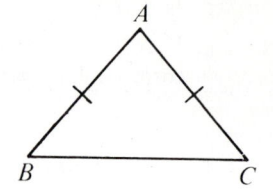

The base angles of $\triangle ABC$ appear to be congruent. Can you show that they must be congruent? You can if you can show that they are *corresponding parts of congruent triangles*.

Let $\overrightarrow{AD}$ be the bisector of $\angle BAC$. Then $\angle BAD \cong \angle CAD$. Also, $\overline{AD} \cong \overline{AD}$. By SAS,

$$\triangle BAD \cong \triangle CAD.$$

Since $\angle B$ and $\angle C$ are corresponding parts of the congruent triangles,

$$\angle B \cong \angle C.$$

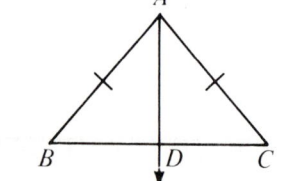

> The base angles of an isosceles triangle are congruent.

EXAMPLE $\triangle RST$ is equilateral. What is the measure of each of its angles?

SOLUTION Since $\overline{RS} \cong \overline{RT}, m\angle S = m\angle T$.
Since $\overline{ST} \cong \overline{SR}, m\angle T = m\angle R$.
$m\angle R + m\angle S + m\angle T = 180$ (Why?)
$m\angle T + m\angle T + m\angle T = 180$ (By substitution.)
$3(m\angle T) = 180$
$m\angle T = 60$
$m\angle R = 60, m\angle S = 60, m\angle T = 60$

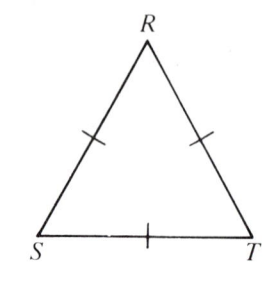

Notice that every equilateral triangle is equiangular.

ORAL EXERCISES

Congruent sides are marked. Which numbered angles are congruent?

1.

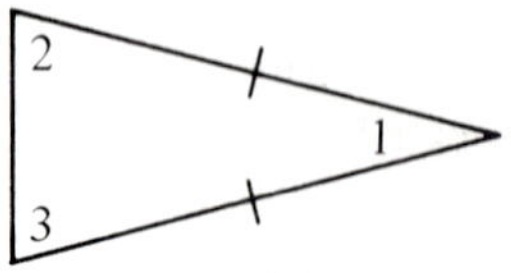

2.

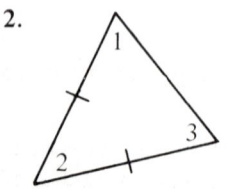

3.

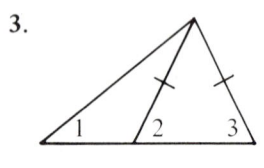

186   CHAPTER 6

4. 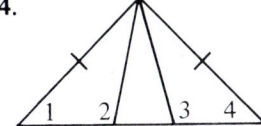
5. 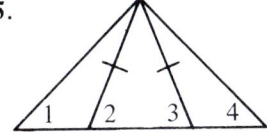
6.

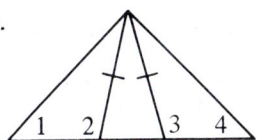

7. Must an isosceles triangle be equilateral?
8. Must an equilateral triangle be isosceles?
9. Can a right triangle be isosceles?
10. Can a right triangle be equilateral?

WRITTEN EXERCISES

Find the measures of the numbered angles.

A 1. 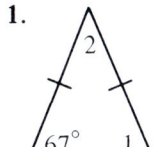
2. 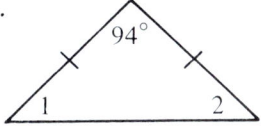
3. 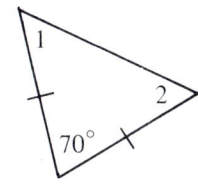
4. 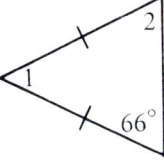
5. 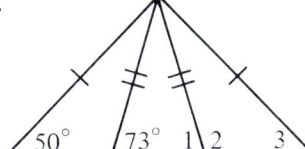
6.

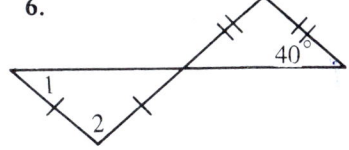

In Exercises 7-12 find the numerical values of the measures of the vertex angle and a base angle of an isosceles triangle.

B

	7.	8.	9.	10.	11.	12.
Vertex angle	50	y	z	$r + 12$	$3s$	$4t - 2$
Base angle	x	50	$2z$	$3r$	$s - 10$	$t + 4$

13. The measure of the vertex angle of an isosceles triangle is 12 more than twice the measure of a base angle of the triangle. Find the measures of all three angles.

14. The measure of a base angle of an isosceles triangle is 6 less than half the measure of the vertex angle of the triangle. Find the measures of all three angles.

CONGRUENT TRIANGLES

15. a. Another way of stating that "The base angles of an isosceles triangle are congruent" is "If two sides of a triangle are congruent, then the angles opposite those sides are congruent."
    Write the converse of this conditional.
   b. Given: $m\angle B = m\angle C$ and $\overrightarrow{AD}$ is the bisector of $\angle BAC$. What triangles would you show are congruent in order to show that $\overline{AB}$ is congruent to $\overline{AC}$?
   c. What parts of these triangles do you know are congruent?
   d. What angle is congruent to $\angle BDA$? Why?
   e. What reason would you give for the triangles you listed in (b) being congruent?
   f. Must $\overline{AB}$ and $\overline{AC}$ be congruent? Explain.
   g. Is the converse you wrote in (a) true?

16. What is the converse of "Every equilateral triangle is equiangular"? Do you think that this converse is true? Explain. (*Hint:* Refer to Exercise 15.)

C 17. Explain why $\triangle ABE$ is congruent to $\triangle DCE$.

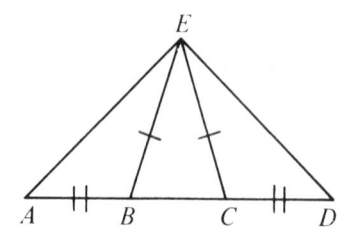

18. Explain why $\triangle WXV$ is congruent to $\triangle ZYV$.

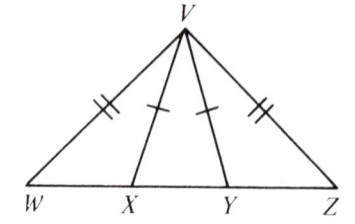

Section 6-4 PARALLELOGRAMS

In this section you will see how congruent triangles can be used to derive certain properties of special quadrilaterals called *parallelograms*.

If both pairs of opposite sides of a quadrilateral are contained in parallel lines, the quadrilateral is called a *parallelogram*. For simplicity, we say that the sides themselves are parallel. Thus, quadrilateral ABCD is a parallelogram if and only if

$$\overline{AB} \parallel \overline{DC} \quad \text{and} \quad \overline{AD} \parallel \overline{BC}.$$

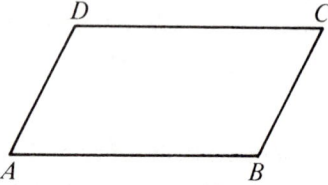

The opposite sides and opposite angles of the parallelogram shown *appear* to be congruent. We will show that these parts must be congruent in any parallelogram by showing that they are corresponding parts of congruent triangles.

$\overline{AC}$ is a diagonal of parallelogram ABCD. It is also a transversal of two different pairs of parallel line segments. Notice that two different pairs of alternate interior angles are shown.

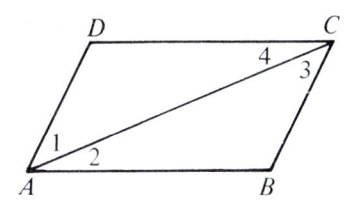

$\overline{AD} \parallel \overline{BC}$ implies that $m\angle 1 = m\angle 3$, which implies that $\angle 1 \cong \angle 3$.
$\overline{DC} \parallel \overline{AB}$ implies that $m\angle 4 = m\angle 2$, which implies that $\angle 4 \cong \angle 2$.
Since $\overline{AC} \cong \overline{AC}$,
$$\triangle ADC \cong \triangle CBA \text{ by ASA.}$$

> A diagonal of a parallelogram separates the parallelogram into two congruent triangles.

Notice that $\overline{DC}$ and $\overline{BA}$ are corresponding parts of congruent triangles, as are $\overline{DA}$ and $\overline{BC}$. Thus,
$$\overline{DC} \cong \overline{BA} \quad \text{and} \quad \overline{DA} \cong \overline{BC}.$$

> Opposite sides of a parallelogram are congruent.

Another way of stating that the opposite sides of a parallelogram are congruent is:

If a quadrilateral is a parallelogram, then both pairs of opposite sides are congruent.

The converse of this statement is also true, as you will see in Exercise 24.

> If both pairs of opposite sides of a quadrilateral are congruent, then the quadrilateral is a parallelogram.

Diagonal $\overline{AC}$ separates parallelogram ABCD into congruent triangles ABC and CDA that have $\angle ABC$ and $\angle CDA$ as corresponding parts. Diagonal $\overline{BD}$ separates the parallelogram into congruent triangles DAB and BCD that have $\angle DAB$ and $\angle BCD$ as corresponding parts. Thus,
$$\angle ABC \cong \angle CDA \quad \text{and} \quad \angle DAB \cong \angle BCD.$$

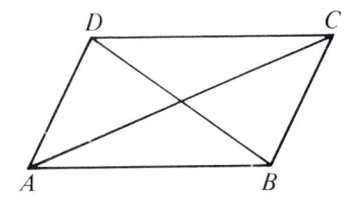

> Opposite angles of a parallelogram are congruent.

CONGRUENT TRIANGLES   189

Suppose you know that the sides $\overline{AD}$ and $\overline{BC}$ of quadrilateral ABCD are both parallel and congruent. That is,

$$\overline{AD} \parallel \overline{BC} \quad \text{and} \quad \overline{AD} \cong \overline{BC}.$$

Do you think that this is enough information to show that ABCD *must* be a parallelogram?

Since $\overline{AD} \parallel \overline{BC}$, you know that $\angle 1 \cong \angle 3$.
$\overline{AC} \cong \overline{AC}$ and $\overline{AD} \cong \overline{BC}$
Therefore, $\triangle ADC \cong \triangle CBA$. (Why?)
Therefore, $\angle 4 \cong \angle 2$. (Why?)
Therefore, $\overline{DC} \parallel \overline{AB}$. (Why?)
Therefore, ABCD is a parallelogram. (Why?)

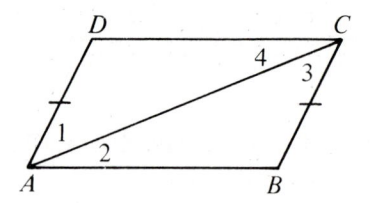

> If two sides of a quadrilateral are both parallel and congruent, then the quadrilateral is a parallelogram.

ORAL EXERCISES

Quadrilateral ABCD is a parallelogram.

1. Name two congruent triangles shown.
2. Name two pairs of congruent line segments.
3. Name four pairs of congruent angles.

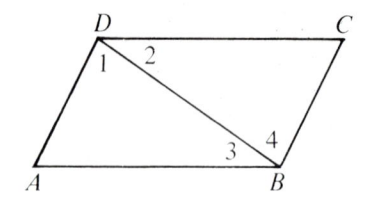

In Exercises 4–9 can you tell from the given information that quadrilateral WXYZ is a parallelogram? Explain.

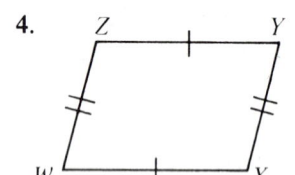

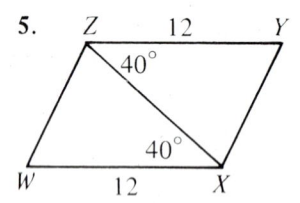

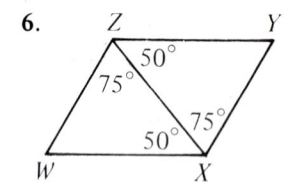

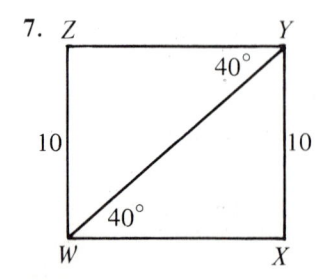

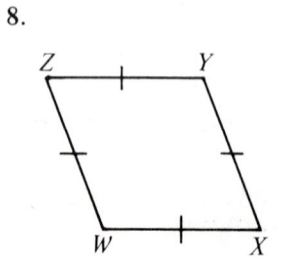

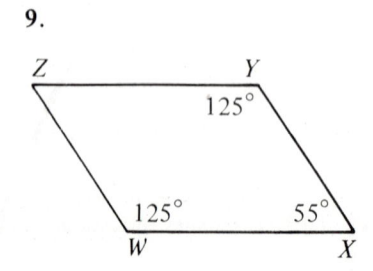

4.

5.

6.

7.

8.

9.

190   CHAPTER 6

WRITTEN EXERCISES

In Exercises 1-6 can you tell from the given information that quadrilateral *JKLM* must be a parallelogram?

A
1. $\overline{JK} \parallel \overline{ML}$
2. $\overline{JK} \cong \overline{ML}$
3. $\overline{JK} \parallel \overline{ML}, \overline{JM} \parallel \overline{KL}$
4. $\overline{JM} \cong \overline{KL}, \overline{JM} \parallel \overline{KL}$
5. $\overline{JK} \cong \overline{ML}, \overline{JK} \parallel \overline{ML}$
6. $\overline{JK} \cong \overline{ML}, \overline{JM} \parallel \overline{KL}$

7. If *WXYZ* is a parallelogram such that $m\angle W = 115$, find the measures of $\angle X, \angle Y,$ and $\angle Z$.

8. Quadrilateral *ABCD* is a parallelogram. Find the measures of the numbered angles.

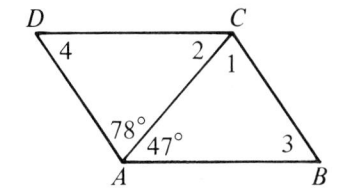

In Exercises 9-11 each quadrilateral is a parallelogram. Find the numerical value of *x*.

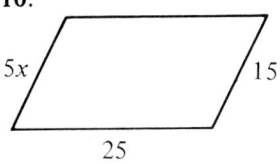

9. (sides: 18, 12, 3x)
10. (sides: 5x, 15, 25)
11. (sides: 2x + 7, 13, 19)

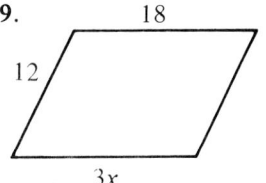

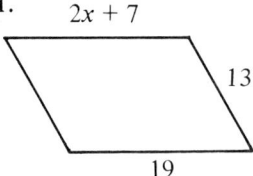

For what values of *x* and *y* is each quadrilateral a parallelogram?

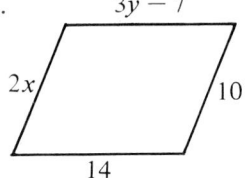

12. (sides: 9x, 28, 4y, 36)
13. (sides: 3y − 7, 2x, 10, 14)
14. (sides: 12y − 38, 50, 3x + 5, 8y − 4)

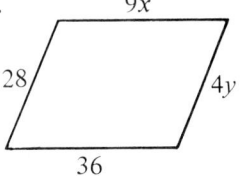

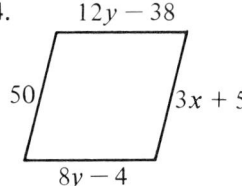

In Exercises 15-19 classify each statement as true or false. If the statement is true, give a reason (or reasons) to support your answer. If the statement is false, draw a diagram that shows why.

B
15. If a diagonal of a quadrilateral separates the quadrilateral into two congruent triangles, then the quadrilateral is a parallelogram.
16. If a quadrilateral is equilateral, then it is a parallelogram.
17. If two opposite sides of a quadrilateral are parallel and two opposite sides are congruent, then the quadrilateral is a parallelogram.
18. If two pairs of sides of a quadrilateral are congruent, then the quadrilateral is a parallelogram.
19. If all pairs of non-opposite angles of a quadrilateral are supplementary, then the quadrilateral is a parallelogram.

In Exercises 20 and 21 the lengths of the sides of each quadrilateral are given in terms of a variable. Is there a value of the variable for which the quadrilateral is a parallelogram? If there is, find the length of each side. If there isn't, explain.

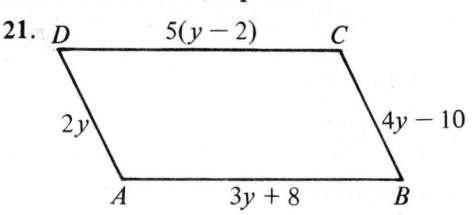

20. Sides: $MJ = x+1$, $ML = 3x$, $JK = x+8$, $LK = 2x-3$

21. Sides: $DC = 5(y-2)$, $DA = 2y$, $CB = 4y-10$, $AB = 3y+8$

C 22. a. Write as a conditional statement: Opposite angles of a parallelogram are congruent.
   b. Write the converse of your answer to part (a).
   c. Is this converse true or false? Explain. (*Hint*: You may use the fact that the sum of the measures of the angles of a quadrilateral is 360, as shown in the example on page 164.)

In Exercises 23 and 24 replace each "why" with the correct reason.

23. Given: $\overleftrightarrow{ML} \parallel \overleftrightarrow{JK}$, $\overleftrightarrow{MJ} \perp \overleftrightarrow{JK}$, $\overleftrightarrow{LK} \perp \overleftrightarrow{JK}$
    Show: $\overline{MJ} \cong \overline{LK}$
    (1) $\angle MJK$ and $\angle LKN$ are right angles. (Why?)
    (2) $\overline{MJ} \parallel \overline{LK}$ (Why?)
    (3) $\overleftrightarrow{ML} \parallel \overleftrightarrow{JK}$ (Why?)
    (4) $JKLM$ is a parallelogram. (Why?)
    (5) $\overline{MJ} \cong \overline{LK}$ (Why?)

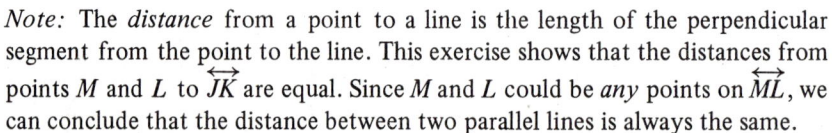

    Note: The *distance* from a point to a line is the length of the perpendicular segment from the point to the line. This exercise shows that the distances from points M and L to $\overleftrightarrow{JK}$ are equal. Since M and L could be *any* points on $\overleftrightarrow{ML}$, we can conclude that the distance between two parallel lines is always the same.

24. If both pairs of opposite sides of a quadrilateral are congruent, then the quadrilateral is a parallelogram.

    Given: In quadrilateral $ABCD$, $\overline{AB} \cong \overline{CD}$ and $\overline{BC} \cong \overline{DA}$.
    Show: $ABCD$ is a parallelogram.

    (1) $\overline{AB} \cong \overline{CD}$, $\overline{BC} \cong \overline{DA}$ (Given)
    (2) $\overline{AC} \cong \overline{AC}$ (A segment is congruent to itself.)
    (3) $\triangle ABC \cong \triangle CDA$ (Why?)
    (4) $\angle CAB \cong \angle ACD$ (Why?)
    (5) $\overline{AB} \parallel \overline{DC}$ (Why?)
    (6) $ABCD$ is a parallelogram. (Why?)

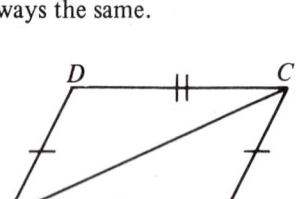

192   CHAPTER 6

Section 6-5 TRAPEZOIDS AND SPECIAL PARALLELOGRAMS

A *trapezoid* is a quadrilateral with exactly one pair of parallel sides. If its nonparallel sides are congruent, the trapezoid is *isosceles*.

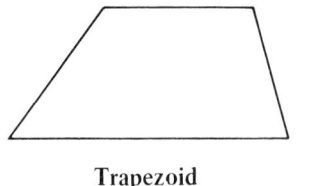

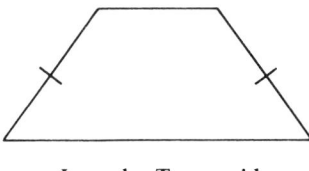

Trapezoid

Isosceles Trapezoid

A *rectangle* is a parallelogram with four right angles.
A *rhombus* is a parallelogram with four congruent sides.
A *square* is a parallelogram with four right angles and four congruent sides.

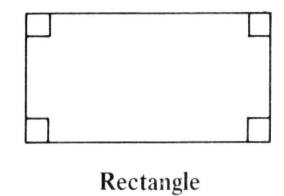

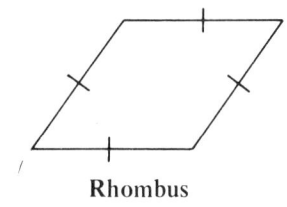

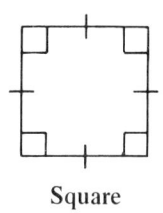

Rectangle

Rhombus

Square

Examine the diagonals of parallelogram *ABCD*, rectangle *JKLM*, and rhombus *RSTW*.

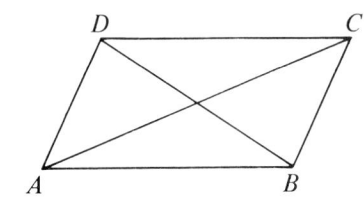

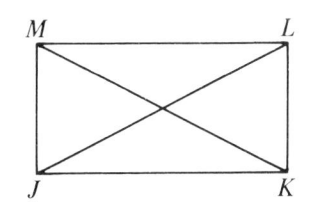

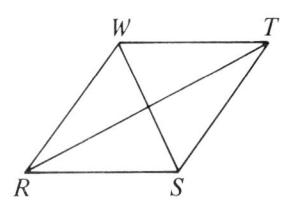

The diagonals of rectangle *JKLM appear* to be congruent, and the diagonals of rhombus *RSTW appear* to be perpendicular. The diagonals of parallelogram *ABCD* are neither congruent nor perpendicular.

Must the diagonals of a rectangle be congruent? If the diagonals of a parallelogram are congruent, must the parallelogram be a rectangle? Must the diagonals of a rhombus be perpendicular? If the diagonals of a parallelogram are perpendicular, must the parallelogram be a rhombus? You will investigate these questions in the text and exercises of this section.

To show that the diagonals $\overline{MK}$ and $\overline{LJ}$ of rectangle *JKLM* are congruent, you would like to find two congruent triangles that have $\overline{MK}$ and $\overline{LJ}$ as corresponding sides. On the following page you will see that it is possible to show that $\triangle MJK$ is congruent to $\triangle LKJ$.

CONGRUENT TRIANGLES 193

$\overline{MJ} \cong \overline{LK}$ (A rectangle is a parallelogram. Opposite sides of a parallelogram are congruent.)

$\overline{JK} \cong \overline{JK}$ (A line segment is congruent to itself.)

∠MJK and ∠LKJ are right angles. (A rectangle has four right angles.)

$m\angle MJK = 90$, $m\angle LKJ = 90$ (A right angle has measure 90.)

∠MJK ≅ ∠LKJ (Two angles are congruent if they have equal measures.)

△MJK ≅ △LKJ (SAS)

$\overline{MK} \cong \overline{LJ}$ (Corresponding parts of congruent triangles are congruent.)

> The diagonals of a rectangle are congruent.

In Exercise 21 you will show that the following is true.

> The diagonals of a rhombus are perpendicular.

If a parallelogram is neither a rectangle nor a rhombus, its diagonals are neither congruent nor perpendicular. However, the diagonals of a parallelogram do have an interesting property, as we show below.

Suppose that ABCD is a parallelogram. Since the opposite sides of a parallelogram are congruent,

$$\overline{AD} \cong \overline{BC}.$$

Since the opposite sides of a parallelogram are parallel,

∠1 ≅ ∠2 and ∠3 ≅ ∠4.

By ASA, △ADE ≅ △CBE.

Since they are pairs of corresponding sides of congruent triangles,

$$\overline{AE} \cong \overline{CE} \quad \text{and} \quad \overline{DE} \cong \overline{BE}.$$

Thus, E is the midpoint of $\overline{AC}$ and $\overline{BD}$.

When two line segments intersect at the midpoint of each segment, they are said to *bisect* each other.

> The diagonals of a parallelogram bisect each other.

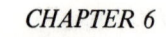

CHAPTER 6

ORAL EXERCISES

Classify each statement as true or false.
1. Every rhombus is a parallelogram.
2. Every quadrilateral is a parallelogram.
3. Every trapezoid is a parallelogram.
4. Every square is a rhombus.
5. The diagonals of a rectangle must be congruent.
6. The diagonals of a parallelogram must bisect each other.

WRITTEN EXERCISES

Classify each statement as true or false.

A
1. Every parallelogram is a rhombus.
2. Every square is a rectangle.
3. Every rhombus is a square.
4. Every square is a quadrilateral.
5. Every rectangle is a parallelogram.
6. The diagonals of a rectangle must bisect each other.
7. The diagonals of a parallelogram must be congruent.
8. A rhombus is an equilateral quadrilateral.
9. A rectangle is an equiangular quadrilateral.
10. Every quadrilateral is either a parallelogram or a trapezoid.
11. A diagonal of a rhombus separates the rhombus into two congruent triangles.
12. If both pairs of opposite sides of a quadrilateral are congruent, then the quadrilateral is a rectangle.
13. Opposite sides of an isosceles trapezoid are congruent.
14. Opposite angles of a rhombus are congruent.

In Exercises 15-18 draw the two line segments described. When you connect the endpoints of the segments, what kind of quadrilateral is formed?

B
15. $\overline{AC}$ and $\overline{BD}$ are noncongruent segments that bisect each other but are not perpendicular.
16. $\overline{AC}$ and $\overline{BD}$ are congruent segments that bisect each other but are not perpendicular.
17. $\overline{AC}$ and $\overline{BD}$ are noncongruent segments that are perpendicular and bisect each other.
18. $\overline{AC}$ and $\overline{BD}$ are congruent segments that are perpendicular and bisect each other.

19. a. What is the converse of "If a quadrilateral is a rectangle, then its diagonals are congruent"?
   b. Draw a diagram to show that the converse is false. (*Hint:* Begin with two congruent segments that do not bisect each other.)

20. a. What is the converse of "If a quadrilateral is a rhombus, then its diagonals are perpendicular"?
   b. Draw a diagram to show that the converse is false.

In Exercises 21 and 22 replace each "why" with the correct reason.

C 21. Show: The diagonals of rhombus *RSTW* are perpendicular.
   (1) $\overline{RS} \cong \overline{RW}$ (Why?)
   (2) $\overline{RZ} \cong \overline{RZ}$ (A segment is congruent to itself.)
   (3) $\overline{ZS} \cong \overline{ZW}$ (A rhombus is a parallelogram. The diagonals of a parallelogram bisect each other.)
   (4) $\triangle RZS \cong \triangle RZW$ (Why?)
   (5) $\angle RZS \cong \angle RZW$ (Why?)
   (6) $\angle RZS$ and $\angle RZW$ are right angles.   ($\angle RZS$ and $\angle RZW$ are both congruent and supplementary, and thus are right angles.)
   (7) $\overline{RT} \perp \overline{SW}$ (Why?)

22. Show: If the diagonals of quadrilateral *ABCD* bisect each other, then the quadrilateral is a parallelogram.
   (1) $\overline{AE} \cong \overline{CE}, \overline{DE} \cong \overline{BE}$ (Given)
   (2) $\angle AED \cong \angle CEB, \angle AEB \cong \angle CED$ (Why?)
   (3) $\triangle AED \cong \triangle CEB, \triangle AEB \cong \triangle CED$ (Why?)
   (4) $\overline{AD} \cong \overline{CB}, \overline{AB} \cong \overline{CD}$ (Why?)
   (5) *ABCD* is a parallelogram.   (Why?)

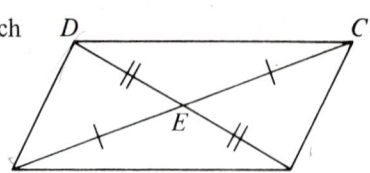

SELF-TEST 2

Find the measures of the numbered angles.

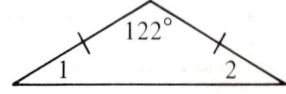

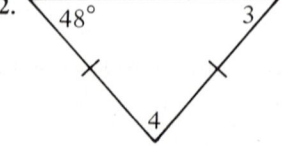

1. [triangle with 122° at top, angles 1 and 2 at base]

2. [triangle with 48° at top, angles 3 and 4]

Section 6-3

3. The measures of the base angles of an isosceles triangle are represented by $3x + 19$ and $8x - 11$. Find the numerical values of the measures of the vertex angle and a base angle of the triangle.

4. Quadrilateral WXYZ is a parallelogram. Fill in the blanks.
   a. △WXY ≅ △ _?_
   b. ∠X ≅ ∠ _?_
   c. ∠1 ≅ ∠ _?_
   d. $\overline{XY}$ ≅ _?_

Section 6-4

5. List three ways you can show that a given quadrilateral must be a parallelogram.

6. Quadrilateral ABCD is a parallelogram. Find the numerical value of x.

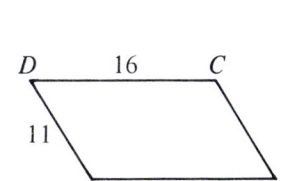

Classify each statement as true or false.

7. Every square is a rectangle and every square is a rhombus.

Section 6-5

8. A trapezoid may have one pair of congruent opposite sides.

9. The diagonals of a rectangle must be perpendicular.

10. Draw a diagram to show that the diagonals of a quadrilateral do not always bisect each other.

11. For which of the following quadrilaterals is it true that the diagonals are always congruent: a parallelogram, a rectangle, a rhombus, a square?

Section 6-6 AREAS OF RECTANGLES AND PARALLELOGRAMS

Two important measures connected with any polygon are its perimeter and its area. In Section 5-7 you studied perimeter. In this section you will begin your study of area with the:
- area of a rectangle
- area of a parallelogram

In rectangle ABCD the length, AB, is 6 units, and the width, BC, is 3 units. You can divide the region enclosed by the rectangle into squares 1 unit on a side, called *square units*. Notice that there are 6 squares along $\overline{AB}$. Since there are three rows of squares, the region contains 3 × 6, or 18, square units. We say that this number, 18, is a measure of the *area* of the rectangle:

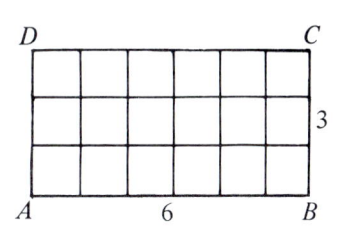

area ABCD = 18 square units

For every rectangle it is true that:

area = length × width

or

$A = l \times w = lw$

CONGRUENT TRIANGLES 197

> The area of a rectangle is the product of its length and width.
>
> $$A = lw$$

To use this formula, be sure that both dimensions, l and w, are expressed in the same unit.

EXAMPLE   What is the area of a rectangle 15 cm long and 1 m wide?

SOLUTION   1 m = 100 cm

$$A = lw = 15 \cdot 100 = 1500 \text{ (cm}^2\text{)}$$

More general terms for the length and width of a rectangle are *base* and *height*, respectively. If we let b stand for the base and h for the height, we may write $A = lw$ as

$$A = bh.$$

The following discussion suggests why $A = bh$ may be used to find the area of *any* parallelogram.

Any side of a parallelogram may be thought of as its *base*. (The term *base* is also used to refer to *base length*.) An *altitude* to that base is a segment perpendicular to the line containing the base from a point on the opposite side. The length of the altitude is the *height*. All altitudes to the same base are congruent and hence have the same length (see Exercise 23 on page 192). In the figure, $\overline{AB}$ is a base and $\overline{DE}$ is an altitude to it.

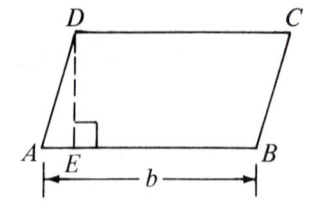

Start with any parallelogram *WXYZ*. Draw the congruent altitudes $\overline{XN}$ and $\overline{WM}$ to form rectangle *WXNM*. Notice that triangles *XYN* and *WZM* appear to be congruent. You will be asked to prove this in Exercise 39 on page 202. Since these triangles are congruent, they must have equal areas. Thus,

$$\text{area } \triangle XYN = \text{area } \triangle WZM$$

area *WXNZ* + area $\triangle XYN$ = area *WXNZ* + area $\triangle WZM$

and so   area *WXYZ* = area *WXNM*

That is, the product bh gives the area of both the rectangle *WXNM* and the parallelogram *WXYZ*.

> The area of a parallelogram is the product of the length of a base and the length of an altitude to that base.
>
> $$A = bh$$

EXAMPLE *QRST* is a parallelogram with base and height as shown. What is its area?

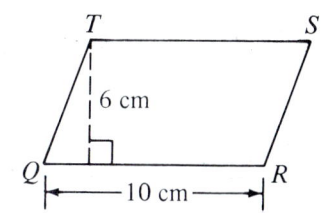

SOLUTION $A = bh$
$b = 10, h = 6$
$A = 10 \cdot 6 = 60 \text{ (cm}^2\text{)}$

ORAL EXERCISES

For each rectangle, give: **(a)** the perimeter, **(b)** the area.

1. 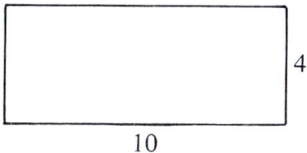 10 by 4

2. 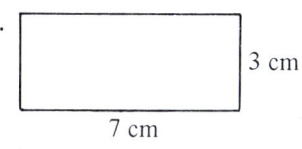 7 cm by 3 cm

3. 4 m by 8 m

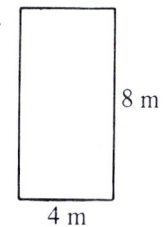

4. In Exercises 1–3, what formulas did you use?

Give the perimeter of each parallelogram.

5. 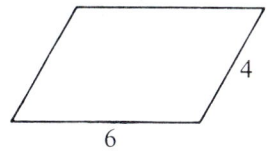 sides 6 and 4

6. sides 1 cm and 5 cm

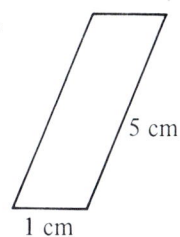

7. 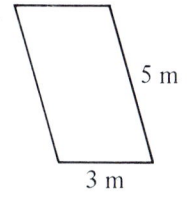 sides 3 m and 5 m

State the values of *b* and *h*, then use the formula $A = bh$ to find the area of each parallelogram.

8. 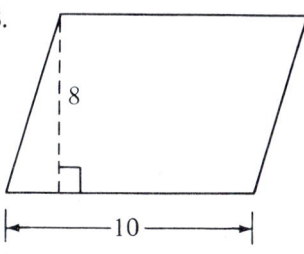 base 10, height 8

9. 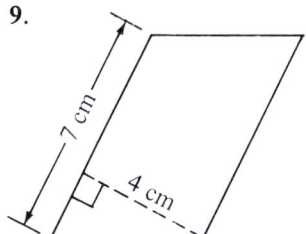 base 4 cm, height 7 cm

10. base 5, height 10

CONGRUENT TRIANGLES   199

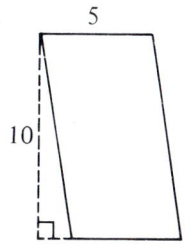

WRITTEN EXERCISES

A 1. A rectangle is 15 units long and 14 units wide. What is its area?

2. The length of a rectangle is 12 units and its perimeter is 36 units. What is its width?

Find the area of the rectangle with the given length and width. Give results to the nearest tenth.

3. $l = 20, w = 15$
4. $l = 15, w = 25$
5. $l = 28, w = 43$
6. $l = 0.5$ m, $w = 25$ m
7. $l = 2.6$ cm, $w = 1.3$ cm
8. $l = 11.5$ cm, $w = 3.7$ cm

Find the area of the parallelogram with the given base and height. Give results to the nearest tenth.

9. $b = 13, h = 22$
10. $b = 28, h = 32$
11. $b = 32, h = 16$
12. $b = 0.5$ m, $h = 16$ m
13. $b = 16.3$ cm, $h = 12.1$ cm
14. $b = 10.9$ cm, $h = 12.4$ cm

Complete the table for each rectangle.

	15.	16.	17.	18.
Length	40 cm	?	23.5 m	?
Width	25 cm	15 cm	?	17.5 cm
Perimeter	?	?	?	38.6 cm
Area	?	225 cm²	385.4 m²	?

Given parallelogram $ABCD$ with base $\overline{AB}$ and altitude $\overline{DX}$. Complete the table.

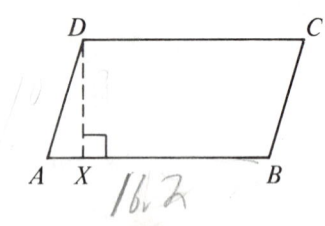

	AB	AD	DX	Perimeter ABCD	Area ABCD
19.	13 cm	10 cm	9 cm	?	?
20.	?	13 cm	12 cm	?	204 cm²
21.	16.2 cm	?	?	44.6 cm	72.9 cm²
22.	?	?	3.5 m	39.6 m	51.8 m²

For each parallelogram, find: (a) the perimeter and (b) the area.

23.

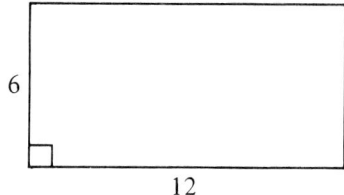

24.

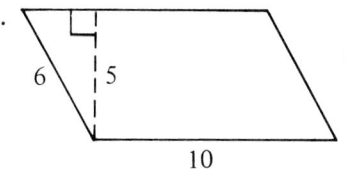

25.

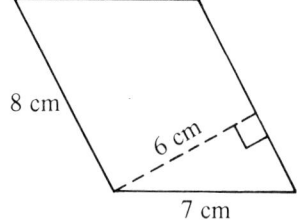

26.

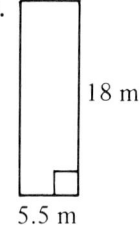

B 27. Find the area of a rectangle 1 m long and 35 cm wide.

28. Find the area of a parallelogram with base 100 cm and height 2 m.

29. a. Rectangle *ABCD* is 15 cm long and 5 cm wide. Rectangle *WXYZ* is 30 cm long and 5 cm wide. Find their areas.
   b. What happens to the area of a rectangle when its length is doubled and the width remains the same?
   c. What happens to the area of a rectangle when both the length and the width are doubled?

30. The length of the base of a parallelogram equals that of a rectangle. The length of the altitude to that base of the parallelogram equals the width of the rectangle. Compare the areas of the two quadrilaterals.

31. A square has sides 30 cm long. What is its area?

32. The perimeter of a square is 32 cm. What is its area?

Find the area and perimeter of the following figures. (You may assume segments to be parallel or perpendicular if they appear to be so.)

33.

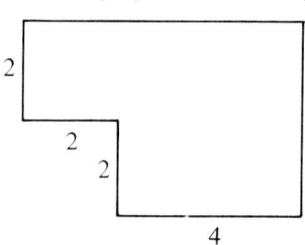

34.

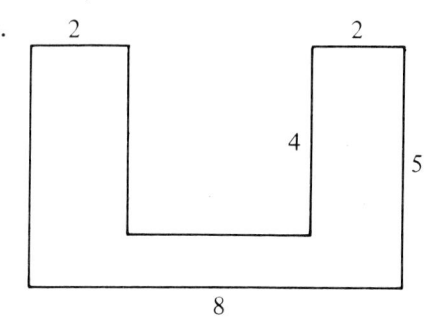

CONGRUENT TRIANGLES

Find the area of the following figures. (You may assume segments to be parallel or perpendicular if they appear to be so.)

35.

36.

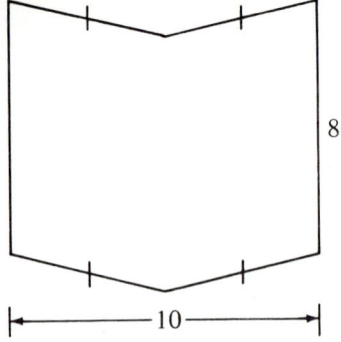

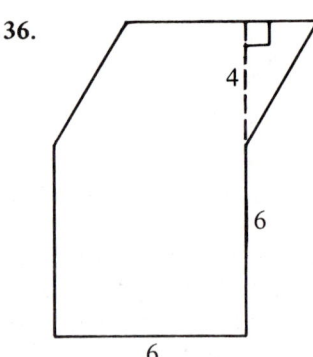

C 37. If the length of a rectangle is twice the width, and the perimeter is 36 cm, find (a) the lengths of the sides and (b) the area.

38. If the length of one side of a rectangle is 3 more than that of another side, and the perimeter is 38, find the area.

In Exercise 39 replace each "why" with the correct reason.

39. Given: $WXYZ$ is a parallelogram.
$\overline{WM}$ and $\overline{XN}$ are both perpendicular to $\overline{MY}$.
Show: $\triangle WZM \cong \triangle XYN$

(1) $\angle WMZ$ and $\angle XNY$ are right angles. (Why?)
(2) $\angle WMZ \cong \angle XNY$ (The measure of a right angle is 90. Congruent angles are angles whose measures are equal.)
(3) $\overline{WZ} \parallel \overline{XY}$ (Why?)
(4) $m\angle WZM = m\angle XYN$ (Two lines cut by a transversal are parallel if and only if corresponding angles have equal measures.)
(5) $\angle WZM \cong \angle XYN$ (Why?)
(6) $\overline{WZ} \cong \overline{XY}$ (Why?)
(7) $\triangle WZM \cong \triangle XYN$ (AAS. See Ex. 26, page 184.)

Section 6-7 AREAS OF TRIANGLES AND TRAPEZOIDS

Now that you know how to find the area of a parallelogram, you can understand the two formulas developed in this section for the:
- area of a triangle
- area of a trapezoid

Any side of a triangle may be regarded as its *base*. The *altitude* to that base is the segment from the opposite vertex that is perpendicular to the line containing the base.

CHAPTER 6

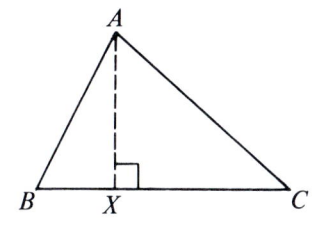

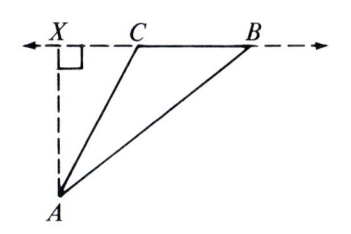

In each triangle above, $\overline{AX}$ is the altitude to base $\overline{BC}$ from vertex A.

To develop a formula for the area of a triangle, start with $\triangle ABC$ having base $\overline{AB}$ and altitude $\overline{CD}$ to that base. Draw a line through C parallel to $\overline{AB}$ and a line through B parallel to $\overline{AC}$. Let E be the point where these lines intersect. $ABEC$ is a parallelogram. Diagonal $\overline{BC}$ divides $ABEC$ into two congruent triangles. This means that:

$$\text{area } \triangle ABC = \text{area } \triangle ECB$$

But: area $\triangle ABC$ + area $\triangle ECB$ = area $ABEC$

$$2(\text{area } \triangle ABC) = \text{area } ABEC$$
$$2(\text{area } \triangle ABC) = bh$$
$$\text{area } \triangle ABC = \frac{1}{2}bh$$

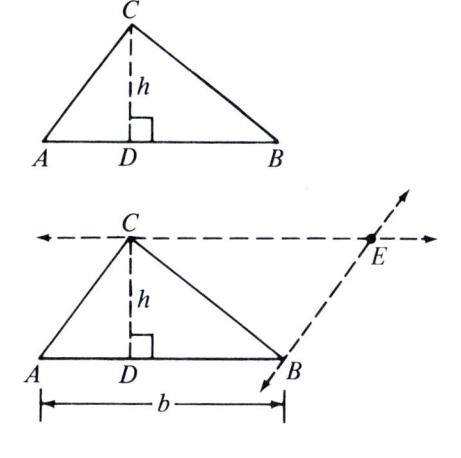

The area of a triangle is half the product of the length of a base and the length of the altitude to that base.

$$A = \frac{1}{2}bh$$

EXAMPLE A triangle has base 13 m and height 6 m. Find its area.

SOLUTION $A = \frac{1}{2}bh$

$A = \frac{1}{2} \cdot 13 \cdot 6 = 39$ (m²)

The parallel sides of a trapezoid are called its *bases*. In the trapezoid $ABCD$ shown, the bases are $\overline{AB}$ and $\overline{CD}$. An *altitude* of a trapezoid is a line segment connecting the two bases and perpendicular to them. $\overline{XY}$ is an altitude of $ABCD$.

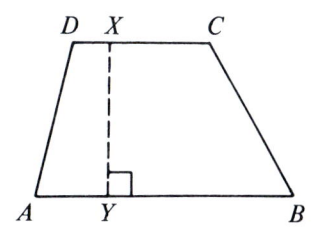

CONGRUENT TRIANGLES

By separating a trapezoid into two triangles, we can develop a formula for its area. Let b_1 be the length of $\overline{AB}$ and b_2 be the length of $\overline{CD}$. Let h be the length of $\overline{DE}$. Draw $\overleftrightarrow{BF}$ parallel to $\overleftrightarrow{ED}$ and intersecting $\overleftrightarrow{DC}$ at F. Then $\overline{BF} \cong \overline{DE}$ (why?) and $\angle DEB \cong \angle BFD$ (why?). So $BF = h$ and $\overline{BF}$ is an altitude of $\triangle BCD$.

$$\text{area } ABCD = \text{area } \triangle ABD + \text{area } \triangle BCD$$
$$= \frac{1}{2}b_1 h + \frac{1}{2}b_2 h$$
$$= \frac{1}{2}h(b_1 + b_2)$$

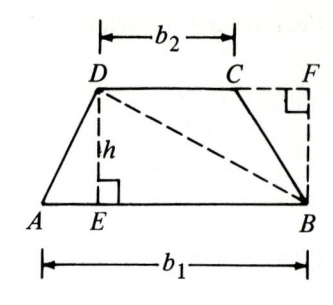

The area of a trapezoid is half the product of the length of an altitude and the sum of the lengths of its bases.

$$A = \frac{1}{2}h(b_1 + b_2)$$

EXAMPLE In trapezoid $ABCD$, $AB = 7$ cm, $DC = 4$ cm, $DE = 3$ cm. Find the area of $ABCD$.

SOLUTION
$A = \frac{1}{2}h(b_1 + b_2)$
$h = 3, b_1 = 7, b_2 = 4$
$A = \frac{1}{2} \cdot 3(4 + 7)$
$A = \frac{3}{2} \cdot 11 = \frac{33}{2} = 16.5$ (cm²)

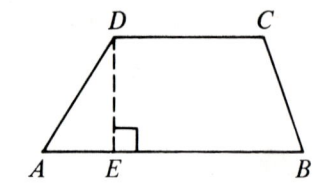

ORAL EXERCISES

1. You are given $\triangle ABC$ and the length of $\overline{AB}$. What other measurement do you need in order to apply the formula $A = \frac{1}{2}bh$?

2. In the formula $A = \frac{1}{2}h(b_1 + b_2)$, what do b_1 and b_2 represent?

3. If you consider $\overline{BC}$ the base of $\triangle ABC$, what segment is its altitude?

4. If you consider $\overline{AB}$ the base of $\triangle ABC$, what two lengths would you need in order to find its area?

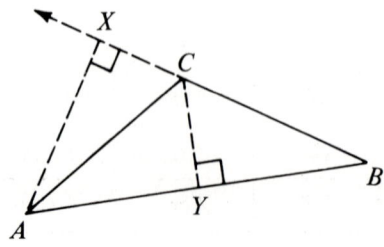

Find the area of each triangle.

5. (triangle: base 6 cm, height 4 cm)

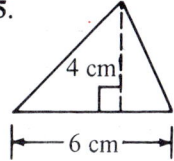

6. (right triangle: legs 6 cm and 8 cm)

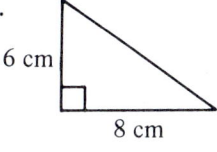

7. (triangle: height 4 cm, side 10 cm)

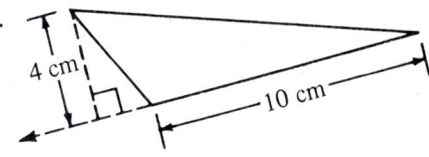

Find the area of each trapezoid.

8. (trapezoid: parallel sides 6 and 10, height 4)

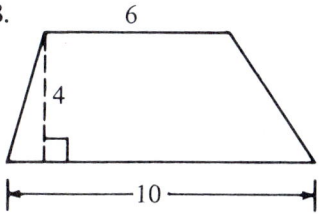

9. (trapezoid: parallel sides 6 and 4, base 2)

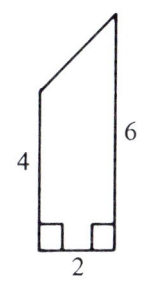

WRITTEN EXERCISES

The dimensions given for Exercises 1-4 refer to the figure. Find the area of △ABC.

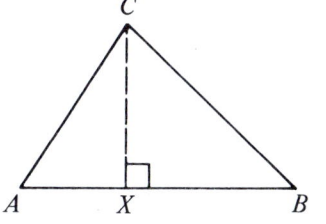

A

	1.	2.	3.	4.
AB	16 cm	14.8 cm	10.5 m	172 cm
CX	8 cm	7.5 cm	7.4 m	84 cm

Find the perimeter and area of each triangle.

5. (triangle with sides 15 cm, 25 cm, 39 cm, and base segment 16 cm)

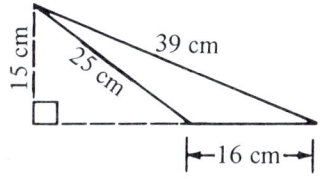

6. (triangle: top 25 m, sides 15 m and 20 m, height 12 m)

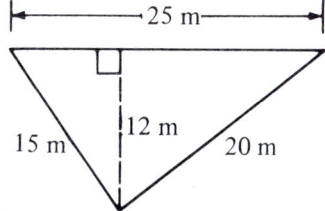

The dimensions given for Exercises 7-10 refer to the figure. Find the area of trapezoid QRST.

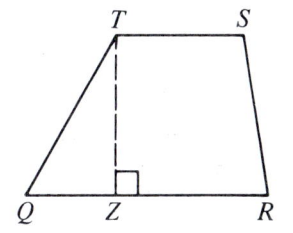

	7.	8.	9.	10.
QR	9 cm	37 cm	9.5 m	61.5 cm
TS	7 cm	15 cm	4.5 m	40.5 cm
TZ	3 cm	10 cm	7.0 m	21.0 cm

CONGRUENT TRIANGLES 205

Find the perimeter and area of each trapezoid.

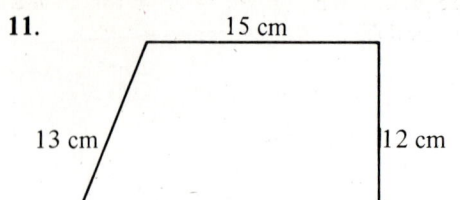

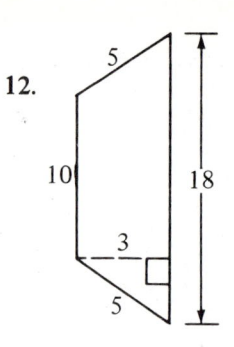

11. (trapezoid with top 15 cm, left side 13 cm, right side 12 cm, bottom 20 cm)

12. (trapezoid with segments 5, 10, 3, 5, and height 18)

EXAMPLE A triangle has area 45 m² and height 9 m. Find its base.

SOLUTION $A = \frac{1}{2}bh$

$45 = \frac{1}{2} \cdot b \cdot 9$

$90 = 9b$

$10 \text{ (m)} = b$

Refer to the figure and complete the table.

	XY	AZ	area △XYZ
B 13.	12 cm	?	240 cm²
14.	?	32 cm	512 cm²
15.	?	12 cm	144 cm²
16.	24 m	?	786 m²

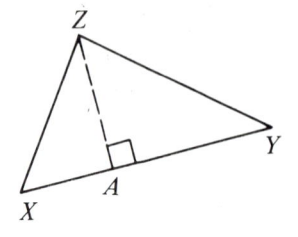

17. Show two methods of finding the area of the trapezoid.

18. Find the perimeter of the trapezoid.

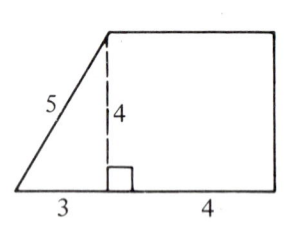

(trapezoid with sides 5, 4, and bases 3 and 4)

19. Find the perimeter and area of the figure.

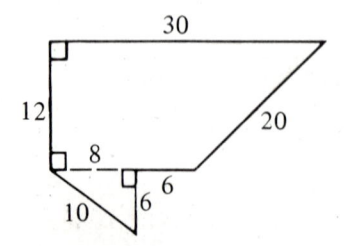

(figure with measurements 30, 12, 20, 8, 6, 6, 10)

20. Find the area of the shaded region.

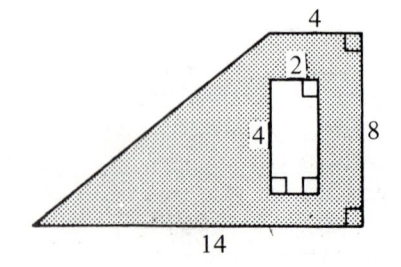

(shaded trapezoid with measurements 4, 2, 4, 8, 14 and an interior rectangle)

206 CHAPTER 6

C 21. a. What is the area of the triangle?
   b. Find h.

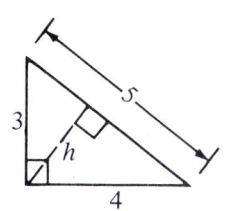

22. You are given $\triangle ABC$. If $AC = 12$, $BX = 5$, and $AB = 6$, find CY.

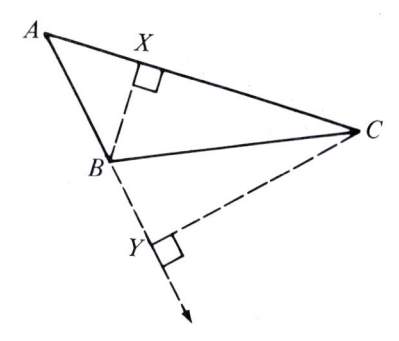

Section 6-8 AREAS OF PRISMS AND PYRAMIDS

You are now ready to apply your knowledge of area to two familiar three-dimensional figures. In this section you will study:
- prisms
- pyramids
- lateral areas and total areas of prisms and pyramids

A *prism* has two bases that are congruent polygons lying in parallel planes (planes that do not intersect). Prisms are named by the shape of their bases, as shown below.

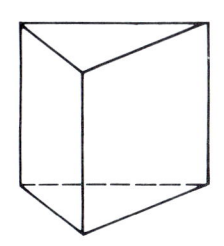

Triangular Prism   **Rectangular Prism**   **Hexagonal Prism**

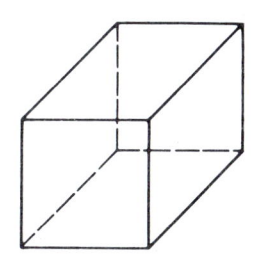

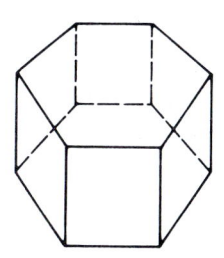

The bases of a prism are joined by *lateral faces* that are parallelograms. In a *right prism* each lateral face is a rectangle.

The segments connecting the corresponding vertices of the two bases are called *lateral edges*. In a right prism the *height* of the prism equals the length of a lateral edge.

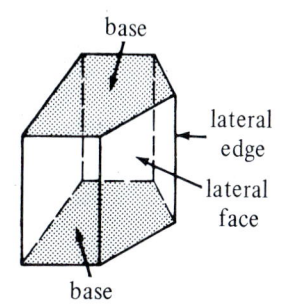

CONGRUENT TRIANGLES

A *cube* is a rectangular prism in which all six faces are congruent squares. The intersection of two adjacent faces is called an *edge* of a cube.

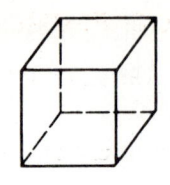

The *lateral area* of a prism is the sum of the areas of its lateral faces. As you might guess, the *total area* of a prism is the sum of the lateral area and the areas of the two bases. (The total area is sometimes called the *surface area*.)

EXAMPLE Find **(a)** the lateral area and **(b)** the total area of a right square prism in which the length of a lateral edge is 9 and the length of each side of the square base is 4.

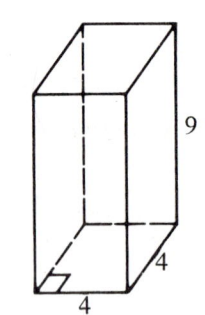

SOLUTION **(a)** area of one lateral face = 9 · 4 = 36

Since there are 4 lateral faces, the lateral area = 4 · 36 = 144 (square units)

(b) area of each base = 4 · 4 = 16

total area of prism = 16 + 16 + 144 = 176 (square units)

Closely related to a prism is a *pyramid*. A pyramid, however, has just one base, which is a polygon. Each of the lateral faces of a pyramid is a triangle. The point where they meet is the *vertex* of the pyramid. The perpendicular segment from the vertex to the plane containing the base is the *altitude* of the pyramid. Its length is the *height* of the pyramid. The base of the pyramid shown is a pentagon. The vertex of the pyramid is V and the altitude is $\overline{VX}$.

A *regular pyramid* is one whose base is a regular polygon and whose lateral edges are all congruent. In a regular pyramid the altitudes of the lateral faces all have the same length, called the *slant height* (l) of the pyramid. For the pyramid shown, $l = VY$.

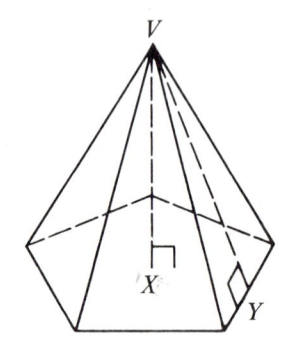

Lateral area of pyramid = sum of areas of lateral faces
Total area of pyramid = lateral area + base area

EXAMPLE Find **(a)** the lateral area and **(b)** the total area of the regular square pyramid pictured.

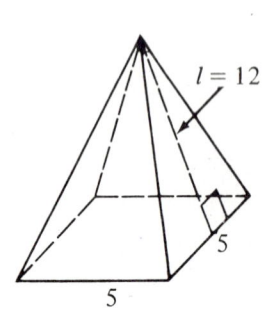

SOLUTION **(a)** Each lateral face is a triangle with $b = 5$, $h = 12$, and area $= \frac{1}{2}bh$.

Lateral area $= 4 \cdot \left(\frac{1}{2} \cdot 5 \cdot 12\right)$
$= 120$ (square units)

(b) Area of base = 5 · 5 = 25 (square units)
Total area = 120 + 25 = 145 (square units)

ORAL EXERCISES

1. If *ABCD* and *WXYZ* are the bases of the prism, name its lateral faces.
2. How many faces are there in a rectangular prism?

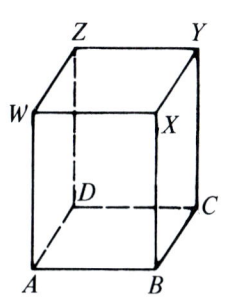

Exs. 1, 2

3. What is the total number of edges of a cube?
4. Name the edges of the cube shown.
5. Compare the lengths of the edges of a cube.

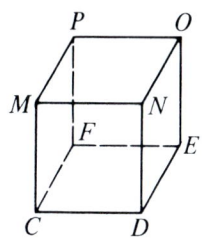

Exs. 3-5

6. Can the two bases of a prism have different shapes?
7. What is the shape of the lateral faces of a right prism?
8. How many lateral faces does a triangular pyramid have?
9. What is the shape of the lateral faces of any pyramid?
10. What is the shape of the lateral faces of any regular pyramid?

WRITTEN EXERCISES

In Exercises 1-4 find **(a)** the lateral area and **(b)** the total area of the right rectangular prisms with l, w, and h as given.

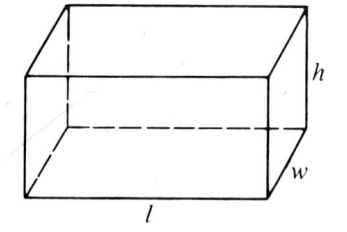

A
1. $l = 4, w = 2, h = 5$
2. $l = 8$ cm, $w = 6$ cm, $h = 12$ cm
3. $l = 9$ cm, $w = 11$ cm, $h = 4$ cm
4. $l = 15, w = 12, h = 20$

In Exercises 5-8 find the lateral area and total area of the regular square pyramid with s and l as given.

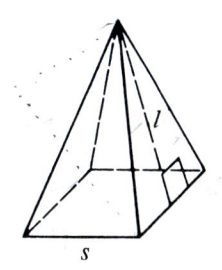

5. $s = 6, l = 5$
6. $s = 10$ cm, $l = 13$ cm
7. $s = 12$ cm, $l = 10$ cm
8. $s = 48, l = 25$

CONGRUENT TRIANGLES

In Exercises 9-14 find the lateral area of the right prism described.

	Shape of Base	Length(s) of Sides of Base	Height
9.	Parallelogram	6 and 8	12
10.	Parallelogram	4 cm and 7 cm	10 cm
11.	Regular Pentagon	4 cm	9 cm
12.	Regular Pentagon	8	15
13.	Regular Hexagon	10	12
14.	Regular Decagon	12 cm	14 cm

In Exercises 15-20 find the lateral area of the regular pyramid from the data given.

	Shape of Base	Length of Sides of Base	Slant Height
15.	Square	4 cm	11 cm
16.	Square	8	6
17.	Regular Pentagon	6 cm	10 cm
18.	Regular Hexagon	10	10
19.	Regular Octagon	12 cm	20 cm
20.	Regular Decagon	8	14

B 21. Show that the lateral area of a right prism is the product of the height of the prism and the perimeter of a base.

22. Show that the lateral area of a regular pyramid is half the product of the perimeter of the base and the slant height of the pyramid. (*Hint*: Let n represent the number of sides in the base.)

23. What is the total area of a cube whose edges are 5 cm long?

24. What is the total area of a cube whose edges are 10 cm long?

25. What happens to the total area of a cube when you double the length of an edge?

26. A kite has the shape of a right regular hexagonal prism. Each side of a base is 40 cm long and the height of the prism is 120 cm. What is the area of the material covering the lateral faces?

27. A steeple in the shape of a regular square pyramid has a slant height of 8.1 m. Each side of the base is 2 m long. What is the lateral area of the steeple?

C 28. A three-tiered wedding cake, rectangular in shape, is ready to be frosted. The tiers are baked separately and arranged in layers before frosting begins. What is the area to be frosted, given the following data?

     First tier: 40 cm by 30 cm by 6 cm high
     Second tier: 30 cm by 20 cm by 5 cm high
     Third tier: 16 cm by 12 cm by 4 cm high

29. The center strip of the avenue leading to the Great Pyramid is planted with shrubs clipped in the shape of regular pyramids. Assume that the shrubs have a square base 60 cm on a side and a slant height of approximately 67 cm. If a worker is to clip 36 of these shrubs, what is the area to be clipped?

Section 6-9 VOLUMES OF PRISMS AND PYRAMIDS

Now that you have learned how to find the areas of various polygons, you can apply your knowledge to finding the:
- volume of a prism
- volume of a pyramid

As you know, a cube is a prism all of whose faces are congruent squares. A cube whose edges measure 1 unit is called a *unit cube*. Just as we used unit squares to measure area, we can use unit cubes to measure volume.

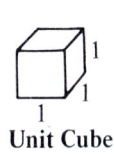

Unit Cube

In the rectangular prism pictured, there are 3 · 2 · 4, or 24, unit cubes. We say that the *volume* of the prism is 24 cubic units. We multiplied the area of the base, (3 · 2), by the height, 4.

If we let B stand for the area of the base and h for the height, then a formula for the volume V of the prism is $V = Bh$.

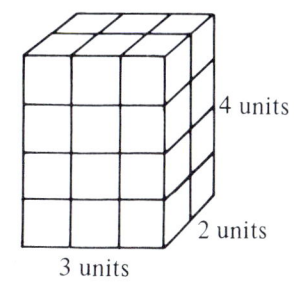

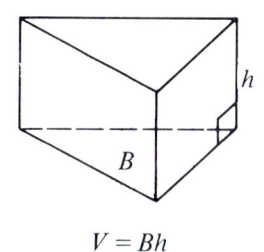

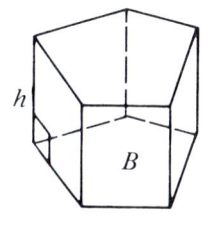

$V = Bh$

$V = Bh$

The volume of a prism is the product of the area of a base and the height.

$$V = Bh$$

CONGRUENT TRIANGLES **211**

EXAMPLE What is the volume of a prism whose base is a rectangle 8 cm long and 5 cm wide and whose height is 10 cm?

SOLUTION Area of base: $B = 5 \cdot 8 = 40$ (cm²)

Volume: $V = Bh = 40 \cdot 10 = 400$ (cm³)

An interesting fact is that the volume of a pyramid is exactly one third that of a prism having an equal base area and the same height.

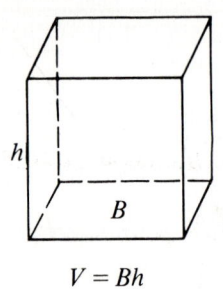

$V = Bh$

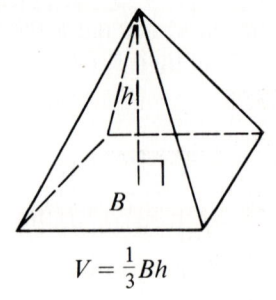

$V = \frac{1}{3}Bh$

The volume of a pyramid is one third the product of the area of the base and the height.

$$V = \frac{1}{3}Bh$$

EXAMPLE A pyramid 21 cm high has a base in the shape of the right triangle shown at the right. What is the volume of the pyramid?

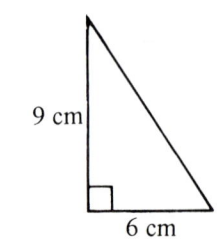

SOLUTION Area of base: $B = \frac{1}{2} \cdot 6 \cdot 9 = 27$ (cm²)

Volume: $V = \frac{1}{3}Bh = \frac{1}{3} \cdot 27 \cdot 21 = 189$ (cm³)

ORAL EXERCISES

1. A prism and a pyramid have equal volumes. Must their bases be congruent?
2. How would you find the volume of your classroom?
3. What is the volume of a pyramid 3 units high whose base has an area of 10 square units?
4. If you knew the volume and height of a prism, how would you find the area of a base?

5. If you knew the volume and height of a pyramid, how would you find the area of the base?

6. A prism and a pyramid have bases equal in area and equal heights. What is the relationship between their volumes?

7. How would you have to change the height of the pyramid in Exercise 6 to make the volumes of the pyramid and prism equal?

WRITTEN EXERCISES

A 1. Find the volume of the square prism.

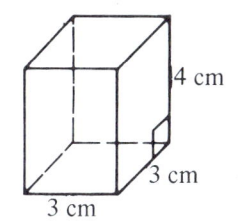

2. Find the volume of the square pyramid.

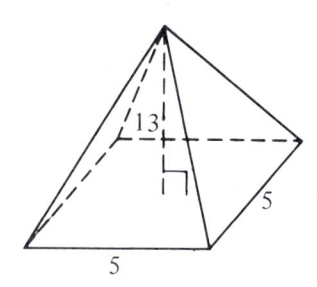

In Exercises 3-6, l and w represent the dimensions of the base of a rectangular prism and h represents its height. Compute its volume, giving results to the nearest tenth of a unit.

3. $l = 5$ cm, $w = 4$ cm, $h = 10$ cm

4. $l = 16$ cm, $w = 10$ cm, $h = 20$ cm

5. $l = 4.5$ cm, $w = 3.4$ cm, $h = 6.5$ cm

6. $l = 12.8$ m, $w = 10$ m, $h = 6.2$ m

Exercises 7-10 give the area of the base of a pyramid and its height. Compute its volume.

7. $B = 30$ cm^2, $h = 15$ cm

8. $B = 96$ cm^2, $h = 37$ cm

9. $B = 5.7$ m^2, $h = 2$ m

10. $B = 29$ m^2, $h = 3.3$ m

11. What is the volume of a cube whose edges are 5 cm long?

12. What is the volume of a cube whose edges are 10 cm long?

B 13. What is the effect on the volume of a cube when you double the length of each side?

14. Find the area of the base of a square prism having a volume of 150 cm^3 and height 6 cm.

15. The volume of a pyramid is 84 cm^3. Its base has an area of 12 cm^2. What is the height of the pyramid?

16. A paperweight in the shape of a pyramid has a square base 4 cm on a side. Its volume is 48 cm^3. What is its height?

CONGRUENT TRIANGLES

C 17. A water trough is in the shape of a right trapezoidal prism, as shown. The bases of the trapezoid are 2.5 m and 3 m and its height is 1 m. If the length of the trough is 5 m, how many cubic meters of water does it hold?

Ex. 17

Ex. 18

18. A metal cube 9 cm high has a hole cut through it in the shape of a right triangular prism whose bases have $b = 3$ cm and $h = 4$ cm. How much metal remains in the cube after the hole has been cut?

19. If the shaded pyramid is removed from the rectangular prism shown, what is the volume of the remaining figure?

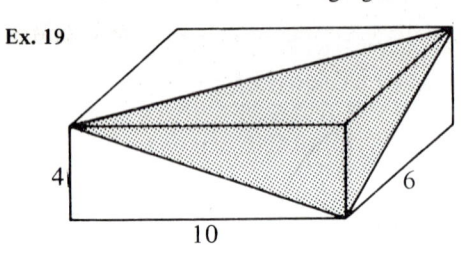

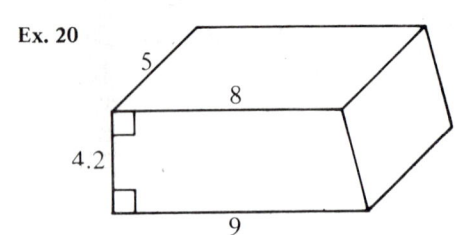

Ex. 19

Ex. 20

20. A small lamp base is in the shape of a right trapezoidal prism. Dimensions shown are in centimeters. What is the volume of the lamp base?

SELF-TEST 3

Find **(a)** the perimeter and **(b)** the area of each parallelogram.

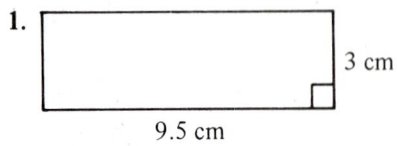

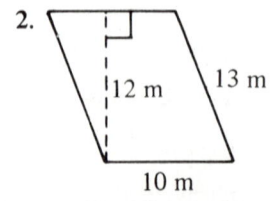

Section 6-6

1.

2.

3. A square with sides 12 cm long has the same area as a rectangle with length 18 cm. Find the width of the rectangle.

Find **(a)** the perimeter and **(b)** the area of the triangle and trapezoid shown.

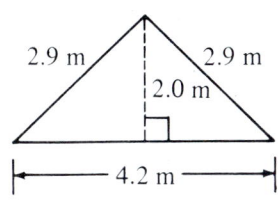

4.

5.

Section 6-7

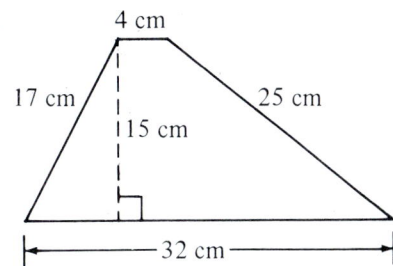

In Exercises 6 and 7, find **(a)** the lateral area and **(b)** the total area of the figure described.

6. A right rectangular prism with length 7 m, width 5.5 m, and height 4 m

Section 6-8

7. A regular square pyramid in which the slant height is 10 cm and each side of the base is 15 cm long

8. The base of a regular pyramid is an octagon with sides 5 m long. If the slant height is 4 m, find the lateral area.

Find the volume of the figure described.

9. A right square prism whose base has edges 7 cm long and whose height is 9 cm

Section 6-9

10. A rectangular pyramid whose base is 12 m long and 8 m wide and whose height is 5.2 m

Computer Exercises

1. Copy and RUN the following program for the values of L and W given below.

```
10  PRINT "THIS PROGRAM WILL FIND THE PERIMETER AND THE AREA"
20  PRINT "    OF A RECTANGLE, GIVEN ITS LENGTH AND WIDTH."
30  PRINT
40  PRINT "INPUT LENGTH AND WIDTH: L, W";
50  INPUT L,W
60  PRINT "PERIMETER =";2*L+2*W
70  PRINT "AREA =";L*W
80  END
```

   a. L = 5, W = 2   b. L = 43, W = 11
   c. L = 54, W = 21   d. L = 8.6, W = 7.2

2. a. Write a program that will find the area of a triangle.
   b. Write a program that will find the area of a trapezoid.

CONGRUENT TRIANGLES

3. a. Complete lines 60 and 70 in the following program to find the surface area and volume of a right rectangular prism, given its length, width, and height.

```
10  PRINT "THIS PROGRAM WILL FIND THE SURFACE AREA"
20  PRINT "    AND VOLUME OF A RIGHT RECTANGULAR PRISM."
30  PRINT
40  PRINT "WHAT ARE THE LENGTH, WIDTH, HEIGHT: L, W, H";
50  INPUT L,W,H
60  LET A=
70  LET V=
80  PRINT "SURFACE AREA=";A;"    VOLUME=";V
90  END
```

b. RUN the program for several combinations of L, W, and H.

4. In your work with compass and straightedge constructions, you may have noticed that the sum of the lengths of any two sides of a triangle must be greater than the length of the third side. (See Ex. 15, page 177.) Copy and RUN the following program for the values of A, B, and C given below.

```
10   PRINT "THIS PROGRAM WILL DETERMINE WHETHER OR NOT"
20   PRINT "    A TRIANGLE CAN BE CONSTRUCTED WITH"
30   PRINT "    SIDES OF THE GIVEN LENGTHS."
40   PRINT
50   PRINT "INPUT THREE POSITIVE NUMBERS: A, B, C";
60   INPUT A,B,C
70   IF A+B <= C THEN 120
80   IF B+C <= A THEN 120
90   IF C+A <= B THEN 120
100  PRINT "THERE IS A TRIANGLE WITH SIDES OF LENGTHS";
110  GOTO 130
120  PRINT "THERE IS NO TRIANGLE WITH SIDES OF LENGTHS";
130  PRINT A;",";B;",";C;"."
140  END
```

a. $A = 3, B = 4, C = 5$ b. $A = 2, B = 3, C = 5$

c. $A = 8, B = 10, C = 4$ d. $A = 3, B = 8, C = 4$

CHAPTER REVIEW

Draw a large triangle like $\triangle XYZ$.

1. Construct a $\triangle ABC$ so that $\angle A \cong \angle X, \angle B \cong \angle Y$, and $\overline{AB} \cong \overline{XY}$.

2. Construct a $\triangle RST$ so that $\overline{RS} \cong \overline{YZ}, \overline{ST} \cong \overline{YZ}$, and $\overline{TR} \cong \overline{YZ}$.

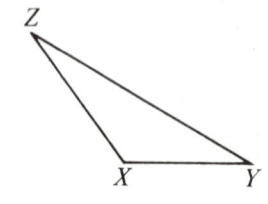

Section 6-1

Name two congruent triangles and tell why they are congruent (SSS, SAS, ASA).

3. *Section 6-2*

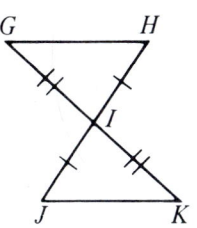

4.

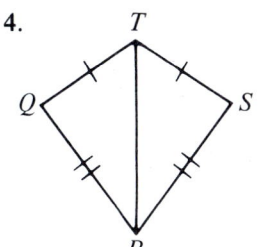

5.

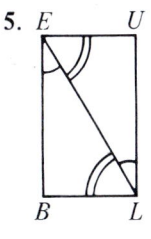

Find the numerical value of the measure of each angle.

6. *Section 6-3*

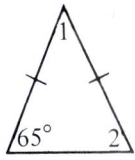

7.

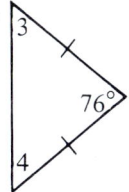

8.

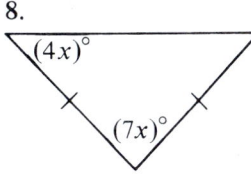

Quadrilateral *ABCD* is a parallelogram with $AB = 10$ and $m\angle B = 80$. Complete each statement, if possible. If not, write "not possible."

9. $m\angle D = $ __?__   **10.** $BC = $ __?__   **11.** $m\angle A = $ __?__   *Section 6-4*

12. In the diagram, $m\angle 1 = m\angle 2$ and $m\angle 3 = m\angle 4$. Must the quadrilateral be a parallelogram? Explain.

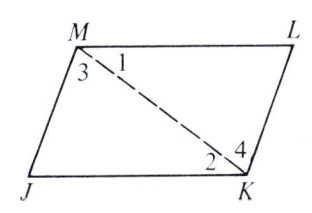

Complete each statement with *always*, *sometimes*, or *never*.

13. A rhombus is __?__ a parallelogram.   *Section 6-5*

14. The diagonals of a rectangle are __?__ perpendicular.

15. The diagonals of a trapezoid are __?__ congruent.

Find the area of each parallelogram.

16. *Section 6-6*

13 cm
15 cm

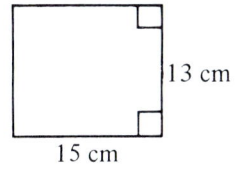

17.

1.2 m   1.4 m
2.4 m

CONGRUENT TRIANGLES   217

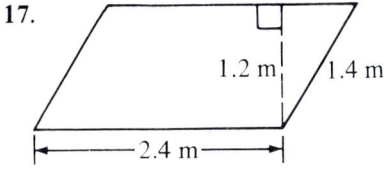

Find the area.

18.

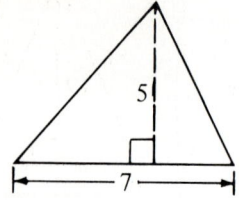

19. 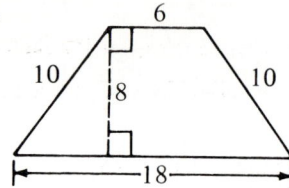 Section 6-7

In Exercises 20 and 21 a right rectangular prism and a regular square pyramid are shown. Find **(a)** the lateral area and **(b)** the total area of each.

20.

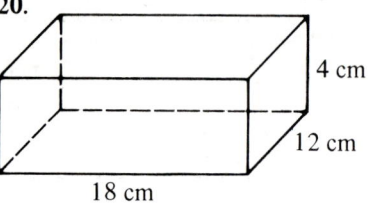

21. 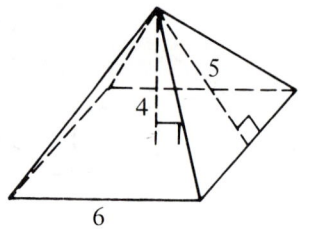 Section 6-8

22. What is the total area of a cube whose edges are 6 cm long?

23. Find the volume of the prism in Exercise 20.   Section 6-9
24. Find the volume of the pyramid in Exercise 21.

CHAPTER TEST

Begin by drawing line segments and angles like those below.

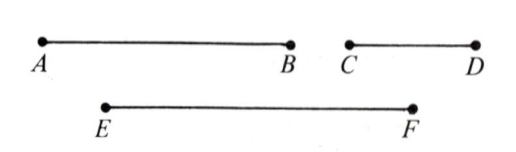

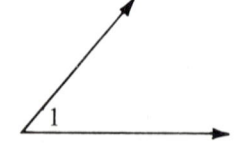

1. Construct a triangle whose sides are congruent to $\overline{AB}$, $\overline{CD}$, and $\overline{EF}$.
2. Construct an isosceles triangle with vertex angle congruent to $\angle 1$ and with two sides congruent to $\overline{AB}$.

Refer to isosceles △RST.

3. Name two pairs of congruent angles. Justify each answer.
4. To prove that △RXS ≅ △TXS by SSS, what additional information would you need?
5. Find the perimeter and the area of △RST.

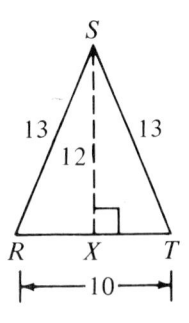

Find the numerical values of x and y. (The quadrilateral shown in Exercise 8 is a parallelogram.)

6.

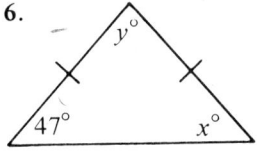

7.

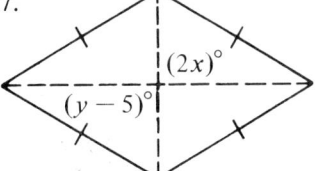

8.

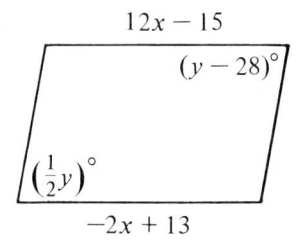

Classify each statement as true or false.

9. The SSA property may be used to prove that two triangles are congruent.
10. The diagonals of every trapezoid bisect each other.
11. An equilateral triangle is also equiangular.
12. Some trapezoids are parallelograms.
13. In quadrilateral WXYZ, if $\overline{WX} \cong \overline{YZ}$ and $\overline{XY} \parallel \overline{WZ}$, then WXYZ must be a parallelogram.
14. The diagonals of a rhombus must be congruent.
15. If a parallelogram has at least one right angle, then the parallelogram must be a rectangle.
16. In quadrilateral ABCD, if $\overline{AB} \cong \overline{CD}$ and $\overline{AD} \cong \overline{BC}$, then ABCD must be a parallelogram.
17. A square is a regular quadrilateral.

In Exercises 18-20 a rectangle, a parallelogram, and a trapezoid are shown. Find (a) the perimeter and (b) the area of each.

18.

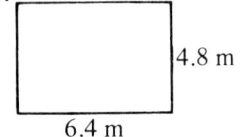

19.

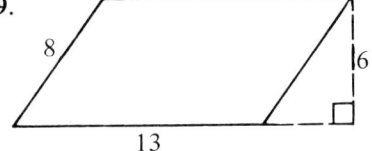

20.

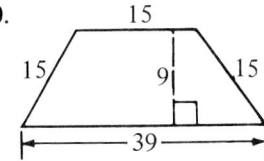

CONGRUENT TRIANGLES

In Exercises 21-23 a cube, a right rectangular prism, and a regular square pyramid are shown. Find (a) the lateral area, (b) the total area, and (c) the volume of each.

21.

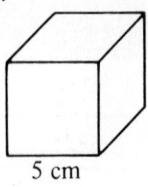

5 cm

22.

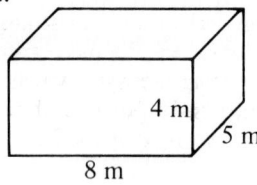

4 m, 5 m, 8 m

23.

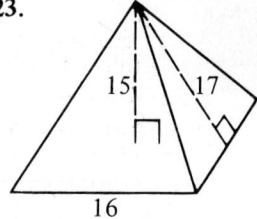

15, 17, 16

24. Quadrilateral *MNOP* is a parallelogram. Write the reason that justifies each statement.
   a. $\overline{PZ} \cong \overline{NZ}$ and $\overline{MZ} \cong \overline{OZ}$
   b. $\overline{PM} \cong \overline{NO}$
   c. $\triangle PZM \cong \triangle NZO$

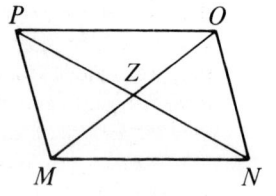

Application — Baseball and Mathematics

Baseball is as much a game of numbers as it is of hot dogs, crowded stands, and trying to catch foul balls. Player statistics are the most visible part of this baseball math. A lot of fans follow players' batting averages, which are updated after every time at bat, as closely as do the managers. If they like, they can also keep count of every other success or failure of their favorite players, from an **RBI** (runs batted in) count to errors in the field.

Did you know how important a role geometry plays in baseball? Think about it. Have you ever seen a home plate that looked like an octagon or a baseball diamond that was a rectangle? How about a bat that was 6 inches thick and 5 feet long? You have not seen any of those things because most aspects of baseball are carefully regulated. Home base is always a pentagon, the distance between all the bases is 90 feet, and bats cannot be more than 2¾ inches thick.

Everyone who has ever watched a baseball game knows how few times even the best players get hits (just look at the statistics!). But do you know why hitting a baseball is the hardest feat in sports? Advanced mathematics, through analysis of the ball's curve and other mechanical forces, helps us understand the problem.

Good pitchers can throw balls at 80 to 100 miles per hour. That sounds hard enough, but the ball is not even traveling in a straight line. Since the pitcher throws from a mound, the ball travels toward the plate at a downward angle. Even worse, as the ball spins, it begins to wobble. No wonder .300 hitters — ones who have missed the ball 70% of the time — can still hold their heads up in batting practice. The diagram below shows some of the mathematics and physics involved in what looks like a very simple act.

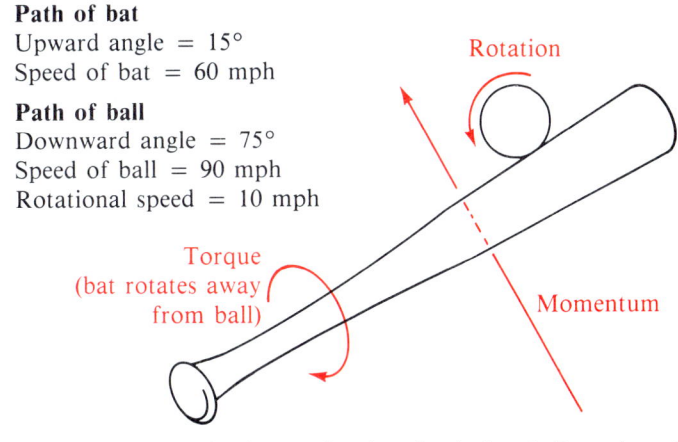

Path of bat
Upward angle = 15°
Speed of bat = 60 mph

Path of ball
Downward angle = 75°
Speed of ball = 90 mph
Rotational speed = 10 mph

Torque
(bat rotates away
from ball)

Rotation

Momentum

The computer has made the number keeping in baseball much easier. All those averages and totals can now be made in split seconds and kept easily accessible for many years. The computer, though, has still another mathematical role in baseball. Computers with graphics screens and attached cameras can be used to analyze each player's swing motions. The computer can find the point in the swing where the player wastes motion or fails to use arm and back power to the best advantage. The coach's eye is no longer the best critic.

CONGRUENT TRIANGLES

VOCABULARY REVIEW

Be sure that you understand the meaning of these terms:

congruent triangles, p. 173
construction, p. 173
isosceles triangle, p. 186
parallelogram, p. 188
trapezoid, p. 193
isosceles trapezoid, p. 193
rectangle, p. 193
rhombus, p. 193
square, p. 193

bisector of a line segment, p. 194
altitude, p. 198
prism, p. 207
cube, p. 208
lateral area, p. 208
total area, p. 208
pyramid, p. 208
regular pyramid, p. 208
slant height, p. 208

MIXED REVIEW

Solve if the domain of x is {positive integers}.

1. $2(5 + x) = 4$
2. $1 - 2x \leqslant -5$

3. If the sides of a triangle have the same lengths as the sides of another triangle, must the triangles be congruent? Explain.

4. Show that $p \rightarrow \sim q$ and $\sim p \vee \sim q$ are logically equivalent.

5. One angle of a right triangle has measure 37. Find the measures of the other two angles.

6. A rectangle has length 12 cm and an area of at least 84 cm². At least how great is its perimeter?

7. A regular square pyramid has base length s and height h.
    a. Find a formula for the volume V in terms of s and h.
    b. Solve your formula for h.

8. Give the negation of "$-12 \div (-4) = -3$" and the truth value of the negation.

9. In the diagram at the right, $a \parallel b$. Find the measures of $\angle 1$ and $\angle 2$.

10. Is {negative integers} a well-defined set? Is it specified by the roster method or by a rule?

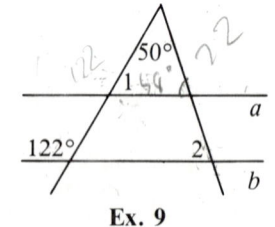

Solve.

11. $\frac{1}{2}x - 1 = -3$
12. $2(3 - y) = -2y$

Ex. 9

13. Graph the solution set of $-3 \leqslant y + 2 < -1$.

14. Consecutive sides of a parallelogram have lengths 60, $5x$, $y - 20$, and 40. Find the values of x and y.

15. How many planes contain a given triangle?

Simplify.

16. $\dfrac{2}{3} - \dfrac{13}{15}$

17. $\dfrac{2}{3} \div \left(-\dfrac{13}{15}\right)$

18. Evaluate $(2r - s^2)t$ when $r = 5$, $s = 2$, and $t = -1$.

19. Points P and Q have coordinates -15 and -8 on a number line. Find PQ.

20. Twelve years ago Melissa was two-thirds as old as she is now. How old is she now?

21. Translate into symbols: The product of 2 and x is less than or equal to the square of 3.

22. Draw two segments, $\overline{MN}$ and $\overline{RS}$, and an acute angle, $\angle A$. Construct a $\triangle XYZ$ so that $\overline{XY} \cong \overline{MN}$, $\overline{YZ} \cong \overline{RS}$, and $\angle XYZ \cong \angle A$.

23. A parallelogram with a right angle is best described as a __?__.

24. The measure of a supplement of an angle is six times the measure of a complement of the angle. Find the measure of the angle.

In Exercises 25 and 26, find the area of the given figure.

25. A square with perimeter 36

26. $\triangle ABC$ if $ABCD$ is a rectangle with $AB = 10$ and $AD = 9$

27. Find the lateral area of a regular pyramid with slant height 8 cm if the base is an equilateral triangle with sides of 12 cm.

28. Find the total area and the volume of a right rectangular prism with length 8, width 6, and height 2.5.

In Exercises 29 and 30, let p be "$2^4 = 32$," q be "$\dfrac{-1}{-1} = 1$," and r be "$\dfrac{0.04}{-0.2} = -0.02$." Give the truth value of each statement.

29. $(\sim p \wedge q) \rightarrow r$

30. $\sim p \leftrightarrow r$

31. Write the inverse of "If $x + 0 = x$, then $x \neq -x$" and give the truth value of the inverse.

32. In $\triangle ABC$, the measure of $\angle B$ is twice the measure of $\angle A$, and the measure of $\angle C$ is 30 less than that of $\angle B$. Find the numerical measure of each angle.

CONGRUENT TRIANGLES

To solve certain types of problems involving the area of a rectangular object, such as a swimming pool, you must first solve a quadratic equation. Solving quadratic equations by factoring and applying this method to problem solving are discussed in this chapter. The pool shown is on the Spanish island Ibiza.

Polynomials and Factoring

7

Can you solve the following problem?

The world's largest swimming pool is a saltwater pool in Casablanca, Morocco. Its area is 36,000 m². If the length is 30 m more than six times the width, what are the dimensions of the pool?

Area = 36,000 m²   w

6w + 30

$$w(6w + 30) = 36,000$$
$$6w^2 + 30w = 36,000$$

In Chapter 3 you learned how to solve equations such as those used in problems involving perimeters. However, the methods you learned in Chapter 3 cannot be used to solve equations, such as $6w^2 + 30w = 36,000$, which involve area.

The algebraic expressions w, $6w + 30$, and $6w^2 + 30w$ are *polynomials*. In this chapter you will use the distributive law to multiply polynomials like w and $6w + 30$. You will also learn how to solve equations such as $6w^2 + 30w = 36,000$.

Section 7-1 EXPONENTS AND FACTORS

In this section you will learn about:
- factors
- powers
- exponents
- bases

When two or more numbers are multiplied, each number is called a *factor* of the product. We call a factored form of a product a *factorization*. Following are some examples of factorizations of integers.

EXAMPLES	Integer	Factorization	Integer	Factorization
	4	2 · 2	18	2 · 3 · 3
	−4	−2 · 2	18	6 · 3
	15	3 · 5	−42	−7 · 6

225

Factorizations of integers sometimes involve fractions. For example, 18 can be factored as $\frac{1}{3} \cdot 54$. In this book, however, when we talk about the factors of an integer, we mean the factors that appear in factorizations containing *only* integers.

EXAMPLE

$12 = 1 \cdot 12 \qquad 12 = 2 \cdot 6 \qquad 12 = 3 \cdot 4$
$12 = (-1)(-12) \qquad 12 = (-2)(-6) \qquad 12 = (-3)(-4)$

The factors of 12 are $-12, -6, -4, -3, -2, -1, 1, 2, 3, 4, 6,$ and 12.

A number that is a product of equal factors is called a *power*.

EXAMPLES

$2 \cdot 2$, or 2^2, is the second power of 2. $\qquad 2^2 = 4$
$2 \cdot 2 \cdot 2$, or 2^3, is the third power of 2. $\qquad 2^3 = 8$
$2 \cdot 2 \cdot 2 \cdot 2$, or 2^4, is the fourth power of 2. $\qquad 2^4 = 16$
$\underbrace{2 \cdot 2 \cdot \ldots \cdot 2}_{n \text{ factors}}$, or 2^n, is the *n*th power of 2.

We also say that 2^1 is the first power of 2. We may read 2^4 as "two to the fourth power." Other powers may be read in a similar manner. Second and third powers may be read in other ways as well. For example, 2^2 may be read as "two squared" or "the square of two" and 2^3 may be read as "two cubed" or "the cube of two."

If x is a real number and n is a positive integer, then x^n is the product in which x is used as a factor n times. The number n is called the *exponent*, and x is called the *base*.

EXAMPLES

$3^4 = 3 \cdot 3 \cdot 3 \cdot 3 = 81 \quad$ (3 is the base and 4 is the exponent.)
$2^5 = 2 \cdot 2 \cdot 2 \cdot 2 \cdot 2 = 32 \quad$ (2 is the base and 5 is the exponent.)
$(-5)^2 = (-5)(-5) = 25 \quad$ (-5 is the base and 2 is the exponent.)
$a^3 = a \cdot a \cdot a$
$3x^2y^4 = 3 \cdot x \cdot x \cdot y \cdot y \cdot y \cdot y$

ORAL EXERCISES

Name the base and the exponent. Tell what number the expression represents.

EXAMPLE $(-1)^3$

SOLUTION The base is -1 and the exponent is 3. $(-1)^3 = (-1)(-1)(-1) = -1$

1. 3^2
2. 5
3. $(-4)^2$
4. $(-3)^3$

5. $(-2)^4$ 6. 1^7 7. 4^3 8. $(-6)^2$

Read as a power.

EXAMPLE $(-2)(-2)(-2)(-2)(-2)(-2)(-2)$

SOLUTION The seventh power of -2 (or -2 to the seventh power)

9. $2 \cdot 2 \cdot 2 \cdot 2$ 10. $n \cdot n \cdot n \cdot n \cdot n$ 11. $(-5)(-5)$ 12. $9 \cdot 9 \cdot 9 \cdot 9 \cdot 9 \cdot 9$

Name the positive factors of the integer.

13. 6   14. 5   15. -4 16. 16
17. 2   18. 10   19. 13   20. -12

21. Each side of a square is s units long. Express the area of the square in terms of s.
22. Each side of a cube is x units long. Express the volume of the cube in terms of x.

WRITTEN EXERCISES

Write using exponents.

A 1. $(-3)(-3)(-3)$ 2. $7 \cdot 7 \cdot 7 \cdot 7 \cdot 7$ 3. $x \cdot x$ 4. $(-y)(-y)(-y)$
5. $2 \cdot z \cdot z$ 6. $5 \cdot a \cdot a \cdot a \cdot a$ 7. $11 \cdot b \cdot b \cdot c$ 8. $3 \cdot a \cdot a \cdot b \cdot b \cdot b$

Write the number in simplest form.

9. 3^5 10. 2^7 11. $(-8)^3$ 12. $(-6)^4$
13. 10^6 14. $(-4)^4$ 15. $2^3 \cdot 3^2$ 16. $3^4 \cdot 5^2$
17. $\left(-\frac{1}{3}\right)^2$ 18. $\left(-\frac{1}{5}\right)^3$ 19. $(0.2)^2$ 20. $(0.5)^3$

Find the factors of the integer.

21. 9   22. -14 23. -16 24. 20
25. 15   26. 18   27. 17   28. -27

Express the area using exponents.

B 29. square, side $4x$

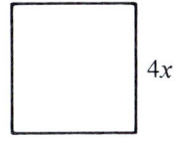

30. rectangle, sides $\frac{y}{2}$ and y

31. triangle, base $6b$, height b

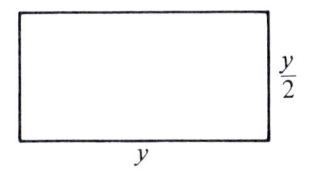

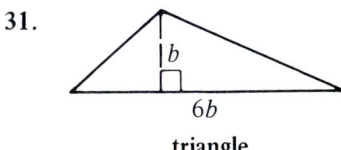

POLYNOMIALS AND FACTORING   227

Express the volume using exponents.

32. cube, $2x$

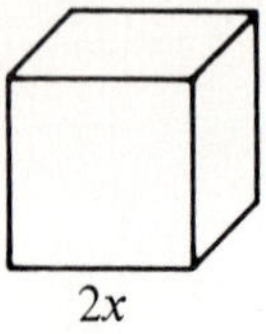

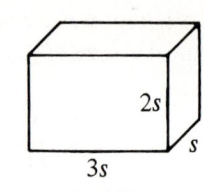

33. rectangular prism, $3s$, $2s$, s

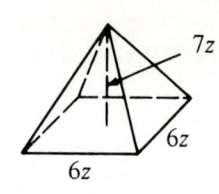

34. square pyramid, $6z$, $6z$, $7z$

35. What is the area of a rectangle if its length is $8d$ and its width is $3d$?

36. What is the volume of a rectangular prism that has a length of $12n$, a width of $5n$, and a height of $2n$?

Classify the statement as true or false.

37. $(-7)^2 + 24^2 = 25^2$

38. $(-1)^{49}(-y)^5 = (-y)^5$

C 39. $3^3 \cdot (-4)^4 \div 2^6 = 216$

40. $12^2 \cdot \left(-\frac{1}{2}\right)^3 \div \left(\frac{1}{2}\right)^2 = -72$

Section 7-2 INTRODUCTION TO POLYNOMIALS

In this section you will learn about polynomials and how to simplify them. You will study:
- monomials, binomials, and trinomials
- the degree of a polynomial
- similar terms
- the standard form of a polynomial

A *monomial* is a number, a variable, or a product of a number and powers of one or more variables. The numerical part of a monomial is called its *numerical coefficient*, or simply its *coefficient*.

EXAMPLES

Monomial	Numerical coefficient
$3xy$	3
$-7xy^2$	-7
ab (or $1 \cdot ab$)	1
$-z^5$ (or $-1 \cdot z^5$)	-1

The expression $\frac{x}{3}$ is a monomial because $\frac{x}{3} = x \div 3$, which is the same as $x \cdot \frac{1}{3}$, or $\frac{1}{3}x$. However, an expression such as $\frac{3}{x}$ that has a variable in the denominator is not considered to be a monomial.

228 CHAPTER 7

A *polynomial* is a monomial or a sum of monomials. Since subtracting a monomial is the same as adding its opposite, a difference of monomials is also a polynomial. Each monomial in a polynomial is called a *term* of the polynomial.

EXAMPLES

Polynomials	Terms
$3x^2 + 2x$	$3x^2, 2x$
$9b^3 - ab$	$9b^3, -ab$
-12	-12
$x^5 - 2x^4 + 4x^2 - x + 6$	$x^5, -2x^4, 4x^2, -x, 6$

Some polynomials have special names determined by the number of their terms. Polynomials with one term are called *monomials*, those with two terms are called *binomials*, and those with three terms are called *trinomials*.

EXAMPLES

Monomials: $5, 6a, -\frac{1}{3}x^3$

Binomials: $x + y, 3t + 7, a - 5m^2n^3$

Trinomials: $x + 2y + 3z^2, 5b^2 - 2b + 3$

Two or more terms are *similar terms*, or *like terms*, if they are identical or if only their numerical coefficients are different.

EXAMPLES

Similar Terms	Not Similar Terms
$3xy$ and $-5xy$	$-7xy^2$ and $-7x^2y$
$4t^2, -11t^2$, and $\frac{5}{4}t^2$	$6c$ and $7d$
$z^5, -5z^5$, and $3z^5$	t^2 and t
$2ab^2$ and $7b^2a$	

A polynomial is *simplified*, or is in *simplest form*, if no two of its terms are similar. You can use the distributive property to simplify a polynomial with similar terms.

EXAMPLE Simplify $3xy - 7xy + 2xy$.

SOLUTION $3xy - 7xy + 2xy = (3 - 7 + 2)xy$
$= -2xy$

EXAMPLE Simplify $6ab^2 + 2bc - 5b^2a$.

SOLUTION $6ab^2 + 2bc - 5b^2a = 6ab^2 + 2bc - 5ab^2$
$= 6ab^2 - 5ab^2 + 2bc$
$= (6 - 5)ab^2 + 2bc$
$= ab^2 + 2bc$

The *degree of a term* is the sum of the exponents of the variables in the term. Since $-5cd^2 = -5c^1d^2$, the degree of $-5cd^2$ is $1 + 2$, or 3. A number such as 2 has degree 0. The monomial 0 has no degree.

The *degree of a polynomial* is the same as that of the term of greatest degree after the polynomial has been simplified.

EXAMPLES

$5x^2 + 3x^3 - 7$ has a degree of 3.
$x^3y^2 + 4xy + 4$ has a degree of 5.
$z^2 - z^2 + 5z$, or $5z$, has a degree of 1.

When working with a polynomial in one variable, it is often convenient to write the polynomial so that its exponents are arranged in descending order. A polynomial written in this manner is said to be in *standard form*.

EXAMPLES

$5x^2 - x + x^3 + 7$ in standard form is $x^3 + 5x^2 - x + 7$.
$y^3 - 5 - 7y^6 + y^8$ in standard form is $y^8 - 7y^6 + y^3 - 5$.

ORAL EXERCISES

Tell whether the expression is a monomial, binomial, trinomial, or none of these. If the expression is a polynomial, name its degree.

1. $3r^3s$
2. $2y^2 - y + 5$
3. $-\dfrac{5}{t}$
4. $8 - 7z^3$
5. $4ab^2 - 3a^2b + a^3 + 8$
6. 0
7. $d + d^4$
8. $\dfrac{k}{2}$
9. $\dfrac{15y}{x^2}$
10. $-m + n - mn + m^2n$

Name the coefficient.

11. $2x^3y^4$
12. $-\dfrac{1}{8}p$
13. $1.5q^6$
14. $\dfrac{yz}{5}$
15. a^3b^2c

Are the given terms similar? Why or why not?

16. $5x, -11x$
17. $-3y, \dfrac{1}{2}y$
18. $3a^2, -4a^3$
19. $8bc, 8cb$
20. $xyz^2, -8xyz^2$
21. $18jk^3, -9j^3k$
22. $rst, -rst^2$
23. $\dfrac{2}{3}x^2zy^3, 6y^3x^2z$

Name the degree of each polynomial. Is the polynomial in standard form? If not, read it in standard form.

24. $5x^3 - x^2 + 3x - 7$
25. $8y + y^7 - 3y^2 + y^{10}$
26. $2 + n^3 - 3n^4 - 7n$
27. $z^2 - z^4 + 1$
28. $-3t^3 - t^2 + 11 + t$
29. $d + 7d^5 - 2d^8 + 12$

WRITTEN EXERCISES

Is the expression a monomial, a binomial, a trinomial, or none of these? If the expression is a polynomial, name its degree.

A 1. $9y$ 2. $4x - 7$ 3. $5y^2 + 10$ 4. $-12ab^2$ 5. $x^3 y$ 6. $-m - n$
   7. -1 8. 0 9. $-10r^6 s^2$ 10. $m + n^2$ 11. $3ab^2 c$ 12. $7 + y$
   13. $3x^2 + 4x - 9x^3$ 14. $-c^2 d - 5c^3 d^2 + 6c^4 d^3$ 15. $4mn - 3n^2 + 9$
   16. $\dfrac{11}{s^2} - \dfrac{t^2}{s^2}$ 17. $\dfrac{2y^3 z}{5} + \dfrac{9y^5 z^2}{10} - \dfrac{8y^7 z^2}{3}$ 18. $\dfrac{a + 2bc^2}{a}$

Select the similar terms, if any.

19. $m, 2m^2, -3m^2, 4m^3$
20. $8y^2, -y, -4y^4, \dfrac{y}{3}$
21. $8x, 9x, -10x^2, 5x^2$
22. $-z^5, z, -6z^3, \dfrac{1}{9}z$
23. $-jk^2, 7j^2 k, -8j^2 k, 4j^2 k^2$
24. $13xyz^2, -3x^2 yz, 7xy^2 z, 2x^2 y^2 z^2$

Write the polynomial in standard form.

25. $2x + 1 + 5x^2$
26. $4 - y^3 + 8y^2 - y$
27. $\dfrac{d^4}{2} - \dfrac{d^2}{3} + 9$
28. $-\dfrac{3}{2}x^2 + \dfrac{3}{2} - \dfrac{23}{4}x + 2x^3$
29. $-4 - 4.5m + 5m^3$
30. $17n - 17n^2 + 3n^3 - 4$

Simplify and write in standard form if possible. Give the degree of the polynomial and also the coefficient of the term of greatest degree.

31. $-2r^5 + r^2 - 3$
32. $t - 3t^2$
33. $4x - 2x + 3x^2$
34. $6y + 2y + 5y^2 - 8y^2$
35. $14b^2 + 6b - 14b^2 + 2$
36. $-11r^3 + 2r^2 - 5r^3 - 2r^2$
37. $5z^2 - 15z^5 + 25z^5 - 5z^5$
38. $-7x^3 + 8 + 2x^3 - x + 6$

B 39. $ab - b^2 + 3ab - a^2$
40. $5mn + 6m^2 n^2 + 2mn$
41. $3x^2 y + 9x^2 - 7y^2 - 4x^2 y$
42. $2st^2 - 10s^2 + 6t^2 s$
43. $-7b^2 c^2 + 2bc + 3cb - c^2 b^2$
44. $4x^4 y^3 - 2x^4 - x^4 y^4 + 3y^3 x^4$

45. Write two different polynomials of degree 3.
46. Write two different polynomials of degree 4.
47. Write a trinomial of degree 2.
48. Write a binomial of degree 7.

POLYNOMIALS AND FACTORING

Section 7-3 ADDITION AND SUBTRACTION OF POLYNOMIALS

In the last section the distributive property was used to add similar terms. When you add polynomials you use the commutative, associative, and distributive properties.

> To add two polynomials, you add similar terms.

EXAMPLES

$$(2x^2 + 3x) + (5x^2 + 7x) = (2x^2 + 5x^2) + (3x + 7x)$$
$$= 7x^2 + 10x$$

$$(-5y + 7) + (y^2 + 8y + 4) = y^2 + (-5y + 8y) + (7 + 4)$$
$$= y^2 + 3y + 11$$

$$(-2rst^2 + 9rs) + (8r^2st + 3rs) = -2rst^2 + 8r^2st + (9rs + 3rs)$$
$$= -2rst^2 + 8r^2st + 12rs$$

If you wanted to subtract one polynomial from another, how would you do it? Recall that subtracting a number is the same as adding its opposite.

EXAMPLE Find the opposite of $x^4 + 2x - 5$.

SOLUTION The opposite of $x^4 + 2x - 5$ may be written as $-(x^4 + 2x - 5)$.
$$-(x^4 + 2x - 5) = -1(x^4 + 2x - 5)$$
$$= (-1 \cdot x^4) + (-1 \cdot 2x) + (-1)(-5)$$
$$= -x^4 - 2x + 5$$

> To subtract one polynomial from another, you replace each term of the polynomial you are subtracting by its opposite. Then you add similar terms.

EXAMPLE
$$(3x^4 - 6x^2 + 5) - (x^4 + 2x - 5) = (3x^4 - 6x^2 + 5) + (-x^4 - 2x + 5)$$
$$= (3x^4 - x^4) - 6x^2 - 2x + (5 + 5)$$
$$= 2x^4 - 6x^2 - 2x + 10$$

ORAL EXERCISES

Add the two given polynomials.

1. $3a + 5, -7a$
2. $-8b, -4b + 9$
3. $z^2 + 5, 3z^2 - 12$
4. $-t^3 + 7, 6t^3 - t$
5. $n^3 + n + 5, 4n^3 - 3n^2 + 7$
6. $5x^2 + 3x - 1, -5x^2 - 7x + 2$
7. $-s^2 + s - 9, 11s - 3$
8. $4y + 3, -y^2 - 7$

9–16. In each of Exercises 1–8, subtract the second polynomial from the first.

WRITTEN EXERCISES

Add.

A
1. $(3x + 2) + (-7x + 3)$
2. $(17s^2 + 14) + (-5s^2 - 14)$
3. $(y^2 + 3y + 7) + (-y^2 - 11y)$
4. $(-5k^2 + k - 3) + (11k^2 + 7k - 8)$
5. $(s + 2) + (3s - 5) + (-s^2 + 2s)$
6. $(-t^2 + 3) + (3t^2 - 2t) + (-9t - 7)$
7. $(6ab + 7a) + (-8ab - b)$
8. $(-wz + 3wz^2) + (9w^2z - 7wz^2 + wz)$
9. $(r^2 - 9rs - s^2) + (5s^2 - 3rs - 7r^2)$
10. $(ac^2d^2 + 4cd - 1) + (-3cd + 1)$

Subtract.

11. $(17c + 5) - (7c - 2)$
12. $(4f - 3) - (5f + 7)$
13. $(-3z^2 + z + 13) - (6z^2 + 7z)$
14. $(x^3 + 3x^2 - 2x) - (9x^2 + 5x - 8)$
15. $(11y^2 - y^4 + 8) - (19 + 5y - y^2)$
16. $(-h - h^2 + h^3) - (-h - h^2 + 2h^3)$
17. $(3n + 5x - 10nx^2) - (2nx^2 - n + 11x)$
18. $(-6y^4z^2 - 3y^2z^2 + 17yz) - (-2y^4z^2 + 3y^2z^2)$
19. $(s^2t^2 - st - 12) - (14s^2t^2 + 9st + 1)$
20. $(16a^4b^2 - 9) - (3a^4b^2 - 2a^2b - 16)$

Simplify.

B
21. $(z^2 + 3z - 7) + (3z^2 - 5z + 2) - (11z^2 - 8z - 3)$
22. $(5k^2 + 3k - 2) - (8k^2 + 7k + 2) + (10k^2 + 9k - 5)$
23. $\left(\frac{1}{7}s^4 - \frac{6}{7}s^2 + \frac{3}{7}\right) - \left(\frac{3}{7}s^4 + \frac{2}{7}s^2 - \frac{5}{7}\right)$
24. $\left(\frac{3}{4}x^4 + \frac{1}{8}x^2 - \frac{3}{16}\right) + \left(\frac{5}{4}x^2 + \frac{7}{8}x + \frac{5}{16}\right)$
25. $(0.7y^3 - 0.3y^2 - 0.6) - (0.2y^2 - 0.5y + 0.9)$
26. $(1.2z^2 - 0.5z + 0.1) + (-0.3z^2 + 2.3z - 0.8)$

Solve.

27. $(5x + 7) - (3x + 2) = 11$
28. $(-4y - 1) + (y - 6) = -4$
29. $(-8t + 3) + (4t - 10) = -9$
30. $(r^2 - 2r) - (r^2 + 8r) = 20$
31. $(11a - 5) - (8a - 6) = 1$
32. $(-d^3 - 4) + (d^3 - 1) = 0$
33. $(z + 7) + (-5 - 3z) + (-2) = -2z$
34. $(5.1q - 1.9) - (4.6q - 3.3) = 1.6$
35. $(7k - 2) - (3 - 2k) = 16$
36. $(-3 + 11m) - (4 + 2m) + (6 - 3m) = 17$
37. $\left(\frac{4}{5}s + \frac{3}{5}\right) - \left(\frac{3}{5}s + \frac{2}{5}\right) = \frac{2}{5}$
38. $\left(\frac{7}{8}p + \frac{3}{8}\right) + \left(\frac{1}{8}p - \frac{5}{8}\right) = \frac{3}{8}$

POLYNOMIALS AND FACTORING

To solve an equation containing fractions, such as $\frac{a}{2} + \frac{a}{6} = 2$, first multiply both sides of the equation by a number that will eliminate the fractions. The easiest number to use is the lowest common multiple (LCM) of the denominators.

EXAMPLE Solve $\frac{a}{6} + \frac{a}{9} = 10$.

SOLUTION Method 2 may be used if you do not recognize that 18 is the LCM of 6 and 9.

Method 1	Method 2
LCM = 18	$6 \cdot 9 = 54$
$18\left(\frac{a}{6} + \frac{a}{9}\right) = 18 \cdot 10$	$54\left(\frac{a}{6} + \frac{a}{9}\right) = 54 \cdot 10$
$18\left(\frac{a}{6}\right) + 18\left(\frac{a}{9}\right) = 180$	$54\left(\frac{a}{6}\right) + 54\left(\frac{a}{9}\right) = 540$
$3a + 2a = 180$	$9a + 6a = 540$
$5a = 180$	$15a = 540$
$a = 36$	$a = 36$

39. $\dfrac{x}{2} + \dfrac{x}{6} = 3$ 40. $\dfrac{r}{5} + \dfrac{r}{10} = 6$ 41. $\dfrac{s}{3} - \dfrac{s}{7} = 4$ 42. $\dfrac{t}{2} - \dfrac{t}{5} = 1$

43. $\dfrac{z}{4} + \dfrac{z}{6} = 1$ 44. $\dfrac{h}{6} - \dfrac{h}{8} = 3$ 45. $\dfrac{b}{9} - \dfrac{b}{12} = 7$ 46. $\dfrac{c}{12} - \dfrac{c}{15} = 4$

Section 7-4 MULTIPLICATION OF MONOMIALS

Which of the expressions below do you think has the greatest value?

$$10^{50} \cdot 10^{50}, \quad 2^{100} \cdot 5^{100}, \quad (100{,}000)^{20}$$

The rules of exponents that you will study in this section will help you to see that the three expressions are equal!

In this section you will learn:
- three rules of exponents
- how to multiply monomials

The examples below show how to multiply two powers with the *same base*.

EXAMPLES $2^3 \cdot 2^2 = (2 \cdot 2 \cdot 2)(2 \cdot 2) = 2^5$
$a^2 \cdot a^4 = (a \cdot a)(a \cdot a \cdot a \cdot a) = a^6$

If r and s are positive integers, then $a^r \cdot a^s = a^{r+s}$.

In other words, when you multiply two powers with the *same base*, you add the exponents.

The examples below lead to the rule for finding a power of a product.

EXAMPLES

$$(2x)^3 = (2x)(2x)(2x) = (2 \cdot 2 \cdot 2)(x \cdot x \cdot x) = 2^3 x^3 = 8x^3$$
$$(yz)^4 = (yz)(yz)(yz)(yz) = (y \cdot y \cdot y \cdot y)(z \cdot z \cdot z \cdot z) = y^4 z^4$$

If r is a positive integer, then $(ab)^r = a^r \cdot b^r$.

The next examples will show you how to find the power of a power.

EXAMPLES

$$(5^2)^3 = (5^2)(5^2)(5^2) = (5 \cdot 5)(5 \cdot 5)(5 \cdot 5) = 5^6$$
$$(x^4)^2 = x^4 \cdot x^4 = (x \cdot x \cdot x \cdot x)(x \cdot x \cdot x \cdot x) = x^8$$

If r and s are positive integers, then $(a^r)^s = a^{rs}$.

In other words, to find the power of a power, you *multiply* the exponents.

Now you can apply the three laws of exponents to answer the question at the beginning of this section.

$$10^{50} \cdot 10^{50} = 10^{50+50} = 10^{100}$$
$$2^{100} \cdot 5^{100} = (2 \cdot 5)^{100} = 10^{100}$$
$$(100{,}000)^{20} = (10^5)^{20} = 10^{100}$$

The three expressions are equal.

The examples below show how the rules of exponents can be used to multiply monomials.

EXAMPLES

$$(2x)(3x^2) = (2 \cdot 3)(x \cdot x^2) = 6x^3$$
$$(-5y^2)(7y^3) = (-5 \cdot 7)(y^2 \cdot y^3) = -35y^5$$
$$(-6abc)(-3a^2bc^5) = (-6)(-3)(a \cdot a^2)(b \cdot b)(c \cdot c^5) = 18a^3b^2c^6$$
$$(-4z^2)^3 = (-4)^3(z^2)^3 = -64z^6$$

To multiply monomials, you multiply the coefficients and use the rules of exponents to simplify the variable part of the product.

POLYNOMIALS AND FACTORING

ORAL EXERCISES

Simplify.

1. $a^7 \cdot a^3$
2. $(-b)^3 (-b)^5$
3. $(3x)^2$
4. $(-2y)^4$
5. $(z^7)^3$
6. $(-m^2 n)^3$

Multiply.

7. $(3x)(5x)$
8. $(-2y)(7y^2)$
9. $(-3a^7)(-4a^2 b)$
10. $(2xyz)(-5x^2 yz^2)$
11. $(6rs^2 t)(8r^2 st)$
12. $\left(\frac{1}{2}j^2 k\right)(2j^4 k^3)$

WRITTEN EXERCISES

Find the square of the given monomial.

A 1. $3a$
2. $-4b^2$
3. $-2c^3$
4. $5d^4$
5. $10ab$
6. $x^2 y$
7. $0.2jk^2$
8. $\frac{1}{3}a^2 bc^3$

Simplify.

9. $a^4 \cdot a^9$
10. $b^5 \cdot b^3$
11. $c(-c)$
12. $(-d^3)(-d^2)$
13. $(-y)(-y)^6$
14. $(x^3)^{10}$
15. $(ab^2)^5$
16. $(-xy^2 z)^4$

Multiply.

17. $(4a^2)(3a^3)$
18. $(-2b^4)(5b^6)$
19. $(14c^5)(-3c^7)$
20. $(-9d^{10})(-5d^5)$
21. $(8rs)(3r^2 s^3)$
22. $(3x^2 y)(-11xy^3)$
23. $(2a^2 bc^3)(-5ab^2 c)$
24. $(7x^2 y^2 z)(-3x^5 yz^4)$
25. $(-8c^4 d^2)(-10c^2 d^7)$
26. $(-3xyz^3)(-39x^2 y^5 z^{11})$
27. $(-2rst^2)(18rt^9)$
28. $(14x^7 y)(8y^2 z^4)$

B 29. $a^2 \cdot a^{11} \cdot a^3$
30. $(-b)^5 (-b)^7 (-b)$
31. $(-2cd)(4cd^2)(3c^5 d^3)$
32. $(5rs^2)(-7r^4 s)(-3r^5 s^2)$
33. $(x^3 y^2 z)(x^4 z^2)(y^3 z^5)$
34. $(-a^4 c^5)(-b^2 c^6)(-a^3 b^4)$

Simplify.

35. $(r^2 s^3)^5 (r^3 s^2)^2$
36. $(-w^4 t^5)^3 (-w^3 t^4)^4$
37. $(6a^3 b^3)^2 - (3a^2 b^2)^3$
38. $(4x^2 y^4)^3 - (32x^2 y^7)(2x^4 y^5)$
39. $\left((2x)^2\right)^3 + (2x^2)^3$
40. $(5cd^2)(3cd)^2 + (8cd^2)(-6c^2 d^2)$

236 CHAPTER 7

Solve for x.

EXAMPLE $7^x \cdot 7^3 = 7^9$
SOLUTION $7^{x+3} = 7^9$
$x + 3 = 9$
$x = 6$

C 41. $5^x \cdot 5^{10} = 5^{15}$
42. $2^x \cdot 2^x = 2^{16}$
43. $3^{5x} \cdot 3^{3x} = 3^{32}$
44. $b^{2x} \cdot b^5 = b^{3x}$
45. $6^{12} = 2^x \cdot 3^x$
46. $(a^5)^{10} = a^x$
47. $9^x = 3^{36}$
48. $4^{10} = 2^x$
49. $a^x \cdot a^{x-2} \cdot a^{x+2} = a^{12}$
50. $y^{2x} \cdot y^{3x-4} \cdot y^6 = y^{12}$

SELF-TEST 1

Simplify.

1. 3^4
2. $(-2)^5$ Section 7-1
3. $5 \cdot 5 \cdot 5 \cdot x \cdot x$
4. $(-4)(-4)(-4)(-4) \cdot y \cdot y \cdot y$
5. $6t - 4t + 3t$
6. $4z - z^2 + 6z + 3z^2$ Section 7-2

Write in standard form.

7. $s^2 + 2s^3 - 5 + 3s$
8. $p^5 - p^7 + 2p^2 - p^{10}$

Add or subtract.

9. $(3t + 4) + (4t - 7)$
10. $(z^2 - 3z + 2) + (3z^2 + 5z - 7)$ Section 7-3
11. $(16d - 3) - (12d + 5)$
12. $(3q^3 + 2q - 5) - (q^3 + 4q^2 - 7)$

Simplify.

13. $d^3 \cdot d^5$
14. $(-x^2yz^3)^3$ Section 7-4
15. $(7rt^2)(4r^3t)$
16. $(2xy^2)(3x^2y)(4xy)$

Section 7-5 MULTIPLICATION OF POLYNOMIALS

In the last section you learned how to multiply two monomials. In this section you will use the distributive property to multiply:
- a polynomial by a monomial
- a polynomial by a polynomial

How would you multiply $2y^3$ and $6y^2 + 3y + 1$? You know that $2y^3 \cdot 6y^2 = 12y^5$, $2y^3 \cdot 3y = 6y^4$, and $2y^3 \cdot 1 = 2y^3$. Thus,

$$2y^3(6y^2 + 3y + 1) = (2y^3 \cdot 6y^2) + (2y^3 \cdot 3y) + (2y^3 \cdot 1)$$
$$= 12y^5 + 6y^4 + 2y^3$$

POLYNOMIALS AND FACTORING

> To multiply a polynomial by a monomial, you multiply each term of the polynomial by the monomial. Then you add the products.

EXAMPLE Multiply $3x + 4$ and $2x$.
SOLUTION $(3x + 4)2x = 3x \cdot 2x + 4 \cdot 2x$
$= 6x^2 + 8x$

EXAMPLE Multiply $3ab^2$ and $5a^2 + 3ab - 4b^2$.
SOLUTION $3ab^2(5a^2 + 3ab - 4b^2) = (3ab^2)(5a^2) + (3ab^2)(3ab) + (3ab^2)(-4b^2)$
$= 15a^3b^2 + 9a^2b^3 - 12ab^4$

To multiply two polynomials, you apply the distributive property twice, as shown in the first two lines of the example below.

EXAMPLE $(3x + 4)(2x + 5) = 3x(2x + 5) + 4(2x + 5)$
$= 3x \cdot 2x + 3x \cdot 5 + 4 \cdot 2x + 4 \cdot 5$
$= 6x^2 + 15x + 8x + 20$
$= 6x^2 + 23x + 20$

> To multiply a polynomial by a polynomial, you multiply each term in the first polynomial by each term in the second polynomial. Then you add the products.

A convenient method of multiplying polynomials is to use a vertical form as shown in the following example.

EXAMPLE Multiply $6x - 2$ and $x^3 + 3x - 4$.
SOLUTION
$$\begin{array}{r} x^3 + 0x^2 + 3x - 4 \\ 6x - 2 \\ \hline \end{array}$$

$6x(x^3 + 3x - 4) \longrightarrow 6x^4 + 0x^3 + 18x^2 - 24x$
$-2(x^3 + 3x - 4) \longrightarrow - 2x^3 - 0x^2 - 6x + 8$
$ 6x^4 - 2x^3 + 18x^2 - 30x + 8$

ORAL EXERCISES

Multiply.

1. $x(x^2 + 5)$
2. $-2y(y^3 - 3)$
3. $(t^2 - t - 7)6t$
4. $(z^3 - 5z^2 + z)z^2$
5. $(3d^4 - 5d^2 + 1)4d$
6. $-c^2(c^3 - c^2 + 6)$
7. $2ab(a + 3)$
8. $5x(x^2 + xy)$
9. $-2m^2p(mp - 3mp^2)$

WRITTEN EXERCISES

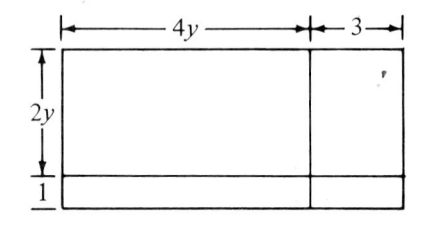

Multiply.

A
1. $11x(13x^2 + 5)$
2. $9y(8y^3 - 7)$
3. $3z^2(9z^2 - 7z + 2)$
4. $10a^2(9a^2 - 8a + 7)$
5. $(b^2 + 3b - 5)(-2b^2)$
6. $(c^3 - 7c + 3)5c^5$
7. $4r^2s^2(2r^2 - 3rs + 6s^2)$
8. $(x^4 - 9x^2y^2 - 4y^4)(-3xy^2)$
9. $(-3a^3b^3 + 2a^2b^2 - 9ab)4a^3b^2$
10. $-4c^2d^3(2c^5d^5 - c^3d^3 + 5cd)$
11. $(x + 3)(x + 5)$
12. $(y + 9)(y + 2)$
13. $2(r - 5)(r - 1)$
14. $3(t - 12)(t - 5)$
15. $(s + 6)(s - 6)$
16. $(z - 9)(z + 9)$
17. $(3x - 2)(5x + 4)$
18. $(6b + 2)(b - 7)$
19. $(3t + 2)^2$
20. $(8v - 6)^2$

B
21. $(x^2 + x + 1)(2x - 5)$
22. $(2y - 3)(3y^3 + 2y^2 - 7)$
23. $(4c^2 + 9)(2c^2 - 15)$
24. $(3t^2 - 16t)(3t^2 - 2t)$
25. $(a^3 - 3a^2b - ab^2)(a - 2b)$
26. $(x - y)(x^2 + 2xy - y^2)$
27. $(a + b)(2a^2 + 3ab - 4b^2)$
28. $(4r^4 - r^2s^2 + 9s^4)(r^2 - s^2)$
29. $(-5z^2 - z + 2)(2z^2 + z - 2)$
30. $(j^2 + 3j - 5)(3j^2 - j + 2)$

Find the perimeter of the figure.

31. A square whose side has length $3x + 2$
32. A square whose side has length $x - 4$
33. A rectangle with sides of lengths $2y + 4$ and $y + 7$
34. A rectangle with sides of lengths $3d - 5$ and $4d + 2$

Find the area of the figure.

35. A square whose side has length $\frac{1}{2}s - 3$
36. A rectangle of length $8z - 7$ and width $3z + 2$
37. A triangle with base $2x - 1$ and height $4x - 6$
38. A trapezoid with bases $3x - \frac{1}{2}$ and $5x + \frac{3}{2}$, and height $4x$

Represent the area of each large rectangle as the product of two polynomials.

C 39.

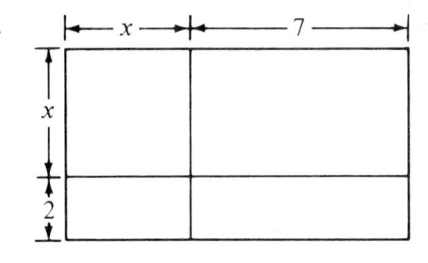

40.

POLYNOMIALS AND FACTORING   239

Multiply.

41. $(k + 2)(k + 3)(k - 5)$
42. $(2z + 1)(z - 7)(3z - 2)$
43. $(s^2 - 5)(s^2 + 5)(s + 3)$
44. $(2r^2 + 5)(r + 7)(2r^2 - 5)$
45. $(c - 2d)^3$
46. $(4d + 5)^3$

Section 7-6 DIVISION OF POLYNOMIALS BY MONOMIALS

In this section you will learn:
- another rule of exponents
- how to divide a monomial by another monomial
- how to divide a polynomial by a monomial

In Section 7-4 you learned that to multiply two powers with the same base you add their exponents. You might guess that to divide a power by another power with the same base you would subtract exponents. But in some cases this would result in a negative exponent or an exponent equal to zero, and so far you know how to use only positive exponents. Thus you cannot simply subtract exponents. The examples below suggest rules for dividing two powers with the same base.

EXAMPLES

$$\frac{7^5}{7^2} = \frac{7 \cdot 7 \cdot 7 \cdot 7 \cdot 7}{7 \cdot 7} = 7^3$$

$$\frac{7^5}{7^5} = \frac{7 \cdot 7 \cdot 7 \cdot 7 \cdot 7}{7 \cdot 7 \cdot 7 \cdot 7 \cdot 7} = 1$$

$$\frac{7^2}{7^5} = \frac{7 \cdot 7}{7 \cdot 7 \cdot 7 \cdot 7 \cdot 7} = \frac{1}{7^3}$$

If r and s are positive integers and $x \neq 0$, then:

$\dfrac{x^r}{x^s} = x^{r-s}$ if $r > s$; $\dfrac{x^r}{x^s} = 1$ if $r = s$; $\dfrac{x^r}{x^s} = \dfrac{1}{x^{s-r}}$ if $s > r$

The examples below show how this rule of exponents can be used in dividing monomials. (In the examples and exercises of this section you may assume that no divisors are zero.)

EXAMPLES

$4x^5 \div 2x^2 = \dfrac{4x^5}{2x^2}$
$= \dfrac{4}{2} \cdot \dfrac{x^5}{x^2}$
$= 2 \cdot x^{5-2}$
$= 2x^3$

$\dfrac{-16y}{-2y^7} = \dfrac{-16}{-2} \cdot \dfrac{y}{y^7}$
$= 8 \cdot \dfrac{1}{y^{7-1}}$
$= 8 \cdot \dfrac{1}{y^6}$
$= \dfrac{8}{y^6}$

$\dfrac{6rs^2t^2}{-3st^3} = \dfrac{6}{-3} \cdot \dfrac{r}{1} \cdot \dfrac{s^2}{s} \cdot \dfrac{t^2}{t^3}$
$= -2rs \cdot \dfrac{1}{t}$
$= -\dfrac{2rs}{t}$

240   CHAPTER 7

> To divide two monomials, you divide the coefficients and simplify the variable part of the quotient.

How would you divide $4x^5 + 6x^3$ by $2x^2$? Recall that dividing by a nonzero number is the same as multiplying by its reciprocal. Using this fact and the distributive property, you can perform the division as follows:

EXAMPLE Divide $4x^5 + 6x^3$ by $2x^2$.

SOLUTION
$$\frac{4x^5 + 6x^3}{2x^2} = (4x^5 + 6x^3) \cdot \frac{1}{2x^2}$$
$$= \left(4x^5 \cdot \frac{1}{2x^2}\right) + \left(6x^3 \cdot \frac{1}{2x^2}\right)$$
$$= \frac{4x^5}{2x^2} + \frac{6x^3}{2x^2}$$
$$= 2x^3 + 3x$$

> To divide a polynomial by a monomial, you divide each term of the polynomial by the monomial. Then you add the resulting quotients.

EXAMPLE Divide $10y^4 - 4y^3 + 6y^2 + 3y$ by $2y$.

SOLUTION
$$\frac{10y^4 - 4y^3 + 6y^2 + 3y}{2y} = \frac{10y^4}{2y} - \frac{4y^3}{2y} + \frac{6y^2}{2y} + \frac{3y}{2y}$$
$$= 5y^3 - 2y^2 + 3y + \frac{3}{2}$$

ORAL EXERCISES

Divide.

1. $\dfrac{5^8}{5^2}$
2. $\dfrac{7^5}{7^3}$
3. $\dfrac{(-5)^4}{(-5)^3}$
4. $\dfrac{2^7}{2^{12}}$
5. $\dfrac{x^3}{x^{11}}$
6. $\dfrac{t^9}{t^9}$
7. $\dfrac{z^a}{z^b}$ $(a > b)$
8. $\dfrac{z^a}{z^b}$ $(a < b)$
9. $\dfrac{14b^5}{7b^2}$
10. $\dfrac{12t^8}{-3t}$
11. $\dfrac{-21y^2}{-7y^9}$
12. $\dfrac{10x^3}{20x^7}$
13. $\dfrac{8x^2 + 4x^3}{2x}$
14. $\dfrac{10y^3 + 15y^5}{5y^2}$
15. $\dfrac{7s^2t^3 - 5st}{st}$
16. $\dfrac{6m^2g^2 - 27mg}{-3mg}$

POLYNOMIALS AND FACTORING

WRITTEN EXERCISES

Divide.

A 1. $\dfrac{r^{12}}{r^2}$ 2. $\dfrac{s^{18}}{s^{24}}$ 3. $\dfrac{(-t)^{11}}{(-t)^{21}}$ 4. $\dfrac{(-c)^8}{(-c)^3}$

5. $\dfrac{(-3y)^{19}}{(-3y)^{19}}$ 6. $\dfrac{(7k)^5}{(7k)^7}$ 7. $\dfrac{(-5d)^9}{(-5d)^6}$ 8. $\dfrac{(5a)^8}{(5a)^8}$

9. $\dfrac{28x^{12}}{7x^7}$ 10. $\dfrac{11y^{15}}{121y^{11}}$ 11. $\dfrac{88x^3y^2}{-4xy^2}$ 12. $\dfrac{-144x^9y^7}{6x^5y^6}$

13. $\dfrac{-35a^3b^5c^7}{-28a^3b^5c^7}$ 14. $\dfrac{13r^2st^3}{52st}$ 15. $\dfrac{3x^2yt}{-42xt}$ 16. $\dfrac{32d^4e^2f^2}{16d^2e^2f}$

17. $\dfrac{10x^3 + 5x^2 - 15x}{5x}$ 18. $\dfrac{30z^5 - 15z^2 + 6z}{3z}$

19. $\dfrac{-28y^7 + 14y^5 - 7y^3}{7y^2}$ 20. $\dfrac{4b^4 - 16b^3 - 40b^2 + 12b}{-4b}$

21. $\dfrac{r^3k^3 - 5r^2k^2 + 2rk}{rk}$ 22. $\dfrac{j^3k^5 - j^2k^2 + j^2k^4}{j^2k^2}$

23. $\dfrac{16x^3y^2z - 8xy^3z^5 + 24x^5y^2z}{8xyz}$ 24. $\dfrac{-10c^2d^2 + 6c^2d + 2cd^2 - 18c^2d^3}{-2cd^2}$

Simplify.

B 25. $\dfrac{15y + 20y^2}{5y} - \dfrac{28y^3 - 49y^2}{7y^2}$ 26. $\dfrac{8x^4 + 4x^3}{2x} + \dfrac{10x^5 - 5x^6}{5x^3}$

27. $\dfrac{3z^6 - 6z^4}{3z^2} - \dfrac{10z^4 + 5z^7}{5z^3}$ 28. $\dfrac{-27a^7 + 18a^4}{6a^3} + \dfrac{-24a^2 + 4a^9}{8a^5}$

29. $\dfrac{40t^2 - 35t + 5t^3}{-5t} + \dfrac{16t^2 - 8}{8}$ 30. $\dfrac{3r^5 - r^3}{-r} - \dfrac{11r^6 + 33r^8 - 55r^{10}}{11r^4}$

Negative integers and zero are used as exponents in more advanced mathematics and in some scientific applications. These exponents are *defined* so that the rules of exponents can still be applied.

C 31. How would you define 2^0 so that $2^5 \cdot 2^0 = 2^{5+0}$ and $\dfrac{2^5}{2^5} = 2^{5-5}$?

32. How would you define b^0 for $b \neq 0$? (Note: 0^0 is not defined.)

33. How would you define 2^{-3} so that $\dfrac{2^3}{2^6} = 2^{3-6} = 2^{-3}$ and $2^3 \cdot 2^{-3} = 2^{3-3} = 2^0$?

34. How would you define b^{-x} for $b \neq 0$ and $x > 0$?

Section 7-7 COMMON MONOMIAL FACTORS

In this section you will learn about:
- prime factors
- common factors
- greatest common factors
- common monomial factors

You may recall from your earlier studies that any integer greater than 1 that has only itself and 1 as positive factors is called a *prime number*. For example, 2, 5, 11, and 23 are prime numbers. When we express a number as a product of its prime factors, we say we have written its *prime factorization*.

EXAMPLES

$$9 = 3 \cdot 3$$
$$18 = 2 \cdot 3 \cdot 3$$
$$34 = 2 \cdot 17$$
$$45 = 3 \cdot 3 \cdot 5$$

Notice in the examples above that 3 is a *common factor* of 9, 18, and 45. Also, that $3 \cdot 3$, or 9, is another common factor of 9, 18, and 45. Here are some other examples of common factors.

EXAMPLES

$12 = 2 \cdot 2 \cdot 3$ and $20 = 2 \cdot 2 \cdot 5$
2 and 4 are common factors of 12 and 20.

$14x = 2 \cdot 7 \cdot x$ and $7x^2 = 7 \cdot x \cdot x$
7, x, and $7x$ are common factors of $14x$ and $7x^2$.

The examples below show how to find the *greatest* common factor (GCF) of two expressions.

EXAMPLES

$12 = 2 \cdot \boxed{2 \cdot 3}$ and $18 = \boxed{2 \cdot 3} \cdot 3$
The GCF of 12 and 18 is $2 \cdot 3$, or 6.

$90 = \boxed{2 \cdot 3 \cdot 3} \cdot 5$ and $126 = \boxed{2 \cdot 3 \cdot 3} \cdot 7$
The GCF of 90 and 126 is $2 \cdot 3 \cdot 3$, or 18.

$12x^2 = \boxed{2 \cdot 2 \cdot 3 \cdot x} \cdot x$ and $24x = 2 \cdot \boxed{2 \cdot 2 \cdot 3 \cdot x}$
The GCF of $12x^2$ and $24x$ is $2 \cdot 2 \cdot 3 \cdot x$, or $12x$.

In algebra we often wish to *factor a polynomial*, that is, express a polynomial as the product of polynomial factors. To do this, we use the "reverse" of the distributive property. The simplest example occurs when one of the factors is a monomial. We usually find the *greatest* common monomial factor. For polynomials with integral coefficients, we consider only the integral factors of the coefficients.

POLYNOMIALS AND FACTORING

EXAMPLES

$$2x^2 + 4x = 2x(x) + 2x(2)$$
$$= 2x(x + 2)$$
$$12y^3 - 30y^5 = 6y^3(2) - 6y^3(5y^2)$$
$$= 6y^3(2 - 5y^2)$$
$$3t^5 + 12t^4 - 21t^2 = 3t^2(t^3 + 4t^2 - 7)$$

ORAL EXERCISES

Tell whether or not a given number is a prime. If it is not a prime, give its prime factorization.

1. 7
2. 21
3. 17
4. 31
5. 24
6. 19

Find the GCF.

7. 10 and 20
8. 12 and 24
9. $3x$ and $12x^2$
10. $5y^2$ and $10y$
11. $4y^2$ and $7y$
12. $16x^2$ and $8x$
13. $5x^2y$ and $12xy^2$
14. $3yz^2$ and $5y^2z^3$

Find the missing factor.

15. $2x^3 = 2x(?)$
16. $5y^4 = 5y^2(?)$
17. $7z^7 = z^4(?)$
18. $-3t^5 = 3t(?)$
19. $8a^2b = 4ab(?)$
20. $10xyz = 5x(?)$
21. $2x^2 + 4x = 2x(?)$
22. $6y + 3y^3 = 3y(?)$

WRITTEN EXERCISES

Write the prime factorization of the integer.

A 1. 8
2. 44
3. 35
4. 125
5. 144
6. 91

Find the GCF.

7. 16 and 32
8. 20 and 40
9. $12x^2$ and $36x^4$
10. $18y^3$ and $24y^5$
11. $8a^2b^3$ and $6a^4b^2$
12. $4r^2t^5$ and $10r^3t^2$
13. $14x^2yz^3$ and $7xy^2z^2$
14. $5x^3y^2z$ and $15xy^3z^4$

Find the missing factor.

15. $24x^5 = 3x(?)$
16. $36y^7 = 6y^2(?)$
17. $110a^3b^2 = 5ab(?)$
18. $162r^5s^3 = 27r^2s(?)$
19. $8x^3 + 4x^2 = 4x^2(?)$
20. $10y^5 - 5y^3 = 5y^3(?)$
21. $2x^3 + 6x^2 + 4x = 2x(?)$
22. $3y^3 - 6y + 12y^2 = 3y(?)$

Factor each of the following expressions into the product of the greatest common monomial factor and another expression.

23. $8x^2 + 4x^3$
24. $12y^3 + 6y^2$
25. $12a^2b - 4ab^2$
26. $20y^3z - 15y^2z^2$
27. $4x^3 + 2x^2 - 6x$
28. $5y^5 - 10y^2 + 25y$

29. $12a^5 - 3a^2 + 4a^4$

30. $14b^3 + 2b^5 - 7b^4$

31. $a^2x^2 + ax^2 - a^2x$

32. $by^2 - b^2y^3 + b^3y^4$

33. $c^3z^3 + c^2z^2 + cz$

34. $d^2s^2 - d^3s^3 + d^4s^4$

35. $a^2x^3y^4 + a^3x^2y^3 - a^2x^3y^2$

36. $b^2y^3z^3 - b^3y^2z^2 + b^2y^3z^4$

37. $2x^2(4x + 6x^3)$

38. $3y(9y^2 + 6y)$

B 39. If the area of a rectangle is $30a^2$ and its width is 6, what is its length?

40. If the area of a rectangle is $42y^2$ and its length is $6y$, what is its width?

41. If the area of a rectangle is $5x^2 + 15x$ and its length is $5x$, what is its width?

42. If the area of a rectangle is $6y^2 + 24y$ and its width is $3y$, what is its length?

C 43. If the volume of a rectangular prism is $30x^3$, its length is $2x$, and its width is $3x$, what is its height?

44. If the volume of a rectangular prism is $216ab^2$, its length is $6ab$, and its width is $8b$, what is its height?

Section 7-8 MULTIPLYING BINOMIALS

In Section 7-5 you saw how the distributive property can be used to multiply a polynomial by a polynomial.

EXAMPLE
$$(x + 2)(x + 5) = x(x + 5) + 2(x + 5)$$
$$= x \cdot x + x \cdot 5 + 2 \cdot x + 10$$
$$= x^2 + 5x + 2x + 10$$
$$= x^2 + 7x + 10$$

When the two polynomials you are multiplying are binomials of degree 1, it is possible to shorten your work. Try thinking about the multiplication in the way shown below.

The first term in the product of two binomials is the product of the *first* terms of the binomials:

$$(x + 5)(x + 2) = x^2 + \ldots$$

The second term in the product is the sum of the products of the *outer* terms and the *inner* terms:

$$(x + 5)(x + 2) = \ldots + (2x + 5x) + \ldots$$

The third term in the product is the product of the *last* terms:

$$(x + 5)(x + 2) = \ldots + \ldots + 10$$

Thus $(x + 5)(x + 2) = x^2 + 7x + 10$.

POLYNOMIALS AND FACTORING

This method is often called the **FOIL** method. The letters F and L stand for First and Last terms. The letters O and I remind you to add the products of the Outer terms and the Inner terms in order to find the middle term of the product. To use this method, you must watch the signs of the terms carefully.

EXAMPLE Multiply $x + 1$ and $x - 7$.

SOLUTION First term of product: $x \cdot x$, or x^2 F
Second term of product: $-7 \cdot x + 1 \cdot x$, or $-6x$ O + I
Third term of product: $1(-7)$, or -7 L
$(x + 1)(x - 7) = x^2 - 6x - 7$

EXAMPLE Multiply $2x + 7$ and $3x + 4$.

SOLUTION First term of product: $2x \cdot 3x$, or $6x^2$ F
Second term of product: $4 \cdot 2x + 7 \cdot 3x$, or $8x + 21x$, or $29x$ O + I
Third term of product: $7 \cdot 4$, or 28 L
$(2x + 7)(3x + 4) = 6x^2 + 29x + 28$

As you gain experience, you will be able to multiply binomials at sight.

EXAMPLE Multiply $x - 2$ and $x - 5$.

SOLUTION $(x - 2)(x - 5) = x^2 - 7x + 10$

Learning to think about multiplication of binomials in this way will help you to factor polynomials, as you will see in the next section.

ORAL EXERCISES

Complete.

1. $(x + 2)(x + 2) = x^2 + \underline{\ ?\ }x + 4$
2. $(y + 5)(y + 3) = y^2 + 8y + \underline{\ ?\ }$
3. $(z + 6)(z - 2) = z^2 + \underline{\ ?\ }z - 12$
4. $(t + 4)(t - 3) = t^2 + t + \underline{\ ?\ }$
5. $(p - 7)(p + 1) = p^2 + \underline{\ ?\ }p - 7$
6. $(k - 5)(k + 2) = k^2 - 3k + \underline{\ ?\ }$
7. $(n - 4)(n - 1) = n^2 + \underline{\ ?\ }n + 4$
8. $(x - 3)(x - 1) = x^2 - 4x + \underline{\ ?\ }$
9. $(x + 7)(x + 11) = x^2 + \underline{\ ?\ }x + 77$
10. $(y + 9)(y + 10) = y^2 + 19y + \underline{\ ?\ }$
11. $(z + 8)(z - 3) = z^2 + \underline{\ ?\ }z - 24$
12. $(t + 12)(t - 5) = t^2 + 7t + \underline{\ ?\ }$
13. $(s - 13)(s + 4) = s^2 + \underline{\ ?\ }s - 52$
14. $(p - 8)(p + 2) = p^2 - 6p + \underline{\ ?\ }$
15. $(q - 10)(q - 11) = q^2 + \underline{\ ?\ }q + 110$
16. $(d - 6)(d - 12) = d^2 - 18d + \underline{\ ?\ }$
17. $(c - 4)^2 = c^2 + \underline{\ ?\ }c + 16$
18. $(e + 6)(e - 6) = e^2 + \underline{\ ?\ }$

246   CHAPTER 7

WRITTEN EXERCISES

Multiply using the FOIL method.

A
1. $(x + 1)(x + 2)$
2. $(y + 2)(y + 3)$
3. $(z + 2)(z - 1)$
4. $(t - 3)(t + 4)$
5. $(s + 5)(s - 6)$
6. $(x - 7)(x + 2)$
7. $(p - 2)(p - 4)$
8. $(q - 5)(q - 3)$
9. $(b + 3)^2$
10. $(r - 2)^2$
11. $(d + 1)(d - 1)$
12. $(y - 3)(y + 3)$
13. $(x + 3)(x + 9)$
14. $(y + 11)(y + 4)$
15. $(z + 12)(z - 3)$
16. $(t + 10)(t - 7)$
17. $(s - 8)(s - 6)$
18. $(r - 15)(r + 10)$
19. $(k - 9)(k - 7)$
20. $(t - 8)(t - 7)$
21. $(h + 7)^2$
22. $(q - 5)^2$
23. $(x + 10)(x - 10)$
24. $(p - 8)(p + 8)$
25. $(2x + 1)(4x + 2)$
26. $(3y + 3)(2y + 5)$
27. $(z + 5)(5z + 6)$
28. $(2a + 3)(4a + 1)$
29. $(2b + 4)(3b + 1)$
30. $(6p + 2)(p + 7)$
31. $(5q + 2)(7q + 3)$
32. $(8s + 4)(6s + 5)$
33. $(3h + 1)^2$
34. $(5g - 2)^2$
35. $(5d - 3)(2d + 7)$
36. $(6f - 8)(2f + 11)$
37. $(7n - 5)(2n - 4)$
38. $(3m - 6)(8m - 5)$
39. $(8k + 2)(8k - 2)$
40. $(9v + 3)(9v - 3)$
41. $(4z - 3)(4z + 3)$
42. $(7t - 2)(7t + 2)$

Complete.

B
43. $(z + \underline{?})(z - 5) = z^2 + z - 30$
44. $(r + 21)(r - \underline{?}) = r^2 + 19r - 42$
45. $(x + \underline{?})(x + 2) = x^2 + \underline{?}x + 2$
46. $(y + 3)(y + \underline{?}) = y^2 + \underline{?}y + 12$
47. $(s + \underline{?})(s - 8) = s^2 - \underline{?}s - 40$
48. $(t - 10)(t + \underline{?}) = t^2 - \underline{?}t - 30$
49. $(p - \underline{?})(p - 11) = p^2 - \underline{?}p + 66$
50. $(x - 12)(x - \underline{?}) = x^2 - \underline{?}x + 24$
51. $(k + \underline{?})(k - 5) = k^2 + \underline{?}k - 25$
52. $(z - \underline{?})(z - 6) = z^2 - \underline{?}z + 36$

C
53. Find two numbers whose sum is 10 and whose product is 21.
54. Find two numbers whose sum is 13 and whose product is 40.
55. Find two numbers whose sum is 35 and whose product is 250.
56. Find two numbers whose sum is -11 and whose product is 30.
57. Find two numbers whose sum is 0 and whose product is -25.

Complete.

58. $(x + \underline{?})(x + \underline{?}) = x^2 + 8x + 15$
59. $(y + \underline{?})(y + \underline{?}) = y^2 + 15y + 56$
60. $(z + \underline{?})(z - \underline{?}) = z^2 + 2z - 24$
61. $(s - \underline{?})(s - \underline{?}) = s^2 - 17s + 72$
62. $(t + \underline{?})(t + \underline{?}) = t^2 + 6t + 9$
63. $(u - \underline{?})(u - \underline{?}) = u^2 - 10u + 25$
64. $(v + \underline{?})(v - \underline{?}) = v^2 - 16$
65. $(h + \underline{?})(h - \underline{?}) = h^2 - 121$

POLYNOMIALS AND FACTORING

SELF-TEST 2

Multiply.

1. $2x(5x^2 + 3)$
2. $(z^3 - 2z^2 + 5)(-2z)$ Section 7-5
3. $(3k + 4)^2$
4. $(h - 5)(2h + 3)$

Divide.

5. $\dfrac{t^{12}}{t^7}$
6. $\dfrac{(3k)^2}{(3k)^5}$ Section 7-6
7. $\dfrac{65x^2y^4}{5xy^2}$
8. $\dfrac{12y^5 - 16y^2 + 4y^3}{4y}$

Find the positive common factors and the GCF.

9. 24 and 32
10. 15 and 45   Section 7-7

Factor.

11. $4x^3 + 2x^2$
12. $a^2y^2 + ay^2 - a^2y$

Multiply.

13. $(z + 5)(z - 6)$
14. $(s - 2)(s + 2)$ Section 7-8
15. $(3q + 1)^2$
16. $(2t - 3)(3t + 5)$

Section 7-9 FACTORING QUADRATIC POLYNOMIALS

In this section you will learn about:
- quadratic polynomials
- factoring quadratic polynomials

A polynomial that has a degree of 2 and has just one variable is called a *quadratic polynomial*.

EXAMPLES

$$x^2, \quad y^2 - 2, \quad 4m^2 + 3, \quad 10d^2 - 6d + 2, \quad -3b^2$$

The term of degree 2 in a quadratic polynomial is called the *quadratic term*, the term of degree 1 is called the *linear term*, and the numerical term is called the *constant term*.

EXAMPLES

Quadratic polynomial	Quadratic term	Linear term	Constant term
$y^2 - 3y + 7$	y^2	$-3y$	7
$2m^2 - 5$	$2m^2$	none	-5
$-3x^2$	$-3x^2$	none	none

248   CHAPTER 7

Factoring a quadratic polynomial is the "reverse" process of multiplying two binomials. Since
$$(x + 2)(x + 5) = x^2 + 7x + 10,$$
you know that $x^2 + 7x + 10$ can be factored as $(x + 2)(x + 5)$. Notice that the sum of 2 and 5 is 7, and their product is 10.

In order to factor a quadratic polynomial in which the coefficient of the quadratic term is 1, you use the fact that
$$(x + a)(x + b) = x^2 + (a + b)x + ab.$$
That is, you can factor the polynomial if you can find two integers whose sum is the coefficient of the linear term and whose product is the constant term.

EXAMPLE Factor $x^2 + 5x + 6$.

SOLUTION 6 can be factored as follows:
$$6 = 1 \cdot 6 \qquad 6 = (-1)(-6)$$
$$6 = 2 \cdot 3 \qquad 6 = (-2)(-3)$$
Only one choice will give 5 as the sum of the factors. That choice is $6 = 2 \cdot 3$. Therefore,
$$x^2 + 5x + 6 = (x + 2)(x + 3)$$

EXAMPLE Factor $y^2 + 3y - 10$.

SOLUTION -10 can be factored as follows:
$$-10 = 1(-10) \qquad -10 = -1(10)$$
$$-10 = 2(-5) \qquad -10 = -2(5)$$
The factors whose sum is 3 are -2 and 5. Therefore,
$$y^2 + 3y - 10 = (y - 2)(y + 5)$$

Factoring a quadratic polynomial in which the coefficient of the quadratic term is not 1 requires more work, since you must also consider the factors of the quadratic term. The example below illustrates the thought process. As you acquire skill, you will be able to omit many of these steps.

EXAMPLE Factor $2x^2 + 11x + 5$.

SOLUTION Possible factors of $2x^2$: $\quad x \cdot 2x$
Possible factors of 5: $\quad 5 \cdot 1$ or $(-5)(-1)$
Possible factors of $2x^2 + 11x + 5$: $\quad (x + 5)(2x + 1) \qquad (x - 5)(2x - 1)$
$\qquad\qquad\qquad\qquad\qquad\qquad\qquad\quad (x + 1)(2x + 5) \qquad (x - 1)(2x - 5)$

Since the middle term, $+11x$, is positive, we know that neither choice on the right is correct. Of the other choices, only the first one gives $+11x$ for the middle term. Therefore, $2x^2 + 11x + 5 = (x + 5)(2x + 1)$.

ORAL EXERCISES

State the possible factors of the constant term for each of the following quadratic trinomials.

1. $x^2 + 4x + 3$
2. $y^2 + 6y + 5$
3. $z^2 + 2z - 3$
4. $t^2 + 4t - 5$
5. $s^2 - 5s + 6$
6. $c^2 - 5c + 6$
7. $n^2 - 3n + 2$
8. $h^2 + 7h + 12$

9-16. Factor the polynomials in Exercises 1-8.

WRITTEN EXERCISES

Factor.

A 1. $x^2 + 7x + 6$
2. $y^2 + 9y + 8$
3. $z^2 + 6z - 7$
4. $r^2 + 4r - 5$
5. $s^2 + 6s + 8$
6. $t^2 + 9t + 18$
7. $h^2 + 7h - 18$
8. $g^2 + 5g - 24$
9. $p^2 + 4p + 4$
10. $q^2 - 6q + 9$
11. $u^2 - u - 30$
12. $v^2 - 3v - 28$
13. $a^2 - 9a + 20$
14. $d^2 - 10d + 24$
15. $x^2 + 10x + 25$
16. $y^2 - 8y + 16$
17. $z^2 - 9$
18. $c^2 - 25$
19. $m^2 + 7m - 78$
20. $n^2 - 4n - 77$

Complete.

21. $6x^2 + 5x + 1 = (\underline{\;?\;}x + 1)(2x + \underline{\;?\;})$
22. $20y^2 - 19y + 3 = (5y - \underline{\;?\;})(\underline{\;?\;}y - 3)$
23. $12z^2 + 13z + 3 = (\underline{\;?\;}z + 1)(4z + \underline{\;?\;})$
24. $15t^2 - 26t + 8 = (3t - \underline{\;?\;})(\underline{\;?\;}t - 2)$
25. $21p^2 + 29p - 10 = (\underline{\;?\;}p - 2)(3p + \underline{\;?\;})$
26. $2q^2 - 7q - 22 = (q + \underline{\;?\;})(\underline{\;?\;}q - 11)$
27. $9k^2 + 42k + 49 = (\underline{\;?\;}k + 7)^2$
28. $25h^2 + 20h + 4 = (5h + \underline{\;?\;})^2$
29. $9s^2 - 24s + 16 = (\underline{\;?\;}s - 4)^2$
30. $4u^2 - 4u + 1 = (2u - \underline{\;?\;})^2$
31. $4n^2 - 49 = (\underline{\;?\;}n + 7)(2n - \underline{\;?\;})$
32. $9m^2 - 100 = (3m + \underline{\;?\;})(\underline{\;?\;}m - 10)$

EXAMPLE Factor $6y^2 + 3y - 9$.

SOLUTION $\quad 6y^2 + 3y - 9 = 3(2y^2 + y - 3)$
$\qquad\qquad\qquad\qquad = 3(2y + 3)(y - 1)$

Factor.

B 33. $4x^2 + 6x + 2$
34. $12y^2 + 11y + 2$
35. $15z^2 + 26z + 8$
36. $12t^2 + 20t + 3$
37. $6k^2 - k - 1$
38. $20m^2 - 6m - 2$
39. $2n^2 - 4n - 6$
40. $3h^2 - 9h - 12$
41. $10a^2 - 7a + 1$
42. $21c^2 - 10c + 1$
43. $8d^2 - 26d + 15$
44. $18g^2 - 26g + 8$
45. $9p^2 - 1$
46. $16q^2 - 1$
47. $25u^2 - 4$
48. $36v^2 - 49$
49. $9x^2 + 6x + 1$
50. $25y^2 - 10y + 1$
51. $4z^2 + 28z + 49$
52. $16s^2 - 24s + 9$

EXAMPLE Factor $2z^4 - 7z^2 + 5$.

SOLUTION $\quad 2z^4 - 7z^2 + 5 = (2z^2 - 5)(z^2 - 1) = (2z^2 - 5)(z + 1)(z - 1)$

Factor.

C 53. $x^4 + 3x^2 + 2$
54. $6z^4 + 13z^2 + 6$
55. $8t^4 + 22t^2 + 15$
56. $18g^6 - 3g^3 - 10$
57. $21s^8 - 5s^4 - 6$
58. $25t^6 + 10t^3 + 1$
59. $9u^4 - 25$
60. $100v^6 - 36$
61. $y^4 - 6y^2 + 5$
62. $16h^4 - 8h^2 + 1$

Section 7-10 SPECIAL PRODUCTS AND FACTORS

There are some products and factors that follow special patterns which occur quite frequently in mathematics. For example, when you multiply the sum and difference of two terms or numbers, you always arrive at a binomial which is the difference of two squares. In general, the pattern is:

$$(a + b)(a - b) = a^2 - b^2 \quad \text{or} \quad a^2 - b^2 = (a + b)(a - b)$$

POLYNOMIALS AND FACTORING

EXAMPLES

$$(x + 3)(x - 3) = x^2 - 9, \text{ or } x^2 - 3^2$$
$$(2y + 1)(2y - 1) = 4y^2 - 1, \text{ or } (2y)^2 - 1^2$$
$$(50 + 2)(50 - 2) = 50^2 - 2^2$$
$$z^2 - 16 = z^2 - 4^2 = (z + 4)(z - 4)$$
$$9t^2 - 4 = (3t)^2 - 2^2 = (3t + 2)(3t - 2)$$
$$10^2 - 1 = 10^2 - 1^2 = (10 + 1)(10 - 1)$$

This special pattern is useful in developing skill in performing mental multiplications.

EXAMPLES

$$22 \cdot 18 = (20 + 2)(20 - 2) = 20^2 - 2^2 = 400 - 4 = 396$$
$$53 \cdot 47 = (50 + 3)(50 - 3) = 50^2 - 3^2 = 2500 - 9 = 2491$$
$$296 \cdot 304 = (300 - 4)(300 + 4) = 300^2 - 4^2 = 90{,}000 - 16 = 89{,}984$$

Two other special patterns arise when multiplying binomials. One involves the square of a sum of two terms and the other involves the square of a difference of two terms.

$$(a + b)^2 = a^2 + 2ab + b^2 \quad \text{or} \quad a^2 + 2ab + b^2 = (a + b)^2$$
$$(a - b)^2 = a^2 - 2ab + b^2 \quad \text{or} \quad a^2 - 2ab + b^2 = (a - b)^2$$

EXAMPLES

$$(x + 3)^2 = x^2 + 2 \cdot x \cdot 3 + 3^2$$
$$= x^2 + 6x + 9$$
$$(y - 7)^2 = y^2 - 2 \cdot y \cdot 7 + 7^2$$
$$= y^2 - 14y + 49$$
$$n^2 + 10n + 25 = n^2 + 2 \cdot n \cdot 5 + 5^2$$
$$= (n + 5)^2$$
$$t^2 - 18t + 81 = t^2 - 2 \cdot t \cdot 9 + 9^2$$
$$= (t - 9)^2$$

These special patterns can be helpful in squaring numbers.

EXAMPLES

$$23^2 = (20 + 3)^2 \qquad\qquad 57^2 = (60 - 3)^2$$
$$= 20^2 + 2 \cdot 20 \cdot 3 + 3^2 \qquad = 60^2 - 2 \cdot 60 \cdot 3 + 3^2$$
$$= 400 + 120 + 9 \qquad\qquad = 3600 - 360 + 9$$
$$= 529 \qquad\qquad\qquad\quad = 3249$$

ORAL EXERCISES

Multiply using the technique shown in the examples at the bottom of page 252, as indicated in Exercises 1, 2, 7, and 8.

1. $19 \cdot 21 = (20 - 1)(20 + 1)$
2. $33 \cdot 27 = (30 + 3)(30 - 3)$
3. $18 \cdot 22$
4. $26 \cdot 14$
5. $35 \cdot 25$
6. $41 \cdot 59$
7. $15^2 = (10 + 5)^2$
8. $28^2 = (30 - 2)^2$
9. 19^2
10. 26^2
11. 44^2
12. 38^2

Factor.

13. $x^2 - 1$
14. $d^2 - 9$
15. $a^2 - 4$
16. $t^2 - 100$
17. $4s^2 - 1$
18. $16r^2 - 9$
19. $y^2 + 2y + 1$
20. $x^2 + 4x + 4$
21. $z^2 - 2z + 1$
22. $p^2 - 4p + 4$
23. $q^2 - 6q + 9$
24. $h^2 - 10h + 25$
25. $d^2 + 8d + 16$
26. $k^2 + 14k + 49$

WRITTEN EXERCISES

Factor.

A
1. $x^2 - 64$
2. $y^2 - 25$
3. $z^2 - 81$
4. $4t^2 - 9$
5. $16q^2 - 1$
6. $100g^2 - 49$
7. $144h^2 - 121$
8. $81k^2 - 400$
9. $z^2 + 20z + 100$
10. $s^2 + 12s + 36$
11. $p^2 - 18p + 81$
12. $q^2 - 16q + 64$
13. $k^2 - 22k + 121$
14. $d^2 - 30d + 225$

Multiply.

15. $(x + 2)(x - 2)$
16. $(y - 5)(y + 5)$
17. $(2z + 1)(2z - 1)$
18. $(4q - 3)(4q + 3)$
19. $(5t - 4)(5t + 4)$
20. $(8p - 7)(8p + 7)$
21. $(x + 4)^2$
22. $(y - 5)^2$
23. $(z + 3)^2$
24. $(s - 7)^2$
25. $(5g + 2)^2$
26. $(8s - 3)^2$
27. $(5h + 11)^2$
28. $(7n - 6)^2$

POLYNOMIALS AND FACTORING

EXAMPLE Factor $3x^2 - 12$.

SOLUTION $3x^2 - 12 = 3(x^2 - 4)$
$= 3(x + 2)(x - 2)$

Factor.

B 29. $3x^2 - 75$ 30. $2d^2 - 162$
31. $6y^2 - 54$ 32. $4t^2 - 64$
33. $20n^2 - 45$ 34. $72x^2 - 128$
35. $3s^2 + 6s + 3$ 36. $4x^2 - 24x + 36$
37. $5n^2 + 40n + 80$ 38. $13n^2 - 52n + 52$
39. $9s^2 + 54s + 81$ 40. $15s^2 - 90s + 135$

C 41. $4y^2 + 4y + 1$ 42. $25x^2 + 20x + 4$
43. $144y^2 - 240y + 100$ 44. $x^2y^4 + 6xy^2 + 9$

Section 7-11 SOLVING QUADRATIC EQUATIONS

In Chapter 3 you learned how to solve equations of degree one. Such equations are called *linear equations*. In order to solve equations of degree two, called *quadratic equations*, you need to learn some new techniques.

EXAMPLES

Linear equations	*Quadratic equations*
$x = 6$	$x^2 + 4x + 4 = 0$
$y + 3 = 4$	$4y^2 - 13 = 3$
$2t - 5 = 0$	$s^2 = 2s$

Linear and quadratic equations fit the following general forms, where a, b, and c are real numbers and $a \neq 0$.

Linear equation: $ax + b = 0$

Quadratic equation: $ax^2 + bx + c = 0$

A member of the solution set of a quadratic equation is called a *root* of the equation. Some quadratic equations can be solved by factoring. The basic tool for solving such quadratic equations is the zero-product property of real numbers.

For all real numbers a and b, $ab = 0$ if and only if $a = 0$ or $b = 0$.

To solve a quadratic equation by factoring, you proceed as follows:
1. Write the quadratic equation in the form $ax^2 + bx + c = 0$, in which the right side of the equation is zero.
2. Factor the left side of the equation.
3. Set each factor equal to zero and solve the resulting linear equations.
4. Check the roots in the original quadratic equation.

EXAMPLE Solve $x^2 - x = 6$.

SOLUTION
1. Write in the form $ax^2 + bx + c = 0$: $x^2 - x - 6 = 0$
2. Factor: $x^2 - x - 6 = 0$ in factored form is
$$(x - 3)(x + 2) = 0.$$
3. Set each factor equal to zero and solve:
$$x - 3 = 0 \quad \text{or} \quad x + 2 = 0$$
$$x = 3 \qquad\qquad x = -2$$
4. Check the solutions:

$3^2 - 3 \;?\; 6 \qquad\qquad (-2)^2 - (-2) \;?\; 6$
$9 - 3 \;?\; 6 \qquad\qquad\quad 4 + 2 \;?\; 6$
$6 = 6 \qquad\qquad\qquad\quad 6 = 6$

The solution set of $x^2 - x = 6$ is $\{-2, 3\}$.

EXAMPLE Solve $2x^2 + 11x + 5 = 0$

SOLUTION
1. $2x^2 + 11x + 5 = 0$
2. $(x + 5)(2x + 1) = 0$
3. $x + 5 = 0 \quad \text{or} \quad 2x + 1 = 0$
$\quad\; x = -5 \quad \text{or} \quad\;\; 2x = -1$
$\qquad\qquad\qquad\qquad\qquad x = -\dfrac{1}{2}$
4. $2(-5)^2 + 11(-5) + 5 \;?\; 0 \qquad 2\left(-\dfrac{1}{2}\right)^2 + 11\left(-\dfrac{1}{2}\right) + 5 \;?\; 0$
$\qquad\quad 50 - 55 + 5 = 0 \qquad\qquad\quad \dfrac{1}{2} - \dfrac{11}{2} + 5 = 0$

The solution set of $2x^2 + 11x + 5 = 0$ is $\left\{-5, -\dfrac{1}{2}\right\}$.

In order to find the dimensions of the swimming pool described at the beginning of this chapter, you need to solve the equation $6w^2 + 30w = 36{,}000$. First write the equation in the form $6w^2 + 30w - 36{,}000 = 0$. Although $6w^2 + 30w - 36{,}000$ can be factored, it might take you a very long time to discover that
$$6w^2 + 30w - 36{,}000 = 6(w + 80)(w - 75),$$
and therefore the width of the pool is 75 m and the length is 480 m.

In your future study of algebra, you will develop a formula that will enable you to solve any quadratic equation quickly.

POLYNOMIALS AND FACTORING

ORAL EXERCISES

What is the solution set for the *factored* quadratic equation?

1. $(x - 1)(x - 2) = 0$
2. $y(y - 3) = 0$
3. $(r + 3)(r - 4) = 0$
4. $(s - 5)(s + 8) = 0$
5. $(t + 2)(t + 6) = 0$
6. $(q + 4)(q + 11) = 0$
7. $2(p - 2)(p + 3) = 0$
8. $3(h - 1)(h + 3) = 0$

WRITTEN EXERCISES

Solve.

A
1. $x^2 - 2x + 1 = 0$
2. $y^2 - 4y + 4 = 0$
3. $z^2 - 1 = 0$
4. $h^2 - 16 = 0$
5. $x^2 - 3x + 2 = 0$
6. $y^2 - 3y = 0$
7. $r^2 + r - 12 = 0$
8. $s^2 - 3s - 40 = 0$
9. $n^2 + 9n + 14 = 0$
10. $m^2 + 14m + 40 = 0$
11. $h^2 - h - 6 = 0$
12. $t^2 + t - 2 = 0$
13. $p^2 + 3p = 10$
14. $q^2 - 8q = 9$
15. $c^2 - 5c + 6 = 0$
16. $d^2 - 8d + 12 = 0$

B
17. $6a^2 + 13a + 6 = 0$
18. $12d^2 + 26d + 10 = 0$
19. $30x^2 + 28x + 6 = 0$
20. $12y^2 + 57y + 63 = 0$
21. $6a^2 + 5a - 6 = 0$
22. $12d^2 - 14d - 10 = 0$
23. $8y^2 - 26y + 15 = 0$
24. $10z^2 - 23z + 12 = 0$
25. $x^2 - 2x = 168$
26. $y^2 + 4y = 396$
27. $p^2 - 34p + 285 = 0$
28. $q^2 - 38q + 336 = 0$
29. $50k^2 - 20k = 6$
30. $48h^2 - 12h = 6$
31. $n^2 - 90n + 2000 = 0$
32. $w^2 + 5w - 6000 = 0$

Write a quadratic equation with the given solution set. Use x for the variable.

EXAMPLE $\{-4, 7\}$

SOLUTION
$x = -4$ or $x = 7$
$x + 4 = 0$ or $x - 7 = 0$
$(x + 4)(x - 7) = 0$
$x^2 - 3x - 28 = 0$

33. $\{1, 2\}$
34. $\{3, 4\}$
35. $\{-1, 5\}$
36. $\{-3, 2\}$
37. $\{-7, 6\}$
38. $\{-2, 8\}$
39. $\{-9, -4\}$
40. $\{-10, -3\}$

EXAMPLE Solve $x^3 + x^2 - 2x = 0$.

SOLUTION
1. $x^3 + x^2 - 2x = 0$
2. $x(x^2 + x - 2) = 0$
   $x(x + 2)(x - 1) = 0$
3. $x = 0$ or $x + 2 = 0$ or $x - 1 = 0$
   $\phantom{x = 0 \text{ or }}$ $x = -2$ $\phantom{\text{ or }}$ $x = 1$
4. $0^3 + 0^2 - 2 \cdot 0 = 0$ $(-2)^3 + (-2)^2 - 2(-2) = 0$ $1^3 + 1^2 - 2 \cdot 1 = 0$

The solution set of $x^3 + x^2 - 2x = 0$ is $\{-2, 0, 1\}$.

Solve.

C
41. $x^3 + x^2 - 6x = 0$
42. $y^3 - 2y^2 - 15y = 0$
43. $z^3 + 2z^2 = 24z$
44. $h^3 = 8h^2 - 15h$
45. $t^4 - 5t^2 + 4 = 0$
46. $p^4 - 10p^2 + 9 = 0$
47. $16x^4 - 8x^2 + 1 = 0$
48. $4x^4 - 17x^2 + 4 = 0$

Section 7-12 SOLVING PROBLEMS WITH QUADRATIC EQUATIONS

In this section you will learn how to solve problems involving quadratic equations. Remember that you should always check a possible solution to make sure it fulfills the requirements stated in the problem.

EXAMPLE The length of a rectangle is 2 m greater than the width and the area is 63 m². What are the dimensions of the rectangle?

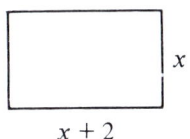

DISCUSSION The basic facts are:
(1) The length is 2 m greater than the width.
(2) The area is 63 m².
The information that isn't known is the length and width of the rectangle.

SOLUTION Let x = the width (in meters).
Then $x + 2$ = the length (in meters).
Since the length times the width equals the area, which is 63 m², we can write:
$$x(x + 2) = 63$$
$$x^2 + 2x = 63$$
$$x^2 + 2x - 63 = 0$$
$$(x + 9)(x - 7) = 0$$
Then $x + 9 = 0$ or $x - 7 = 0$
$\phantom{\text{Then }}$ $x = -9$ or $x = 7$

Check   Since the width cannot be a negative number, we discard -9. If the width is 7 m, then the length, $x + 2$, is 9 m and the area is 7 m × 9 m, or 63 m². This checks with the conditions of the problem. Thus the dimensions of the rectangle are 7 m and 9 m.

EXAMPLE The product of two consecutive positive integers is 20. What are the numbers?

SOLUTION Let n = the smaller integer.
$n + 1$ = the next consecutive integer.
Then $n(n + 1) = 20$
$n^2 + n = 20$
$n^2 + n - 20 = 0$
$(n + 5)(n - 4) = 0$
$n + 5 = 0$ or $n - 4 = 0$
$n = -5$ $n = 4$
$n + 1 = -4$ $n + 1 = 5$

Check   Since the problem asks for positive integers, we discard -5. If 4 is the smaller integer, then $4 + 1$, or 5, is the next consecutive integer. Their product, $4 \cdot 5$, is 20. This checks with the conditions of the problem. Thus the numbers are 4 and 5.

EXAMPLE Jack plans to make a pan by cutting squares from each corner of a 14 cm by 16 cm sheet of tin and then folding up the sides. The bottom of the pan is to have an area of 120 cm². What is the length of the side of each square that Jack should cut out?

SOLUTION Let s = length of side of each square (in centimeters).
Then $16 - 2s$ = length of pan (in centimeters).
$14 - 2s$ = width of pan (in centimeters).
Then $(16 - 2s)(14 - 2s) = 120$
$224 - 60s + 4s^2 = 120$
$4s^2 - 60s + 104 = 0$
$s^2 - 15s + 26 = 0$
$(s - 13)(s - 2) = 0$
Then $s - 13 = 0$ or $s - 2 = 0$
$s = 13$ $s = 2$

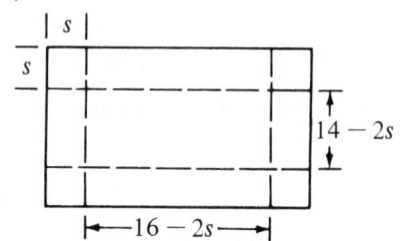

Check   It is impossible to cut four squares, each having a side 13 cm long, from a 14 cm by 16 cm sheet, so we discard 13. If the side of each square is 2 cm long, then the area of the bottom of the pan will be $(16 - 2 \cdot 2)(14 - 2 \cdot 2) = 12 \cdot 10$, or 120 cm². This checks with the conditions of the problem. Thus the length of the side of each square should be 2 cm.

ORAL EXERCISES

1. Why is it important that you check your answer to a problem with the facts of the original problem?

Using the variable x, tell how you would represent the information asked for.

2. The product of two consecutive numbers is 90. What are the numbers?
3. The length of a rectangle is 15 m longer than the width. The area is 100 m². What are the dimensions of the rectangle?
4. The area of a square is 36 m². What is the length of each side?
5. The sum of a number plus its square is 56. What is the number?

Make up a problem dealing with the given topic that you think can be solved with a quadratic equation.

6. a square and its area
7. a rectangle and its area
8. consecutive integers

WRITTEN EXERCISES

Solve by writing a quadratic equation. Check your answer.

A
1. The product of two consecutive positive integers is 42. Find the integers.
2. The product of two consecutive positive integers is 110. Find the integers.
3. The product of two consecutive negative integers is 56. Find the integers.
4. The product of two consecutive negative integers is 132. Find the integers.
5. The product of two consecutive positive even integers is 48. Find the integers.
6. The product of two consecutive negative even integers is 224. Find the integers.
7. The product of two consecutive positive odd integers is 35. Find the integers.
8. The product of two consecutive negative odd integers is 143. Find the integers.
9. The length of a rectangle is 2 m greater than the width, and the area is 35 m². Find the dimensions of the rectangle.
10. The length of a rectangle is 5 m greater than the width, and the area is 24 m². Find the dimensions of the rectangle.
11. The width of a rectangle is 3 m less than the length. If the area is 88 m², find the dimensions of the rectangle.
12. If the width of a rectangle is 4 m less than the length and the area is 96 m², find the dimensions of the rectangle.
13. Mary is five years older than Beth. The product of their ages is 150. How old is Beth?
14. Jack is four years younger than Charles. The product of their ages is 357. How old is Jack?

POLYNOMIALS AND FACTORING

B 15. The sum of the squares of two consecutive positive integers is 85. What are the integers?

16. The sum of the squares of three consecutive positive integers is 110. What are the integers?

17. The sum of the squares of two consecutive positive even integers is 340. What are the integers?

18. The sum of the squares of two consecutive negative odd integers is 514. What are the integers?

19. Four squares are cut from the corners of a 20 cm by 24 cm rectangular sheet of tin. The sides are then bent up to make a pan. What is the length of each side of the squares if the area of the bottom of the pan is to be 192 cm²?

20. Repeat Exercise 19, but use a 16 cm by 20 cm rectangular sheet of tin.

C 21. The perimeter of a rectangle is 18 m and the area is 20 m². What are the dimensions of the rectangle?

22. Find three consecutive positive even integers such that the product of the two smaller exceeds the largest by 16.

23. If the second of three consecutive positive integers is added to the product of the first and third, the result is 109. Find the integers.

24. If the first of three consecutive positive integers is added to the product of the second and third, the result is 194. Find the integers.

SELF-TEST 3

Factor.

1. $x^2 + 4x + 3$
2. $y^2 + 3y - 10$ Section 7-9
3. $8h^2 + 2h - 1$
4. $10s^2 + 13s - 3$
5. $t^2 - 25$
6. $k^2 + 6k + 9$ Section 7-10

Multiply.

7. $(x + 2)(x - 2)$
8. $(3g - 5)^2$

Solve.

9. $m^2 + m - 12 = 0$
10. $t^2 - 9t + 20 = 0$ Section 7-11
11. $3h^2 - 14h - 5 = 0$
12. $6z^2 + z - 35 = 0$

Solve.

13. The product of two consecutive negative integers is 72. Find the integers.   Section 7-12

14. The sum of the squares of two consecutive positive odd integers is 130. Find the integers.

Computer Exercises

1. **a.** This program provides practice in finding the GCF of two positive integers. Copy and RUN the program.

```
10   READ A,B
20   PRINT "WHAT IS THE GCF OF";A;" AND";B;
30   INPUT X
40   FOR F=A TO 1 STEP -1
50   IF INT(A/F) <> A/F THEN 70
60   IF INT(B/F)=B/F THEN 80
70   NEXT F
80   IF X=F THEN 110
90   PRINT "SORRY, THE GCF IS";F
100  GOTO 10
110  PRINT "RIGHT!"
120  GOTO 10
130  DATA 12,18,11,36,54,15,84,120,8,32,300,65
140  END
```

b. To provide more practice, type a new line 130 listing different pairs of positive integers.

c. INT(Q) is defined to be the greatest integer less than or equal to Q. For example, INT(2.3) = 2, INT(3.9) = 3, and INT(12/2) = 6. The INT function can be used to test whether or not one number is a factor of another number.

$$\text{INT}(12/2) = 12/2, \text{ so 2 is a factor of 12.}$$
$$\text{INT}(12/5) \ne 12/5, \text{ so 5 is not a factor of 12.}$$

Write a program that determines whether or not one number is a factor of the other when you INPUT any two positive integers.

2. Copy and RUN the following program for the products given below.

```
10   PRINT "THIS PROGRAM WILL FIND THE COEFFICIENTS OF"
20   PRINT "    THE PRODUCT OF (AX + B) AND (CX + D)."
30   PRINT
40   PRINT "INPUT A, B, C, D";
50   INPUT A,B,C,D
60   PRINT "THE COEFFICIENTS OF THE PRODUCT ARE:"
70   PRINT "   ";A*C;"  ,  ";A*D+B*C;"  ,  ";B*D
80   END
```

a. $(x + 2)(x + 3)$ b. $(x + 7)(x + 7)$ c. $(2x + 5)(x + 6)$

d. $(x + 8)(3x + 5)$ e. $(3x - 4)(5x - 2)$ f. $(7x + 1)(3x - 6)$

POLYNOMIALS AND FACTORING

CHAPTER REVIEW

1. Write using exponents: $2 \cdot y \cdot y \cdot y \cdot y$ — Section 7-1

2. Simplify: $\left(-\frac{2}{3}\right)^3$

3. Find the positive factors of 30.

4. State the degree of the polynomial $2x^3 - 3xy + 7x$ and the coefficient of the term of greatest degree. — Section 7-2

Simplify and write in standard form.

5. $3t - 5t^5 + 3t^3 + 6 - 2t$
6. $3m - 2n + m - m^2 + 2n$

In Exercises 7-8, (a) add the two polynomials; (b) subtract the second polynomial from the first.

7. $7r^2 + 6r - 4,\ 8r^2 + 2$
8. $4a^2b^4 - 4ab^2 + 1,\ 10 - 3a^2b^4$ — Section 7-3

Multiply or divide.

9. $(-g)^8(-g)^2$
10. $(3x^2y^3)^2$
11. $(-c^3d)(6b^2cd^2)$ — Section 7-4

12. $-2n(3n^2 - 2n)$
13. $4(y+2)(y-3)$
14. $(2a-7)^2$ — Section 7-5

15. $\dfrac{(8gh)^{14}}{(8gh)^{16}}$
16. $\dfrac{-6r^7s^4}{42r^3s^2}$
17. $\dfrac{2w^6 - w^4 + 3w^2}{-w^3}$ — Section 7-6

18. Write the prime factorization of 135. — Section 7-7
19. Find the GCF of $12a^3b^5$ and $20a^3b^2$.

Factor.

20. $40x^5y^2 + 16xy^3$
21. $36b^3 + 12b^6 - 18b^2$

Multiply.

22. $(z+7)(z+5)$
23. $(4h-3)^2$
24. $(3d+2)(3d-5)$ — Section 7-8

Factor.

25. $s^2 - 12s - 45$
26. $8k^2 - 10k + 3$ — Section 7-9

27. $100m^2 - 49$
28. $q^2 - 26q + 169$ — Section 7-10

Solve.

29. $b^2 + 18b + 81 = 0$
30. $10r^2 - r = 2$ — Section 7-11

Solve.

31. The product of two consecutive negative even integers is 80. Find the integers. — Section 7-12

32. The length of a rectangle is three times the width. The area of the rectangle is 108 m². Find the dimensions of the rectangle.

CHAPTER TEST

1. Write the prime factorization of 168.
2. Write $3z^2 - 5z^4 + 15 - 3z$ in standard form. State the degree of the polynomial and the coefficient of the term of greatest degree.

Simplify.

3. $(-3w^2)(-w^3)$
4. $7a^3 - 9a + a^5 - 2a^3 + a$
5. $(d^4 - d - 12) - (d^2 + 6d - 16)$
6. $\dfrac{-56d^2e^3f^5}{-8d^4e^3f^2}$
7. $(13x - 4 + 12y) + (-7x + 7 - y)$
8. $(-n^4 \cdot n^2)^3$
9. $(-7t^2 + 4t - 10)(-2t^5)$
10. $(4y - 1)(3y + 5)$
11. $(-2ab^2)^2(8a^3b)$
12. $(m - 7)^2$
13. $(3v + 8)(3v - 8)$
14. $\dfrac{84y^8 - 60y^6 - 36y^3}{6y^3}$

Find the GCF.

15. 28 and 70
16. $24m^2n$ and $12m^3p$

Factor.

17. $x^2 + 6x$
18. $v^2 - 12v + 11$
19. $m^2 - 64$
20. $k^2 + 8k + 16$
21. $12s^2 - s - 6$
22. $u^2 + 3u - 54$

Find the solution set of each equation.

23. $m^2 + 7m = 0$
24. $r^2 = 10r - 25$
25. $5j^2 + 33j - 56 = 0$
26. $2x^2 - 72 = 0$

Solve.

27. The sum of two consecutive positive integers is 11 less than their product. Find the integers.
28. Each side of a square is $(3x - 1)$ cm long. The area of the square is 36 cm². Find the value of x.

CUMULATIVE REVIEW: CHAPTERS 1–7

Indicate the best answer by writing the appropriate letter.

1. If $t = 8$, then an angle whose measure is $9t + 18$ is what kind of angle?
   a. acute b. obtuse c. right d. straight

2. The graph at the right is the graph of the solution set of which inequality?
   a. $2 \leqslant -2z < 8$ b. $-2 < 3z - 5 \leqslant 7$
   c. $6 \geqslant -z + 2 > 1$ d. $1 > 3z + 4 \geqslant -8$

3. Simplify: $(-3)^2 \cdot \left(-\dfrac{1}{3}\right)^3$
   a. $-\dfrac{1}{9}$ b. $-\dfrac{1}{3}$ c. -3 d. -1

4. Simplify: $(a^2 b)^3 \div a^4$
   a. $a^{10} b^3$ b. ab^3 c. $\dfrac{b^3}{a^2}$ d. $a^2 b^3$

5. If $7k$ and $8k$ represent the measures of two complementary angles, find the value of k.
   a. 6 b. 90 c. 42 d. 12

6. The measures of the angles of a triangle are represented by $4x + 8$, $5x - 3$, and $-x + 63$. Find the measure of the largest angle.
   a. 64 b. 67 c. 14 d. 49

7. The inequality $x - \dfrac{3}{4} < \dfrac{1}{2}$ is equivalent to:
   a. $x < \dfrac{5}{4}$ b. $x > \dfrac{1}{4}$ c. $x < \dfrac{2}{3}$ d. $x < -\dfrac{1}{4}$

8. The equation $(x^2 - 3x) - (x^2 + x - 4) = 0$ is equivalent to:
   a. $-4x - 4 = 0$ b. $-4x + 4 = 0$ c. $-2x + 4 = 0$ d. $-2x - 4 = 0$

9. Express in lowest terms: $-\dfrac{3}{4} \div \dfrac{9}{8}$
   a. $-\dfrac{24}{36}$ b. $-\dfrac{15}{8}$ c. $-\dfrac{27}{32}$ d. $-\dfrac{2}{3}$

10. If $\triangle TOP \cong \triangle BIG$, which statement must be true?
    a. $\overline{TP} \cong \overline{GB}$ b. $m\angle T = m\angle G$ c. $\overline{OP} \cong \overline{BI}$ d. $\overline{LO} \cong \overline{LG}$

11. The degree of the polynomial $z^3 + 8z - z^3 + 5$ is:
    a. 8 b. 3 c. 1 d. 0

12. Multiply: $(8s - 7)(8s + 7)$
    a. $64s^2 + 49$ b. $64s^2 - 49$ c. $16s$ d. $64s^2 - 112s - 49$

13. Complete: If two parallel lines are cut by a transversal, then interior angles on the same side of the transversal must be:
    a. complementary b. supplementary c. congruent d. vertical

14. Multiply: $(7r - 5)(3r + 4)$
    a. $21r^2 + 13r - 20$
    b. $21r^2 - 13r - 20$
    c. $21r^2 - 43r - 20$
    d. $21r^2 + 43r - 20$

15. Which property is *not* always true for a parallelogram $ABCD$?
    a. $\angle A \cong \angle C$
    b. $\overline{BC} \cong \overline{AD}$
    c. $AC = BD$
    d. $\triangle ABC \cong \triangle CDA$

16. Find the area of the parallelogram shown.
    a. 26 cm²
    b. 16 cm²
    c. 40 cm²
    d. 32 cm²

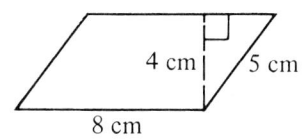

17. Which statement does *not* guarantee that $a \parallel b$?
    a. $m\angle 3 = m\angle 6$
    b. $m\angle 4 = m\angle 8$
    c. $\angle 5 \cong \angle 8$
    d. $\angle 2 \cong \angle 7$

18. Find the solution set of $2x = 3x$ if the domain of x is {integers}.
    a. $\left\{\dfrac{2}{3}\right\}$
    b. $\{0, 1\}$
    c. $\emptyset$
    d. $\{0\}$

19. Find the area of the trapezoid shown.
    a. 144 m²
    b. 56 m²
    c. 288 m²
    d. 192 m²

20. Find the roots of $8x^2 - 2x = 15$.
    a. $-\dfrac{5}{4}, \dfrac{3}{2}$
    b. $\dfrac{5}{4}, -\dfrac{3}{2}$
    c. $\dfrac{3}{4}, -\dfrac{5}{2}$
    d. $-\dfrac{3}{4}, \dfrac{5}{2}$

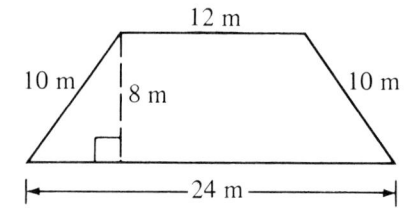

21. The measure of the vertex angle of an isosceles triangle is 54. Find the measures of the other two angles.
    a. 54, 72
    b. 63, 63
    c. 60, 66
    d. 65, 65

22. Which statement allows you to conclude that quadrilaterial $RSTU$ must be a parallelogram?
    a. $m\angle R + m\angle S = 180$
    b. $\overline{RS} \cong \overline{TU}$ and $\overline{ST} \parallel \overline{RU}$
    c. $\overline{RT} \cong \overline{SU}$ and $\angle S \cong \angle U$
    d. $\overline{RS} \cong \overline{TU}$ and $\overline{ST} \cong \overline{RU}$

23. Solve for a in terms of b, c, and x: $bx + a = c$
    a. $c - bx$
    b. $c + bx$
    c. $\dfrac{c - a}{b}, b \neq 0$
    d. $\dfrac{c - bx}{b}, b \neq 0$

24. Refer to the figure at the right. What method can you use to show that $\triangle ABC \cong \triangle FBD$?
    a. SSS
    b. SAS
    c. SSA
    d. ASA

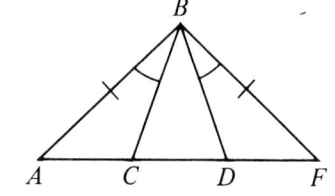

POLYNOMIALS AND FACTORING

25. The converse of $p \to \sim q$ is:
    a. $\sim q \to p$ b. $\sim p \to q$ c. $q \to \sim p$ d. $q \to p$

26. Let p represent "The quadrilateral contains exactly one pair of parallel sides," and let q represent "The quadrilateral contains at least one right angle." Which of the following is true if the quadrilateral is a rectangle?
    a. $p \wedge q$ b. p c. $\sim q$ d. $p \vee q$

27. If each side of a square is multiplied by 3, then the area of the square is multiplied by:
    a. $\frac{1}{9}$ b. $\frac{1}{3}$ c. 3   d. 9

28. Find the product of $4a^3$ and $9a^4$.
    a. $13a^7$ b. $13a^{12}$
    c. $36a^7$ d. $36a^{12}$

29. Find the lateral area of the regular square pyramid shown.
    a. 4320 cm²   b. 1476 cm²
    c. 1800 cm²   d. 324 cm²

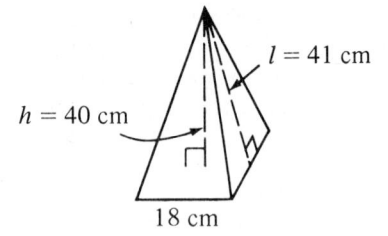

    $h = 40$ cm, $l = 41$ cm, 18 cm

30. The perimeter of a square is represented by $16x + 8$. Express the length of one side of the square in terms of x.
    a. $8x$ b. $4x$ c. $8x + 4$ d. $4x + 2$

31. Subtract $(3x - 2)$ from $(7 - 2x)$.
    a. $5x - 9$ b. $x - 5$ c. $x - 9$ d. $-5x + 9$

32. A right square prism is 4 units high. If its volume is 256 cubic units, find the length of each edge of the base.
    a. 8 units   b. 16 units   c. 64 units   d. $42\frac{2}{3}$ units

33. Solve: $-2(4 - z) \geq 14$
    a. $\{z : z \leq 11\}$ b. $\{z : z \geq 11\}$ c. $\{z : z \leq -11\}$ d. $\{z : z \geq -3\}$

34. Find the solution set of $-\frac{1}{2}(18r - 12) = 3(2 - 3r)$.
    a. {whole numbers}   b. {real numbers}   c. $\emptyset$ d. $\{0\}$

35. If $p \to q$ is false, which of the following must be false?
    a. $p \wedge \sim q$ b. $p \vee \sim q$ c. $p \wedge q$ d. $p \vee q$

36. How many members does {the odd whole numbers between -4 and 4} have?
    a. 7   b. 2   c. 4   d. 9

37. The inequality $9 - 8p \leq -2p + 33$ is equivalent to:
    a. $p \geq 4$ b. $p \leq 4$ c. $p \geq -4$ d. $p \leq -4$

38. If r represents "Today is Friday" and s represents "We have pizza for lunch," the inverse of "If today is not Friday, then we don't have pizza for lunch" can be written in symbolic form as
    a. $\sim s \to \sim r$ b. $\sim r \to \sim s$ c. $s \to r$ d. $r \to s$

CHAPTER 7

Biographical Note

Ada Byron Lovelace

Ada Byron Lovelace was the daughter of the English poet Lord Byron. Born in 1815, she married before she was twenty and died when she was only thirty-six. Her only published work was an annotated translation of a description of someone else's invention. Why, then, do we remember her as a contributor to the development of mathematical thinking?

The invention that Lovelace wrote about happened to be Charles Babbage's Analytical Engine, a forerunner of the computer. Because of her clear description of the method of giving information and instructions to the machine, Lovelace is now called the first computer programmer. A computer language named after her, called Ada, has been developed for a United States government agency. A general-purpose language based on Pascal, it is intended as a replacement for some of the specialized languages now in use.

Although Lovelace was a poet's daughter, her absorbing interest from childhood to the end of her life was not literature but mathematics. Even after she entered fashionable London society, she continued her studies. Fascinated by Babbage's first invention, the Difference Engine, she enthusiastically offered to put her mind at his disposal. Babbage accepted her help, and she worked with him for eighteen years.

During this time Babbage conceived the idea of a more advanced machine, the Analytical Engine. This machine was designed to get its instructions from punched cards and to include a memory. Although the Analytical Engine was never completed, Babbage's plans attracted considerable attention. In 1842, an Italian engineer wrote an article about the machine. It was translated by Lovelace, with detailed annotations of her own. One of her comments shows how clearly she understood the limitations as well as the power of a calculating machine: "The Analytical Engine has no pretensions whatever to originate anything. It can do whatever we know how to order it to perform. It can follow analysis, but it has no power of anticipating any analytical revelations or truths."

POLYNOMIALS AND FACTORING

VOCABULARY REVIEW

Be sure that you understand the meaning of these terms:

factorization, p. 225
power, p. 226
monomial, p. 228
coefficient, p. 228
polynomial, p. 229
binomial, p. 229
trinomial, p. 229
similar (like) terms, p. 229
simplified polynomial, p. 229

degree of a polynomial, p. 230
standard form of a polynomial, p. 230
prime factorization, p. 243
greatest common factor (GCF), p. 243
factor a polynomial, p. 243
quadratic polynomial, p. 248
linear equation, p. 254
quadratic equation, p. 254
root of an equation, p. 254

MIXED REVIEW

Multiply.

1. $(-3rs^3)^3(r^2s)$
2. $(5a - 1)(3a + 2)$
3. $(-2)^4 \cdot (-1)^3$

4. Which statement is logically equivalent to $\sim p \rightarrow \sim q$, $\sim q \vee p$ or $\sim p \vee q$?

5. A parking meter contains $28.80 in quarters and dimes. If it contains twice as many dimes as quarters, how many dimes does it contain?

6. Graph: {counting numbers}

Solve.

7. $9x^2 = 4$
8. $7y - 2 = 2y + \dfrac{1}{2}$
9. $6 - 0.3z \leqslant -9$
10. $-9 < 5t + 1 < 6$
11. $m^2 - 4m = -4$
12. $-\dfrac{20}{21} = -\dfrac{2}{7}d$

13. If $\triangle ABC \cong \triangle RST$, list the pairs of congruent corresponding parts.

14. Find the total area and the volume of a regular square pyramid with height 4, slant height 5, and base edge 6.

Exercises 15 and 16 refer to the diagram at the right, in which $r \parallel s$.

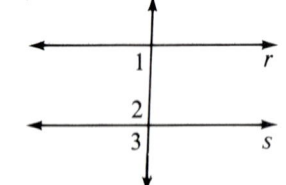

15. If $m\angle 1 = 17x + 3$ and $m\angle 2 = 20x - 8$, find the value of x.

16. If $m\angle 1 = 7y - 4$ and $m\angle 3 = 8y - 16$, find the value of y.

Exs. 15, 16

17. In $\triangle DEF$, $\overline{DE} \cong \overline{EF} \cong \overline{DF}$. Find the measure of $\angle E$.

18. In quadrilateral $WXYZ$, $\overline{WX} \cong \overline{YZ}$. Must $WXYZ$ be a parallelogram if $\overline{WX} \parallel \overline{YZ}$? if $\overline{XY} \parallel \overline{WZ}$? if $\overline{XY} \cong \overline{WZ}$?

19. The sum of the squares of two consecutive negative even integers is 164. Find the integers.

20. Find the area of a trapezoid with bases of 8 and 10 and height 9.

21. If *WXYZ* is a rhombus, must $\overline{WY}$ and $\overline{XZ}$ bisect each other? be congruent? be perpendicular?

22. Construct a truth table for the converse of $p \rightarrow q$.

23. If the domain of x is $\{-2, -1, 0, 1, 2\}$, find the solution set of $4 - \frac{3}{2}x \geq 2$.

Factor.

24. $-3x^3 + 6x$ 25. $y^2 - 9y + 8$ 26. $5z^2 + 7z - 6$

27. Draw a large triangle. Construct a triangle congruent to it. Name the method you used.

28. Each base of a right prism is a regular pentagon with sides of length 2. If the height of the prism is 6, find the lateral area.

29. If two planes intersect, describe their intersection.

30. The sum of three consecutive integers is less than 20. Find the greatest possible values for the three numbers.

Find the value of x.

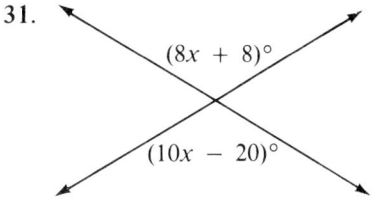

31.

$(8x + 8)°$

$(10x - 20)°$

32.

$x°$

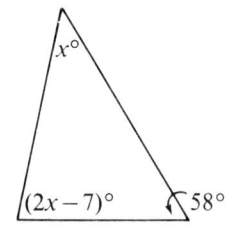

$(2x - 7)°$ $58°$

Translate each expression into symbols.

33. In twenty years, Marty's age will be three times what it is now.

34. The value in cents of $3x - 4$ quarters.

35. y is between -7 and 14.

36. A base angle of an isosceles triangle has measure 52. Find the measure of each of the other two angles.

37. Graph the solution set: $-3 \leq -2t + 5 < 7$.

POLYNOMIALS AND FACTORING

The square bricks in the photograph have been arranged in a circular pattern. The length of a side of a square with area 2 and the ratio of the circumference of a circle to its diameter are two well-known irrational numbers.

Rational and Irrational Numbers   8

Can you find the length of a side of each square shown below?

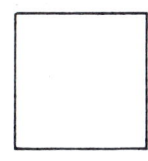

Area = 4

Area = $\frac{1}{4}$

Area = 2

Since $2 \cdot 2 = 4$, the length of each side of the square on the left is 2.
Since $\frac{1}{2} \cdot \frac{1}{2} = \frac{1}{4}$, the length of each side of the smallest square is $\frac{1}{2}$.
There is no integer or rational number whose product with itself is 2. However, there is such a real number, as you will discover in this chapter.

Section 8-1 WRITING RATIONAL NUMBERS AS DECIMALS

In this section you will learn about:
- terminating decimals
- repeating decimals
- repetends

In Chapter 1 you learned that a rational number is a number that can be expressed as the quotient of two integers.

EXAMPLES OF RATIONAL NUMBERS

$$3, \quad 0, \quad \frac{1}{9}, \quad -1\frac{1}{8}, \quad 0.5, \quad \frac{56}{13}$$

Each of the examples given on the preceding page can be expressed as the quotient of integers.

$$3 = \frac{3}{1}, \qquad 0 = \frac{0}{2}, \qquad -1\frac{1}{8} = \frac{-9}{8}, \qquad 0.5 = \frac{1}{2}$$

Sometimes you may need to express a rational number as a decimal. To do so, simply perform the indicated division.

EXAMPLE Express $\frac{8}{25}$ as a decimal.

SOLUTION
$$\begin{array}{r} 0.32 \\ 25\overline{)8.00} \\ \underline{7\ 5} \\ 50 \\ \underline{50} \\ 0 \end{array}$$

$$\frac{8}{25} = 0.32$$

EXAMPLE Express $2\frac{1}{4}$ as a decimal.

SOLUTION $2\frac{1}{4} = \frac{9}{4}$

$$\begin{array}{r} 2.25 \\ 4\overline{)9.00} \\ \underline{8} \\ 1\ 0 \\ \underline{8} \\ 20 \\ \underline{20} \\ 0 \end{array}$$

$$2\frac{1}{4} = 2.25$$

When there is eventually a remainder of 0, as above, the decimal is called a *terminating decimal*.

If the division does not result in a 0 remainder, it goes on indefinitely, and eventually a sequence of digits is repeated, as in the examples below. Such nonterminating decimals are called *repeating decimals*. The digits that repeat are known as the *repetend*.

EXAMPLE Express $\frac{2}{3}$ as a decimal.

SOLUTION
$$\begin{array}{r} 0.666 \\ 3\overline{)2.000} \\ \underline{1\ 8} \\ 20 \\ \underline{18} \\ 20 \\ \underline{18} \\ 2 \end{array}$$

$$\frac{2}{3} = 0.666\ldots$$

EXAMPLE Express $\frac{35}{22}$ as a decimal.

SOLUTION
$$\begin{array}{r} 1.59090 \\ 22\overline{)35.00000} \\ \underline{22} \\ 13\ 0 \\ \underline{11\ 0} \\ 2\ 00 \\ \underline{1\ 98} \\ 200 \\ \underline{198} \\ 2 \end{array}$$

$$\frac{35}{22} = 1.59090\ldots$$

The repetends in the examples just given are 6 and 90. One way to indicate a repetend is to use a bar over the digits that repeat:

$$\frac{2}{3} = 0.\overline{6} \qquad \frac{39}{22} = 1.5\overline{90}$$

> Every rational number can be expressed as either a terminating or a repeating decimal.

ORAL EXERCISES

Match each rational number with its decimal representation.

1. $\frac{3}{10}$ a. 0.03
2. $\frac{3}{100}$ b. 0.003
3. $\frac{3}{1000}$ c. 0.3

4. $\frac{61}{10}$ a. 0.0061
5. $\frac{61}{100}$ b. 6.1
6. $\frac{61}{10,000}$ c. 0.61

7. $\frac{2}{5}$ a. 0.125
8. $\frac{2}{500}$ b. 0.04
9. $\frac{1}{8}$ c. 0.4
10. $\frac{1}{80}$ d. 0.0125
         e. 0.004

11. $\frac{1}{3}$ a. $3.\overline{3}$
12. $\frac{1}{30}$ b. 0.3
13. $\frac{1}{300}$ c. $0.0\overline{3}$
14. $\frac{10}{3}$ d. $0.\overline{3}$
         e. $0.00\overline{3}$

Complete.

15. Since $\frac{7}{10} = 0.7$, $\frac{7}{1000} = \underline{\ ?\ }$.

16. Since $\frac{3}{40} = 0.075$, $\frac{3}{400} = \underline{\ ?\ }$.

17. Since $\frac{1}{6} = 0.1\overline{6}$, $\frac{1}{60} = \underline{\ ?\ }$.

18. Since $\frac{5}{33} = 0.\overline{15}$, $\frac{5}{330} = \underline{\ ?\ }$.

WRITTEN EXERCISES

Express as a quotient of integers.

A   1. 32   2. 0   3. $-2\frac{1}{2}$ 4. 5   5. 0.4   6. 3.2

RATIONAL AND IRRATIONAL NUMBERS

Write each rational number as a decimal.

7. $\frac{1}{4}$ 8. $\frac{3}{4}$ 9. $-\frac{7}{10}$ 10. $\frac{29}{10}$ 11. $-\frac{13}{8}$ 12. $-\frac{11}{40}$

13. $3\frac{1}{5}$ 14. $2\frac{3}{20}$ 15. $-\frac{2}{9}$ 16. $\frac{5}{3}$ 17. $-\frac{7}{11}$ 18. $\frac{20}{33}$

19. $\frac{8}{27}$ 20. $-\frac{19}{37}$ 21. $\frac{14}{33}$ 22. $\frac{3}{55}$ 23. $\frac{5}{22}$ 24. $\frac{11}{24}$

B 25. $\frac{1}{64}$ 26. $\frac{1}{101}$ 27. $\frac{50}{101}$ 28. $\frac{3}{7}$ 29. $\frac{1}{13}$ 30. $\frac{2}{39}$

Change all fractions to decimals. Then arrange the numbers in order from least to greatest.

31. $\frac{78}{100}$, $0.\overline{78}$, $0.7\overline{8}$ 32. $\frac{44}{100}$, $\frac{11}{20}$, $0.\overline{4}$ 33. $9.\overline{39}$, $0.\overline{93}$, $0.9\overline{39}$

34. $0.\overline{5}$, $\frac{1}{2}$, 0.556 35. $\frac{3}{4}$, 3.4, $0.\overline{34}$ 36. $\frac{16}{99}$, $\frac{4}{25}$, $\frac{1}{6}$

Section 8-2 CHANGING FROM DECIMAL TO FRACTIONAL FORM

In this section you will learn:
- how to change terminating decimals to fractional form
- how to change repeating decimals to fractional form

In earlier courses you learned how to read decimals and change them to fractions with a power of 10 as the denominator. Some examples are given below.

$$0.3 = \text{three tenths} = \frac{3}{10}$$

$$0.25 = \text{twenty-five hundredths} = \frac{25}{100}$$

$$0.005 = \text{five thousandths} = \frac{5}{1000}$$

When you need to convert a terminating decimal to its fractional form, simply follow the plan suggested by the examples above. The denominator is always a power of 10.

EXAMPLE Express as a fraction in lowest terms: (a) 0.05, (b) 1.004, (c) 0.0006

SOLUTION (a) $0.05 = \frac{5}{100} = \frac{1}{20}$

(b) $1.004 = \frac{1004}{1000} = \frac{251}{250}$

(c) $0.0006 = \frac{6}{10,000} = \frac{3}{5000}$

To change a repeating decimal to fractional form you may use the following method.

EXAMPLE

Express $0.\overline{123}$ as a fraction in lowest terms.

METHOD

1. Let N equal the decimal.
2. Multiply both sides by that power of 10 whose exponent is the number of digits in the repetend.
3. Subtract the first equation from the second.
4. Solve for N.

SOLUTION

1. $N = 0.\overline{123}$
2. (There are 3 digits in 123, so the multiplier will be 10^3, or 1000.)
   $1000 N = 123.\overline{123}$
3. $1000 N = 123.\overline{123}$
   $ 1 N = 0.123$
   $\overline{999 N = 123}$
4. $N = \dfrac{123}{999}$, or $\dfrac{41}{333}$

   $0.\overline{123} = \dfrac{41}{333}$

EXAMPLE Express $0.2\overline{47}$ as a fraction in lowest terms.

SOLUTION
1. $N = 0.2\overline{47}$
2. $100N = 24.7\overline{47}$
3. $1N = 0.2\overline{47}$
   $\overline{99N = 24.5}$
4. $990N = 245$

   $N = \dfrac{245}{990}$, or $\dfrac{49}{198}$

   $0.2\overline{47} = \dfrac{49}{198}$

(Since the repetend, 47, has 2 digits, the multiplier in Step 2 is 10^2, or 100.)

It is important to note the following:

> All terminating decimals and all repeating decimals can be expressed as the quotient of two integers.

ORAL EXERCISES

Read each of these decimals.

1. 0.3
2. 3.4
3. 0.01
4. 4.05
5. 0.005
6. 10.101
7. 0.0001
8. 5.0531

RATIONAL AND IRRATIONAL NUMBERS

What power of 10 is each number?

9. 100
10. 10,000
11. 10
12. 1000

When the decimal is written as a fraction, the denominator will be what power of 10?

13. 0.35
14. 0.01
15. 0.0009
16. 0.016
17. 0.19
18. 0.617
19. 0.7104
20. 0.7

WRITTEN EXERCISES

Express as a fraction in lowest terms.

A
1. 0.8
2. 0.4
3. 0.12
4. 0.26
5. 0.214
6. 0.617
7. $0.\overline{1}$
8. $2.\overline{8}$
9. $0.\overline{13}$
10. $5.\overline{63}$
11. $0.\overline{312}$
12. $0.\overline{147}$

B
13. $0.5\overline{7}$
14. $0.6\overline{23}$
15. $0.0\overline{9}$
16. $0.8\overline{97}$
17. $0.12\overline{15}$
18. $0.\overline{1215}$
19. $3(0.\overline{17})$
20. $(0.\overline{6})^2$

Replace the __?__ with <, =, or > to form a true sentence.

C
21. $0.\overline{3}$ __?__ $\frac{3}{8}$
22. $4(0.\overline{6})$ __?__ $2.\overline{4}$
23. $0.\overline{7} + 0.\overline{9}$ __?__ $1.\overline{6}$
24. $\frac{1}{0.\overline{5}}$ __?__ 1.8
25. 1 __?__ $0.\overline{63} + 0.\overline{36}$
26. $11(0.\overline{35})$ __?__ $3.\overline{53}$

SELF-TEST 1

Write each rational number as a decimal.

1. $\frac{5}{16}$
2. $\frac{1}{15}$
3. $\frac{2}{33}$

Section 8-1

Express as a fraction in lowest terms.

4. 0.325
5. $0.\overline{45}$
6. $0.2\overline{8}$

Section 8-2

Section 8-3 RATIONAL SQUARE ROOTS

In this section you will learn:
- the meaning of square root
- how to find the square roots of a perfect square

The inverse of squaring a number is finding its *square root*. Both 4^2 and $(-4)^2$ equal 16, so 16 has two square roots, 4 and −4. You can use a *radical sign*, $\sqrt{}$, to indicate the positive, or *principal*, square root:

$$\sqrt{16} = 4$$

276 CHAPTER 8

A negative sign before the radical sign indicates the negative square root:
$$-\sqrt{16} = -4$$

The number written under the radical sign, such as 16 above, is called the *radicand*. The entire expression, $\sqrt{16}$ or $-\sqrt{16}$, is called a *radical*.

Only nonnegative numbers have square roots that are real numbers. (Recall that the square of a real number cannot be negative.) The number zero has only one square root, which is 0.

Some numbers have square roots that are rational. These numbers are called *perfect squares*. You can find the square roots of a perfect square, either by sight or by factoring.

EXAMPLES

Perfect Square	Principal Square Root		Reason
81	9	$\sqrt{81} = 9$ because	$9 \cdot 9 = 81$
$\frac{4}{9}$	$\frac{2}{3}$	$\sqrt{\frac{4}{9}} = \frac{2}{3}$ because	$\frac{2}{3} \cdot \frac{2}{3} = \frac{4}{9}$
0.64	0.8	$\sqrt{0.64} = 0.8$ because	$0.8 \cdot 0.8 = 0.64$

Factoring is used to find the indicated square root in the following examples.

EXAMPLE Find $\sqrt{5625}$.

SOLUTION
$$5625 = 5 \cdot 1125$$
$$= 5 \cdot 5 \cdot 225$$
$$= 5 \cdot 5 \cdot 15 \cdot 15$$
$$= (5 \cdot 15) \cdot (5 \cdot 15)$$
$$= 75 \cdot 75$$
$$\sqrt{5625} = \sqrt{75 \cdot 75} = 75$$

EXAMPLE Find $-\sqrt{1521}$.

SOLUTION
$$1521 = 3 \cdot 507$$
$$= 3 \cdot 3 \cdot 169$$
$$= 3 \cdot 3 \cdot 13 \cdot 13$$
$$= (3 \cdot 13) \cdot (3 \cdot 13)$$
$$= 39 \cdot 39$$
$$-\sqrt{1521} = -\sqrt{39 \cdot 39} = -39$$

Numbers that are *not* perfect squares have square roots that are *not* rational numbers. In the next section you will learn more about these square roots.

ORAL EXERCISES

State the two square roots of each number.

1. 25
2. 4
3. 9
4. 64
5. 100
6. 36

State the principal square root.

7. 25
8. 49
9. $\frac{1}{4}$
10. $\frac{1}{81}$
11. $\frac{1}{100}$
12. $\frac{16}{25}$

Find the indicated square root.

13. $\sqrt{100}$
14. $-\sqrt{49}$
15. $-\sqrt{36}$
16. $\sqrt{121}$
17. $\sqrt{\frac{1}{16}}$
18. $-\sqrt{\frac{25}{36}}$

RATIONAL AND IRRATIONAL NUMBERS

True or false?

19. 20 is a perfect square.
20. 100 is a perfect square.
21. $\sqrt{(-3)^2} = -3$
22. In $\sqrt{64}$ the radicand is 64.
23. $\sqrt{49}$ indicates only one square root of 49.
24. 0 has no square root.

WRITTEN EXERCISES

Each radicand is a perfect square. Find the indicated square root.

A
1. $\sqrt{225}$
2. $\sqrt{625}$
3. $-\sqrt{196}$
4. $-\sqrt{484}$
5. $-\sqrt{1225}$
6. $\sqrt{441}$
7. $\sqrt{144}$
8. $-\sqrt{400}$
9. $-\sqrt{3025}$
10. $-\sqrt{2025}$
11. $\sqrt{4900}$
12. $\sqrt{2401}$
13. $-\sqrt{729}$
14. $\sqrt{1024}$
15. $\sqrt{1296}$
16. $-\sqrt{3600}$
17. $\sqrt{361}$
18. $-\sqrt{529}$
19. $-\sqrt{\frac{16}{81}}$
20. $\sqrt{\frac{64}{121}}$
21. $\sqrt{\frac{256}{225}}$
22. $-\sqrt{\frac{81}{10{,}000}}$
23. $-\sqrt{\frac{49}{900}}$
24. $-\sqrt{\frac{169}{6400}}$

B
25. $\frac{2}{3}\sqrt{324}$
26. $\frac{1}{4}\sqrt{576}$
27. $-\sqrt{961}$
28. $\sqrt{841}$
29. $-\sqrt{3721}$
30. $-\sqrt{2809}$
31. $\sqrt{7225}$
32. $-\sqrt{5476}$
33. $\sqrt{0.01}$
34. $\sqrt{0.16}$
35. $-\sqrt{0.0049}$
36. $\sqrt{0.0225}$

37. The areas of the two smaller squares shown are 25 cm² and 9 cm².
  a. Find the values of x and y.
  b. What is the area of the shaded portion of the large square?
  c. What is the area of the large square?
  d. What is the length of each side of the large square?

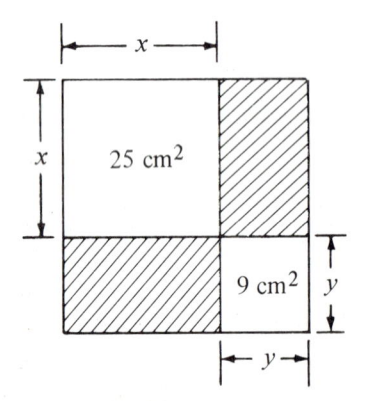

In Exercises 38-45, assume that all variables represent positive numbers. Find each indicated square root and then simplify.

C
38. $\sqrt{x^2}$
39. $-\sqrt{4x^2y^2}$
40. $-\sqrt{64y^4}$
41. $-\sqrt{225a^2}$
42. $\frac{1}{4}\sqrt{16x^2}$
43. $x\sqrt{49x^2}$
44. $\frac{2}{3}\sqrt{9x^6}$
45. $\frac{\sqrt{16a^2}}{\sqrt{25b^2}}$

46. The large square at the right has an area of 576 cm². The unshaded square in the middle has an area of 144 cm². What is the area of the shaded portion of the large square?

The *cube root* of a given number is the number whose cube is the given number. The radical sign $\sqrt[3]{}$ indicates a cube root.

EXAMPLES $\sqrt[3]{27} = 3$ since $3^3 = 3 \cdot 3 \cdot 3 = 27$.
$\sqrt[3]{-27} = -3$ since $(-3)^3 = (-3)(-3)(-3) = -27$.

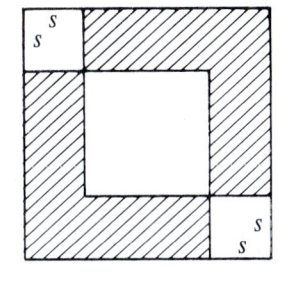

Find the cube root.

47. $\sqrt[3]{-8}$ 48. $\sqrt[3]{125}$ 49. $\sqrt[3]{64}$ 50. $\sqrt[3]{-1}$

51. If the volume of a cube is 1000 cm³, what is the length of each side of the cube?

Section 8-4 IRRATIONAL SQUARE ROOTS

In this section you will learn:
- the meaning of the term "irrational number"
- how to use a square root table
- a method for finding a decimal approximation of an irrational square root

In Chapter 1 you learned that every point on a number line is paired with a real number. So far you have worked with just one kind of real number, a rational number. Now you will learn about another kind of real number, an *irrational number*. The rational numbers together with the irrational numbers make up the entire set of real numbers.

Unlike rational numbers, irrational numbers *cannot* be expressed as the quotient of two integers; however, they can be approximated by decimals. Decimals for irrational numbers continue indefinitely without terminating or repeating.

> All real numbers can be expressed as decimals, and all decimals are real numbers.

EXAMPLES OF IRRATIONAL NUMBERS

0.121121112 ... ⎫
0.2468101214 ... ⎭ These are examples of decimals that exhibit patterns that do not repeat.

π (pi) Approximations for this familiar number include 3.14, 3.1416, and 3.14159. Computers have enabled mathematicians to approximate the value of π to a million decimal places.

$\sqrt{5} \approx 2.2361$ ⎫
$\sqrt{45} \approx 6.7082$ ⎬ Many square roots are irrational numbers.
$\sqrt{2352} \approx 48.4974$ ⎭ These have been rounded to four decimal places.
(The symbol "$\approx$" means "is approximately equal to.")

RATIONAL AND IRRATIONAL NUMBERS

The easiest way to find a *decimal approximation* for an irrational square root is to use a calculator or a table of square roots such as the one on page 281. A little practice will enable you to use this table quickly.

EXAMPLE Find $\sqrt{2}$ from the table on page 281.

SOLUTION Locate 2 in the column headed "Number," or "N," then move to the right and find 1.414 in the column headed "Positive Square Root," or "$\sqrt{N}$." $\sqrt{2} \approx 1.414$ rounded to three decimal places.

EXAMPLE Find $\sqrt{59}$ in the table on page 281 and round it to tenths.

SOLUTION From the table, $\sqrt{59} \approx 7.681$. Rounded to tenths, $\sqrt{59} \approx 7.7$.

It is a good idea to learn at least one method for approximating square roots. The one below is known as the divide-and-average method. As you will see by studying the example, you just keep averaging pairs of factors until a good approximation is reached.

EXAMPLE Find an approximation for $\sqrt{59}$ to the nearest thousandth.

METHOD

1. Make a good guess for a divisor that will approximate the quotient.

2. Divide the radicand by the divisor chosen in Step 1. Carry the quotient to two more digits than those in the divisor.

3. Find the average of the divisor and quotient. (The average of the divisor and quotient is their sum divided by 2.)

4. Use the average as a new divisor of the radicand.

SOLUTION

1. 59 lies between 7^2 and 8^2, or 49 and 64. It is nearer to 64 than to 49, so we take 7.6 as our first divisor.

2. $59 \div 7.6 \approx 7.763$

3. $\dfrac{7.6 + 7.763}{2} = \dfrac{15.363}{2} \approx 7.681$

4. $59 \div 7.681 \approx 7.68129$
   Since the divisor and quotient agree to three decimal places, $\sqrt{59} \approx 7.681$ to the nearest thousandth.

In this method, the approximation obtained is accurate to at least as many digits as match in the divisor and quotient. If you had been asked to find an approximation for $\sqrt{59}$ to the nearest hundredth, you could have stopped dividing 59 by 7.681 when you obtained the quotient 7.68.

SQUARE ROOTS OF INTEGERS FROM 1 TO 100

Decimal approximations for irrational square roots are rounded to three decimal places.

Number N	Positive Square Root $\sqrt{N}$	Number N	Positive Square Root $\sqrt{N}$	Number N	Positive Square Root $\sqrt{N}$	Number N	Positive Square Root $\sqrt{N}$
1	1	26	5.099	51	7.141	76	8.718
2	1.414	27	5.196	52	7.211	77	8.775
3	1.732	28	5.292	53	7.280	78	8.832
4	2	29	5.385	54	7.348	79	8.888
5	2.236	30	5.477	55	7.416	80	8.944
6	2.449	31	5.568	56	7.483	81	9
7	2.646	32	5.657	57	7.550	82	9.055
8	2.828	33	5.745	58	7.616	83	9.110
9	3	34	5.831	59	7.681	84	9.165
10	3.162	35	5.916	60	7.746	85	9.220
11	3.317	36	6	61	7.810	86	9.274
12	3.464	37	6.083	62	7.874	87	9.327
13	3.606	38	6.164	63	7.937	88	9.381
14	3.742	39	6.245	64	8	89	9.434
15	3.873	40	6.325	65	8.062	90	9.487
16	4	41	6.403	66	8.124	91	9.539
17	4.123	42	6.481	67	8.185	92	9.592
18	4.243	43	6.557	68	8.246	93	9.644
19	4.359	44	6.633	69	8.307	94	9.695
20	4.472	45	6.708	70	8.367	95	9.747
21	4.583	46	6.782	71	8.426	96	9.798
22	4.690	47	6.856	72	8.485	97	9.849
23	4.796	48	6.928	73	8.544	98	9.899
24	4.899	49	7	74	8.602	99	9.950
25	5	50	7.071	75	8.660	100	10

ORAL EXERCISES

Does the expression stand for a rational or an irrational number? Explain.

1. $\sqrt{17}$
2. $\sqrt{10}$
3. $\sqrt{11}$
4. $\sqrt{\frac{24}{6}}$
5. $\sqrt{7} - \sqrt{7}$
6. $\sqrt{3.3}$
7. $\sqrt{81}$
8. $\sqrt{10-4}$
9. $5.49\overline{2}$
10. $0.575575557\ldots$
11. 27.623
12. $10.1100111000\ldots$

WRITTEN EXERCISES

Use the table on page 281 to find an approximation for the square root to the nearest hundredth.

A
1. $\sqrt{21}$
2. $\sqrt{19}$
3. $\sqrt{82}$
4. $-\sqrt{15}$
5. $-\sqrt{54}$
6. $\sqrt{27}$
7. $-\sqrt{74}$
8. $-\sqrt{90}$
9. $\sqrt{24}$
10. $-\sqrt{79}$
11. $\sqrt{29}$
12. $\sqrt{80}$

Use the divide-and-average method to find an approximation for the square root to the nearest hundredth.

B
13. $\sqrt{39}$
14. $-\sqrt{129}$
15. $-\sqrt{228}$
16. $\sqrt{524}$
17. $-\sqrt{624}$
18. $-\sqrt{52.8}$
19. $\sqrt{19.7}$
20. $\sqrt{76.4}$

Rewrite the numbers in order from least to greatest.

C
21. $\sqrt{54}$, 7.33, 7.35
22. 8.74, 8.72, $\sqrt{76}$
23. 13.36, $\sqrt{179}$, 13.34
24. $-12.83, -12.85, -\sqrt{165}$
25. 21.23, $\sqrt{450}$, 21.22
26. $-27.99, -\sqrt{782}, -27.97$

A display shelf is $\frac{4}{9}$ as wide as it is long and has an area of 1296 cm². An interior decorator plans to cover it with square tiles. The tiles available have the areas given in Exercises 27–34. Is it possible to cover the shelf completely with tiles of the given area? If so, how many tiles would be needed? Only whole tiles are to be used.

27. 4 cm²
28. 9 cm²
29. 16 cm²
30. 36 cm²
31. 81 cm²
32. 144 cm²
33. 324 cm²
34. 1296 cm²

SELF-TEST 2

Each radicand is a perfect square. Find the indicated square root.

1. $\sqrt{900}$
2. $-\sqrt{4225}$
3. $\sqrt{784}$

Section 8-3

Use the table on page 281 to find an approximation for the square root to the nearest hundredth.

4. $-\sqrt{50}$
5. $\sqrt{98}$
6. $-\sqrt{67}$

Section 8-4

7. Use the divide-and-average method to find an approximation for $\sqrt{200}$ to the nearest hundredth.

Section 8-5 MULTIPLICATION AND DIVISION OF RADICALS

In this section you will learn how to:
- multiply square-root radicals
- divide square-root radicals

Sometimes you need to multiply or divide radicals. It is easy when the radicands are perfect squares.

$$\sqrt{4} \cdot \sqrt{9} = 2 \cdot 3 = 6$$

You arrive at the same result, 6, if you write

$$\sqrt{4} \cdot \sqrt{9} = \sqrt{4 \cdot 9} = \sqrt{36} = 6.$$

You can use the latter method even if the numbers are irrational. For example,

$$\sqrt{5} \cdot \sqrt{2} = \sqrt{10}$$

Likewise, for division, you can write

$$\frac{\sqrt{4}}{\sqrt{9}} = \frac{2}{3}$$

or

$$\frac{\sqrt{4}}{\sqrt{9}} = \sqrt{\frac{4}{9}} = \frac{2}{3}$$

In a similar manner,

$$\frac{\sqrt{2}}{\sqrt{5}} = \sqrt{\frac{2}{5}}$$

If a and b represent any two nonnegative real numbers, then:

$$\sqrt{a} \cdot \sqrt{b} = \sqrt{ab}$$

and

$$\frac{\sqrt{a}}{\sqrt{b}} = \sqrt{\frac{a}{b}} \quad \text{where } b \neq 0.$$

EXAMPLES

$$\sqrt{5} \cdot \sqrt{11} = \sqrt{5 \cdot 11} = \sqrt{55}$$
$$(\sqrt{7})^2 = \sqrt{7} \cdot \sqrt{7} = \sqrt{7 \cdot 7} = \sqrt{49} = 7$$
$$\sqrt{2.7} \cdot \sqrt{0.3} = \sqrt{(2.7)(0.3)} = \sqrt{0.81} = 0.9$$
$$\frac{\sqrt{69}}{\sqrt{3}} = \sqrt{\frac{69}{3}} = \sqrt{23}$$

If the numbers contain both rational and irrational factors, we simply use the commutative and associative properties as illustrated on the following page.

EXAMPLES

$$2\sqrt{7} \cdot 3\sqrt{5} = 2 \cdot 3 \cdot \sqrt{7} \cdot \sqrt{5}$$
$$= 6 \cdot \sqrt{7 \cdot 5}$$
$$= 6\sqrt{35}$$

$$\frac{20\sqrt{21}}{4\sqrt{3}} = \frac{20}{4} \cdot \frac{\sqrt{21}}{\sqrt{3}}$$
$$= 5 \cdot \sqrt{\frac{21}{3}}$$
$$= 5\sqrt{7}$$

$$(3\sqrt{2})^2 = 3\sqrt{2} \cdot 3\sqrt{2}$$
$$= 3 \cdot 3 \cdot \sqrt{2} \cdot \sqrt{2}$$
$$= 9 \cdot \sqrt{2 \cdot 2} = 9 \cdot \sqrt{4} = 9 \cdot 2 = 18$$

ORAL EXERCISES

Perform the indicated operations. Leave irrational numbers in radical form.

1. $\sqrt{4} \cdot \sqrt{16}$
2. $\sqrt{5} \cdot \sqrt{3}$
3. $\sqrt{11} \cdot \sqrt{7}$
4. $(\sqrt{10})^2$
5. $(-\sqrt{3})^2$
6. $(\sqrt{5})^2$
7. $\frac{\sqrt{14}}{\sqrt{2}}$
8. $\frac{\sqrt{150}}{\sqrt{5}}$
9. $\frac{\sqrt{12}}{\sqrt{3}}$
10. $4\sqrt{2} \cdot 2\sqrt{5}$
11. $6\sqrt{7} \cdot 5\sqrt{2}$
12. $9\sqrt{6}(-2\sqrt{5})$
13. $\frac{24\sqrt{8}}{6\sqrt{2}}$
14. $\frac{-12\sqrt{22}}{-4\sqrt{11}}$
15. $\frac{56\sqrt{42}}{-8\sqrt{6}}$

WRITTEN EXERCISES

Perform the indicated operations. Leave irrational numbers in radical form.

A
1. $\sqrt{3} \cdot \sqrt{7}$
2. $\sqrt{13} \cdot \sqrt{5}$
3. $4\sqrt{2} \cdot 6\sqrt{3}$
4. $7\sqrt{3} \cdot 9\sqrt{11}$
5. $-15\sqrt{7} \cdot 4\sqrt{5}$
6. $(-8\sqrt{2})(-6\sqrt{11})$
7. $\frac{\sqrt{42}}{\sqrt{2}}$
8. $\frac{\sqrt{39}}{\sqrt{3}}$
9. $\frac{28\sqrt{30}}{4\sqrt{15}}$
10. $\frac{54\sqrt{28}}{6\sqrt{2}}$
11. $\frac{-72\sqrt{121}}{-8\sqrt{11}}$
12. $\frac{36\sqrt{56}}{-4\sqrt{8}}$
13. $4\sqrt{12} \cdot 5\sqrt{3}$
14. $8\sqrt{20} \cdot 2\sqrt{5}$
15. $\frac{9\sqrt{98}}{3\sqrt{2}}$
16. $\frac{8\sqrt{75}}{2\sqrt{3}}$
17. $\frac{2}{3}\sqrt{\frac{6}{7}} \cdot 6\sqrt{\frac{7}{6}}$
18. $\frac{5}{7}\sqrt{\frac{16}{5}} \cdot 14\sqrt{\frac{5}{4}}$
19. $(\sqrt{15})^2$
20. $(-\sqrt{23})^2$
21. $(3\sqrt{5})^2$
22. $(4\sqrt{8})^2$
23. $3(-\sqrt{5})^2$
24. $4(\sqrt{8})^2$

B 25. $\sqrt{0.3} \cdot \sqrt{1.2}$ 26. $\sqrt{0.8} \cdot \sqrt{0.2}$ 27. $\sqrt{3.2} \cdot \sqrt{0.2}$
   28. $\sqrt{0.9} \cdot \sqrt{0.4}$ 29. $\dfrac{\sqrt{4.5}}{\sqrt{0.5}}$ 30. $\dfrac{\sqrt{0.064}}{\sqrt{0.4}}$
   31. $\sqrt{2} \cdot \sqrt{3} \cdot \sqrt{5}$ 32. $-\sqrt{3} \cdot \sqrt{6} \cdot \sqrt{2}$ 33. $\sqrt{8} \cdot \sqrt{10} \cdot \sqrt{5}$
   34. $(-2\sqrt{5})(3\sqrt{6})(-\sqrt{7})$ 35. $(\sqrt{5})^4$ 36. $(3\sqrt{2})^4$

37. a. Square A has an area of 42 cm². What is the length of one of its sides?
    b. Square B has twice the area of Square A. What is the length of one of its sides?
    c. The length of a side of Square C is twice the length of a side of Square A. What is the area of Square C?
    d. Which has the greater area, Square B or Square C? How many times as great?
    e. How many times greater in area is Square C than Square A?

Multiply. Assume that all variables represent nonnegative real numbers.

C 38. $\sqrt{xy} \cdot \sqrt{x^3y}$ 39. $\sqrt{ab} \cdot \sqrt{a^3b^3}$ 40. $-4\sqrt{x^3y} \cdot 3\sqrt{xy^3}$
   41. $(a\sqrt{a^3})^2$ 42. $\sqrt{x} \cdot \sqrt{y} \cdot \sqrt{xy}$ 43. $\sqrt{ab} \cdot \sqrt{a^3} \cdot \sqrt{b^5}$

Section 8-6 SIMPLIFYING RADICALS

You know what is meant by the direction "Express $\dfrac{6}{8}$ in simplest form." Similarly, it is often best to express a square-root radical in its simplest form. To do so, you must be sure to meet these three requirements:

1. Leave no perfect square factor (other than 1) under a radical sign.
2. Leave no fraction under a radical sign.
3. Leave no radical in a denominator.

EXAMPLES

$$\sqrt{27} = \sqrt{3^2 \cdot 3} = 3\sqrt{3} \qquad \sqrt{588} = \sqrt{2 \cdot 2 \cdot 3 \cdot 7 \cdot 7} = 2 \cdot 7\sqrt{3}, \text{ or } 14\sqrt{3}$$

$$\sqrt{\dfrac{1}{49}} = \dfrac{1}{7} \qquad \sqrt{\dfrac{15}{4}} = \dfrac{1}{2}\sqrt{15}$$

$$\dfrac{2}{\sqrt{5}} = \dfrac{2}{\sqrt{5}} \cdot \dfrac{\sqrt{5}}{\sqrt{5}} = \dfrac{2\sqrt{5}}{5}, \text{ or } \dfrac{2}{5}\sqrt{5}$$

The last example uses the identity property for multiplication, since multiplying by $\dfrac{\sqrt{5}}{\sqrt{5}}$ is equivalent to multiplying by 1. The process of eliminating the radical in a denominator, as in this example, is called *rationalizing the denominator.*

RATIONAL AND IRRATIONAL NUMBERS

How would you simplify $\dfrac{\sqrt{3}}{\sqrt{8}}$? You could multiply by $\dfrac{\sqrt{8}}{\sqrt{8}}$. However, there is an easier way to rationalize the denominator if you recognize that 16 is the least perfect square of which 8 is a factor.

EXAMPLE Express $\dfrac{\sqrt{3}}{\sqrt{8}}$ in simplest form.

SOLUTION $\dfrac{\sqrt{3}}{\sqrt{8}} = \dfrac{\sqrt{3}}{\sqrt{8}} \cdot \dfrac{\sqrt{2}}{\sqrt{2}} = \dfrac{\sqrt{6}}{\sqrt{16}} = \dfrac{1}{4}\sqrt{6}$

ORAL EXERCISES

Express in simplest form.

1. $\sqrt{12}$
2. $\sqrt{45}$
3. $-\sqrt{50}$
4. $-\sqrt{28}$
5. $3\sqrt{18}$
6. $\dfrac{1}{\sqrt{7}}$
7. $\sqrt{\dfrac{11}{16}}$
8. $\sqrt{\dfrac{8}{25}}$

WRITTEN EXERCISES

Express in simplest form.

A
1. $\sqrt{20}$
2. $\sqrt{63}$
3. $-5\sqrt{44}$
4. $6\sqrt{24}$
5. $9\sqrt{40}$
6. $-10\sqrt{52}$
7. $2\sqrt{99}$
8. $-\sqrt{72}$
9. $\sqrt{75}$
10. $\sqrt{275}$
11. $\dfrac{5}{7}\sqrt{98}$
12. $\dfrac{2}{3}\sqrt{153}$
13. $\dfrac{1}{4}\sqrt{128}$
14. $\dfrac{1}{7}\sqrt{245}$
15. $\dfrac{3\sqrt{56}}{4}$
16. $\dfrac{2\sqrt{117}}{3}$
17. $-\sqrt{\dfrac{7}{25}}$
18. $\sqrt{\dfrac{3}{49}}$
19. $\sqrt{\dfrac{32}{81}}$
20. $-\sqrt{\dfrac{125}{144}}$
21. $\dfrac{5}{\sqrt{3}}$
22. $\dfrac{5}{\sqrt{5}}$
23. $\sqrt{\dfrac{2}{7}}$
24. $\sqrt{\dfrac{3}{2}}$

B
25. $-\dfrac{3}{\sqrt{27}}$
26. $-\dfrac{4}{\sqrt{18}}$
27. $\dfrac{\sqrt{5}}{\sqrt{135}}$
28. $\dfrac{\sqrt{3}}{\sqrt{24}}$
29. $\sqrt{5} \cdot \sqrt{10}$
30. $\sqrt{33} \cdot \sqrt{3}$
31. $4\sqrt{15} \cdot 3\sqrt{3}$
32. $2\sqrt{10} \cdot 5\sqrt{2}$

Express in simplest form. Then use the table on page 281 to find an approximation for each square root to the nearest hundredth.

33. $\sqrt{212}$ 34. $-\sqrt{198}$ 35. $\sqrt{125}$ 36. $\sqrt{200}$ 37. $-\sqrt{156}$ 38. $\sqrt{752}$

Express in simplest form. Assume that all variables represent positive integers.

C 39. $\sqrt{49x^3}$ 40. $\sqrt{9x^9}$ 41. $\sqrt{8x^8}$ 42. $\sqrt{12a^7b^6}$ 43. $\sqrt{18b^2c^3}$ 44. $\sqrt{25x^9y^{12}}$

Section 8-7 ADDITION AND SUBTRACTION OF RADICALS

In this section you will learn how to:
- recognize like square-root radicals
- simplify radical expressions involving addition and subtraction

You will recall that $2a + 5a$ can be simplified by use of the distributive property: $2a + 5a = (2 + 5)a = 7a$. Similarly, you can simplify a radical expression where *like* radicals, radicals with the same radicand, occur:

$$2\sqrt{3} + 5\sqrt{3} = (2+5)\sqrt{3} = (7)\sqrt{3}, \text{ or } 7\sqrt{3}$$

Sometimes you can combine terms by first simplifying the radicals, as shown below.

$$\sqrt{18} + \sqrt{50} = \sqrt{9 \cdot 2} + \sqrt{25 \cdot 2}$$
$$= \sqrt{9} \cdot \sqrt{2} + \sqrt{25} \cdot \sqrt{2}$$
$$= 3\sqrt{2} + 5\sqrt{2} = 8\sqrt{2}$$

Unless the terms contain like radicals they cannot be combined. For example, the expression $\sqrt{3} + \sqrt{5}$ must be left in that form.

EXAMPLES

$\sqrt{50} - \sqrt{18} = 5\sqrt{2} - 3\sqrt{2}$
$\phantom{\sqrt{50} - \sqrt{18} } = 2\sqrt{2}$

$\sqrt{75} + \sqrt{125} = 5\sqrt{3} + 5\sqrt{5}$

$2x^2\sqrt{5} - 11x^2\sqrt{5} = -9x^2\sqrt{5}$

$3\sqrt{24} - \sqrt{\dfrac{2}{3}} = 3\sqrt{4 \cdot 6} - \dfrac{\sqrt{2}}{\sqrt{3}}$
$\phantom{3\sqrt{24} - \sqrt{\dfrac{2}{3}}} = 3(2\sqrt{6}) - \left(\dfrac{\sqrt{2}}{\sqrt{3}} \cdot \dfrac{\sqrt{3}}{\sqrt{3}}\right)$
$\phantom{3\sqrt{24} - \sqrt{\dfrac{2}{3}}} = 6\sqrt{6} - \dfrac{1}{3}\sqrt{6}$
$\phantom{3\sqrt{24} - \sqrt{\dfrac{2}{3}}} = \left(6 - \dfrac{1}{3}\right)\sqrt{6}$
$\phantom{3\sqrt{24} - \sqrt{\dfrac{2}{3}}} = \dfrac{17}{3}\sqrt{6}$

ORAL EXERCISES

In each exercise choose two terms that could be combined after they are simplified.

1. $3\sqrt{5}, 2\sqrt{7}, 8\sqrt{7}$ 2. $4\sqrt{2}, 5\sqrt{8}, \sqrt{3}$
3. $\sqrt{10}, \sqrt{20}, \sqrt{40}$ 4. $2x\sqrt{3}, x^2\sqrt{5}, x\sqrt{27}$

RATIONAL AND IRRATIONAL NUMBERS

Simplify.

5. $2\sqrt{11} + 5\sqrt{11}$
6. $-8\sqrt{2} + 3\sqrt{2}$
7. $\frac{8}{3}\sqrt{3} - \frac{2}{3}\sqrt{3}$
8. $21\sqrt{5} - 21\sqrt{5}$

WRITTEN EXERCISES

Simplify. Assume that all variables represent positive integers.

A
1. $9\sqrt{7} + 5\sqrt{7}$
2. $15\sqrt{14} - 10\sqrt{14}$
3. $-13\sqrt{5} - 8\sqrt{5}$
4. $-9\sqrt{6} + 11\sqrt{6}$
5. $14\sqrt{3} - 19\sqrt{3}$
6. $-21\sqrt{2} - 2\sqrt{2}$
7. $8\sqrt{15} + \sqrt{15}$
8. $\sqrt{7} + 2\sqrt{7}$
9. $\sqrt{3} + \sqrt{3}$
10. $\frac{2}{5}\sqrt{7} + \frac{3}{5}\sqrt{7}$
11. $\frac{11}{3}\sqrt{13} - \frac{2}{3}\sqrt{13}$
12. $5\sqrt{2} + 3\sqrt{50}$
13. $8\sqrt{63} - 5\sqrt{7}$
14. $2\sqrt{27} + 2\sqrt{300}$
15. $5\sqrt{48} - 4\sqrt{75}$

B
16. $4\sqrt{2} + \sqrt{2} - \sqrt{8}$
17. $7\sqrt{3} - \sqrt{3} + \sqrt{27}$
18. $\sqrt{\frac{1}{3}} + \frac{2\sqrt{3}}{3}$
19. $\frac{2}{\sqrt{5}} + \sqrt{20}$
20. $\sqrt{72} + \frac{1}{\sqrt{8}}$
21. $\sqrt{\frac{2}{3}} + \sqrt{\frac{3}{2}}$
22. $\sqrt{2}(2\sqrt{3} + 3\sqrt{3})$
23. $3\sqrt{7}(11\sqrt{3} + 4\sqrt{3})$
24. $\frac{1}{3}\sqrt{6}(11\sqrt{2} - 2\sqrt{2})$
25. $\sqrt{2}(\sqrt{8} + \sqrt{2})$
26. $2\sqrt{5}(\sqrt{8} + \sqrt{18})$
27. $5\sqrt{2}(\sqrt{2} + \sqrt{32})$

C
28. $2x\sqrt{3} + 9x\sqrt{3}$
29. $3x\sqrt{2} + 7\sqrt{2x^2}$
30. $7\sqrt{x^3} - 12\sqrt{x}$

Solve for x. The domain of x is the set of nonnegative real numbers.
31. $x\sqrt{2} + \sqrt{32} = 5\sqrt{2x^2} + \sqrt{18}$
32. $\sqrt{10} + 10x\sqrt{40} = 3\sqrt{10} + 3x\sqrt{90}$
33. $\sqrt{75} + 3\sqrt{3x^2} = \sqrt{27} + 8x\sqrt{3}$

SELF-TEST 3

Simplify.

1. $(-2\sqrt{5})(-3\sqrt{13})$
2. $\frac{8\sqrt{45}}{-4\sqrt{5}}$
3. $(-\sqrt{12})^2$ Section 8-5

4. $\sqrt{245}$
5. $-\sqrt{\frac{27}{64}}$
6. $\frac{5}{\sqrt{11}}$ Section 8-6

7. $7\sqrt{7} + 13\sqrt{7}$
8. $\sqrt{45} - \sqrt{20}$
9. $4\sqrt{27} - 3\sqrt{75}$ Section 8-7

Section 8-8 BINOMIALS CONTAINING RADICALS

Radicals often occur in binomial expressions. In this section you will learn:
- how to multiply two binomials containing radicals
- the meaning of the term "conjugate pairs"
- how to rationalize a binomial denominator containing radicals

In Chapter 7 you learned how to multiply two binomials. You can use the same patterns when radicals are involved.

1. $(a + b)(c + d) = ac + ad + bc + bd$
   $(2 + \sqrt{5})(3 + \sqrt{2}) = 6 + 2\sqrt{2} + 3\sqrt{5} + \sqrt{10}$
2. $(a + b)^2 = a^2 + 2ab + b^2$
   $(3 + \sqrt{5})^2 = 9 + 2 \cdot 3 \cdot \sqrt{5} + 5 = 14 + 6\sqrt{5}$
3. $(a - b)^2 = a^2 - 2ab + b^2$
   $(\sqrt{3} - 4)^2 = 3 - 2 \cdot \sqrt{3} \cdot 4 + 16 = 19 - 8\sqrt{3}$
4. $(a + b)(a - b) = a^2 - b^2$
   $(2 + \sqrt{5})(2 - \sqrt{5}) = 2^2 - (\sqrt{5})^2 = 4 - 5 = -1$

Binomials such as $2 + \sqrt{5}$ and $2 - \sqrt{5}$ are called *conjugate pairs*. In general, $\sqrt{x} + \sqrt{y}$ and $\sqrt{x} - \sqrt{y}$ are *conjugates* of each other. You can use the fact that
$$(\sqrt{x} + \sqrt{y})(\sqrt{x} - \sqrt{y}) = x - y$$
to *rationalize* a denominator that is a binomial containing a radical.

EXAMPLE Simplify $\dfrac{6}{3 - \sqrt{13}}$.

SOLUTION The conjugate of the denominator of $\dfrac{6}{3 - \sqrt{13}}$ is $3 + \sqrt{13}$. So,

$$\frac{6}{3 - \sqrt{13}} = \frac{6}{3 - \sqrt{13}} \cdot \frac{3 + \sqrt{13}}{3 + \sqrt{13}} = \frac{6(3 + \sqrt{13})}{9 - 13} = \frac{6(3 + \sqrt{13})}{-4}$$

$$= \frac{-3(3 + \sqrt{13})}{2} = \frac{-9 - 3\sqrt{13}}{2}$$

EXAMPLE Simplify $\dfrac{\sqrt{6}}{\sqrt{2} - \sqrt{3}}$.

SOLUTION $\dfrac{\sqrt{6}}{\sqrt{2} - \sqrt{3}} = \dfrac{\sqrt{6}}{\sqrt{2} - \sqrt{3}} \cdot \dfrac{\sqrt{2} + \sqrt{3}}{\sqrt{2} + \sqrt{3}} = \dfrac{\sqrt{12} + \sqrt{18}}{2 - 3} = \dfrac{2\sqrt{3} + 3\sqrt{2}}{-1}$

$= -2\sqrt{3} - 3\sqrt{2}$

RATIONAL AND IRRATIONAL NUMBERS

ORAL EXERCISES

State the conjugate.

1. $4 - \sqrt{11}$
2. $-7 - \sqrt{3}$
3. $2\sqrt{5} + 8$
4. $x + 3\sqrt{6}$
5. $\sqrt{11} + \sqrt{2}$
6. $5\sqrt{7} - 9\sqrt{10}$
7. $\sqrt{13} + 2$
8. $3\sqrt{7} - 26$
9. $2\sqrt{3} - \sqrt{6}$

WRITTEN EXERCISES

State which of the four patterns on page 289 you will use. Then simplify.

A
1. $(7 - \sqrt{5})^2$
2. $(\sqrt{11} - 2)^2$
3. $(7 + \sqrt{5})(7 - \sqrt{5})$
4. $(\sqrt{11} - 2)(\sqrt{11} + 2)$
5. $(\sqrt{3} + \sqrt{2})(\sqrt{5} + \sqrt{6})$
6. $(3\sqrt{2} + 4\sqrt{5})(3\sqrt{2} - 4\sqrt{5})$
7. $(\sqrt{6} + 2)^2$
8. $(x + \sqrt{11})^2$
9. $(x + \sqrt{3})(x - \sqrt{3})$

Simplify. Rationalize denominators.

10. $\dfrac{5}{4 - \sqrt{3}}$
11. $\dfrac{3}{\sqrt{13} + 2}$
12. $\dfrac{1}{\sqrt{11} - 1}$
13. $\dfrac{11}{6 - \sqrt{3}}$
14. $\dfrac{3}{4 + \sqrt{7}}$
15. $\dfrac{5}{\sqrt{10} - 2}$

B
16. $(3\sqrt{2} - 5\sqrt{7})^2$
17. $(2\sqrt{6} + 5\sqrt{2})^2$
18. $(2\sqrt{2} + 1)(3\sqrt{3} + 5)$
19. $(2\sqrt{3} + 6\sqrt{2})^2$
20. $(3\sqrt{5} - 1)(2\sqrt{3} + 6)$
21. $(3\sqrt{2} - 5\sqrt{7})(3\sqrt{2} + 5\sqrt{7})$
22. $\dfrac{23}{4\sqrt{6} - 7}$
23. $\dfrac{21}{5\sqrt{6} - 3\sqrt{3}}$
24. $\dfrac{4}{4\sqrt{7} + 2\sqrt{5}}$

C
25. Express as the product of two binomials: $x^2 - 3$.
26. The conjugate of $x + \sqrt{15}$ equals the reciprocal of $x + \sqrt{15}$. Find the positive value of x.
27. Solve for x and express the result in simplest form: $3x + x\sqrt{2} = 1$.

Section 8-9 THE QUADRATIC FORMULA

You learned in Chapter 7 how to solve some quadratic equations by factoring.

$$x^2 - 3x - 4 = 0$$
$$(x - 4)(x + 1) = 0$$
$$x - 4 = 0 \quad \text{or} \quad x + 1 = 0$$
$$x = 4 \quad \text{or} \quad x = -1$$

However, you do not need to limit yourself to this method. There is a formula that works for any quadratic equation.

Recall that the standard form of a quadratic equation is $ax^2 + bx + c = 0$, where a and b are coefficients of x^2 and x respectively, c is the constant term, and $a \neq 0$.

The following formula, called the *quadratic formula,* may be used *provided* the equation is written in the standard form.

$$x = \frac{-b \pm \sqrt{b^2 - 4ac}}{2a}$$

The sign "±" is read "plus or minus" and represents *both* square roots. Thus, the formula is equivalent to the following disjunction.

$$x = \frac{-b + \sqrt{b^2 - 4ac}}{2a} \quad \text{or} \quad x = \frac{-b - \sqrt{b^2 - 4ac}}{2a}$$

If $b^2 - 4ac \geq 0$, then the solutions, or roots, of the equation are real numbers.

Let's see if the quadratic formula gives the same solutions for $x^2 - 3x - 4 = 0$ as we obtained by factoring. Notice that:

$$a = 1 \quad \text{(coefficient of } x^2\text{)}$$
$$b = -3 \quad \text{(coefficient of } x\text{)}$$
$$c = -4 \quad \text{(constant term)}$$

Substituting, we get:

$$x = \frac{-(-3) \pm \sqrt{(-3)^2 - 4 \cdot 1 \cdot (-4)}}{2 \cdot 1}$$
$$= \frac{3 \pm \sqrt{9 + 16}}{2}$$
$$= \frac{3 \pm \sqrt{25}}{2}$$

Thus, $x = \dfrac{3 + \sqrt{25}}{2} = \dfrac{3 + 5}{2} = 4 \quad \text{or} \quad x = \dfrac{3 - \sqrt{25}}{2} = \dfrac{3 - 5}{2} = -1.$

The roots are the same as those arrived at by the factoring method, 4 and −1.

EXAMPLE Solve $2x^2 + 5x = 3$.

SOLUTION In standard form, the equation is $2x^2 + 5x - 3 = 0$.
Then $a = 2$, $b = 5$, and $c = -3$.

Then
$$x = \frac{-5 \pm \sqrt{5^2 - 4 \cdot 2 \cdot (-3)}}{2 \cdot 2}$$
$$= \frac{-5 \pm \sqrt{25 + 24}}{4}$$
$$= \frac{-5 \pm \sqrt{49}}{4}$$
$$= \frac{-5 \pm 7}{4}$$

$x = \dfrac{-5 + 7}{4} = \dfrac{1}{2} \quad \text{or} \quad x = \dfrac{-5 - 7}{4} = -3$

The solution set is $\left\{\dfrac{1}{2}, -3\right\}$.

EXAMPLE Solve $x^2 + 11x + 1 = 0$.

SOLUTION $a = 1, b = 11, c = 1$

$$x = \frac{-11 \pm \sqrt{11^2 - 4 \cdot 1 \cdot 1}}{2 \cdot 1}$$

$$= \frac{-11 \pm \sqrt{121 - 4}}{2}$$

$$= \frac{-11 \pm \sqrt{117}}{2}$$

$$= \frac{-11 \pm 3\sqrt{13}}{2}$$

The solution set is $\left\{ \dfrac{-11 + 3\sqrt{13}}{2}, \dfrac{-11 - 3\sqrt{13}}{2} \right\}$.

ORAL EXERCISES

State the standard form of the equation.

1. $3x^2 + 4x = 5$
2. $10x^2 - 2x = 15$
3. $5x^2 - 12 = 100x$

State the values of *a*, *b*, and *c* you would use when applying the quadratic formula.

4. $x^2 + 13x + 40 = 0$
5. $4x^2 + 11x = 0$
6. $5y^2 - 14 = 0$
7. $x^2 + 27 = 12x$
8. $9x^2 + 17x + 4 = 2x^2$
9. $6x^2 = 7$

WRITTEN EXERCISES

Solve by the quadratic formula. Check by the factoring method.

A
1. $x^2 - 7x + 12 = 0$
2. $y^2 - y - 2 = 0$
3. $6x^2 - 17x - 3 = 0$
4. $2z^2 - 7z - 4 = 0$
5. $15y^2 + 7y - 2 = 0$
6. $3x^2 - 25x - 18 = 0$

Express in standard form. Then solve by the quadratic formula.

7. $x^2 + 3x = 4$
8. $9x^2 - 6x = -1$
9. $2y^2 + 3y = 5$
10. $3x^2 - 8 = 2x$
11. $7w^2 + 47w = 14$
12. $x^2 - 15 = 10$

Solve.

B 13. $x^2 - x - 10 = 0$
14. $y^2 + 3y - 8 = 0$
15. $5y^2 + 3y = 1$
16. $2z^2 + z - 5 = 0$
17. $7x^2 - 4x - 1 = 0$
18. $y^2 + 8y - 4 = 0$

Solve. Then use the table on page 281 to approximate each root to the nearest hundredth.

C 19. $y^2 - 8y + 1 = 0$
20. $x^2 + 3x - 5 = 0$
21. $4y^2 - y = 1$
22. $3x^2 - 3x - 1 = 0$

SELF-TEST 4

Simplify.

1. $(\sqrt{5} - 2)^2$
2. $\dfrac{7}{\sqrt{11} + 2}$
3. $(2\sqrt{3} + 1)(\sqrt{2} - 1)$ Section 8-8

Use the quadratic formula to solve.

4. $2x^2 - 3x - 2 = 0$
5. $y^2 + 8y = -4$ Section 8-9

Computer Exercises

1. This program rounds a given number N to the nearest integer. Copy and RUN the program for several values of N.

```
10  INPUT N
20  PRINT "WHEN ROUNDED TO THE NEAREST INTEGER,"
30  PRINT "    "N;" IS";INT(N+.5);"."
40  END
```

This program uses the INT function, which was described on page 261, to round N to the nearest integer. If N=5.2, for example, then:

$$INT(N + 0.5) = INT(5.2 + 0.5)$$
$$= INT(5.7)$$
$$= 5$$

If N=5.6, then:

$$INT(N + 0.5) = INT(5.6 + 0.5)$$
$$= INT(6.1)$$
$$= 6$$

To round a number to the nearest tenth, use INT(10∗N + 0.5)/10. If N=4.37, for example, then:

$$INT(10*N + 0.5)/10 = INT(43.7 + 0.5)/10$$
$$= INT(44.2)/10$$
$$= 44/10$$
$$= 4.4$$

RATIONAL AND IRRATIONAL NUMBERS

2. This program computes, to the nearest hundredth, square roots of positive real numbers, using the divide-and-average method. Lines 70 and 80 test whether the quotient and divisor agree to the nearest thousandth. Line 120 rounds the final answer to the nearest hundredth using the method described in Exercise 1.

Copy and RUN the program.

```
10    PRINT "   N","SQUARE RT N (NEAREST 100TH)"
20    READ N
30    IF N=.99999 THEN 150
40    IF N<0 THEN 20
50    LET Q=N/2
60    LET D=(2+Q)/2
70    IF Q-D >= .0001 THEN 90
80    IF D-Q<.0001 THEN 120
90    LET Q=N/D
100   LET D=(D+Q)/2
110   GOTO 70
120   PRINT N,INT(100*D+.5)/100
130   GOTO 20
140   DATA 36,92,75,169,684,2728,256,.00143,.99999
150   END
```

3. a. SQR(X) is defined to be the positive square root of X. For example, SQR(42.7) ≈ 6.5345, SQR(32/2) = 4, but SQR(−6) is not defined and will produce an error statement. Write a program using SQR to compute the square roots of the numbers in the DATA statement in the program in Exercise 2.

   b. Revise the program you wrote for part (a) so that each square root is rounded to the nearest thousandth.

CHAPTER REVIEW

Express as a decimal.

1. $\dfrac{7}{12}$ 2. $\dfrac{13}{125}$ 3. $-\dfrac{67}{66}$ Section 8-1

Express as a fraction in lowest terms.

4. 0.064 5. $1.\overline{297}$ 6. $0.34\overline{5}$ Section 8-2

Each radicand is a perfect square. Simplify.

7. $-\sqrt{8100}$ 8. $-\sqrt{\dfrac{49}{324}}$ 9. $\dfrac{1}{2}\sqrt{676}$ Section 8-3

Use the table on page 281 to approximate the square root to the nearest hundredth.

10. $\sqrt{23}$ 11. $-\sqrt{89}$ Section 8-4

12. Use the divide-and-average method to approximate $\sqrt{127}$ to the nearest hundredth.

Simplify.

13. $\left(\frac{1}{3}\sqrt{2}\right)(-3\sqrt{50})$ 14. $\frac{-15\sqrt{91}}{-5\sqrt{7}}$ 15. $(4\sqrt{3})^2$ Section 8-5

16. $-\sqrt{252}$ 17. $\sqrt{\frac{27}{100}}$ 18. $\frac{6}{\sqrt{3}}$ Section 8-6

19. $\sqrt{11} - 5\sqrt{11}$ 20. $\sqrt{28} + \frac{1}{2}\sqrt{63}$ 21. $3\sqrt{32} - 4\sqrt{18}$ Section 8-7

22. $(6-4\sqrt{2})(6+4\sqrt{2})$ 23. $\frac{4}{\sqrt{15}-3}$ 24. $(5\sqrt{5}-\sqrt{7})^2$ Section 8-8

Solve by the quadratic formula.

25. $8x^2 = 2x + 3$ 26. $4y^2 + 8y - 1 = 0$ Section 8-9

CHAPTER TEST

Express each fraction as a decimal. Then name the greater fraction.

1. $\frac{131}{200}, \frac{11}{18}$ 2. $-\frac{10}{27}, -\frac{307}{1000}$ 3. $\frac{68}{75}, \frac{29}{32}$

Express as a fraction in lowest terms.

4. 2.612 5. $0.\overline{57}$ 6. $0.4\overline{91}$

Simplify

7. $\sqrt{\frac{40{,}000}{9}}$ 8. $\frac{5}{6}\sqrt{216}$ 9. $7\sqrt{3}\left(-6\sqrt{\frac{2}{27}}\right)$

10. $-3\sqrt{5} + 12\sqrt{5}$ 11. $(8+\sqrt{7})(5-\sqrt{7})$ 12. $-\sqrt{1764}$

13. $\sqrt{\frac{1}{12}} + \frac{1}{3}\sqrt{3}$ 14. $\frac{9}{7+\sqrt{13}}$ 15. $\sqrt{\frac{7}{18}}$

16. $\frac{1}{2}\sqrt{96} - \sqrt{54}$ 17. $\frac{-72\sqrt{45}}{24\sqrt{5}}$ 18. $9\sqrt{180}$

19. $(3\sqrt{10} - 2\sqrt{2})^2$ 20. $\sqrt{1.44}$ 21. $10(-\sqrt{13})^2$

RATIONAL AND IRRATIONAL NUMBERS

Express in simplest form. Then use the table on page 281 to approximate each square root to the nearest hundredth.

22. $\sqrt{61}$ **23.** $-\sqrt{\dfrac{13}{9}}$ **24.** $\sqrt{104}$

25. Use the divide-and-average method to approximate $\sqrt{239}$ to the nearest hundredth.

Solve by the quadratic formula.

26. $6w^2 = 29w - 9$ **27.** $z^2 - 6z - 1 = 0$

Careers — Automobile Mechanic

Very few people are patient about car trouble. Fears of countless trips to the shop and huge repair bills are aroused the first time a car sputters or fails to start. Ignorance about how an internal combustion engine works makes many people feel helpless.

The growth of suburban areas during the last few decades has brought good automobile mechanics into great demand. A mechanic's most valuable skill is the ability to quickly and accurately diagnose problems. Some mechanics handle everything that can go wrong with a car, while others handle only particular types of problems. Possible specializations include automatic transmission mechanics, who work on hydraulic pumps and gear trains, and tune-up mechanics, who adjust or replace such things as spark plugs and distributor points.

Computers are becoming common in the repair and replacement of auto parts, such as in fine tuning or wheel balancing. Other testing and repair equipment with which mechanics become familiar on the job include motor analyzers, spark plug testers, pneumatic wrenches, lathes, and grinding machines.

Most mechanics learn their skills on the job, though courses are available from some high schools and many vocational or technical schools. A three- or four-year apprenticeship program, which can be arranged through an auto dealer or repair shop, is recommended. The amount of training you will need depends on whether you want to learn a difficult specialty, such as automatic transmission repair.

Certification by the National Institute for Automotive Service Excellence is the accepted proof of a mechanic's skill.

The number of mechanics in a shop usually ranges from one to five, though it can reach one hundred. Employers include automobile dealers, gas stations, department stores with automobile service facilities, and automobile leasing companies.

VOCABULARY REVIEW

Be sure that you understand the meaning of these terms:

terminating decimal, p. 272
repeating decimal, p. 272
square root, p. 276
radical sign, p. 276
principal square root, p. 276
radicand, p. 227

radical, p. 277
perfect square, p. 277
irrational number, p. 279
rationalizing the denominator, p. 285
like radicals, p. 287
conjugate pairs, p. 289

MIXED REVIEW

Graph the solution set of $x + 2 < -1$ if the domain of x is the given set.

1. {integers}
2. {real numbers}

If $p, q,$ and r are all false, find the truth value of each statement.

3. $(p \wedge q) \leftrightarrow r$
4. $(p \vee \sim q) \rightarrow r$

5. Sketch four points that are not coplanar.

6. If $m\angle 1 = t + 10$, then a complement of $\angle 1$ has measure $\underline{}$ and a supplement of $\angle 1$ has measure $\underline{}$. (Give answers in terms of t.)

Factor.

7. $4x^2 - 100$
8. $2y^2 + 5y - 12$
9. $n^2 - 2n - 35$

Simplify.

10. $\dfrac{\sqrt{56}}{\sqrt{6}}$
11. $\sqrt{175} - 2\sqrt{63}$
12. $\dfrac{3}{\sqrt{2} + 1}$

13. Solve $V = \dfrac{1}{3}Bh$ for h.

14. A right rectangular prism has length 7, width 4, and total area 166. Find its volume.

15. In $\triangle RST$, $\overline{RS} \cong \overline{ST}$ and $m\angle R = 42$. Find the measures of $\angle S$ and $\angle T$.

16. Find the GCF of $72r^3s^2t^7$ and $-54rs^2$.

Solve.

17. $-5 \geqslant -\dfrac{3}{2}n$
18. $0.08w = -4$
19. $9d^2 + 6d + 1 = 0$
20. $4s^2 - 11s + 6 = 0$
21. $z^2 + 4z = -2$
22. $-(x + 1) = 8 - x$

298   CHAPTER 8

23. Must the triangles shown in the diagram be congruent? If so, give a reason (SSS, SAS, or ASA).

24. Are the diagonals of ABCD necessarily congruent if ABCD is a parallelogram? a rectangle? a rhombus? an isosceles trapezoid?

25. Construct a truth table for the disjunction $\sim p \vee q$.

26. a. The opposite angles of a parallelogram are __?__.
    b. Two consecutive angles of a parallelogram are __?__.

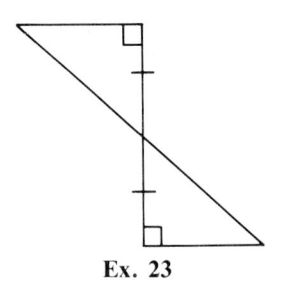

Ex. 23

27. When the length of a certain rectangle is decreased by 2 cm and the width is increased by 2 cm, a square is formed. If the sum of the areas of the rectangle and the square is 124 cm², find the dimensions of the rectangle.

Multiply or divide.

28. $(\sqrt{3} + \sqrt{6})^2$

29. $(x + \sqrt{7})(x - \sqrt{7})$

30. $\dfrac{15x^7}{(3xy^2)^2}$

31. $\dfrac{18v^{10} - 81v^4 + 99v}{3v}$

32. Find the area of the figure shown in the diagram.

33. Simplify the expression and give its degree:
$(8x - 5x^2 + 2) - (4x^2 + 10 - 3x^3)$

34. Express $1.2\overline{34}$ as a fraction in lowest terms.

35. Use the divide-and-average method to approximate $\sqrt{141}$ to the nearest tenth.

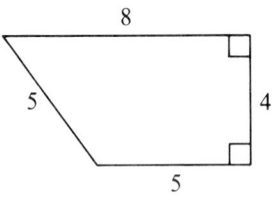

Ex. 32

36. Use a protractor to draw an ∠A of measure 130. Then construct an angle congruent to ∠A.

Exercises 37 and 38 refer to the diagram at the right.

37. Write the biconditional that justifies the statement "If $a \parallel b$, then $m\angle 1 = m\angle 2$."

38. If $a \parallel b$, $m\angle 1 = x^2 + 3$, and $m\angle 2 = 12x - 8$, find x. Then find the numerical measure of the angles.

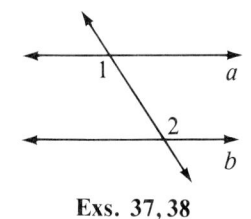

Exs. 37, 38

39. If the measures of the angles of a triangle are $2x - 3$, $3x - 3$, and $4x - 3$, find the numerical measure of each angle.

40. If one pad of paper costs 90 cents, at most how many pads of paper can be bought for $10?

RATIONAL AND IRRATIONAL NUMBERS

An important application of the concepts of similarity and proportion is in making scale models and scale drawings. The scale model in the photograph shows the Harborplace market, World Trade Center, and inner harbor in Baltimore, Maryland.

Ratio and Similarity

9

Timco Enterprises uses a crate with dimensions $l \times w \times h$ to ship its product. In order to save on crating costs, Timco would like to double the volume of the crate. Suppose a crate with dimensions $2l \times 2w \times 2h$ is produced. Do you think that the volume will be double that of the original crate?

In this chapter you will learn that the larger crate is *similar* to the smaller one, and that its volume is eight times that of the smaller one.

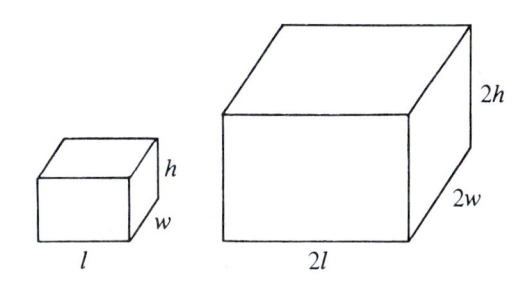

Section 9-1 RATIOS

We constantly need to make comparisons in our daily lives, whether we are shopping for a car with good gas mileage or deciding between two products in a supermarket. In this section you will learn how to make comparisons using *ratios*.

EXAMPLE Joanne pays $300 of her $1200 monthly pay check for rent. Roger earns $800 monthly and pays $200 for rent. What part of each person's salary goes for rent?

SOLUTION $\frac{300}{1200}$, or $\frac{1}{4}$, of Joanne's pay check is used for rent. Roger pays $\frac{200}{800}$, which is also $\frac{1}{4}$ of his pay check.

We say that the *ratio* of rent to salary is 1 to 4 in each case. The *ratio* of one number to another is the quotient of the first number divided by the second.

Notice that in order to compare the two ratios, we wrote each in simplest form as $\frac{1}{4}$. Ratios are ordinarily expressed in simplest form.

As the following example shows, it is important that the numbers being compared are expressed in terms of the same unit.

EXAMPLE Pearl is 1.5 m tall and Leo is 165 cm tall. What is the ratio of their heights?

SOLUTION 1.5 m is 150 cm. The ratio of Pearl's height to Leo's height is $\frac{150}{165}$, or $\frac{10}{11}$.

The ratio of a to b is sometimes written as $a:b$. This form is especially useful for comparing more than two numbers.

EXAMPLE Find the ratio of the lengths of the sides of $\triangle RST$.

SOLUTION The ratio $RS:ST:RT$ is $4:2\sqrt{5}:6$.
Since 2 is the factor common to each number in the ratio, the simplest form of the ratio is $2:\sqrt{5}:3$.

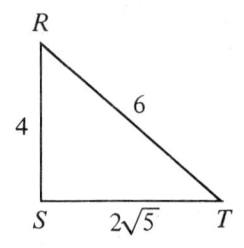

To find two numbers that are in a given ratio it is often convenient to express the numbers as multiples of x.

EXAMPLE The measures of an angle and its supplement are in the ratio $7:8$. Find the measure of each angle.

SOLUTION Let $7x$ and $8x$ be the measures of the angle and its supplement.
$$7x + 8x = 180$$
$$15x = 180$$
$$x = 12$$
The measure of the angle is $7x$, or 84.
The measure of its supplement is $8x$, or 96.

Check $84:96 = 7:8$ $84 + 96 = 180$

ORAL EXERCISES

Express each ratio in simplest form.

1. 9 eggs to 3 dozen eggs
2. 7.2 kg to 36 g
3. 6 dollars to 25 cents
4. 48 min to 4 h
5. $AB:AC$
6. $AC:AB$
7. $BC:AC$
8. $AB:BC:AC$
9. $m\angle C:m\angle A$
10. $m\angle A:m\angle B$
11. $m\angle C:m\angle B$
12. $m\angle C:m\angle A:m\angle B$

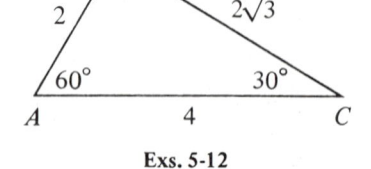

Exs. 5-12

13. $DE:EF$
14. $EF:DE$
15. $DE:DF$
16. $EF:DF$

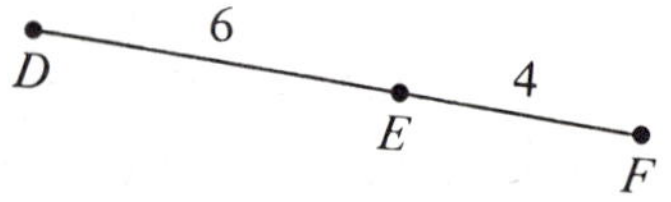

17. Represent in terms of x three numbers that are in the ratio $2:3:5$.

WRITTEN EXERCISES

Express each ratio in simplest form.

A
1. $AB:BC$
2. $BC:CD$
3. $CD:AB$
4. $AB:AC$
5. $BC:BD$
6. $BC:AD$
7. $CD:BD:AD$
8. $AB:BC:CD$

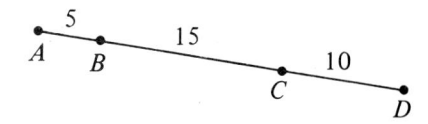

9. A tennis camp has 36 instructors and 126 students. What is the ratio of instructors to students?

10. A rectangle is 2.5 cm wide and 10 cm long. What is the ratio of the width to the length?

11. There are 18 females in a class and 15 males. What is the ratio of females to the total number of students in the class?

12. Of the 400 tickets sold to a play, 80 were $5 tickets and 200 were $3 tickets. The remainder were $2 tickets. Express as a ratio the number of $5 tickets, $3 tickets, and $2 tickets.

13. In a recent school election, Rita beat Chris by a ratio of $4:3$. If 1428 students voted for either Rita or Chris, how many votes did each receive?

14. John has quarters and dimes, in the ratio of $2:3$. If he has 200 coins in all, how many of each kind does he have?

15. The measures of two complementary angles are in the ratio of $3:2$. Find the measure of each angle.

16. The ratio of the measures of the two acute angles of a right triangle is $2:7$. Find the measure of each angle.

17. 2000 tickets were sold for a concert of "The What." If the ratio of reserved seats sold to unreserved seats was $3:5$, how many tickets were for reserved seats?

18. Two positive numbers whose ratio is $7:5$ have a difference of 28. Find the numbers.

19. Richie told his friends that he had a dream about a very unusual fish whose tail, body, and head were in the ratio $3:1:2$. If he was sure that the fish was 60 cm long, how long was the head?

20. The ratio of the measures of the three angles of a triangle is $2:5:8$. What is the measure of each angle of the triangle?

21. Three numbers that are in the ratio $4:9:2$ have a sum of -90. Find the numbers.

22. The ratio of the numbers of A, B, and C exercises in one section of a textbook is $6:3:1$. If there are 40 exercises in all, how many are B exercises?

23. What is the ratio of the length of one side of a square to its perimeter?

24. What is the ratio of the length of one side of a regular octagon to its perimeter?

RATIO AND SIMILARITY

B 25. Ellen has nickels and dimes in the ratio of $2:3$. If the total value of the coins is $3.60, how many coins of each kind does she have?

26. Sven Carlson wants to cut a pipe 42 m long into three sections so that each section is twice as long as the preceding one. How long should each section be?

27. The ratio of the lengths of two adjacent sides of a rectangle is $3:8$. The perimeter of the rectangle is 88 cm. Find the dimensions of the rectangle.

28. The ratio of the measure of one of the base angles of an isosceles triangle to that of the vertex angle is $7:4$. Find the measure of the vertex angle.

29. The ratio of the measure of the vertex angle of an isosceles triangle to that of one of the base angles is $2:1$. Find the measure of the vertex angle.

30. If $\triangle ABC$ is equilateral, what is the ratio of the measures of $\angle B$ and $\angle C$?

31. Find the ratio of the length of a rectangle to its width if the width is 14 mm and the perimeter is 72 mm.

32. Find the lengths of the sides of an isosceles triangle if the ratio of the length of one of the congruent sides to the length of the third side is $3:1$ and the perimeter is 28 mm.

33. Pat typed a page containing 250 words in 4 min and made 4 mistakes. Compute the speed as a ratio of integers using the equation:

$$\text{speed} = \frac{\text{number of words} - \text{mistakes}}{\text{time}}$$

C 34. The ratio of the number of 8-track tapes Alice has to those Brian has is $5:3$. What part of those Alice has should she give to Brian so that each will have the same number of tapes?

35. In $\triangle ABC$, the measures of $\angle A$ and $\angle B$ are in the ratio $1:2$. If the measure of $\angle C$ is 12 less than the measure of $\angle A$, find the measure of each angle.

Section 9-2 PROPORTIONS

An equation such as

$$\frac{12}{8} = \frac{3}{2} \quad \text{or} \quad 6:9 = 10:15$$

which states that two ratios are equal is called a *proportion*. The second proportion above is read as "6 is to 9 as 10 is to 15."

The first and fourth *terms* of a proportion are called the *extremes* of the proportion. The second and third terms are its *means*.

First → 12 = 3 ← Third
Second → 8   2 ← Fourth

12 and 2 are the extremes.
8 and 3 are the means.

extremes
↓   ↓
6 : 9 = 10 : 15
↑   ↑
means

Sometimes you will know three terms of a proportion and want to find the value of the unknown term. In order to do this, you use the fact that when you multiply both sides of the general proportion

$$\frac{a}{b} = \frac{c}{d} \qquad (b \neq 0, d \neq 0)$$

by the common denominator, bd, you obtain an equivalent equation.

$$\left(\frac{a}{b}\right)bd = \left(\frac{c}{d}\right)bd$$
$$ad = bc$$

In a proportion, the product of the extremes is equal to the product of the means.

EXAMPLE   Solve: a. $\dfrac{14}{9} = \dfrac{21}{y}$ b. $\dfrac{2x-5}{x} = \dfrac{3}{4}$

SOLUTION   a. $14y = 9 \cdot 21$ b. $(2x-5)4 = x \cdot 3$
                 $14y = 189$ $8x - 20 = 3x$
                   $y = 13.5$ $8x - 3x = 20$
                                               $5x = 20$
                                               $x = 4$

At any particular time, the ratios of heights of objects to the lengths of their shadows are equal. We say that the height is *directly proportional* to the length of the shadow.

EXAMPLE   A person who is 1.6 m tall casts a shadow 2 m long. At the same time, the length of the shadow of a nearby house is 15 m. Find the height of the house.

SOLUTION   Let h be the height of the house.

$$\frac{1.6}{2} = \frac{h}{15}$$
$$1.6 \cdot 15 = 2h$$
$$24 = 2h$$
$$12 = h \qquad \text{The house is 12 m high.}$$

RATIO AND SIMILARITY

The property tax on a house is *directly proportional* to the valuation of the house.

EXAMPLE If the property tax on a house valued at $50,000 is $1100, find the property tax on a neighboring house valued at: a. $40,000 b. $62,000

SOLUTION a. Let x be the tax.

$$\frac{x}{40{,}000} = \frac{1100}{50{,}000}$$

$$\frac{x}{40{,}000} = \frac{11}{500}$$

$$500x = 440{,}000$$

$$x = 880$$

Tax is $880.

b. Let y be the tax.

$$\frac{y}{62{,}000} = \frac{1100}{50{,}000}$$

$$\frac{y}{62{,}000} = \frac{11}{500}$$

$$500y = 682{,}000$$

$$y = 1364$$

Tax is $1364.

ORAL EXERCISES

Choose the ratio that is *not* equal to the others.

1. $\frac{1}{2}, \frac{2}{4}, \frac{3}{6}, \frac{4}{7}$

2. $\frac{5}{15}, \frac{2}{10}, \frac{1}{3}, \frac{4}{12}$

3. $\frac{8}{9}, \frac{9}{8}, \frac{27}{24}, \frac{18}{16}$

4. $8:5,\ 10:16,\ 4:2.5,\ 24:15$

Complete.

5. If $\frac{2}{9} = \frac{6}{a}$, then $2a = \underline{\ ?\ }$.

6. If $\frac{b}{14} = \frac{5}{6}$, then $6b = \underline{\ ?\ }$.

7. If $\frac{3}{4} = \frac{c}{12}$, then $4c = \underline{\ ?\ }$.

8. If $\frac{4}{d} = \frac{16}{18}$, then $16d = \underline{\ ?\ }$.

WRITTEN EXERCISES

Solve.

A 1. $\frac{x}{8} = \frac{21}{24}$

2. $\frac{7}{y} = \frac{49}{63}$

3. $\frac{9}{12} = \frac{a}{20}$

4. $\frac{12}{18} = \frac{10}{b}$

5. $\frac{6}{7} = \frac{21}{r}$

6. $\frac{s}{26} = \frac{10}{16}$

7. $\frac{x+7}{30} = \frac{x-3}{5}$

8. $\frac{12}{27} = \frac{z-2}{2z-3}$

In Exercises 9–12, $\frac{AB}{BD} = \frac{AC}{CE}$.

9. $AB = 7,\ BD = 14,\ AC = 5.$ Find CE.
10. $BD = 9,\ AC = 8,\ CE = 12.$ Find AB.
11. $AB = 4,\ AD = 10,\ CE = 7.5.$ Find AC.
12. $AB = 12,\ AC = 10,\ AE = 25.$ Find AD.

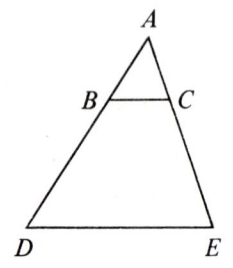

13. If the sales tax on a $7000 car is $490, find the sales tax on a $5000 car.
14. A woman 175 cm tall is only 7 cm high in a photograph. If her son is 5 cm high in the photograph, find the son's actual height.
15. A bronze alloy contains 176 g of copper and 24 g of tin. Kevin wants to make this same alloy using 120 g of tin. How much copper should he use?
16. If a basketball player who is 2 m tall casts a shadow 1.6 m long, find the height of the post holding the basket if it casts a shadow 2.8 m long.

B 17. The house that the Chans are building is drawn on a blueprint in which 1 cm represents 15 cm.
   a. How long is the living room if it is 28 cm long on the blueprint?
   b. What is the length of the dining room on the blueprint if the actual length will be 3.6 m?
18. On a certain map, 1 cm represents a distance of 120 km.
   a. What is the distance between two cities that are 5.2 cm apart on the map?
   b. If two cities are 270 km apart, what is the distance between them on the map?
19. The manager of radio station JAZZ wants the ratio of commercial time to total air time to be 2 to 15. How many minutes will be used for commercial breaks over a 3-hour period?
20. Carlos is 5 times as old as his sister Theresa. In 10 years, the ratio of their ages will be $5:3$. Find the present age of each.

Solve if the domain is the set of positive real numbers.

21. $\dfrac{4}{x} = \dfrac{x}{16}$ 22. $\dfrac{z}{2\sqrt{5}} = \dfrac{2\sqrt{5}}{10}$ 23. $\dfrac{6}{k} = \dfrac{k}{18}$

Solve.

24. $\dfrac{x+1}{x+3} = \dfrac{x-1}{x+2}$ 25. $\dfrac{x+2}{x-2} = \dfrac{3x+6}{3x+4}$ 26. $\dfrac{x+1}{x-1} = \dfrac{2x+2}{x+4}$

C 27. Chris mixed red and yellow paint in the ratio of $3:4$ instead of $4:3$ when she prepared 21 L of orange paint. How many liters of red paint should she add to obtain the shade she wants?

Section 9-3 PERCENTS

A *percent* is a special type of ratio. In this section you will explore:
- the meaning of percent
- its use in problem solving

Percents are ratios in which a number *r* is compared to 100.

EXAMPLES $\frac{37}{100} = 37\%$ $\frac{16.2}{100} = 16.2\%$ $\frac{125}{100} = 125\%$

There are two ways to solve problems involving percents. The first method involves the proportion

$$\frac{r}{100} = \frac{P}{B}$$

where *r* is the number being compared to 100 and *P* is the *part* being compared to the *whole*, *B*.

EXAMPLE   Find 75% of 24.

SOLUTION   You want to find the part, *P*.

$$\frac{75}{100} = \frac{P}{24}$$
$$\frac{3}{4} = \frac{P}{24}$$
$$72 = 4P$$
$$18 = P \qquad 75\% \text{ of } 24 \text{ is } 18.$$

EXAMPLE   10.8 is 24% of what number?

SOLUTION   You want to find the whole, *B*.

$$\frac{24}{100} = \frac{10.8}{B}$$
$$24B = 1080$$
$$B = \frac{1080}{24} = 45 \qquad 10.8 \text{ is } 24\% \text{ of } 45.$$

EXAMPLE   What percent of 25 is 30?

SOLUTION   You want to find *r*.

$$\frac{r}{100} = \frac{30}{25}$$
$$\frac{r}{100} = \frac{6}{5}$$
$$5r = 600$$
$$r = 120 \qquad 30 \text{ is } 120\% \text{ of } 25.$$

The other method of solving such problems makes use of the formula

$$RB = P$$

where *R* is the percent (usually written as a decimal), *B* is the whole, and *P* is the part.

EXAMPLE   Find 36% of 68.

SOLUTION $0.36(68) = P$
$24.48 = P$
24.48 is 36% of 68.

EXAMPLE   What percent of 16 is 19?

SOLUTION $R(16) = 19$
$R = 1.1875$
19 is 118.75% of 16.

Most problems involving percents can be solved by either method.

EXAMPLE   Each month Jan Bodeis is paid a base salary of $1200 plus a commission of 6% of her sales. In April she sold $12,000 worth of merchandise. What were her earnings in April?

SOLUTION   First find Jan Bodeis's commission earnings.

$$\frac{6}{100} = \frac{P}{12{,}000}$$
$$6(12{,}000) = 100P$$
$$72{,}000 = 100P$$
$$720 = P \quad \text{\$720 earned by commission}$$

To find the total, add the base salary to the commission.

$$\$1200 + \$720 = \$1920$$

Jan Bodeis earned $1920 in April.

EXAMPLE   The Longhill School expects a 12% decrease in enrollment next year. There are 475 students enrolled this year. How many are expected next year?

SOLUTION   To find the enrollment for next year, you first find the decrease in enrollment.

$$\frac{12}{100} = \frac{P}{475}$$
$$12 \cdot 475 = 100P$$
$$5700 = 100P$$
$$57 = P$$

To find next year's enrollment, you subtract.

$$475 - 57 = 418$$

Longhill School's enrollment for next year is expected to be 418.

EXAMPLE   Megan Daly's shares in Pedaling Cycles were worth $3000 last year. They are now worth $3540. Find the percent of increase in their value.

SOLUTION   First find the amount of increase in value.

$$\$3540 - \$3000 = \$540$$

Then use the amount of increase to determine the percent of increase.

$$R(3000) = 540$$
$$R = \frac{540}{3000}$$
$$R = 0.18 = 18\%$$

Check   Since $0.18(3000) = 540$, 540 is 18% of 3000. $\$3000 + \$540 = \$3540$

The following type of problem is most easily solved using the formula $RB = P$.

EXAMPLE Ron Dodd invested $1500, part in Allied Computers with a 9% return and the rest in Rotech, Inc., at 6%. He earned the same amount of money from each investment. How much did he invest in each company?

SOLUTION Let x = investment in Allied Computers.
Then $1500 - x$ = investment in Rotech, Inc.
Since the two investments yielded the same earnings, we get the following equation.
$$0.09(x) = 0.06(1500 - x)$$
$$0.09(x) = 90 - 0.06x$$
$$0.15x = 90$$
$$x = 600$$
$$1500 - x = 900$$

Check Do these answers meet the conditions of the problem? The earnings from both investments must be equal.
Allied Computers $0.09(600) = 54$
Rotech, Inc. $0.06(900) = 54$

They are the same, so Ron Dodd invested $600 in Allied Computers and $900 in Rotech, Inc.

ORAL EXERCISES

State the proportion you would use to solve each exercise.

1. Find 35% of 440.
2. 64% of what number is 80?
3. What percent of 180 is 99?
4. 72% of what number is 162?
5. Find 250% of 48.
6. What percent of 75 is 51?
7. Find 18% of 124.
8. What percent of 25 is 45?
9. 18% of what number is 63?
10. Find 5.4% of 65.
11. What percent of 56 is 19.6?
12. 125% of what number is 105?

13–24. For Exercises 1–12, state an equation of the form $RB = P$ you would use to solve the exercise.

WRITTEN EXERCISES

A 1–12. Solve Oral Exercises 1–12.

13. If it takes Karen 4 h to mow a lawn, what percent of the job can she do in 1 h?
14. If Ray can paint a room in 5 h, what percent of the room can he paint in 3 h?
15. In a recent year, the minimum age for a regular driver's license was 16 in 30 of the 50 states in the United States. In 7 states the minimum age was lower. What percent of the states allowed a 16-year-old to have a regular driver's license?

16. In a certain year, 25.4% of the 74,500,000 United States households with television sets watched M*A*S*H. About how many households watched this program?

17. Mark Shark has insured his boat for $23,800, which is 85% of its value. What is the value of the boat?

18. In 1969, the chemicals and minerals in the body of an average-sized human were worth $1.00. Ten years later, the same body was worth $7.43. Calculate the percent increase in the body's worth in the ten-year period.

19. The Tallman City Volunteer Fire Department has raised 78% of the goal for this year's fund drive. The volunteers have $12,480 now. How much money do they hope to raise?

20. On a unit test in biology, there were 13 A's, 14 B's, 19 C's, 11 D's, and 8 F's. What percent of the students got A's?

21. Lui Lang makes $365 a week. His deductions are 29% for federal taxes and Social Security, 1% for union dues, and 5% for state taxes. What is his take-home pay each week?

22. The Edmunds have their house insured for $68,000. It would cost $85,000 to rebuild it today. By what percent should they increase their coverage to meet today's costs?

23. If 15,000 cm^3 of air contain 11,700 cm^3 of nitrogen, calculate the percent of nitrogen in the air.

24. In the 1976 presidential election, Jimmy Carter received 297 electoral votes and Gerald Ford 240. Calculate, to the nearest tenth of a percent, the percent of these electoral votes Jimmy Carter received.

B 25. Stanley's company pays him 15% commission on all sales over $5000. If his sales this week were $8000, how much did he earn as commission?

26. A *quorum* is the number of people needed to conduct business at a meeting of an organization. Find the number needed for a quorum in each of the following instances.
   a. A quorum is a *majority* (the least integer greater than 50%). The club has 100 members.
   b. A quorum is a majority. The club has 77 members.
   c. A quorum is 75%. The club has 259 members.

27. Montvale Motors is selling the new Scarp car for 10% off the regular $5000 sticker price. There is an additional reduction of 5% of the sale price if you pay cash. How much will you pay for the car if
   a. you don't pay cash?
   b. you pay cash under the given terms?
   c. you receive instead a straight 15% discount for cash?

28. Susan Madison invested $4500, part at 8% and the rest at 10%. If she earned the same amount from each investment, how much did she invest at each rate?

29. Hank Charles invested $1000, part at 9% and the rest at 7%. He made $86 in all on his investments. How much did he invest at each rate?

30. Tri-state Electric returns $5\frac{1}{2}$% on an investment and NOW Telephone returns 8%. Eugenia invested $750 more in the telephone company than in the electric company. She made twice as much from the telephone company as from the electric company. How much did she invest in each utility?

31. The Surnee Foundation made 4% on its investment in the Theater Troupe and lost 6% of its investment in Poetry Classics. The foundation's investment in Theater Troupe was twice that in Poetry Classics. If the total profit was $120, how much was invested in Poetry Classics?

32. A 40% discount is offered by Today's Designers Company to stores buying its men's suits. The Miele Shop bought a shipment of $200 suits from Today's Designers and plans to sell them at 40% above cost. What is the price of these suits at the Miele Shop?

33. What is 35% of the number which is 145% of 36?

EXAMPLE How many milliliters of 45% antifreeze solution should be added to a 15% antifreeze solution to obtain 750 mL of 35% antifreeze solution?

SOLUTION Let x = number of milliliters of 45% solution.
Then $750 - x$ = number of milliliters of 15% solution.

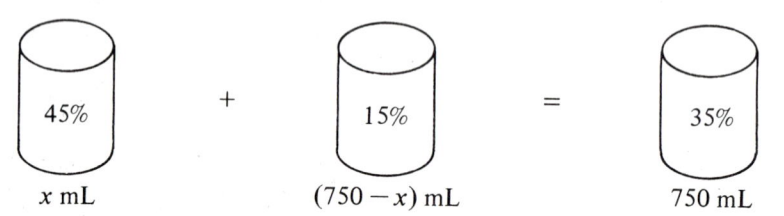

[45%] x mL + [15%] $(750 - x)$ mL = [35%] 750 mL

The amount of antifreeze in the two solutions is equal to the total amount in the final solution.

$0.45(x) + 0.15(750 - x) = 0.35(750)$ (To eliminate decimals, multiply
$45x + 15(750 - x) = 35(750)$ each side by 100.)
$45x + 11{,}250 - 15x = 26{,}250$
$30x = 15{,}000$
$x = 500$

Check 750 mL of 35% antifreeze solution are needed.
$750 - 500 = 250$
$500(0.45) + 250(0.15) = 225 + 37.5$
$= 262.5$
$= 0.35(750)$

Therefore 500 mL of 45% antifreeze solution is required.

34. How many milliliters of a 10% solution of acid should Joyce add to a 60% solution to produce 150 mL of a 30% solution?

35. John Collins has 60 L of a solution which is 40% alcohol. How many liters of a solution that is 80% alcohol should he add to produce a solution that is 50% alcohol?

C 36. How much water should be added to 20 mL of a solution that is 30% alcohol to produce a solution that is 12% alcohol?

37. To the nearest gram, how much pure salt should Pedro add to 1.5 kg of a 15% saline solution to increase its salt content to 31%?

SELF-TEST 1

1. The Circleville Cinema has 250 seats on the main floor. There are 75 balcony seats. What is the ratio of balcony seats to main-floor seats?   Section 9-1

2. Two numbers whose ratio is 8:5 have a sum of 52. Find the numbers.

3. The lengths of the sides of a triangle are in the ratio 2:3:4. The perimeter of the triangle is 108. Find the lengths of the three sides of the triangle.

4. For their last 228 games the Monumett Squires have a 7:5 win-to-loss ratio. How many games of the last 228 have they won?

Solve.

5. $\dfrac{x}{21} = \dfrac{16}{12}$ 6. $\dfrac{x-4}{8} = \dfrac{x+9}{10}$ Section 9-2

7. Cheryl's car traveled 390 km on 36 L of gas. How far can it travel on 48 L of gas?

8. McNulty's Auto Body Shop can paint 15 cars in 12 hours. How many hours would the shop need to paint 35 cars?

9. Find 36% of 650.   10. 6% of what number is 45?   Section 9-3

11. There are 1250 students at the Wilbur E. Cox High School. A recent survey of the students found that 66% of them buy their lunches at school. How many students buy their lunches at the Wilbur E. Cox High School?

12. A winter storm caused $900 worth of damage to the Duncans' home. Their insurance policy requires them to pay $250 of any claim they submit. To the nearest tenth of a percent, how much of the damage will be paid for by the insurance company?

RATIO AND SIMILARITY

Section 9-4 SIMILAR POLYGONS

If you used a copying machine to reduce the polygon shown on the left below, you might obtain the polygon shown on the right. Although these polygons do not have the same size, they do have the same shape. We have labeled the polygons and measured each side so that you can compare the figures.

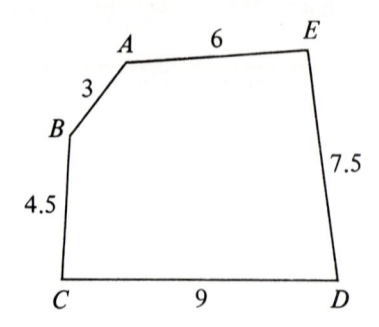

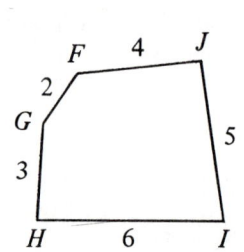

Since polygons *ABCDE* and *FGHIJ* have the same shape, their corresponding angles are congruent:

$$\angle A \cong \angle F \quad \angle B \cong \angle G \quad \angle C \cong \angle H \quad \angle D \cong \angle I \quad \angle E \cong \angle J$$

Notice that the ratio of the length of each side of polygon *ABCDE* to that of the corresponding side of polygon *FGHIJ* is 3:2.

$$\frac{AB}{FG} = \frac{3}{2} \quad \frac{BC}{GH} = \frac{4.5}{3} \quad \frac{CD}{HI} = \frac{9}{6} \quad \frac{DE}{IJ} = \frac{7.5}{5} \quad \frac{EA}{JF} = \frac{6}{4}$$

$$= \frac{45}{30} \qquad = \frac{3}{2} \qquad = \frac{75}{50} \qquad = \frac{3}{2}$$

$$= \frac{3}{2} \qquad\qquad\qquad = \frac{3}{2}$$

Since the lengths of corresponding sides have the same ratio, we say that they are *in proportion*. We write the following *extended proportion*.

$$\frac{AB}{FG} = \frac{BC}{GH} = \frac{CD}{HI} = \frac{DE}{IJ} = \frac{EA}{JF}$$

Polygons that have the same shape, such as *ABCDE* and *FGHIJ*, are said to be *similar*. The symbol ~ is used for "is similar to."

polygon *ABCDE* ~ polygon *FGHIJ*

Two polygons are *similar* if their consecutive vertices can be paired so that:
1. Corresponding angles are congruent.
2. Lengths of corresponding sides are in proportion.

EXAMPLE In the diagram at the top of page 315, quadrilateral *JKLM* ~ quadrilateral *PQRS*. Find the measure of ∠*J* and the length of $\overline{QR}$.

314   CHAPTER 9

SOLUTION Since corresponding angles are congruent,
$\angle J \cong \angle P$. Thus, $m\angle J = 130$.

Because lengths of corresponding sides are in proportion:

$$\frac{KL}{QR} = \frac{JK}{PQ}$$
$$\frac{6}{QR} = \frac{4}{5}$$
$$6 \cdot 5 = QR \cdot 4$$
$$\frac{30}{4} = QR$$
$$7.5 = QR$$

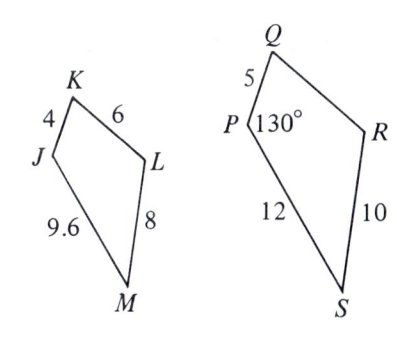

In the diagram, $\triangle ABC \sim \triangle DEF$. Thus, the lengths of corresponding sides are in porportion:

$$\frac{AB}{DE} = \frac{BC}{EF} = \frac{AC}{DF} = \frac{1}{3}$$

What do you think is the ratio of the perimeters of the triangles?

Perimeter of $\triangle ABC = 9$ Perimeter of $\triangle DEF = 27$

The ratio of the perimeters is $\frac{9}{27}$, or $\frac{1}{3}$.

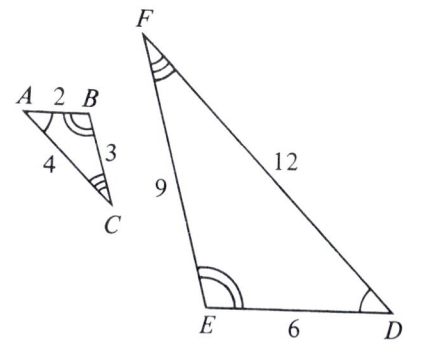

Thus, the ratio of the perimeters of $\triangle ABC$ and $\triangle DEF$ is the same as the ratio of the lengths of the corresponding sides. In fact, this is true for *all* pairs of similar polygons.

If two polygons are similar, the ratio of their perimeters is equal to the ratio of the lengths of any two corresponding sides.

EXAMPLE In the diagram, quadrilateral $GHIJ \sim$ quadrilateral $KLMN$.
Find the perimeter, P, of quadrilateral $KLMN$.

SOLUTION Since side $\overline{HI}$ corresponds to side $\overline{LM}$, the ratio of the lengths of any two corresponding sides is $3:5$. The perimeters have the same ratio:

$$\frac{\text{Perimeter of } GHIJ}{\text{Perimeter of } KLMN} = \frac{3}{5}$$
$$\frac{2+3+4.5+7}{P} = \frac{3}{5}$$
$$(16.5)(5) = 3P$$
$$82.5 = 3P$$
$$27.5 = P$$

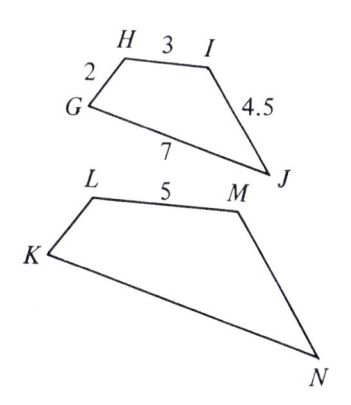

RATIO AND SIMILARITY

From the definition of similar polygons, you know that two triangles are similar if three pairs of corresponding angles are congruent and the ratios of the lengths of three pairs of corresponding sides are equal. However, as in the case of congruent triangles, you do not actually need all this information to show that two triangles are *similar*.

> If two angles of one triangle are congruent to two angles of a second triangle, then the triangles are similar. (AA)

EXAMPLE If $\angle B \cong \angle C$, show that $\triangle ABE \sim \triangle DCE$.

SOLUTION
$\angle B \cong \angle C$ (Given)
$\angle AEB \cong \angle DEC$ (Vertical angles are congruent.)
$\triangle ABE \sim \triangle DCE$ (AA)

EXAMPLE To find the height of a utility pole, Ludwig stands so that the top of his shadow lines up with the top of the pole's shadow. Ludwig is 1.6 m tall, his shadow is 2 m long, and the pole's shadow is 8 m long. How high is the pole?

SOLUTION $\triangle ABE \sim \triangle CDE$ because
$\angle ABE \cong \angle CDE$ and $\angle AEB \cong \angle CED$.
Therefore, $\dfrac{AB}{CD} = \dfrac{BE}{DE}$.

You know that $CD = 1.6$, $BE = 8$, and $DE = 2$.

$$\dfrac{AB}{1.6} = \dfrac{8}{2}$$

$$\dfrac{AB}{1.6} = 4$$

$$AB = 4(1.6) = 6.4$$

The pole is 6.4 m high.

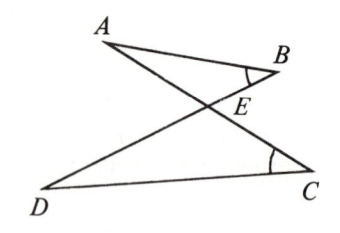

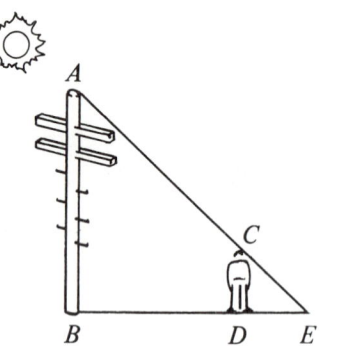

ORAL EXERCISES

Classify each statement as true or false.

1. All squares are similar.
2. If two angles of one quadrilateral are congruent to two angles of a second quadrilateral, then the quadrilaterals must be similar.
3. If two polygons are congruent, then they are similar.
4. All equilateral triangles are similar.

5. a. State which triangles shown below are similar. Give a reason for each answer.
  b. State an extended proportion for each pair of similar triangles.

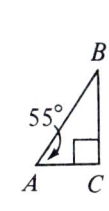

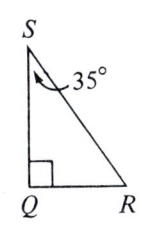

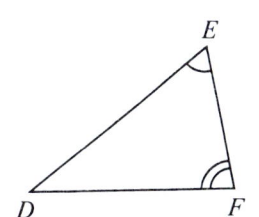

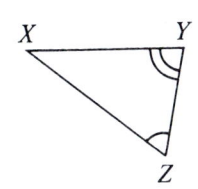

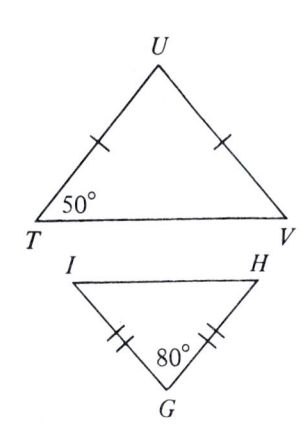

WRITTEN EXERCISES

Exercises 1–4 refer to the diagram at the right, in which $\triangle DEF \sim \triangle GHI$.

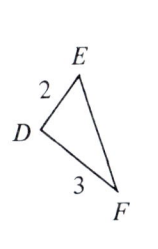

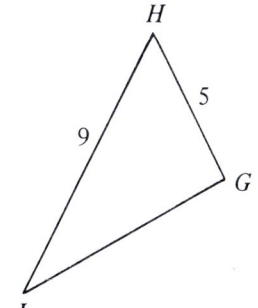

A 1. Determine $\dfrac{DE}{GH}$.

2. Determine $\dfrac{EF}{HI}$.

3. Find EF.

4. Find the perimeter of $\triangle GHI$.

Exercises 5–11 refer to the figure, in which $\overline{ST} \parallel \overline{PQ}$.

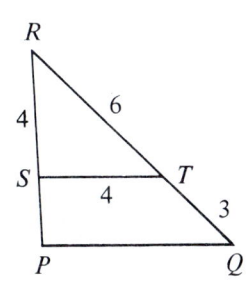

5. Complete: $\angle P \cong \angle \underline{\ ?\ }$

6. Complete: $\angle RTS \cong \angle \underline{\ ?\ }$

7. Name two similar triangles.

8. Find $\dfrac{RT}{RQ}$.

9. Find PQ.

10. Find RP.

11. Find the perimeter of $\triangle RPQ$.

RATIO AND SIMILARITY   317

Exercises 12-14 refer to the figure, in which $\overline{DE} \parallel \overline{BC}$.

12. Find $\dfrac{AD}{AB}$.

13. If $DE = x$ and $BC = x + 3$, find the value of x.

14. If $AD = 15$, find DB.

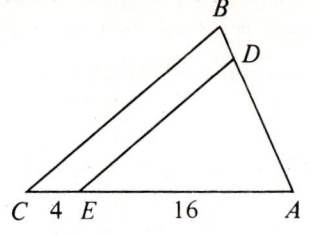

15. The two quadrilaterals on the right are similar. Find the values of w, x, y, and z.

16. The lengths of the sides of a triangle are 16, 24, and 20. If the length of the longest side of a similar triangle is 8, find the length of its shortest side.

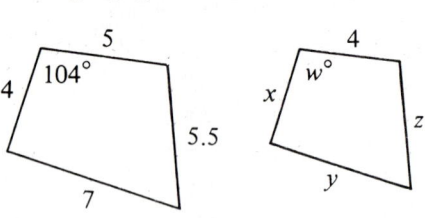

17. The lengths of the sides of a triangle are 4, 5, and 7. If the length of the shortest side of a similar triangle is 6, find the length of its longest side.

18. The lengths of the sides of a triangle are 15, 18, and 24. If the length of the shortest side of a similar triangle is 10, find its perimeter.

19. The lengths of the sides of a triangle are 7, 12, and 9. If the perimeter of a similar triangle is 14, find the length of the longest side.

Exercises 20-23 refer to the figure, in which $\overline{QR} \perp \overline{RU}$ and $\overline{QS} \perp \overline{ST}$.

B 20. Find $\dfrac{QR}{QS}$.

21. Find ST.

22. Find the area of $\triangle QST$.

23. Find the area of trapezoid $RSTU$.

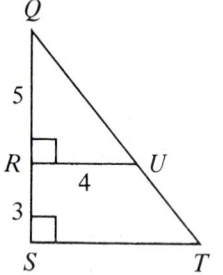

24. The diagram below illustrates a means of finding the width of a river. Use the given measurements to find the value of x.

Ex. 24

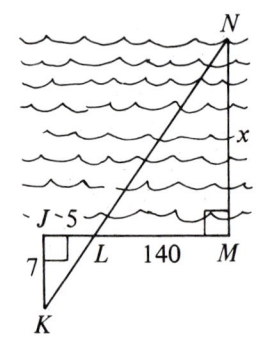

Ex. 25

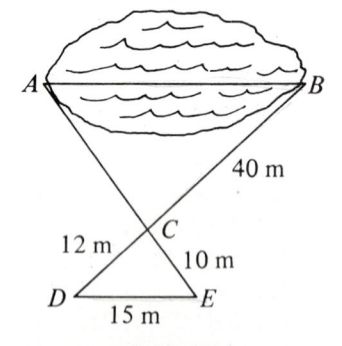

25. In the figure, $\angle A \cong \angle E$. Use the given measurements to find the length of the pond.

318   CHAPTER 9

26. A tree casts a shadow 64 m long at the same time a person 1.8 m tall casts a shadow 8 m long. Find the height of the tree.

EXAMPLE $\triangle JKL \sim \triangle JMN$
$JN = 8$, $JL = 5$, $KM = 2$.
Find JK.

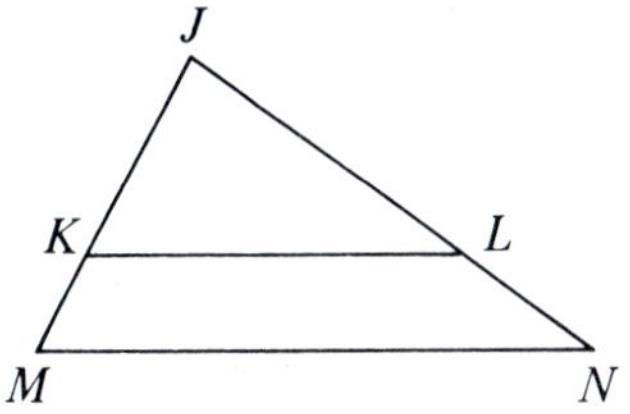

SOLUTION Since $\triangle JKL \sim \triangle JMN$,
$$\frac{JL}{JN} = \frac{JK}{JM}.$$
Let $JK = x$.
Then $JM = x + 2$.
$$\frac{5}{8} = \frac{x}{x+2}$$
$5(x + 2) = 8x$
$5x + 10 = 8x$
$10 = 3x$
$3\frac{1}{3} = x$ That is, $JK = 3\frac{1}{3}$.

In Exercises 27–30, $\triangle ABC \sim \triangle ADE$.

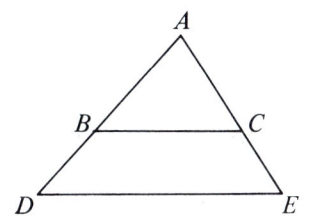

27. $AD = 10$, $AB = 6$, $CE = 5$. Find AC.
28. $AC = 8$, $CE = 2$, $BD = 1.5$. Find AB.
29. $AE = 10$, $CE = 3$, $DE = 12$. Find BC.
30. $DE = 10$, $BC = 8$, $CE = 4$. Find AE.

31. Replace each "why" with the correct reason.

Given: $\triangle ABC \sim \triangle QRS$.
$\overline{BX}$ is an altitude to $\overline{AC}$.
$\overline{RY}$ is an altitude to $\overline{QS}$.
Show: Corresponding altitudes of similar triangles are in the same ratio as corresponding sides.

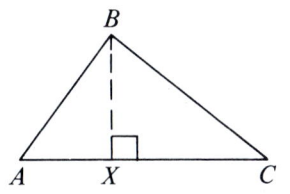

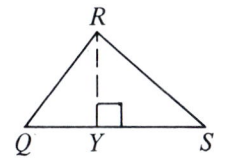

(1) $\triangle ABC \sim \triangle QRS$ (Given)
(2) $\angle C \cong \angle S$ (Why?)
(3) $\angle BXC \cong \angle RYS$ (Why?)
(4) $\triangle BXC \sim \triangle RYS$ (Why?)
(5) $\dfrac{BX}{RY} = \dfrac{BC}{RS}$ (Why?)

RATIO AND SIMILARITY

32. Find the perimeter of quadrilateral ABCD.

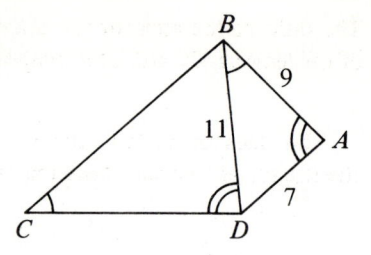

33. If △FGH ~ △JKL, m∠F = 70, and m∠L = 40, find $\dfrac{KL}{JL}$.

C 34. Replace each "why" with the correct reason.
Given: $\overline{CD}$ is the altitude to the hypotenuse of right △ABC.
Show: △ACD ~ △ABC
 (1) $\overline{CD}$ is the altitude to $\overline{AB}$. (Given)
 (2) $\overline{CD} \perp \overline{AB}$ (Definition of altitude)
 (3) ∠ADC and ∠ACB are right angles. (Why?)
 (4) ∠ADC ≅ ∠ACB (Why?)
 (5) ∠CAD ≅ ∠BAC (Why?)
 (6) △ACD ~ △ABC (Why?)

35. Use an argument like the one in Exercise 34 to show that △CBD ~ △ABC.
36. Use the results of Exercise 34 to show that $(AC)^2 = AB \cdot AD$.
37. Use the results of Exercise 35 to show that $(BC)^2 = AB \cdot BD$.

Section 9-5 AREAS AND VOLUMES OF SIMILAR FIGURES

What is the ratio of the areas of the two squares shown below?

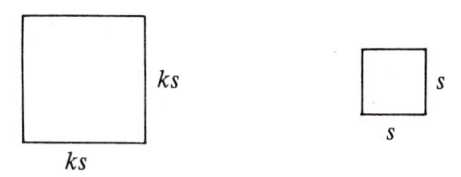

You can see immediately that the ratio of the length of each side of the smaller square to the length of the corresponding side of the larger square is $s : ks$, or $1 : k$. You might guess that the ratio of the areas is also $1 : k$. Let's check to see if this is so.

Area of smaller square: $A_1 = s^2$
Area of larger square: $A_2 = (ks)^2 = k^2 s^2$

Ratio of the areas:
$$\dfrac{A_1}{A_2} = \dfrac{s^2}{k^2 s^2}$$
$$\dfrac{A_1}{A_2} = \dfrac{1}{k^2} = \left(\dfrac{1}{k}\right)^2$$

320   CHAPTER 9

The ratio of the areas of the squares is $1:k^2$, which equals the *square* of the ratio of the lengths of a pair of corresponding sides.

How does the ratio of the areas of two similar triangles compare to the ratio of the lengths of two corresponding sides? In the figure below, $\triangle ABC \sim \triangle DEF$.

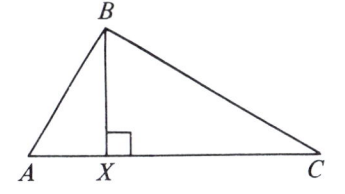

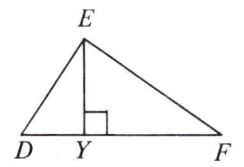

From Exercise 31 on page 319, you know that the lengths of corresponding altitudes of the triangles are in the same ratio as the lengths of any two corresponding sides.

$$\frac{BX}{EY} = \frac{AC}{DF}$$

$$\text{Area of } \triangle ABC = \frac{1}{2}(AC)(BX)$$

$$\text{Area of } \triangle DEF = \frac{1}{2}(DF)(EY)$$

$$\frac{\text{Area of } \triangle ABC}{\text{Area of } \triangle DEF} = \frac{\frac{1}{2}(AC)(BX)}{\frac{1}{2}(DF)(EY)}$$

$$= \frac{AC}{DF} \cdot \frac{BX}{EY}$$

$$= \frac{AC}{DF} \cdot \frac{AC}{DF}$$

$$= \left(\frac{AC}{DF}\right)^2$$

Since $\dfrac{AC}{DF} = \dfrac{AB}{DE} = \dfrac{BC}{EF}$, you can write:

$$\frac{\text{Area } \triangle ABC}{\text{Area } \triangle DEF} = \left(\frac{AC}{DF}\right)^2 = \left(\frac{AB}{DE}\right)^2 = \left(\frac{BC}{EF}\right)^2$$

In general, the following is true for any two similar polygons.

The ratio of the areas of two similar polygons is equal to the square of the ratio of the lengths of any two corresponding sides.

RATIO AND SIMILARITY

EXAMPLE Polygon ABCDE ~ Polygon QRSTU, BC = 4, and RS = 7.
If the area of polygon ABCDE is 48, find the area of polygon QRSTU.

SOLUTION $\dfrac{\text{area of polygon } ABCDE}{\text{area of polygon } QRSTU} = \left(\dfrac{BC}{RS}\right)^2$

$$\frac{48}{x} = \left(\frac{4}{7}\right)^2$$

$$\frac{48}{x} = \frac{16}{49}$$

$$16x = 48 \cdot 49$$

$$x = \frac{48 \cdot 49}{16}$$

$$x = 3 \cdot 49$$

$$x = 147$$

The area of polygon QRSTU is 147.

Until now you have studied similarity as it relates to plane figures only, but there are *similar solids* as well. Two solids are similar if they have the same shape. For example, all cubes are similar.

Consider the two cubes shown below. How does the ratio of their volumes compare to the ratio of the lengths of the edges?

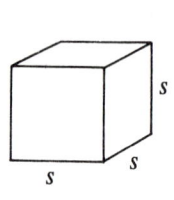

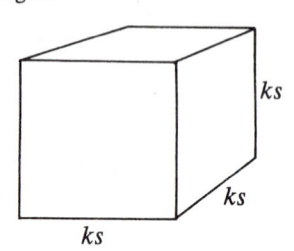

Volume of smaller cube: $V_1 = s^3$
Volume of larger cube: $V_2 = (ks)^3 = k^3 s^3$

$$\frac{V_1}{V_2} = \frac{s^3}{k^3 s^3} = \frac{1}{k^3} = \left(\frac{1}{k}\right)^3$$

The ratio of the volumes of the two cubes is equal to the cube of the ratio of the lengths of their edges.

The following is true for all similar solids, including prisms and pyramids.

> The ratio of the volumes of two similar solids is equal to the cube of the ratio of the lengths of two corresponding segments.

EXAMPLE The length of each edge of a rectangular right prism is doubled. How is the volume affected?

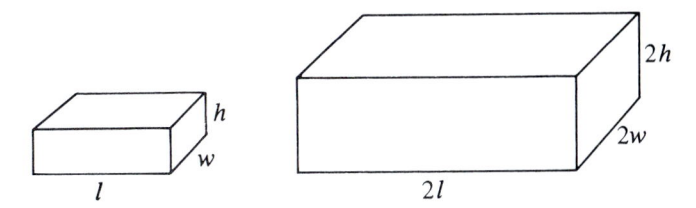

SOLUTION The ratio of the lengths of two corresponding segments is $1:2$. Therefore, the ratio of the volumes of the solids is $\left(\frac{1}{2}\right)^3$, or $\frac{1}{8}$. If the length of each edge is doubled, then the volume is multiplied by 8.

ORAL EXERCISES

Find the ratio of the areas of two similar triangles in which the ratio of the lengths of two corresponding sides is:

1. $2:3$
2. $5:11$
3. $7:4$
4. $\sqrt{2}:1$

Find the ratio of the volumes of two similar solids in which the ratio of the lengths of two corresponding segments is:

5. $2:3$
6. $5:4$
7. $3:10$
8. $\sqrt{2}:1$

Find the ratio of the lengths of a pair of corresponding sides of two similar triangles if the ratio of their areas is:

9. $9:25$
10. $81:16$
11. $49:100$
12. $1:3$

WRITTEN EXERCISES

A
1. Two equilateral triangles have sides of lengths 12 cm and 18 cm. Find the ratio of their areas.
2. Two triangles are similar. The longest side of one triangle is five times as long as the corresponding side of the second triangle. How many times greater is the area of the larger triangle than the area of the smaller triangle?
3. The ratio of the areas of two similar polygons is $49:144$. Find the ratio of their perimeters.
4. Two similar quadrilaterals have areas 14 cm² and 59 cm². Find the ratio of their perimeters.

RATIO AND SIMILARITY

Exercises 5-8 refer to the figure at the right in which $\triangle ABC \sim \triangle DEF$.

5. If $AB = 12$ and $DE = 9$, find the ratio of the areas of $\triangle ABC$ and $\triangle DEF$.

6. If $AC = 24$ and $DF = 12$, what fractional part of the area of $\triangle ABC$ is the area of $\triangle DEF$?

7. If the area of $\triangle ABC$ is 36 cm², the area of $\triangle DEF$ is 16 cm², and $\overline{AC}$ is 12 cm long, find DF.

8. If $AB = 18$, $DE = 16$, and the area of $\triangle DEF = 96$, find the area of $\triangle ABC$.

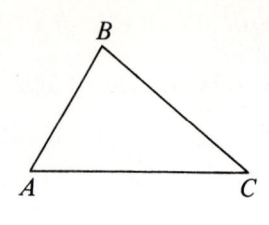

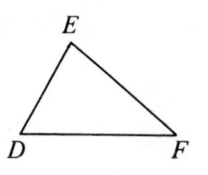

9. Two rectangular prisms are similar and the length of one edge of the larger prism is three times the length of the corresponding edge of the smaller prism. Find the ratio of their volumes.

10. If each edge of a pyramid is quadrupled in length, by what number is the volume multiplied?

11. Two similar right rectangular prisms have corresponding edges whose lengths are 16 cm and 12 cm. If the volume of the smaller prism is 81 cm³, find the volume of the larger prism.

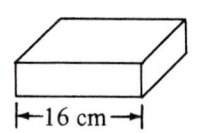

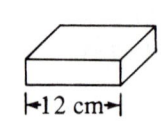

12. In two similar triangular pyramids the lengths of two corresponding edges are 24 cm and 8 cm. If the volume of the larger pyramid is 243 cm³, find the volume of the smaller pyramid.

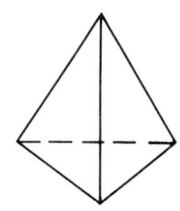

In Exercises 13-16, $\triangle ABC \sim \triangle ADE$.

B 13. If $AB = 3$, $BD = 4$, and the area of $\triangle ABC = 18$, find the area of $\triangle ADE$.

14. If the area of $\triangle ABC$ is 108, the area of $\triangle ADE$ is 147, and $AC = 12$, find CE.

15. If $AB:BD = 2:3$, and the area of $\triangle ABC$ is 15, find the area of $\triangle ADE$.

16. If $BC = 4\sqrt{3}$, $DE = 8$, and the area of $\triangle ADE$ is 28, find the area of $\triangle ABC$.

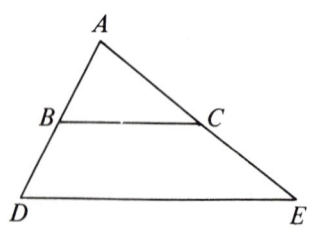

Exs. 13-16

17. A photographer wants to enlarge a square print so that its area will be doubled. How many times as long as the original will each side of the enlarged print be?

18. The area of one equilateral triangle is three times the area of a smaller equilateral triangle of side 8 cm. What is the length of a side of the larger triangle?

19. In the figure, $\overline{RS} \parallel \overline{TU}$.
    If $QR = 12$, $RT = 8$, and the area of $\triangle QRS = 108$, find:
    a. the area of $\triangle QTU$.
    b. the area of trapezoid $RTUS$.

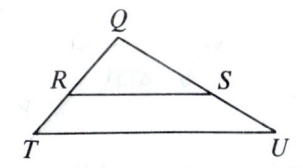

C 20. The area of the larger of two equilateral triangles is 25% greater than the area of the smaller. If the length of one side of the larger triangle is 15 cm, find the length of a side of the smaller triangle.

21. The difference between the lengths of the sides of two equilateral triangles is 12 cm. If the ratio of their areas is $1:2$, find the length of a side of the smaller triangle.

22. The capacity of a column to support mass is directly proportional to the area of its cross section. If the larger of two similar columns is three times as high as the smaller column, what is the ratio of the masses they can support?

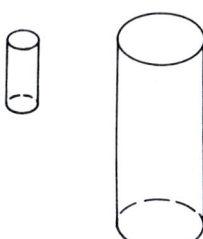

SELF-TEST 2

Exercises 1 and 2 refer to the figure, in which $\overline{LN} \parallel \overline{KJ}$.

1. Find MJ.
2. Find the perimeter of $\triangle LMN$.

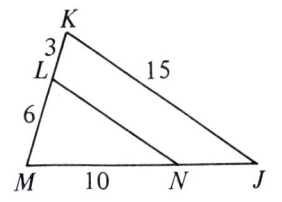

Section 9-4

3. The lengths of the sides of a triangle are 15, 20, and 28. If the longest side of a similar triangle has length 12, find the perimeter of the smaller triangle.

4. The ratio of the lengths of two corresponding sides of two similar polygons is $5:11$. Find the ratio of their areas.

Section 9-5

5. The perimeters of two similar triangles are 18 and 42. Find the ratio of their areas.

6. The length of each edge of one pyramid is 5 times the length of each edge of a similar pyramid. Find the ratio of their volumes.

7. The volume of the larger of two similar right rectangular prisms is 480 cm³. The heights of the prisms are 9 cm and 12 cm, respectively. Find the volume of the smaller prism.

Section 9-6 THE PYTHAGOREAN THEOREM

In this section you will learn:
- an important relationship involving the sides of a right triangle
- how to solve problems using this relationship

The side opposite the right angle of a right triangle is called the *hypotenuse* of the triangle. The other two sides are called *legs*. In the figure, the hypotenuse of $\triangle ABC$ is $\overline{AB}$, and c is its length. The legs are $\overline{AC}$ and $\overline{BC}$, with lengths b and a.

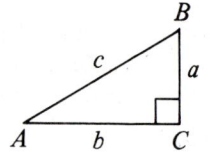

What is the relationship between a, b, and c? You can use the fact that the area of $\triangle ABC$ is $\frac{1}{2}ab$ to investigate this question. Extend the legs of the triangle to form two sides of a square with sides of length $a + b$, as shown on the left below.

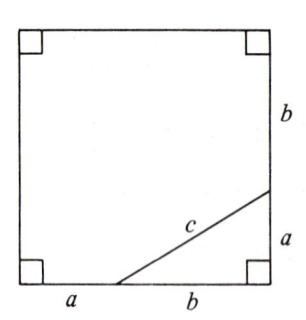

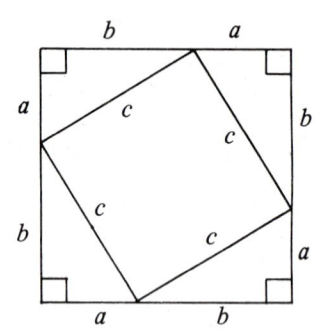

Then draw three more triangles whose legs have lengths a and b, as shown on the right. By SAS, the four right triangles are congruent. Thus, the length of each hypotenuse is c. As you will show in Oral Exercise 16, each angle of the inner quadrilateral is a right angle. Thus, that quadrilateral is a square with area c^2.

Area of larger square = Area of smaller square + Areas of four triangles

$$(a + b)^2 = c^2 + 4\left(\frac{1}{2}ab\right)$$
$$a^2 + 2ab + b^2 = c^2 + 2ab$$
$$a^2 + b^2 = c^2$$

> In a right triangle, the square of the length of the hypotenuse is equal to the sum of the squares of the lengths of the legs.

This relationship has been known for thousands of years. Because many believe that the Greek philosopher and mathematician Pythagoras was the first to show that the statement is true, it is known as the Pythagorean Theorem.

EXAMPLE Find the length of the diagonal of a rectangle whose sides have lengths 7 and 24.

SOLUTION
$d^2 = 24^2 + 7^2$
$= 576 + 49$
$= 625$

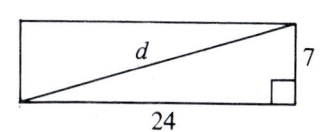

Since the length must be a positive number, $d = 25$.

EXAMPLE A ladder 4 m long leans against a building. If the foot of the ladder is 1 m from the building, how high up the building does the ladder reach?

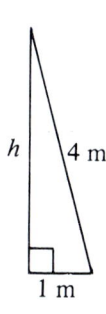

SOLUTION
$h^2 + 1^2 = 4$
$h^2 + 1 = 16$
$h^2 = 15$
$h = \sqrt{15}$

To the nearest tenth of a meter, the top of the ladder is 3.9 m from the ground.

The converse of the Pythagorean Theorem is also true.

> If the square of the length of one side of a triangle is equal to the sum of the squares of the lengths of the other two sides, then the triangle is a right triangle.

EXAMPLE △PQR has sides of length 8, 15, and 17, as shown. Show that ∠P is a complement of ∠Q.

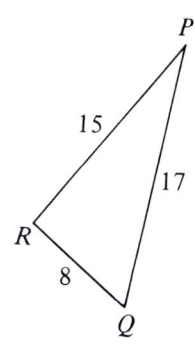

SOLUTION
$(RQ)^2 + (PR)^2 = 8^2 + 15^2$ $(PQ)^2 = 17^2$
$ = 64 + 225$ $ = 289$
$ = 289$
$(RQ)^2 + (PR)^2 = (PQ)^2$

By the converse of the Pythagorean Theorem, △PQR is a right triangle. $\overline{PQ}$ is the hypotenuse, and ∠R is the right angle. Because the two acute angles of a right triangle are complementary, ∠P is the complement of ∠Q.

ORAL EXERCISES

For each right triangle, tell which variable represents the length of the hypotenuse.

1. 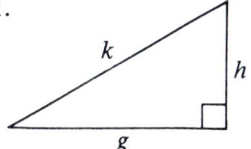
2. 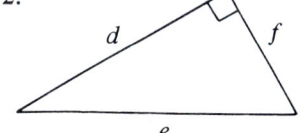
3.

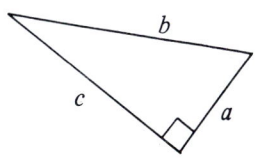

RATIO AND SIMILARITY

State the equation you would use to solve each of the following.

4.

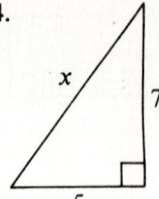

5.

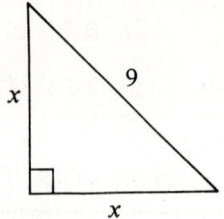

6.

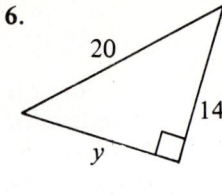

What is the square of each of the following?

7. 11

8. $\sqrt{3}$

9. $3\sqrt{5}$

What is the principal square root of each of the following? Give irrational numbers in simplest radical form.

10. 144

11. 50

12. 12

Is the triangle whose sides have the given lengths a right triangle?

13. 6, 8, 10

14. $\sqrt{2}, \sqrt{3}, \sqrt{5}$

15. 5, 13, 12

16. a. Why are ∠3 and ∠4 complementary?
    b. Why are ∠1 and ∠4 congruent?
    c. $m\angle 1 + m\angle 3 = \underline{\quad?\quad}$
    d. $m\angle 2 = \underline{\quad?\quad}$

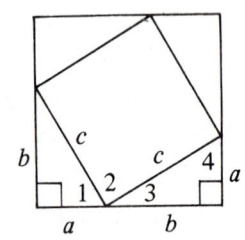

WRITTEN EXERCISES

Find the value of x in each right triangle. Give irrational answers in simplest radical form.

A 1.

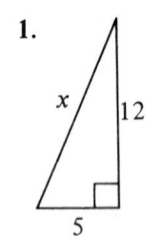

2.

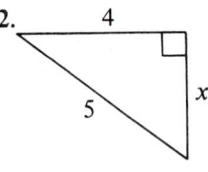

3.

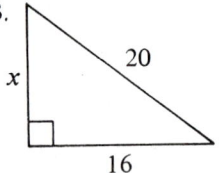

4.

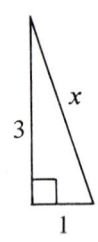

5.

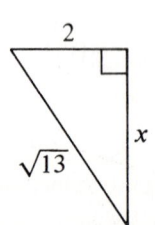

6.

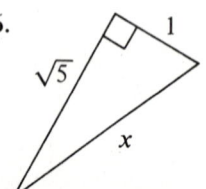

7.

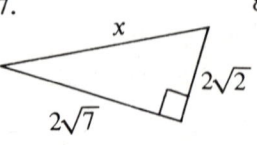

8.

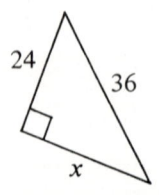

328   CHAPTER 9

Is the triangle whose sides have the given lengths a right triangle?

9. 2, 3, 5
10. $\frac{3}{5}, \frac{4}{5}, 1$
11. 8, 17, 15
12. 13, 15, 17
13. 5, 5, 5√2
14. 7, 14, 7√3
15. 2√5, √2, 3√2
16. 4, 4√5, 4√6

For Exercises 17 and 18, use the table on page 253.

17. To the nearest tenth of a meter, find the length of each diagonal of a rectangle with length 5 m and width 2 m.

18. To the nearest tenth of a centimeter, find the width of a rectangle whose length is 6 cm and whose diagonal is 8 cm long.

19. The ratio of the lengths of the sides of a right triangle is 3:4:5. Find the length of each leg if the length of the hypotenuse is:
    a. 10
    b. 15
    c. 50
    d. 2

20. The ratio of the lengths of the sides of a right triangle is 5:12:13. Find the lengths of the other sides if the length of the shortest side is:
    a. 10
    b. 15
    c. 50
    d. 1

B 21. If the lengths of the two longest sides of a right triangle are 3 and 4, find the length of the shortest side.

22. If the lengths of the two longest sides of a right triangle are 24 and 26, find the area of the triangle.

Exercises 23–28 refer to rhombus $PQRS$ whose diagonals intersect in point T. (Recall that the diagonals of any parallelogram bisect each other and that the diagonals of a rhombus are perpendicular.)

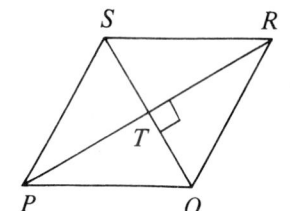

23. If $TR = 5$ and $ST = \sqrt{11}$, find SR.
24. If $PQ = 8$ and $TQ = 4$, find PT.
25. If PT is twice the length of ST and $SP = 4\sqrt{5}$, find ST.
26. If $PT:TQ = 3:2$ and $PQ = 2\sqrt{13}$, find PT.
27. If $SR = 6$ and $SQ = 2\sqrt{11}$, find PR.
28. If the perimeter of rhombus $PQRS$ is 32 and $SQ = 2\sqrt{15}$, find PR.

29. Write the Pythagorean Theorem and its converse as a biconditional.

30. In the figure shown, $\overline{AB}, \overline{BC}, \overline{CD}, \overline{DE}, \overline{EF},$ and $\overline{FG}$ each have length 1. Complete.
    a. $AC = \underline{\ ?\ }$
    b. $AD = \underline{\ ?\ }$
    c. $AE = \underline{\ ?\ }$
    d. $AF = \underline{\ ?\ }$
    e. $AG = \underline{\ ?\ }$

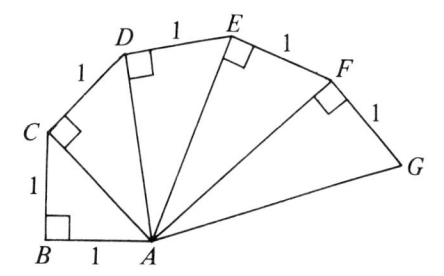

Exercises 31 and 32 refer to the square pyramid shown, with altitude $\overline{EF}$ and slant height EG. Point F is the intersection of diagonals $\overline{AC}$ and $\overline{BD}$. G is the midpoint of $\overline{BC}$.

31. If $AB = 6$ and $EF = 4$, complete.
    a. $FG = $ _?_
    b. $EG = $ _?_
    c. $GC = $ _?_
    d. $EC = $ _?_
    e. Volume of the pyramid = _?_

32. If $AB = 10$ and $EB = 13$, complete.
    a. $BG = $ _?_
    b. $EG = $ _?_
    c. $FG = $ _?_
    d. $EF = $ _?_
    e. Lateral area of the pyramid = _?_
    f. Total area of the pyramid = _?_

EXAMPLE   Find the lengths of the sides of the triangle shown.

SOLUTION
$(3x + 5)^2 = (2x + 1)^2 + (2x + 8)^2$
$9x^2 + 30x + 25 = 4x^2 + 4x + 1 + 4x^2 + 32x + 64$
$9x^2 + 30x + 25 = 8x^2 + 36x + 65$
$9x^2 - 8x^2 + 30x - 36x + 25 - 65 = 0$
$x^2 - 6x - 40 = 0$
$(x - 10)(x + 4) = 0$
$x - 10 = 0$ or $x + 4 = 0$
    $x = 10$ or $x = -4$

Since $2x + 1$ must be positive, we reject $x = -4$.
$2x + 1 = 21$ $2x + 8 = 28$ $3x + 5 = 35$

Find the lengths of the sides of each triangle.

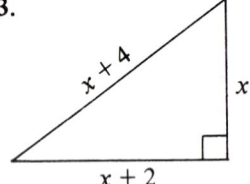

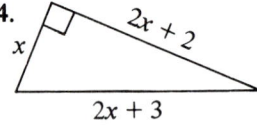

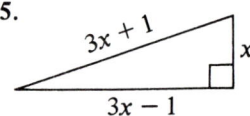

33. (triangle with sides $x+4$, x, $x+2$)

34. (right triangle with legs x, $2x+3$ and hypotenuse $2x+2$)

35. (right triangle with sides $3x+1$, x, $3x-1$)

36. The length of the hypotenuse of a right triangle is 3 mm longer than one leg and 6 mm longer than the other. Find the length of the hypotenuse.

37. In ancient times the Chinese used the Pythagorean Theorem to solve the following problem: A bamboo plant 2 m long was bent so that the top of the bamboo touched the ground 1 m from its base. At what height did the break occur?

C   **38.** The perimeter of a right triangle is 90 cm. The length of the hypotenuse is 41 cm. Find the length of the longer leg.

39. The perimeter of a right triangle is 80 mm. The length of the hypotenuse is 34 mm. Find the length of the shorter leg.

40. The Babylonians considered this problem: A ladder is flush against a wall. When it slides down the wall 30 cm, it slides out 90 cm. What is the length of the ladder?

41. Another way to show that the Pythagorean Theorem is true is based on similarity. (See Exercises 34-37 of Section 9-4.)
    Given: Right $\triangle ABC$, with $\overline{CD}$ the altitude to the hypotenuse.
    Show: $(AB)^2 = (AC)^2 + (BC)^2$

$AB \cdot AD = (AC)^2$	(Exercise 36, page 288)
$AB \cdot BD = (BC)^2$	(Exercise 37, page 288)
$\underline{?} = (AC)^2 + (BC)^2$	(Addition)
$AB(\underline{?} + \underline{?}) = (AC)^2 + (BC)^2$	(Distributive property)
$AB(\underline{?}) = (AC)^2 + (BC)^2$	($AD + BD = AB$)

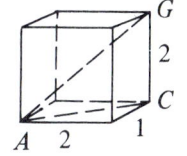

42. The rectangular prism shown has length 2, width 1, and height 2. Complete.
    a. $AC =$ __?__
    b. $AG =$ __?__
    $\overline{AG}$ is called a *diagonal* of the prism.

43. What is the length of a diagonal of a cube whose edges each have length 2?

44. What is the length of the longest stick that can be placed inside a box that is 3 cm wide, 4 cm long, and 12 cm high? (Disregard the width of the stick.)

45. What is the length of the longest stick that can be placed inside a box that is 32 cm long, 24 cm wide, and 9 cm high?

46. Another ancient Chinese problem is the following one: The length of each side of a square-shaped pond is 10 m. A reed growing from the center of the pond is 1 m higher than the surface of the water. When it is pulled to a corner of the pond, it just touches the surface. Find the depth of the pond.

Section 9-7 SPECIAL RIGHT TRIANGLES

Certain right triangles occur frequently in geometric figures. In this section you will learn about the ratio of the lengths of the sides of each:
- 45°-45°-90° triangle
- 30°-60°-90° triangle

Since the acute angles of an isosceles right triangle are both congruent (why?) and complementary, the measure of each is 45. For this reason, an isosceles right triangle is often called a 45°-45°-90° triangle. To find the relationship between the lengths of the sides of a 45°-45°-90° triangle, we use the Pythagorean Theorem, as shown on the next page.

EXAMPLE   If each leg of a 45°-45°-90° triangle has length 1, find the length of the hypotenuse.

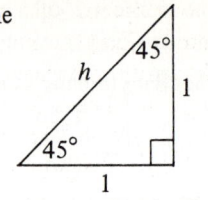

SOLUTION $h^2 = 1^2 + 1^2$
$= 1 + 1$
$= 2$
$h = \sqrt{2}$

Notice that the ratio of the lengths of the sides of the triangle in the example is $1:1:\sqrt{2}$. Because every 45°-45°-90° triangle is similar to the triangle in the example, we have the following:

> In a 45°-45°-90° triangle, the length of the hypotenuse is equal to the length of a leg multiplied by $\sqrt{2}$.

EXAMPLE   Find the value of x in each figure.

a.

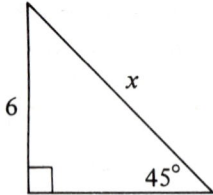

b.

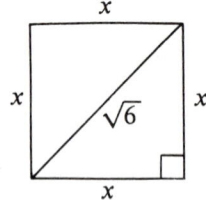

SOLUTION   All triangles shown are 45°-45°-90° triangles.

a. $x = 6\sqrt{2}$

b. $x\sqrt{2} = \sqrt{6}$
$x = \dfrac{\sqrt{6}}{\sqrt{2}}$
$x = \sqrt{3}$

EXAMPLE   The ratio of the perimeters of two isosceles right triangles is 2:3. If the length of each leg of the smaller triangle is 14, find the length of the hypotenuse of the larger triangle.

SOLUTION   To find the length, z, of the hypotenuse of the larger triangle, first find the length, y, of each leg of the larger triangle.

$\dfrac{2}{3} = \dfrac{14}{y}$
$2 \cdot y = 3 \cdot 14$
$2y = 42$
$y = 21$
Thus, $z = 21\sqrt{2}$.

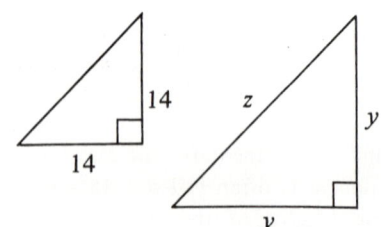

Another special right triangle is a 30°-60°-90° triangle. To find the relationship between the lengths of the sides of a 30°-60°-90° triangle, consider an equilateral triangle whose sides have length 2. Notice that a bisector of an angle of this triangle divides the equilateral triangle into two congruent 30°-60°-90° triangles. The shorter legs of each of these 30°-60°-90° triangles are congruent and each has length 1. To find the length, x, of the longer leg, use the Pythagorean Theorem:

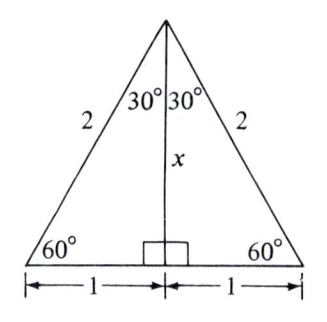

$$2^2 = 1^2 + x^2$$
$$4 = 1 + x^2$$
$$3 = x^2$$
$$\sqrt{3} = x$$

Thus, the sides of each 30°-60°-90° triangle shown are in the ratio $1 : 2 : \sqrt{3}$, with the shorter leg opposite the 30° angle and the longer leg opposite the 60° angle. Since these triangles are similar to all 30°-60°-90° triangles, the following is true.

In a 30°-60°-90° triangle, the hypotenuse is twice as long as the shorter leg. The length of the longer leg is equal to the length of the shorter leg multiplied by $\sqrt{3}$.

EXAMPLE In each triangle the length of one side is given. Find the lengths of the other sides.

a.

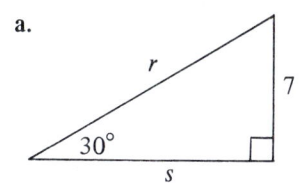

b.

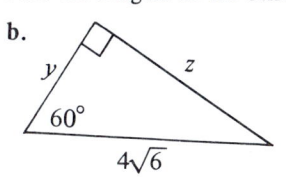

SOLUTION Each triangle is a 30°-60°-90° triangle.

a. $r = 2 \cdot 7 = 14$
$s = 7\sqrt{3}$

b. $4\sqrt{6} = 2y$
$2\sqrt{6} = y$
$z = (2\sqrt{6})\sqrt{3} = 2\sqrt{18} = 6\sqrt{2}$

EXAMPLE A ladder leans against a wall so that it makes a 30° angle with the top of the wall. If the ladder is 10 m long, find the height of the wall.

SOLUTION To find the height, h, of the wall, we first find the distance, b, of the foot of the ladder from the bottom of the wall.

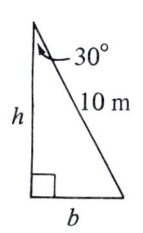

$2b = 10$
$b = 5$ (m) $h = 5\sqrt{3}$ (m)
$h \approx 8.66$ m

RATIO AND SIMILARITY

ORAL EXERCISES

1. Find the length of the shorter leg of a 30°-60°-90° triangle if the length of the hypotenuse is:
   a. 40
   b. 19
   c. 3.2
   d. $\sqrt{7}$

2. Find the length of the hypotenuse of a 30°-60°-90° triangle if the length of the shorter leg is:
   a. 12
   b. $5\frac{1}{2}$
   c. 5.6
   d. $3\sqrt{2}$

3. Find the length of the longer leg of a 30°-60°-90° triangle if the length of the shorter leg is:
   a. 7
   b. 1
   c. $\sqrt{3}$
   d. $\sqrt{5}$

4. Find the length of the longer leg of a 30°-60°-90° triangle if the length of the hypotenuse is:
   a. 14
   b. 1
   c. 5
   d. $\sqrt{3}$

5. Find the length of a diagonal of a square if the length of each side is:
   a. 3
   b. $\sqrt{5}$
   c. $\sqrt{2}$
   d. $3\sqrt{2}$

6. Find the length of each leg of a 45°-45°-90° triangle if the length of the hypotenuse is:
   a. $\sqrt{2}$
   b. $3\sqrt{2}$
   c. $\sqrt{6}$
   d. 1

WRITTEN EXERCISES

Complete.

A

	1.	2.	3.	4.	5.	6.	7.	8.
JK	2	5	$4\sqrt{2}$	$\sqrt{6}$	?	?	?	?
LK	?	?	?	?	$2\sqrt{2}$	$\sqrt{10}$	4	9

	9.	10.	11.	12.	13.	14.	15.	16.
RS	5	$\frac{1}{2}$	?	?	?	?	$\sqrt{6}$	?
ST	?	?	?	?	$4\sqrt{3}$	6	?	$\sqrt{6}$
RT	?	?	26	7	?	?	?	?

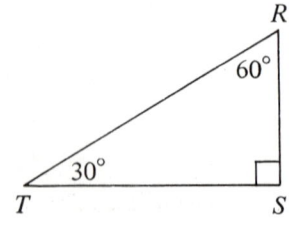

17. Find the length of a diagonal of a square whose perimeter is 60.
18. Find the perimeter of a square if the length of its diagonal is $5\sqrt{2}$.

In Exercises 19 and 20, $ABCD$ is a rectangle and $m\angle CAB = 30$.

19. If $BC = 12$, find the area of $\triangle ABC$.
20. If $AD = 8$, find the area of rectangle $ABCD$.

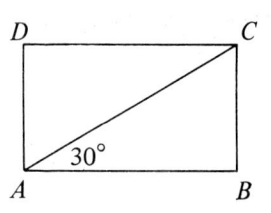

B 21. The ratio of the perimeters of two isosceles right triangles is $3:4$. If the length of the hypotenuse of the larger triangle is $12\sqrt{2}$, find the length of each leg of the smaller triangle.

22. The ratio of the perimeters of two isosceles right triangles is $2:5$. If the length of each leg of the smaller triangle is 7, find the length of the hypotenuse of the larger triangle.

In Exercises 23 and 24, the diagonals of square $JKLM$ intersect at point O. (Recall that the diagonals of any parallelogram bisect each other and that the diagonals of a rhombus are perpendicular.)

23. If $JK = 6$, find JO.
24. If $JO = 9\sqrt{2}$, find KL.

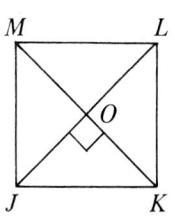

In Exercises 25-30, the diagonals of rhombus $ABCD$ intersect at point O and $m\angle ABC = 120$.

25. What is the measure of $\angle AOB$?
26. What is the measure of $\angle ABO$?
27. If $AC = 10\sqrt{3}$, find BC.
28. If $CD = 14$, find BD.
29. If $BD = 22$, find AC.
30. If $AC = 36\sqrt{3}$, find BO.

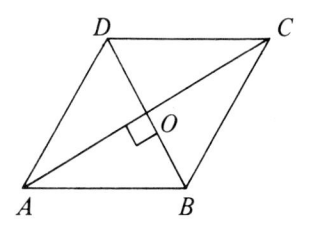

31. $\triangle RST$ and $\triangle XYZ$ are equilateral triangles whose sides have lengths 1 and 3 respectively.
    a. What is the length of an altitude of $\triangle RST$?
    b. What is the area of $\triangle RST$?
    c. What is the length of an altitude of $\triangle XYZ$?
    d. What is the area of $\triangle XYZ$?
    e. What is the ratio of the area of $\triangle RST$ to the area of $\triangle XYZ$?

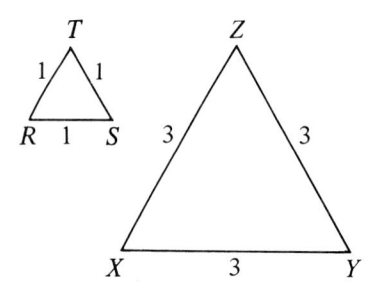

32. a. What is the length of an altitude of an equilateral triangle whose sides have length s?
    b. What is the area of an equilateral triangle whose sides have length s?

Exercises 33-37 refer to the figure.

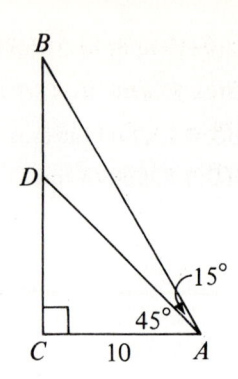

C 33. Find *BD*.

34. Find the perimeter of △*ABC*.

35. Find the perimeter of △*ABD*.

36. Find the area of △*ABD*.

37. Find $\dfrac{AB}{AD}$.

Exercises 38-40 refer to the figure, with *GH* = 6.

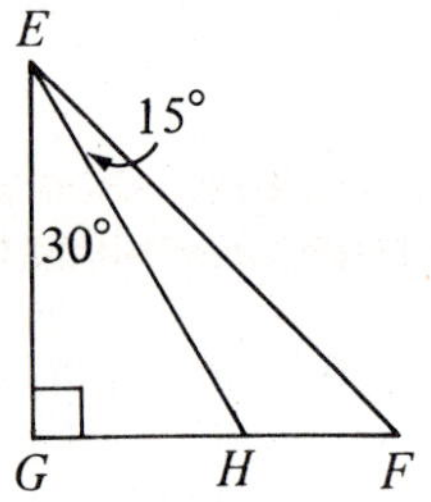

38. Find the area of △*EHG*.

39. Find the area of △*EFG*.

40. Find the area of △*EHF*.

41. The length of both the diagonal of a square and the shorter diagonal of a rhombus is 12 cm. If the measure of one of the angles of the rhombus is 120, find the ratio of the length of each side of the square to the length of each side of the rhombus.

SELF-TEST 3

In Exercises 1 and 2, find the value of *x*. Give irrational answers in simplest radical form.

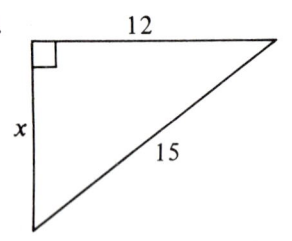

1.

2.

Section 9-6

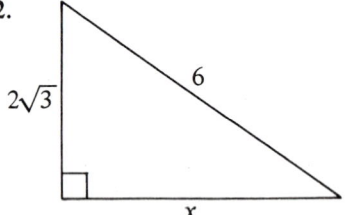

3. Is a triangle whose sides have lengths $2\sqrt{5}$, $3\sqrt{2}$, and 6 a right triangle?

4. In simplest radical form, what is the length of a diagonal of a rectangle with width 8 and length 16?

Exercises 5-7 refer to △*BMT*.

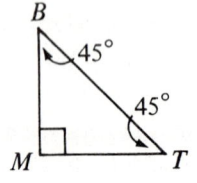

5. If *BM* = 8, find *BT*.

6. If *BT* = $7\sqrt{2}$, find *MT*.

7. If *BT* = 24, find *BM*.

Section 9-7

Exercises 8–10 refer to $\triangle CHR$.

8. If $CH = 6$, find CR.
9. If $CR = 14$, find HR.
10. If $HR = 18$, find CR.

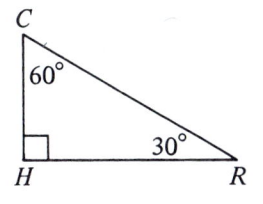

Computer Exercises

1. a. In BASIC you can represent a variable by a letter followed by a number. For example, C1, C2, and J7 are variables. Note that C1 and C2 are different variables. In the following program, the perimeters, P1 and P2, of two similar polygons are given in line 90. The lengths of the sides of the first polygon are given in line 100. The program computes and prints out the lengths of the corresponding sides of the second polygon.

    This program also illustrates a useful way to *exit* from a loop. An inappropriate number (in this case −1, which cannot be the length of a side of a polygon) is inserted as the last entry in the DATA statement (line 100). The program reads (line 50) each entry in the DATA statement and tests it (line 60) to see if it matches the predetermined number. When the match is made the program exits from the loop.

    Copy and RUN the program.

    ```
    10   READ P1,P2
    20   PRINT "SINCE THE RATIO OF THE PERIMETERS IS"
    30   PRINT P2/P1;",";" THE LENGTHS OF THE CORRESPONDING"
    40   PRINT "SIDES OF THE SECOND POLYGON ARE"
    50   READ S
    60   IF S=-1 THEN 110
    70   PRINT (P2/P1)*S;" ";
    80   GOTO 50
    90   DATA 43,64.5
    100  DATA 6,8,2,7,8,5,3,4,-1
    110  END
    ```

    b. Note that the sum of the entries (excluding −1) in line 100 must equal the first entry in line 90. Find the sum of the numbers in the print-out. Explain the significance of this sum.

    c. RUN the program using perimeters of 20 and 25. Use 3, 5, 4, 1, and 7 as the lengths of the sides of the first polygon.

2. a. A *Pythagorean triple* is any three positive integers that could represent the lengths of the sides of a right triangle. For example, (3, 4, 5) is a Pythagorean triple because $3^2 + 4^2 = 5^2$.

   A *primitive* Pythagorean triple is any Pythagorean triple whose numbers do not have a common factor. Some examples of primitive Pythagorean triples are (3, 4, 5), (5, 12, 13), and (8, 15, 17). The triple (6, 8, 10) is not primitive because 2 is a factor of each member.

   The following program generates 28 Pythagorean triples. Copy and RUN the program.

```
10   PRINT "THIS PROGRAM PRINTS PYTHAGOREAN TRIPLES."
20   PRINT
30   FOR M=2 TO 8
40   FOR N=1 TO M-1
50   PRINT M^2-N^2,2*M*N,M^2+N^2
60   NEXT N
70   NEXT M
80   END
```

   b. In your print-out, cross out the triples that are not primitive. You now have a list of primitive Pythagorean triples.

   c. If you wish to make a longer list of Pythagorean triples, increase the upper limit for M in line 30 and RUN the program again.

3. a. Write a program that computes and prints out the lengths of the legs of a 30°-60°-90° triangle when you INPUT the length of the hypotenuse.

   b. Revise your program so that it will print out the lengths of the shorter leg and the hypotenuse when you INPUT the length of the longer leg.

CHAPTER REVIEW

Express each ratio in simplest form.

1. 765 mm to 3 m   Section 9-1

2. 12 cm to 8 cm to 2 cm

3. Two negative numbers are in the ratio 5:2. Their difference is −21. Find the numbers.

Solve.

4. $\dfrac{a}{a+6} = \dfrac{7}{10}$ 5. $\dfrac{3k+1}{12} = \dfrac{2k+2}{9}$ Section 9-2

6. Six kilograms of aluminum chloride contain 1.5 kg of aluminum and 4.5 kg of chlorine. How much chlorine does 9 kg of aluminum chloride contain?

338   CHAPTER 9

7. Find 135% of 640.   Section 9-3

8. A $280 camera was on sale for $229.60. Find the percent of discount.

9. What amount of money, invested for one year at 8.5% annual interest, earns $306?

10. Refer to the diagram at the right.   Section 9-4
    a. Name two similar triangles.
    b. Write an extended proportion, using the lengths of the sides of the triangles.

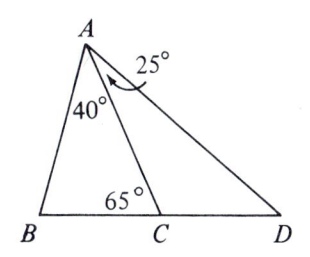

11. The lengths of the sides of a quadrilateral are 9, 10, 12, and 15. The longest side of a similar quadrilateral has length 9. Find the perimeter of the smaller quadrilateral.

12. Two similar triangles have perimeters of 18 cm and 30 cm. Find the ratio of their areas.   Section 9-5

13. In the figure, $\triangle RXY \sim \triangle RST$. The area of $\triangle RXY$ is 24 cm² and the area of $\triangle RST$ is 54 cm². If $\overline{XY}$ is 10 cm long, find ST.

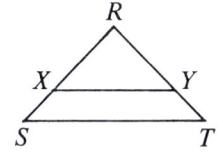

14. Two similar regular pyramids have slant heights of 2 and 8. The volume of the smaller pyramid is 10. Find the volume of the larger pyramid.

Is the triangle whose sides have the given lengths a right triangle?

15. 18, $2\sqrt{19}$, 20   16. 53, 28, 45   Section 9-6

17. If the legs of a right triangle are 18 cm long and 24 cm long, find the perimeter of the triangle.

Find the values of x and y.

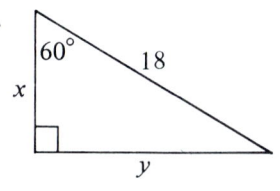

18.   19.   Section 9-7

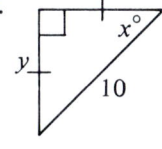

20.

RATIO AND SIMILARITY   339

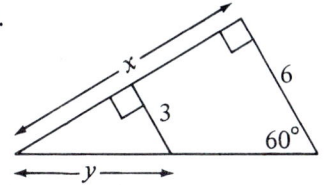

CHAPTER TEST

Find each ratio or length in simplest form.

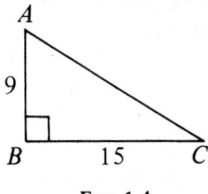

1. $AB:BC$
2. AC

3. If $\triangle XYZ \sim \triangle ABC$ and $XY = 3\sqrt{5}$, find $\dfrac{\text{perimeter of } \triangle XYZ}{\text{perimeter of } \triangle ABC}$.

4. A triangle similar to $\triangle ABC$ has area 750. Find the length of the longer leg of this triangle.

Exs. 1-4

Solve.

5. $\dfrac{n+5}{3n+1} = \dfrac{1}{2}$

6. $\dfrac{12}{5t-2} = \dfrac{10}{4t-1}$

7. What is 35% of 280?
8. 96% of what number is 36?
9. What percent of 400 is 3?

Each diagram shows similar triangles. Find the values of x and y.

10. 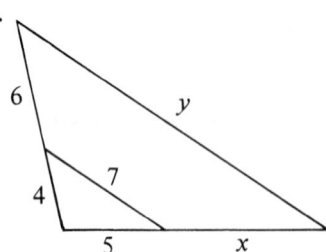
11. 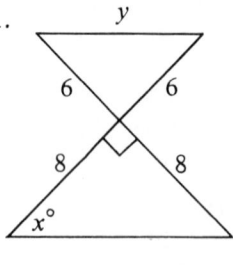
12.

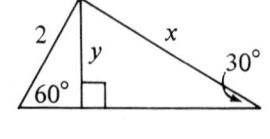

13. Find the ratio of the areas of the similar triangles shown in Exercise 10.

Is the triangle whose sides have the given lengths a right triangle?

14. 0.8, 1.5, 1.7
15. $2\sqrt{3}$, $3\sqrt{2}$, $5\sqrt{6}$

16. An equilateral triangle has an altitude $7\sqrt{3}$ cm long. Find the perimeter of the triangle.

17. A survey found that of 600 customers shopping in a supermarket, 282 redeemed manufacturers' coupons to save money. What percent did *not* use coupons?

18. One week, the ratio of the amount of money Patrick spent to the amount he saved was 29:7. If his weekly take-home pay is $270, how much did he save that week?

19. If the length of each edge of a cube is divided by 4, how is the volume of the cube affected?

20. A child who is 110 cm tall casts a shadow 132 cm long. At the same time, the child's dog casts a shadow 51 cm long. Find the height of the dog.

Application — Using a Credit Card

Banks, oil companies, and stores issue credit cards for the convenience of their customers. Credit card receipts are a good record of expenses, and having "plastic money" means you do not have to carry around as much cash. Using a credit card responsibly, however, is not as simple as writing your name on the receipt. You have to remember what your credit limit is, because as soon as you go over that limit, many stores will not accept this form of payment for your purchases. You have to keep your credit card in a safe place and if it is lost or stolen, report it to the company that issued it as soon as possible.

You will receive a statement of your purchases from the company that issued the card. If you do not pay the balance due within a specified period, you will have to pay a finance charge. Finance charges are based on a periodic rate, such as 1.65% or 1.5% per month. One method of calculating a finance charge is to multiply the periodic rate by the previous balance.

Previous-Balance Method

finance charge = periodic rate × previous balance

new balance = previous balance + finance charge + new purchases − (payments and credits)

Emily Chamberlain has a charge account with a store whose finance charge is 1.5% of the previous balance. Find her new balance with the information given.

PREVIOUS BALANCE	TOTAL PURCHASES	PAYMENTS	FINANCE CHARGE
$60.43	$28.62	$15.00	$0.91

1. Find the previous balance. $60.43
2. Find the finance charge.
   1.5% of $60.43 = $0.90645 $ 0.91
3. Find the new purchases. $28.62
4. Find the new balance.
   $60.43 + $0.91 + $28.62 − $15.00 $74.96

Exercise

Using the information above, find Emily's new balance if the periodic rate is 1.65%.

There are other methods of calculating a finance charge, and different methods can make a real difference in your bill. Be sure you understand how the company calculates its finance charge when you use a credit card.

RATIO AND SIMILARITY

VOCABULARY REVIEW

Be sure that you understand the meaning of these terms:

ratio, p. 301
proportion, p. 304
extremes, p. 305
means, p. 305
percent, p. 308

similar polygons, p. 314
similar solids, p. 332
hypotenuse, p. 326
leg of a right triangle, p. 326

MIXED REVIEW

1. If the figure in the diagram is such that $\overline{BE} \parallel \overline{CD}$, $AE = 10.2$, $AD = 17$, and $BC = 6$, find AB.

2. If all the sides of a quadrilateral are congruent, then the quadrilateral *must* be a ___?___, *may* be a ___?___, and *cannot* be a ___?___.

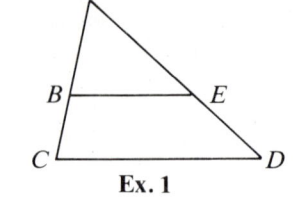

3. Use a truth table to show that $\sim(p \vee q) \rightarrow \sim p$ is a tautology.

4. If A and B have coordinates -3 and 13 on a number line, find the coordinate of the midpoint of $\overline{AB}$.

Ex. 1

5. Draw a large $\triangle XYZ$. Then construct $\triangle RST$ so that $\angle R \cong \angle X$, $\overline{RS} \cong \overline{XY}$, and $\angle S \cong \angle Y$. Why must it be that $\triangle XYZ \cong \triangle RST$?

Factor.

6. $81x^2 - 1$

7. $y^2 - 2y - 15$

8. If a right rectangular prism has length $t + 1$, width t, and height $t - 2$, give the volume of the prism as a polynomial in standard form.

Simplify.

9. $\sqrt{\dfrac{4}{3}} + \sqrt{\dfrac{3}{4}}$

10. $-\dfrac{2}{3}\sqrt{396}$

11. Find the ratio of the length of the longer leg to the length of the hypotenuse of a $30°$-$60°$-$90°$ triangle. (*Hint:* Let the shorter leg have length 1.)

12. Use the quadratic formula to solve $3x^2 - 5x = -1$.

Graph the solution set.

13. $-\dfrac{2}{3}y \geqslant 4$

14. $3.4 > k - 1.5 > -0.5$

15. How much do you earn when you invest $1500 at 8.5% interest for one year?

16. A regular square pyramid has base edges 12, height 8, and slant height 10. Find its volume.

17. Find the factors of 33.

18. a. Write the converse of "If an integer is even, then it is divisible by 2."
    b. Give the truth value of the converse.
19. The measure of a complement of ∠1 is less than 40. What can you conclude about the measure of ∠1?
20. In the diagram at the right, if $\overrightarrow{PQ} \parallel \overrightarrow{RS}$, $m\angle SPQ = 60$, and $m\angle SPR = 40$, find $m\angle R$.

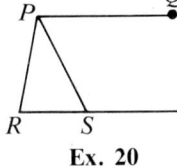

Ex. 20

21. If $a^2 + ab = 4$, solve for b.
22. In $\triangle JKL$, $\overline{JK} \perp \overline{JL}$, $JK = 6$ and $JL = 8$. Find KL.

Express as a decimal.

23. $\dfrac{23}{25}$

24. $\dfrac{7}{15}$

Multiply.

25. $(2 - 3\sqrt{5})^2$

26. $(a \cdot a^3)^2 \cdot (2a)^3$

27. The height of a parallelogram is 3 cm less than twice the length of the base. If the area is 54 cm², find the height.

Solve.

28. $x^2 = 225$

29. $5y^2 - 8y + 3 = 0$

30. $4z - 12 < \dfrac{4}{3}$

31. $9z^2 - 12z + 4 = 0$

32. $-\dfrac{5}{4}x = 10 - x$

33. $2(y - 1) > 4 - 2y$

34. When two lines intersect, the measure of one of the angles formed is 37. What are the measures of the other three angles?
35. Must any three points be collinear? coplanar?
36. If the domain of x is {0, 1, 2}, find a value of x that makes $x^3 = x$ true and a value that makes $x^3 = x$ false.

Find the value of x.

37.

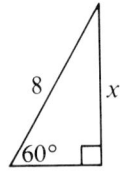

38.

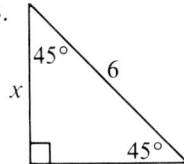

39.

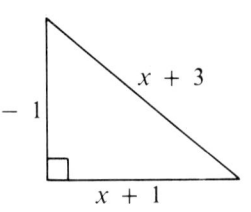

RATIO AND SIMILARITY

The streets shown crisscrossing Dupont Circle in Washington, D.C., in the aerial photograph above suggest secants to a circle. Circular arcs, central angles, diameters, and radii can also be seen. In this chapter you will study properties of these and other geometric figures related to circles.

Geometry of the Circle 10

Have you ever stopped to wonder why large fruits such as melons and pumpkins grow on the ground while smaller fruits like peaches and apples grow on trees? As you will see in this chapter, the volume of the fruit is directly proportional to the cube of its radius. But the strength of a tree limb or stalk is directly proportional to the square of its radius. Therefore, if a tree were to produce pumpkins it would have to have very, very thick and heavy branches.

Section 10-1 BASIC DEFINITIONS

You probably are familiar with many of the terms presented in this section. Which of the following do you recognize?
- circle
- radius
- chord
- diameter
- secant
- tangent
- congruent circles

A *circle* with *center O* is the set of all points in a plane at a given distance from point *O* in that plane. A *radius* (plural *radii*) of a circle is a line segment whose endpoints are the center of the circle and a point on the circle.

You may name a circle by its center. The figure shows ⊙*O* (read "circle *O*") with radius $\overline{OA}$. Can you explain why all radii of a circle are congruent?

The word radius may also refer to the length of a radius. If, for example, $OA = 3$, you would say that ⊙*O* has a radius of 3.

Any segment whose endpoints are points on a circle is called a *chord* of the circle. In ⊙*O*, $\overline{AB}$ and $\overline{CD}$ are chords.

If a chord contains the center of a circle, then it is a *diameter* of the circle. $\overline{AB}$ is a diameter, but $\overline{CD}$ is not. The word diameter may also refer to the length of that segment. Thus, if $AB = 6$, you would say that ⊙*O* has a diameter of 6. Notice that the length of a diameter is twice the length of a radius.

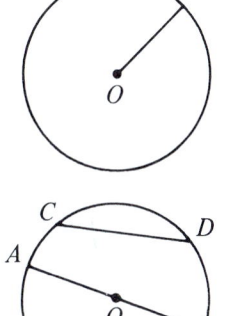

A *secant* is any line that contains a chord. $\overleftrightarrow{PQ}$ and $\overleftrightarrow{RS}$ are secants.

A *tangent* to a circle is a line that lies in the same plane as the circle and intersects the circle in exactly one point. $\overleftrightarrow{TU}$ is a tangent to $\odot O$. T is the *point of tangency*.

A segment or ray that is part of the tangent line and contains the point of tangency is also called a tangent. $\overrightarrow{TU}$ and $\overline{TU}$ are tangents.

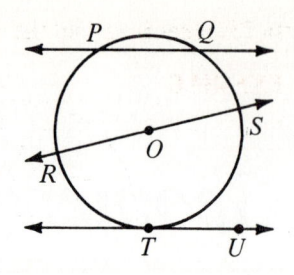

Two circles are *congruent* if they have congruent radii. Thus, if $\overline{AB} \cong \overline{QR}$, then $\odot A$ is congruent to $\odot Q$.

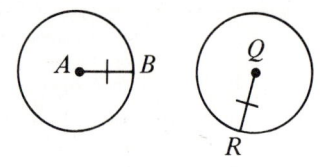

ORAL EXERCISES

Exercises 1–3 refer to $\odot O$ shown at the right.

1. a. Name two chords.
   b. Name a secant.
   c. Name a diameter.
   d. Name three tangents.
   e. Name two radii.
2. How many diameters can be drawn through F? through O?
3. If $OD = 4$, find: a. OA b. OC c. OB d. AD
4. State the diameter of a circle if its radius is 5; 8; 0.5; 10; s.
5. State the radius of a circle if its diameter is 6; 18; 5; 1; t.

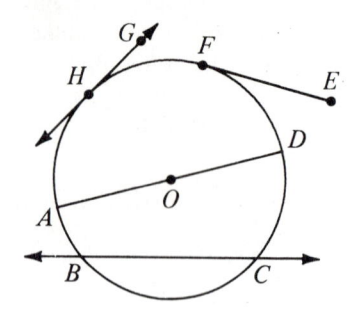

WRITTEN EXERCISES

A   1. Refer to $\odot O$ shown at the right. What term best describes each set of points listed below?
     a. $\overline{ST}$ b. $\overline{UR}$
     c. $\overleftrightarrow{UR}$ d. $\overleftrightarrow{QR}$
     e. $\overline{OS}$ f. $\overleftrightarrow{ST}$
     g. $\overline{UT}$ h. $\overline{TO}$

In Exercises 2–7 tell whether the statement is true or false.

2. A secant can intersect a circle in more than two points.
3. A diameter is a chord.
4. A radius is a chord.
5. A tangent may pass through the center of a circle.
6. All chords through the center of a circle are congruent.
7. Two circles are congruent if they have congruent diameters.

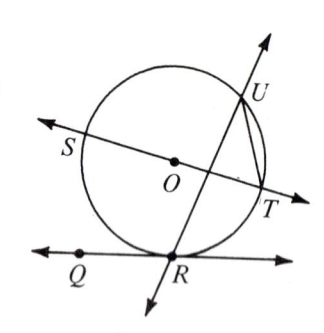

346   CHAPTER 10

In Exercises 8-11 find the values of x, y, and z for the given ⊙O.

EXAMPLE

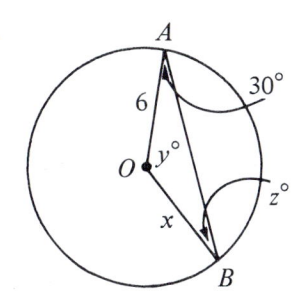

SOLUTION $\overline{OA} \cong \overline{OB}$, so $x = 6$.
$\overline{OA} \cong \overline{OB}$, so $\angle OAB \cong \angle OBA$.
Therefore, $z = 30$.
$180 = m\angle OAB + m\angle OBA + m\angle AOB$
$180 = 30 + 30 + y$
$120 = y$

8.

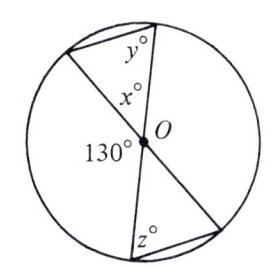

9.

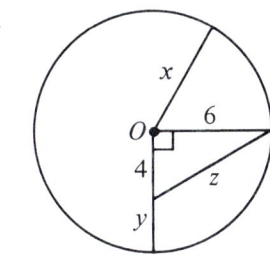

10.

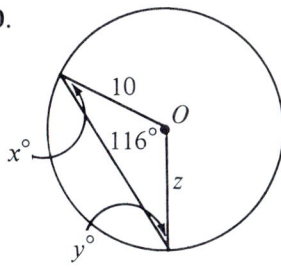

11.

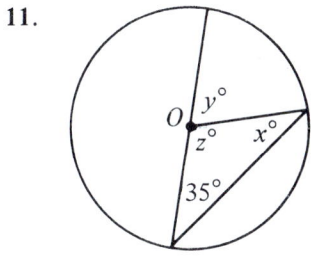

12. In the figure, find AB.

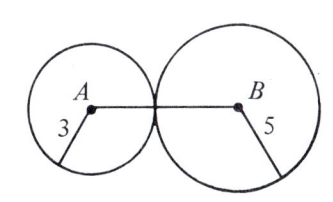

In Exercises 13 and 14 find the values of x, y, and z for the given ⊙P.

B 13.

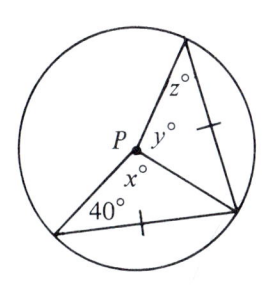

14.

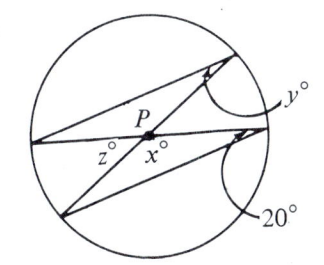

GEOMETRY OF THE CIRCLE   347

In Exercises 15 and 16 find the values of x, y, and z for the given $\odot P$.

15.

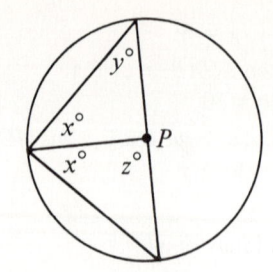

16.

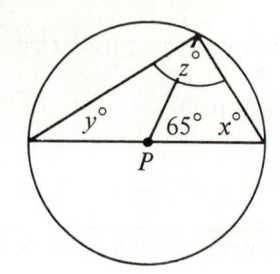

In Exercises 17 and 18 fill in the blanks and replace each "why" with the correct reason.

C 17. Given: $\overline{OA} \perp \overline{BC}$
Show: $\overline{BM} \cong \overline{CM}$

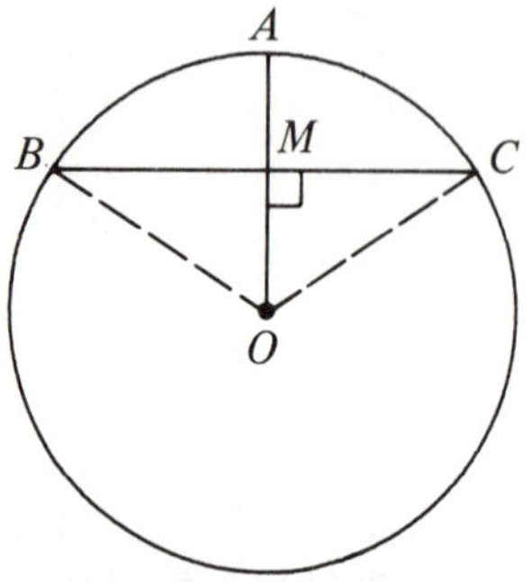

Draw $\overline{OB}$ and $\overline{OC}$.
(1) $\overline{OA} \perp \overline{BC}$ (Given)
(2) $\angle OMB$ and $\angle OMC$ are right angles. (Why?)
(3) $\angle OMB \cong \angle OMC$ (Why?)
(4) $\overline{OB} \cong \overline{OC}$ (Why?)
(5) $\angle MBO \cong \angle MCO$ (Why?)
(6) $\triangle \underline{\ ?\ } \cong \triangle \underline{\ ?\ }$ (AAS. See Exercise 26, page 184.)
(7) $\overline{BM} \cong \overline{CM}$ (Why?)

Note that this exercise shows that if a radius is perpendicular to a chord, then the radius bisects the chord.

18. Given: $\overline{BM} \cong \overline{CM}$
$\overline{BC}$ is not a diameter.
Show: $\overline{OA} \perp \overline{BC}$

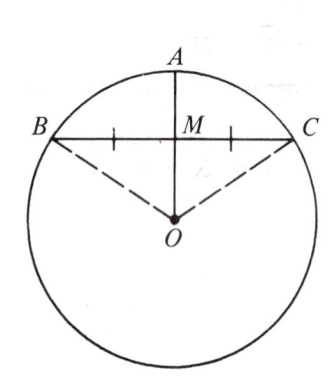

Draw $\overline{OB}$ and $\overline{OC}$.
(1) $\overline{BM} \cong \overline{CM}$ (Why?)
(2) $\overline{OM} \cong \overline{OM}$ (Why?)
(3) $\overline{OB} \cong \overline{OC}$ (Why?)
(4) $\triangle \underline{\ ?\ } \cong \triangle \underline{\ ?\ }$ (Why?)
(5) $\angle OMB \cong \angle OMC$ (Why?)
(6) $m\angle OMB + m\angle OMC = 180$
(Definition of supplementary angles)
(7) $2m\angle OMB = 180$ (Substitution)
(8) $m\angle OMB = 90$ (Dividing both sides by 2)
(9) $\overline{OA} \perp \overline{BC}$ (Why?)

Note that this exercise shows that if a radius bisects a chord that is not a diameter, then the radius is perpendicular to the chord.

348   CHAPTER 10

Section 10-2 CENTRAL ANGLES AND ARCS OF CIRCLES

This section introduces some new concepts, including:
- arc
- minor arc, major arc, semicircle
- central angle
- measure of an arc
- congruent arcs

In the figures below, the solid curves illustrate the three different kinds of *arcs* of a circle.

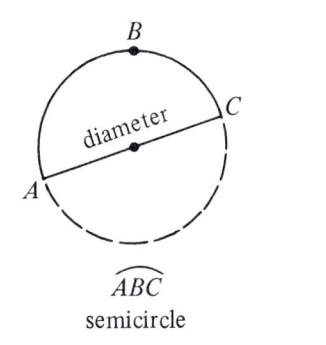

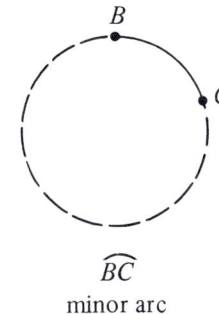

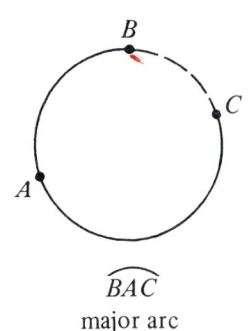

$\widehat{ABC}$
semicircle

$\widehat{BC}$
minor arc

$\widehat{BAC}$
major arc

Naming an arc by its endpoints can be confusing. In the figures above, B and C are the endpoints of both a minor arc and a major arc. To avoid ambiguity, you may assume that an arc named only by its endpoints is a minor arc.

A *central angle* of a circle is an angle whose vertex is the center of the circle. $\angle MON$ is a central angle.

The *measure of a minor arc* is the measure of its central angle.

$$m\,\widehat{MN} = m\angle MON$$
$$m\,\widehat{MN} = 75$$

The *measure of a major arc* is 360 minus the measure of the corresponding minor arc.

$$m\,\widehat{MRN} = 360 - m\,\widehat{MN}$$
$$= 360 - 75$$
$$= 285$$

The *measure of a semicircle* is 180.

$$m\,\widehat{ABC} = 180 \qquad m\,\widehat{CDA} = 180$$

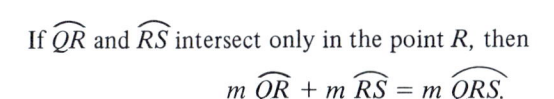

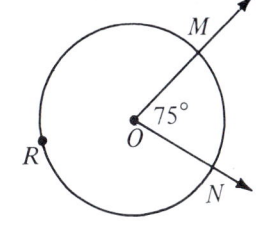

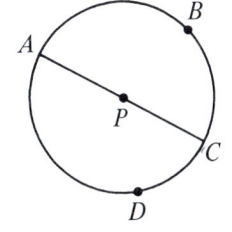

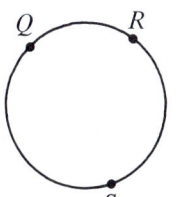

If $\widehat{QR}$ and $\widehat{RS}$ intersect only in the point R, then

$$m\,\widehat{QR} + m\,\widehat{RS} = m\,\widehat{QRS}.$$

GEOMETRY OF THE CIRCLE

EXAMPLE Given ⊙O, find the values of x, y, and z.

SOLUTION $m \widehat{AB} = m\angle AOB$
$x = 60$

$\widehat{ABC}$ is a semicircle; $m \widehat{ABC} = 180$.
$m \widehat{ABC} = m \widehat{AB} + m \widehat{BC}$
$180 = 60 + y$
$120 = y$

$\angle BOC$ and $\angle BOA$ are supplementary.
$z + 60 = 180$
$z = 120$

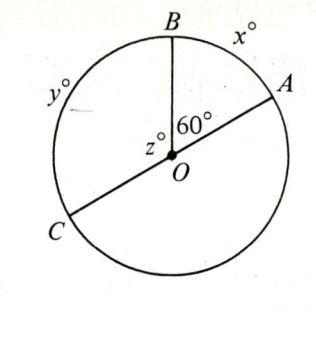

If two or more arcs of the same circle or congruent circles have equal measures, they are called *congruent arcs*.

EXAMPLES

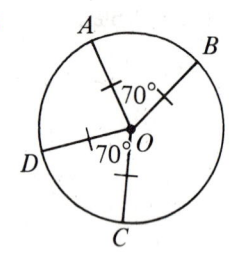

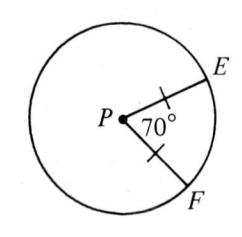

$\widehat{AB}$, $\widehat{CD}$, and $\widehat{EF}$ are congruent arcs.

ORAL EXERCISES

Exercises 1–8 refer to ⊙O at the right.

1. Name: a. three central angles
   b. three minor arcs
   c. three major arcs

2. If $m\angle AOB = 70$, what is $m \widehat{AB}$?

3. If $m \widehat{AC} = 165$, what is $m\angle AOC$?

4. If $m \widehat{CAB} = 220$, name an arc whose measure is 140.

5. If $m\angle BOC$ is 120, name:
   a. an arc whose measure is 120   b. an arc whose measure is 240

6. If $\angle AOC \cong \angle BOC$, name two congruent arcs.

7. Complete.
   a. $m \widehat{AB} + m \widehat{BC} = m$?   b. $m \widehat{ACB} - m \widehat{BC} = m$?
   c. $m \widehat{AC} + m \widehat{BC} = m$?   d. Does $m \widehat{AB} + m \widehat{BC} = m \widehat{AC}$?

8. If $\widehat{AC} \cong \widehat{BC}$ and $m \widehat{AC} = 140$, find $m \widehat{AB}$.

WRITTEN EXERCISES

A 1. If P is the center of the circle whose semicircle is shown, find the measures of the numbered angles.

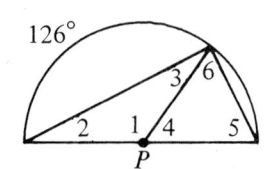

2. P is the center of both circles pictured at the right. If $m\angle BPA = 72$ and $m\angle QPS = 93$, find each of the following.
   a. $m\stackrel{\frown}{AB}$ 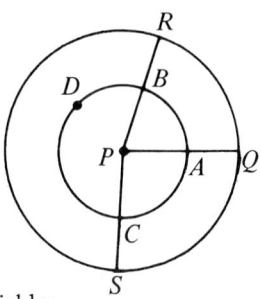
   b. $m\stackrel{\frown}{QR}$
   c. $m\stackrel{\frown}{AC}$
   d. $m\stackrel{\frown}{QS}$
   e. $m\stackrel{\frown}{BC}$
   f. $m\stackrel{\frown}{BDC}$
   g. Are $\stackrel{\frown}{AB}$ and $\stackrel{\frown}{QR}$ congruent arcs?

In Exercises 3–11, O is the center of the circle. Find the values of the variables.

3.

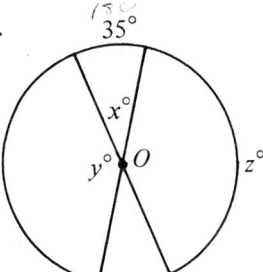

4.

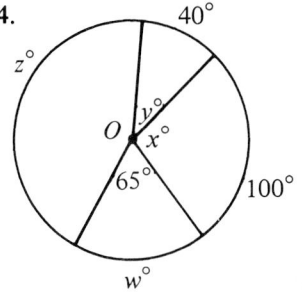

5.

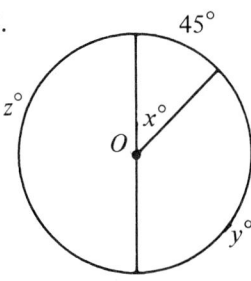

6.

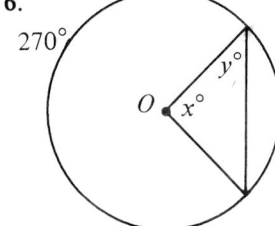

7.

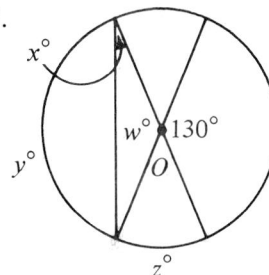

8.

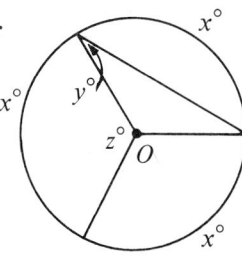

9.

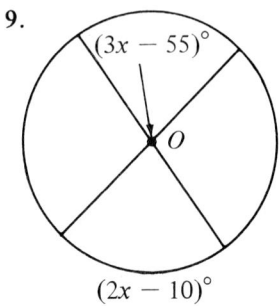

10.

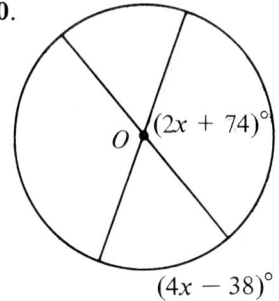

11.

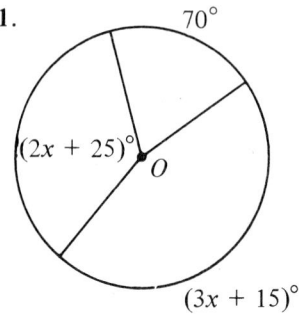

GEOMETRY OF THE CIRCLE 351

In Exercises 12-14, O is the center of the circle. Find the values of the variables.

B 12.

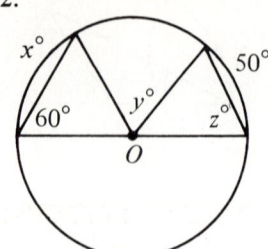

13.

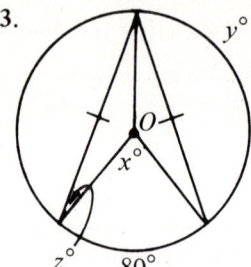

14.

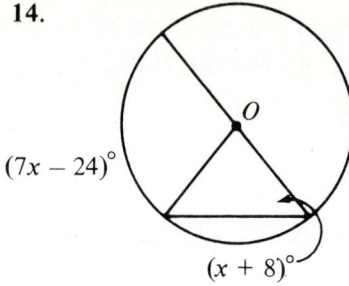

15. If $\angle ROQ$ is a central angle of a circle of radius 7, and $m\angle ROQ = 60$, find RQ.

16. If $\angle AOB$ is a central angle of a circle of radius 4, and $AB = 4\sqrt{2}$, find $m\angle AOB$.

17. Points A and B on a circle determine a major arc and a minor arc. If the measure of the major arc is 18 more than 5 times the measure of the minor arc, find the measure of the minor arc.

C 18. Fill in the blanks and replace each "why" with the correct reason.
Given: $\odot O$, with $\overline{AB} \cong \overline{CD}$
Show: $\overset{\frown}{AB} \cong \overset{\frown}{CD}$
Draw $\overline{OA}, \overline{OB}, \overline{OC},$ and $\overline{OD}$.
(1) $\overline{AB} \cong \overline{CD}$ (Why?)
(2) $\overline{OA} \cong \overline{OC}; \overline{OB} \cong \overline{OD}$ (Why?)
(3) $\triangle AOB \cong \triangle COD$ (Why?)
(4) $\angle\underline{\ ?\ } \cong \angle\underline{\ ?\ }$ (Why?)
(5) $\overset{\frown}{AB} \cong \overset{\frown}{CD}$ (Why?)

This shows that congruent chords cut off congruent arcs.

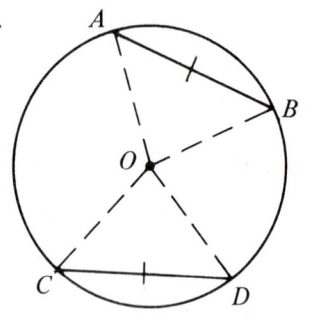

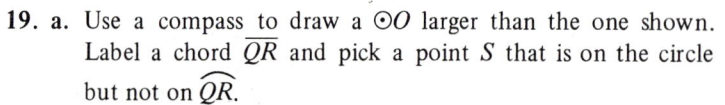

19. a. Use a compass to draw a $\odot O$ larger than the one shown. Label a chord $\overline{QR}$ and pick a point S that is on the circle but not on $\overset{\frown}{QR}$.
   b. Use a protractor to measure $\angle QOR$ and $\angle QSR$.
   c. How are $m\angle QOR$ and $m\angle QSR$ related?
   d. Check your results with two other circles.

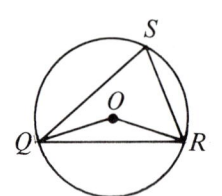

SELF-TEST 1

Complete.

1. A secant of $\odot O$ is __?__ .
2. $\overline{OB}$ is a __?__ of $\odot O$.
3. $\overline{FA}$ is a __?__ of $\odot O$.

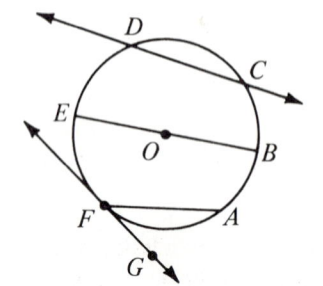

Section 10-1

352   CHAPTER 10

4. Find the values of x y, and z in ⊙P shown at the right.

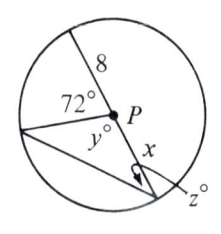

In Exercises 5-8 refer to ⊙O at the right.

5. If $m\widehat{RS} = 58$, then $m\widehat{RUS} = $? .

6. If $m\angle ROU = 114$, then $m\widehat{RU} = $? .

7. If $m\widehat{RU} = 126$, then $m\angle OUR = $? .

8. If $m\widehat{RS} = 58$ and $m\angle ROU = 2x + 12$, find the value of x.

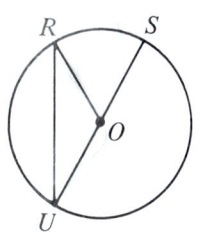

Section 10-2

Section 10-3 INSCRIBED ANGLES AND INTERCEPTED ARCS

Now you will learn more about the relationships between angles and arcs of a circle. After completing this section you will be able to identify:
- an inscribed angle
- an intercepted arc
- a polygon inscribed in a circle
- a circle circumscribed about a polygon

In the figures below, each angle *intercepts* the solid arc. Each angle is *inscribed* in the dashed arc.

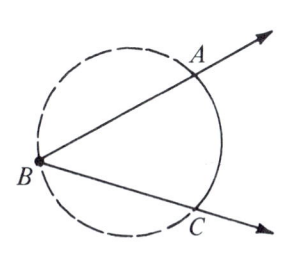

$\angle ABC$ intercepts $\widehat{AC}$.
$\angle ABC$ is inscribed in $\widehat{ABC}$.

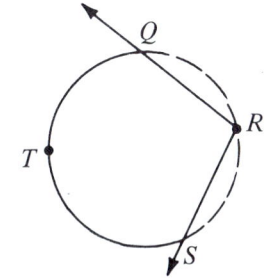

$\angle QRS$ intercepts $\widehat{QTS}$.
$\angle QRS$ is inscribed in $\widehat{QS}$.

An angle *intercepts* an arc if
(1) each side of the angle contains an endpoint of the arc;
(2) all other points of the arc lie in the interior of the angle.

GEOMETRY OF THE CIRCLE   353

An angle can intercept two arcs as shown.

∠1 intercepts $\widehat{AB}$ and $\widehat{CD}$.

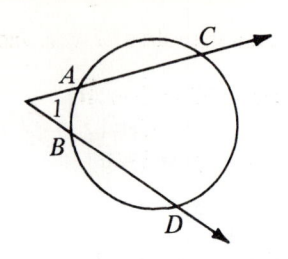

An angle is *inscribed* in an arc if
(1) its vertex lies on the arc, but is not an endpoint of the arc;
(2) its sides contain the endpoints of the arc.

| The measure of an inscribed angle is half the measure of its intercepted arc. |

To show this, you must consider the three cases illustrated below, where O is the center of the circle.

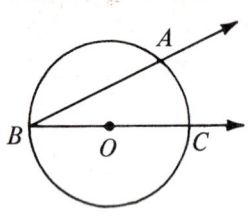

Case I
Point O lies on ∠ABC.

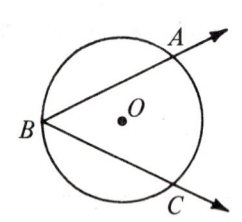

Case II
Point O is in the interior of ∠ABC.

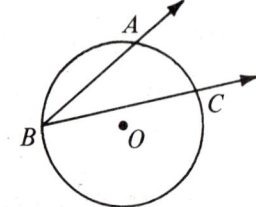

Case III
Point O is in the exterior of ∠ABC.

The reasoning for Case I is given below. Case II and Case III are left as Exercises 15 and 16 on page 358.

Given: ∠ABC is inscribed in ⊙O. O lies on $\vec{BC}$.

Show: $m\angle ABC = \frac{1}{2} m\widehat{AC}$

Begin by drawing $\overline{OA}$.
$\overline{OB} \cong \overline{OA}$ (All radii of a circle are congruent.)
∠ABC ≅ ∠1 (Why?)
$m\angle ABC = m\angle 1$ (Why?)
$m\angle 1 + m\angle ABC = m\angle AOC$ (The measure of an exterior angle of a triangle equals the sum of the measures of the two remote interior angles.)
$2m\angle ABC = m\angle AOC$ (Substitution)
$m\widehat{AC} = m\angle AOC$ (The measure of a minor arc equals the measure of its central angle.)
$2m\angle ABC = m\widehat{AC}$ (Substitution)
$m\angle ABC = \frac{1}{2} m\widehat{AC}$ (Dividing both sides by 2)

EXAMPLE Find the values of x, y, and z in $\odot O$.

SOLUTION

$m\angle ABC = \dfrac{1}{2} m\stackrel{\frown}{AC}$ $\qquad m\angle CBD = \dfrac{1}{2} m\stackrel{\frown}{CD}$

$x = \dfrac{1}{2}(40) \qquad\qquad 35 = \dfrac{1}{2}y$

$x = 20 \qquad\qquad\qquad 70 = y$

$\stackrel{\frown}{CAB}$ is a semicircle.

$m\stackrel{\frown}{CAB} = 180;\ m\stackrel{\frown}{CAB} = m\stackrel{\frown}{CA} + m\stackrel{\frown}{AB}$

$\qquad\qquad\qquad\quad 180 = 40 + z$

$\qquad\qquad\qquad\quad 140 = z$

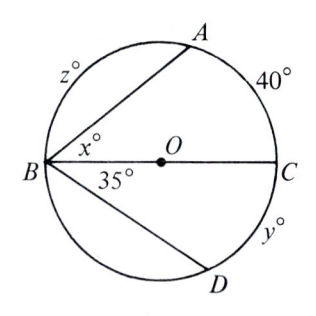

An angle inscribed in a semicircle is a right angle.

Clearly, $\angle 1$ intercepts a semicircle. The measure of a semicircle is 180 and so $m\angle 1 = \dfrac{1}{2}(180) = 90$.

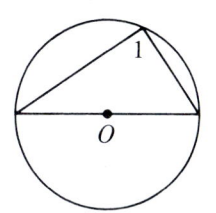

Inscribed angles that intercept the same arc are congruent.

The inscribed angles have the same measure and therefore are congruent.

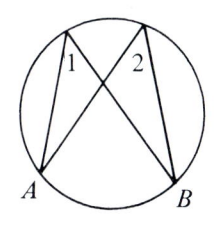

$$m\angle 1 = \dfrac{1}{2} m\stackrel{\frown}{AB} = m\angle 2$$

A polygon is *inscribed* in a circle and a circle is *circumscribed* about the polygon if each vertex of the polygon lies on the circle.

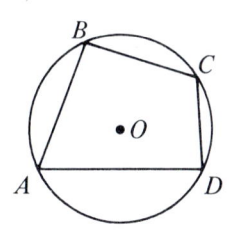

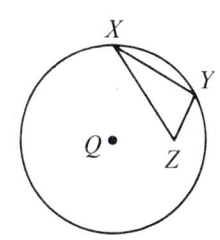

Quadrilateral $ABCD$ is inscribed in $\odot O$. $\triangle XYZ$ is not inscribed in $\odot Q$. $\odot O$ is circumscribed about quadrilateral $ABCD$, but $\odot Q$ is not circumscribed about $\triangle XYZ$.

GEOMETRY OF THE CIRCLE

In Exercise 17 you will be asked to show why the following is true.

> If a quadrilateral is inscribed in a circle, then opposite angles are supplementary.

EXAMPLE

$\angle A$ and $\angle C$ are supplementary.
$\angle B$ and $\angle D$ are supplementary.

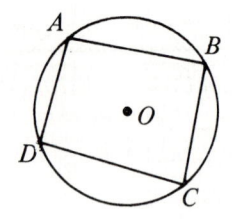

ORAL EXERCISES

In Exercises 1-6 name all inscribed angles and intercepted arcs (if there are any).

1. 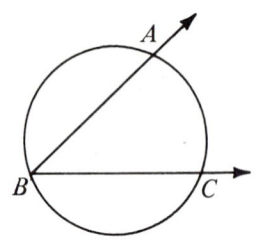
2. 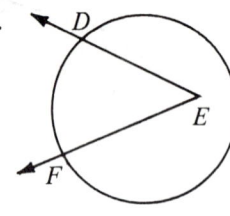
3. 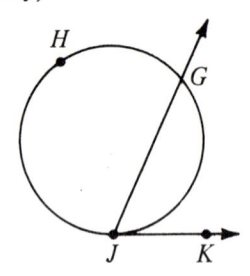
4. 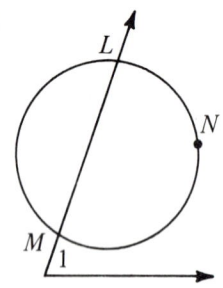
5. 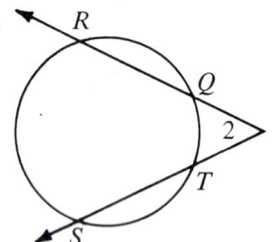
6.

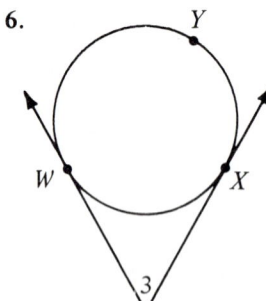

7. Refer to the figure at the right.
   a. If $m \overset{\frown}{AB} = 50$, then $m\angle 1 = \underline{\ ?\ }$.
   b. If $m\angle 1 = 42$, then $m \overset{\frown}{AB} = \underline{\ ?\ }$.
   c. If $m \overset{\frown}{ACB} = 210$, then $m\angle 1 = \underline{\ ?\ }$.
   d. If $m\angle 1 = 36$, then $m \overset{\frown}{ACB} = \underline{\ ?\ }$.

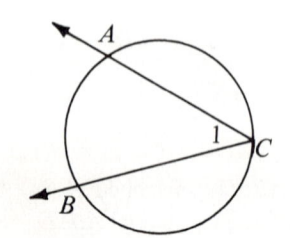

8. Refer to ⊙P at the right.
   a. m∠1 = __?__
   b. m∠2 = __?__
   c. m∠3 = __?__

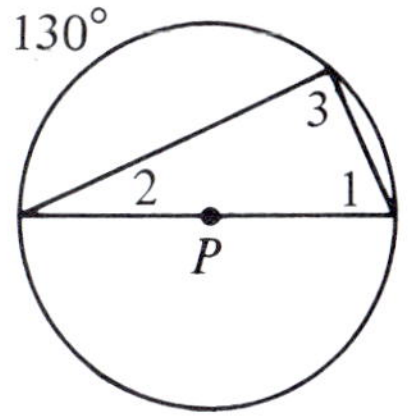

9. Refer to the figure at the right, in which m $\widehat{AB}$ = 100.
   a. m∠1 = __?__
   b. m∠2 = __?__
   c. m∠3 = __?__
   d. m∠4 = __?__

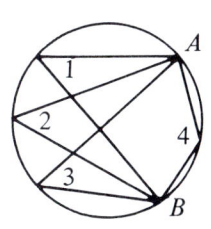

WRITTEN EXERCISES

In Exercises 1–6 find the measures of the numbered angles. O is the center of the circle.

A 1.

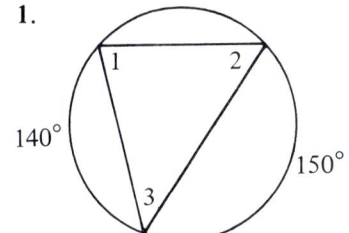

2.

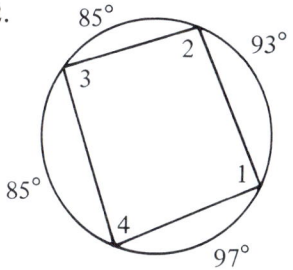

3.

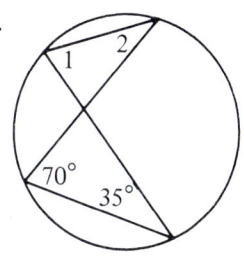

4.

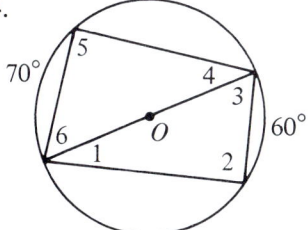

5.

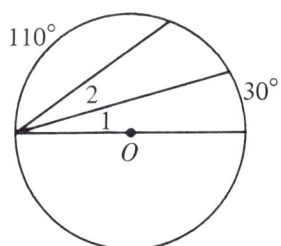

6.

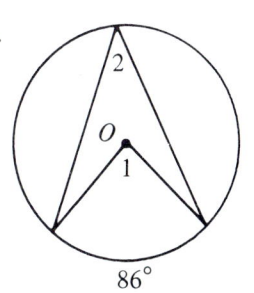

In Exercises 7–9 find the values of the variables. O is the center of the circle.

7.

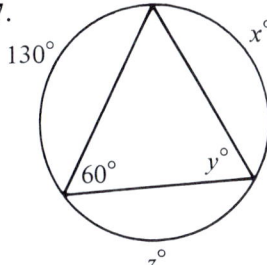

8.

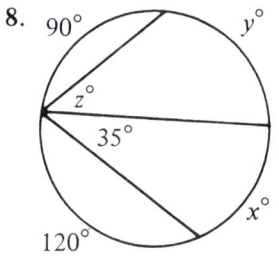

9.

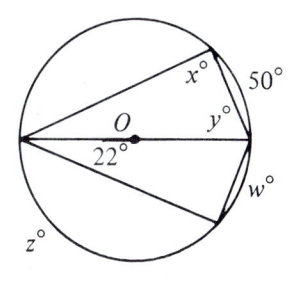

GEOMETRY OF THE CIRCLE

10. a. If $\overline{AB} \parallel \overline{CD}$, what must be true of $\angle 1$ and $\angle 2$? Why?
    b. Which two arcs must be congruent? Why?
    c. Complete: If chords $\overline{AB}$ and $\overline{CD}$ are parallel, then $\widehat{AC}$ and $\widehat{BD}$ are __?__.

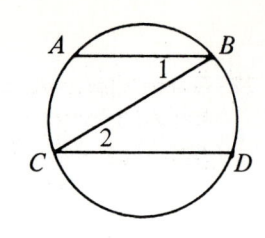

Use the results of Exercise 10 to find the value of x in Exercises 11–13. O is the center of the circle and $\overline{AB} \parallel \overline{CD}$.

11. 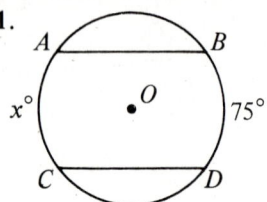
12. 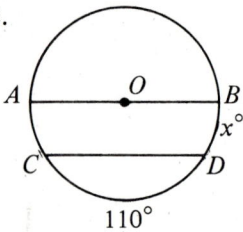
13.

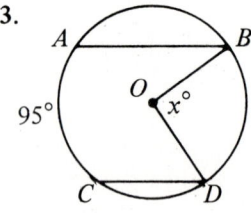

B 14. If $\widehat{QS} \cong \widehat{RT}$, explain why $\overline{QR} \parallel \overline{ST}$.

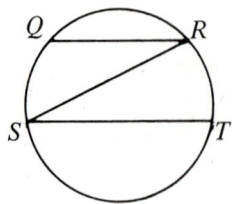

15. Show that the measure of an inscribed angle is half the measure of its intercepted arc for Case II on page 354 by replacing each "why" with the correct reason.

Given: $\angle ABC$ is inscribed in $\odot O$.
O lies in the interior of $\angle ABC$.

Show: $m\angle ABC = \frac{1}{2} m\widehat{AC}$

Draw $\overrightarrow{BO}$. Label the point where $\overrightarrow{BO}$ intersects $\odot O$ as E. 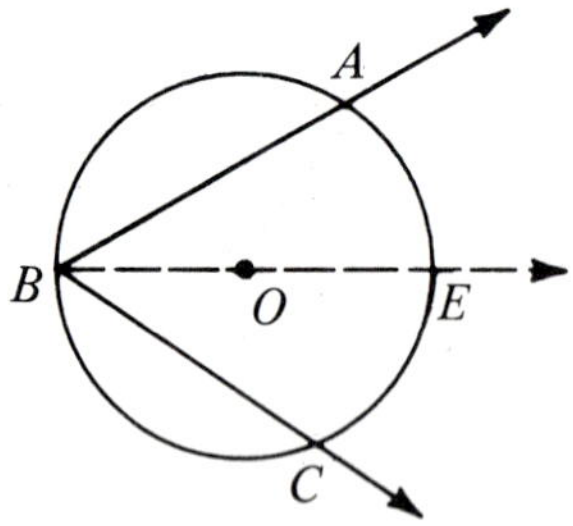
(1) $m\angle ABC = m\angle ABE + m\angle EBC$
(2) $m\angle ABE = \frac{1}{2} m\widehat{AE}$ (Why?)
(3) $m\angle EBC = \frac{1}{2} m\widehat{EC}$ (Why?)
(4) $m\angle ABC = \frac{1}{2} m\widehat{AE} + \frac{1}{2} m\widehat{EC}$ (Why?)
(5) $m\angle ABC = \frac{1}{2} (m\widehat{AE} + m\widehat{EC})$ (Why?)
(6) $m\angle ABC = \frac{1}{2} m\widehat{AC}$ (Substitution)

16. Show that the measure of an inscribed angle is half the measure of its intercepted arc for Case III shown on page 354.

358   CHAPTER 10

17. Given: Quadrilateral $ABCD$ is inscribed in $\odot O$.
    Explain why $m\angle A + m\angle C = 180$.
    (*Hint:* Let $m\stackrel{\frown}{BCD} = x$. Express $m\angle A$ and $m\angle C$ in terms of x.)

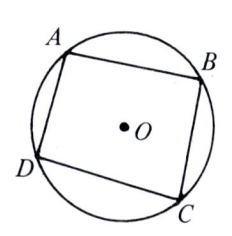

C 18. If $m\angle Q = 3x + 10$, $m\angle R = 3x + 20$, and $m\angle S = 2x + 5$, find the numerical measure of $\angle T$.

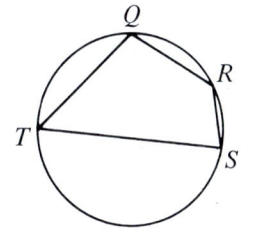

19. A, B, and C are three points on circle O. If $m\angle ABC = 50$ and $m\angle BOA = 132$, find $m\angle BAC$ and $m\angle ACB$. (*Hint:* There are two sets of answers.)

Section 10-4 OTHER ANGLES AND INTERCEPTED ARCS

This section completes your study of the relationships between different kinds of angles and arcs of a circle.

> The measure of an angle formed by a chord and a tangent is half the measure of the intercepted arc.

EXAMPLES

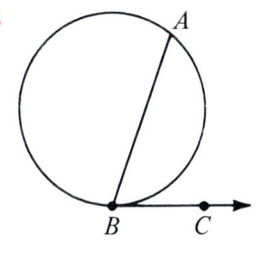

$m\angle ABC = \dfrac{1}{2} m\stackrel{\frown}{AB}$

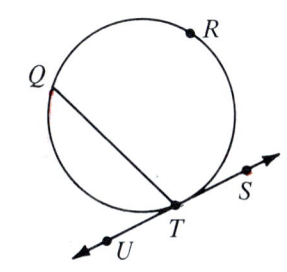

$m\angle UTQ = \dfrac{1}{2} m\stackrel{\frown}{TQ}$

$m\angle STQ = \dfrac{1}{2} m\stackrel{\frown}{QRT}$

GEOMETRY OF THE CIRCLE

> The measure of an angle formed by two secants intersecting in the interior of a circle is half the sum of the measures of the arcs intercepted by the angle and its vertical angle.

Given: Secants $\overleftrightarrow{AB}$ and $\overleftrightarrow{QR}$ intersect at X.
Show: $m\angle 1 = \frac{1}{2}(m\widehat{AQ} + m\widehat{RB})$

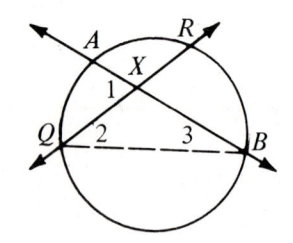

Draw $\overline{QB}$.

$m\angle 1 = m\angle 2 + m\angle 3$	(The measure of an exterior angle of a triangle is equal to the sum of the measures of the two remote interior angles.)
$m\angle 2 = \frac{1}{2} m\widehat{RB}$	(Why?)
$m\angle 3 = \frac{1}{2} m\widehat{AQ}$	(Why?)
$m\angle 1 = \frac{1}{2} m\widehat{RB} + \frac{1}{2} m\widehat{AQ}$	(Substitution)
$m\angle 1 = \frac{1}{2}(m\widehat{RB} + m\widehat{AQ})$	(Why?)

EXAMPLE Find $m\angle 1$.

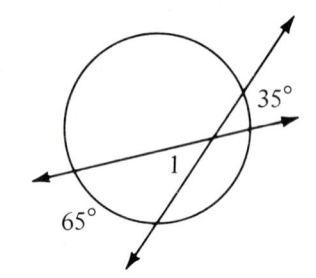

SOLUTION $m\angle 1 = \frac{1}{2}(65 + 35)$
$= 50$

> The measure of an angle formed by two secants intersecting in the exterior of a circle is half the difference of the measures of the intercepted arcs.

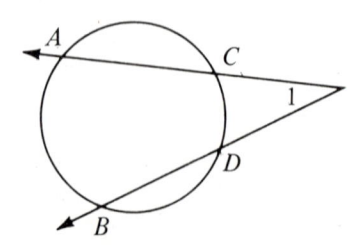

Given: Secants $\overleftrightarrow{AC}$ and $\overleftrightarrow{BD}$
Show: $m\angle 1 = \frac{1}{2}(m\widehat{AB} - m\widehat{CD})$

You will be asked to show that this is true in Exercise 22, page 365.

EXAMPLE Find $m\angle 1$.

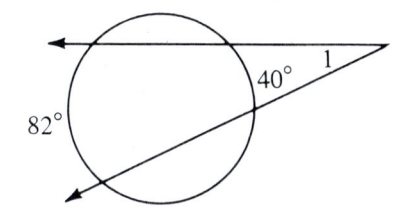

SOLUTION $m\angle 1 = \dfrac{1}{2}(82 - 40)$

$m\angle 1 = 21$

EXAMPLE Find the value of x.

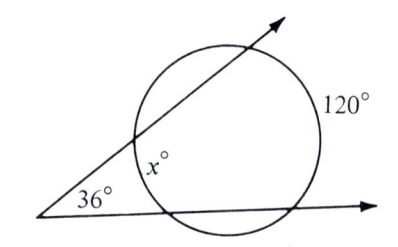

SOLUTION $36 = \dfrac{1}{2}(120 - x)$

$72 = 120 - x$
$-48 = -x$
$48 = x$

The measure of an angle formed by two tangents or by a tangent and a secant intersecting in the exterior of a circle is half the difference of the measures of the intercepted arcs.

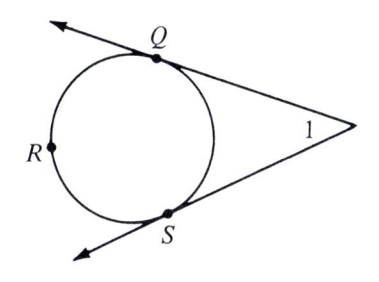

$m\angle 1 = \dfrac{1}{2}(m\overarc{QRS} - m\overarc{QS})$ $m\angle 2 = \dfrac{1}{2}(m\overarc{AC} - m\overarc{AB})$

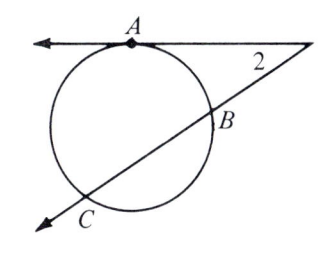

EXAMPLE If $m\overarc{AB} = 128$, find $m\angle 1$.

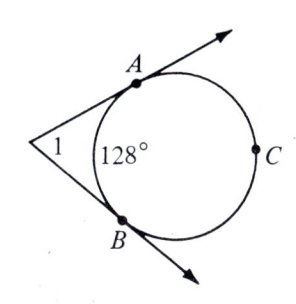

SOLUTION $m\angle 1 = \dfrac{1}{2}(m\overarc{ACB} - m\overarc{AB})$

$m\overarc{ACB} = 232$ (Why?)

$m\angle 1 = \dfrac{1}{2}(232 - 128)$

$m\angle 1 = 52$

GEOMETRY OF THE CIRCLE

A polygon is *circumscribed* about a circle and a circle is *inscribed* in the polygon if each side of the polygon is tangent to the circle.

In the figure on the left below, the pentagon is circumscribed about the circle. In the figure on the right, the quadrilateral is *not* circumscribed about the circle.

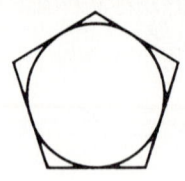

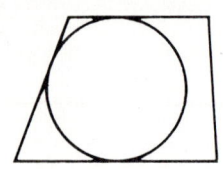

EXAMPLE The triangle is circumscribed about the circle. Find $m\overset{\frown}{BC}$.

SOLUTION Let $m\overset{\frown}{BC} = x$. Then $m\overset{\frown}{CAB} = 360 - x$.

$$64 = \frac{1}{2}(m\overset{\frown}{CAB} - m\overset{\frown}{BC})$$

$$64 = \frac{1}{2}[(360 - x) - x]$$

$$64 = \frac{1}{2}(360 - 2x)$$

$$64 = 180 - x$$

$$x = 116$$

$$m\overset{\frown}{BC} = 116$$

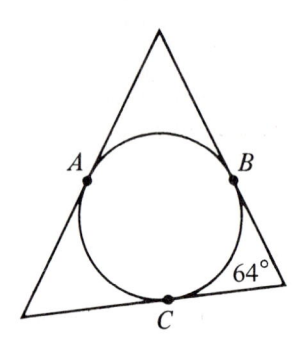

ORAL EXERCISES

Find the measure of each numbered angle.

1.
160°

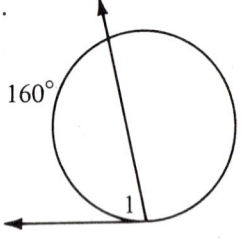

2.
30°
140°

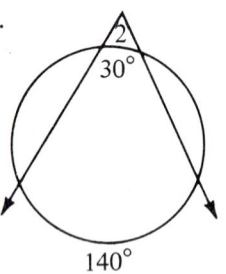

3.
150°
120°

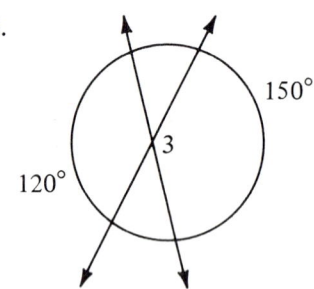

4.
60°
154°

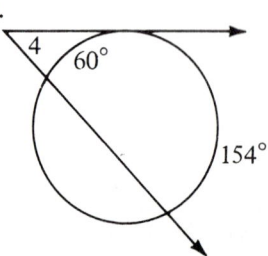

5.
110°

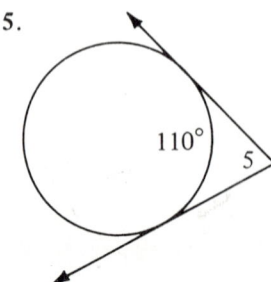

6.
60°
64°

362 CHAPTER 10

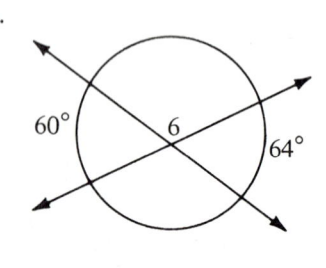

Complete.

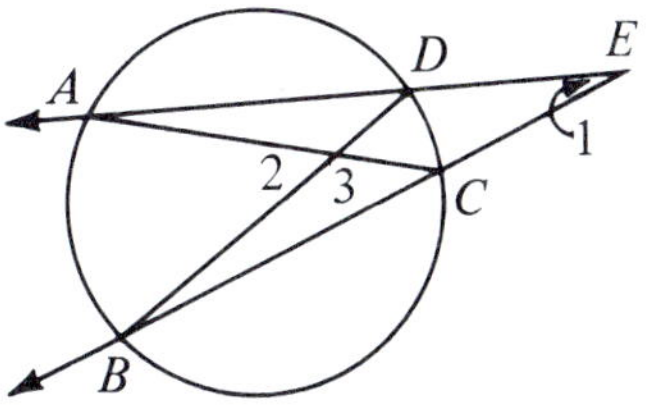

7. $m\angle 1 = \frac{1}{2}(m\underline{\ ?\ } - m\underline{\ ?\ })$

8. $m\angle 2 = \frac{1}{2}(m\underline{\ ?\ } + m\underline{\ ?\ })$

9. $m\angle 3 = \frac{1}{2}(m\underline{\ ?\ } + m\underline{\ ?\ })$

10. $m\widehat{AB} + m\widehat{CD} = 2m\angle \underline{\ ?\ }$

11. If $m\widehat{AD} + m\widehat{BC} = 240$, then $m\angle 3 = \underline{\ ?\ }$.

12. If $m\widehat{AB} = 85$ and $m\widehat{CD} = 23$, then $m\angle 1 = \underline{\ ?\ }$ and $m\angle 2 = \underline{\ ?\ }$.

WRITTEN EXERCISES

A 1. Find the measures of the numbered angles if:
$\overleftrightarrow{AE}$ is tangent to the circle,
$m\widehat{BC} = 120$,
$m\widehat{BE} = 50$,
$m\widehat{CD} = 20$.

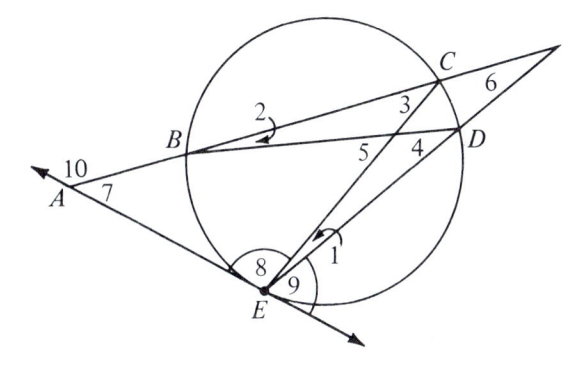

Find the value of x. You may assume that segments, rays, or lines that appear to be tangents are tangents.

2.

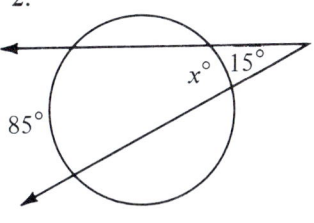

3.

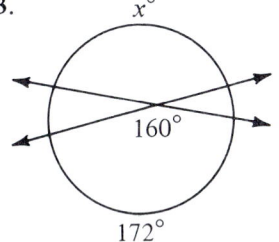

4.

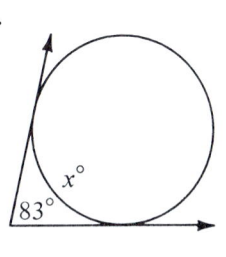

5.

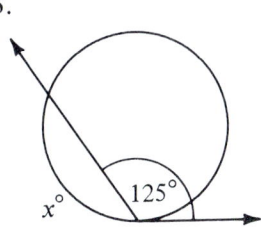

6.

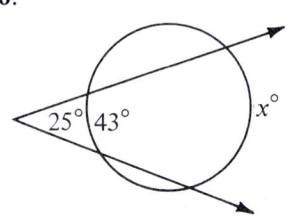

7.

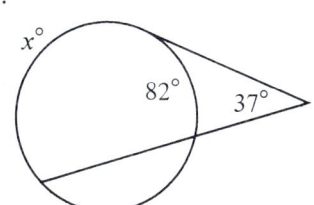

GEOMETRY OF THE CIRCLE

8. In the figure, the triangle is circumscribed about the circle and $m\angle 1 = 52$. Find $m\widehat{QS}$.

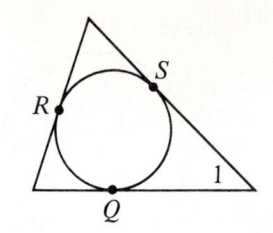

9. In the figure the triangle is circumscribed about the circle and $m\widehat{QS} = 142$. Find $m\angle 1$.

10. Two secants intersecting in a point A in the exterior of a circle intercept arcs of 87° and 31°. Find the measure of $\angle A$.

11. The angle formed by a chord and a tangent has measure 127. Find the measure of the intercepted arc.

12. $\overline{AB}$ is a diameter. $\overline{CD}$ is a tangent. $m\widehat{BD} = 72$. Find $m\widehat{AD}$, $m\angle C$, $m\angle CDB$, and $m\angle CBD$.

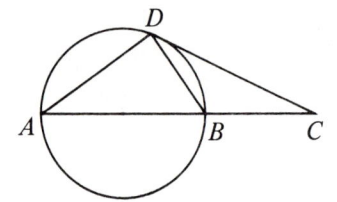

B 13. In the figure, $\overrightarrow{BA}$ and $\overrightarrow{BC}$ are tangent to the circle at A and C respectively.
   a. Explain why $\angle BCA \cong \angle BAC$.
   b. If $\angle BCA \cong \angle BAC$, what must be true of $\overline{BA}$ and $\overline{BC}$? Why?
   c. If $m\angle B = 60$, what kind of triangle is $\triangle ABC$?

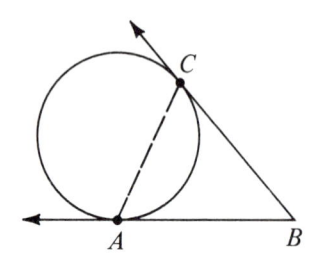

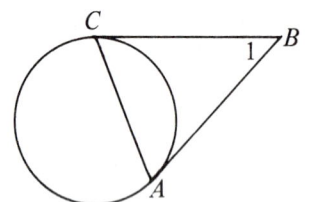

14. If $\overline{BC}$ and $\overline{BA}$ are tangents and $\overline{AC} \cong \overline{BC}$, find $m\angle 1$.

Find the value of x. You may assume that rays that appear to be tangent are tangent.

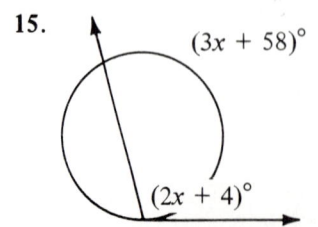

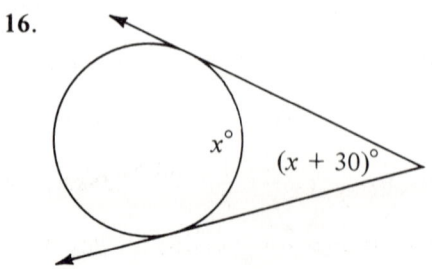

15. $(3x + 58)°$, $(2x + 4)°$

16. $x°$, $(x + 30)°$

364   CHAPTER 10

17. [Figure: circle with chords intersecting inside, arcs labeled $(4x+15)°$, $82°$, $(2x+17)°$]

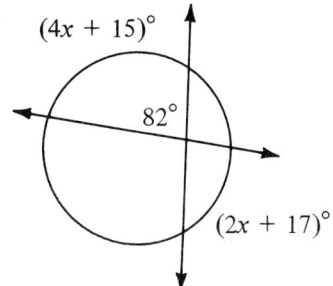

18. 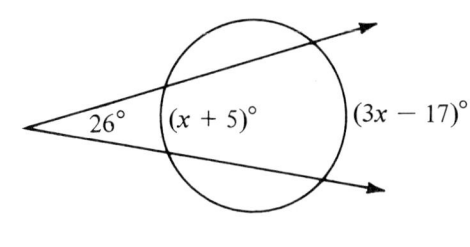 [Figure: two secants from external point, $26°$, $(x+5)°$, $(3x-17)°$]

19. 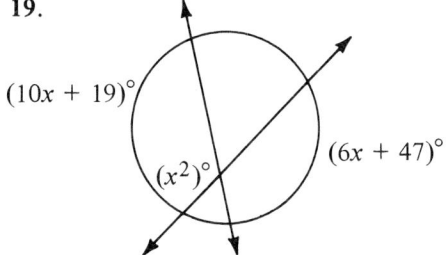 [Figure: two chords/secants intersecting, $(10x+19)°$, $(x^2)°$, $(6x+47)°$]

20. 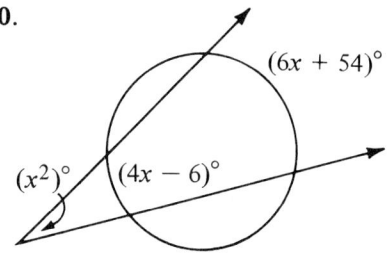 [Figure: two secants from external point, $(6x+54)°$, $(x^2)°$, $(4x-6)°$]

21. $\overline{AB}$, $\overline{BC}$, and $\overline{CA}$ are tangent to the circle at P, Q, and R, respectively. If $m\angle B = 38$, and $m\angle RPQ = 46$, find $m\angle A$, $m\angle C$, $m\angle PRQ$, and $m\angle RQP$.

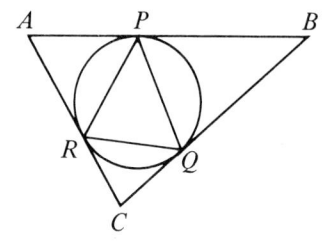

22. Replace each "why" with the correct reason to show that the measure of an angle formed by two secants intersecting in the exterior of a circle is half the difference of the measures of the intercepted arcs.

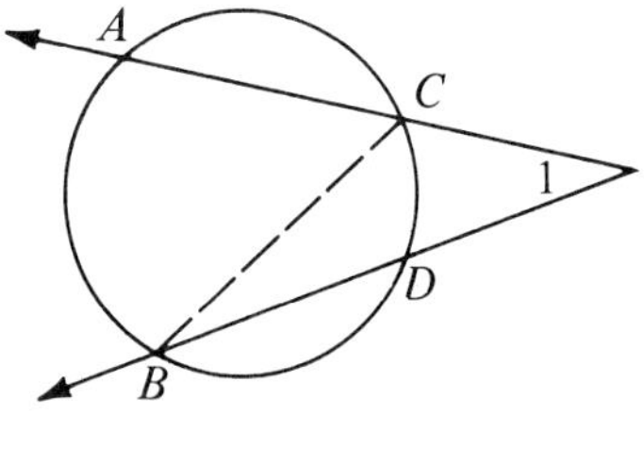

Given: Secants $\overleftrightarrow{AC}$ and $\overleftrightarrow{BD}$

Show: $m\angle 1 = \dfrac{1}{2}(m\widehat{AB} - m\widehat{CD})$

Draw $\overline{CB}$.

(1) $m\angle ACB = m\angle 1 + m\angle CBD$ (Why?)
(2) $m\angle 1 = m\angle ACB - m\angle CBD$ (Addition Property of Equivalent Equations)
(3) $m\angle ACB = \dfrac{1}{2} m\widehat{AB}$ (Why?)
(4) $m\angle CBD = \dfrac{1}{2} m\widehat{CD}$ (Why?)
(5) $m\angle 1 = \dfrac{1}{2} m\widehat{AB} - \dfrac{1}{2} m\widehat{CD}$ (Substitution Principle)
(6) $m\angle 1 = \dfrac{1}{2}(m\widehat{AB} - m\widehat{CD})$ (Why?)

GEOMETRY OF THE CIRCLE

SELF-TEST 2

Exercises 1 and 2 refer to ⊙O at the right. Complete.

1. If $m \widehat{KE} = 86$, then $m\angle 1 = \underline{\ ?\ }$.

2. If $m \widehat{KI} = 62$, then $m\angle 2 = \underline{\ ?\ }$.

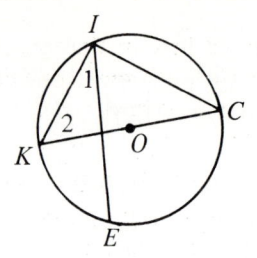

Section 10-3

In Exercises 3-4, P is the center of the circle and $m \widehat{RST} = 132$.

3. Find $m\angle RST$.
4. Find $m\angle RSU$.

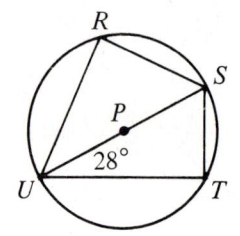

5. Find the measure of the numbered angles if O is the center of the circle, $m \widehat{BC} = 77$, and $m \widehat{DC} = 53$.

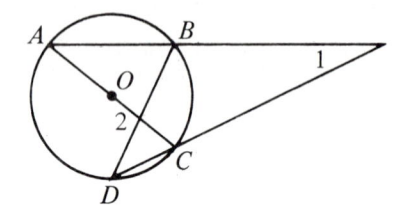

Section 10-4

6. $\overleftrightarrow{XY}$ is a tangent to ⊙P and $m \widehat{WX} = 84$. Find $m\angle WXY$.

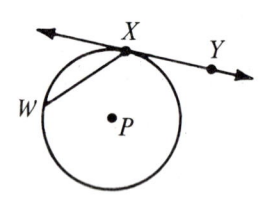

In Exercises 7-8, $\overrightarrow{KR}$ and $\overrightarrow{KN}$ are tangent to ⊙P.

7. If $m \widehat{RN} = 158$, find $m\angle RKN$.

8. If $m \widehat{RM} = 116$ and $m \widehat{MQ} = 170$, find $m\angle RKM$.

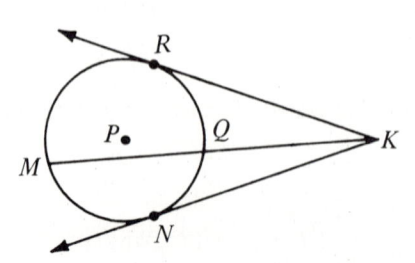

366   CHAPTER 10

Section 10-5 CIRCUMFERENCE AND AREA OF A CIRCLE

Just as you learned how to find the perimeters and areas of certain polygons, in this section you will learn how to calculate the:
- circumference of a circle
- area of a circle

The distance around a circle is called the *circumference* of the circle. To measure the distance around a circle you might wrap a string once around the circle, then pull the string tight and measure it.

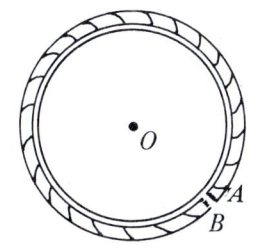

Try this yourself. Use a ruler to measure the diameter of a coffee can, for example. Then wrap a piece of string around the can and measure the string. Express the ratio of the length of the string to the diameter of the can as a decimal. Is your answer between 3.1 and 3.2?

Archimedes (287-212 B.C.) showed that the ratio of the circumference, C, of any circle to its diameter, d, is less than $3\frac{1}{7}$ but greater than $3\frac{10}{71}$. Since the ratio $\frac{C}{d}$ is the same for all circles, mathematicians have given it a special name, π (read "pi").

$$\pi = \frac{\text{circumference}}{\text{diameter}}$$

The number π is irrational. Some useful approximations for π are given below.

$$\pi \approx 3.14 \qquad \pi \approx 3.14159 \qquad \pi \approx \frac{22}{7}$$

The circumference C of a circle of diameter d and radius r is given by the formulas

$$C = \pi d \quad \text{and} \quad C = 2\pi r.$$

EXAMPLE Find the circumference of a circle of radius 7 cm. (Use $\pi \approx \frac{22}{7}$.)

SOLUTION
$C = 2\pi r$
$C \approx 2\left(\frac{22}{7}\right)7$
$C \approx 44$ (cm)

GEOMETRY OF THE CIRCLE

EXAMPLE Find the diameter of a circle with circumference 94.2 m. (Use $\pi \approx 3.14$.)

SOLUTION
$C = \pi d$
$94.2 \approx 3.14d$
$9420 \approx 314d$
$30 \approx d$
$d \approx 30$ (m)

How would you find the area of a circle? You might cut it up into several congruent wedge-shaped pieces.

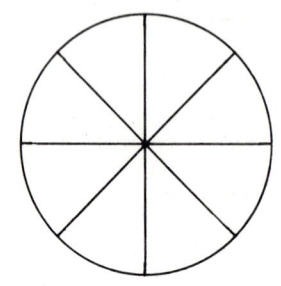

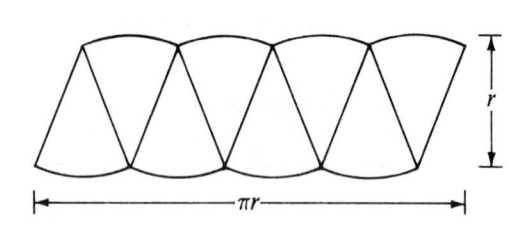

If you arrange the pieces as shown on the right, you will get a figure that looks roughly like a parallelogram. The height of this "parallelogram" is r and the base is approximately πr. Applying the formula for the area of a parallelogram, $A = bh$, you find that the area is approximately $(\pi r)r$, or πr^2.

The area of a circle of radius r is given by the formula
$$A = \pi r^2.$$

EXAMPLE Find the area of a circle of radius 5 mm. (Use $\pi \approx 3.14$.)

SOLUTION $A = \pi r^2$
$A \approx (3.14)(5^2)$
$A \approx 78.5$ (mm²)

EXAMPLE Find the area of a circle of diameter 70 cm. (Use $\pi \approx \frac{22}{7}$.)

SOLUTION If $d = 70$, then $r = 35$.
$A = \pi r^2$
$A \approx \left(\frac{22}{7}\right)(35)^2$
$A \approx 3850$ (cm²)

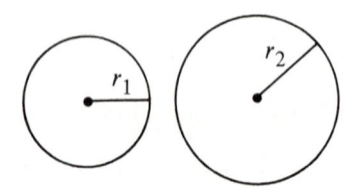

How does the ratio of the circumferences of two circles compare to the ratio of their radii? The circumference, C_1, of the first circle is $2\pi r_1$. The circumference, C_2, of the second circle is $2\pi r_2$. Thus, $\dfrac{C_1}{C_2} = \dfrac{2\pi r_1}{2\pi r_2} = \dfrac{r_1}{r_2}$.

The ratio of the circumferences of two circles equals the ratio of their radii:
$$\frac{C_1}{C_2} = \frac{r_1}{r_2}$$

Now consider the ratio of the areas of two circles. If $A_1 = \pi r_1^2$ and $A_2 = \pi r_2^2$, then $\dfrac{A_1}{A_2} = \dfrac{\pi r_1^2}{\pi r_2^2} = \dfrac{r_1^2}{r_2^2} = \left(\dfrac{r_1}{r_2}\right)^2$.

The ratio of the areas of two circles equals the square of the ratio of their radii:
$$\frac{A_1}{A_2} = \left(\frac{r_1}{r_2}\right)^2$$

EXAMPLE If the radius of a circle is tripled, by what number is the area multiplied?

SOLUTION Let A_1 represent the area of the smaller circle and A_2 represent the area of the larger circle. The ratio of the radii of the two circles is $1:3$. You know that the ratio of the areas of two circles is the square of the ratio of their radii. Thus

$$\frac{A_1}{A_2} = \left(\frac{1}{3}\right)^2 = \frac{1}{9}, \text{ or } A_2 = 9A_1.$$

Therefore, if the radius is tripled, then the area is multiplied by 9.

If you are asked to find the *exact* circumference or area of a circle, leave your answer in terms of π.

EXAMPLE Find the exact circumference and the exact area of a circle of radius 2.5.

SOLUTION
$C = 2\pi r$
$ = 2\pi(2.5)$
$ = 5\pi$

$A = \pi r^2$
$ = \pi(2.5)^2$
$ = 6.25\pi$

ORAL EXERCISES

1. Find the exact circumference of a circle with the given radius.

   a. 2 cm   b. 11 m   c. $\dfrac{1}{2}$ d. 1.2 km

2. Find the exact area of the circles described in Exercise 1.

3. Use $\pi \approx \dfrac{22}{7}$ to find the circumference of a circle with the given diameter.

   a. 7 cm   b. 14 m   c. 42 m   d. 2.8 mm

GEOMETRY OF THE CIRCLE

4. Use $\pi \approx 3.14$ to find the area of a circle with the given radius.
   a. 1 cm
   b. 10 m
   c. 2 mm
   d. 3

5. Find the radius of a circle with the given circumference.
   a. 12π m
   b. 3.2π cm
   c. 428π
   d. 2π mm

6. By what value is the circumference of a circle multiplied if the radius is multiplied by the given value?
   a. 2
   b. 10
   c. $\frac{1}{4}$
   d. n

7. By what value is the area of a circle multiplied if the radius is multiplied by the given value?
   a. 2
   b. 10
   c. $\frac{1}{4}$
   d. n

WRITTEN EXERCISES

In Exercises 1–4 find the exact circumference and the exact area of a circle with the given radius.

A 1. $3\sqrt{2}$ cm
2. 2.8 m
3. $12k$ mm
4. 2π km

In Exercises 5–7 find the circumference and the area of each circle O.

5. Use $\pi \approx 3.14$. 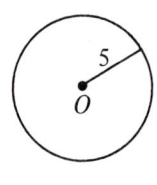
6. Use $\pi \approx \frac{22}{7}$. 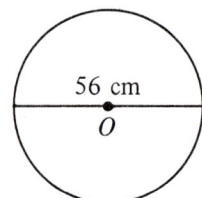
7. Use $\pi \approx \frac{22}{7}$.

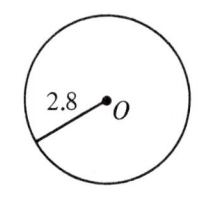

8. A circle has a circumference of 208π. Find its diameter.

9. Jane is roller skating around a circular rink of radius 4.5 m. If she skates 12 times around the rink, how far will she have traveled? (Use $\pi \approx 3.14$. Give your answer to the nearest tenth of a meter.)

10. The circumference of a circle is increased from 40π cm to 80π cm. By how many centimeters is the radius increased?

11. The area of one circle is 36 times the area of a smaller circle. How do their radii compare?

12. The ratio of the radii of two circles is 3:2. Find the ratio of their circumferences and the ratio of their areas.

B 13. The area of a semicircle is 24π. Find the radius of the semicircle in simplest radical form.

14. The diameter of Ted's bicycle wheels is 56 cm. How many rotations does a wheel complete when Ted bicycles 36 km? (Use $\pi \approx 3.14$. Give your answer to the nearest hundred.)

15. The circumferences of two circles are 12π and 36π. Find the ratio of the areas of the circles.

16. A square with perimeter 16 cm is circumscribed about a circle. Find the exact area of the circle.

17. The figure shows two circles with the same center, O.
    a. Find the exact area of the larger circle.
    b. Find the exact area of the smaller circle.
    c. Find the exact area of the shaded region.

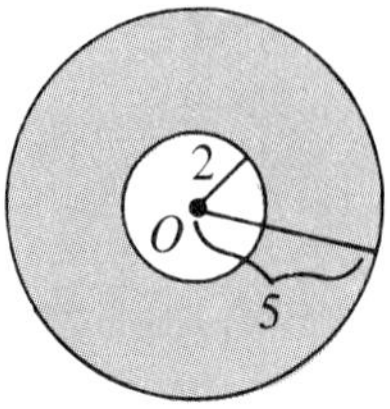

In Exercises 18–21 find the exact area of the shaded region.

18.

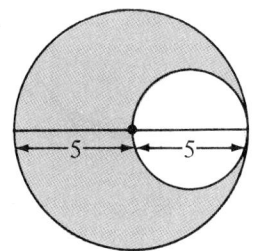

19.

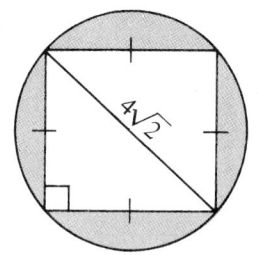

20.

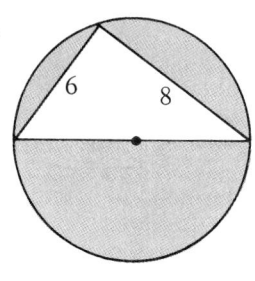

21.

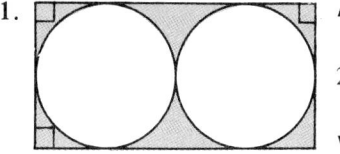

C 22. The circle is circumscribed about a regular hexagon with perimeter 36 cm. Find the exact area of the shaded region.

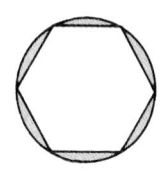

23. Triangle ABC is an equilateral triangle inscribed in $\odot O$. The perimeter of $\triangle ABC$ is 12 cm.
    a. Find the area of $\triangle ABC$.
    b. Find the radius of $\odot O$. (*Hint:* Consider a 30°-60°-90° triangle whose hypotenuse is $\overline{OC}$.)
    c. Find the exact area of the shaded region.

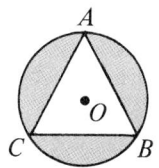

24. A circle has circumference 64π cm. Find the area of an equilateral triangle inscribed in it.

GEOMETRY OF THE CIRCLE 371

Section 10-6 SURFACE AREAS AND VOLUMES OF SPHERES

Now you are ready to study the three-dimensional counterpart of a circle. In this section you will learn about:
- spheres
- hemispheres
- surface areas and volumes of spheres

A *sphere* with center P is the set of all points in space at a given distance from a given point P. Many of the terms used for circles are also used for spheres.

$\overline{PA}$ and $\overline{PB}$ are radii.
$\overline{AB}$ is a diameter.

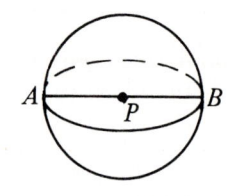

The three-dimensional version of a semicircle is a *hemisphere*.

The surface area, S.A., of a sphere of radius r is given by the formula
$$\text{S.A.} = 4\pi r^2.$$

EXAMPLE A sphere has diameter 10 cm. Find its surface area.

SOLUTION If $d = 10$, then $r = 5$.
$$\text{S.A.} = 4\pi r^2$$
$$= 4\pi(5^2)$$
$$= 100\pi \text{ (cm}^2)$$

How can you find a formula for the volume of a sphere? You may use an argument similar to the one in Section 10-5 for deriving the formula for the area of a circle. Think of a sphere as being divided into many "pyramids," each with the center of the sphere as vertex. (Of course none of these shapes could really be a pyramid because their bases would be curved. This is only an approximation.) The approximate volume of each of these "pyramids" can be found by using the formula $V = \frac{1}{3}Bh$. The height of the pyramid is approximately r, the radius of the sphere.

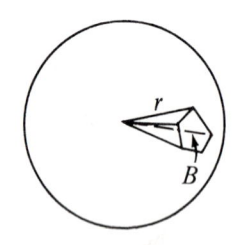

$$V \approx \frac{1}{3}Br$$

To find the volume of the sphere, you would add the volumes of all the "pyramids."

$$V \approx \frac{1}{3}B_1 r + \frac{1}{3}B_2 r + \cdots$$
$$\approx \frac{1}{3}r(B_1 + B_2 + \cdots)$$

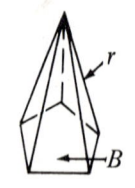

372 CHAPTER 10

Notice, however, that $B_1 + B_2 + \cdots$ would equal the surface area of the sphere.

$$V \approx \frac{1}{3}r(4\pi r^2)$$

$$V \approx \frac{4}{3}\pi r^3$$

The volume, V, of a sphere of radius r is given by the formula

$$V = \frac{4}{3}\pi r^3.$$

EXAMPLE   A sphere has a radius of 6 m. Find its volume.

SOLUTION $V = \frac{4}{3}\pi r^3$

$= \frac{4}{3}\pi(6^3)$

$= \frac{4}{3}\pi \cdot 216$

$= 4\pi \cdot 72$

$= 288\pi$ The volume is 288π m^3.

How does the ratio of the surface areas of two spheres compare to the ratio of their radii?

The surface area of the large sphere shown is $4\pi r_1^2$, and the surface area of the smaller sphere is $4\pi r_2^2$.

The ratio of the surface areas, $\dfrac{4\pi r_1^2}{4\pi r_2^2}$, simplifies to $\left(\dfrac{r_1}{r_2}\right)^2$.

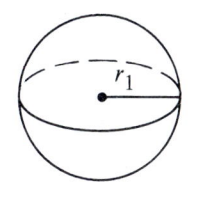

The ratio of the surface areas of two spheres equals the square of the ratio of their radii:

$$\frac{(S.A.)_1}{(S.A.)_2} = \left(\frac{r_1}{r_2}\right)^2$$

EXAMPLE   The surface area of one sphere is 36π cm^2. Find the surface area of a sphere whose radius is twice the radius of the first sphere.

SOLUTION   The ratio of the radii of the two spheres is $1:2$.
The surface area of the smaller sphere is 36π cm^2.

$\dfrac{36\pi}{x} = \left(\dfrac{1}{2}\right)^2$

$\dfrac{36\pi}{x} = \dfrac{1}{4}$

$x = 144\pi$ The surface area of the second sphere is 144π cm^2.

GEOMETRY OF THE CIRCLE

In Exercise 20 on page 375 you will be asked to show that the following is true.

> The ratio of the volumes of two spheres equals the cube of the ratio of their radii:
> $$\frac{V_1}{V_2} = \left(\frac{r_1}{r_2}\right)^3$$

EXAMPLE   The radius of one sphere is 3 times the radius of a smaller sphere. If the volume of the smaller sphere is 4 m³, find the volume of the larger sphere.

SOLUTION   The ratio of the radii is 1:3. The volume of the smaller sphere is 4 m³.

$$\frac{4}{V} = \left(\frac{1}{3}\right)^3$$

$$\frac{4}{V} = \frac{1}{27}$$

$$V = 4 \cdot 27$$
$$= 108$$

The volume of the larger sphere is 108 m³.

ORAL EXERCISES

Find the exact surface area and volume of a sphere with the given radius.

1. 10   2. 6 m   3. $\frac{1}{2}$ 4. $3x$ cm

Find the exact surface area and volume of a sphere with the given diameter.

5. 30 cm   6. 2   7. 0.6   8. $3x$ m

WRITTEN EXERCISES

Complete for a sphere of radius r, diameter d, surface area S.A., and volume V. Leave your answers in terms of π.

	r	d	S.A.	V
A 1.	3 cm	?	?	?
2.	?	18 m	?	?
3.	1.2 m	?	?	?
4.	?	0.8	?	?
5.	?	?	144π m²	?
6.	?	?	25π	?
7.	$\sqrt{2}$	?	?	?
8.	?	$2\sqrt{3}$	?	?
9.	?	?	4π	?

374   CHAPTER 10

10. Find the exact volume of the hemisphere shown at the right.
11. The radius of a sphere is doubled. How is its volume affected?
12. The radius of a sphere is tripled. How is its surface area affected?

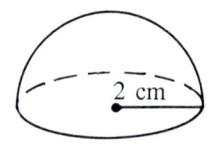

B 13. A plane passing through the center of a sphere intercepts a circle with area 3.6π cm². Find the surface area of the sphere.

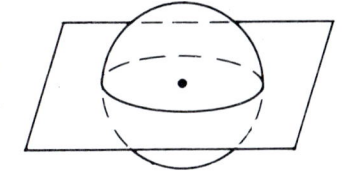

14. The ratio of the surface areas of two spheres is 16:25. Find the ratio of (a) their radii; (b) their volumes.
15. One sphere has a diameter 3 times that of a smaller sphere. What is the ratio of their surface areas?
16. A sphere is inscribed in a cube whose edges are 10 units long. Find the surface area and volume of the sphere.

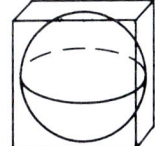

Exercises 17 and 18 refer to the solid spherical shell shown at the right.
17. If the radius of the inner sphere is 3 and the radius of the outer sphere is 4, find the volume of the shell.
18. If the diameter of the inner sphere is 8 cm and the thickness of the shell is 3 cm, find the volume of the shell.

C 19. The sum of the volumes of two spheres is $\frac{1120}{3}\pi$. If the ratio of their surface areas is 4:9, find the volume of each sphere.
20. Show that the ratio of the volumes of two spheres equals the cube of the ratio of their radii.

Section 10-7 SURFACE AREAS AND VOLUMES OF CYLINDERS AND CONES

You are now ready to apply your knowledge of the circumference and area of a circle to other three-dimensional figures. In this section you will learn about:
- cylinders
- cones
- lateral area, total area, and volume of cylinders and cones

GEOMETRY OF THE CIRCLE

A *right circular cylinder* is like a right prism except that its bases are congruent circles instead of congruent polygons.

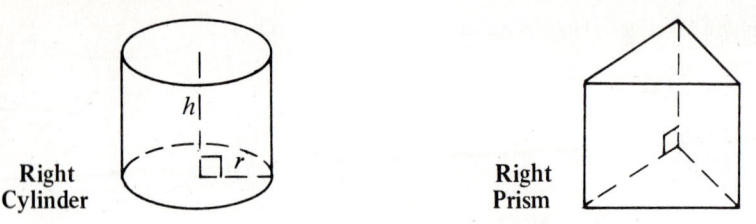

Right Cylinder

Right Prism

The *altitude* of a right cylinder is the segment whose endpoints are the centers of the bases. The length of the altitude is the *height* of the cylinder.

A *right circular cone* is like a right pyramid except that its base is a circle instead of a polygon. The *altitude* of a right circular cone is the segment whose endpoints are the vertex of the cone and the center of the base.

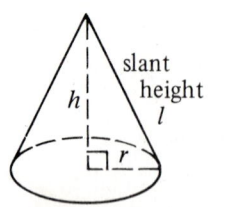

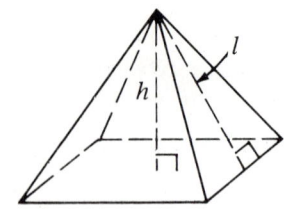

When we refer to either a cylinder or a cone in this book we will mean a right circular cylinder or a right circular cone.

The formulas for lateral area, total area, and volume of cylinders and cones are much like those for prisms and pyramids.

Cylinder L.A. $= 2\pi rh$	Cone L.A. $= \pi rl$
T.A. $=$ L.A. $+ 2B = 2\pi rh + 2\pi r^2$	T.A. $=$ L.A. $+ B = \pi rl + \pi r^2$
$V = Bh = \pi r^2 h$	$V = \frac{1}{3}Bh = \frac{1}{3}\pi r^2 h$

EXAMPLE Find the lateral area, total area, and volume of a cylinder with a radius of 3 cm and a height of 8 cm.

SOLUTION
L.A. $= 2\pi rh$
$\quad\quad = 2\pi(3)(8)$
$\quad\quad = 48\pi$
L.A. $= 48\pi$ cm²

T.A. $=$ L.A. $+ 2\pi r^2$
$\quad\quad = 48\pi + 2\pi(3^2)$
$\quad\quad = 66\pi$
T.A. $= 66\pi$ cm²

$V = \pi r^2 h$
$\quad = \pi(3^2)(8)$
$\quad = 72\pi$
$V = 72\pi$ cm³

376   CHAPTER 10

EXAMPLE The volume of the cone shown at the right is 100π m³. Find the height and the slant height of the cone.

SOLUTION
$$V = \frac{1}{3}\pi r^2 h$$
$$100\pi = \frac{1}{3}\pi(25)h$$
$$300\pi = 25\pi h$$
$$h = 12 \text{ m}$$

By the Pythagorean Theorem, $r^2 + h^2 = l^2$.
$(5)^2 + (12)^2 = l^2$, and $l = 13$ m.

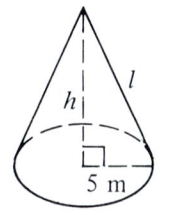

ORAL EXERCISES

Find the exact lateral area, total area, and volume of each cylinder.
1. $r = 4$ cm, $h = 10$ cm
2. $r = 2$ m, $h = 9$ m
3. $r = 5$ mm, $h = 4$ mm

Complete the table for the cone.

	r	h	l	L.A.	T.A.	V
4.	3 cm	4 cm	?	?	?	?
5.	8 m	?	17 m	?	?	?
6.	?	12	13	?	?	?

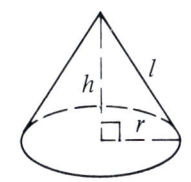

7. If rectangle $ABCD$ is revolved around side $\overline{AD}$ it generates a cylinder.
   a. What is the radius of the cylinder?
   b. What is the height of the cylinder?
   c. If $AB = 2$ and $BC = 5$, find the volume of the cylinder.
8. If rectangle $ABCD$ were revolved around side $\overline{CD}$, would it generate a cylinder with the same volume as in Exercise 7?

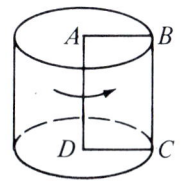

9. Suppose that right $\triangle QRS$ is revolved around $\overline{QS}$. What kind of figure is generated?

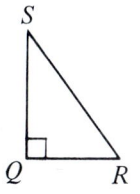

GEOMETRY OF THE CIRCLE

WRITTEN EXERCISES

Copy and complete the table for the cylinder.

A

	h	r	L.A.	T.A.	V
1.	3	$\sqrt{2}$	?	?	?
2.	?	5	?	?	50π
3.	4	?	24π	?	?
4.	a	2a	?	?	?
5.	3	?	?	?	192π

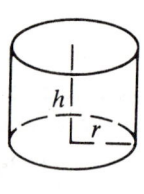

Find the lateral area, total area, and volume for each cone.

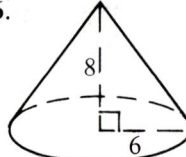

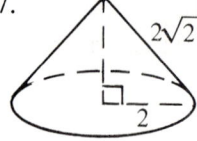

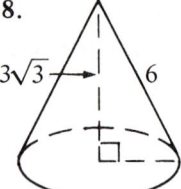

6. [cone with height 8, radius 6]

7. [cone with height $2\sqrt{2}$, radius 2]

8. [cone with slant $3\sqrt{3}$, radius 6]

9. A cone has height 2 cm and radius 2 cm. How many conefuls of sand would it take to fill a sphere of radius 2 cm?

10. A cone and a cylinder have the same height and radius. How many times greater is the volume of the cylinder than the volume of the cone?

11. Brand X soup comes in a cylindrical can with a radius of 4 cm and a height of 10 cm. Brand Y soup is packaged in a cylindrical can of radius 3 cm and height of 16 cm. Each can costs 23¢. Which gives you more for your money?

12. A cone with volume 300π cm³ has a height of 10 cm. Find the radius and slant height of the cone.

B 13. What happens to the volume of a cylinder if the height is doubled while the radius remains the same?

14. What happens to the volume of a cylinder if the radius is tripled while the height remains the same?

15. A cone is inscribed in a regular square pyramid. The length of each edge of the base of the pyramid is 10 cm and the slant height of the pyramid is 13 cm. Find the volume of the cone.

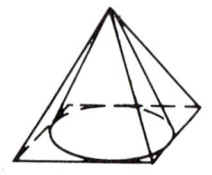

16. Find the radius of a cone if the slant height of the cone is 4 cm longer than the radius, and the lateral area of the cone is 12π cm².

378   CHAPTER 10

17. A cylindrical hole of radius 4 cm is drilled through a right triangular prism. If the base of the prism is an equilateral triangle 10 cm on a side and the height of the prism is 10 cm, find the volume of the resulting figure.

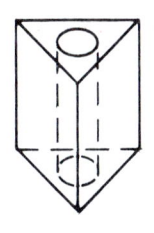

C 18. The number of cubic centimeters in the volume of a certain cone equals the number of square centimeters in the lateral area of the cone. If the height of the cone is 5 cm, find the radius of the cone.

19. Find the radius and height of a cylinder if the height is 3 less than the radius and the total area is 70π.

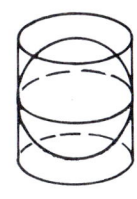

20. A sphere is inscribed in a cylinder. Show that the volume of the sphere is two thirds the volume of the cylinder.

21. A metal cylinder of radius r and height h is melted down and recast as a cone of radius r. Find the height of the cone.

SELF-TEST 3

1. Find the exact circumference and the exact area of a circle with radius 8 m. Section 10-5

2. The circumference of a standard bicycle tire is 68.58 cm. Using $\pi \approx 3.14$, find the radius of the tire to the nearest tenth of a centimeter.

3. The radii of two circles are 18 and 24. Find the ratio of their circumferences and the ratio of their areas.

4. Find the radius of a circle whose area is 2500π cm^2.

5. Find the exact surface area of a sphere with diameter 48 cm. Section 10-6

6. Find the exact volume of a sphere with radius 15 cm.

7. The ratio of the radii of two spheres is 3:5. The surface area of the smaller sphere is 144π. Find the surface area of the larger sphere.

In Exercises 8-9, find the exact lateral area, total area, and volume of the figure shown.

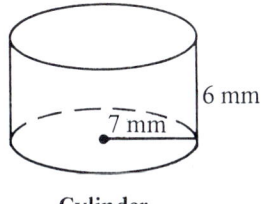

8. Section 10-7

Cylinder (6 mm height, 7 mm radius)

9. Cone ($2\sqrt{10}$ slant, radius 3)

GEOMETRY OF THE CIRCLE

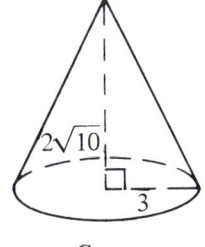

Computer Exercises

1. The circumference of a circle with radius 2 is 4π. In the diagram at the right, the length of $\overset{\frown}{AE}$ is one fourth the circumference of a circle with radius 2. Therefore, the length of $\overset{\frown}{AE}$ is $\frac{1}{4}(4\pi)$, or π.

   The horizontal radius, $\overline{OE}$, has been subdivided into 4 congruent parts. Line segments perpendicular to $\overline{OE}$ at the dividing points intersect $\overset{\frown}{AE}$ at B, C, and D. The following program uses the sum of the lengths of the chords $\overline{AB}$, $\overline{BC}$, $\overline{CD}$, and $\overline{DE}$ as an approximation for π. That is,

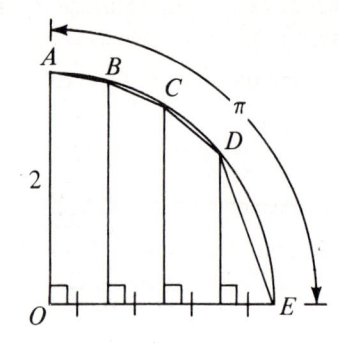

   $$\pi \approx AB + BC + CD + DE.$$

   N represents the number of subdivisions of $\overline{OE}$. The larger N is, the better the approximation for π.

   Copy and RUN the program for several values of N. Use powers of 2 as values of N.

   ```
   10   INPUT N
   20   LET H1=2
   30   FOR I=1 TO N
   40   LET D=D+2/N
   50   LET H2=SQR(4-D^2)
   60   LET L=L+SQR((H1-H2)^2+(2/N)^2)
   70   LET H1=H2
   80   NEXT I
   90   PRINT "PI IS APPROXIMATELY";L
   100  END
   ```

2. Use $\pi \approx 3.14159$.

   a. Write a program to find the volume and surface area of a sphere given its radius.

   b. Write a program to find the area of a circle given its circumference.

CHAPTER REVIEW

Match each term with the appropriate set of points in the right-hand column. *Each letter may be used only once.*

1. radius
2. tangent
3. secant
4. diameter
5. chord

a. $\overleftrightarrow{AB}$
b. $\overleftrightarrow{CF}$
c. $\overrightarrow{OC}$
d. $\overline{OB}$
e. $\overline{OA}$
f. $\overline{BC}$
g. $\overline{DE}$
h. $\overline{CF}$

Section 10-1

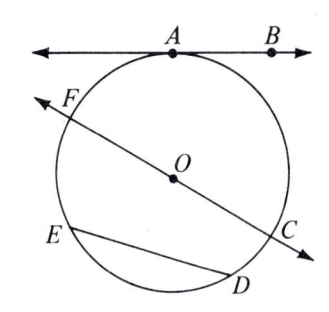

In Exercises 6–11, O is the center of the circle. Find the values of the variables. You may assume that lines that appear to be tangents *are* tangents.

6. Section 10-2

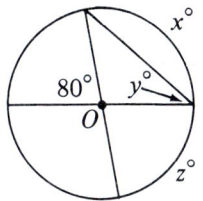

7.

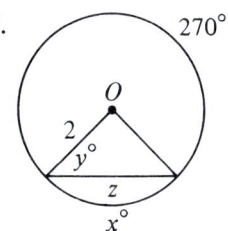

8. Section 10-3

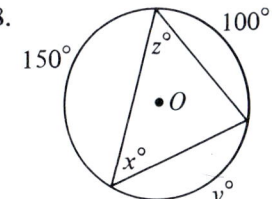

9.

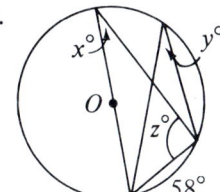

10. Section 10-4

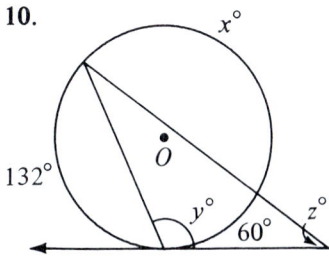

11.

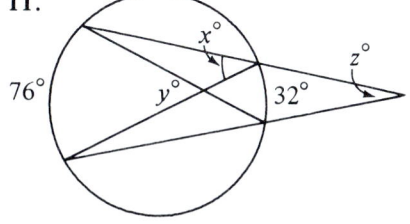

12. Find the circumference and area of a circle with radius 10 cm. (Use $\pi \approx 3.14$.)   Section 10-5
13. Find the diameter of a circle whose area is 121π m².
14. The area of one circle is 49 times the area of a smaller circle. What is the ratio of their radii?

GEOMETRY OF THE CIRCLE

15. Find the exact surface area and volume of a sphere with radius $\sqrt{3}$. Section 10-6

16. The radius of a sphere is divided by 2. How is its volume affected?

Find the exact lateral area, total area, and volume of each figure.

17. 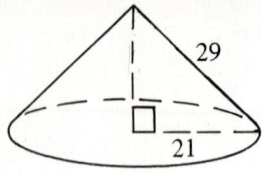 18. 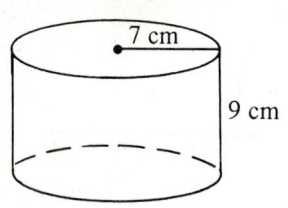 Section 10-7

19. A cone with volume 96π cm³ has a radius of 6 cm. Find the height of the cone.

CHAPTER TEST

Exercises 1-9 refer to $\odot O$ shown at the right. State whether the statement is true or false. You may assume that lines that appear to be tangents *are* tangents.

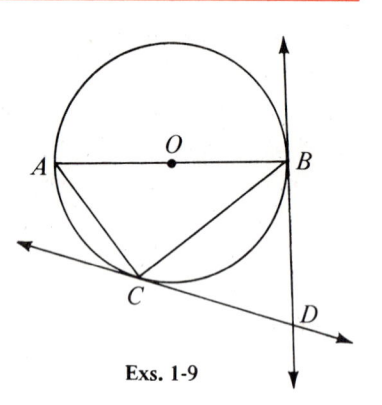

1. If $AB = 15$, then $OC = 7\frac{1}{2}$.
2. $\overline{BC}$ is a secant of $\odot O$.
3. $m\angle ABC = 2\, m\widehat{AC}$
4. $m\angle ABD = m\angle ACB$
5. If $m\widehat{BC} = x$, then $m\widehat{BAC} = 180 - x$.
6. One and only one tangent line can be drawn to $\odot O$ through point A.
7. $m\angle BOC = \frac{1}{2} m\widehat{BC}$
8. $\triangle ABC$ is inscribed in $\odot O$.
9. If $m\widehat{BC} = 70$, then $m\angle BDC = 100$.

Exs. 1-9

In Exercises 10-13, find the values of x and y for each $\odot O$. You may assume that lines that appear to be tangents are tangents.

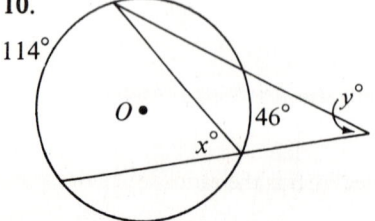

10. 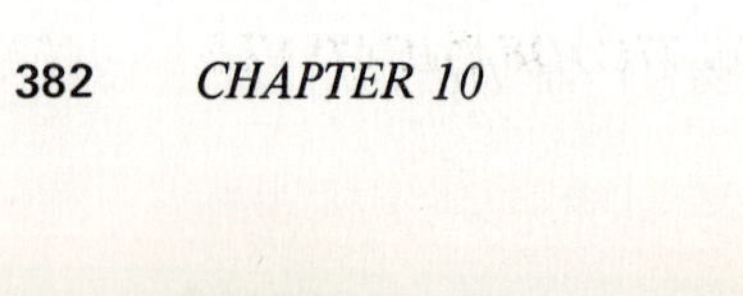 11.

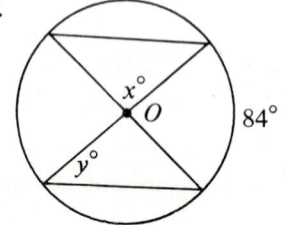

382   CHAPTER 10

12.

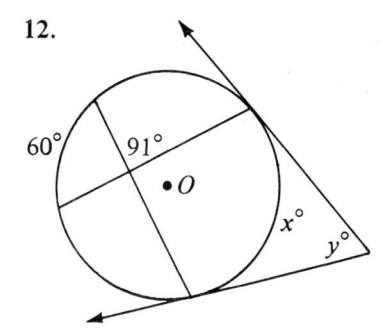

13.

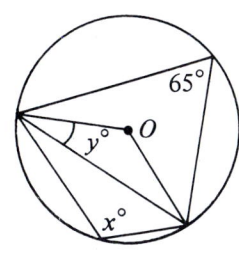

14. Find the exact circumference and area of a circle whose diameter is 9.2 cm.
15. The ratio of the radii of two spheres is 2:5. The volume of the smaller sphere is 48π mm³. Find the volume of the larger sphere.
16. A cylinder has a radius of 8.4 cm and a height of 5 cm. Find its exact lateral area, total area, and volume.
17. The areas of two circles are 128π and 50π. What is the ratio of their radii?
18. A cone has a radius of 9 and a height of 12. Find its exact lateral area, total area, and volume.
19. Find the exact surface area and volume of a sphere with radius 6 cm.
20. The surface area of a sphere is equal to the lateral area of a cylinder. If each radius is 5 cm, find the height of the cylinder.

CUMULATIVE REVIEW: CHAPTERS 1-10

Indicate the best answer by writing the appropriate letter.

1. If the length of one leg of a right triangle is 8 and the length of the hypotenuse is 12, find the length of the other leg.
   a. $8\sqrt{3}$ b. $4\sqrt{5}$ c. $4\sqrt{13}$ d. 4

2. Refer to the diagram. Which statement need *not* be true?
   a. $m\widehat{AD} = m\widehat{DC}$ b. $\triangle ABE \sim \triangle DCE$
   c. $\angle B \cong \angle C$ d. $\dfrac{AB}{DC} = \dfrac{BE}{CE}$

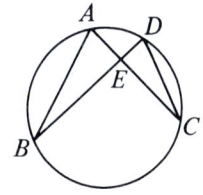

3. The inverse of $p \rightarrow q$ is:
   a. $\sim q \rightarrow \sim p$ b. $\sim p \rightarrow q$ c. $\sim p \rightarrow \sim q$ d. $q \rightarrow p$

4. Complete: $6z^2 + 7z - 3 = (3z \underline{\ ?\ })(2z \underline{\ ?\ })$
   a. $+1; -3$ b. $-1; +3$ c. $+3; -1$ d. $-3; +1$

5. Solve: $4y - 7 = 13$
   a. $\dfrac{3}{2}$ b. 5   c. $-\dfrac{3}{2}$ d. -5

6. Simplify: $(-39 + 12) \div (-9)^2$
   a. $\frac{1}{3}$
   b. $-\frac{1}{3}$
   c. 3
   d. -3

7. Simplify: $\dfrac{2\sqrt{6} \cdot \sqrt{3}}{\sqrt{2}}$
   a. $2\sqrt{3}$
   b. $2\sqrt{9}$
   c. $\dfrac{2\sqrt{18}}{\sqrt{2}}$
   d. 6

8. Let p represent: $x + 3 \leq 0$
   Let q represent: $2x < -10$
   When $x = -4$, which of the following is true?
   a. $p \lor q$
   b. $p \land q$
   c. $p \rightarrow q$
   d. $\sim p \land q$

9. To the nearest tenth, what percent of 40 is 27?
   a. 0.7%
   b. 1.5%
   c. 148.1%
   d. 67.5%

10. Multiply: $(3r - 7)(2r + 5)$
    a. $6r^2 - r - 35$
    b. $6r^2 - 29r - 35$
    c. $6r^2 + r - 35$
    d. $6r^2 + 29r - 35$

11. The ratio of the lengths of a pair of corresponding sides of two similar triangles is $5:7$. Find the ratio of the areas of the triangles.
    a. $10:49$
    b. $5:7$
    c. $125:343$
    d. $25:49$

12. If $\widehat{ABC}$ is a semicircle of $\odot O$, what kind of triangle must $\triangle ABC$ be?
    a. equilateral
    b. isosceles
    c. right
    d. obtuse

13. Evaluate $2a\left(\dfrac{b-c}{d}\right)$ if $a = 1.5$, $b = -8$, $c = -2$, and $d = -3$.
    a. 6
    b. -6
    c. 10
    d. -60

14. Simplify: $\dfrac{4}{\sqrt{2}}$
    a. $2\sqrt{2}$
    b. $4\sqrt{2}$
    c. 2
    d. $\sqrt{2}$

15. If quadrilateral $ABCD$ is a parallelogram, which property *must* be true?
    a. $\angle A$ and $\angle B$ are supplementary.
    b. $\triangle ABC \cong \triangle BCD$
    c. $\angle A \cong \angle B$
    d. $\overline{AC} \cong \overline{BD}$

16. Simplify: $\sqrt{\dfrac{7}{20}}$
    a. $\dfrac{\sqrt{35}}{10}$
    b. $\dfrac{\sqrt{140}}{20}$
    c. $\dfrac{\sqrt{7}}{2\sqrt{5}}$
    d. $\dfrac{1}{2}\sqrt{\dfrac{7}{5}}$

17. The sum of the measures of a complement and a supplement of an angle is 220. Find the measure of the angle.
    a. 65
    b. 155
    c. 55
    d. 25

18. If 80% of a number is 28, what is the number?
    a. 22.4
    b. 140
    c. 35
    d. 3.5

19. You can show that $\triangle RTS \cong \triangle RTU$ by:
    a. SSS
    b. AA
    c. SSA
    d. ASA

20. Alfreda paid $51 for an automobile tune-up. If the charge for parts was $7 less than the charge for labor, find the charge for labor.
   a. $29   b. $27   c. $25   d. $22

21. Simplify: $\sqrt{175} + \sqrt{7} - \sqrt{28}$
   a. $3\sqrt{7}$ b. $4\sqrt{7}$ c. $9\sqrt{7}$ d. $22\sqrt{7}$

22. Solve: $0.5x + 0.2 = 1.7$
   a. 3.8   b. 0.3   c. 0.38   d. 3

23. Which statement is *not always* true when two parallel lines are cut by a transversal?
   a. Alternate exterior angles are congruent.
   b. Alternate interior angles are supplementary.
   c. Corresponding angles are congruent.
   d. Interior angles on the same side of the transversal are supplementary.

24. Find the value of x in the diagram.
   a. 110   b. 17   c. 34   d. 55

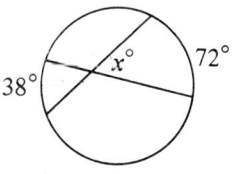

25. Express $\frac{13}{72}$ as a decimal.
   a. $0.1\overline{805}$ b. $0.180\overline{5}$ c. 0.1805 d. $0.18\overline{5}$

26. For which quadrilaterals is it true that the diagonals are *always* perpendicular?
   I. squares   II. rectangles   III. rhombuses
   a. II   b. I and II   c. I and III   d. I, II, and III

27. The surface areas of two spheres are 225π and 100π. What is the ratio of their radii?
   a. 9:4   b. 100:225   c. 8:27   d. 3:2

28. The graph at the right shows the solution set of which inequality?
   a. $-9 \leq 2x - 7 < -3$ b. $10 \geq 8 - x > 9$
   c. $-5 < x - 4 \leq -1$ d. $3 > -3x \geq -6$

29. The ratio of 195 students with blond, brown, or black hair is 3:7:5. How many students have brown hair?
   a. 65   b. 91   c. 13   d. 39

30. Simplify: $(3x^2y)^3(2xy^2)$
   a. $54x^7y^5$ b. $6x^7y^5$ c. $6x^6y^5$ d. $54x^6y^6$

31. Solve: $-5(3x + 2) = 2(7x + 24)$
   a. 2   b. $\frac{1}{2}$ c. -2 d. $-\frac{1}{2}$

32. If the radius of a circle is tripled, then the circumference of the circle is multiplied by:
   a. 6   b. 3   c. 9   d. 27

33. Express $1.7\overline{6}$ as a fraction in lowest terms.
   a. $\dfrac{69}{90}$ b. $\dfrac{159}{90}$ c. $\dfrac{26}{15}$ d. $\dfrac{53}{30}$

34. The lengths of the sides of a pentagon are 6, 3, 8, 5, and 2. If the length of the longest side of a similar pentagon is 4, find the perimeter of the smaller pentagon.
   a. 48   b. 16   c. 12   d. 24

35. The figure shows two circles with the same center O. Find the area of the shaded region.
   a. 24π b. 4π c. 2π d. 49π

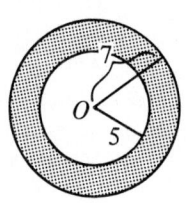

36. Which statement is false?
   a. Two lines that intersect are not parallel.
   b. The sum of the measures of the acute angles of a right triangle is 90.
   c. Two planes may intersect in just one point.
   d. Two angles may be congruent, adjacent, and complementary.

37. A trapezoid has height 10. The lengths of its bases are in the ratio of $2:5$. If the area of the trapezoid is 140, find the length of the longer base.
   a. 10   b. 8   c. 20   d. 14

38. The lengths of the sides of an isosceles triangle are integers. Each of the two congruent sides is twice as long as the third side. If the perimeter of the triangle is at most 38, find the greatest possible length of the shortest side.
   a. 7.6   b. 8   c. 7   d. 14

39. A cylinder has height 7 and volume 448π. Find its radius.
   a. 4   b. 64   c. 8   d. 64π

40. $\overline{XY}$ and $\overline{ZY}$ are tangent to the circle. If $m\stackrel{\frown}{XWZ} = 210$, find $m\angle XYZ$.
   a. 75   b. 85   c. 150   d. 30

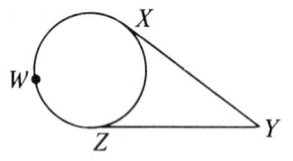

41. The length of a rectangle is 7 cm more than the width. If the length is decreased by 6 cm and the width is increased by 1 cm, the area of the resulting square is 29 cm² less than the area of the original rectangle. Find the area of the original rectangle.
   a. 49 cm²   b. 107 cm²   c. 78 cm²   d. 64 cm²

42. If a triangle has sides of length 8, 10, and 6, then its area
   a. is 40.   b. is 24.   c. is 30.
   d. cannot be determined from the information given.

386   CHAPTER 10

Careers — Meteorologist

Weather forecasts can change plans for a frisbee game or a space shuttle launching. Weather forecasters, or operational meteorologists, use such data as air pressure, temperature, humidity, and wind velocity before predicting cloud coverage and precipitation. Weather satellites and computer analysis have increased the accuracy of these predictions.

Meteorology is the study of the atmospheres of Earth and other planets. Weather forecasting is only one of several specialties in the field. Physical meteorologists, for example, examine the chemical and electrical properties of the atmosphere. They are interested in the transmission of light, sound, and radio waves, as well as the formation of clouds, rain, snow, and other phenomena.

The mathematics these research meteorologists use makes them essentially mathematical physicists. They apply principles of physics to cyclones, thunderstorms, and clouds, as well as to processes such as condensation and convection. Because you cannot very well bring a snowstorm into a laboratory, these researchers study natural phenomena by constructing mathematical models of the atmosphere that are based on physical laws and statistical relationships of weather variables. These models can be used to forecast not only weather but also air pollution potential, which is a measure of the conditions that favor pollutant concentration.

Another meteorological specialty is climatology, which is a study of climate trends, such as wind, rainfall, sunshine, and temperature, in order to determine the general weather pattern for an area. These data can then be used to choose appropriate heating and cooling systems, building designs, and land use.

The National Oceanic and Atmospheric Administration (NOAA) is a large employer of meteorologists. Commercial airlines, weather consulting firms, designers of meteorological instruments, and aerospace firms are other possibilities. Universities and colleges offer research and teaching positions.

A major in meteorology is a requirement for many jobs in the field. An alternative path is to combine a major in a related science or engineering field with courses in meteorology. To succeed in this field, you will need skills in such areas as physics, mathematics, chemistry, statistics, and computer science.

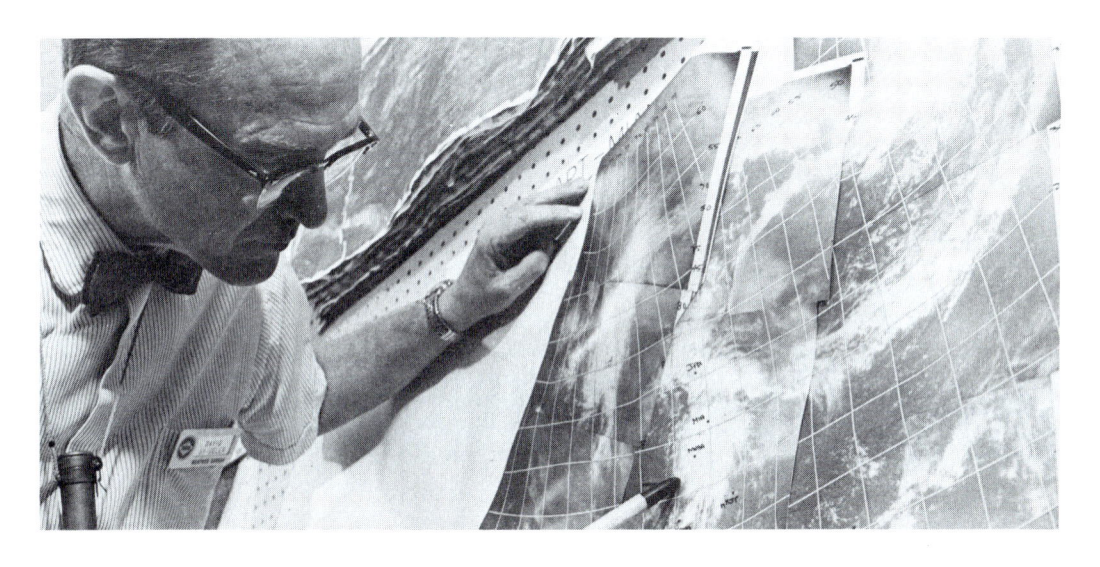

GEOMETRY OF THE CIRCLE

VOCABULARY REVIEW

Be sure that you understand the meaning of these terms:

circle, p. 345
chord, p. 345
diameter, p. 345
secant, p. 346
tangent, p. 346
semicircle, p. 349
minor arc, p. 349
major arc, p. 349
central angle, p. 349
congruent arcs, p. 350

intercepted arc, p. 353
inscribed angle, p. 354
inscribed polygon, p. 355
circumscribed circle, p. 355
circumscribed polygon, p. 362
inscribed circle, p. 362
circumference, p. 367
sphere, p. 372
right circular cylinder, p. 376
right circular cone, p. 376

MIXED REVIEW

1. Is a triangle with sides of lengths 3, $2\sqrt{5}$, and 5 a right triangle? Explain.

2. If $\angle AOB$ is a central angle of measure 60 in a circle of radius 6, find the perimeter and area of $\triangle AOB$.

3. One number is 7 more than another number. If the product of the numbers is 120, find the numbers.

4. Find the area of an isosceles triangle with sides of length 6, 6, and 8. (*Hint:* Draw the altitude to the base.)

5. If a quadrilateral has four congruent angles, must its diagonals be perpendicular? congruent?

6. Two circles have circumferences of π and 4π. If the area of the larger circle is 4π, find the area of the smaller circle.

7. The Korachs' house was worth $72,000 four years ago and is worth $85,500 now. Find the percent of increase in the value of their house.

Simplify.

8. $\sqrt{\dfrac{18}{5}} \cdot \sqrt{\dfrac{5}{2}}$

9. $\dfrac{4}{\sqrt{6}-2}$

10. $(5\sqrt{3}+2)(5\sqrt{3}-2)$

Solve.

11. $-\dfrac{2}{3}x + \dfrac{1}{2} = 4$

12. $2(x-1) > -5-x$

13. $4x^2 - 25 = 0$

14. $8 > -1.6x \geq -2$

15. $6x^2 - 19x + 10 = 0$

16. $x^2 - x = 1$

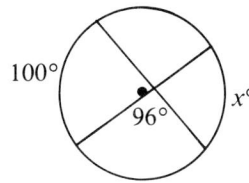

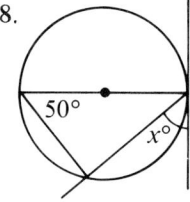

Find the value of x.

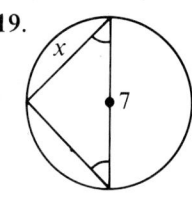

17. [circle with intersecting chords, arcs 100° and 96°, angle $x°$]

18. [circle with inscribed angle, arc 50°, angle $x°$]

19. [circle with inscribed right triangle, angle $x°$, side 7]

20. Jan is one-third as old as her cousin. In 8 years she will be three-fifths as old. How old is each now?

21. At Rent-A-Car you pay $14 a day and 8¢ per kilometer to rent a car. If you want to pay no more than $20 to rent a car for one day, find the maximum distance you can travel.

22. Multiply: a. $2x^2(3xy^3)^2$ b. $(x - 2y)(x^2 - 3xy + 2y^2)$

23. A regular square pyramid has base edges of 6 and volume 60. Find its slant height.

24. Write the prime factorization of 540.

25. Use the divide-and-average method to approximate $-\sqrt{281}$ to the nearest tenth.

26. $m\angle 1 = 3x - 9$ and $m\angle 2 = x + 5$. Find the value of x if $\angle 1$ and $\angle 2$ are:
    a. supplements   b. vertical angles

27. A triangle has angles with measures in the ratio 4:7:9. Find the numerical measure of each angle.

28. In $\odot O$, $m\stackrel{\frown}{AB} = 90$ and $OA = 5$. Find AB.

29. Which has the greater volume, a sphere with radius x or a cylinder with radius x and height x?

30. If p, q, and r are all false, find the truth value of $(p \wedge \sim q) \rightarrow (\sim q \vee r)$.

31. Write the formula for the volume of a cone. Then solve for h.

32. Write $2.\overline{72}$ as a fraction in lowest terms.

Exercises 33 and 34 refer to the diagram at the right.

33. If $\overline{AB} \parallel \overline{CD}$, explain why the triangles must be similar.

34. If $\angle A \cong \angle C$, $AX = 10$, $XC = 6$, and $BX = 9$, find:
    a. XD b. the ratio of the areas of $\triangle ABX$ and $\triangle XCD$.

35. What number is 2.5% of 3?

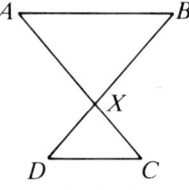

Exs. 33, 34

People planning to attend an important outdoor event are vitally interested in the probability of rain. As you will learn in this chapter (page 418), since "rain" and "no rain" are complementary events, the sum of their probabilities is one.

Probability 11

Did you ever draw straws to see who plays first in a game? This is one way of giving players equal chances of being chosen to go first. It is an example of a *random experiment,* a topic in probability which you will study in this chapter.

Probability helps us to answer questions such as Who is most likely to win? What will probably happen? Which has the better chance? It plays an important role in our lives.

Section 11-1 ONE-STAGE EXPERIMENTS

A one-stage experiment is an experiment in which one of several possible *outcomes* occurs. The set of all possible outcomes is called the *sample space.*

In the first example below, the experiment of tossing a coin has two possible outcomes, heads or tails, making a sample space {Heads, Tails}, or simply {H, T}.

When you toss a coin, you cannot predict with certainty what the outcome will be, heads or tails. Such experiments, in which you cannot predict the outcome, are called *random* experiments. Rolling a die (*die* is the singular of *dice*) is another example of a random experiment.

EXAMPLES   *Experiment*   *Sample Space*

Tossing a coin   {H, T}

Rolling one die   {1, 2, 3, 4, 5, 6}

Choosing a day of the week   {Su, M, Tu, W, Th, F, Sa}

In each of the first two experiments on the preceding page, you assume that the outcomes are equally likely to occur. That is, you assume that the coin being tossed is perfectly symmetrical and is just as likely to land heads up as tails up. In the same way you assume that the die being rolled is just as likely to show any one of the outcomes 1, 2, 3, 4, 5, or 6 as any other. Unless otherwise stated, we shall assume that all experiments in this chapter have equally likely outcomes. How might you perform the experiment of choosing a day of the week so that the outcomes are equally likely?

If you tossed a coin over and over again, you would expect, in the long run, that it would come up heads half the time and tails half the time. We say that the probability of heads is $\frac{1}{2}$ and the probability of tails is also $\frac{1}{2}$. The *probability* of an outcome is a fraction for the proportion of times we expect that outcome to occur. We will show these probabilities in two ways, in a table and in a tree diagram.

EXAMPLES

Experiment — *Table* — *Tree Diagram*

Tossing a coin

Outcome	H	T
Probability	$\frac{1}{2}$	$\frac{1}{2}$

Rolling one die

Outcome	1	2	3	4	5	6
Probability	$\frac{1}{6}$	$\frac{1}{6}$	$\frac{1}{6}$	$\frac{1}{6}$	$\frac{1}{6}$	$\frac{1}{6}$

Choosing a day of the week

Outcome	Su	M	Tu	W	Th	F	Sa
Probability	$\frac{1}{7}$	$\frac{1}{7}$	$\frac{1}{7}$	$\frac{1}{7}$	$\frac{1}{7}$	$\frac{1}{7}$	$\frac{1}{7}$

392   CHAPTER 11

In each of the preceding experiments, the outcomes were equally likely and thus the probabilities for the outcomes were the same. This is not always the case. Suppose that you have five cards whose backs are identical but whose faces have the following colors:

| red | green | red | blue | red |

The cards are shuffled and placed face down. The table and tree diagram showing the probabilities for choosing a card of a certain color are shown below:

Table

Outcome	red	blue	green
Probability	$\frac{3}{5}$	$\frac{1}{5}$	$\frac{1}{5}$

Tree Diagram

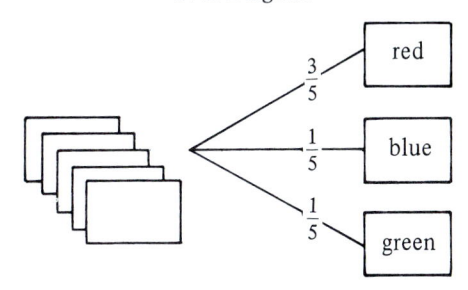

Since there are 3 red cards, the probability of choosing a red card is 3 out of 5 or $\frac{3}{5}$.

Notice that the probability of choosing a red card, $\frac{3}{5}$, is three times as great as the probability of choosing a blue card, $\frac{1}{5}$, or a green card, $\frac{1}{5}$.

ORAL EXERCISES

Give the sample space for each one-stage experiment. For Exercises 1-8, discuss how you might set up each experiment so that it would be performed in a random manner.

1. Choosing a letter of the alphabet
2. Choosing a month of the year
3. Choosing an integer from 1 to 15 including both 1 and 15
4. Choosing an integer between 7 and 8
5. Choosing one of the positive integers less than 2001
6. Choosing one of the negative integers greater than -7
7. Tossing a two-headed coin
8. Tossing a two-tailed coin

PROBABILITY 393

Give the sample space for each one-stage experiment.

9. Spinning this spinner:

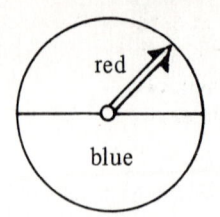

10. Spinning this spinner:

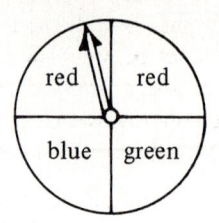

Note: In spinning experiments you always assume that the spinner never stops on a line.

WRITTEN EXERCISES

A 1–10. Name the probability for each of the outcomes in Oral Exercises 1–10.

Give the sample space and the probability for each of the outcomes for the following one-stage experiments. Assume all experiments are done randomly.

11. Picking a card from: purple, purple, yellow, orange

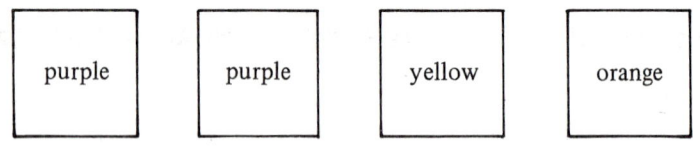

12. Picking a card from: red, red, blue, green, red, blue

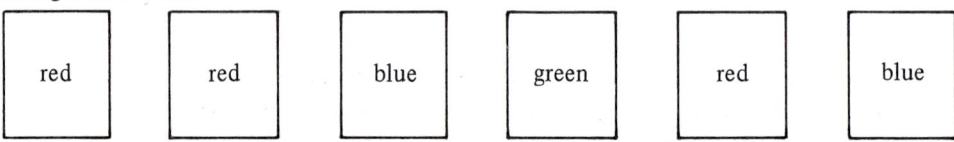

13. Spinning: brown, gray, yellow (arrow on yellow)

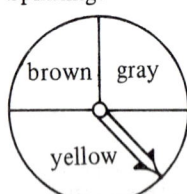

14. Spinning: green, violet

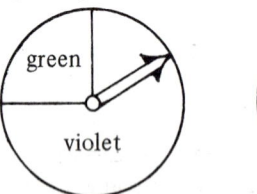

15. Spinning: 5¢, 10¢, 1¢, 25¢

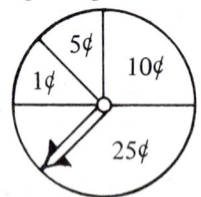

16. Spinning: $1, $2, $10, $5, $20

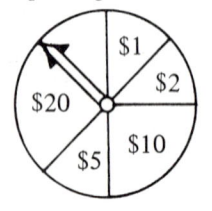

17. Picking a card from an ordinary deck of playing cards
18. Picking a red card from an ordinary deck of playing cards
19. Picking a jack from an ordinary deck of playing cards
20. Picking a black ten from an ordinary deck of playing cards
21. Picking a diamond from an ordinary deck of playing cards

Section 11-2 EVENTS AND THEIR PROBABILITIES

Getting heads when tossing a coin, rolling an even number with a die, choosing a day of the week beginning with the letter S are all examples of *events*. An *event* is associated with any set of one or more outcomes of the sample space for a given experiment. We can assign probabilities to events by adding up the probabilities of all the outcomes for that event. If we use the letter A to stand for some event, then we write the probability of A as $P(A)$.

EXAMPLES

Event	Sample Space	Set of Outcomes for the Event	Probability of Event
Event H = Getting heads when tossing a coin	$\{H, T\}$	$\{H\}$	$P(H) = \dfrac{1}{2}$
Event E = Rolling an even number with one die	$\{1, 2, 3, 4, 5, 6\}$	$\{2, 4, 6\}$	$P(E) = P(2) + P(4) + P(6)$ $= \dfrac{1}{6} + \dfrac{1}{6} + \dfrac{1}{6}$ $= \dfrac{3}{6} = \dfrac{1}{2}$
Event S = Choosing a day of the week beginning with S from all the days of the week	$\{\text{Su, M, Tu, W, Th, F, Sa}\}$	$\{\text{Su, Sa}\}$	$P(S) = P(\text{Su}) + P(\text{Sa})$ $= \dfrac{1}{7} + \dfrac{1}{7} = \dfrac{2}{7}$

Sometimes an event may consist of two or more sets of outcomes. In that case, care must be taken to determine whether any of the sets of outcomes overlap. Study the following example.

EXAMPLE When rolling a die, what is the probability of getting a number divisible by:
(a) *both* 2 and 3?   (b) *either* 2 or 3?

SOLUTION The sample space is $\{1, 2, 3, 4, 5, 6\}$.
The set of outcomes "divisible by 2" is $\{2, 4, 6\}$.
The set of outcomes "divisible by 3" is $\{3, 6\}$.

(a) We inspect these sets of outcomes to find any outcomes which appear in both of them. There is only one such outcome, 6. Thus the probability of the event "divisible by both 2 and 3" is $P(6) = \dfrac{1}{6}$.

(b) By inspection we see that there are four outcomes that appear in either one set or the other, forming the set: $\{2, 3, 4, \text{and } 6\}$. (Notice that 6 appears in both sets, but is counted only once.) Thus the probability of the event "divisible by either 2 or 3" is $P(2) + P(3) + P(4) + P(6) = \dfrac{4}{6}$, or $\dfrac{2}{3}$.

ORAL EXERCISES

Give the sample space for the experiment and the set of outcomes for the event.

1. Getting tails when tossing a coin
2. Getting the number 3 when rolling a die
3. Getting an odd number when rolling a die
4. Getting a number greater than 4 when rolling a die
5. Choosing a day of the week ending in y from all the days of the week
6. Choosing a month beginning with J from all the months of the year

WRITTEN EXERCISES

A 1-6. Give the probability of the events described in Oral Exercises 1-6.

Give the sample space for the experiment, the set of outcomes for the event, and the probability of the event.

7. Getting at least a 2 when rolling a die
8. Getting at most a 5 when rolling a die
9. Getting a prime number when rolling a die
10. Choosing a month of the year with 8 letters from all the months of the year
11. Choosing a month of the year beginning with M or with A from all the months of the year
12. Picking a 7 from an ordinary deck of cards
13. Picking a club from an ordinary deck of cards
14. Picking a heart or club from an ordinary deck of cards
15. Picking a face card (jack, queen, or king) from an ordinary deck of cards

What is the probability that the spinner shown stops on

16. an even number?
17. an odd number?
18. a number less than 4?
19. a number greater than 2?
20. a number divisible by 3?
21. a number divisible by 6?
22. a prime number?
23. a perfect square?

B 24. a number divisible by both 2 and 3?
25. a number divisible by either 2 or 3?
26. a number both greater than 3 and divisible by 5?
27. a number either greater than 3 or divisible by 5?

What is the probability that a card drawn from an ordinary deck of cards is

28. black and a 3?
29. red or a face card (jack, queen, or king)?

30. red and less than 10? (Count the ace as 1.)
31. a club or a jack?
32. a diamond and a 9?
33. the king of spades or black?

Section 11-3 MANY-STAGE EXPERIMENTS

Rolling two dice, tossing a coin three times, rolling a die then tossing a coin, are all examples of experiments that have more than one stage. Such experiments are called *many-stage experiments*. As with a one-stage experiment, the sample space for a many-stage experiment is the set of all the possible outcomes for the experiment. The following examples show some different techniques for listing the outcomes in the sample space of a many-stage experiment.

EXAMPLES

Experiment		Sample space
Tossing two coins	Use HH to show the outcome that both coins land heads up, HT for the outcome that the first coin lands heads up and the second coin lands tails up, and so on.	{HH, HT, TH, TT}
Rolling a die, then tossing a coin	Trace along each branch, starting with the left one. The first outcome listed is 1H, then 1T, 2H, and so on.	{1H, 1T, 2H, 2T, 3H, 3T, 4H, 4T, 5H, 5T, 6H, 6T}

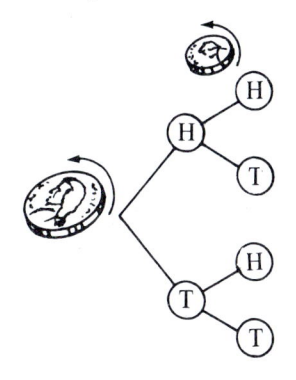

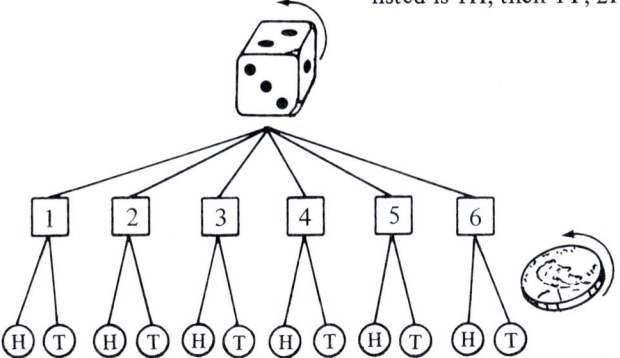

PROBABILITY 397

Experiment: Rolling a red die and a green die

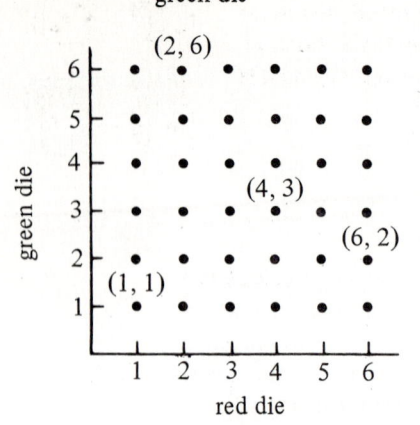

The dot diagram helps to organize the outcomes. In each number pair below, the first number is that shown on the red die and the second number is that shown on the green die. Thus (2, 6) stands for a red 2 and a green 6.

Sample space

$\{(1, 1), (1, 2), (1, 3), (1, 4), (1, 5), (1, 6),$
$(2, 1), (2, 2), (2, 3), (2, 4), (2, 5), (2, 6),$
$(3, 1), (3, 2), (3, 3), (3, 4), (3, 5), (3, 6),$
$(4, 1), (4, 2), (4, 3), (4, 4), (4, 5), (4, 6),$
$(5, 1), (5, 2), (5, 3), (5, 4), (5, 5), (5, 6),$
$(6, 1), (6, 2), (6, 3), (6, 4), (6, 5), (6, 6)\}$

If the outcomes of each *stage* of a many-stage experiment are equally likely, then the outcomes of the many-stage experiment *itself* are equally likely and we can assign them equal probabilities. For the first two experiments at the beginning of the section, we have the following:

Experiment

Tossing two coins

Table

Outcome	HH	HT	TH	TT
Probability	$\frac{1}{4}$	$\frac{1}{4}$	$\frac{1}{4}$	$\frac{1}{4}$

Rolling a die, then tossing a coin

Outcome	1H	1T	2H	2T	3H	3T	4H	4T	5H	5T	6H	6T
Probability	$\frac{1}{12}$	$\frac{1}{12}$	$\frac{1}{12}$	$\frac{1}{12}$	$\frac{1}{12}$	$\frac{1}{12}$	$\frac{1}{12}$	$\frac{1}{12}$	$\frac{1}{12}$	$\frac{1}{12}$	$\frac{1}{12}$	$\frac{1}{12}$

We can assign probabilities to events of many-stage experiments just as we did with one-stage experiments.

EXAMPLE What is the probability for the event that one and only one head shows when two coins are tossed?

SOLUTION From the table for this experiment shown just above, we can see that there are two outcomes with one and only one head showing, HT and TH, each with a probability of $\frac{1}{4}$. Thus,

$$P(\text{one and only one head}) = \frac{1}{4} + \frac{1}{4} = \frac{1}{2}$$

When two dice are rolled, you usually look for the sum of the two numbers displayed. We call the sum the *event*. Thus when (1, 1) is rolled, the event is 2, and when (6, 3) is rolled, the event is 9. The following table shows all such events and their probabilities. Note that because there are 36 possible outcomes, the probability of any one of them occurring is $\frac{1}{36}$.

Experiment: Finding the sum when two dice are rolled.

Event	Outcomes for the Event	Probability
2	{(1, 1)}	$\frac{1}{36}$
3	{(1, 2), (2, 1)}	$\frac{2}{36}$
4	{(1, 3), (2, 2), (3, 1)}	$\frac{3}{36}$
5	{(1, 4), (2, 3), (3, 2), (4, 1)}	$\frac{4}{36}$
6	{(1, 5), (2, 4), (3, 3), (4, 2), (5, 1)}	$\frac{5}{36}$
7	{(1, 6), (2, 5), (3, 4), (4, 3), (5, 2), (6, 1)}	$\frac{6}{36}$
8	{(2, 6), (3, 5), (4, 4), (5, 3), (6, 2)}	$\frac{5}{36}$
9	{(3, 6), (4, 5), (5, 4), (6, 3)}	$\frac{4}{36}$
10	{(4, 6), (5, 5), (6, 4)}	$\frac{3}{36}$
11	{(5, 6), (6, 5)}	$\frac{2}{36}$
12	{(6, 6)}	$\frac{1}{36}$

The following table summarizes this information.

Event	2	3	4	5	6	7	8	9	10	11	12
Probability	$\frac{1}{36}$	$\frac{2}{36}$	$\frac{3}{36}$	$\frac{4}{36}$	$\frac{5}{36}$	$\frac{6}{36}$	$\frac{5}{36}$	$\frac{4}{36}$	$\frac{3}{36}$	$\frac{2}{36}$	$\frac{1}{36}$

This table can be used to find the probability for other events that occur when a pair of dice are rolled.

EXAMPLE What is the probability of getting a 7 or an 11?

SOLUTION $P(7 \text{ or } 11) = P(7) + P(11)$
$$= \frac{6}{36} + \frac{2}{36}$$
$$= \frac{8}{36} = \frac{2}{9}$$

EXAMPLE What is the probability of getting a 2, a 3, or a 12 when two dice are rolled?

SOLUTION $P(2, 3, \text{ or } 12) = P(2) + P(3) + P(12)$
$$= \frac{1}{36} + \frac{2}{36} + \frac{1}{36}$$
$$= \frac{4}{36} = \frac{1}{9}$$

ORAL EXERCISES

Give the sample space for the many-stage experiment.

1. Tossing one coin twice
2. Tossing a penny and a dime
3. Two spins of this spinner: 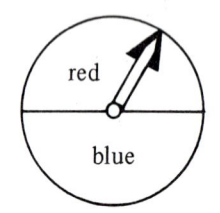
4. Two spins of this spinner: 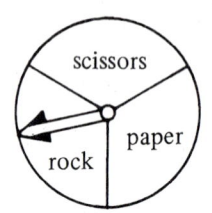
5. Tossing a coin and then spinning this spinner:

For the experiment of rolling two dice and finding the sum, tell the probability of the event. Use the table in the text.

6. 6	7. 3	8. 10	9. 4
10. 5	11. 7	12. 4 or 6	13. 8 or 10
14. 2, 3, or 4	15. 6, 7, or 8	16. 5, 9, or 10	17. 8, 11, or 12

WRITTEN EXERCISES

A 1–5. Give the probability for each outcome of the many-stage experiments in Oral Exercises 1–5.

List the sample space for the many-stage experiment and give the probability for each outcome. You may find it helpful to draw a tree or dot diagram.

6. One spin of this spinner [long/short spinner], followed by one spin of this spinner [hop/skip/jump spinner]

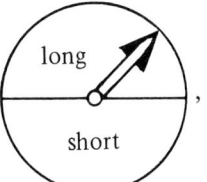

7. Spinning this spinner [red/blue/green/yellow spinner], 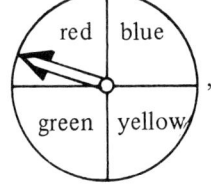 and then tossing a coin

8. Tossing a coin and then rolling a die

What is the probability for the given event when two coins are tossed?

9. The coins show the same face.
10. The coins show different faces.
11. At least one coin shows heads.
12. Neither coin shows tails.

For the experiment of finding the sum when two dice are rolled, what is the probability of the given event?

13. An odd number
14. An even number
15. A prime number
16. A number that is not prime
17. A number less than 7
18. A number equal to or greater than 9
19. A factor of 12
20. A number divisible by 3

List the sample space for the many-stage experiment and give the probability for each outcome.

B 21. Tossing three coins

22. Two spins of [do/re spinner] followed by one spin of [fa/mi/sol spinner]

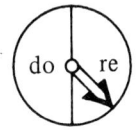

PROBABILITY   401

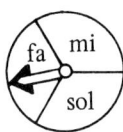

List the sample space for the many-stage experiment and give the probability for each outcome.

23. One spin of (do, re, mi) followed by one spin of (fa, sol) followed by one spin of (do, la, ti)

24. One spin of (country, rock, rhythm) followed by one spin of (western, roll, blues) followed by one spin of (band, singer)

What is the probability for the given event when three coins are tossed? (Refer to the sample space for Exercise 21.)

25. Only one coin shows tails.
26. There are only two heads.
27. There are at least two heads.
28. The coins all show the same face.
29. There is at least one tail.
30. There is no more than one tail.

Two spins are made on the spinner at the right. What is the probability that

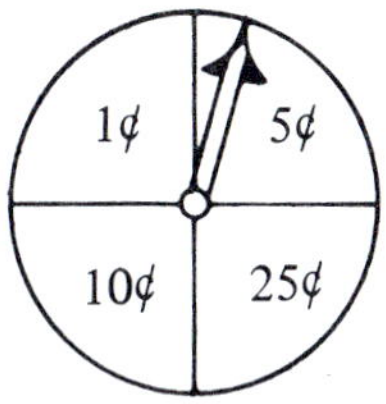

31. the sum of the two spins is 15¢?
32. the first spin is greater in value than the second spin?
33. the sum of the two spins is greater than 20¢?
34. the sum of the two spins is less than 35¢?

What is the probability for the given event when two dice are rolled and the sum of the numbers displayed is found?

C 35. The sum is an odd number or a number divisible by 3.
36. The sum is an even number or a number divisible by 3.
37. The sum is a prime number or a number equal to or greater than 7.
38. The sum is a factor of 12 or a perfect square.

SELF-TEST 1

Give the sample space and the probability for each of the outcomes for the following one-stage experiments. Assume all experiments are done randomly.

1. Choosing a season of the year from all the seasons of the year
2. Choosing an integer from 2 to 9, including both 2 and 9
3. Choosing an odd integer between 110 and 114
4. Spinning the spinner at the right

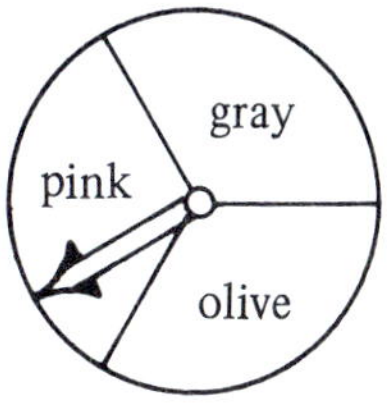

Section 11-1

Give the probability for each of the following events.

5. Getting at least a three when rolling a die
6. Picking a ten from an ordinary deck of cards
7. Getting an even number with the spinner at the right
8. Getting a perfect square with the spinner at the right

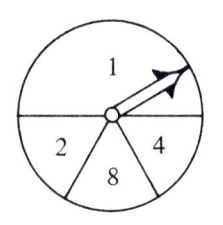

Section 11-2

Give the probability for each of the following events.

9. Getting a ten when rolling two dice
10. Getting at least a ten when rolling two dice
11. Getting three heads when three coins are tossed
12. Getting at least one head when three coins are tossed

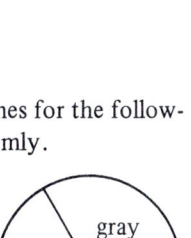

Section 11-3

PROBABILITY 403

Section 11-4 THE COUNTING PRINCIPLE

In order to compute the probability of an event of a many-stage experiment, an easy way is needed to find the number of outcomes for the experiment. Tree diagrams are useful, but only when the number of outcomes is small. However, the following two examples of simple many-stage experiments introduce a different way of counting the number of outcomes of any many-stage experiment.

EXAMPLE Draw a tree diagram for the two-stage experiment of rolling a die and then tossing a coin. Find out how many outcomes are in the sample space.

SOLUTION

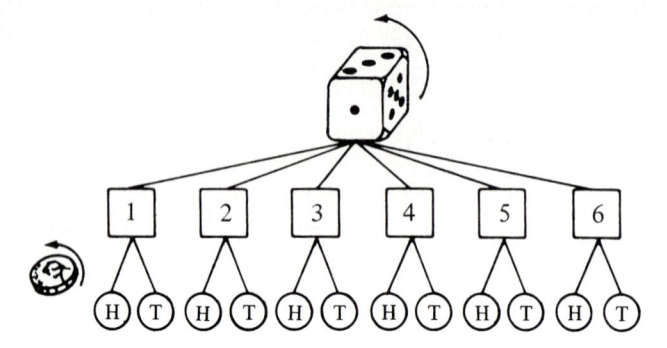

The sample space is { 1H, 1T, 2H, 2T, 3H, 3T, 4H, 4T, 5H, 5T, 6H, 6T }. By counting we find that there are 12 outcomes. Another way of finding the number of outcomes is to notice that for each of the 6 outcomes obtained by rolling a die, 2 outcomes are obtained by tossing a coin. Thus there are 6 × 2, or 12, outcomes for the two-stage experiment.

EXAMPLE Draw a tree diagram for the three-stage experiment of tossing a coin three times. Find out how many outcomes there are in the sample space.

SOLUTION

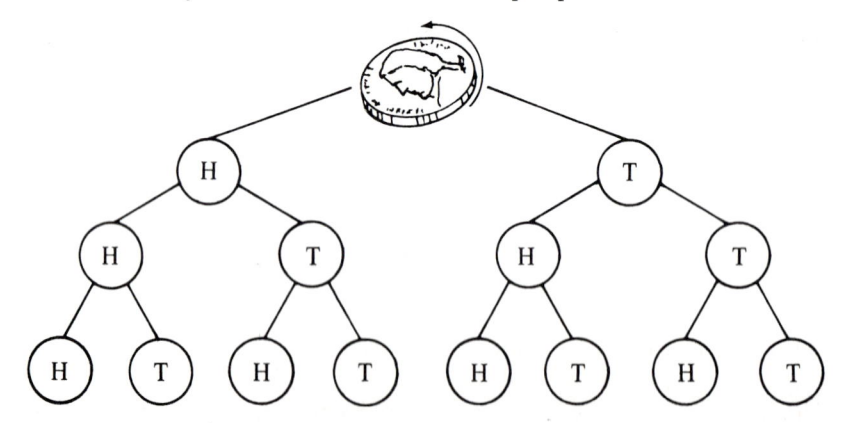

The sample space is { HHH, HHT, HTH, HTT, THH, THT, TTH, TTT }. By counting we find that there are 8 outcomes. However, notice also that the first toss has 2 outcomes, the first two tosses have 2 × 2, or 4 outcomes, and the three tosses have 2 × 2 × 2, or 8 outcomes.

These examples suggest the following rule, called the Counting Principle.

> To find the total number of outcomes for a many-stage experiment, multiply together the number of outcomes for each stage of the experiment.

EXAMPLE Mary wishes to drive from San Francisco to Boston by way of Denver and Chicago. She has a choice of four routes between San Francisco and Denver, of three routes between Denver and Chicago, and of two routes between Chicago and Boston. How many different ways can Mary drive from San Francisco to Boston?

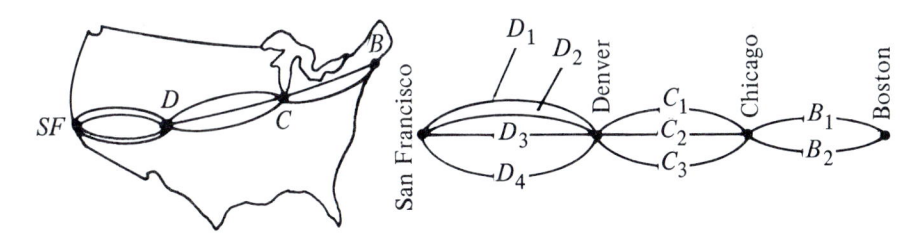

SOLUTION There are:

4 choices for the first stage,	4 · ___ · ___
3 choices for the second stage,	4 · 3 · ___
2 choices for the third stage,	4 · 3 · 2

Using the counting principle, there are 4 · 3 · 2 = 24 ways for Mary to make the trip from San Francisco to Boston.

EXAMPLE In how many different orders can we arrange the letters A, B, and C? What are they?

SOLUTION We have

3 different choices for the first letter:	3 · ___ · ___
2 different choices for the second letter:	3 · 2 · ___
and 1 choice for the third letter:	3 · 2 · 1

Thus there are 3 · 2 · 1 = 6 different orders. They are ABC, ACB, BAC, BCA, CAB, and CBA.

EXAMPLE How many 3-digit numbers can be formed from the digits 2, 4, 5, and 8 if repetition of digits is not allowed?

SOLUTION There are 4 choices for the first digit, 3 choices for the second digit, and 2 choices for the third digit. Thus there are 4 · 3 · 2 = 24 numbers with 3 different digits that can be formed. (For example, 245, 248, 458, and so on.)

EXAMPLE If one of the 3-digit numbers formed in the preceding example is selected at random, what is the probability that it is even?

SOLUTION For the number selected to be even, the units' digit must be 2, 4, or 8. Thus there are 3 choices for the units' digit. This leaves 3 choices for a second digit and 2 choices for the third digit. There are 3 · 3 · 2 = 18 even numbers among the 24 3-digit numbers formed. Thus the probability that the number selected is even is

$$P(\text{even}) = \frac{18}{24} = \frac{3}{4}$$

EXAMPLE How many 3-digit numbers can be formed from the digits 2, 4, 5, and 8 if repetition of digits is allowed?

SOLUTION There are 4 choices for each of the three digits. Thus there are 4 · 4 · 4 = 64 3-digit numbers that can be formed if repetition is allowed.

EXAMPLE If one of the 3-digit numbers formed in the preceding example is selected at random, what is the probability that it is less than 500?

SOLUTION For the number to be less than 500, the hundreds' digit must be either 2 or 4. Thus there are 2 choices for the hundreds' digit and 4 choices for each of the other two digits. There are 2 · 4 · 4 = 32 numbers less than 500 among the 64 3-digit numbers formed. The probability that the number selected is less than 500 is

$$P(\text{less than } 500) = \frac{32}{64} = \frac{1}{2}$$

ORAL EXERCISES

Using the diagram at the right below, tell how many different paths you may take in traveling between the given points.

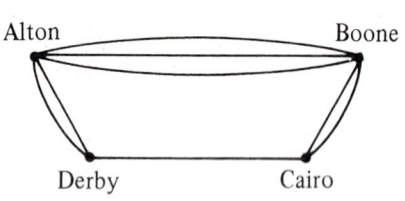

1. From Alton directly to Boone
2. From Boone directly to Cairo
3. From Alton to Cairo by way of Boone
4. From Boone to Derby by way of Cairo
5. From Boone to Alton by way of Cairo and Derby
6. From Boone to Derby by way of Alton
7. From Cairo directly to Derby
8. From Cairo to Derby by way of Boone and Alton

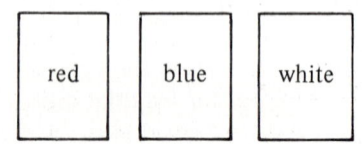

9. In how many ways can you arrange a red card, a blue card, and a white card side by side?

10. How many different 2-letter pairs can you form from the letters in the word INTO?

11. How many different 2-digit numbers can you form from the digits 1, 3, 4, 5, and 8 without repetition of digits?

12. How many different 2-digit numbers can you form from the digits 2, 5, and 8 if repetition of digits is allowed?

WRITTEN EXERCISES

A 1. How many different arrangements can be made of the letters W, X, Y, and Z?

2. List the arrangements in Exercise 1 in which X is the first letter.

3. In how many different ways may 6 runners competing in a marathon race finish? Assume there are no ties.

4. Seven candidates are running for the office of mayor of Space City. Their names are to be listed on the ballot at random. How many different arrangements are possible?

5. There are 5 different hiking paths to the top of Mount Seaview. In how many different ways may one climb and descend the mountain using different paths for the climb and descent?

6. A dog, a cat, a canary, a hamster, a turtle, a pig, and a parrot are entered in a pet show. In how many different ways may they be awarded first, second, and third prizes?

For Exercises 7-14, use the numbers 1, 2, 4, 6, and 8 as digits.

B 7. How many 3-digit numbers are there if repetition of digits is not allowed?

8. How many of these 3-digit numbers are odd?

9. How many of these 3-digit numbers are even?

10. How many of these 3-digit numbers are larger than 350?

11. How many of these 3-digit numbers are larger than 525?

12. How many of these 3-digit numbers are less than 742?

13. How many of these 3-digit numbers have 4 as a units' digit?

14. How many of these 3-digit numbers have either 1 or 8 as a tens' digit?

A computer is programmed to select at random and print a 3-digit number using 1, 2, 4, 6, and 8 as digits, with repetition of digits not allowed. What is the probability that the number is

15. odd? (Refer to Exercise 8.)

16. even? (Refer to Exercise 9.)

17. larger than 350? (Refer to Exercise 10.)

18. larger than 525? (Refer to Exercise 11.)

19. less than 742? (Refer to Exercise 12.)

20. a number having 4 as a units' digit? (Refer to Exercise 13.)

21. a number having either 1 or 8 as a tens' digit? (Refer to Exercise 14.)

Exercises 22–27 refer to all possible 5-digit numbers with digits not repeated. Allow for arrangements in which the first digit is zero.

C 22. How many 5-digit numbers are there?

23. How many are even?

24. How many are odd?

25. How many end in 7?

26. How many are multiples of 5?

27. How many have 3 as the second digit?

A computer is programmed to select at random and print a 5-digit number as described in the direction line for Exercises 22–27. What is the probability that the number is

28. even? (Refer to Exercise 23.)

29. odd? (Refer to Exercise 24.)

30. a number ending in 7? (Refer to Exercise 25.)

31. a multiple of 5? (Refer to Exercise 26.)

32. a number having 3 as the thousands' digit? (Refer to Exercise 27.)

Section 11-5 PERMUTATIONS

In this section you will learn about:
- permutations
- factorial numbers

In the preceding section you learned that you can arrange the three letters A, B, and C in six different orders: ABC, ACB, BAC, BCA, CAB, CBA. We call each of these arrangements a *permutation* of the letters A, B, and C. A *permutation* of some elements is an ordered arrangement of the elements.

Using the counting principle you can see that the number of permutations of:

$$3 \text{ elements taken 3 at a time is } 3 \cdot 2 \cdot 1 = 6$$
$$4 \text{ elements taken 4 at a time is } 4 \cdot 3 \cdot 2 \cdot 1 = 24$$
$$5 \text{ elements taken 5 at a time is } 5 \cdot 4 \cdot 3 \cdot 2 \cdot 1 = 120$$

and so on.

We use the symbol 3! to stand for $3 \cdot 2 \cdot 1$. We read 3! as "factorial three" or "three factorial." Similarly,

$$4! = 4 \cdot 3 \cdot 2 \cdot 1$$
$$8! = 8 \cdot 7 \cdot 6 \cdot 5 \cdot 4 \cdot 3 \cdot 2 \cdot 1$$

In general,

$n! = n(n-1)(n-2) \ldots (2)(1)$
$1! = 1 \qquad\qquad 0! = 1 \quad$ (by special definition)

This leads us to the following rule:

> The number of permutations of n things taken n at a time is $n!$.

EXAMPLE In how many different ways can 5 different butterfly specimens be displayed in a row on a shelf?

SOLUTION There are 5! ways.
$5! = 5 \cdot 4 \cdot 3 \cdot 2 \cdot 1 = 120$
The butterfly specimens may be displayed in 120 different ways.

EXAMPLE In how many different ways can the letters in the word JUPITER be arranged?

SOLUTION There are 7 letters in the word JUPITER.
$7! = 7 \cdot 6 \cdot 5 \cdot 4 \cdot 3 \cdot 2 \cdot 1 = 5040$
Thus there are 5040 different ways of arranging the letters in the word JUPITER.

In the preceding section we used the counting principle to find the number of permutations in which only some of the elements are chosen from a group of elements.

We shall use the symbol $_nP_r$ to stand for the number of permutations of n different things taken r at a time. For example,

$_3P_2$ = number of permutations of 3 things taken 2 at a time

$_{10}P_4$ = number of permutations of 10 things taken 4 at a time

and so on.

EXAMPLE Find $_4P_2$. Illustrate the permutations by using the letters A, B, C, and D.

SOLUTION $_4P_2$ stands for the number of permutations of 4 elements taken 2 at a time. Since there are 4 choices for the first element and 3 choices for the second element, $_4P_2 = 4 \cdot 3 = 12$.

Using A, B, C, and D, the 12 permutations for arranging 2 letters are:

AB　　AC　　AD　　BA　　BC　　BD
CA　　CB　　CD　　DA　　DB　　DC

The two examples on the next page illustrate the following rule:

> If $_nP_r$ stands for the number of permutations of n things taken r at a time, then:
> $$_nP_r = \frac{n!}{(n-r)!}$$

PROBABILITY　　409

EXAMPLE Find the number of permutations of the letters in the word HOUND taken 3 at a time by **(a)** the counting principle and **(b)** the rule $_nP_r = \dfrac{n!}{(n-r)!}$.

SOLUTION **(a)** There are 5 letters in the word HOUND. By the counting principle, there are $5 \times 4 \times 3 = 60$ permutations.

(b) By the rule, $_nP_r = {_5P_3} = \dfrac{5!}{(5-3)!}$
$$= \dfrac{5!}{2!} = \dfrac{5 \cdot 4 \cdot 3 \cdot 2 \cdot 1}{2 \cdot 1}$$
$$= 5 \cdot 4 \cdot 3 = 60$$

Again, there are 60 permutations.

EXAMPLE In how many ways may 4 different record albums be distributed among 12 members of a music club so that each member receiving an album gets at most one album?

SOLUTION The 1st album may go to any one of the 12 members, the 2nd album to any one of the 11 remaining members, the 3rd album to any one of the 10 remaining members, and the 4th album to any one of the 9 remaining members. Thus the number of different ways of distributing the albums amounts to the permutation $_{12}P_4$.

$$_{12}P_4 = \dfrac{12!}{(12-4)!} = \dfrac{12!}{8!} = 12 \cdot 11 \cdot 10 \cdot 9 = 11{,}880$$

There are 11,880 ways the 4 albums may be distributed among the 12 members.

ORAL EXERCISES

Give the answer to each of Exercises 1–4 in the form $_nP_r$.

EXAMPLE How many two-letter permutations can you form from ten different letters?
SOLUTION $_{10}P_2$

1. How many three-letter permutations can you form from 6 different letters?
2. How many five-letter permutations can you form from 8 different letters?
3. How many six-letter permutations can you form from 12 different letters?
4. How many 12-letter permutations can you form from 15 different letters?

Evaluate each of the following.

5. $_3P_1$ 6. $_4P_1$ 7. $_3P_2$ 8. $_4P_2$ 9. $_5P_2$ 10. $_7P_2$

WRITTEN EXERCISES

Evaluate.

A 1. $_6P_1$ 2. $_{12}P_1$ 3. $_6P_6$ 4. $_5P_3$ 5. $_{10}P_3$ 6. $_8P_4$
7. $_6P_5$ 8. $_{10}P_4$ 9. $_9P_4$ 10. $_8P_8$ 11. $_{100}P_2$ 12. $_{100}P_5$

13. How many 3-letter permutations can you form from the letters of the word SPANIEL?

14. How many 5-letter permutations can you form from the letters of the word SEQUOIA?

15. How many 6-letter permutations can you form from the letters of the word LABYRINTHS?

16. How many 8-letter permutations can you form from the letters of the word TWICKENHAM?

Exercises 17-22 refer to a class of 18 students made up of 10 females and 8 males.

B 17. In how many ways can a slate of officers (class president, class secretary, and class treasurer) be selected? (Assume that a student may not hold more than one office.)

18. How many of these slates will be all female?

19. How many of these slates will be all male?

20. How many of these slates will have a female as class president?

21. How many of these slates will have a male as secretary?

22. How many of these slates will contain at least one male?

23-27. For each of Exercises 18-22, find the probability that the slate described would be chosen if the choice were made at random.

Exercises 28-32 refer to the following situation. A company that produces canvas tote bags uses 3 colors of canvas for each bag. The bag itself is one color, the handles are a second color, and a side pocket is a third color. Canvas is available in the colors violet, turquoise, magenta, brown, raspberry, rust, cream, olive, and blue. How many different colored bags are possible if

28. all possible color combinations are used?

29. the bag itself must be brown but the handles and side pocket may be any of the other colors?

30. the side pocket may be only rust or raspberry but the bag itself and handles may be any of the other colors?

31. the handles must be olive, the side pocket must be magenta, but the bag itself may be any of the other colors?

32. the bag itself must be cream or turquoise, the handles must be blue, but the side pocket may be any of the other colors?

C 33. How many 2-letter permutations can you form from the letters of the word TREE? (Assume that you cannot distinguish between the E's.)

34. How many 2-letter permutations can you form from the letters of the word CIVIC? (Assume that you cannot distinguish between the I's or between the C's.)

PROBABILITY   411

SELF-TEST 2

1. How many different arrangements can be made with the letters R, S, and T? *Section 11-4*
2. In how many different ways can 6 people sit in 6 adjacent seats?

For Exercises 3 and 4, use the numbers 1, 3, 5, 7, 8, and 9 as digits.

3. How many 3-digit numbers are there if repetition of digits is not allowed?
4. How many of these 3-digit numbers are even?

Evaluate.

5. $_7P_2$ 6. $_{50}P_3$ *Section 11-5*

7. How many 3-letter permutations can you form from the letters of the word COMPANY?
8. How many 5-letter permutations can you form from the letters of the word NETWORKS?

Section 11-6 PROBABILITIES OF MANY-STAGE EXPERIMENTS

So far we have assigned probabilities to outcomes of many-stage experiments when the outcomes were equally likely. We should also like to assign probabilities when the outcomes are not equally likely. We can use the following general rule for either situation.

> The probability of any outcome in the sample space of a many-stage experiment is the product of all probabilities along the path that represents that outcome on the tree diagram.

EXAMPLE An urn contains 3 white balls and 2 black balls. A ball is chosen at random, its color is noted, and the ball is replaced. Then a second ball is chosen and its color noted. What are the outcomes and their probabilities?

SOLUTION We can draw a tree diagram:

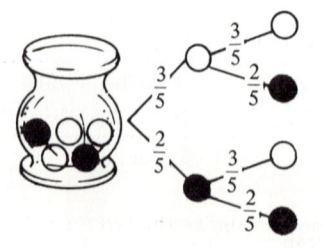

The outcomes are: ○○, ○●, ●○, ●●. (The symbol ○●, for example, means that the first ball chosen is white and the second ball chosen is black.)

The probabilities are:

$$P(○○) = \frac{3}{5} \cdot \frac{3}{5} = \frac{9}{25} \qquad P(○●) = \frac{3}{5} \cdot \frac{2}{5} = \frac{6}{25}$$

$$P(●○) = \frac{2}{5} \cdot \frac{3}{5} = \frac{6}{25} \qquad P(●●) = \frac{2}{5} \cdot \frac{2}{5} = \frac{4}{25}$$

Shown in a table:

Outcome	○○	○●	●○	●●
Probability	$\frac{9}{25}$	$\frac{6}{25}$	$\frac{6}{25}$	$\frac{4}{25}$

Note that the sum of these probabilities is 1. That is

$$\frac{9}{25} + \frac{6}{25} + \frac{6}{25} + \frac{4}{25} = \frac{25}{25} = 1.$$

EXAMPLE The spinner shown is spun twice. What are the outcomes and their probabilities?

SOLUTION We can draw a tree diagram:

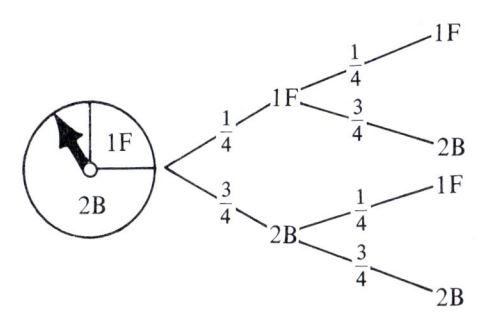

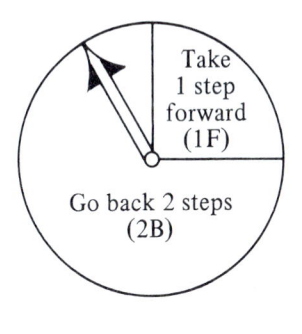

The outcomes are: (1F, 1F), (1F, 2B), (2B, 1F), (2B, 2B).

The probabilities are:
$P(1F, 1F) = \frac{1}{4} \cdot \frac{1}{4} = \frac{1}{16}$

$P(1F, 2B) = \frac{1}{4} \cdot \frac{3}{4} = \frac{3}{16}$

$P(2B, 1F) = \frac{3}{4} \cdot \frac{1}{4} = \frac{3}{16}$

$P(2B, 2B) = \frac{3}{4} \cdot \frac{3}{4} = \frac{9}{16}$

In table form:

Outcome	(1F, 1F)	(1F, 2B)	(2B, 1F)	(2B, 2B)
Probability	$\frac{1}{16}$	$\frac{3}{16}$	$\frac{3}{16}$	$\frac{9}{16}$

Suppose in this last example you wished to find the probability that your final position was one step back from your starting position. This occurs with either outcome (1F, 2B) or (2B, 1F). Thus the probability of ending one step back is

$$P(1F, 2B) + P(2B, 1F) = \frac{3}{16} + \frac{3}{16} = \frac{6}{16}.$$

PROBABILITY

EXAMPLE A card is drawn at random from an ordinary deck of cards. It is returned to the deck, which is then shuffled. A second card is drawn. What is the probability that both cards are jacks?

SOLUTION Since there are 4 jacks in the deck of 52 cards, the probability of drawing a jack at each draw is $\frac{4}{52}$.

$$P(\text{two jacks}) = \frac{4}{52} \cdot \frac{4}{52} = \frac{16}{2704} = \frac{1}{169}$$

ORAL EXERCISES

The tree diagram below shows the outcomes that occur when drawing two balls at random from the urn shown at the right. The first ball is replaced and the balls are mixed before the second ball is drawn.

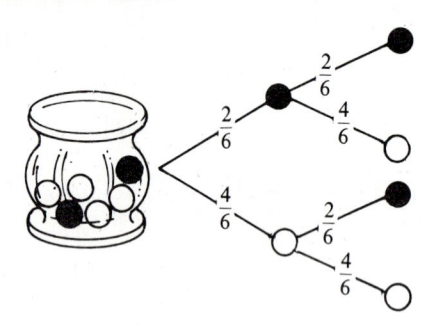

What is the probability of the given outcome?

1. BB   2. BW   3. WB   4. WW

WRITTEN EXERCISES

Exercises 1-6 refer to an experiment in which two balls are drawn, with replacement, from the urn shown at the right.

A 1. Draw a tree diagram and find the probability for each outcome.
2. What is the probability that one, and no more than one, white ball is drawn?
3. What is the probability that at least one white ball is drawn?
4. What is the probability that one, and no more than one, black ball is drawn?
5. What is the probability that at least one black ball is drawn?
6. What is the probability that both balls are the same color?

7-12. Repeat Exercises 1-6 but refer instead to the urn shown at the right.

Exercises 13-18 refer to the experiment in which the spinner below is spun twice.

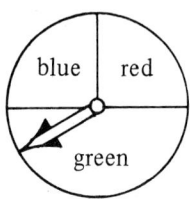

13. Draw a tree diagram for the experiment and find the probability for each outcome.

What is the probability that the spinner stops

14. first on red and then on blue?
15. first on red and then on green?
16. once on red and once on blue?
17. on green once and no more than once?
18. on green at least once?

Draw a tree diagram and find the probability for each possible outcome for the experiment described.

B 19. Draw three balls, with replacement, from the urn shown below.

20. Draw three balls, with replacement, from the urn shown below.

A card is drawn at random from an ordinary deck of cards. It is returned to the deck, which is then shuffled. A second card is drawn. What is the probability that

21. both cards are the king of diamonds?
22. both cards are hearts?
23. both cards are red?
24. both cards are the same?
25. both cards are not the same?
26. the first card is an ace and the second card is a 5?
27. the first card is a 7 and the second is a diamond?
28. the first card is red and the second card is a heart?

PROBABILITY 415

On the game board shown at the right, the forward direction (F) is clockwise and the backward direction (B) is counterclockwise. After one spin of the spinner shown below the game board, what is the probability of

29. winning a bicycle?
30. getting a $15 parking ticket?
31. being taxed $50?
32. winning $100?
33. winning a skateboard?
34. winning a camera?

B F
← →

Get a $15 parking ticket	START	Win $100
Win a camera		Win a bicycle
Be taxed $50		Be fined $100
Win a calculator	Go to jail	Win a skateboard

Playing the same game, after two spins what is the probability of

C
35. winning a calculator?
37. landing back on START?
39. winning a skateboard?
41. going to jail?
43. getting a $15 parking ticket?

36. being fined $100?
38. winning a bicycle?
40. being taxed $50?
42. winning a camera?
44. winning $100?

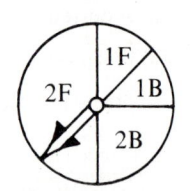

1F = 1 step forward
1B = 1 step backward

Section 11-7 GENERAL RULES ABOUT PROBABILITY

Would you bet that the sun will rise in the west tomorrow? Of course not, because this is an impossible event. On the other hand, would you bet that spring will follow winter next year? Yes! This event is certain to happen.

In this section, you will consider probabilities for:
- events that cannot occur
- events that are certain to occur
- complementary events

You know that the sample space of all possible outcomes when you roll a die is $\{1, 2, 3, 4, 5, 6\}$. In this experiment, the outcomes are equally likely to occur and the sum of their probabilities, $\frac{1}{6} + \frac{1}{6} + \frac{1}{6} + \frac{1}{6} + \frac{1}{6} + \frac{1}{6}$, is equal to 1. In general, this is true for all experiments whether or not the outcomes are equally likely. That is:

> The sum of the probabilities of all possible outcomes of an experiment is equal to 1.

EXAMPLE

For the spinner shown at the right, the probability of getting each color shown is:

$$P(\text{red}) = \frac{1}{4} \qquad P(\text{blue}) = \frac{1}{4} \qquad P(\text{white}) = \frac{1}{2}$$

$$P(\text{red}) + P(\text{blue}) + P(\text{white}) = \frac{1}{4} + \frac{1}{4} + \frac{1}{2}$$
$$= 1$$

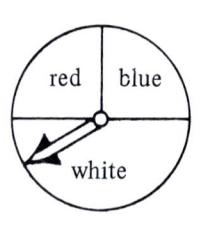

What is the probability that you will get a number greater than 6 when you roll a die? This event is impossible and a probability of 0 is assigned to it.

The probability of an event that cannot occur is 0.

EXAMPLES

        Getting a fraction when you roll a die.
        Getting 3 heads when you toss 2 pennies.
        Finding a 4-dollar bill in your wallet.

On the other hand, what is the probability that you will get a positive integer less than 7 when you roll a die? This event is certain to occur and a probability of 1 is assigned to it.

The probability of an event that is certain to occur is 1.

EXAMPLES

        Getting either a head or a tail when you toss a coin.
        Getting either an odd or an even number when you roll a die.
        Getting wet when you go for a swim.

Since the possibility of an event occurring runs the range from not occurring at all, to occurring some of the time, to being certain to occur, we have the following rule:

The probability of an event A occurring is either 0, a fraction between 0 and 1, or 1. That is,

$$0 \leqslant P(A) \leqslant 1.$$

Suppose you wanted to find the probability of not getting a 5 when rolling a die. Since the probability of getting a 5 is $\frac{1}{6}$, it seems reasonable that the probability of not getting a 5 is $\frac{5}{6}$. The event "getting a 5" is called the *complement* of "not getting a 5." Also, "getting a 5" and "not getting a 5" are called *complementary events*. In general, if A is an event with a certain set of outcomes, the event that contains all the outcomes not in A is called the *complement* of A and is written A'. A and A' are said to be *complementary events*.

EXAMPLES

Experiment: Rolling One Die

		Set of Outcomes	Probability
Event:	Rolling a 4	$\{4\}$	$\frac{1}{6}$
Complementary Event:	Not rolling a 4	$\{1, 2, 3, 5, 6\}$	$\frac{5}{6}$
Event:	Rolling a number less than 5	$\{1, 2, 3, 4\}$	$\frac{4}{6}$
Complementary Event:	Rolling a number equal to or greater than 5	$\{5, 6\}$	$\frac{2}{6}$
Event:	Rolling an even number	$\{2, 4, 6\}$	$\frac{3}{6}$
Complementary Event:	Rolling an odd number	$\{1, 3, 5\}$	$\frac{3}{6}$

Notice that in each of these examples the sum of the probability of an event and the probability of its complementary event is 1. This suggests the following rule:

> The probability of an event A added to the probability of the complementary event A' equals 1. That is,
> $$P(A) + P(A') = 1, \quad \text{or} \quad P(A') = 1 - P(A).$$

EXAMPLES

If the probability that it will rain tomorrow is 0.4, then the probability that it will not rain is $1 - 0.4$, or 0.6.

If the probability of drawing a diamond at random from an ordinary deck of cards is $\frac{1}{4}$, then the probability of not drawing a diamond is $1 - \frac{1}{4}$, or $\frac{3}{4}$.

If the probability of winning a game is 0.48, then the probability of losing, assuming no ties are possible, is $1 - 0.48$, or 0.52.

ORAL EXERCISES

Give the probability of each event.
1. Getting a 7 when rolling one die.
2. Getting a sum of 13 when rolling a pair of dice.
3. Getting heads when tossing a two-headed coin.
4. Getting a number from 1 to 6, including 1 and 6, when rolling one die.
5. Running a kilometer in less than 5 seconds.
6. Getting at least one head or one tail when tossing 3 coins.
7. Choosing a day of the week with at least 6 letters.
8. Finding a 20-cent piece in a piggy bank.

List the outcome(s) for the event and for its complement.
9. Tossing a coin and getting heads.
10. Tossing two coins and getting two heads.
11. Rolling one die and getting a 5.
12. Choosing a day of the week with 7 letters from all the days of the week.
13. Rolling one die and getting a 1 or a 6.
14. Picking a month of the year beginning with the letter X from all the months of the year.
15-20. Give the probability for each event and for its complement in Exercises 9-14.

WRITTEN EXERCISES

List or describe the outcome(s) for the event and for its complement.

A
1. Getting at least one tail when tossing two coins.
2. Getting two heads and no more than two heads when tossing three coins.
3. Getting heads and an even number when tossing a coin and rolling a die.
4. Getting tails or an odd number when tossing a coin and rolling a die.
5. Rolling a sum of 2 or 12 with a pair of dice.
6. Rolling a sum of 7 or 11 with a pair of dice.

7-12. For each of Exercises 1-6, find the probability of the event and of its complement.

One card is drawn from an ordinary deck of playing cards. What is the probability that the card is not

13. the queen of diamonds?
14. a spade?
15. a 9?
16. a red king?
17. a black 2, 3, or 4?
18. the ten, jack, queen, king, or ace of clubs?

PROBABILITY 419

For Exercises 19-22, two drawings are made with replacement from the urn at the right. Find the probability of each event and of its complement.

B 19. Drawing two white balls.

20. Drawing two black balls.

21. Drawing first a white ball and then a black ball.

22. Drawing one white ball and one black ball.

For Exercises 23-26, three drawings are made with replacement from the urn at the right. Find the probability of each event and of its complement.

23. Drawing three white balls.

24. Drawing three black balls.

25. Drawing black for the first ball and white for the other two.

26. Drawing at least two white balls.

SELF-TEST 3

Two balls are drawn, with replacement, from the urn shown.

1. Draw a tree diagram and find the probability for each outcome.   Section 11-6

2. What is the probability that one, and only one, white ball is drawn?

3. What is the probability that at least one white ball is drawn?

4. What is the probability that at most one white ball is drawn?

List or describe the outcomes for the event and for its complement. Find the probability of the event and of its complement.

5. Choosing a day of the week with 8 letters from all the days of the week   Section 11-7

6. Choosing a day of the week with at least 8 letters from all the days of the week

7. Getting at least one head when tossing two coins

8. Getting a sum of 6 or 8 with a pair of dice

Section 11-8 DRAWING WITH AND WITHOUT REPLACEMENT

Experiments such as those involving drawing balls from an urn or cards from a deck of cards which have been discussed so far have all been performed with replacement. That is, the item drawn has been replaced in, or returned to, its original group, the items have been mixed or shuffled, and then another drawing has been made. In this method of drawing, the probabilities for selecting specific items remain the same at every stage.

    Now consider experiments in which drawings are made without replacement. This means that an item drawn is not returned to its original group before another drawing is made. In this method of drawing, the probabilities of specific items change at each stage.

EXAMPLE What is the probability of drawing two hearts from an ordinary deck of playing cards   (a) with replacement?   (b) without replacement?

SOLUTION (a) Since there are 13 hearts in a deck of 52 cards, the probability of drawing a heart on each draw is $\frac{13}{52}$, or $\frac{1}{4}$.

With replacement,
$$P(\text{two hearts}) = \frac{1}{4} \cdot \frac{1}{4} = \frac{1}{16}.$$

Expressed as a decimal, $\frac{1}{16} = 0.0625$.

(b) The probability of drawing a heart on the first draw is $\frac{13}{52}$, or $\frac{1}{4}$. If the first card drawn is a heart and is not replaced, there remain for the second draw 51 cards of which 12 are hearts. Thus the probability of getting a heart on the second draw is $\frac{12}{51}$, or $\frac{4}{17}$.

Thus, without replacement,
$$P(\text{two hearts}) = \frac{1}{4} \cdot \frac{4}{17} = \frac{1}{17}.$$

Expressed as a decimal, $\frac{1}{17} \approx 0.0588$.

Notice that your chances for drawing two hearts are just a little better when drawing with replacement.

EXAMPLE An urn contains 5 white balls and 3 black balls. Two balls are drawn from it at random. Find all the outcomes and their probabilities when the drawings are made
(a) with replacement.
(b) without replacement.

SOLUTION (a) We can draw a tree diagram and from it, determine the probabilities.

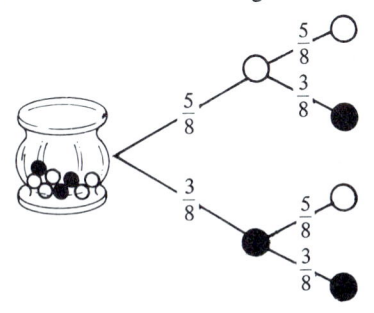

$$P(\circ\circ) = \frac{5}{8} \times \frac{5}{8} = \frac{25}{64}$$

$$P(\circ\bullet) = \frac{5}{8} \times \frac{3}{8} = \frac{15}{64}$$

$$P(\bullet\circ) = \frac{3}{8} \times \frac{5}{8} = \frac{15}{64}$$

$$P(\bullet\bullet) = \frac{3}{8} \times \frac{3}{8} = \frac{9}{64}$$

This information can be shown in a table:

Note that $\frac{25}{64} + \frac{15}{64} + \frac{15}{64} + \frac{9}{64} = 1$.

Outcome	○○	○●	●○	●●
Probability	$\frac{25}{64}$	$\frac{15}{64}$	$\frac{15}{64}$	$\frac{9}{64}$

PROBABILITY   421

(b) When the drawings are made without replacement, the probabilities change for the second draw. For example, the probability that the first ball drawn is white is $\frac{5}{8}$. If a white ball *is* draw on the first draw, there remain 7 balls in the urn, 4 of which are white. So the probability of the ball also being white on the second draw is $\frac{4}{7}$. This is shown by the top branch of the tree diagram below.

Be sure you understand how the probability for the second draw for each of the other outcomes is assigned.

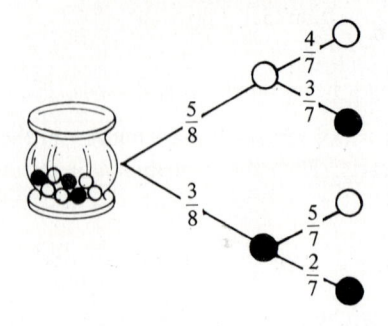

$$P(\circ\circ) = \frac{5}{8} \times \frac{4}{7} = \frac{20}{56}$$

$$P(\circ\bullet) = \frac{5}{8} \times \frac{3}{7} = \frac{15}{56}$$

$$P(\bullet\circ) = \frac{3}{8} \times \frac{5}{7} = \frac{15}{56}$$

$$P(\bullet\bullet) = \frac{3}{8} \times \frac{2}{7} = \frac{6}{56}$$

In table form:

Outcome	○○	○●	●○	●●
Probability	$\frac{20}{56}$	$\frac{15}{56}$	$\frac{15}{56}$	$\frac{6}{56}$

Note that $\frac{20}{56} + \frac{15}{56} + \frac{15}{56} + \frac{6}{56} = 1$.

ORAL EXERCISES

In each exercise, the two given cards are drawn without replacement from an ordinary deck of cards. Describe the deck of cards after each drawing.

EXAMPLE   Two kings

SOLUTION   After the first drawing there are 51 cards left of which 3 are kings. After the second drawing there are 50 cards left of which 2 are kings.

1. Two sevens
2. Two clubs
3. Two red cards
4. Two black aces
5. First a queen, second a king
6. First a red queen, second a red king

422   CHAPTER 11

In each exercise, two balls are drawn from an urn without replacement. Tell how many balls are left in the urn and how many there are of each color after each drawing.

	Balls in urn originally	First drawing	Second drawing
7.	3 white, 2 black	white	black
8.	3 white, 2 black	black	white
9.	4 white, 2 black	white	white
10.	4 white, 2 black	black	black
11.	5 white, 4 black	white	black
12.	5 white, 4 black	white	white

WRITTEN EXERCISES

Find the probability that the two given cards are drawn from an ordinary deck of cards both with and without replacement.

A   1. Two queens   2. Two fives

3. Two clubs   4. Two black cards

5. Two red tens   6. First a black ten, second a red ten

7. First an ace, then a two   8. First the ace of hearts, then the two of spades

9. Neither card is an ace   10. Neither card is a diamond

11. Neither card is red   12. Neither card is a jack or a king

Two balls are drawn from an urn containing 3 white and 4 black balls. Find the probability that the two given balls are drawn both with and without replacement.

13. Two black balls   14. Two white balls

15. First ball white, second ball black   16. First ball black, second ball white

17. One, and only one, ball is white   18. One, and only one, ball is black

Two balls are drawn from an urn containing 3 red, 4 green, and 5 yellow balls. Find the probability that the two given balls are drawn both with and without replacement.

B   19. Two red balls   20. Two green balls

21. Two yellow balls   22. First ball red, second ball green

23. First ball red, second ball yellow   24. First ball green, second ball red

25. First ball green, second ball yellow   26. First ball yellow, second ball red

27. First ball yellow, second ball green   28. One, and only one, red ball

29. One, and only one, green ball   30. One, and only one, yellow ball

PROBABILITY

11-9 EMPIRICAL PROBABILITY

By *empirical probability* we mean a probability based on actual experience or observation. You have already accepted the fact that the probability of obtaining heads when a coin is tossed is $\frac{1}{2}$, or 0.5. Do you really believe it? What does it actually mean? The following experiment should help you understand the idea better.

EXPERIMENT Each of 5 students was given a coin and asked to toss it 20 times and count the number of heads observed. Rosa's results were recorded first and the relative frequency of heads was calculated, as shown in the table below. Next, Arnie's results were combined with Rosa's and the relative frequency of heads was calculated for the combined data. The table continues in this fashion.

Student	No. of Heads	Total No. of Heads	Total No. of Tosses	Fraction, or Relative Frequency, of Heads
Rosa	8	8	20	$\frac{8}{20} = 0.400$
Arnie	7	15	40	$\frac{15}{40} = 0.375$
Dom	14	29	60	$\frac{29}{60} \approx 0.483$
Mimi	12	41	80	$\frac{41}{80} \approx 0.513$
Freda	10	51	100	$\frac{51}{100} = 0.510$

Notice that as more and more tosses are made, the fraction of heads appearing (that is, the relative frequency of heads) varies up or down but generally appears to be getting closer and closer to 0.5. This is shown more clearly in the graph below.

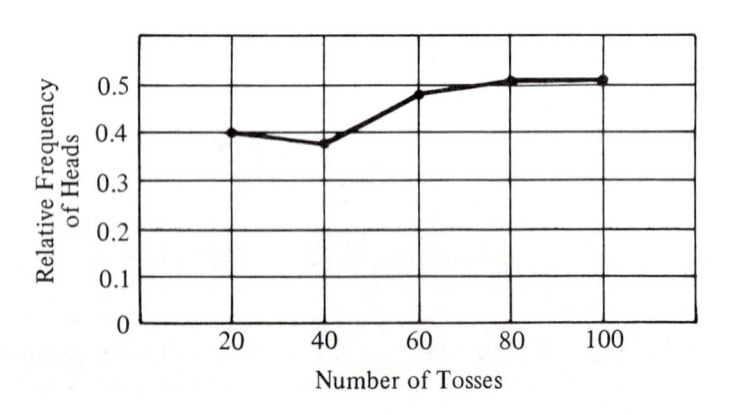

If this experiment were continued, and many, many coin tosses were made, the graph would probably look something like the following:

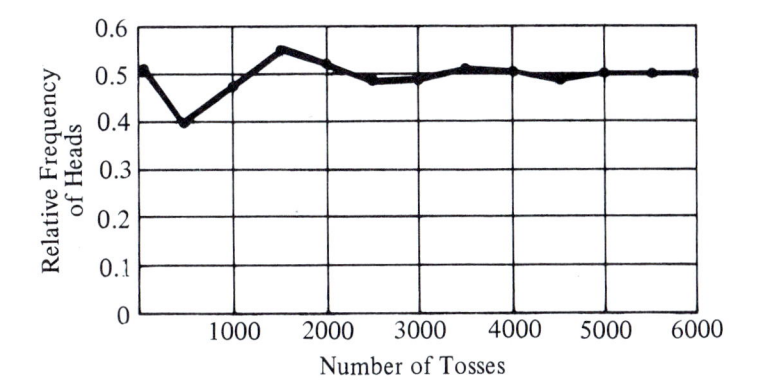

Saying that the probability of obtaining heads is 0.5 means that the relative frequency of heads stabilizes at, or gets fairly close to, 0.5 as the number of tosses becomes greater. It does *not* mean that you will get *exactly* 3000 heads in 6000 tosses.

You can accept the fact that the probability of getting heads when tossing a coin is 0.5 without actually running an experiment. However, there are other activities for which the probability can be determined *only* by actual experimentation. For example, if you toss a thumbtack it will land either point up or point down; but it is impossible to predict the probability of either outcome without experimenting. That is, in order to approximate the probability of landing point up (or point down) you have to actually toss the thumbtack and record the number of times it lands point up (or point down) and the total number of times tossed. Of course, the more times you toss the thumbtack, the more closely you will approximate the actual probabilities.

EXPERIMENT A student wishes to determine the probability that a certain kind of thumbtack will land point up when tossed. She performs an experiment in which each trial consists of tossing 5 such thumbtacks 10 times and recording the number that land point up. The results of the first two tosses are illustrated below.

Point up: 3   Point up: 1

Since 5 tacks are tossed together each time, a trial that consists of 10 tosses is equivalent to tossing 1 thumbtack 50 times. The student made 10 such trials. The results are shown in the table on the following page. Notice that, after the first trial, the results of each trial are combined with the preceding results before a new relative frequency is calculated.

PROBABILITY   425

Trial	No. with Point Up	Total No. with Point Up	Total No. of Tacks Tossed	Relative Frequency of No. with Point Up
1	13	13	50	$\frac{13}{50} = 0.260$
2	22	35	100	$\frac{35}{100} = 0.350$
3	20	55	150	$\frac{55}{150} \approx 0.367$
4	23	78	200	$\frac{78}{200} = 0.390$
5	22	100	250	$\frac{100}{250} = 0.400$
6	19	119	300	$\frac{119}{300} \approx 0.397$
7	23	142	350	$\frac{142}{350} \approx 0.406$
8	24	166	400	$\frac{166}{400} = 0.415$
9	19	185	450	$\frac{185}{450} \approx 0.411$
10	24	209	500	$\frac{209}{500} = 0.418$

The student concluded that the probability that the kind of thumbtack she was using would land point up when tossed was approximately 0.4.

ORAL EXERCISES

1. You suspect that a coin has been tampered with so that it is not as equally likely to land heads up as tails up. Describe an experiment to test your theory and determine approximately the probability of each outcome.

2. You suspect that the probabilities of rolling a 1, 2, 3, 4, 5, or 6 with a certain die are not equally likely. How would you determine approximately the probability of each outcome?

3. A bag holds 10,000 marbles all the same size but of three different colors, red, white, and blue. Describe how you might determine approximately the number of marbles of each color without actually counting them.

4. You plan to base an estimate of the total number of giraffes on a game preserve in Africa on the numbers of giraffes in sample groups. How might your methods of selecting sample groups affect the reliability of your estimate?

5. A factory is polluting a river by dumping its waste products into it. You plan to set up an experiment and take samples of the river water to find out what is in the waste products and what their concentrations are. Discuss what you might have to take into account in setting up your experiment so that your results will be reasonably valid.

WRITTEN EXERCISES

A 1. Toss a coin 10 times and count the number of tails. Record your data in a table like the one for the experiment on page 426. Calculate the fraction of tails appearing and repeat the experiment until you have performed 5 trials. How close do you come to the theoretical probability of getting tails?

2. Toss two coins together 20 times and count the number of times you get 0, 1, or 2 heads. Copy the table below and record your data in it. Find the theoretical probability for each outcome and compare it with the value you have found experimentally.

No. of Heads	Tally	Total No.	Fraction
0			
1			
2			

3. Roll a die 20 times and count the number of times the outcome is odd or even. Calculate the theoretical probability of each outcome and compare it with the value you have found experimentally.

4. Repeat Exercise 3 using a pair of dice and finding the sum.

5. Repeat Exercise 2 using three coins. What are the probabilities for getting 0, 1, 2, or 3 heads?

6. Do you think that the vowels a, e, i, o, and u are used with the same relative frequency? Devise an experiment to check your claim and carry it out.

7. Do you think that the digits 0, 1, 2, . . . , 9 are used with the same relative frequency in telephone numbers? Devise an experiment to check your claim and carry it out.

8. From several different colors of construction paper, cut about 30 squares, each with sides 2 cm long. Vary the number of squares for the different colors. Put your squares in a paper bag. Without looking, draw a square from the paper bag and then observe its color. Replace the square in the bag and shake the bag up. Perform this activity 50 times, keeping a record of the colors drawn. Based on your data, what fraction of the total number of squares would you expect the squares of each color to be? Calculate the fraction for each color by actually counting the squares. How do your experimental results compare with your actual results?

SELF-TEST 4

Find the probability that the two given cards are drawn from an ordinary deck of cards both with and without replacement.

1. Two aces   Section 11-8
2. Two clubs
3. First a black seven, then a red three
4. Neither card is a king.

What do you think the empirical probabilities for each of the following events should approach?

5. Getting tails when tossing a fair coin   Section 11-9
6. Rolling a 2, 3, or 5 with one die
7. Rolling an odd number with a pair of dice
8. Having a telephone number that ends in an even digit

Computer Exercises

1. Copy and RUN the following program.

```
10 FOR I=1 TO 5
20 READ X
30 PRINT X
40 NEXT I
50 DATA 53461.92,534619.2,5346192,53461920,534619200
60 END
```

2. Notice that the computer prints out 5.34619E+07 instead of 53461920. This is an example of *scientific notation* in BASIC. The number 5.34619E+07 means 5.34619×10^7, which is an approximation for 53,461,920. Compare the rest of the output with the other entries in line 50. Note that the number following the "E" states how many places to the left of its original location the decimal point has been moved.

   Each of the following numbers is written in BASIC scientific notation. Write out each number, placing the decimal point in the correct location.

   a. 4.63863E+03   b. 5.1091E+01

   c. 9.38842E+05   d. 1.4697E+08

3. a. The following program computes N!. Copy and RUN the program using several values of N.

```
10  INPUT N
20  LET P=1
30  FOR I=2 TO N
40  LET P=P*I
50  NEXT I
60  PRINT N;"!=";P
70  END
```

   b. What is the largest value of N you can use without requiring the computer to write N! in scientific notation?

CHAPTER REVIEW

Give the sample space and the probability for each of the outcomes for the following one-stage experiments. Assume that all experiments are done randomly.

1. Choosing the winner from four finalists, Claire, Li-Hui, Beth, and Ivan — Section 11-1
2. Choosing an even integer between −1 and 9 from all the integers between −1 and 9, including −1 and 9
3. Picking a card from a deck containing four green cards, three white cards, and five yellow cards

Give the probability for each event.

4. Getting at most a 3 when rolling a die — Section 11-2
5. Getting a vowel when a letter is chosen from the word EXPERIMENT
6. Picking a 2 or a 6 from an ordinary deck of cards
7. Getting at least a sum of 4 when rolling two dice — Section 11-3
8. Getting a prime number and tails when rolling a die and then tossing a coin
9. Getting one odd number and one even number when spinning the spinner at the right twice

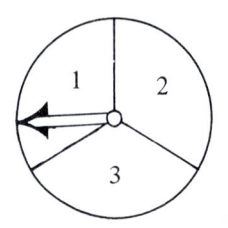

PROBABILITY 429

For Exercises 10-12, use the letters A, B, C, D, and E.

10. How many different three-letter arrangements can be made if repetition of letters is not allowed?
    Section 11-4

11. Repeat Exercise 10 if repetition is allowed.

12. How many of the arrangements in Exercise 10 begin with a vowel?

13. Evaluate $_{12}P_4$.
    Section 11-5

14. How many 6-letter permutations can you form from the letters M, N, O, P, Q, and R?

15. How many 5-letter permutations can you form from the letters of the word CHEMISTRY?

Two balls are drawn, with replacement, from the urn shown at the right.

16. Draw a tree diagram and find the probability for each outcome.
    Section 11-6

17. What is the probability that both balls are the same color?

18. What is the probability that at least one ball is black?

List or describe the outcomes for the event and for its complement. Find the probability of the event and of its complement.

19. Getting two, and only two, heads when tossing three coins
    Section 11-7

20. Getting no more than three tails when tossing three coins

21. Choosing a day of the week beginning with a vowel from all the days of the week

Two balls are drawn from the urn shown for Exercises 16-18. Find the probability that the two given balls are drawn both with and without replacement.

22. Two black balls   23. One, and only one, white ball   *Section 11-8*

24. First ball black, second ball white

What do you think the empirical probability for each event should approach?

25. Getting a 2 when spinning the spinner shown in Exercise 9
    Section 11-9

26. Getting two tails when tossing two fair coins

27. A quality-control engineer wishes to evaluate the batteries produced in a factory. Describe an experiment the engineer could use to determine the probability that a randomly selected battery is defective.

430   CHAPTER 11

CHAPTER TEST

Give the sample space and the probability for each of the outcomes for the following one-stage experiments. Assume all experiments are done randomly.

1. Choosing a ball from a bag containing three green balls, seven orange balls, and two yellow balls
2. Spinning:

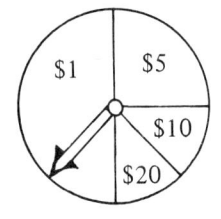

3. Two balls are drawn with replacement from the urn at the right. Draw a tree diagram and find the probability of each outcome.

Evaluate

4. $_9P_3$ 5. $_6P_4$

6. How many 2-letter permutations can you form from 7 different letters?

Find the probability of each event. In Exercises 7–10, also find the probability of the complement of the event.

7. Winning $1 or $10 when spinning the spinner above
8. Getting a sum greater than 9 when rolling two dice
9. Choosing an even integer between 4 and 6
10. Choosing a 30-day month from all the months of the year
11. Winning at most $6 when spinning the spinner above twice
12. Getting one, and only one, white ball when two balls are drawn, with replacement, from the urn shown above
13. Getting at least one black ball when two balls are drawn, without replacement, from the urn shown above
14. Getting a heart, a diamond, a spade, or a club when picking a card from an ordinary deck of cards
15. Getting at least a 3 when rolling a die
16. Getting either two or three heads when tossing three coins
17. Getting an ace and then a black four when picking two cards, with replacement, from an ordinary deck of cards
18. Getting one club and then one five when picking two cards, with replacement, from an ordinary deck of cards

19. In how many different ways may you choose a dinner if there are four choices for the main course, three choices of vegetables, and three choices of potato? (Assume you must select a main course, a vegetable, and a potato.)

20. In how many different ways may four people be lined up for a photograph?

21. In tossing a coin 50 times, Art noticed that the coin landed heads up 28 times. Is it likely that the coin is a fair one? Explain.

22. Describe a method that a newspaper editor could use to predict the probability of winning for each of three candidates running for mayor.

In Exercises 23-25, use the digits 1, 3, 5, 7, and 8.

23. How many different 5-digit numbers can you form if repetition of digits is not allowed?   if repetition is allowed?

24. How many different 3-digit numbers can you form if repetition is not allowed? How many of these are even?

25. How many 4-digit permutations can you form?

Biographical Note

Benjamin Banneker

Imagine that you have never seen a clock, but you have decided to make one. All you have at your disposal is your math book, a copy of Newton's *Principia*, a picture of a clock, an old pocket watch, some woodworking tools, and some pieces of wood. Could you do it?

Benjamin Banneker (1731-1806) did just that. In 1761, with only a few years' training in a Friends' school, he made all the necessary calculations. He hand-carved all the gears, and when he had finished he had constructed the first clock ever made in the Colonies. It kept good time for forty years.

Banneker also taught himself astronomy. George Ellicott, a neighbor of Banneker's in Maryland, lent him some books on astronomy, a telescope, and some drafting instruments. Using these, Banneker predicted the solar eclipse of 1789 more accurately than two well-known mathematicians of the time were able to do.

This interest led him into further study of the motions of celestial bodies. He computed tables and prepared a series of almanacs that were published over a ten-year period. The almanacs included predictions of eclipses, calculations for weather forecasts, tide tables, phases of the moon, and times of sunrise and sunset.

When Major Andrew Ellicott went to survey the ten-mile square of land that was to become Washington, D.C., Banneker went with him and helped him set up the base lines and boundaries for the new capital.

Banneker sent some of his tables to Thomas Jefferson, then Secretary of State, and a correspondence developed between the two men. Banneker, the son of freed slaves, challenged Jefferson's position on slavery. The resulting letters were later published in pamphlet form and aroused much interest.

Banneker's manuscript notes and journal, preserved after his death in 1806, have enabled modern scientists to examine his work in some detail. Their studies have supported Banneker's reputation as an excellent self-taught mathematician and amateur astronomer.

PROBABILITY

VOCABULARY REVIEW

Be sure that you understand the meaning of these terms:

outcome, p. 391
sample space, p. 391
random experiment, p. 391
probability, p. 392
event, p. 395

many-stage experiment, p. 397
permutation, p. 408
factorial, p. 408
complementary events, p. 418
empirical probability, p. 424

MIXED REVIEW

1. A cone has radius 5 and slant height 13. Find its total area and its volume.

2. Equal amounts of money were invested at 9.5% and at 7%. The earnings at the higher rate were $100 more than those at the lower rate. How much was invested at each rate?

3. Show that $\sim(p \vee \sim q)$ and $\sim p \wedge q$ are logically equivalent.

4. When the radius of a given circle was increased by 3 cm, the area was increased by 39π cm^2. What was the original radius?

Solve if the domain of x is {positive integers}.

5. $5\left(3 - \dfrac{x}{2}\right) = 2 - 9x$

6. $-0.28x > -3.5$

A deck of cards consists of three red cards, two green cards, and one yellow card. Find the probability that the two given cards are drawn both with and without replacement.

7. At least one card is red.

8. Neither card is red.

9. In $\odot O$, shown in the diagram, $m \stackrel{\frown}{YZ} = 60$ and $YZ = 2$. Find $m\angle Z$ and XZ.

10. In how many different ways can 7 books be placed on a bookshelf?

11. Factor: a. $x^2 - 8x - 48$ b. $2y^2 + 25y + 12$

12. The length of a rectangle is one less than twice the width. If the perimeter is 40, find the area. (Give a numerical answer.)

13. A circle has circumference 20π. Find its area.

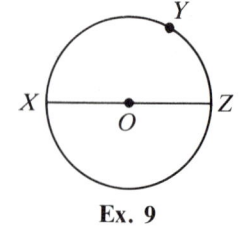

Ex. 9

Find the value of x. (In Exercise 14, $\overline{RS} \parallel \overline{TU}$.)

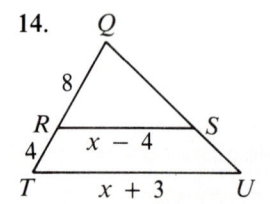

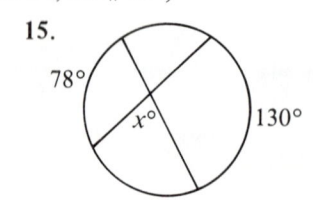

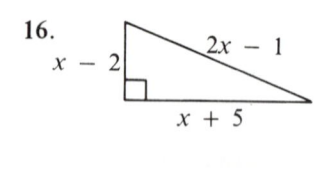

14. Triangle with Q at top, R and S on sides with $QR = 8$, $RT = 4$, $RS = x - 4$, $TU = x + 3$.

15. Circle with chords; angles $78°$, $x°$, $130°$.

16. Right triangle with legs $x - 2$ and $x + 5$, hypotenuse $2x - 1$.

434   CHAPTER 11

17. Give the probability of picking a heart or a 7 from an ordinary deck of cards.
18. An isosceles right triangle has a hypotenuse of length 10. Find the length of a leg.

Simplify.

19. $(7 - 5\sqrt{3})^2$
20. $\frac{7}{5}\sqrt{875}$
21. $\frac{2}{1 - \sqrt{3}}$

22. Two parallel lines, m and n, are cut by a transversal, t. If two interior angles on the same side of t are congruent, what can you conclude about m and t? Explain.
23. A regular pyramid has a base in the shape of an equilateral triangle. If each base edge is 8 cm and the slant height is 10 cm, find the lateral area of the pyramid.
24. If you randomly choose a 2-digit positive integer, find the probability that it is less than 70.
25. How many 5-letter permutations can you form from the letters of the word MAJORITY?
26. A cylinder has radius 3 and height 4. Find its total area and volume.
27. An equilateral triangle has a base of 10 cm. Find its height and its area.
28. If $m\angle 1 = 48$, find the ratio in simplest form of the measures of a complement and a supplement of $\angle 1$.
29. Must two given points be collinear? coplanar?
30. Graph the solution set of $-5 > 2x - 9 \geq -17$.
31. Write the negation of $-1(-1) = -1$ and give the truth value of the negation.
32. Name the property illustrated by the statement "$0 \cdot (5 \cdot 1) = (0 \cdot 5) \cdot 1$."
33. The ratio of the areas of two similar rectangles is $64:9$. Find the ratio of their widths.
34. In $\odot P$, $m \stackrel{\frown}{RS} = 121$. Find $m\angle RPS$.
35. Find the probability of getting a 2 or a 4 when rolling two dice.

In this chapter, you will learn techniques for organizing and presenting data. For example (Exercises 10-14, page 445), the lengths of suspension bridges in North America can be displayed in a histogram or a frequency polygon. The San Francisco-Oakland Bay Bridge is shown above.

Statistics 12

Information that a computer uses can be stored on silicon chips called the computer's *memory*. The world production of silicon chips with a memory of approximately 64,000 storage elements totaled 20,000,000 in a recent year. It is predicted that within five years chips with a memory of approximately 256,000 storage elements will be in standard use.

These statements are all examples of *statistics*, or numerical data. The word *statistics* is also used to name the area of mathematics that deals with the collection, analysis, and interpretation of facts called *data*. You will learn more about statistics in this chapter.

Section 12-1 INTRODUCTION TO STATISTICS

Often data are not easily analyzed in the form in which they are first collected. In this section you will learn techniques for compiling and organizing data so that you can more easily draw conclusions from them.

Statistics is divided into two branches. *Descriptive statistics* is concerned with analyzing or summarizing data that have been collected. *Inferential statistics* is concerned with making predictions and drawing conclusions based on incomplete information.

Often the incomplete information is in the form of a *sample*, as shown in the example that follows. In this example, the data are organized by arranging them in a frequency table. Notice that tallies have been used to count the data.

EXAMPLE   For a class project, some students bought 20 packages of Tasty Almonds and counted the number of almonds in each package. These 20 packages are a sample of all packages produced by the Tasty Almond Company. The numbers of almonds they counted were:

17, 20, 20, 15, 18, 17, 18, 18, 18, 19,
17, 18, 18, 19, 20, 18, 17, 18, 17, 20

Next, they used tallies to count the number of packages containing each quantity of almonds and made a *frequency table,* as shown on the following page.

Number of almonds per package	Tally	Frequency
15	/	1
16		0
17	ℍ	5
18	ℍ ///	8
19	//	2
20	////	4

From the table you can see that there are 2 packages that contain 19 almonds. We say that the *frequency* of packages containing 19 almonds is 2.

If you were to buy a package of Tasty Almonds on your way home, how many almonds would you expect to get? Is it likely that you would get 25 almonds? Probably not. It seems more likely that you would get 17 to 20 almonds.

When large numbers of data are collected they are usually grouped into *intervals*.

EXAMPLE The following data are the heights in centimeters of all the 50 entering freshmen at Central College.

```
163  150  171  142  171  183  152  154  157  197
165  172  177  183  189  179  195  175  176  166
162  181  165  161  187  153  181  148  152  183
152  172  142  171  155  154  151  167  182  156
168  189  157  187  159  188  156  151  185  157
```

Searching through the data we find that the shortest height is 142 cm and the tallest is 197 cm. We shall arrange the data in intervals of 140 cm through 149 cm, 150 cm through 159 cm, and so on, as shown in the frequency table below.

Heights in Centimeters of Freshmen at Central College

Interval	Tally	Frequency
140-149	///	3
150-159	ℍ ℍ ℍ /	16
160-169	ℍ ///	8
170-179	ℍ ////	9
180-189	ℍ ℍ //	12
190-199	//	2

Notice that most of the entering freshmen are in the intervals 150-159 and 180-189. Can you guess why? Is the average height of males and females the same?

From the frequency table you can also see that 3 + 16, or 19, of the freshmen are less than 160 cm tall, that 3 + 16 + 8, or 27, are less than 170 cm tall, and so on. You can display this kind of information more usefully in a *cumulative frequency table*, as shown below.

Heights in Centimeters of Freshmen at Central College

Interval	Frequency	Cumulative Frequency
140–149	3	3
150–159	16	19
160–169	8	27
170–179	9	36
180–189	12	48
190–199	2	50

From this table you can see directly, for example, that 36 of the freshmen are less than 180 cm tall, and 48 are less than 190 cm tall.

ORAL EXERCISES

The following table represents the number of movies 100 students at North High School attended last year.

Interval	Frequency	Cumulative Frequency
0–2	27	27
3–5	12	39
6–8	32	71
9–11	18	89
over 11	11	100

1. How many intervals does this table have?
2. How many students saw 3–5 movies?
3. How many students saw fewer than 6 movies?
4. How many students saw 9–11 movies?
5. How many students saw at most 11 movies?
6. How many students saw more than 11 movies?
7. Which interval has the greatest frequency? What is this frequency?
8. Which interval has the least frequency? What is this frequency?
9. How many students saw more than 2 movies but fewer than 9?

STATISTICS

The table at the right gives the yearly rainfall in centimeters for Waterville over the last 50 years.

Interval	Frequency	Cumulative Frequency
0–19	2	2
20–39	3	5
40–59	5	10
60–79	28	38
80–99	7	45
100–119	4	49
120–139	1	50

10. In how many years was the rainfall greater than 39 cm and less than 60 cm?

11. In how many years was the rainfall greater than 119 cm and less than 140 cm?

12. In how many years was the rainfall less than 40 cm?

13. In how many years was the rainfall less than 120 cm?

14. In how many years was the rainfall at most 59 cm?

15. In how many years was the rainfall more than 39 cm but less than 80 cm?

16. In how many years was the rainfall at most 99 cm?

17. In how many years was the rainfall at most 79 cm?

WRITTEN EXERCISES

For Exercises 1–5, use the following scores Pat made playing golf:

83, 85, 95, 90, 72, 104, 92, 92, 86, 93,
91, 95, 97, 101, 95, 99, 86, 95, 90, 91

Use intervals of 70–75, 76–81, 82–87, 88–93, 94–99, and 100–105.

A 1. Make a frequency table and a cumulative frequency table.

2. How many times did Pat score below 94?

3. How many times did Pat score above 87?

4. How many times did Pat score at least 88?

5. In what interval is Pat most likely to score?

For Exercises 6–10, use the following data which are the times in seconds of 30 students running the 100-meter dash.

12.8	12.6	11.0	14.9	11.3	13.7	13.8	12.1	13.1	13.2
12.2	12.3	10.8	13.0	13.9	14.1	16.1	12.9	12.8	11.6
15.2	14.7	13.7	14.1	15.2	12.7	14.2	11.7	15.4	13.0

B 6. Round the times to the nearest second and make a frequency table and a cumulative frequency table.

7. How many students finished in less than 12 s?

8. How many students finished in less than 14 s?

9. How many students took more than 14 s to run the dash?
10. What was the time to the nearest second of the student who came in 10th?

Section 12-2 CHARTS AND GRAPHS

In this section you will learn to construct:
- histograms and cumulative histograms
- frequency polygons and cumulative frequency polygons

A *histogram* or bar chart is a drawing that helps you visualize the information presented in a frequency table. A *cumulative* histogram helps you visualize the information in a cumulative frequency table.

EXAMPLE In order to pass the written test given at the end of a first-aid course, a person must answer at least 45 of the 50 questions correctly. The following table gives the scores of 30 people who passed the test.

Score	Frequency	Cumulative Frequency
45	2	2
46	2	4
47	5	9
48	10	19
49	5	24
50	6	30

The data in this frequency table are graphed in the histogram at the left below. The cumulative frequency histogram at the right below is a graph of the data in the cumulative frequency table.

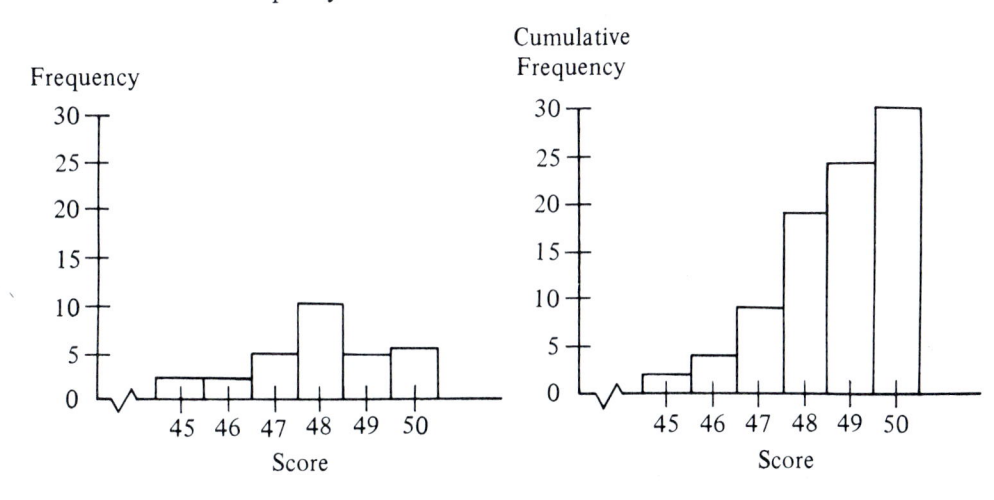

Frequency Histogram

Cumulative Frequency Histogram

STATISTICS

In the following example, histograms are shown for data that have been grouped in intervals. Notice that an endpoint common to two successive intervals is to be included in the interval on the right. For example, the second vertical line in the frequency histogram intersects the horizontal axis at 20.

EXAMPLE The frequency and cumulative frequency table below shows the ages of 35 students enrolled in a graduate course at State University. The data are also presented in a histogram and a cumulative frequency histogram.

Interval	Frequency	Cumulative Frequency
10–19	1	1
20–29	17	18
30–39	10	28
40–49	5	33
50–59	0	33
60–69	2	35

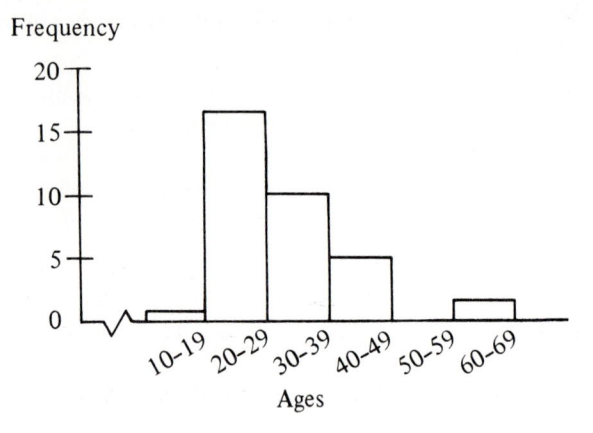

Frequency Histogram

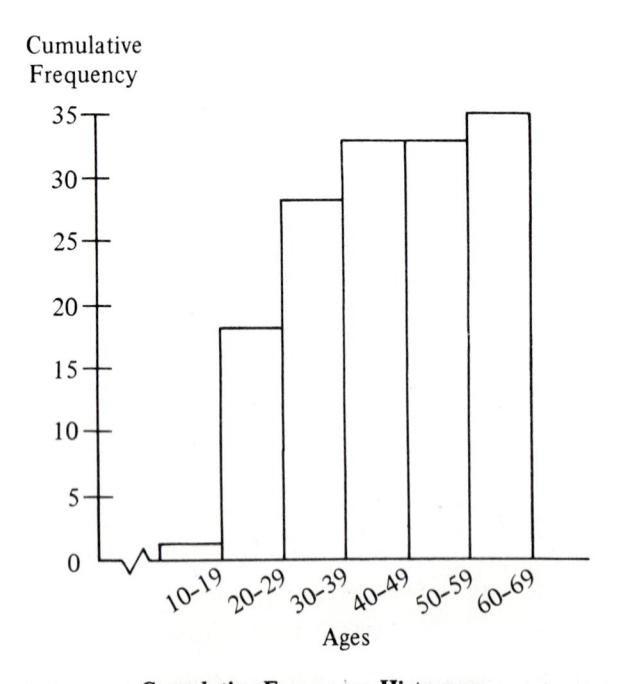

Cumulative Frequency Histogram

442   CHAPTER 12

A *frequency polygon* is a line graph made by connecting the midpoints of the tops of consecutive bars of a frequency histogram. In the diagram at the right, the dashed line is the frequency polygon for the frequency histogram in the preceding example. Notice that the frequency polygon has been drawn to join the horizontal axis both to the left and to the right of the histogram.

Frequency Polygon (dashed line)

In the same way a *cumulative frequency polygon* can be drawn for a cumulative frequency histogram. Usually, however, a cumulative frequency polygon is drawn by connecting the right-hand endpoints of the tops of consecutive bars rather than their midpoints.

In the diagram below, the dashed line is the cumulative frequency polygon for the cumulative frequency histogram in the preceding example. Notice that it is connected to the horizontal axis at the left of the histogram but not at the right.

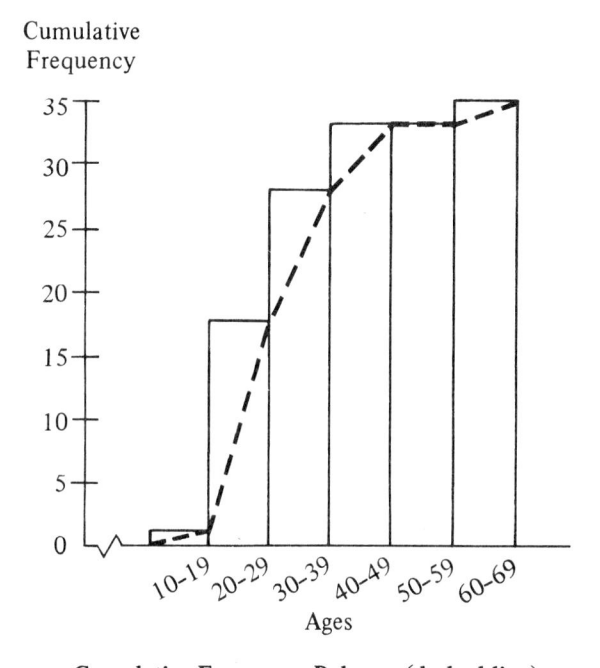

Cumulative Frequency Polygon (dashed line)

STATISTICS 443

ORAL EXERCISES

The histogram at the right shows the number of times Blue Bay Drawbridge opened each day to let boats pass through over a two-week period.

1. On how many days did the bridge open 3 times?
2. What was the greatest number of times it opened on any one day?
3. What was the fewest number of openings on any one day?
4. Was there any day when the bridge didn't open during the two-week period?

The cumulative frequency histogram for the Blue Bay Drawbridge data given above is shown at the right. On how many days did the drawbridge open

5. fewer than 5 times?
6. fewer than 6 times?
7. fewer than 7 times?
8. fewer than 8 times?

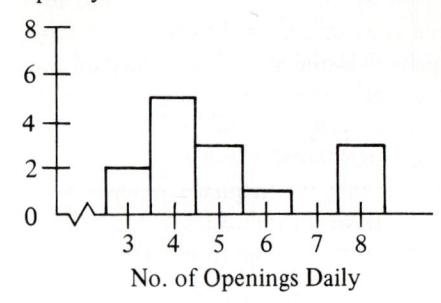

WRITTEN EXERCISES

Use the following data for Exercises 1–5. These are the average daily temperatures in degrees Celsius for a two-week period in Sun City.

26.9, 24.1, 20.3, 28.6, 27.2, 22.9, 23.2,
27.0, 28.3, 19.8, 22.1, 24.4, 25.8, 23.7

A 1. Round the temperatures to the nearest degree and make a frequency table and a cumulative frequency table.
2. Construct a frequency histogram.
3. Construct a frequency polygon.
4. Construct a cumulative frequency histogram.
5. Construct a cumulative frequency polygon.

The table at the top of page 445 shows the number of takeoffs and landings at 15 airports during a recent year. Using the information given in the table, construct the following graphs.

6. A frequency histogram
7. A frequency polygon
8. A cumulative frequency histogram
9. A cumulative frequency polygon

444   CHAPTER 12

Interval	Frequency	Cumulative Frequency
350,000–399,000	7	7
400,000–449,000	1	8
450,000–499,000	3	11
500,000–549,000	1	12
550,000–599,000	2	14
600,000–649,000	0	14
650,000–699,000	1	15

Following are the lengths to the nearest meter of the longest spans of 15 suspension bridges in North America. Use these data for Exercises 10-14.

1299, 1280, 1158, 1067, 853,
847, 704, 701, 668, 655,
610, 564, 549, 533, 497

B 10. Make a frequency table and a cumulative frequency table using intervals of 400-499 m, 500-599 m, and so on.

11. Construct a histogram.

12. Construct a frequency polygon.

13. Construct a cumulative frequency histogram.

14. Construct a cumulative frequency polygon.

SELF-TEST 1

The following table shows the numbers of records owned by 100 students at East High School.

Interval	Frequency	Cumulative Frequency
0–5	6	6
6–11	32	38
12–17	13	51
18–23	22	73
24–29	27	100

1. How many students owned 18-23 records?

2. How many students owned 17 or fewer records?

3. How many students owned 18 or more records?

4. How many students owned fewer than 24 records?

Section 12-1

Use the information in the table above to construct the following graphs.

5. A histogram

6. A frequency polygon

7. A cumulative frequency histogram

8. A cumulative frequency polygon

Section 12-2

STATISTICS

Section 12-3 MEASURES OF CENTRAL TENDENCY

In this section you will learn about the following measures of central tendency:
- arithmetic average or mean
- median
- mode

The frequency histograms discussed in the last section provide a lot of useful information about data in the form of an easy-to-read chart or graph. A *measure of central tendency*, on the other hand, is a single number that describes the given data.

The most widely used measure of central tendency is the *arithmetic average*, or *mean*. The *mean* for a group of numbers is found by adding all the numbers in the group and then dividing that sum by the number of numbers in the group.

EXAMPLES

Group of numbers: 82, 66, 86, 98

Mean: $\frac{82 + 66 + 86 + 98}{4} = \frac{332}{4} = 83$

Group of numbers: 27, 31, 28, 33, 26, 35, 30

Mean: $\frac{27 + 31 + 28 + 33 + 26 + 35 + 30}{7} = \frac{210}{7} = 30$

Group of numbers: 1, −4, −6, 3

Mean: $\frac{1 + (-4) + (-6) + 3}{4} = \frac{-6}{4} = -1.5$

Another measure of central tendency is the *median*. The *median* for a group of numbers is the middle number after the numbers have been arranged, or ranked, in either increasing or decreasing order. If the group consists of an odd number of numbers, the median is the middle number after ranking. If the group consists of an even number of numbers, the median is taken to be the mean of the two middle numbers after ranking.

EXAMPLES

Group of numbers: 87, 95, 91
Ranked numbers: 87, 91, 95
Median: 91

Group of numbers: 24, 30, 20, 14
Ranked numbers: 14, 20, 24, 30
Median: $\frac{20 + 24}{2} = \frac{44}{2} = 22$

Group of numbers: 1, −4, −6, 3
Ranked numbers: −6, −4, 1, 3
Median: $\frac{-4 + 1}{2} = \frac{-3}{2} = -1.5$

A third measure of central tendency is the *mode*. The *mode* for a group of numbers is the number that appears most frequently. If no number is repeated, we say that there is no mode. It is possible for a group of numbers to have more than one mode.

EXAMPLES Group of numbers: 87, 92, 95, 87
Mode: 87

Group of numbers: 1, 0, 2, 0, 3, 0, 4, 0
Mode: 0

Group of numbers: 16, 23, 19, 25, 12, 85
This group has no mode.

Group of numbers: 15, 15, 6, 3, 8, 3, 15, 9, 9, 3
This group has two modes, 3 and 15.

The mode is useful when dealing with data that are not numerical, such as color of eyes. With such data it is not possible to calculate a median or average.

It is often useful to calculate more than one measure of central tendency for a group of numbers. For example, the mean is influenced by extremely high or extremely low numbers while the median is not.

EXAMPLE The following data are a sample of yearly incomes:
$10,000, $10,000, $10,000, $18,000, $22,000, $80,000

Mean = $25,000
Median = $14,000
Mode = $10,000

Do you see that in this case the median, $14,000, gives a better picture of how the salaries are distributed than does the mean, $25,000, or the mode, $10,000?

EXAMPLE Sam's five marks in Film-Making for the marking period were 100, 75, 100, 60, 70.
Mean = 81   Median = 75   Mode = 100

Which measure of central tendency seems to be the fairest in this case?

ORAL EXERCISES

Give the mean, median, and each mode (if there is one) for each of the following groups of numbers.

1. 4, 4, 4
2. 2, 2, 4, 4
3. 1, 2, 3, 4, 5
4. 50, 49, 1, 0
5. 0, 0, 0, 1, 2, 3
6. −1, −2, −3, −4, −5
7. 1, 3, 5
8. 6, 4, 2
9. 1, 2, 87

WRITTEN EXERCISES

Find the mean, median, and each mode (if there is any) for each of the following groups of numbers.

A 1. 2, 1, 7, 4, 1
2. 80, 70, 90
3. 67, 60, 68
4. 83, 83, 81, 81, 82
5. 15, 14, 13, 14, 15
6. 22, 21, 25, 22, 27
7. 62, 83, 91, 85, 99, 75, 82, 85, 90, 72
8. −12, 15, 13, −10, −9, 17, 6, −2, 8, 13
9. $\frac{1}{2}, \frac{1}{3}, \frac{1}{6}$
10. $\frac{1}{2}, \frac{1}{4}, \frac{1}{8}, \frac{1}{8}$
11. 12.2, 11.1, 15.7, 25.2, 31.4, 8.7, 1.2, 19.7, 8.9, 9.5
12. 0.25, 0.37, 0.42, 0.12, 0.18
13. −0.12, 0.25, −0.08, 0.32, 0.12
14. −0.001, −0.002, −0.005, −0.002

B 15. The average of the numbers 20, 28, 30, and x is 31. Find the value of x.
16. The average of the numbers 88, 63, 70, 85, 78, and y is 78. Find the value of y.

For each of the groups of data described, indicate whether you think the mean would be larger than the median, the median would be larger than the mean, or both mean and median would be about the same.

17. annual incomes for all people in Canada
18. masses of all adult males in the United States
19. ages of all teenagers in New York City
20. all golf scores for non-professional players
21. heights of all high-school students in Toronto
22. the lengths of all full-grown porpoises

Construct a group of five numbers satisfying each of the following conditions.

23. The average is 10.
24. The mode is 12.
25. The median is 63.
26. The median is 12.5.

C 27. The mean is 25 and the median is 20.
28. The mean is 20 and the median is 25.
29. The mean is 80 and the median is 50.
30. The mean is 50, the median is 40, and the mode is 30.

Section 12-4 QUARTILES AND PERCENTILES

Shown below are the scores 16 students received on a math quiz. They are ranked in order from least to greatest.

$$\left| \begin{array}{cccccccc} 60 & 62 & 62 & 65 & 70 & 74 & 74 & 76 \end{array} \right| \begin{array}{cccccccc} 78 & 82 & 85 & 94 & 96 & 98 & 98 & 99 \end{array} \right|$$

$$\longleftarrow 50\% \longrightarrow | \longleftarrow 50\% \longrightarrow$$

Median

$$\text{Median} = \frac{76 + 78}{2} = 77$$

As you can see, the median separates the scores into two equal groups. In the problem above, there is an even number of scores. In such cases, one half, or 50%, of the scores are less than the median. The other half, or 50%, of the scores are greater than the median.

For practical purposes, data are often separated into four equal groups after being ranked in order so that each of the groups contains 25% of the scores. The numbers that separate the groups are called *quartiles*. There are three quartiles, called the *first, second,* and *third* quartiles, for any set of scores.

EXAMPLE Find the first, second, and third quartiles for the math quiz scores given above.

SOLUTION Separate the ranked scores into two equal groups as shown above, using the median as a divider. Then calculate the medians of these groups. The original median and the medians of these two groups divide the original group into four equal groups as shown below.

$$\left| \begin{array}{cccc} 60 & 62 & 62 & 65 \end{array} \right| \begin{array}{cccc} 70 & 74 & 74 & 76 \end{array} \right| \begin{array}{cccc} 78 & 82 & 85 & 94 \end{array} \right| \begin{array}{cccc} 96 & 98 & 98 & 99 \end{array} \right|$$

$$\longleftarrow 25\% \longrightarrow | \longleftarrow 25\% \longrightarrow | \longleftarrow 25\% \longrightarrow | \longleftarrow 25\% \longrightarrow$$

First Quartile Second Quartile Third Quartile

$$\text{First Quartile} = \frac{65 + 70}{2} = 67.5$$

$$\text{Second Quartile} = \frac{76 + 78}{2} = 77$$

$$\text{Third Quartile} = \frac{94 + 96}{2} = 95$$

Notice that the second quartile is always the same as the median. The first quartile is also sometimes called the *lower quartile,* and the third quartile is sometimes called the *upper quartile.*

STATISTICS

Quartiles are generally used when statisticians deal with a large amount of data. In this case, you can use a cumulative frequency polygon to estimate the quartiles.

EXAMPLE Using the data from Section 12-1, construct a cumulative frequency polygon for the heights of 50 freshmen at Central College. Then use this polygon to estimate the quartiles for the data.

SOLUTION The cumulative frequency table for the data in Section 12-1 is repeated at the right. From this table we can construct a cumulative frequency histogram and cumulative frequency polygon as shown below.

Interval	Cumulative Frequency
140–149	3
150–159	19
160–169	27
170–179	36
180–189	48
190–199	50

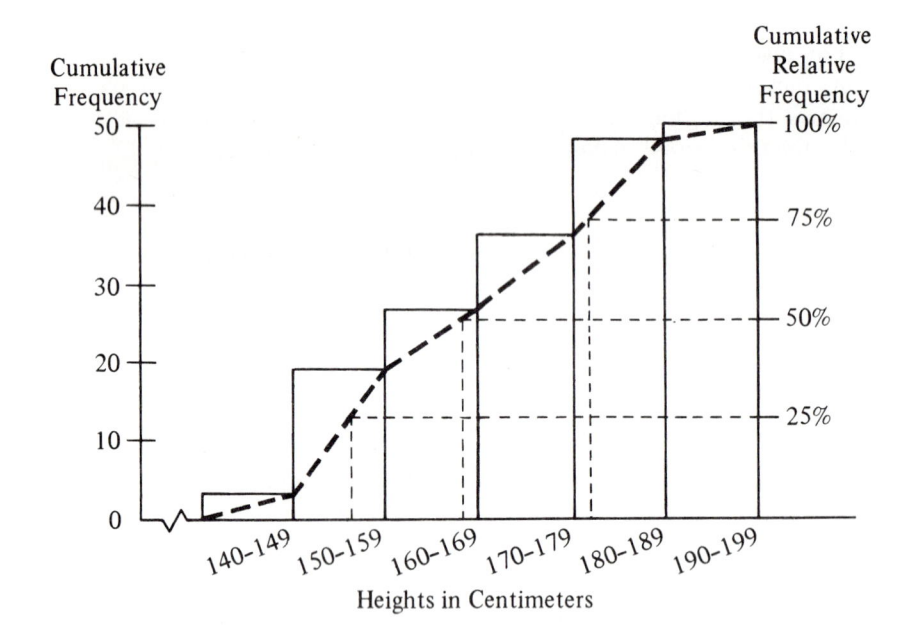

The dashed line is the cumulative frequency polygon line. The height of the right-hand bar of the histogram represents all, or 100%, of the scores. We mark off the bar into four equal segments and label the three boundary points we find as 25%, 50%, and 75%. These are called measures of *cumulative relative frequency*.

To estimate the first quartile, start from the 25% mark on the bar and draw a horizontal line to the left until it meets the cumulative frequency polygon. From this point draw a vertical line down until it intersects the horizontal axis. Since

450   CHAPTER 12

this point of intersection is a little more than halfway along the 150-159 interval, we estimate the first quartile as 156 cm. We estimate the second and third quartiles in the same way, starting with the 50% mark and the 75% mark, respectively.

We arrive at the following estimates:

$$\text{First quartile} \approx 156 \text{ cm}$$
$$\text{Second quartile} \approx 168 \text{ cm}$$
$$\text{Third quartile} \approx 182 \text{ cm}$$

Thus we can estimate that 50% of the entering freshmen are less than 168 cm tall, and 75% are less than 182 cm tall. To check these estimates you could go back to the original data, arrange them in order, separate them into four equal groups, and calculate the quartiles.

A measure called a *percentile* is often used to deal with a large collection of scores. In this case the scores are separated into 100 equal groups and the points of separation mark the percentiles.

The scores made by 500 students on a chemistry test are shown by the cumulative histogram and the cumulative frequency polygon below.

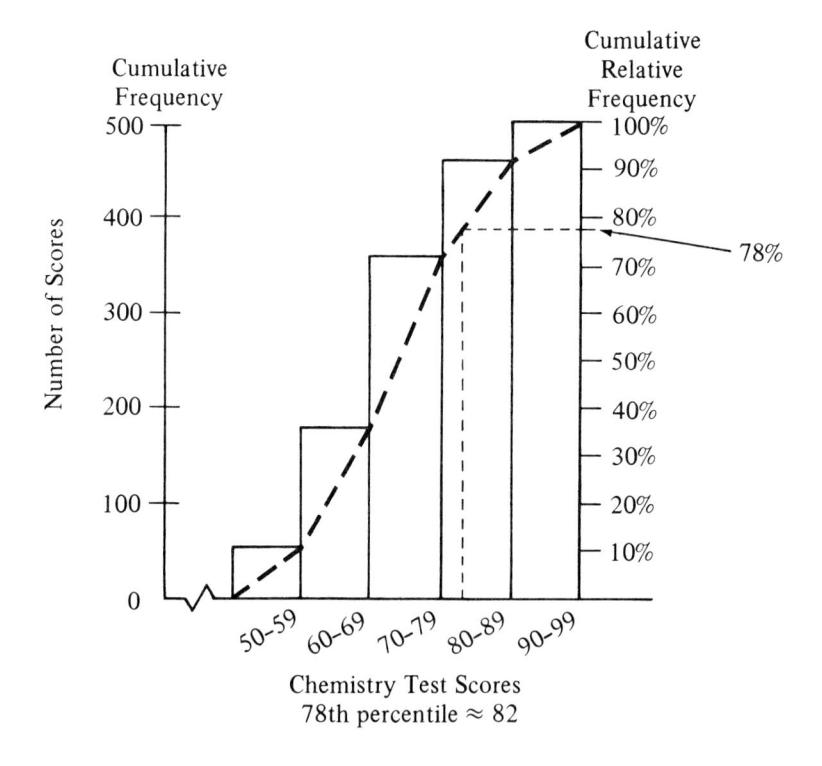

Chemistry Test Scores
78th percentile ≈ 82

The graph indicates that the 78th percentile is approximately a score of 82. This means that about 78% of the students had a score of 82 or less, and about 22% had a score greater than 82.

ORAL EXERCISES

1. What percentile corresponds to the first quartile? the second quartile? the third quartile?

2. Give two other names for the second quartile.

3. a. What information is given by the statement that the 43rd percentile of a group of data is 623?
   b. In part (a), what percent of the data are above 623?

Find the first, second, and third quartiles for each of the following groups of data.

4. 2, 4, 6, 8

5. 1, 3, 5, 7, 9, 11, 13, 15

WRITTEN EXERCISES

Find the first, second, and third quartile for each of the following groups of data.

A 1. 82, 72, 74, 70

2. 87, 80, 80, 85, 84, 88, 83, 82

In Exercises 3 and 4, use the given cumulative frequency polygon to estimate, to the nearest integer, the first, second, and third quartiles.

3. Cumulative Frequency

4. Cumulative Frequency

452   CHAPTER 12

Use the cumulative frequency polygon to estimate which test score corresponds to each of the following.

5. The 70th percentile
6. The 40th percentile
7. The 85th percentile
8. The median
9. The first quartile
10. The third quartile

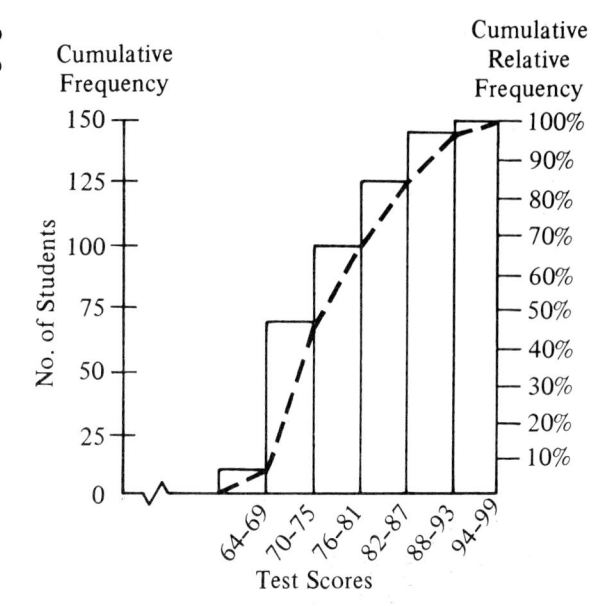

Some tests are being conducted on top of Old Smokey to see if it would be a good location for a windmill. For a period of two weeks the average hourly wind speed in kilometers per hour was recorded. The cumulative frequency polygon is shown below. Use it for Exercises 11–21 on the next page.

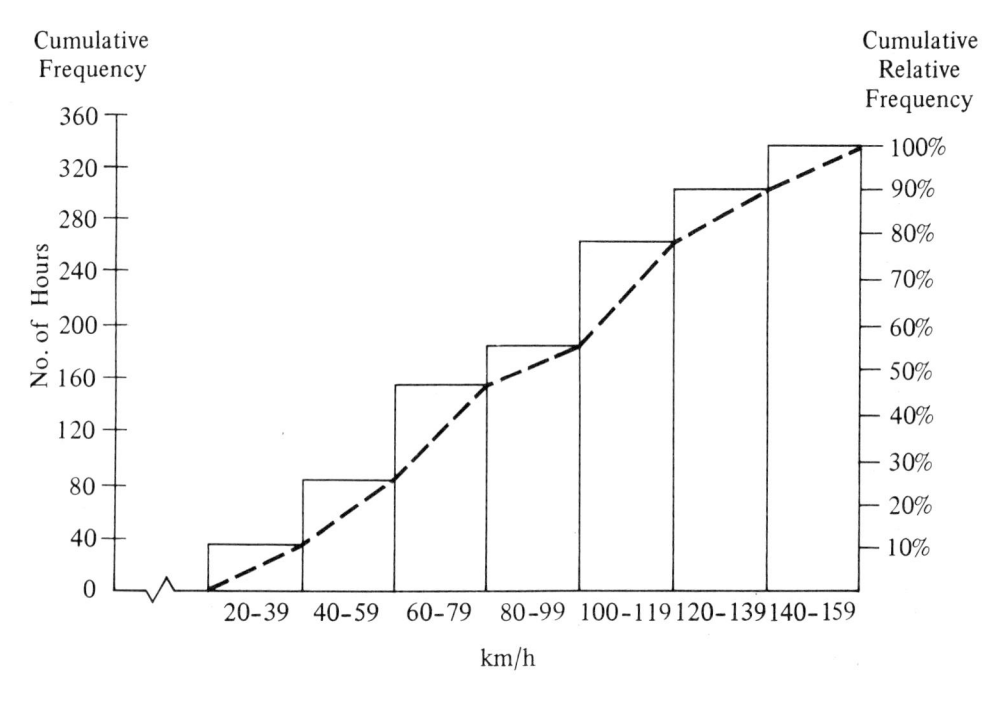

STATISTICS 453

Use the cumulative frequency polygon on page 453 to estimate the following.

B 11. The wind speed at the 30th percentile
12. The wind speed at the 65th percentile
13. The wind speed at the 10th percentile
14. The median wind speed
15. The wind speed at the first quartile
16. The wind speed at the third quartile
17. The percent of the hours that the wind speed was under 130 km/h during the two-week period
18. The percent of the hours that the wind speed was above 80 km/h
19. The percent of the hours that the wind speed was between 50 and 100 km/h
20. The number of hours during which the wind speed was below the 65th percentile
21. The number of hours during which the wind speed was above the 70th percentile

Use the frequency table to construct the cumulative frequency polygon and estimate the following.

C 22. The first quartile
23. The median
24. The third quartile
25. The 95th percentile
26. The 12th percentile
27. The 85th percentile

Interval	Frequency
10–19	40
20–29	65
30–39	75
40–49	15
50–59	5

SELF-TEST 2

Find the mean, median, and each mode (if there is any) for the following groups of scores.

Section 12-3

1. 3, 2, 1, 2, 2
2. 40, 44, 30
3. 20, 15, 12, 20, 16, 13, 30, 22, 18, 14
4. $\frac{4}{5}, \frac{2}{5}, \frac{5}{5}, \frac{1}{5}, \frac{3}{5}$

Use the cumulative frequency polygon to estimate the following.

Section 12-4

5. The 80th percentile
6. The 10th percentile
7. The 30th percentile
8. The first quartile
9. The second quartile
10. The third quartile

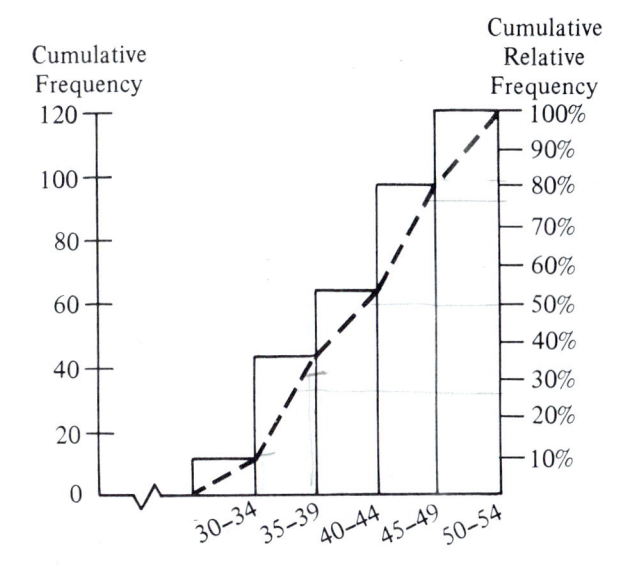

Section 12-5 SAMPLING AND SIMULATION

In this section you will learn about:
- simple random samples
- random number tables
- simulation

Manufacturers are concerned with maintaining control of the quality of the goods they produce by testing or inspecting them. But when goods are produced in large quantities, it is often impractical to inspect or test each item. For this reason, manufacturers usually rely on taking a sample of some of the items for testing or inspecting. In order that the sample be representative of all the items produced, its members must be chosen with care. A method is used so that all the items produced have the same probability of being chosen for the sample. Such samples are called *simple random samples*.

EXAMPLE Devise a method for choosing a simple random sample of 5 soccer balls from a set of 30 soccer balls.

SOLUTION Number the soccer balls from 1 to 30. Take 30 identical slips of paper and number them from 1 to 30. Put the slips in a bag and shuffle them. Without looking, reach into the bag and draw out 5 slips. Read the numbers from the slips and choose for your sample the 5 soccer balls with the same numbers.

STATISTICS 455

Can you see that this procedure of choosing simple random samples could be tedious and time consuming if you were asked to pick, say, a random sample of 10 students from an entire student body of 2379? Statisticians have devised a more scientific and convenient way to choose simple random samples by using *random number tables*. A small portion of such a table appears below.

Columns

1	2	3	4	5	6	7	8	9	10
81439	87143	37357	40597	76569	94255	65709	59813	33116	27042
98091	46151	12606	63635	13837	80501	38663	56081	50848	94268
53132	97120	74805	28139	15571	69375	63622	57645	64842	20723
75873	79270	61575	09576	71963	50852	30227	63849	56777	80040
19508	97337	12568	54544	99918	27268	90823	07923	17137	36782
31873	14687	20029	26078	80979	08431	82008	95979	69653	16772
34502	07678	30970	06837	31293	62400	17608	35036	62450	44736
91373	94895	36632	73400	52962	56678	50498	03158	80708	92557
63975	21678	47056	81082	00134	85054	81533	84692	88429	35748
48940	94031	70043	46710	11539	32803	84097	35410	43968	16930

The following examples illustrate how to use such a table to choose a simple random sample.

EXAMPLE Ten students are to be chosen at random from the 100 members of the Sports Club as winners of bicycles. Use the random number table above to choose the students.

SOLUTION Assign a number from 00, 01, 02, and so on through 99 to each student. Decide on a method of picking 2-digit numbers from the random number table. We will use the first column and read off the 2 left-hand digits going down the column. Part of the column is shown at the right with the chosen numbers underlined. If a number is repeated, we shall use it only once. We continue until we have 10 numbers.

In this way we obtain the numbers 81, 98, 53, 75, 19, 31, 34, 91, 63, and 48. Thus the 10 students with those numbers assigned will win the 10 bicycles.

<u>81</u>439
<u>98</u>091
<u>53</u>132
<u>75</u>873
.....

In the example above, we could have chosen several alternative methods for picking 2-digit numbers. For example, we could have started at the left of the top row and moved to the right, obtaining the numbers 81, 43, 98, 71, and so on. When using a random number table it is important that before you look at the table you decide where to start and what method to use for picking the desired numbers.

EXAMPLE Use the random number table to choose a sample of 10 students from a student body of 2379.

SOLUTION Assign a number from 0001, 0002, and so on through 2379 to each student. You could decide to start at the top of Column 5 and read off the 4 left-hand digits going down the column. If you come to a number larger than 2379, ignore it. Thus your sample will be the 10 students numbered 1383, 1557, 0013, 1153, 0843, 1760, 0792, 0315, 1713, and 2072 if you use Columns 5, 6, 7, 8, 9, and 10 in order.

Computers are programmed to generate random numbers. For more information, see the Computer Exercises on pages 462–464.

Random number tables can also be used to *simulate,* or imitate, experiments.

EXAMPLE How would you use a random number table to simulate the experiment of tossing a coin and determining the approximate probability of heads?

SOLUTION Let's associate heads with an even number and tails with an odd number. Then we shall choose a block of 100 digits and count the even digits in it. Say we choose Columns 9 and 10, which have 100 digits. By counting we find there are 57 even digits.

$$\text{Approximate probability of heads} = \frac{57}{100}$$

ORAL EXERCISES

1. Will every digit (0–9) appear exactly the same number of times in a random number table? Explain.
2. Does each digit occur at regular intervals in a random number table? Why or why not?
3. a. How many digits are there in each row of the random number table on page 456?
   b. On the average, how many 4's would you expect to find in each row?
   c. How many 4's are there in the first row? the second row? the third row?
   d. Considering your answer to part (b), do your answers to part (c) seem reasonable?
4. Repeat Exercise 3, parts (b), (c), and (d), for the digit 7.

WRITTEN EXERCISES

Use the random number table in this section to choose each of the following samples.

A 1. A sample of 5 batteries from a lot of 100
2. A sample of 10 concert-goers from an audience of 1000
3. A sample of 5 calculators from a lot of 500
4. A sample of 15 ballpoint pens from a lot of 5200

STATISTICS

Tell how you would use the random number table to simulate the following experiments and determine the approximate probability for the events described.

B 5. Obtaining one, and only one, head when two coins are tossed

6. Obtaining one or two heads when two coins are tossed

7. Rolling a 5 with one die

8. Rolling a sum of 7 with a pair of dice

C 9. Obtaining a head and an even number when a coin and a die are tossed together

Section 12-6 MATHEMATICAL EXPECTATION

In this section you will see the strong relationship between statistics and probability by investigating *mathematical expectation,* or *expected value.*

Decisions in business, industry, and everyday life are often based on predictions of what will happen in the future. In this respect it is useful to calculate a quantity called an *expected value.* We will first show how to calculate the expected values of some of the probability experiments you studied in Chapter 11.

The *expected value, E,* of an experiment is the sum of the products of each outcome times its corresponding probability. It tells you what result is obtained, on average, when an experiment is repeated over and over again.

EXAMPLE   What is the expected value, E, when you toss one die?

SOLUTION   The table at the right shows the outcomes and probabilities for the toss of a die.

Outcomes	1	2	3	4	5	6
Probabilities	$\frac{1}{6}$	$\frac{1}{6}$	$\frac{1}{6}$	$\frac{1}{6}$	$\frac{1}{6}$	$\frac{1}{6}$

$$E = 1 \cdot \frac{1}{6} + 2 \cdot \frac{1}{6} + 3 \cdot \frac{1}{6} + 4 \cdot \frac{1}{6} + 5 \cdot \frac{1}{6} + 6 \cdot \frac{1}{6} = \frac{21}{6} = 3.5$$

This means that if you toss a die over and over, record the number observed each time, and then find the average, that average will be close to 3.5.

EXAMPLE   What is the expected number of heads when you toss two coins?

SOLUTION   The table at the right shows the outcomes and their probabilities.

Outcomes	0 Head	1 Head	2 Heads
Probabilities	$\frac{1}{4}$	$\frac{1}{2}$	$\frac{1}{4}$

$$E = 0 \cdot \frac{1}{4} + 1 \cdot \frac{1}{2} + 2 \cdot \frac{1}{4} = 1$$

On the average, you can expect to observe one head when you toss two coins.

You can also use expected value to predict how well you might succeed in playing a game.

EXAMPLE You play a game with a friend by tossing a coin. If a head is tossed you pay your friend $2, and if a tail is tossed your friend pays you $1. In the long run, how do you expect to fare?

SOLUTION The outcomes and probabilities are shown in the table at the right. The negative 2 means that you lose $2.

Outcomes	−2	1
Probabilities	$\frac{1}{2}$	$\frac{1}{2}$

$$E = -2 \cdot \frac{1}{2} + 1 \cdot \frac{1}{2} = -1 + \frac{1}{2} = -\frac{1}{2}$$

The expected value is $-\frac{1}{2}$ dollar, or −50¢. This means that, over the long run, you should expect to lose an average of 50¢ each time you play.

In the game above, the expected value is a negative number, meaning that you lose. If the expected value is 0, the game is said to be *fair*. Can you invent a fair game?

ORAL EXERCISES

EXAMPLE Five cards numbered 1 through 5 are placed in a hat. You close your eyes and pick two cards. If the outcome is the product of the numbers on the cards, state the possible outcomes.

SOLUTION You may wish to list all the possible products.

$1 \cdot 2 = 2$ $1 \cdot 3 = 3$ $1 \cdot 4 = 4$ $1 \cdot 5 = 5$
$2 \cdot 1 = 2$ $2 \cdot 3 = 6$ $2 \cdot 4 = 8$ $2 \cdot 5 = 10$
$3 \cdot 1 = 3$ $3 \cdot 2 = 6$ $3 \cdot 4 = 12$ $3 \cdot 5 = 15$
$4 \cdot 1 = 4$ $4 \cdot 2 = 8$ $4 \cdot 3 = 12$ $4 \cdot 5 = 20$
$5 \cdot 1 = 5$ $5 \cdot 2 = 10$ $5 \cdot 3 = 15$ $5 \cdot 4 = 20$

The possible outcomes are 2, 3, 4, 5, 6, 8, 10, 12, 15, and 20.

In each of the following experiments, state the possible outcomes.

1. You toss two coins. The outcome is the number of tails that come up.
2. You roll a die.
3. You roll two dice. The outcome is the sum of the numbers on the faces of the dice.
4. You roll two dice. The outcome is the product of the numbers on the faces of the dice.
5. You pay $.50 each time you play a game and you receive $10 each time you win.

STATISTICS

Find the expected value.

6.
Outcomes	1	−1
Probabilities	$\frac{1}{2}$	$\frac{1}{2}$

7.
Outcomes	1	2	3	4
Probabilities	$\frac{1}{4}$	$\frac{1}{4}$	$\frac{1}{4}$	$\frac{1}{4}$

8. Must the expected value of an experiment be an outcome of the experiment? Explain.

WRITTEN EXERCISES

Find the expected value.

A 1.
Outcomes	1	2	3	4
Probabilities	0.1	0.2	0.3	0.4

2.
Outcomes	3000	−6000	9000
Probabilities	$\frac{2}{3}$	$\frac{1}{6}$	$\frac{1}{6}$

3.
Outcomes	−5	−4	0	2
Probabilities	$\frac{1}{12}$	$\frac{1}{12}$	$\frac{1}{2}$	$\frac{1}{3}$

4.
Outcomes	$\frac{1}{2}$	$\frac{3}{8}$	$\frac{1}{8}$
Probabilities	$\frac{1}{8}$	$\frac{3}{8}$	$\frac{1}{2}$

5. a. Find the probability of each outcome in Oral Exercise 1.
   b. Find the expected value.

6. a. Complete the table for the spinner game at the right.

Outcomes	−5	10	30	−50
Probabilities	$\frac{1}{2}$	?	$\frac{1}{8}$	?

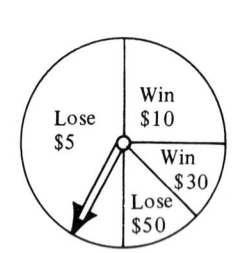

   b. Find the expected value.

Find the expected value for each of the following spinner games.

B 7. Win $100, Lose $20, Lose $50, Win $30

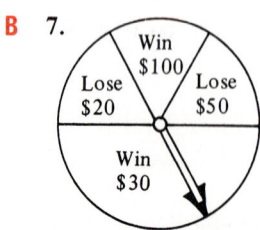

8. Win $1000, Lose $1500, Win $200, Lose $800

9. Lose $7500, Win $10,000, Lose $2500, Win $8000

460 CHAPTER 12

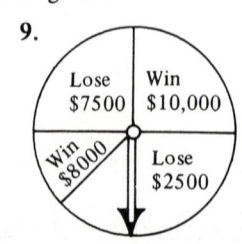

10. a. Find the probability of each outcome in Oral Exercise 3.
    b. Find the expected value.
11. a. Find the probability of each outcome in Oral Exercise 4.
    b. Find the expected value.
12. Find the expected value for the number of heads when 3 coins are tossed.

C 13. You pay 50¢ to choose an item from a grab bag. In the grab bag there are 8 items worth 25¢ each, 6 items worth 50¢ each, 4 items worth 75¢ each, and 2 items worth $1 each. What is the expected value of the item you pick? Do you think 50¢ is a reasonable charge for the grab bag?

14. There is a $\frac{3}{4}$ probability that a dog of a certain breed has a curly tail and a $\frac{1}{4}$ probability that the tail is straight. Puppies with curly tails sell for $200 and puppies with straight tails sell for $300. What is the expected value of the sales price for one of these puppies?

SELF-TEST 3

Use the random number table on page 456 to choose each of the following samples.

1. A sample of 8 digital watches from a lot of 100
2. A sample of 10 bolts from a lot of 500

Section 12-5

Describe how you would use the random number table on page 456 to simulate the following experiments and determine the approximate probability for the events described.

3. Rolling a 2, 4, or 6 with one die
4. Obtaining one, and only one, head when three coins are tossed

Find the expected value for each of the following.

5.

Outcomes	1	2	3	4
Probabilities	0.4	0.3	0.2	0.1

Section 12-6

6.

Outcomes	2	1	0	−1	−2
Probabilities	$\frac{1}{5}$	$\frac{1}{5}$	$\frac{1}{5}$	$\frac{1}{5}$	$\frac{1}{5}$

STATISTICS

Computer Exercises

1. In most versions of BASIC, the function RND(0) generates a random number between 0 and 1.

   Copy and RUN the following program twice and compare your results.

   ```
   10   FOR P=1 TO 15
   20   PRINT RND(0),
   30   NEXT P
   40   END
   ```

   If both outputs are the same, add the statement
   ```
   5 RANDOMIZE
   ```
   and RUN the program again. If you still get the same output, try changing line 5 to read:
   ```
   5 LET X=RND(-1)
   ```
   If either of these additions was necessary, you will have to include that statement in any program involving random numbers if you wish to get different results for different runs of the program.

2. Change line 20 in the previous program to
   ```
   20 PRINT 5*RND(0),
   ```
   and RUN it again. Notice that the computer now generates random numbers between 0 and 5.

3. a. To generate random integers from 0 to 4, change line 20 to:
   ```
   20 PRINT INT(5*RND(0)),
   ```
   b. To generate random integers from 1 to 5, try:
   ```
   20 PRINT INT(5*RND(0))+1,
   ```
   c. Write a program that generates 20 random integers from 1 to 8.

   d. Write a program that generates 20 random integers from 2 to 9.

   Consult your user's manual or your teacher to determine whether your computer has a simpler method for generating random integers. Many computers do.

4. Computers are excellent at simulating experiments. The following program uses random numbers to simulate an experiment in which a die is tossed 50 times.

Copy the program and RUN it several times.

```
10   FOR I=1 TO 50
20   LET X=INT(6*RND(0))+1
30   PRINT X;
40   LET F[X]=F[X]+1
50   NEXT I
60   PRINT
70   PRINT
80   PRINT "NUMBER","FREQUENCY","CUMULATIVE FREQUENCY"
90   PRINT
100  LET C[1]=F[1]
110  PRINT 1,F[1],C[1]
120  FOR X=2 TO 6
130  LET C[X]=F[X]+C[X-1]
140  PRINT X,F[X],C[X]
150  NEXT X
160  END
```

5. a. The following program computes the median and the mean of N numbers, where $N \geq 1$. In the DATA statement the numbers must be listed in order, followed by .9999.

Copy and RUN the following program.

```
10   LET I=1
20   READ X[I]
30   IF X[I]=.9999 THEN 70
40   LET S=S+X[I]
50   LET I=I+1
60   GOTO 20
70   LET I=I-1
80   PRINT "THE MEAN IS ";S/I
90   IF I/2=INT(I/2) THEN 120
100  PRINT "THE MEDIAN ";X[(I+1)/2]
110  GOTO 140
120  PRINT "THE MEDIAN IS ";(X[I/2]+X[I/2+1])/2
130  DATA -2000,-6,-2,1,7,26,41,117,.9999
140  END
```

b. Note how great the difference is between the mean and the median in the program in part (a). Change the data in line 120 so that the mean and the median of the eight numbers will be the same.

6. a. The programs in Exercises 4 and 5 use *subscripted variables.* In Chapter 9 you saw examples of subscripted variables such as C_1 and C_2. In BASIC a subscripted variable is a letter followed by an algebraic expression, called the *subscript,* in parentheses. The value of the subscript must be a positive integer. A[1], P[7], M[I], and X[3∗J + 4] are examples of subscripted variables.

Copy and RUN the following program.

```
10   LET A[1]=5
20   LET A[2]=7
30   LET A[3]=A[1]+A[2]
40   PRINT A[1];A[2];A[3]
50   END
```

The program above contains three subscripted variables, A[1], A[2], and A[3]. Note that A[1] is a completely different variable from A[2].

Do not confuse the value of a subscripted variable with its subscript. In the program above, the *value* of A[1] is 5. The *subscript* of A[1] is 1.

b. Copy and RUN the following program.

```
10   FOR I=1 TO 5
20   READ A[I]
30   NEXT I
40   FOR X=5 TO 1 STEP -1
50   PRINT A[X];
60   NEXT X
70   DATA 4,-7,-2,6,9
80   END
```

CHAPTER REVIEW

A car's gasoline consumption, in kilometers per liter, was computed ten times. Use these data, given below, in Exercises 1-8.

10, 14, 12, 14, 11, 13, 10, 14, 13, 11

1. Make a cumulative frequency table. Section 12-1
2. How many times was the gasoline consumption at most 12 km/L?
3. How many times was it less than 11 km/L?
4. Construct a histogram. Section 12-2
5. Construct a frequency polygon.
6. Construct a cumulative frequency histogram.
7. Construct a cumulative frequency polygon.
8. Find the mean, median, and each mode. Section 12-3

Find the mean, median, and each mode (if there is any) for the following groups of numbers.

9. 7, 8, 2, 9, 6, 5, 9, 2
10. 0.92, 0.53, 0.27, 0.46, 0.72

Use the cumulative frequency polygon below to estimate each of the following.

11. The first quartile
12. The second quartile
13. The third quartile
14. The 15th percentile

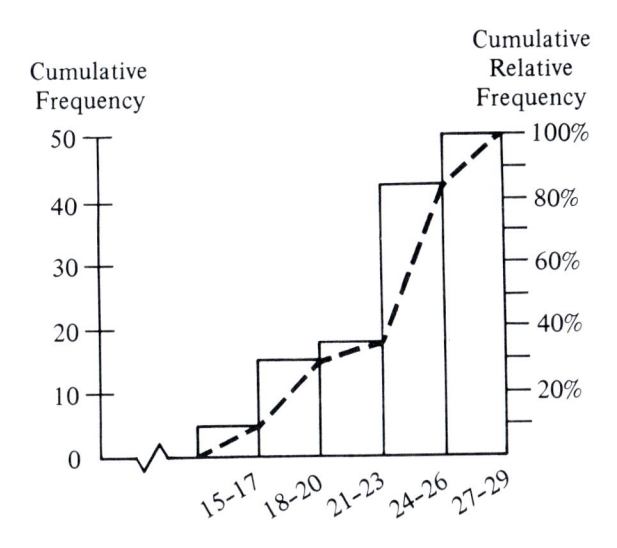

Section 12-4

Use the random number table in Section 12-5 to select each sample.

15. A sample of 5 transistors from a lot of 2000
16. A sample of 8 registered voters in a town with 8000 registered voters
17. Tell how you would use the random number table in Section 12-5 to approximate the probability of obtaining a 2 when selecting a random digit from 0 to 9.

Section 12-5

Find the expected value.

18.
Outcomes	1	2	3	4
Probabilities	$\frac{1}{2}$	$\frac{1}{8}$	$\frac{1}{4}$	$\frac{1}{8}$

Section 12-6

19.
Outcomes	−1	0	1	2
Probabilities	0.1	0.2	0.3	0.4

STATISTICS 465

CHAPTER TEST

The following table shows the approximate heights to the nearest 10 meters of the 20 highest mountain peaks in Antarctica.

Interval	Frequency	Cumulative Frequency
4000–4190	7	7
4200–4390	6	13
4400–4590	2	15
4600–4790	2	17
4800–4990	2	19
5000–5190	1	20

1. Which interval has the greatest frequency?
2. How many mountains are less than 4800 m high?
3. Construct a cumulative frequency histogram.
4. Construct a cumulative frequency polygon.
5. Use your diagram from Exercise 4 to estimate:
   a. the third quartile   b. the first quartile   c. the 90th percentile

The following data show the numbers of cameras sold at the Photo Shoppe during twelve business days.

$$3, 2, 3, 3, 4, 1, 3, 5, 4, 1, 3, 4$$

6. Find the mean, median, and mode.
7. Construct a frequency histogram.
8. Construct a frequency polygon.

Use the random number table in Section 12-5 to

9. choose a random sample of 20 newly minted coins from a collection of 3000.
10. choose a random sample of 8 wrist watches from a lot of 1500.
11. tell how you would approximate the probability of rolling at least one 6 with a pair of dice.

Find the mean, median, and each mode (if there is any) for each group of numbers.

12. 1, 1, 5, 5
13. 13, −10, 3, −2, −5, −7, 1

Find the expected value.

14.
Outcomes	1	2	3	4
Probabilities	$\frac{1}{6}$	$\frac{1}{6}$	$\frac{1}{6}$	$\frac{1}{2}$

15.
Outcomes	1	−2	3	−4
Probabilities	$\frac{1}{4}$	$\frac{1}{8}$	$\frac{1}{2}$	$\frac{1}{8}$

Careers — Statistician

Suppose you were asked to conduct a survey at your school to find out how satisfied the students are with the extracurricular activities now offered and to predict their reaction to a new sport. You would have to choose a group of students to question (the sample), decide what you need to ask them, and interpret the answers you received. Or suppose your principal handed you three computer tapes full of data about the families of students at your school over the last fifty years. Your task would be to sift through all the information and find trends that would help your principal in planning.

These are the kinds of jobs that statisticians do, but usually on a much larger scale. Statisticians must feel at home with numbers — and lots of them. They use mathematical techniques and formulas to analyze data that they have either gathered themselves or been asked to interpret. All the little bits of information must be distilled into a form, such as a trend or a percentage, that decision makers can use. The work of statisticians results in forecasts of change, evaluations of the success of programs, or quality control tests for manufactured products.

The kinds of data that statisticians gather and evaluate are expanding because the number of fields that have recognized the value of statistical analysis is increasing. Workers in all types of fields — from economics to education, from the natural sciences to the physical sciences, from history to law — generate masses of data that they need help interpreting. An important growth area for statistics lies in analyzing the effects of pollution and toxic substances on the environment and on human health.

Statisticians work for private industry, especially manufacturing, finance, and insurance companies, and for governmental offices. An increasing number of statisticians become consultants. An advantage of this choice is the chance to be on the pioneering edge of many different fields.

A major in statistics or a minor in statistics with a major in an applied field of your choice (such as a natural science or economics) is a good way to prepare for a career in this field. You will find a graduate degree in mathematics or statistics necessary for administrative, consulting, or teaching positions. Because of the importance of the computer in gathering and storing large quantities of data, computer proficiency is as essential as a good mathematics background.

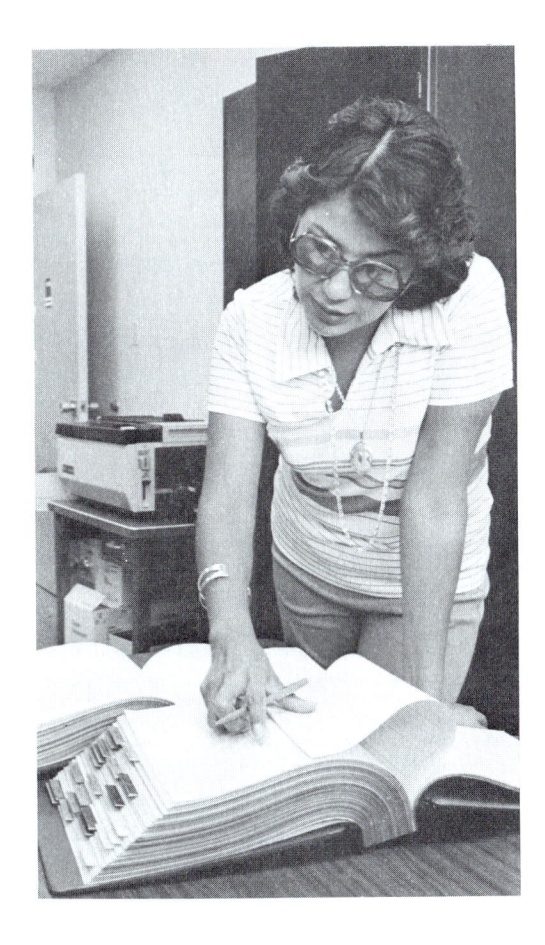

VOCABULARY REVIEW

Be sure that you understand the meaning of these terms:

statistics, p. 437
data, p. 437
sample, p. 437
frequency table, p. 437
cumulative frequency table, p. 439
histogram, p. 441
frequency polygon, p. 443

mean (arithmetic average), p. 446
median, p. 446
mode, p. 447
quartile, p. 449
percentile, p. 451
simple random samples, p. 455
expected value, p. 458

MIXED REVIEW

1. How many different 3-digit numbers can be formed from the digits 1, 3, 5, 7, and 9 if no digit is repeated?

2. Repeat Exercise 1 if repetition is allowed.

3. A rectangle is 6 cm wide and 8 cm long. How long is each diagonal?

4. Refer to the figure at the right. Write the reason that justifies each statement.
   a. $\triangle AEB \cong \triangle CED$ and $\triangle AED \cong \triangle CEB$.
   b. $\overline{AB} \cong \overline{CD}$ and $\overline{AD} \cong \overline{BC}$.
   c. $ABCD$ is a parallelogram.

5. Find the mean, the median, and each mode (if any) for the data: 10, 12, 11, 12, 11, 12

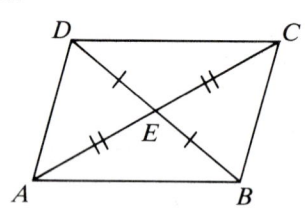

Find the values of x, y, and z. (In Exercise 7, $\overrightarrow{BA} \parallel \overrightarrow{CD}$.)

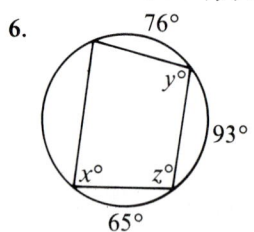

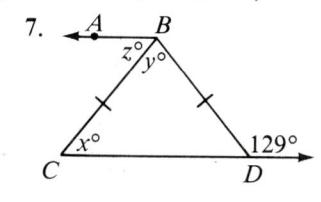

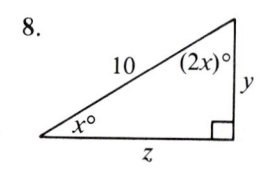

6. [circle with inscribed quadrilateral: 76°, y°, 93°, z°, 65°, x°]

7. [triangle with A, B at top, z°, y°, x°, 129° at D, C]

8. [right triangle with 10, (2x)°, y, x°, z]

9. Graph the solution set: $0 < 2 - y < 3$

10. Find the radius of a sphere with surface area 36π cm².

11. Find the probability of getting a number greater than 8 when rolling two dice.

Exercises 12 and 13 refer to the spinner at the right.

12. Find the expected value when you spin the spinner once.

13. Suppose that you spin the spinner twice. What is the probability that you get two odd numbers? at least one even number?

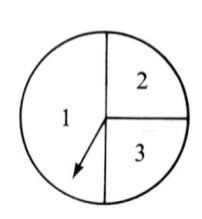

468   CHAPTER 12

Luke's quiz scores were 8, 8, 10, 10, 10, 9, 7, and 10.

14. Make a frequency table and a cumulative frequency table.
15. Construct a frequency polygon.
16. Construct a cumulative frequency histogram.
17. Find the first, second, and third quartiles.
18. A poll found that 52% of the registered voters plan to vote for Candidate B. If there are 20,000 registered voters, how many are expected not to vote for Candidate B?
19. What is the probability of drawing 3 red aces from a deck of cards if cards are not replaced? What is the probability of the complement of this event?
20. Simplify: a. $(\sqrt{2} - \sqrt{6})^2$ b. $\dfrac{6}{\sqrt{2} - \sqrt{6}}$
21. A right rectangular prism has height 3, width 5, and total area 126. Find its volume.

Solve.

22. $\dfrac{2}{3}x - 7 = \dfrac{5}{9}$
23. $2x^2 + 2x = 1$
24. $8z^2 = 6z + 9$

25. In the figure at the right, $m\stackrel{\frown}{RS} = 2x + 5$, $m\stackrel{\frown}{TV} = x - 1$, and $m\angle U = 35$. Find the value of x.
26. A cone has radius 5 and height 10. Find its lateral area.
27. Construct a truth table for $q \rightarrow (p \wedge q)$.
28. A square has a diagonal of length $2\sqrt{14}$. Find the perimeter and the area of the square.

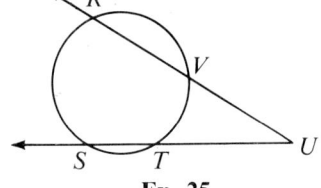

Ex. 25

29. The radius of a circle is an integer. If the circumference is greater than 15π, what is the smallest possible radius for the circle?
30. An urn contains 5 white balls and 7 green balls. If 2 balls are drawn without replacement, find the probability that the balls are the same color.
31. Describe how you could use a random number table to estimate the probability of getting 3 heads when 3 coins are tossed together.

The familiar idea of *steepness*, such as the steepness of a road or a ski slope, is closely related to the mathematical concept of *slope*. In this chapter, you will learn how to find the slope of a line and how to use slope in writing an equation of a line.

Lines and Their Equations   13

When planning the construction of a road or a railroad track, for example, an engineer must take into account the *gradient* of the road. The gradient is the ratio of the height gained to the horizontal distance traveled. The figure below (which is not drawn to scale) illustrates a road with a gradient of $\frac{1}{50}$, or 2%. This means that for every 100 km covered horizontally there is a gain of 2 km in height.

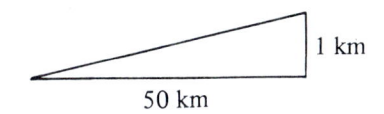

The gradient of a hill is an important factor in the construction of roads, especially with regard to the climbing and braking power of the vehicles using them. You know that a car must use more energy to climb a hill than to travel along a flat stretch of land. But equally important is the effect of a downward slope on the speed of the vehicle. If a car were to begin from rest at the top of the incline illustrated above and roll down without its brakes being applied, then when it reached the bottom of the incline it would be traveling at a rate of 500 km/h!

Just as the steepness of a hill is described by its gradient, the steepness of a line is described by its *slope*. You will learn more about lines and their slopes in this chapter.

Section 13-1 A RECTANGULAR COORDINATE SYSTEM

Until now you have studied geometry without regard to the exact location of lines and points in a plane. Since it is often useful to specify *where* a particular line or point lies in relation to some fixed central point, in this section you will learn how to:
- locate a point by using a rectangular coordinate system
- find the distance between two points on the same horizontal or vertical line

In Section 1-2 you learned how to construct a number line. A *rectangular coordinate system*, or *coordinate plane*, consists of two perpendicular number lines that intersect at their origins. The point of intersection is called the *origin O* of the system.

The number lines are the *coordinate axes*. The horizontal coordinate axis is called the *x-axis*. The vertical one is the *y-axis*. Every point that is not on a coordinate axis lies in one of the four *quadrants* shown.

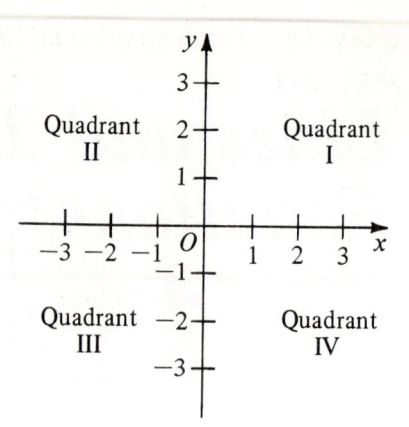

Just as every point on a number line corresponds to exactly one real number, every point in the coordinate plane corresponds to two real numbers, called an *ordered pair*. For example, the point P that is 2 units to the right of the origin and 3 units above it corresponds to the ordered pair (2, 3).

The *x-coordinate*, or *abscissa*, of point P is 2. The *y-coordinate*, or *ordinate*, is 3. When you write an ordered pair, the *x*-coordinate is always given first.

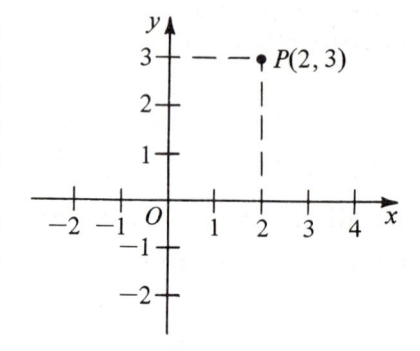

EXAMPLE State the coordinates of points A, B, C, and D.

SOLUTION The coordinates of A are $(3, 2)$.
The coordinates of B are $(-2, 4)$.
The coordinates of C are $(-3, -2)$.
The coordinates of D are $\left(1\frac{1}{2}, -3\right)$.

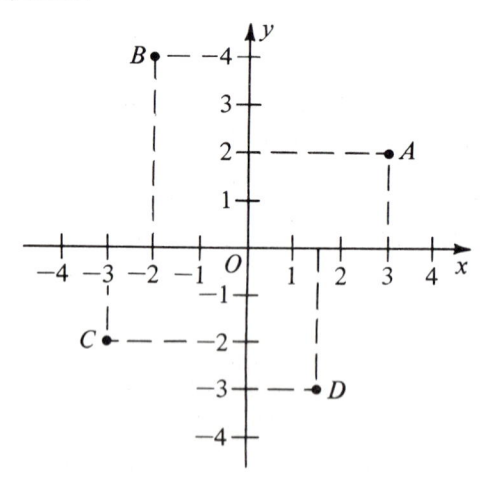

For convenience, we will sometimes refer to point $A(3, 2)$ or simply to $(3, 2)$ when we mean the point with those coordinates.

EXAMPLE Graph the points $E(5, 0)$, $F(0, 4)$, and $G(-3, -2)$.

SOLUTION

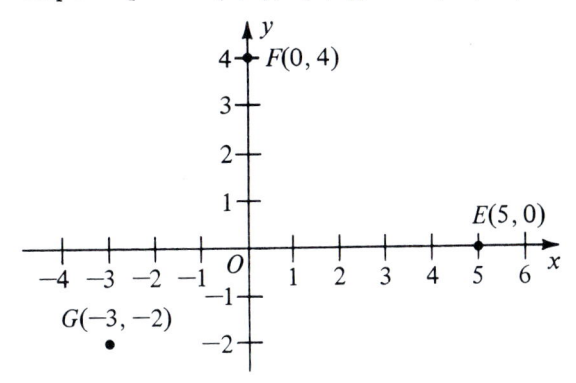

Notice that if the y-coordinate of a point is 0, the point is on the x-axis. If the x-coordinate of a point is 0, the point is on the y-axis.

To find the distance between two points with the same y-coordinate, either count units or subtract the lesser of the two x-coordinates from the greater.

EXAMPLE Find the distance between $(3, 2)$ and $(8, 2)$.

SOLUTION

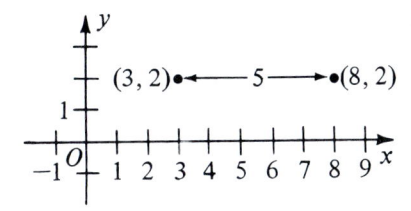

Distance $= 8 - 3 = 5$

To find the distance between two points with the same x-coordinate, either count units or subtract the lesser of the two y-coordinates from the greater.

EXAMPLE Find the distance between $(2, -3)$ and $(2, 4)$.

SOLUTION

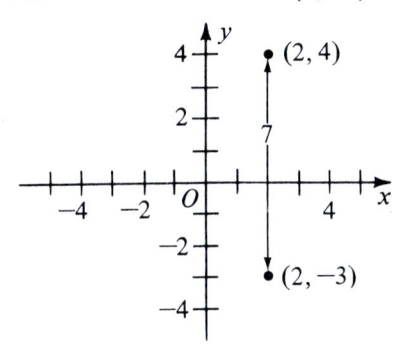

Distance $= 4 - (-3) = 4 + 3 = 7$

LINES AND THEIR EQUATIONS

ORAL EXERCISES

1. Name the coordinates of each point shown.

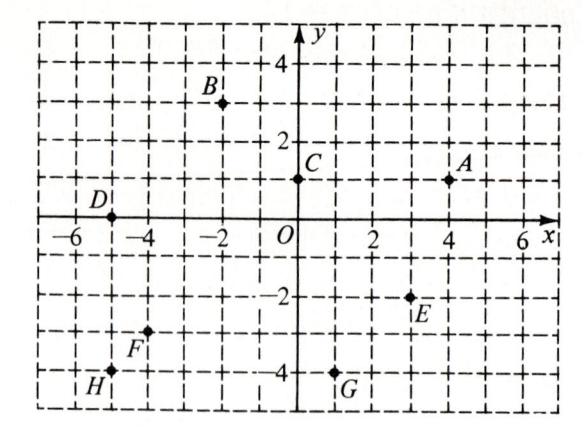

Exercises 2-12 refer to the following points:

$I(3, 7)$ $J(-2, 5)$ $K(-4, 0)$
$L(5, -2)$ $M(3, 6)$ $N(0, -2)$

2. Name a point whose abscissa is -2.
3. Name a point whose ordinate is 5.
4. Name a point on the x-axis.
5. Name a point on the y-axis.
6. Name two points in Quadrant I.
7. Name a point in Quadrant II.
8. Name a point in Quadrant IV.
9. Name two points on the same horizontal line.
10. Name two points on the same vertical line.
11. Name the point that is farthest above the x-axis.
12. Name the point that is farthest to the left of the y-axis.

Ex. 1

WRITTEN EXERCISES

A 1. Give the coordinates of each point shown. If a coordinate is not an integer, estimate it to the nearest half unit.

Graph each of the following points.

2. $G(3, 7)$
3. $H(-2, 5)$
4. $I(6, -4)$
5. $J(-5, -1)$
6. $K(0, 4)$
7. $L(-5, 0)$
8. $M\left(-\frac{5}{2}, 1\right)$
9. $N\left(3, -\frac{7}{2}\right)$
10. $P\left(-\frac{1}{2}, \frac{3}{2}\right)$
11. $Q(0.5, 6)$
12. $R(-2, -2.5)$
13. $S(1.5, -5)$

Ex. 1

Find the distance between each pair of points.

14. $(9, 6), (2, 6)$
15. $(-7, 2), (-7, 5)$
16. $(5, 0), (-4, 0)$
17. $(0, -3), (0, 7)$
18. $(3, -12), (3, -7)$
19. $(-1, -3), (-9, -3)$

474   CHAPTER 13

Graph the four points given. What special kind of quadrilateral do you think *JKLM* is? (Be as specific as possible.)

B 20. $J(0, 3), K(6, 0), L(5, -2), M(-1, 1)$ 21. $J(-3, 0), K(0, 1), L(4, -1), M(-5, -4)$

Sketch rectangle *ABCD* with the given vertices and find its area.

22. $A(-2, -3), B(5, -3), C(5, 4), D(-2, 4)$ 23. $A(6, -1), B(6, 8), C(3, 8), D(3, -1)$

Sketch $\triangle DEF$ with the given vertices and find its area.

24. $D(-3, 1), E(4, 1), F(4, 7)$ 25. $D(2, -5), E(2, 3), F(-4, 3)$
26. $D(1, -2), E(7, -2), F(3, 6)$ 27. $D(-3, 5), E(-3, 1), F(6, 2)$

Sketch parallelogram *ABCD* with the given vertices and find its area.

28. $A(-3, -1), B(0, 5), C(4, 5), D(1, -1)$ 29. $A(4, 5), B(5, -1), C(-3, -1), D(-4, 5)$

Graph the given points and find a fourth point so that all four points will be the vertices of a rectangle. Sketch the rectangle.

30. $K(2, 1), L(2, -1), M(5, -1)$ 31. $C(4, 6), F(-4, 6), I(4, -6)$
32. $B(-5, -2), E(-3, 2), H(1, -5)$ 33. $A(0, 0), D(2, -4), G(6, 3)$

Graph the two given points. Then find a third point *T* so that $\triangle RST$ is an isosceles triangle whose base is $\overline{RS}$ and whose height is $2RS$. (*Hint:* First find the midpoint of $\overline{RS}$.)

C 34. $R(4, 0), S(8, 0)$ 35. $R(-7, 2), S(2, 2)$

Graph the two given points. Then find a third point *C* so that $\triangle ABC$ is a right triangle with hypotenuse $\overline{AB}$. Use the Pythagorean Theorem to find the distance between *A* and *B*.

36. $A(1, 1), B(4, 5)$ 37. $A(-2, 2), B(6, 8)$

Section 13-2 LINEAR EQUATIONS IN TWO VARIABLES

You have already graphed on a number line equations and inequalities containing one variable. In this section you will learn how to:
- graph in a plane a linear equation containing two variables
- recognize an equation that defines a line
- find the *x*-intercept and *y*-intercept of a line

The *graph* in a plane of an equation containing *x* and *y* consists of all the points in the plane whose coordinates *satisfy* the equation.

EXAMPLE $(2, -5)$ is on the graph of the equation $3x + y = 1$, since
$$3(2) + (-5) = 6 - 5 = 1.$$
$(-2, 4)$ is *not* on the graph of $3x + y = 1$, since
$$3(-2) + 4 = -6 + 4 \neq 1.$$

One way to graph the equation $3x + y = 1$ is to first make a list, or table, of the coordinates of several points that satisfy the equation.

EXAMPLE Graph $3x + y = 1$.

SOLUTION To make a table, substitute values for x and solve for y:

$$3(-2) + y = 1$$
$$-6 + y = 1$$
$$y = 7$$

Thus, $(-2, 7)$ is a solution of $3x + y = 1$.

x	y	(x, y)
-2	7	$(-2, 7)$
-1	4	$(-1, 4)$
0	1	$(0, 1)$
1	-2	$(1, -2)$
2	-5	$(2, -5)$

The other values shown in the table are found in the same way. Next, graph the points in the table, as shown on the left below. The diagram suggests that all these points lie on a line. In fact, the line shown on the right consists of *all* points that satisfy the equation $3x + y = 1$.

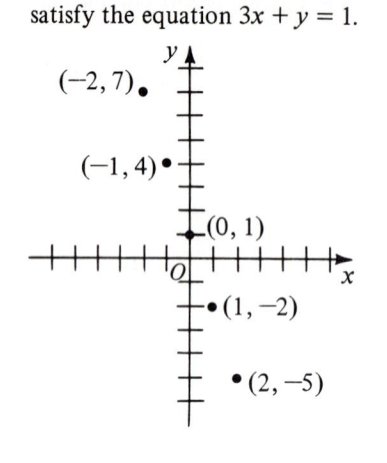

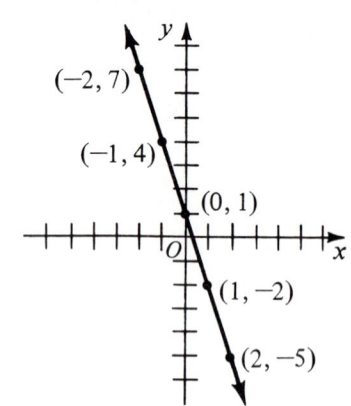

The graph of any equation of the form

$$Ax + By = C,$$

in which A, B, and C are real numbers, with A and B not both 0, is a line. An equation that can be put into this form is called a *linear equation in two variables*.

EXAMPLE Graph the equation $-3x + 2y = 6$.

SOLUTION

x	y
-4	-3
-2	0
0	3
2	6
4	9

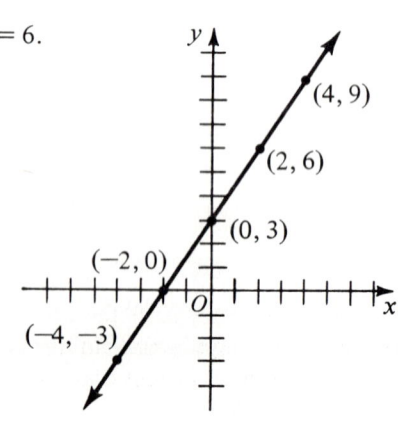

Notice that in the example we chose to substitute even integers for x in order to obtain integers as values of y. Of course, you could substitute other values for x. For example, if you substitute 1 for x, you will find that $\left(1, 4\frac{1}{2}\right)$ satisfies the equation.

EXAMPLE Graph $y = 3$ in the coordinate plane.

SOLUTION This is an equation of a line, with 0 as the coefficient of x. The value of y is 3 no matter what the value of x is. The graph is a line parallel to the x-axis.

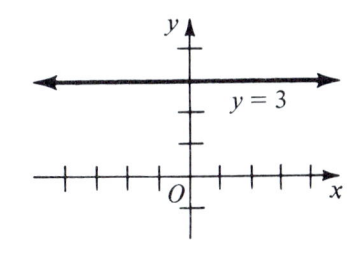

EXAMPLE Graph $5x = -10$ in the coordinate plane.

SOLUTION $5x = -10$ is equivalent to $x = -2$. The graph is a line parallel to the y-axis.

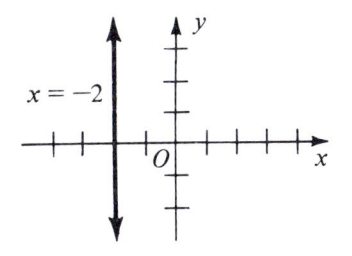

Since two points determine a line (see Section 5-1), you could graph a linear equation in two variables by graphing only *two* points that lie on the line. In practice, it is wise to graph three or four points to guard against arithmetic errors.

The points where a line crosses the x-axis and y-axis are often convenient ones to use when graphing the line. You find the coordinates of the point where the line crosses the x-axis by substituting 0 for y and solving for x. To find the coordinates of the point where the line crosses the y-axis, substitute 0 for x and solve for y.

EXAMPLE Graph $2x - 3y = 6$.

SOLUTION

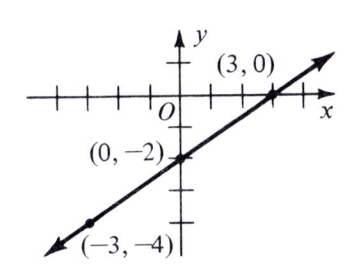

x	y
0	-2
3	0
-3	-4

The line with equation $2x - 3y = 6$ crosses the x-axis at $(3, 0)$. The number 3 is called the *x-intercept* of the line. Since the line crosses the y-axis at $(0, -2)$, the *y-intercept* of the line is -2. In general, if $(a, 0)$ and $(0, b)$ are two points of a line, then a is the x-intercept and b is the y-intercept of the line.

LINES AND THEIR EQUATIONS

ORAL EXERCISES

State whether the given point is on the line whose equation is given.

1. $2x - 5y = 3$; $(4, 1)$
2. $3x + 2y = -4$; $(-2, 5)$
3. $x - 6y = -3$; $(3, -1)$
4. $-x - 3y = 6$; $(6, -4)$
5. $3x - 2y = 0$; $(-2, -3)$
6. $-6x + 2y = 20$; $(-1, 7)$
7. $y = 5$; $(5, -7)$
8. $x = 5$; $(5, -7)$

In Exercises 9–16 the given point lies on the line whose equation is $3x - 2y = 1$. Find the value of k.

9. $(3, k)$
10. $(1, k)$
11. $(-1, k)$
12. $(0, k)$
13. $(k, 7)$
14. $(k, 0)$
15. $(k, -5)$
16. $(k, -11)$

Find (a) the x-intercept and (b) the y-intercept of the line whose equation is given.

17. $5x + 3y = -15$
18. $4x - y = 8$
19. $x - 2y = -10$

State whether the given equation is linear.

20. $-3x + 4y = 2$
21. $xy + x + 3y = 1$
22. $3 = -x + y$
23. $6x = 10$
24. $x^2 + 2y = 1$
25. $y = 2(x - 1)$

WRITTEN EXERCISES

Find the coordinates of at least three points on the graph of each equation, and then graph the equation.

A
1. $x + y = 5$
2. $x - y = 3$
3. $2x - y = 6$
4. $3x + y = 3$
5. $x - 2y = 0$
6. $x + 3y = 6$
7. $3x + 2y = 12$
8. $2x + 3y = 6$
9. $-5x + 2y = 4$
10. $4x - 3y = 0$
11. $x = 2$
12. $y = -1$
13. $2y = 3$
14. $3x = -6$
15. $5x - 3y = -2$
16. $-7x - 2y = 5$

For each line whose equation is given, find the x-intercept and the y-intercept (if they exist).

17. $x + y = 4$
18. $2x - y = 1$
19. $4x - y = 5$
20. $2x + 3y = -6$
21. $x - y = 0$
22. $x = 4y$
23. $y = 4$
24. $x = 3$

Write each equation in the form $Ax + By = C$ and then graph the equation.

B
25. $y = 2(x - 1)$
26. $2y = -3(x + 2)$
27. $y + 1 = -5(x - 1)$
28. $y - 4 = \frac{1}{2}(x - 2)$

Find A or B if the graph of the given linear equation contains the given point.

EXAMPLE $4x + By = 3;\ (2, 1)$

SOLUTION $4(2) + B(1) = 3$
$8 + B = 3$
$B = -5$

29. $Ax + 9y = 7;\ (-2, 1)$
30. $-3x + By = -2;\ (3, 1)$
31. $-2x + By = 4;\ (5, 2)$
32. $Ax - 6y = -7;\ (-5, -3)$
33. $Ax - 3y = 5;\ (2, 0)$
34. $7x + By = 6;\ (0, -4)$
35. $-3x + By = 1;\ (-2, 2)$
36. $Ax - 3y = 8;\ (3, -2)$

Graph each pair of equations on one set of axes. Give the coordinates of the point of intersection of the two graphs.

C 37. $x - 2y = 8$
$x + y = -4$

38. $2x + y = 5$
$x - y = 1$

39. $3x + 2y = -1$
$x + 2y = 1$

40. $2x - 5y = -4$
$-x + 4y = 5$

41. If a line has equation $Ax + By = C$, with $A \neq 0$ and $B \neq 0$, find its x-intercept and y-intercept in terms of A, B, and C.

42. If $A = 0$, does the graph of $Ax + By = C$ have an x-intercept? Explain.

SELF-TEST 1

1. Name the coordinates of each point shown.   Section 13-1

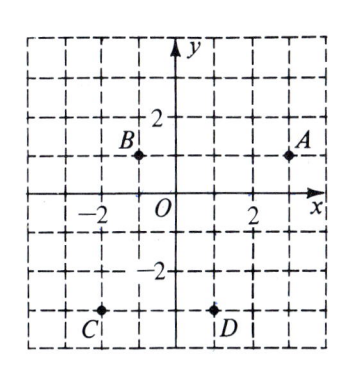

2. Graph the points $E(-4, 1)$ and $F(2, 3)$.
3. Find the distance between $(-1, 3)$ and $(5, 3)$.

Find the coordinates of at least three points on the graph of each equation, and then graph the equation.

4. $2y - x = 4$
5. $y = 4$ Section 13-2
6. $4x - 3y = 6$
7. $3x + 2y = 4$

LINES AND THEIR EQUATIONS

Section 13-3 THE SLOPE OF A LINE

In the figure, $\overleftrightarrow{AB}$ appears to be steeper than $\overleftrightarrow{AD}$, while $\overleftrightarrow{AB}$ and $\overleftrightarrow{CD}$ appear to be parallel. The steepness of a line is described mathematically as the *slope* of the line. In this section you will learn:
- how to find the slope of a line
- how the slopes of parallel lines are related

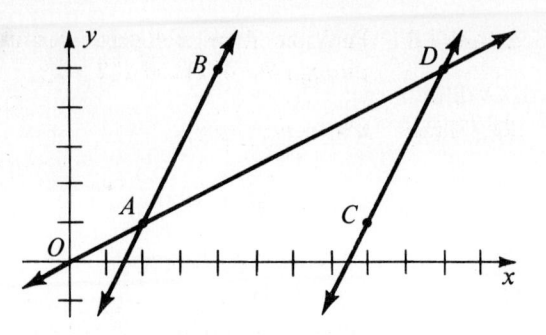

To get from point *A* to point *B*, you move 4 units up and 2 units to the right, as shown on the left below. To get from point *A* to point *D*, you also move 4 units up, but you must move 8 units to the right. The *slope* of each line is the *ratio* of the vertical change to the horizontal change.

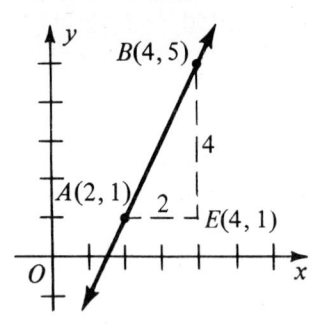

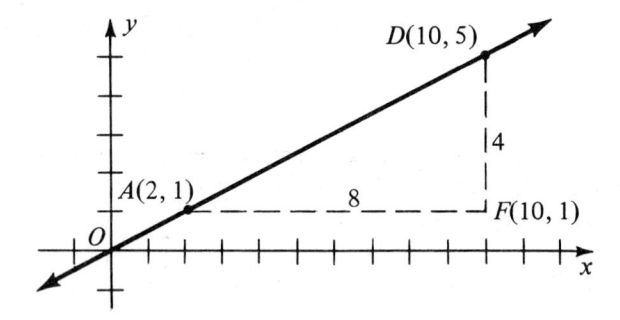

Slope of $\overleftrightarrow{AB} = \dfrac{\text{vertical change}}{\text{horizontal change}}$

$= \dfrac{4}{2}$

$= 2$

Slope of $\overleftrightarrow{AD} = \dfrac{\text{vertical change}}{\text{horizontal change}}$

$= \dfrac{4}{8}$

$= \dfrac{1}{2}$

Notice that the vertical change is the same as the change in the *y*-coordinates and the horizontal change is the same as the change in the *x*-coordinates.

Slope of $\overleftrightarrow{AB} = \dfrac{\text{change in } y\text{-coordinates}}{\text{change in } x\text{-coordinates}}$

$= \dfrac{5-1}{4-2}$

$= \dfrac{4}{2}$

$= 2$

Slope of $\overleftrightarrow{AD} = \dfrac{\text{change in } y\text{-coordinates}}{\text{change in } x\text{-coordinates}}$

$= \dfrac{5-1}{10-2}$

$= \dfrac{4}{8}$

$= \dfrac{1}{2}$

The slopes of both $\overleftrightarrow{AB}$ and $\overleftrightarrow{AD}$ are positive because each line rises from left to right. $\overleftrightarrow{ST}$ in the next example has a negative slope because it falls from left to right.

EXAMPLE Find the slope of the line passing through $S(-4, 1)$ and $T(2, -2)$.

SOLUTION Slope $= \dfrac{-2-1}{2-(-4)}$

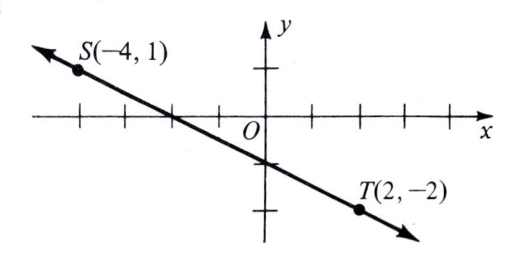

$= \dfrac{-3}{6}$

$= -\dfrac{1}{2}$

In the example we subtracted the *y*-coordinate of *S* from the *y*-coordinate of *T* and subtracted the *x*-coordinate of *S* from the *x*-coordinate of *T*. The slope is the same if the coordinates of *T* are subtracted from those of *S*:

$$\text{Slope} = \dfrac{1-(-2)}{-4-2}$$

$$= \dfrac{3}{-6}$$

$$= -\dfrac{1}{2}$$

The slope of a line does not depend on the two points used to find it. To determine the slope of a line whose equation is given, you may use the coordinates of *any* two points on the line.

EXAMPLE Find the slope of the line whose equation is $3x + y = 1$.

SOLUTION To compute the slope, choose any two points on the line, for example, $(0, 1)$ and $(2, -5)$.

$$\text{Slope} = \dfrac{-5-1}{2-0} = \dfrac{-6}{2} = -3$$

EXAMPLE Find the slopes of the lines with equations **(a)** $y = 1$ and **(b)** $x = 4$.

SOLUTION (a) Two points on the line with equation $y = 1$ are $(0, 1)$ and $(2, 1)$.

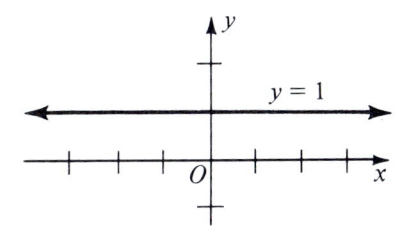

$$\text{Slope} = \dfrac{1-1}{2-0} = \dfrac{0}{2} = 0$$

(b) Two points on the line with equation $x = 4$ are $(4, 6)$ and $(4, 0)$. The change in the *x*-coordinates is 0. Since the denominator of a fraction cannot be 0, the slope of the line is *not defined*.

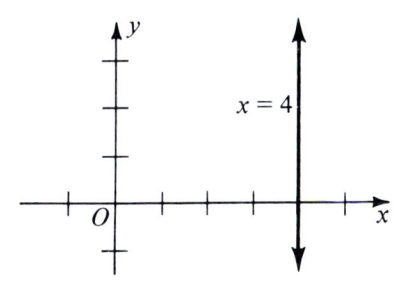

LINES AND THEIR EQUATIONS

Every horizontal line has slope 0, since the change in the *y*-coordinates is 0. The slope of a vertical line is not defined, since the change in the *x*-coordinates is 0, and the denominator of a fraction cannot be 0.

All horizontal lines are parallel to the *x*-axis and to each other. All vertical lines are parallel to the *y*-axis and to each other. To investigate how the slopes of other parallel lines are related, let's look again at the two lines, $\overleftrightarrow{AB}$ and $\overleftrightarrow{CD}$, that were shown at the beginning of this section.

You have seen already that the slope of $\overleftrightarrow{AB}$ is 2. The slope of $\overleftrightarrow{CD}$ is also 2.

Notice that *A* and *C* have been chosen as points with the same *y*-coordinate. Also, *B* and *D* have the same *y*-coordinate. Since $AC = 6$ and $BD = 6$,

$$\overline{AC} \cong \overline{BD}.$$

Because $\overline{AC}$ and $\overline{BD}$ are horizontal, they are parallel. Since two of its sides are both parallel and congruent, quadrilateral *ACDB* is a parallelogram. Thus, $\overleftrightarrow{AB} \parallel \overleftrightarrow{CD}$.

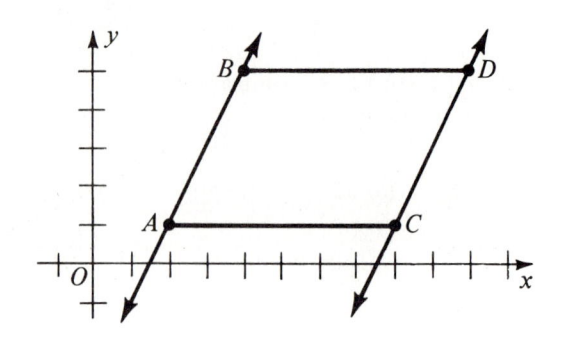

This suggests the following biconditional statement, which is true for *all* nonvertical lines.

> Two nonvertical lines are parallel if and only if they have equal slopes.

ORAL EXERCISES

1. For each line shown, state whether the slope appears to be positive, negative, zero, or not defined.

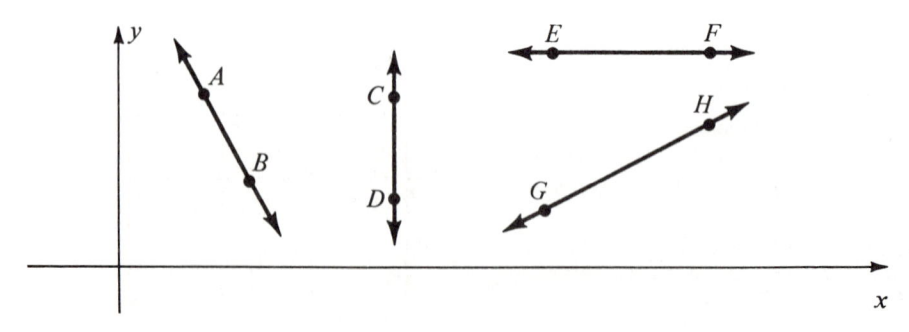

482   CHAPTER 13

For each line shown, find the slope (if it exists).

2. (2, 1), (7, 3)

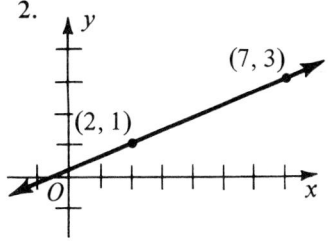

3. (3, 0), (6, 3)

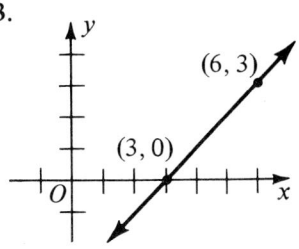

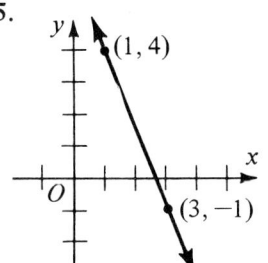

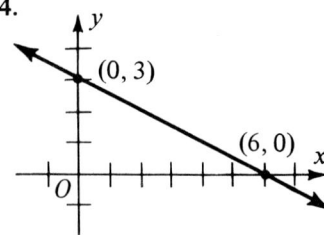

4. (0, 3), (6, 0)

5. (1, 4), (3, −1)

6. (−2, 2), (4, −1)

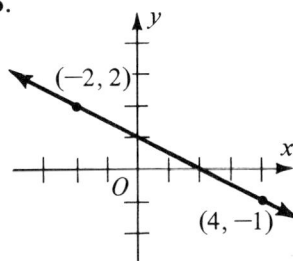

7. (3, 4), (3, −2)

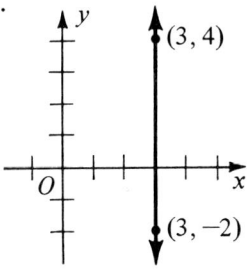

8. (−3, 2), (3, 2)

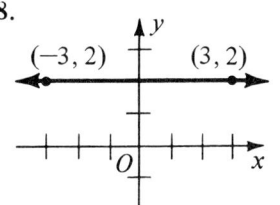

9. (−3, 0), (0, 2)

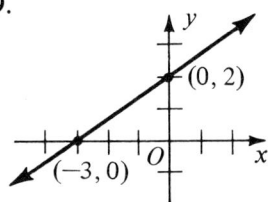

10. (−3, −1), (−1, −2)

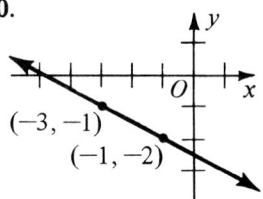

11.

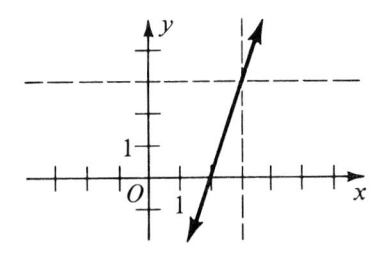

12.

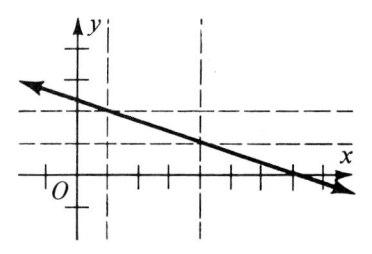

13.

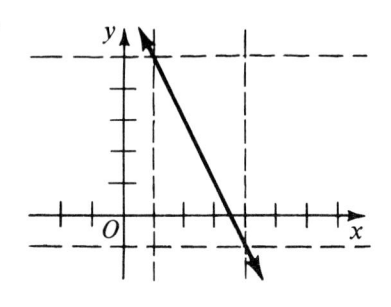

14.

LINES AND THEIR EQUATIONS

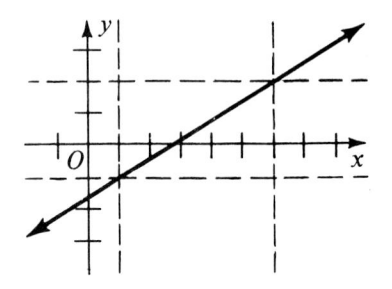

WRITTEN EXERCISES

Find the slope of the line through each pair of points. If the slope is not defined, so state.

A 1. (3, 1), (9, 4)   2. (5, 8), (7, 4)   3. (−1, 5), (3, −7)   4. (−4, −2), (6, 3)
  5. (0, 3), (−4, 0)   6. (−2, 0), (0, −5)   7. (−6, 3), (−6, −1)   8. (−2, 1), (5, 1)

Graph the given points and state whether or not quadrilateral ABCD is a parallelogram. Then use slopes to show that your answer is correct.

9. A(7, 2), B(5, −2), C(1, −3), D(2, 1)   10. A(−3, 0), B(2, −2), C(6, 1), D(1, 3)
11. A(−1, −3), B(1, −2), C(4, 4), D(2, 3)   12. A(1, −2), B(7, 6), C(5, 9), D(2, 5)

Find the slope (if it exists) for each line whose equation is given.

13. $3x + 4y = 7$ 14. $2x + 5y = 6$ 15. $4x - 3y = 1$
16. $x - 6y = 6$ 17. $3x = -2$ 18. $y = \frac{1}{2}$
19. $3x = 2y - 5$ 20. $4y = -x + 2$ 21. $6x = -3(y + 4)$
22. $4(x - 3) = 6y$ 23. $5(x - y) = 1$ 24. $-2(x + y) = 5$

Find the slope of the line through each pair of points.

B 25. $\left(0, \frac{1}{2}\right), \left(3, \frac{1}{2}\right)$ 26. $\left(\frac{1}{2}, 7\right), \left(\frac{3}{2}, 5\right)$ 27. $\left(-\frac{1}{3}, 3\right), \left(\frac{5}{3}, -1\right)$
28. $\left(7, -\frac{7}{2}\right), \left(-5, \frac{1}{2}\right)$ 29. $\left(\frac{1}{4}, \frac{1}{3}\right), \left(\frac{3}{4}, \frac{2}{3}\right)$ 30. $\left(-\frac{1}{2}, 3\right), \left(2, \frac{1}{2}\right)$

31. If (1, 2) and (3, r) are two points on a line with a positive slope, what can you say about r?

32. If (−2, 0) and (1, s) are two points on a line with a negative slope, what can you say about s?

Graph the given points. Then find the slopes of $\overleftrightarrow{AB}$ and $\overleftrightarrow{BC}$ and state whether A, B, and C are collinear.

33. A(2, −5), B(4, −8), C(6, −11)   34. A(−3, −1), B(1, 1), C(4, 3)
35. A(−2, 4), B(1, −1), C(3, −5)   36. A(−1, −4), B(−6, 1), C(−2, −3)

EXAMPLE Graph the line through (3, −1) with slope −2.

SOLUTION A vertical change of 2 corresponds to a horizontal change of −1. Start at (3, −1) and go 2 units up and 1 unit left.
A vertical change of −2 corresponds to a horizontal change of 1. To find another point on the line, start at (3, −1) and go 2 units down and 1 unit right.

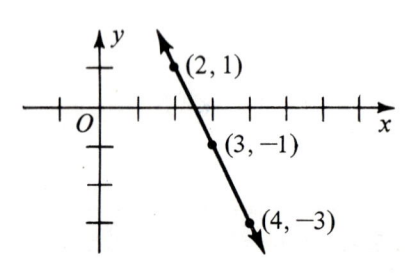

Graph the line with the given slope and passing through the given point. (Refer to the example at the bottom of page 484.)

37. $\frac{4}{5}$; $(0, 0)$ 38. $-\frac{2}{3}$; $(1, 3)$ 39. $-\frac{1}{4}$; $(3, -1)$ 40. $\frac{3}{2}$; $(-2, -1)$

C 41. a. Graph the points $A(-5, 0)$, $B(4, 12)$, and $C(20, 0)$.
  b. Find AB, BC, and AC. (Use the Pythagorean Theorem if necessary.)
  c. What is the measure of $\angle ABC$? Explain.
  d. Find the slopes of $\overleftrightarrow{AB}$ and $\overleftrightarrow{BC}$.
  e. What is the product of the slopes of $\overleftrightarrow{AB}$ and $\overleftrightarrow{BC}$?

42. Repeat Exercise 41 for the points $A(-1, 0)$, $B(0, -2)$, and $C(4, 0)$.

43. Use your results from Exercises 41 and 42 to complete the conjecture:
Two nonvertical lines are __?__ if and only if __?__.

Section 13-4 SLOPE-INTERCEPT FORM

It is often convenient to use an equation of a line that is written in a form other than $Ax + By = C$. For example, $y = 3x + 4$ is an equation of a line since this equation is equivalent to $-3x + y = 4$. Which equation would you prefer to use to find the value of y that corresponds to a given value of x?

EXAMPLE   Graph $y = 3x + 4$.

SOLUTION $y = 3(0) + 4$
       $= 4$
  $y = 3(1) + 4$
       $= 7$
  $y = 3(-1) + 4$
       $= 1$

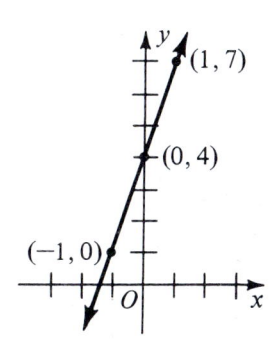

x	y
0	4
1	7
-1	1

Notice in the example that the slope of the line with equation $y = 3x + 4$ is

$$\frac{7 - 4}{1 - 0}, \text{ or } 3,$$

and the y-intercept is 4.

In general, a line with equation

$$y = mx + b$$

has slope m and y-intercept b. For this reason, the form $y = mx + b$ is called the *slope-intercept* form of the equation of a line.

LINES AND THEIR EQUATIONS   485

EXAMPLE Find the slope and y-intercept of the line whose equation is $3x + 2y = 6$.

SOLUTION Rewrite the equation in slope-intercept form:
$$3x + 2y = 6$$
$$2y = -3x + 6$$
$$y = \frac{1}{2}(-3x + 6)$$
$$y = -\frac{3}{2}x + 3$$

Therefore the slope is $-\frac{3}{2}$. The y-intercept is 3.

EXAMPLE Find an equation of the line with slope -4 and passing through the point $(0, -5)$.

SOLUTION Since the slope is -4, we have $m = -4$. Because the y-intercept is -5, we have $b = -5$. Substituting in the equation $y = mx + b$, we find that an equation of the line is $y = -4x - 5$.

EXAMPLE Find an equation of the line with slope 2 and passing through the point $(3, -1)$.

SOLUTION Since the slope is 2, an equation of the line has the form
$$y = 2x + b.$$
To find the value of b, use the fact that the point $(3, -1)$ lies on the line. Substitute 3 for x and -1 for y:
$$-1 = 2(3) + b$$
$$-1 = 6 + b$$
$$-7 = b$$

Thus, an equation is
$$y = 2x - 7.$$

EXAMPLE Find an equation of the line passing through the points $(1, 4)$ and $(-5, 7)$.

SOLUTION First find the slope of the line:
$$\text{slope} = \frac{7 - 4}{-5 - 1} = \frac{3}{-6} = -\frac{1}{2}$$

Therefore $m = -\frac{1}{2}$ and the equation has the form:
$$y = -\frac{1}{2}x + b$$

To find the value of b, substitute for x and y the coordinates of *either* point. If you substitute the coordinates of $(-5, 7)$, you obtain:
$$7 = -\frac{1}{2}(-5) + b$$
$$7 = \frac{5}{2} + b$$
$$\frac{9}{2} = b$$

486 CHAPTER 13

Thus,
$$y = -\frac{1}{2}x + \frac{9}{2}$$
is an equation of the line containing the given points.

EXAMPLE Find an equation of the line with y-intercept -5 and parallel to the line with equation $y = 3x + 2$.

SOLUTION Any line parallel to the line with equation $y = 3x + 2$ has slope 3. An equation of the line with slope 3 and y-intercept -5 is
$$y = 3x - 5.$$

ORAL EXERCISES

State an equation of the line with the given slope and y-intercept.

1. slope -2; y-intercept 5
2. slope $\frac{3}{4}$; y-intercept $\frac{1}{4}$
3. slope 0; y-intercept 3
4. slope -7; y-intercept 0

State the slope and the y-intercept of the line whose equation is given.

5. $y = \frac{1}{3}x + 2$
6. $y = -5x - \frac{1}{2}$
7. $y = -2$
8. $y + 3 = 2x$
9. $3x + y = 4$
10. $2y = 6x - 10$

11. Can the equation of a horizontal line be expressed in slope-intercept form?
12. Can the equation of a vertical line be expressed in slope-intercept form?
13. Describe the relationship of the lines whose equations are $y = -5x + 1$ and $y = -5x - 4$.

WRITTEN EXERCISES

Find the slope and the y-intercept of the line whose equation is given.

A
1. $y = -3x + 2$
2. $y = x + 6$
3. $4y = 8x - 5$
4. $3y = -5x - 9$
5. $2x - y = 5$
6. $3x + 2y = 0$

Write in the form $Ax + By = C$ an equation of the line with the given slope and y-intercept.

7. slope 2; y-intercept -1
8. slope -1; y-intercept -3
9. slope 0; y-intercept -2
10. slope 6; y-intercept 0
11. slope -3; y-intercept 5
12. slope 1; y-intercept 1

Find an equation in slope-intercept form of the line with the given slope and passing through the given point.

13. $3; (0, -4)$
14. $-1; (0, 5)$
15. $-\frac{1}{2}; (0, 3)$
16. $\frac{3}{4}; \left(0, -\frac{1}{4}\right)$

17. $0; (0, 6)$
18. $\frac{5}{4}; (0, 0)$
19. $2; (1, 7)$
20. $3; (-1, 5)$

21. $\frac{1}{2}; (4, -3)$
22. $-\frac{2}{3}; (-6, -1)$
23. $-\frac{3}{2}; (8, -2)$
24. $\frac{2}{5}; (-10, 3)$

Find an equation in slope-intercept form of the line parallel to the given line and passing through the given point.

25. $y = -x + 7; (0, -2)$
26. $y = 3x; (2, -4)$
27. $y = -2x + 1; (-3, 5)$

28. $y = \frac{1}{2}x + 6; (8, -1)$
29. $y = \frac{2}{3}x - \frac{5}{3}; (2, -1)$
30. $y = -\frac{3}{2}x - 1; (3, -4)$

Find an equation in slope-intercept form of the line passing through the two given points.

B 31. $(0, 1), (4, 7)$
32. $(0, -2), (-5, 3)$
33. $(0, -3), (4, -1)$

34. $(0, 2), (-1, -3)$
35. $(3, 6), (-1, 6)$
36. $(-2, -4), (10, -4)$

37. $(2, 5), (6, 7)$
38. $(2, -1), (3, -3)$
39. $(3, -2), (6, 0)$

40. $(-4, 6), (-2, 1)$
41. $(1, 5), (3, 8)$
42. $(-1, 4), (2, 3)$

43. Find an equation in slope-intercept form of the line that passes through the origin and is parallel to the line through $(5, -2)$ and $(-1, 6)$.

44. Find an equation in slope-intercept form of the line that passes through $(3, -4)$ and is parallel to the line through $(-1, 6)$ and $(2, 9)$.

EXAMPLE Graph the equation $y = \frac{3}{2}x + 1$.

SOLUTION Since the y-intercept is 1, the line passes through the point $(0, 1)$. Because the slope is $\frac{3}{2}$, a vertical change of 3 corresponds to a horizontal change of 2. Thus, the point $(2, 4)$ also lies on the line.

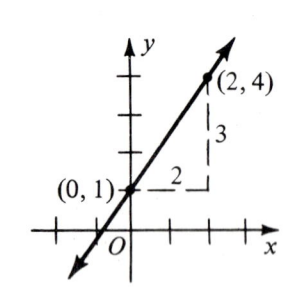

Graph each equation by graphing the y-intercept and then using the slope to determine another point on the line.

45. $y = \frac{1}{2}x - 3$
46. $y = -2x + 7$
47. $y = -\frac{3}{2}x - 1$
48. $y = \frac{3}{4}x + 1$

49. $y = \frac{4}{5}x$
50. $y = -\frac{2}{5}x$
51. $y = -3x + 5$
52. $y = \frac{1}{3}x - 4$

Section 13-5 GRAPHING LINEAR INEQUALITIES IN TWO VARIABLES

In Chapter 4 you learned how to graph on a number line inequalities in one variable. In this section you will learn how to graph in a coordinate plane a linear inequality in two variables.

The line with equation $y = 2$ separates the points in the coordinate plane into three sets. The y-coordinate of each point determines where the point lies.

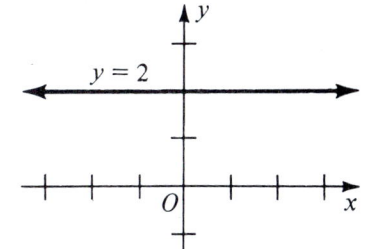

(1) If $y = 2$, the point is on the line.
(2) If $y > 2$, the point is *above* the line.
(3) If $y < 2$, the point is *below* the line.

The points above the line, that is, the points for which $y > 2$, form an *open half-plane*. The half-plane is called open because the *boundary line* with equation $y = 2$ is not included.

To graph the inequality $y > 2$, you shade the part of the coordinate plane lying above the line with equation $y = 2$, as shown on the left below. The boundary line is dashed to indicate that it is not included in the graph.

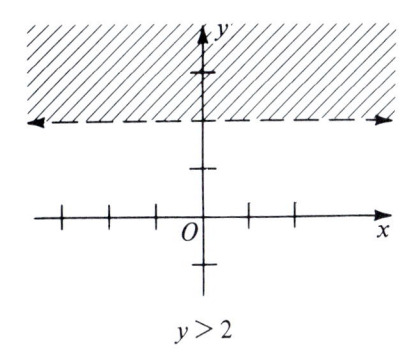

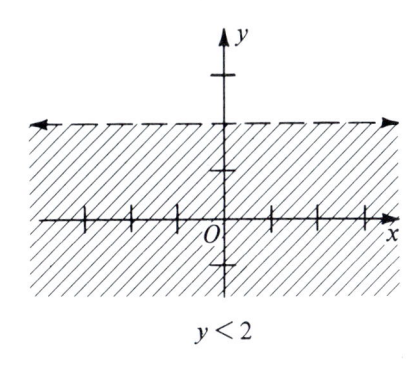

$y > 2$

$y < 2$

The graph in the coordinate plane of the inequality $y < 2$ is shown on the right above.

The graph of an inequality such as $y \geq 2$ includes the boundary line and is called a *closed half-plane*. The boundary line is drawn as a solid line to indicate that it is included in the graph.

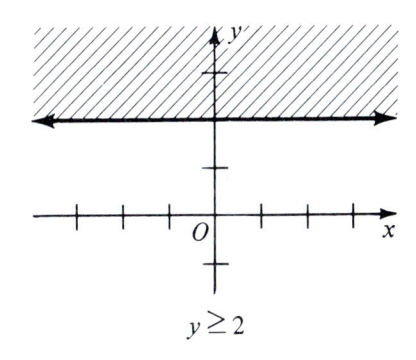

$y \geq 2$

LINES AND THEIR EQUATIONS   489

To graph an inequality whose boundary line is not parallel to the *x*-axis, it is often convenient to use the slope-intercept form of the equation of the boundary line.

EXAMPLE Graph $2y + x \leq 6$.

SOLUTION The equation of the boundary line is $2y + x = 6$. Rewrite the equation in slope-intercept form:

$$y = -\frac{1}{2}x + 3$$

Graph the boundary line as a solid line.

The boundary line divides the plane into 3 parts. For every point in the plane it must be true that its coordinates satisfy either

(1) $y = -\frac{1}{2}x + 3$ or

(2) $y < -\frac{1}{2}x + 3$ or

(3) $y > -\frac{1}{2}x + 3$.

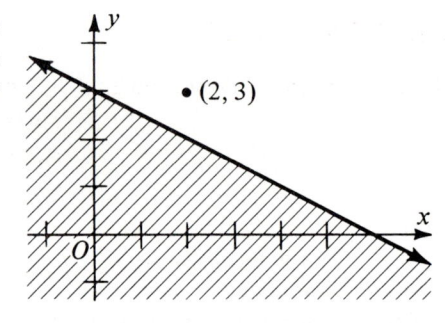

To decide which half-plane you should shade, choose a point not on the boundary line, say (2, 3), and determine which of the two inequalities it satisfies. Because

$$3 > -\frac{1}{2}(2) + 3,$$

the point (2, 3) does *not* satisfy the inequality $y < -\frac{1}{2}x + 3$. Therefore, you shade the half-plane that does *not* contain (2, 3), that is, the half-plane *below* the line.

EXAMPLE Graph $y - 2x > -4$.

SOLUTION Rewrite the inequality as $y > 2x - 4$. Draw the graph of $y = 2x - 4$ as a dashed line. Choose a point, say (0, 0). Since

$$0 > 2(0) - 4,$$

the point (0, 0) does satisfy the inequality $y > 2x - 4$. Therefore you shade the half-plane that contains (0, 0).

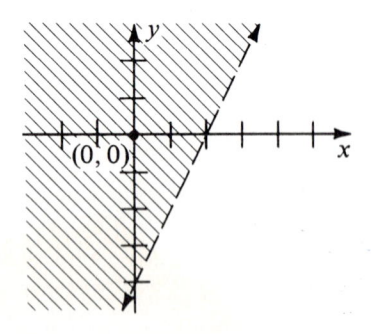

In general, if the boundary line does not contain the origin, then (0, 0) is a convenient point to use to determine which half-plane to shade.

EXAMPLE Graph the inequality $x < 3$ in the coordinate plane.

SOLUTION You cannot write this inequality in slope-intercept form, but it should be clear that the graph consists of all points to the *left* of the line with equation $x = 3$.

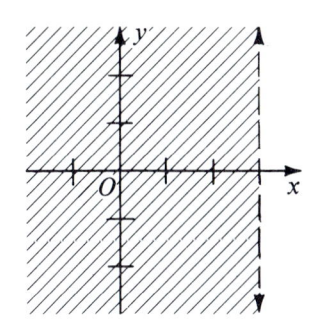

ORAL EXERCISES

State which equation or inequality, A, B, or C, the given ordered pair satisfies.

$$A: y = -2x + 3$$
$$B: y > -2x + 3$$
$$C: y < -2x + 3$$

1. $(0, 0)$	**2.** $(0, 3)$	**3.** $(-1, 2)$	**4.** $(2, 1)$
5. $(1, 1)$	**6.** $(-2, -5)$	**7.** $(6, -4)$	**8.** $(-1, 5)$

a. State whether the graph of the given inequality is an open half-plane or a closed half-plane.

b. State whether you would shade the region *above* or *below* the boundary line when you graph the inequality.

9. $y > 2x - 1$ **10.** $x + y < 2$ **11.** $y \leq \frac{1}{2}x + 3$

12. $-x + 2y \geq 8$ **13.** $x + y < 5$ **14.** $2x + 3y \geq -6$

15. $-x - y \leq 4$ **16.** $x < -2 - y$ **17.** $3x \geq -2y + 4$

WRITTEN EXERCISES

State whether the given ordered pair satisfies the given inequality.

A **1.** $3x + 2y < 5;\ (1, 1)$ **2.** $4x - 5y > 10;\ (2, -1)$
3. $-5x + 2y \geq 12;\ (-3, -1)$ **4.** $-x - 3y \leq 9;\ (-3, -2)$
5. $2y \leq x - 5;\ (0, 5)$ **6.** $y > -3x + 6;\ (-2, 3)$

Graph each inequality.

7. $y \geq 6$ **8.** $y < -1$ **9.** $x \leq 2$

10. $x > -1$ **11.** $y > x + 2$ **12.** $y < x - 2$

13. $y \leq 2x - 1$ **14.** $y \geq \frac{1}{2}x + 3$ **15.** $3x + y < 6$

LINES AND THEIR EQUATIONS

Graph each inequality.

16. $-2x + y \geq 8$
17. $2x - y > 5$
18. $3x - y < 7$
19. $-x - y \geq 4$
20. $-4x + y \leq -2$
21. $-x - 2y \geq 8$
22. $x - 3y < -12$
23. $\frac{1}{3}y > -4$
24. $-x + 2 \leq \frac{1}{2}y$

B 25. $3x + 4y < 6$
26. $-5x + 2y > 8$
27. $4x - 3y \geq 9$
28. $2x - 3y \leq 15$
29. $4x - 5y < 20$
30. $7x - 4y > 30$

Graph the set of points that satisfy both inequalities.

C 31. $x > 0$ and $y > 0$
32. $x < 0$ and $y > 0$
33. $x \geq 2$ and $y < 1$
34. $x < -3$ and $y \geq 4$
35. $-3 \leq x < 2$
36. $2 < y \leq 5$

37. The points in Quadrant II are described by which two inequalities?
38. The points in Quadrant III are described by which two inequalities?

SELF-TEST 2

1. Find the slope of the line through the points $(6, -3)$ and $(-2, 5)$.   Section 13-3
2. Use slopes to determine whether the quadrilateral whose vertices are $A(-2, -5)$, $B(7, -1)$, $C(2, 3)$, and $D(-6, -1)$ is a parallelogram.

Find the slope (if it exists) for each line whose equation is given.

3. $4x - 7y = 28$
4. $6x = -5$

5. Find the slope and y-intercept of the line with equation $5x + 4y = 8$.   Section 13-4

Find an equation in slope-intercept form of the line described.

6. Through point $(6, -1)$ with slope $-\frac{2}{3}$.
7. Through point $(-2, 5)$ and parallel to the line with equation $2y + 3x = 18$.
8. Through points $(-3, -5)$ and $(2, 5)$.

Graph each inequality.

9. $y \leq x - 2$
10. $y > \frac{3}{2}x + 5$ Section 13-5
11. $4x - 5y \leq 10$
12. $4y + 3x < -12$

Computer Exercises

1. The following program prints out the slope-intercept form of the equation of the line determined by two given points (x_1, y_1) and (x_2, y_2).

  a. Note that the program treats $x_1 = x_2$ as a special case. Why is this important? Why must line 30 precede line 40?

```
10    PRINT "INPUT X1,Y1,X2,Y2 IN THAT ORDER.";
20    INPUT X1,Y1,X2,Y2
30    IF X1=X2 THEN 120
40    LET M=(Y2-Y1)/(X2-X1)
50    LET B=Y1-M*X1
60    PRINT "THE EQUATION IN SLOPE-INTERCEPT FORM IS ";
70    PRINT "Y = ";M;"X ";
80    IF B<0 THEN 100
90    PRINT "+";
100   PRINT B
110   GOTO 130
120   PRINT "THE EQUATION IS X = ";X1
130   END
```

  b. Copy and RUN the program using the points $(3, 2)$ and $(1, 0)$.

  c. RUN the program using the points $(-2, 3)$ and $(-2, 4)$.

  d. RUN the program using the points $(1, -5)$ and $(5, -5)$.

2. a. Write a program to print out the slope-intercept form of the equation of a line given the equation in the general form $Ax + By = C$. Begin the program by inputting A, B, and C. RUN the program for the line with equation $3x + 2y = 6$.

  b. Why is it important to test for $B = 0$ in your program in part (a)? If necessary, alter your program in part (a) so that inputting $B = 0$ will not produce an error statement.

3. The following program plots a series of points on a line, given two points, (x_1, y_1) and (x_2, y_2), on the line. If your computer's output device is a video screen, rather than a printer, only a fixed number of lines of output are visible at any moment. When your output is a graph or chart, you will want to use as much of the screen as possible for the final display. Normally, when a program finishes running, the word "READY" and possibly some other symbol are printed beneath the output. To prevent the computer from printing these characters, you must make sure that the program never stops running. To do this, insert a single-statement infinite loop. For the following program use 265 GOTO 265.

LINES AND THEIR EQUATIONS

If you use the method just described, how should you end the program when you have finished inspecting the output on the screen? Pressing the "Break" key or the "CTRL" and "C" keys simultaneously performs this function on some computers. To find out how to terminate a program on your computer, consult your user's manual.

(*Note:* There are four spaces in line 260 between the opening quotation marks and the −6.)

a. Copy and RUN the program using the points (0, 1) and (3, 4).

```
10   PRINT "INPUT X1,Y1,X2,Y2 IN THAT ORDER."
20   INPUT X1,Y1,X2,Y2
30   IF X1=X2 THEN 60
40   LET M=(Y2-Y1)/(X2-X1)
50   LET B=Y1-M*X1
60   FOR Y=6 TO -6 STEP -1
70   PRINT Y;
80   IF Y1 <> Y2 THEN 160
90   IF Y <> Y1 THEN 230
100  PRINT TAB(5);
110  FOR I=1 TO 13
120  PRINT "* ";
130  NEXT I
140  PRINT "Y=";Y
150  GOTO 240
160  IF X1 <> X2 THEN 190
170  LET X=X1
180  GOTO 210
190  LET X=(Y-B)/M
200  IF X<-6 OR X>6 THEN 230
210  PRINT TAB(INT(2*X+17.5));"* (";X;",";Y;")"
220  GOTO 240
230  PRINT
240  NEXT Y
250  PRINT
260  PRINT "    -6-5-4-3-2-1 0 1 2 3 4 5 6"
270  END
```

b. RUN the program using the points (1, −3) and (−2, 3).

c. RUN the program using the points (−2, 4) and (−2, 1).

d. RUN the program using the points (4, 3) and (−6, 3).

CHAPTER REVIEW

1. Give the coordinates of each point shown. *Section 13-1*

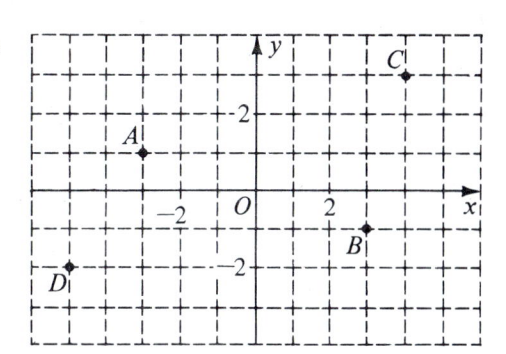

2. Graph each of the following points.
   a. $E(-3, 5)$
   b. $F(2, 4)$
   c. $G(1, -4)$
   d. $H(-2, -3)$

Find the distance between each pair of points.

3. $(3, -2), (3, 8)$
4. $(-4, 2), (-8, 2)$

For each line whose equation is given, find the x-intercept and the y-intercept.

5. $5x - 2y = -10$
6. $3x + 4y = 1$ *Section 13-2*

Find the coordinates of at least three points on the graph of each equation, and then graph the equation.

7. $3x + y = 6$
8. $2x = 3y$
9. $y = -1$
10. $3y - 2x = 12$

Find the slope of the line through each pair of points.

11. $(2, -3), (-2, 4)$
12. $(-2, 5), (7, 8)$
13. $\left(0, \frac{1}{2}\right), (-2, 0)$ *Section 13-3*

14. Given the points $R(0, -2)$, $S(10, -3)$, $T(5, 5)$, and $U(-5, 6)$, is $RSTU$ a parallelogram?

15. Find the slope and y-intercept of the line with equation $2x - 3y = 12$. *Section 13-4*

Find an equation in slope-intercept form for each line described.

16. Through the point $(10, -5)$ and with slope $\frac{2}{5}$

17. Through the points $(-4, 2)$ and $(2, 5)$

18. Through the point $(8, 4)$ and parallel to the line with equation $3x - 4y = 20$

Graph each inequality.

19. $4x - 3 < y$
20. $2y + 3x \leq 6$ *Section 13-5*
21. $y \geq -2$
22. $5y + 10 > 3x$

LINES AND THEIR EQUATIONS

CHAPTER TEST

Graph each of the following points.

1. $A(0, 6)$
2. $B(3, -2)$
3. $C(-4, -1)$

4. Find the distance between the points $(-3, 6)$ and $(-3, -2)$.

Find the coordinates of at least three points on the graph of each equation, and then graph the equation.

5. $x = 2y$
6. $2x - y = 4$
7. $x = 4$
8. $3x + 2y = 18$

Find the slope of the line through each pair of points.

9. $(-4, -1), (2, 6)$
10. $(-3, 2), (-7, 6)$

11. Given the points $W(-5, -2)$, $X(-3, 2)$, $Y(3, 4)$, and $Z(1, 0)$, is $WXYZ$ a parallelogram?

Find the slope (if it exists) for each line whose equation is given.

12. $5x + 4y = 16$
13. $2x = -5$
14. $3(x - 2) = 4y$

Find an equation in slope-intercept form for each line described.

15. Through the point $(-4, 5)$ and with slope $\dfrac{3}{2}$

16. Through the point $(6, -2)$ and parallel to the line with equation $2y + 3x = 10$

17. Through the points $(-4, -1)$ and $(2, 6)$

Graph each inequality.

18. $y \leq -3$
19. $2y - x < 4$
20. $5y - 3x \geq 20$

Application — Line of Best Fit

When statisticians gather data, they look for relationships that can be used to make predictions about future data. For example, assume you are studying the increasing height of the United States population. You have gathered information about people's adolescent and adult heights. There seems to be a clear relationship between these two factors — in other words, that adolescent height is a good predictor of adult height. If this were true, you could make a good prediction now of how tall all 14-year-olds will be when they grow up.

One way of testing your hypothesis is by plotting the data you have gathered on graph paper. You are interested in two factors, a person's height at age 14 and in adulthood, so assign one factor to each axis. You can then plot a pair of heights as a point (x, y) on the graph. The more people you gather data from, the more points you will have on your graph. If there is a relationship between a person's height at age 14 and as an adult, the points will often cluster around a line, called the *line of best fit*. There is a mathematical equation for this line, but you can often draw it quite well "by eye." From this line you can predict values that you did not gather. For example, you can predict how much a teenager who is 156 cm tall will grow. You can also predict how much a teenager who is 170 cm tall will grow by extending your line and the x-axis to the right.

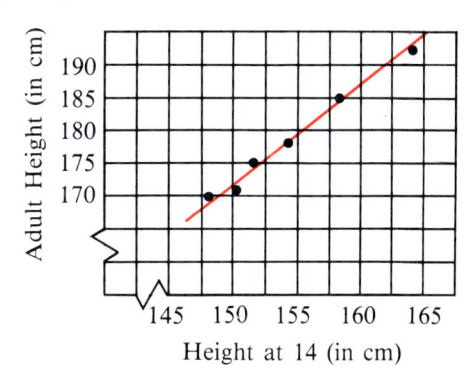

Exercises

You are studying the relationship between people's heart rates and the time they spend exercising.

1. Gather the following data from 20 classmates and members of your family: their heart rate per minute and their average weekly time spent exercising. You can figure out heart rate by finding a person's wrist pulse and counting the number of beats you feel in 60 seconds.

2. For each person, plot a point (x, y) on a graph such that the x-coordinate is the number of hours per week spent exercising and the y-coordinate is the number of beats per minute.

3. On the graph, draw the line that seems to fit the points of the graph most closely. If the plotted points do not cluster about the line, you probably did not find a causal relationship between a person's exercise time and heart rate. If you consider your line a "good fit," try to predict heart rates from exercise times that are not on your graph.

LINES AND THEIR EQUATIONS

VOCABULARY REVIEW

Be sure you understand the meaning of these terms:

coordinate plane, p. 472
origin, p. 472
coordinate axes, p. 472
quadrant, p. 472
ordered pair, p. 472
x-coordinate (abscissa), p. 472
y-coordinate (ordinate), p. 472

graph in a plane, p. 475
linear equation in two variables, p. 476
x-intercept, p. 477
y-intercept, p. 477
slope, p. 480
open half-plane, p. 489
closed half-plane, p. 489

MIXED REVIEW

Find an equation in slope-intercept form for each line.

1. The line through $(-2, 4)$ that is parallel to the line with equation $y = -1$.

2. The line through $(-2, 4)$ and $(7, -8)$.

3. If the measures of the angles of a triangle are in the ratio $2:3:4$, find the measure of the largest angle.

4. The statement $\sim p \rightarrow q$ is false only when p is __?__ and q is __?__.

5. One card is drawn from an ordinary deck. Find the probability that:
   a. the card is not a king or queen   b. the card is a red 7

6. Simplify: a. $\dfrac{\sqrt{3} + \sqrt{2}}{\sqrt{3} - \sqrt{2}}$ b. $\sqrt{48} + \dfrac{1}{\sqrt{12}}$

7. If two sides of a right triangle are each 3 cm long, find the length of the third side.

8. If $\angle AOB$ is a central angle of $\odot O$ with radius 6 and $AB = 6\sqrt{3}$, find the measures of $\angle BAO$ and $\angle AOB$.

9. Two cards are drawn at random from an ordinary deck of cards. What is the probability that both are hearts if the cards are drawn with replacement? without replacement?

10. Given the data 1, 3, 5, 5, 6, 7, 9, 12, find the mode, the median, the mean, and the third quartile.

11. A game consists of tossing two coins. A player who gets two heads wins 15¢, but a player who does not must pay 1¢. How should a player expect to do in the long run?

12. Name the coordinates of three points on the graph of $5x - 3y = 10$ and graph the equation.

13. A guitar that usually costs $105 is on sale for 20% off the regular price. Find the sale price.

14. If $\triangle ABC \sim \triangle RST$, $AB = 10$, $BC = 14$, $AC = 18$, and $TR = 12$, find:
    a. the perimeter of $\triangle RST$
    b. the ratio of the areas of $\triangle ABC$ and $\triangle RST$
15. Find the area of the shaded region in the figure at the right.
16. Solve: a. $25x^2 - 100 = 0$ b. $x^2 - 3x + 1 = 0$
17. Construct a group of five numbers with mean 72.
18. Factor: $7y^2 - 8y - 12$
19. Graph in the coordinate plane: a. $x > -2$ b. $3x - y \leqslant 6$
20. Given the diagram at the right, explain why each statement is true.
    a. $\triangle ABC \cong \triangle ADC$ b. $AB = AD$
21. A true-false quiz consists of four questions. How many different sets of answers are there? (*Hint:* Use the Counting Principle.)
22. The slope of a __?__ line is not defined.
23. Helene kept this record of her running times in minutes for the 5-kilometer run: 20, 19, 18, 19, 19, 20, 19, 18, 17, 18. Make a frequency table, a frequency polygon, and a cumulative frequency histogram for the data.
24. Find the slope of the line with equation $3x + 5y = 8$.
25. How many 4-letter permutations can be formed from the letters of the word MUSICAL?
26. A cone has height 4 and volume 108π. Find its radius.
27. Which is greater, $0.\overline{36}$ or $\dfrac{9}{25}$?
28. Simplify: a. $(9a - 12a^2 + 2) - (a^2 - 5a)$ b. $\dfrac{8b^5 - 4b^3 + 2b^2}{2b^2}$
29. Solve: $5 - 1.2y \geqslant -19$
30. Graph: {even integers}
31. Construct a truth table for $(p \wedge q) \rightarrow \sim p$. Is the statement a tautology, a contradiction, or neither?
32. If B is the midpoint of $\overline{AC}$, $AB = 3x + 5$, and $AC = 8x - 2$, find the numerical value of BC.
33. A rectangle has length 16 and diagonals of length 20. Find its perimeter and its area.
34. A line that intersects a circle in two points is called a __?__.

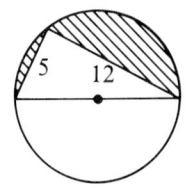

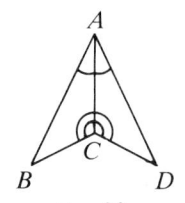

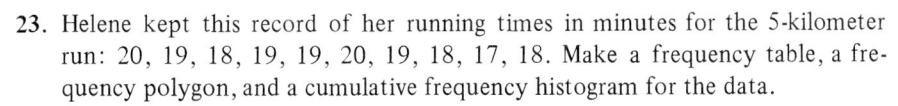

LINES AND THEIR EQUATIONS   499

The technique of *linear programming* involves finding and graphing systems of linear inequalities. Linear programming is used to aid decision-making regarding large-scale industrial processes, such as the manufacture of cars.

Systems of Linear Equations

14

The first known solution of systems of equations such as

$$y = ax - b$$
$$y = cx + d$$

(where a, b, c, and d are positive integers and $a \neq c$) is given in *Chiu Chang Suan Shu*, or *Nine Chapters on the Mathematical Arts*, written before 200 B.C. The general solution is the ordered pair

$$\left(\frac{b+d}{a-c}, \frac{ad+bc}{a-c}\right)$$

In this chapter you will learn how to use graphing, substitution, addition, and subtraction to solve systems of equations like the one above when a, b, c, and d are real numbers.

Section 14-1 USING GRAPHS TO SOLVE PAIRS OF LINEAR EQUATIONS

A *system of linear equations* is the conjunction of two (or more) linear equations. When you have found all the ordered pairs that satisfy every equation of the system you have *solved* the system of equations. As illustrated in the three following examples, a system may have one solution, no solution, or infinitely many solutions.

EXAMPLE Solve the following system by graphing: $x - y = -1$
$3x - y = 1$

SOLUTION To graph the system, first write the equations in slope-intercept form.

$$y = x + 1$$
$$y = 3x - 1$$

The lines intersect in one point. The coordinates of that point satisfy both equations. From the graph you can see that the point of intersection is $(1, 2)$.

Check Check your solution in both equations.

$x - y = -1$ $3x - y = 1$
$1 - 2 = -1$ $3(1) - 2 = 1$

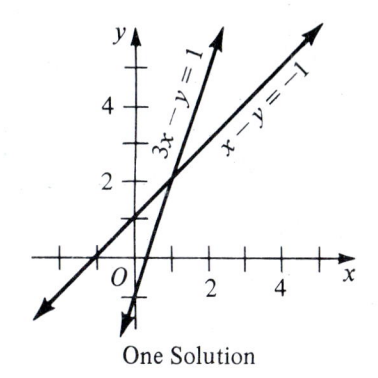

One Solution

As you may recall from Section 13-4, if two lines have equal slopes and different y-intercepts, then they do not intersect, but are parallel. The next example illustrates such a case.

EXAMPLE Solve the following system by graphing.
$$2x - y = 1$$
$$2x - y = 2$$

SOLUTION First write the equations in slope-intercept form.
$$y = 2x - 1$$
$$y = 2x - 2$$

The two lines are parallel since they have the same slope but different y-intercepts. Since the lines do not intersect, there is no ordered pair that satisfies both equations. Thus the solution set is ∅.

No Solution

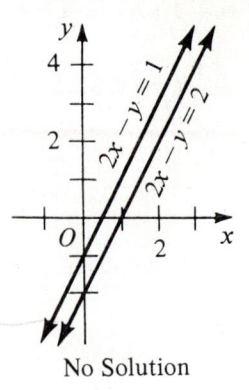

Two equations may represent the same line.

EXAMPLE Solve the following system by graphing.
$$-x + 2y = 1 \quad (1)$$
$$2x - 4y = -2 \quad (2)$$

SOLUTION To graph the system, write the equations in slope-intercept form.

$$-x + 2y = 1 \qquad\qquad 2x - 4y = -2$$
$$2y = x + 1 \qquad\qquad -4y = -2x - 2$$
$$y = \frac{1}{2}x + \frac{1}{2} \qquad\qquad y = \frac{1}{2}x + \frac{1}{2}$$

You do not have to graph these equations to solve the system if you can recognize that equations (1) and (2) are equivalent; multiplying equation (1) by −2 gives you equation (2).

The graphs of equivalent equations coincide. Therefore, there are an infinite number of ordered pairs that satisfy both equations. The solution set can be written as:

$$\{(x, y): -x + 2y = 1\}.$$

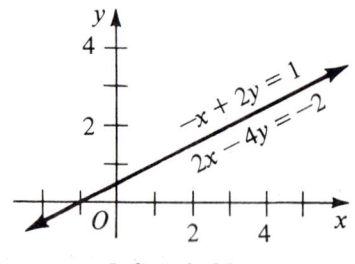

Infinitely Many Solutions

ORAL EXERCISES

Name three ordered pairs that satisfy the given equation.

1. $2x + y = 3$
2. $x - 3y = 6$
3. $y = 5x + 1$
5. $4y + 3x = 28$
6. $5y - 2x = 20$

State the solution of each system.

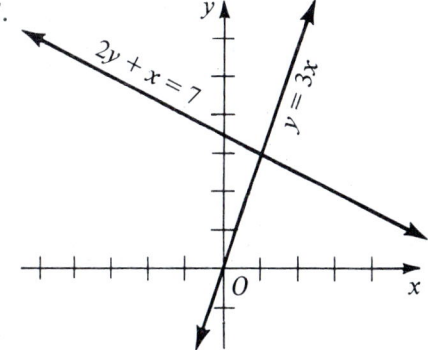

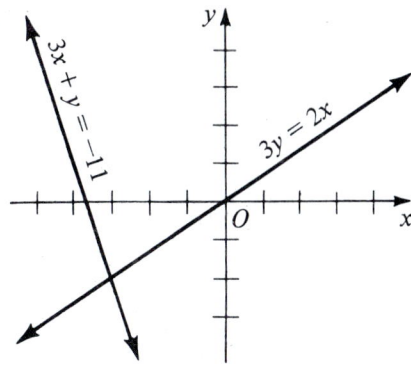

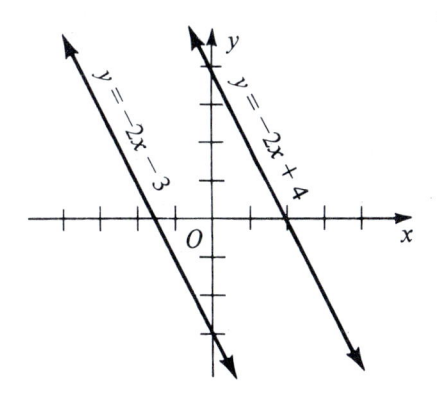

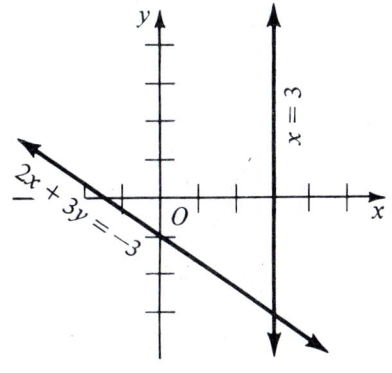

7. [Graph showing lines $2y + x = 7$ and $y = 3x$]

8. [Graph showing lines $3x + y = -11$ and $3y = 2x$]

9. [Graph showing lines $y = -2x - 3$ and $y = -2x + 4$]

10. [Graph showing lines $2x + 3y = -3$ and $x = 3$]

11. $x = 1, y = 2$

12. $x + y = 1$
    $2x + 2y = 2$

WRITTEN EXERCISES

Solve each system by graphing. If the system has exactly one solution, check your answer.

A
1. $x + y = 2$
   $x - y = 0$

2. $y = 2x - 1$
   $y = x - 1$

3. $x - 2y = 2$
   $y + 2x = 9$

4. $3x + y = -2$
   $-3x + y = -2$

5. $3x + 2y = 3$
   $x - y = -4$

6. $x + y = 3$
   $2x + 5y = 0$

7. $-x + 5y = 5$
   $x - 5y = 20$

8. $6x + y = 2$
   $6x + y = 6$

9. $x + 3y = -2$
   $-x - 3y = 2$

10. $2y - 3x = 8$
    $3x + 8 = 2y$

11. $x = 2y + 5$
    $-2x = -y + 2$

12. $x = 3y - 1$
    $-2x = y + 9$

SYSTEMS OF LINEAR EQUATIONS   503

Solve each system by graphing. Estimate the coordinates of the point of intersection to the nearest half unit. Check your solution for reasonableness.

B 13. $y = 2x - 1$
$x + y = 7$

14. $x - 3y = 1$
$2x - y = 1$

15. $2x + y = 9$
$0.5y - 2x = -3$

16. $x - 2y = 8$
$0.5x + y = -5$

17. $\frac{1}{3}x + \frac{1}{2}y = 0$
$\frac{3}{4}x - \frac{1}{3}y = -\frac{7}{2}$

18. $\frac{1}{2}x + \frac{1}{3}y = \frac{2}{3}$
$\frac{3}{4}x + \frac{1}{8}y = 1$

Graph each system of equations on the same set of axes. Find the area of the triangle whose vertices are the points of intersection.

C 19. $y = x + 3$
$y = -x + 5$
$y = 1$

20. $2x = -4$
$2x = y - 8$
$x + y = -4$

Section 14-2 USING THE SUBSTITUTION METHOD

Graphing a system of linear equations frequently gives you at best an approximate solution. Suppose you tried to solve the following system graphically.

$$2x + 3y = 6$$
$$x - 2y = 5$$

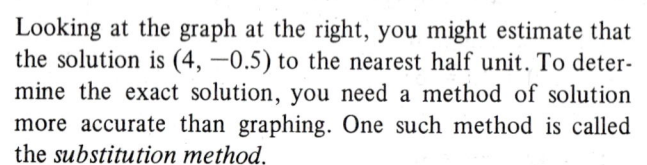

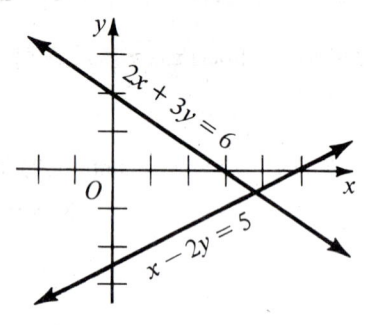

Looking at the graph at the right, you might estimate that the solution is $(4, -0.5)$ to the nearest half unit. To determine the exact solution, you need a method of solution more accurate than graphing. One such method is called the *substitution method*.

To solve a pair of linear equations in two variables by the *substitution method*, you proceed as follows:

1. Solve one equation for one of the variables.
2. Substitute the expression you obtained from Step 1 in the other equation.
3. Solve the resulting equation.
4. Find the corresponding value of the other variable.

EXAMPLE Solve the system using the substitution method.

$$x + 2y = 5 \quad (1)$$
$$3x + 2y = 19 \quad (2)$$

SOLUTION 1. Solve equation (1) for x (since its coefficient is 1).

$$x = 5 - 2y$$

2. Substitute $5 - 2y$ for x in equation (2).

$$3(5 - 2y) + 2y = 19$$

3. Solve the resulting equation.
$$3(5 - 2y) + 2y = 19$$
$$15 - 6y + 2y = 19$$
$$15 - 4y = 19$$
$$-4y = 4$$
$$y = -1$$

4. Find the corresponding value of x.
Substitute -1 for y in either equation, say equation (1).
$$x + 2(-1) = 5$$
$$x - 2 = 5$$
$$x = 7$$

The solution is $(7, -1)$.

Check Check your solution in *both* equations.

$$x + 2y = 5 \qquad\qquad 3x + 2y = 19$$
$$7 + 2(-1) = 7 - 2 \qquad 3(7) + 2(-1) = 21 - 2$$
$$= 5 \qquad\qquad\qquad = 19$$

The following example uses the substitution method to solve the system of equations given at the beginning of this section.

EXAMPLE   Solve the system using the substitution method.
$$2x + 3y = 6 \qquad (1)$$
$$x - 2y = 5 \qquad (2)$$

SOLUTION 1. Solve equation (2) for x. $\quad x = 5 + 2y$

2. Substitute this expression for x in equation (1).
$$2(5 + 2y) + 3y = 6$$

3. Solve.
$$10 + 4y + 3y = 6$$
$$10 + 7y = 6$$
$$7y = -4$$
$$y = -\frac{4}{7}$$

4. Find the corresponding value of x.
$$x - 2\left(-\frac{4}{7}\right) = 5$$
$$x + \frac{8}{7} = 5$$
$$x = 5 - \frac{8}{7} = \frac{35 - 8}{7} = \frac{27}{7}$$

The solution is $\left(\frac{27}{7}, -\frac{4}{7}\right)$.

(Example continued on next page.)

SYSTEMS OF LINEAR EQUATIONS   505

Check $\qquad 2x + 3y = 6 \qquad\qquad\qquad x - 2y = 5$

$$2\left(\frac{27}{7}\right) + 3\left(-\frac{4}{7}\right) = \frac{54}{7} - \frac{12}{7} \qquad \frac{27}{7} - 2\left(-\frac{4}{7}\right) = \frac{27}{7} + \frac{8}{7}$$
$$= \frac{42}{7} \qquad\qquad\qquad\qquad = \frac{35}{7}$$
$$= 6 \qquad\qquad\qquad\qquad\quad = 5$$

The next two examples show that a system may have no solution or an infinite number of solutions. Often you can determine this by inspection without going through all the steps of solving the system.

EXAMPLE Solve the system using substitution.

$$x + 3y = 8 \qquad (1)$$
$$x = 5 - 3y \qquad (2)$$

SOLUTION
1. Equation (2) is already solved for x.
2. Substitute this expression for x in equation (1).
$$(5 - 3y) + 3y = 8$$
3. Solve.
$$5 + 0 = 8$$
$$5 = 8$$

If, in the process of solving, you reach an obvious contradiction such as $5 = 8$, then the system has no solution. You can easily check that the graphs of the two equations are parallel lines.

EXAMPLE Solve the system using substitution.

$$y - 3x = 12 \qquad (1)$$
$$6x - 2y = -24 \qquad (2)$$

SOLUTION
1. Solve equation (1) for y.
$$y = 12 + 3x$$
2. $6x - 2(12 + 3x) = -24$
3. $6x - (24 + 6x) = -24$
$$6x - 24 - 6x = -24$$
$$-24 = -24$$

If, in the process of solving, you reach an identity such as $-24 = -24$ or $0 = 0$, then there are an infinite number of solutions. The solution set of this system is $\{(x, y): y - 3x = 12\}$. The graphs of the two equations are the same line.

ORAL EXERCISES

Solve each equation for the variable indicated.

1. $3x + y = 4$; y
2. $x + 4y = 2$; x
3. $2x = y + 1$; y
4. $x - 3y = 10$; x
5. $y - 2x = -5$; y
6. $5y - x = -1$; x
7. $-2x - y = 12$; y
8. $2x + 4y = 6$; x
9. $3y = 6x - 12$; x

WRITTEN EXERCISES

Use the substitution method to solve each system of equations.

A
1. $y = 2x + 1$
   $x + y = 7$

2. $x + y = 5$
   $x = 1 + y$

3. $2x + 3y = 11$
   $x + y = 6$

4. $2x + y = -1$
   $4x + y = -7$

5. $x + 5y = 1$
   $2x + 10y = 1$

6. $2x - y = 3$
   $4x - 2y = 3$

7. $x = y + 4$
   $y = x - 4$

8. $4x - 8y = 12$
   $x - 2y = 3$

9. $y = x$
   $2y = 5x + 6$

10. $x = -y$
    $5x + 3y = 14$

11. $x = 3y$
    $2(y + 1) = x + 1$

12. $y = -2x$
    $3y = 2(x + 4)$

13. $x + y = 2$
    $3x + 2y = -6$

14. $x + y = 3$
    $2x + 3y = -2$

15. $x - y = 1$
    $5x - 4y = -3$

16. $x - y = 1$
    $6x - 7y = 1$

17. $y - 2x = -1$
    $5x - 2y = -2$

18. $4x - y = 2$
    $3x - 2y = 9$

B
19. $4x + 5y = 3$
    $2x - 3y = 7$

20. $3x - 4y = 7$
    $5x - 2y = -7$

21. $4y - 3x + 10 = 0$
    $5x + 12y + 14 = 0$

22. $3x - 2y + 1 = 0$
    $15x - 7y + 17 = 0$

23. $\frac{1}{3}x + \frac{2}{3}y = \frac{1}{3}$
    $4x - 2y = 9$

24. $\frac{1}{4}x = \frac{1}{2}y$
    $3x + 4y = -15$

Assume that a and b are nonzero constants and that $a \neq c$ and solve the system.

C
25. $x + ay = b$
    $x + 2ay = b$

26. $ax - by = 3a$
    $ax + by = a$

27. $y = ax - b$
    $y = cx + d$

Section 14-3 USING THE ADDITION OR SUBTRACTION METHOD

As you probably discovered in Section 14-2, the substitution method is somewhat difficult to use if neither equation contains a variable with a coefficient of 1. This type of problem is much easier to solve if you use the *addition or subtraction method*.

To solve a pair of linear equations by the *addition or subtraction method*, you proceed as follows:

1. Eliminate one variable by *adding* or *subtracting* the equations.
2. Solve the resulting equation.
3. Substitute the value you obtained from Step 2 in either of the original equations.
4. Find the corresponding value of the other variable.
5. Check your solution in both *original* equations.

SYSTEMS OF LINEAR EQUATIONS

EXAMPLE Solve the system using the addition or subtraction method.
$$-x + y = 1 \quad (1)$$
$$3x - y = 1 \quad (2)$$

SOLUTION *Adding* equations (1) and (2) eliminates the variable y.
$$\begin{array}{r} -x + y = 1 \\ 3x - y = 1 \\ \hline 2x = 2 \\ x = 1 \end{array}$$

Substitute 1 for x in either of the two original equations to find the corresponding value of y.
$$-x + y = 1$$
$$-1 + y = 1$$
$$y = 2$$

The solution is $(1, 2)$. The check is left to you.

Sometimes you will not be able to eliminate either variable by adding or subtracting the *given* equations as you did in the above example. If the coefficients of one variable are not already equals or opposites, multiply both sides of the equation(s) by the appropriate constant(s). As a result, you will have replaced an equation by an equivalent equation. (See Section 3-4.) The new system of equations will have the same solutions as the original system. Then you can add or subtract the new equations.

EXAMPLE Solve the system using the addition or subtraction method.
$$2x + 3y = 2 \quad (1)$$
$$x + 4y = 6 \quad (2)$$

SOLUTION Adding or subtracting these equations will not immediately eliminate one of the variables. To *subtract* the equations, the coefficients of one of the variables must be *equal*. The simplest way to achieve this is to multiply both sides of equation (2) by 2. Then subtract equation (1).

$$2(x + 4y) = 2 \cdot 6 \qquad \begin{array}{r} 2x + 8y = 12 \\ 2x + 3y = 2 \\ \hline 5y = 10 \\ y = 2 \end{array}$$

Substitute 2 for y in equation (2) to find x.
$$x + 4 \cdot 2 = 6$$
$$x + 8 = 6$$
$$x = -2$$

The solution is $(-2, 2)$.

Check Check your solution in both *original* equations.
$$2x + 3y = 2 \qquad\qquad x + 4y = 6$$
$$2(-2) + 3(2) = -4 + 6 = 2 \qquad -2 + 4(2) = -2 + 8 = 6$$

EXAMPLE Solve the system using the addition or subtraction method.

$$2x + 5y = -1 \quad (1)$$
$$3x - y = -10 \quad (2)$$

SOLUTION To *add* the equations, the coefficients of one of the variables must be *opposites*. The simplest way to accomplish this is to multiply both sides of equation (2) by 5. Then add.

$$5(3x - y) = 5(-10)$$

$$\begin{aligned} 2x + 5y &= -1 \\ 15x - 5y &= -50 \\ \hline 17x &= -51 \\ x &= -3 \end{aligned}$$

Substitute -3 for x in equation (2) to find y.

$$\begin{aligned} 3(-3) - y &= -10 \\ -9 - y &= -10 \\ -y &= -1 \\ y &= 1 \end{aligned}$$

The solution is $(-3, 1)$. The check is left to you.

EXAMPLE Solve the system using the addition or subtraction method.

$$2x - 3y = 1 \quad (1)$$
$$-5x + 4y = 8 \quad (2)$$

SOLUTION There are many ways in which you can approach this problem. You have a choice of which variable to eliminate and which method to use, addition or subtraction. Since the signs of the variables in equation (2) are both opposite to those in equation (1), let's use the addition method. Solution I shows how to eliminate the variable x. Solution II shows how to eliminate y. You may use whichever method you prefer.

SOLUTION I

Multiply both sides of equation (1) by 5 and both sides of equation (2) by 2. Then add.

$$5(2x - 3y) = 5 \cdot 1$$
$$2(-5x + 4y) = 2 \cdot 8$$

$$\begin{aligned} 10x - 15y &= 5 \\ -10x + 8y &= 16 \\ \hline -7y &= 21 \\ y &= -3 \end{aligned}$$

$$\begin{aligned} 2x - 3(-3) &= 1 \\ 2x + 9 &= 1 \\ 2x &= -8 \\ x &= -4 \end{aligned}$$

SOLUTION II

Multiply both sides of equation (1) by 4 and both sides of equation (2) by 3. Then add.

$$4(2x - 3y) = 4 \cdot 1$$
$$3(-5x + 4y) = 3 \cdot 8$$

$$\begin{aligned} 8x - 12y &= 4 \\ -15x + 12y &= 24 \\ \hline -7x &= 28 \\ x &= -4 \end{aligned}$$

$$\begin{aligned} -5(-4) + 4y &= 8 \\ 20 + 4y &= 8 \\ 4y &= -12 \\ y &= -3 \end{aligned}$$

The solution is $(-4, -3)$. The check is left to you.

SYSTEMS OF LINEAR EQUATIONS

ORAL EXERCISES

State a number by which you would multiply both sides of one of the equations so that: **(a)** the coefficients of one of the variables would be opposites; **(b)** the coefficients of one of the variables would be equal.

1. $-x + 2y = 1$
   $2x + 3y = 5$
2. $2x + y = 2$
   $3x - 5y = -5$
3. $4x - 3y = 10$
   $2x + 5y = 12$
4. $3x + 2y = 8$
   $-7x + 6y = 5$
5. $2y + 6x - 3 = 0$
   $2x - y + 4 = 0$
6. $3y - 4x + 4 = 0$
   $4y + x - 3 = 0$

Describe what you would do to eliminate: **(a)** the variable x; **(b)** the variable y.

7. $4x + 4y = 3$
   $2x + 6y = 1$
8. $-2x + 2y = 7$
   $3x + 9y = 1$
9. $7x + 5y = 8$
   $3x - 2y = 1$

WRITTEN EXERCISES

Use the addition or subtraction method to solve each system of equations.

A
1. $x - y = 0$
   $x + y = 4$
2. $2x + 3y = 5$
   $4x - 3y = 1$
3. $3x + 2y = 3$
   $3x - y = 3$
4. $3x + 2y = 0$
   $-x + 2y = 8$
5. $4x - 3y = 18$
   $-4x + 5y = -22$
6. $2x - 5y = 3$
   $-x + 5y = 6$
7. $5x + 3y = 5$
   $4x + 3y = 10$
8. $5x + 2y = -7$
   $5x + 4y = 1$
9. $5x + 2y = 3$
   $15x + 7y = -2$
10. $3x + 4y = 4$
    $6x + 7y = 13$
11. $3x + 7y = -12$
    $5x + 14y = -13$
12. $5x - 6y = 92$
    $15x + y = 48$
13. $2x + 3y = -1$
    $5x - 6y = 38$
14. $5x + 12y = 8$
    $2x - 3y = -28$
15. $8x + 3y = 14$
    $-2x + 5y = 8$
16. $-3x + 2y = 7$
    $9x - 5y = -10$
17. $4x + 3y = 15$
    $2x - 9y = -3$
18. $5x + 12y = -21$
    $-10x + 3y = -39$

B
19. $2x + 5y = -1$
    $3x + 4y = 9$
20. $2x - 3y = -13$
    $-3x + 4y = 14$
21. $2x - 5y = 20$
    $3x + 2y = -27$
22. $3x + 4y = 12$
    $2x + 3y = 10$
23. $2x - 3y = 2$
    $6x - 9y = 10$
24. $-5x + 2y = 10$
    $2x + 3y = -23$
25. $6x + 8y = 4$
    $\frac{3}{2}x + 2y = 1$
26. $-3x + 2y = 2$
    $6x - 4y = 1$
27. $5x - 2y = -5$
    $7x - 5y = -51$
28. $17x - 3y = 9$
    $-51x + 9y = -27$
29. $-3x - 5y = 67$
    $4x - 3y = -12$
30. $4y - 3x = 21$
    $3y + 2x = 37$

Assume that a and b are nonzero constants and solve the system.

C
31. $3ax + 3by = b$
    $ax - by = 3b$
32. $ax + by = 3$
    $bx - ay = 0$
33. $bx + ay = 0$
    $-ax + by = 6$

510   CHAPTER 14

SELF-TEST 1

Solve each system of equations by graphing.

1. $y + x = 2$
   $y - x = -4$
2. $2x - y = 1$
   $x + 4 = 2y$
3. $2y - x = 6$
   $2y + x = 2$

Section 14-1

Solve each system of equations by substitution.

4. $x = -3y$
   $3x + 2y = 14$
5. $x + 3y = 2$
   $2x + 6y = 2$
6. $2y - x = 8$
   $y - 3 = x$

Section 14-2

Solve each system of equations by the addition or subtraction method.

7. $3x + 4y = 15$
   $3x - y = 0$
8. $7x - 3y = 6$
   $-4x + 6y = 18$
9. $5x - 3y = 38$
   $4x + 2y = 4$

Section 14-3

Section 14-4 USING SYSTEMS OF EQUATIONS TO SOLVE PROBLEMS

A variety of types of problems may be solved using a system of linear equations. In Chapter 3 you learned how to use one equation in one variable to solve problems. In this section you will learn how to solve problems that can be expressed as two linear equations in two variables.

To solve the word problems presented here, you should follow these steps.

1. Decide what two quantities are unknown and choose two variables to represent them.
2. Translate the information given in the problem into two linear equations in two variables.
3. Solve the equations using either the substitution method or the addition or subtraction method.
4. Check your answer against the conditions of the problem.
5. State the solution.

EXAMPLE The sum of the ages of a father and his daughter is 58 years. The difference in their ages is 26 years. What is the age of each person?

SOLUTION Let f = the father's age.
Let d = the daughter's age.

$$f + d = 58 \quad (1)$$
$$f - d = 26 \quad (2)$$

Add (1) and (2).
$$2f = 84$$
$$f = 42$$

Substitute 42 for f in (1).
$$42 + d = 58$$
$$d = 16$$

Check the solution against the conditions of the problem.
The sum of their ages is 58 years. $42 + 16 = 58$
The difference in their ages is 26 years. $42 - 16 = 26$
Therefore, the father is 42 years old and the daughter is 16 years old.

SYSTEMS OF LINEAR EQUATIONS

EXAMPLE Angelo rows 12 km downstream in 2 h. It takes him 5 h to row the same distance upstream. Find Angelo's rowing rate in still water and the rate of the current.

SOLUTION Let a = Angelo's rowing rate in still water.
Let c = the rate of the current.
Then $a + c$ = the rate at which Angelo travels downstream (with the current),
and $a - c$ = the rate at which Angelo travels upstream (against the current).
A chart may be helpful in organizing the data.

	rate	time	d = rt
downstream	$a + c$	2	$2a + 2c$
upstream	$a - c$	5	$5a - 5c$

$$2a + 2c = 12 \quad (1)$$
$$5a - 5c = 12 \quad (2)$$

Multiply (1) by 5. $10a + 10c = 60$
Multiply (2) by 2. $\underline{10a - 10c = 24}$
Add. $20a = 84$
                        $a = 4.2$

Substitute 4.2 for a in (1). $2(4.2) + 2c = 12$
                                  $8.4 + 2c = 12$
                                  $2c = 3.6$
                                  $c = 1.8$

Check against the conditions of the problem.
Angelo rows 12 km *downstream* in 2 h.
$$(4.2 + 1.8)2 = 6 \cdot 2 = 12$$
Angelo rows 12 km *upstream* in 5 h.
$$(4.2 - 1.8)5 = (2.4)5 = 12$$
Therefore Angelo's rowing rate in still water is 4.2 km/h and the rate of the current is 1.8 km/h.

EXAMPLE How many kilograms of a 30% salt solution and how many kilograms of a 15% salt solution must be mixed together to make 60 kg of a 20% salt solution?

SOLUTION Let x = number of kilograms of 30% salt solution needed.
Let y = number of kilograms of 15% salt solution needed.

$$x + y = 60 \quad (1)$$
$$0.30x + 0.15y = (0.20)60 \quad (2)$$

Multiply (2) by 100. $30x + 15y = 1200$
To simplify the equation you can divide by 15. $2x + y = 80$
Solve (1) for x in terms of y. $x = 60 - y$
Substitute. $2(60 - y) + y = 80$
                                                    $120 - 2y + y = 80$
                                                    $120 - y = 80$
                                                    $y = 40$

512   CHAPTER 14

Substitute in (1).

$$x + 40 = 60$$
$$x = 20$$

Check against the conditions of the problem.
60 kg of solution are needed.

$$20 + 40 = 60$$

(0.20)60 kg of salt are in the resulting solution.

$$(0.30)20 + (0.15)40 = 6 + 6 = 12 = (0.20)(60)$$

20 kg of the 30% solution and 40 kg of the 15% solution must be mixed together.

ORAL EXERCISES

Translate each sentence into an equation in two variables.

1. The sum of two numbers is 40.
2. The width of a rectangle is half the length.
3. Adam is 10 years older than Eva.
4. One number is 8 less than twice another.
5. The value of the nickels and dimes is $2.05.
6. The total interest earned from one investment at 12% and another investment at 8% is $115.
7. Twice Rob's age is 3 years more than Pearl's age.
8. Peanuts worth $4/kg were mixed with almonds worth $7/kg to produce a mixture worth $55.
9. A parking meter contains 47 coins in dimes and quarters only.
10. The perimeter of a rectangle is 42 cm.

WRITTEN EXERCISES

Solve each problem using two equations in two variables.

A 1. The sum of two numbers is 40. Their difference is 14. What are the numbers?
2. The difference between two numbers is 28. The greater is 8 less than twice the lesser. Find the numbers.
3. Adam is 10 years older than Eva. The sum of their ages is 56 years. How old is each person?
4. The sum of Emil's and Rosa's ages is 27 years. Nine years ago Emil was twice as old as Rosa was then. How old is each person?

SYSTEMS OF LINEAR EQUATIONS

5. The width of a rectangle is half the length. The perimeter is 42 cm. Find the dimensions of the rectangle.

6. The perimeter of an equilateral triangle is 6 cm more than the perimeter of a square. The length of a side of the square is 3 cm less than the length of a side of the equilateral triangle. Find the area of the square.

7. Two angles are supplementary. The measure of one is 16 less than 3 times the other. Find the measure of each angle.

8. Two angles are complementary. The measure of one is 9 more than twice the other. Find the measure of each angle.

9. Keith has $265 in $5 bills and $10 bills. He has 37 bills in all. How many of each kind does he have?

10. Heidi has 12 more $5 bills than $2 bills. Altogether she has $137. How many of each kind does she have?

11. Man-sun traveled by train for 3 h and by bus for 1 h to visit his cousin who lived 334 km away. The average rate of the train was 16 km/h faster than the bus. How fast did the train travel?

12. The Twesmey twins left their house at the same time, driving in opposite directions. Nora drove 5 km/h slower than Dora. After 4 h they were 456 km apart. How far did Dora drive?

13. Tickets to the "Potatoes" concert cost $5 and $8. If 960 tickets were sold for a total of $5652, how many tickets were sold at each price?

14. Admission to the Stratham Fair is $4 for adults and $1.50 for children under 12. On Friday 4019 people went to the fair. The gate receipts for that day were $11,568.50. How many adults and how many children attended the fair on Friday?

B 15. How many liters of a 10% alcohol solution and how many liters of a 20% alcohol solution must be mixed together to make 10 L of a 12% alcohol solution?

16. How many kilograms of a 5% salt solution and how many kilograms of a 15% salt solution must be mixed together to make 45 kg of an 8% salt solution?

17. Peanuts worth $4/kg were mixed with almonds worth $7/kg to produce 11.5 kg of nuts worth $55. How many kilograms of each type of nut were used?

18. A grocer carries two brands of coffee, one worth $5.50/kg and the other worth $6.50/kg. How many kilograms of each brand should be mixed together to produce 44 kg of coffee worth $5.75/kg?

19. A parking meter contains 47 coins in dimes and quarters for a total of $5.45. How many of each coin does the meter contain?

20. A beverage machine contains 59 nickels and dimes for a total of $4.55. How many of each coin does the machine contain?

21. A motorboat travels 160 km upstream in 8 h. It travels the same distance downstream in 5 h. What is the rate of the boat in still water?

22. Mimi is one third as old as George. In 5 years she will be one half as old as George. How old is Mimi?

23. Dr. Kaye has invested a total of $8000, part in a 90-day notice account paying 9% interest per year and the rest in a regular savings account paying 6% per year. How much was invested in each account if the total interest earned in one year was $675?

24. Last year Joe Raintree invested $300 more in a stock paying a 6% dividend than he did in a stock paying a 5% dividend. That year he earned $43 more in dividends from the stock paying the higher dividend. How much did he invest in each stock?

C 25. Sarah began her round trip canoe trip from Keene Valley to Keeseville at 8 A.M. She traveled a total of 108 km and returned to Keene Valley at 5 P.M. If she spent all but 1 h of that time traveling on the river, and if the return trip took three times as long as the outbound trip, how fast was the current?

26. The perimeter of a rectangle is 56 cm. If the length were halved and the width doubled, the perimeter would be 58 cm. How long is the original rectangle?

Section 14-5 GRAPHING SYSTEMS OF LINEAR INEQUALITIES

In Section 13-5 you learned that the graph of a simple inequality, such as $x - y > 2$, is an *open half-plane* and that the graph of a compound inequality, such as $x - y \geq 2$, is a *closed half-plane*. Now you will learn how to solve a system of inequalities by graphing more than one inequality on the same set of axes.

EXAMPLE   Solve the following system by graphing.
$$x \geq 4$$
$$y > 1$$

SOLUTION   Graph the line $x = 4$ as a solid line. Shade the half-plane to the right of the line. On the same set of axes graph the line $y = 1$ as a dashed line, and shade the half-plane above the line. The region A where the shadings overlap is the *intersection* of the two half-planes. It consists of the points whose coordinates satisfy *both* inequalities. Thus region A represents the graph of the *solution set* of the given system of linear inequalities. Note that this region contains part of the boundary line $x = 4$. It does not contain any of the boundary line $y = 1$. In particular, it does not contain the point (4, 1).

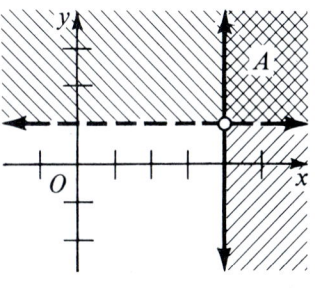

EXAMPLE Solve the following system graphically.
$$x - y \leq 2$$
$$2x + y \geq 4$$

SOLUTION You may find it helpful to rewrite each inequality in slope-intercept form and *then* graph both inequalities on the same set of axes.
$$y \geq x - 2$$
$$y \geq -2x + 4$$

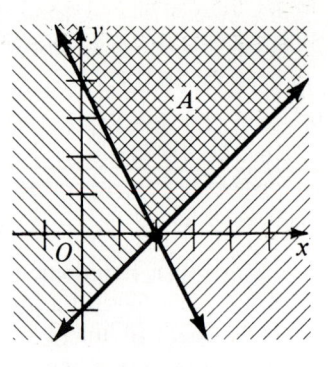

The solution set of the system is represented by A, the intersection of the two half-planes. Note that this region contains part of each boundary line, including the point $(2, 0)$.

EXAMPLE Solve the following system graphically.
$$y > -2$$
$$y \leq x + 1$$
$$y \leq -2x + 1$$

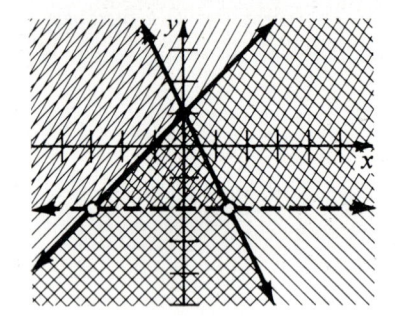

SOLUTION Graph each inequality on the same set of axes. The region common to all the graphs is the graph of the solution set of the given system.

To solve a system of linear inequalities graphically, follow these steps.

1. Graph each inequality on the same set of coordinate axes, shading the appropriate half-planes.
2. Locate the intersection of the half-planes. This region represents the solution set of the system. (If there is no such region, the solution set is the empty set.)

ORAL EXERCISES

Exercises 1–4 refer to the figure at the right, which shows how the lines whose equations are $x = 4$ and $y = -x + 3$ separate the plane into four regions. Name the region that forms the solution set of the given system of inequalities.

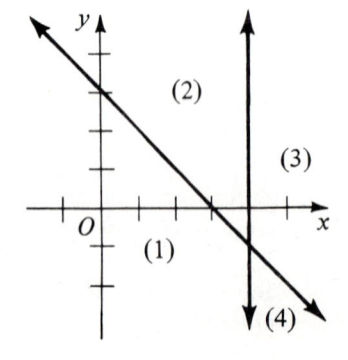

1. $x < 4$
   $y < -x + 3$

2. $x > 4$
   $y < -x + 3$

3. $x > 4$
   $y > -x + 3$

4. $x < 4$
   $y > -x + 3$

Exercises 5-10 refer to the figure at the right, which shows how the lines whose equations are $y = -3$, $y = 2$, and $y = x$ separate the plane into six regions. Name the region or regions that form the solution set of the given system of inequalities.

5. $y < 2$
   $y > -3$

6. $y < x$
   $y < -3$

7. $y > x$
   $y > 2$

8. $y < x$
   $y > 2$

9. $y > x$
   $y > -3$

10. $y < 2$
    $y < x$

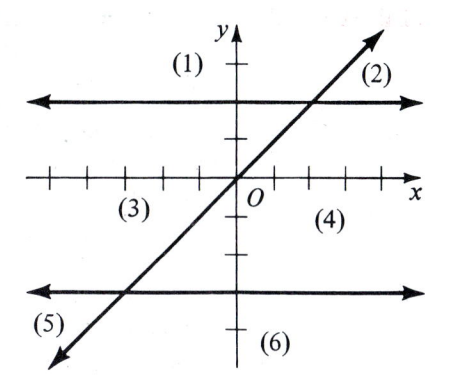

WRITTEN EXERCISES

Solve each system graphically.

A 1. $x > 3$
   $y \leqslant -1$

2. $x \leqslant 4$
   $y < 0$

3. $y < x + 5$
   $y > -2$

4. $y > -x + 1$
   $y \geqslant 1$

5. $y \leqslant 3x + 1$
   $y < x - 1$

6. $2x < y$
   $y < x + 2$

7. $y \leqslant 4x + 3$
   $y < -x - 3$

8. $y \geqslant -2x + 4$
   $y \leqslant x + 1$

9. $y \geqslant 4x$
   $y > -x + 3$

10. $x > y$
    $2x - 3 > y$

11. $x + y \geqslant 1$
    $x - y \leqslant -4$

12. $x - y \leqslant 2$
    $x + y \geqslant -1$

B 13. $x + 2y > 4$
   $x - 3y > 0$

14. $x + 3y > -12$
    $3x - y \geqslant -1$

15. $2y - 3x \leqslant 6$
    $x + 4y \leqslant -2$

16. $x + 2y < 8$
    $2y - 3x < 12$

17. $4x - 5y \leqslant 12$
    $2x + 3y > 6$

18. $3x + 4y \geqslant 8$
    $4x - 2y \geqslant 5$

C 19. $x \geqslant -1$
   $y < 4$
   $x + y \geqslant -2$

20. $y \geqslant -3$
    $x \leqslant 2$
    $x - y > -2$

21. $x > -2$
    $y > x$
    $y < 3$

SELF-TEST 2

Solve each problem using two equations in two variables.

1. The sum of Margaret's and Janet's ages is 24. Four years ago Janet was three times as old as Margaret was then. How old is each person? *Section 14-4*

2. The 4360 m² of the Nordville vegetable gardens are divided into 65 garden plots. The plots are either rectangles 6 m by 12 m or squares 8 m by 8 m. How many plots of each type are there?

(Self-Test continued on next page.)

3. Grove City and Chardon are 492 km apart on Route 297. Miguel left Chardon at 11 A.M. to go to Grove City. At the same time, Preston left Grove City, driving in the direction of Chardon. If Preston was traveling 12 km/h faster than Miguel and they passed each other at 2 P.M., how fast was Miguel traveling?

Solve each system graphically.

4. $y \leqslant 2x$
   $y + 2x > 2$

5. $y - x > 3$
   $y < -x + 1$

6. $2y + x \geqslant 4$
   $y - x \geqslant -3$

Section 14-5

Section 14-6 LINEAR PROGRAMMING (Optional)

Linear programming is a strategy for *maximizing* or *minimizing* a linear expression of the form $ax + by$ (where a and b are real numbers) subject to various *constraints*, or restrictions, expressed as linear inequalities. The following problem illustrates how linear programming may be used.

Due to licensing restrictions, the fluttercraft of the Asteroid Mining Company can mine no more than 600 t (metric tons) of trypton and 500 t of zyptium per week. It takes each fluttercraft 0.2 h to mine a ton of trypton. A ton of zyptium can be mined by a fluttercraft in 0.4 h. Up to 240 h of flight time is available each week. The Asteroid Mining Company makes a profit of $15 per ton of trypton mined and $40 per ton of zyptium. How many tons of each should be mined each week to maximize profit?

Step 1. Determine what quantities are unknown and choose variables to represent them.

In this problem you want to find how many tons of trypton and zyptium should be mined.

Let x = number of tons of trypton.
Let y = number of tons of zyptium.

Step 2. Write the expression to be maximized (or minimized) in the form $ax + by$ (a and b are real numbers; x and y are the variables chosen in Step 1).

$15x$ = weekly profit on trypton
$40y$ = weekly profit on zyptium
$15x + 40y$ = total weekly profit on trypton and zyptium together

Step 3. Using the variables chosen in Step 1, translate the constraints of the problem into inequalities.

$x \geqslant 0$
$y \geqslant 0$ A negative number of tons cannot be mined.

$x \leqslant 600$ No more than 600 t of trypton can be mined.
$y \leqslant 500$ No more than 500 t of zyptium can be mined.
$0.2x + 0.4y \leqslant 240$ Up to 240 h of flight time is available.

Step 4. Graph the inequalities from Step 3 on the same set of coordinate axes and determine the coordinates of the vertices of the region.

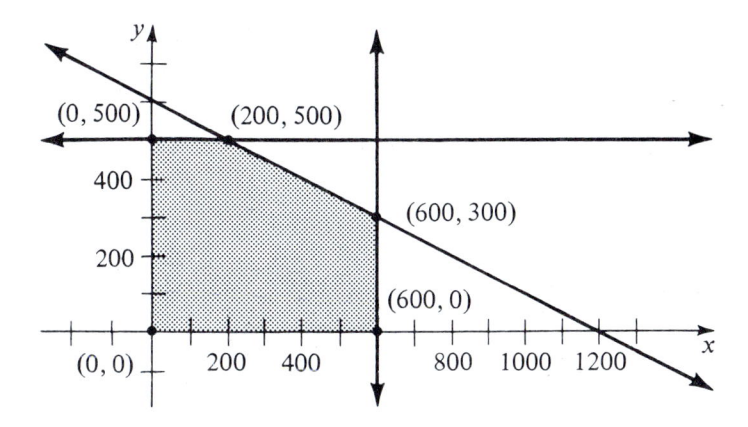

The shaded region in the figure above represents the solution set of the system of constraints.

Now we must determine which of the points in this solution set will maximize profit. Because there are infinitely many points in the solution set, it would be impossible to test each one to see which yields maximum profit. Because the following statement is true, you have to check only a finite number of points.

> If a linear expression of the form $ax + by$ (where a and b are real numbers) has a maximum (or minimum) value over a region determined by the intersection of a finite number of closed half-planes, then the maximum (or minimum) value occurs at a vertex of the region.

Step 5. Test each vertex.

To find the maximum profit, evaluate the expression $15x + 40y$ at each of the vertices of the region.

(0, 0)	$15 \cdot 0 + 40 \cdot 0 = 0$
(0, 500)	$15 \cdot 0 + 40 \cdot 500 = 20{,}000$
(200, 500)	$15 \cdot 200 + 40 \cdot 500 = 23{,}000$
(600, 300)	$15 \cdot 600 + 40 \cdot 300 = 21{,}000$
(600, 0)	$15 \cdot 600 + 40 \cdot 0 = 9000$

Therefore, the Asteroid Mining Company will make a maximum profit of $23,000 when it mines 200 t of trypton and 500 t of zyptium per week.

WRITTEN EXERCISES

For each region graphed below:
a. Find the point that yields the maximum value for the given linear expression.
b. State the maximum value of the expression.
c. Find the point that yields the minimum value for the given expression.
d. State the minimum value of the expression.

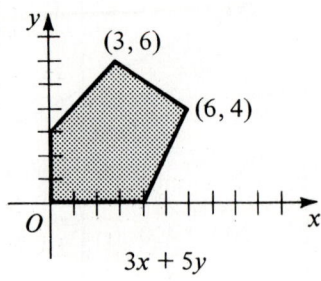

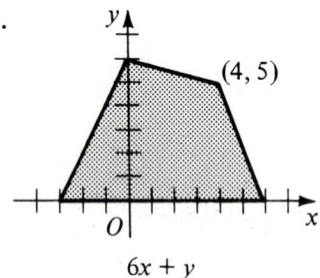

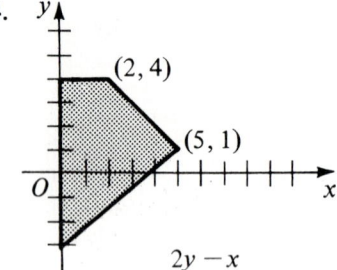

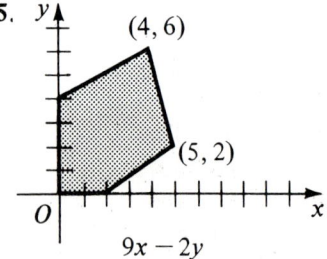

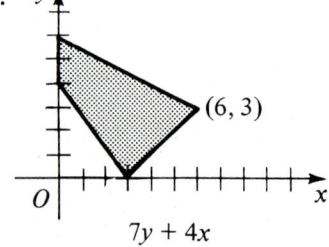

A 1. (4, 5)
$2x + 3y$

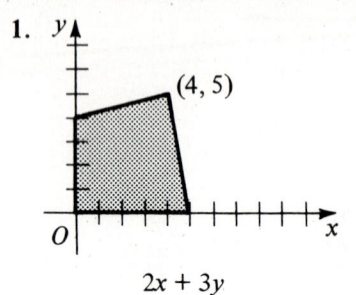

2. (3, 6), (6, 4)
$3x + 5y$

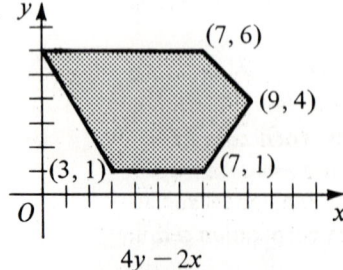

3. (4, 5)
$6x + y$

4. (2, 4), (5, 1)
$2y - x$

5. (4, 6), (5, 2)
$9x - 2y$

6. (6, 3)
$7y + 4x$

7. (7, 6), (9, 4), (3, 1), (7, 1)
$4y - 2x$

8. (2, 6), (5, 4), (6, 2)
$2x - 7y$

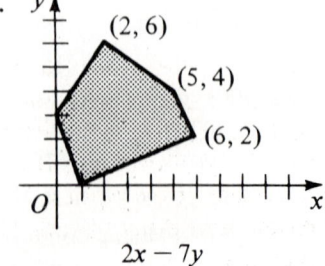

In Exercises 9-14:
a. Graph the solution set of the given inequalities.
b. Determine the coordinates of the vertices of the region graphed.
c. Maximize or minimize the given expression as indicated.

B 9. $x \geq 0$
$y \geq 0$
$x + 2y \leq 4$
Maximize $2x + 7y$.

10. $x \geq 0$
$y \geq 0$
$y \leq -2x + 5$
Maximize $4x + 3y$.

11. $x \geq 0$
$y \leq -x + 5$
$y \geq \frac{2}{3}x$
Maximize $x + 2y$.

12. $x \geq 0$
$y \leq -x + 6$
$y \geq \frac{1}{2}x + 3$
Maximize $4x + 5y$.

13. $x \geq 0$
$y \geq -\frac{1}{2}x + 3$
$3y \geq x - 1$
Minimize $3x + 2y$.

14. $y \geq 3$
$x \geq 2$
$y \geq -\frac{1}{2}x + 6$
$y \leq -\frac{1}{2}x + 9$
Minimize $5x + 6y$.

In Exercises 15-17, use the methods of linear programming to solve the problem.

C 15. The Frangiosa Brothers Landscaping Service estimates that this week it will need at least 80 juniper shrubs and a minimum of 90 yew shrubs. The Bluebell Nursery is currently offering a special sale on a combination of 3 juniper and 2 yew shrubs for $60. Pleasant View Nursery is having a sale, too: 2 juniper and 3 yew shrubs for $75. How many combinations should the Frangiosa Brothers Landscaping Service buy from each nursery to minimize its costs?

16. The Camera Company is having a sale on 135 mm print and slide film. The sale price includes a 20¢ profit on each roll of slide film and a 15¢ profit on each roll of print film. Sales records indicate that the company should have enough film in stock to sell at least 5 rolls of print film for every 3 rolls of slide film. The store currently has 3000 rolls of each type of film on hand. It can order more, and has room to store a maximum of 12,000 rolls. How many rolls of each type of film must the company sell to realize the greatest profit?

17. The Zoom-Aboard Corporation manufactures two types of skateboards, the standard model and the deluxe model. For every deluxe model the corporation manufactures, it can make at most 3 standard models. Total sales are not expected to exceed 20,000 skateboards. It is not likely that more than 8000 deluxe models will be sold. If the profit is $3 on each standard model and $5 on each deluxe model, how many of each type must the corporation sell in order to maximize profits?

Computer Exercises

1. **a.** The ordered pair $\left(\dfrac{CE - BF}{AE - BD}, \dfrac{CD - AF}{BD - AE}\right)$, with $AE \neq BD$, represents the point of intersection of the lines with equations $Ax + By = C$ and $Dx + Ey = F$. The following program computes the coordinates of the point of intersection (if it exists) of these lines. If there is no unique point of intersection, the program branches from line 50 to line 90 to determine whether the lines are parallel ($AF \neq CD$) or whether they coincide ($AF = CD$). Copy and RUN the program for the lines with equations $4x + 3y = 5$ and $3x + 2y = 8$.

```
10   PRINT "INPUT A,B,C";
20   INPUT A,B,C
30   PRINT "INPUT D,E,F";
40   INPUT D,E,F
50   IF A*E=B*D THEN 90
60   PRINT "THE COMMON SOLUTION IS ";
70   PRINT "(";(C*E-B*F)/(A*E-B*D);",";(C*D-A*F)/(B*D-A*E);")"
80   GOTO 140
90   IF A*F=C*D THEN 120
100  PRINT "THERE IS NO COMMON SOLUTION. THE LINES ARE PARALLEL."
110  GOTO 140
120  PRINT "THERE ARE INFINITELY MANY COMMON SOLUTIONS, ";
130  PRINT "THE LINES COINCIDE."
140  END
```

   b. RUN the program for the lines with equations $27x - 92y = 146$ and $45x + 22y = 68$.

   c. RUN the program for the lines with equations $3x + 2y = 5$ and $6x + 4y = 1$. How could you have anticipated your result before using the computer?

   d. RUN the program for the lines with equations $x + 9y = 3$ and $4x + 36y = 12$.

2. **a.** The following program draws the graph of a pair of intersecting lines with equations $Ax + By = C$ and $Dx + Ey = F$. The program does not work if A or D is 0. Copy and RUN the program for the lines with the following equations: $x + y = 2$ (A = 1, B = 1, C = 2) and $x - y = 0$ (D = 1, E = −1, F = 0).

```
10   PRINT "INPUT A,B,C WITH A<>0";
20   INPUT A,B,C
30   PRINT "INPUT D,E,F WITH D<>0";
40   INPUT D,E,F
50   FOR Y=6 TO -6 STEP -1
```

```
60   PRINT Y;
70   LET X1=(C-B*Y)/A
80   LET X2=(F-E*Y)/D
90   IF X1<X2 THEN 190
100  IF X1=X2 THEN 160
110  IF X2<-6 OR X2>6 THEN 130
120  PRINT TAB(INT(2*X2+17.5));"*";
130  IF X1<-6 OR X1>6 THEN 240
140  PRINT TAB(INT(2*X1+17.5));"*"
150  GOTO 250
160  IF X1<-6 OR X2>6 THEN 240
170  PRINT TAB(INT(2*X1+17.5));"* (";X1;",";Y;")"
180  GOTO 250
190  IF X1<-6 OR X1>6 THEN 210
200  PRINT TAB(INT(2*X1+17.5));"*";
210  IF X2<-6 OR X2>6 THEN 240
220  PRINT TAB(INT(2*X2+17.5));"*"
230  GOTO 250
240  PRINT
250  NEXT Y
260  PRINT
270  PRINT "    -6-5-4-3-2-1 0 1 2 3 4 5 6"
280  END
```

b. RUN the program for the lines with equations $2x + y = 1$ and $-2x + y = 3$.

c. The program searches for the point of intersection while looping through integral values of y from 6 to -6. (Lines 50 and 250 define the loop.) The ordered pair (x, y), representing the point of intersection, will be printed only if y is an integer, and if $-6 \leq y \leq 6$ and $-6 \leq x \leq 6$. RUN the program for the lines with equations $x + y = 2$ and $x - 2y = 6$.

d. What is the actual point of intersection of the lines in part (c)?

CHAPTER REVIEW

Solve each system by graphing.

1. $4x - y = 4$
   $3x + 2y = 14$

2. $x + 3y = 0$
   $12 - 9y = 3x$

3. $2x - 3y = 6$
   $5x - 3y = -3$

Section 14-1

Use the substitution method to solve each system.

4. $x = 2y - 3$
   $3y = 2x + 5$

5. $5x + 4y = 22$
   $4x + y = 22$

6. $4x - y = 8$
   $2y - 8x = -16$

Section 14-2

SYSTEMS OF LINEAR EQUATIONS

Use the addition or subtraction method to solve each system.

7. $7x - 4y = -13$
   $3x + 4y = -17$

8. $9x + y = 20$
   $11x + 3y = 28$

9. $2x - 5y = 20$
   $5x + 3y = -12$

Section 14-3

Solve each problem using two equations in two variables.

10. The mean of two numbers is 18. If the difference of the numbers is 14, find the numbers.

Section 14-4

11. How many liters of a 30% acid solution and how many liters of a 25% acid solution must be mixed together to form 20 L of a 28% acid solution?

Solve each system graphically.

12. $y < 3x + 1$
    $y \geqslant -2x - 4$

13. $x - y < 3$
    $3x + 4y < 9$

14. $x \geqslant -y$
    $2x + 5y \leqslant 10$

Section 14-5

For each region, find the maximum value and the minimum value of the expression $2x - 3y$.

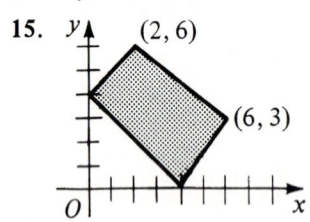

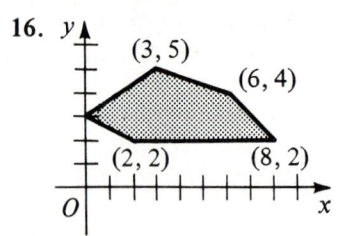

15. (2, 6), (6, 3)

16. (3, 5), (6, 4), (2, 2), (8, 2)

Section 14-6

CHAPTER TEST

Solve each system by graphing.

1. $3x - y = 6$
   $-2x + y = -4$

2. $7x - 3y = 3$
   $y - x = 3$

3. $y = 4x - 4$
   $3y = 12x + 12$

4. $y > -2x + 5$
   $y < x - 1$

5. $x \geqslant -2$
   $3x - 5y \leqslant 15$

6. $2x + y < 9$
   $5x - 6y \geqslant -30$

Solve each system *either* by the substitution method *or* by the addition or subtraction method.

7. $5x + y = -3$
   $13x + 2y = -3$

8. $x - 2y = 3$
   $3x - 2y = -11$

9. $2x - y = 14$
   $3x + 5y = 47$

10. $7x + 3y = 22$
    $8x - 3y = -7$

11. $x + 4y = 4$
    $3x + 8y = 0$

12. $2x - 5y = 9$
    $3x + 2y = -15$

Solve each problem using two equations in two variables.

13. The sum of the ages of a woman and her son is 30. In 24 years her son will be half as old as she will be. How old is each now?

14. A beverage machine contains 112 coins in nickels and dimes only. The value of the coins is $9.40. How many of each coin does the machine contain?

15. (Optional) For the region shown, find the maximum value and the minimum value of the expression $x + 4y$.

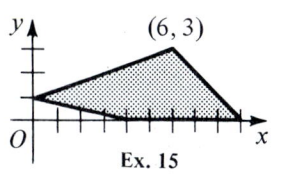

16. (Optional)
    a. Graph the solution set of the system.
    b. Minimize $5x + 2y$ for the region graphed.
    $$x \geq 0$$
    $$y \geq 0$$
    $$3x + y \geq 6$$

Ex. 15

CUMULATIVE REVIEW: CHAPTERS 1-14

Indicate the best answer by writing the appropriate letter.

1. What is the slope of the line whose equation is $6x + 3y = 2$?
   a. -2 b. $\frac{1}{2}$ c. $-\frac{1}{2}$ d. $\frac{2}{3}$

2. If a die is rolled, what is the probability that the outcome is a number less than 6?
   a. $\frac{1}{6}$ b. $\frac{2}{3}$ c. $\frac{5}{6}$ d. 1

3. Simplify: $3 - 6(8 \div 4^2)$
   a. $-\frac{3}{2}$ b. 0   c. -21 d. -12

4. Which ordered pair is a solution of the following system of equations?
   $$4x + y = 5$$
   $$3x - 2y = 12$$
   a. $(0, 5)$ b. $(3, -7)$ c. $(2, -3)$ d. $(-2, -9)$

5. Simplify: $(2\sqrt{3} + \sqrt{7})(2\sqrt{3} - \sqrt{7})$
   a. $4\sqrt{3} - 7$ b. $4\sqrt{3} - 49$ c. 19   d. 5

6. If the length of a diagonal of a square is 6, then the length of a side is:
   a. $3\sqrt{2}$ b. $2\sqrt{3}$ c. $6\sqrt{2}$ d. 6

7. Simplify: $(-3x^2)^3$
   a. $-3x^6$ b. $-3x^5$ c. $-27x^5$ d. $-27x^6$

8. Find the circumference of a circle whose area is 36π.
   a. 324π b. 144π c. 12π d. 6π

9. What is the equation of the line shown?
   a. $x + y = 1$ b. $x - y = 1$
   c. $x + y = -1$ d. $y - x = 1$

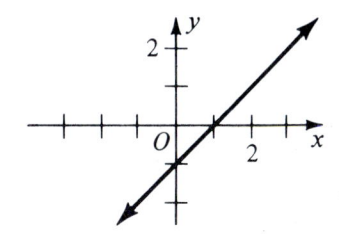

SYSTEMS OF LINEAR EQUATIONS

10. Find the area of the given trapezoid.
    a. 192   b. 40   c. 80   d. 48

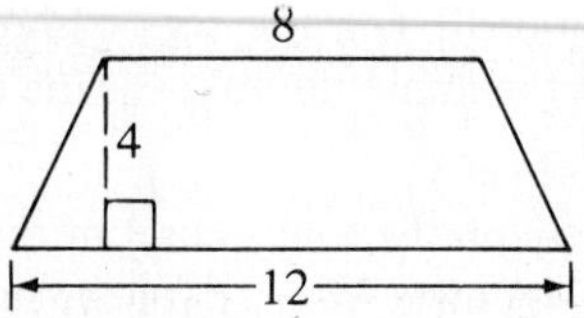

11. Her performance in previous competition indicates that the probability that Elsie will bowl a strike in any given frame is approximately $\frac{3}{11}$. What is the probability that she will not bowl a strike?
    a. $\frac{3}{11}$ b. $-\frac{3}{11}$ c. $\frac{8}{11}$ d. $\frac{11}{3}$

12. If the domain of x is $\{-3, -2, -1, 0, 1, 2, 3\}$, find the solution set of $2x + 5 < 3$.
    a. $\{-1\}$ b. $\{0, 1, 2, 3\}$ c. $\{-3, -2, -1\}$ d. $\{-3, -2\}$

13. Find the value of x in the diagram.
    a. 96   b. 48   c. 32   d. 52

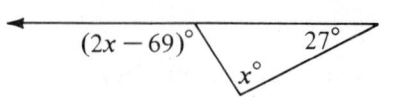

14. The third quartile corresponds to which of the following percentiles?
    a. 25th   b. 33rd   c. 67th   d. 75th

15. What is the solution of the system of equations in the graph?
    a. $(1, -3)$ b. $\left(0, -\frac{3}{2}\right)$
    c. $(-1, 0)$ d. $(-3, 1)$

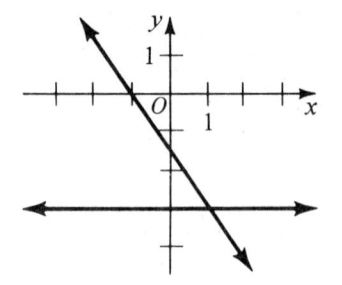

16. The converse of $q \rightarrow p$ is:
    a. $p \rightarrow q$ b. $q \rightarrow p$ c. $\sim p \rightarrow \sim q$ d. $\sim q \rightarrow \sim p$

17. Kelly has $2.20 in dimes and quarters. If the number of dimes she has is 1 more than the number of quarters, how many dimes does she have?
    a. 4   b. 5   c. 6   d. 7

18. The slope of the line containing the points $(3, -4)$ and $(6, -4)$ is:
    a. 0   b. undefined   c. -4 d. 3

19. Harry, Louise, Ned, Terry, and Martha always sit in the front row during class. In how many different ways can the five students sit together?
    a. 5   b. 15   c. 24   d. 120

20. $\overleftrightarrow{CD}$ is tangent to $\odot O$ at C. If $m\angle ACD = 36$, find $m\widehat{ABC}$.
    a. 144   b. 324   c. 288   d. 72

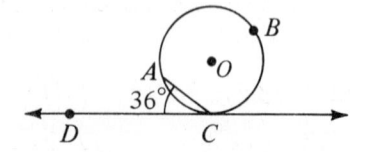

526   CHAPTER 14

21. One factor of $3x^2 - 10x - 8$ is:
   a. $3x + 4$ b. $3x - 2$ c. $3x + 2$ d. $x + 4$

22. The figure is the graph of which inequality?
   a. $y \geq -2$ b. $x \geq -2$
   c. $y > -2$ d. $x > -2$

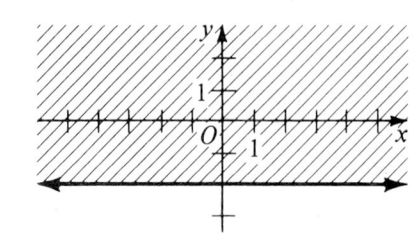

23. Express $3.\overline{42}$ as a fraction in lowest terms.
   a. $\dfrac{154}{45}$ b. $\dfrac{113}{33}$ c. $\dfrac{31}{9}$ d. $\dfrac{38}{11}$

24. The measures of two supplementary angles are in the ratio $2:7$. What is the measure of the larger angle?
   a. 145   b. 70   c. 40   d. 140

Use the following data for Exercises 25-27: $5, 4, 15, 5, 15, 11, 20, 5$

25. Find the median.
   a. 5   b. 8   c. 10   d. 11

26. Find the mean.
   a. 5   b. 8   c. 10   d. 11

27. Find the mode.
   a. 5   b. 8   c. 10   d. 15

28. If $m\angle ACB = 38$, find $m\angle AOB$.
   a. 38   b. 142   c. 76   d. 114

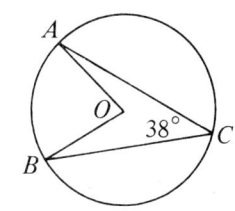

29. The line whose equation is $y = -\dfrac{2}{3}x + 1$ is parallel to the line with which equation?
   a. $3x + 2y = 6$ b. $3x - 2y = 6$ c. $2x - 3y = 6$ d. $2x + 3y = 6$

30. In an experiment a coin is tossed and a card is drawn from a standard deck. What is the probability that a head comes up and a black king is drawn?
   a. $\dfrac{1}{52}$ b. $\dfrac{1}{26}$ c. $\dfrac{1}{13}$ d. $\dfrac{2}{13}$

31. The ratio of the areas of two similar triangles is $9:16$. What is the ratio of the lengths of a pair of corresponding sides?
   a. $\sqrt{3}:2$ b. $9:16$ c. $3:4$ d. $81:256$

32. Simplify: $\dfrac{1}{2}\sqrt{48} - 3\sqrt{12}$
   a. $-4\sqrt{3}$ b. $2\sqrt{3} - 9\sqrt{2}$ c. $-3\sqrt{3}$ d. -15

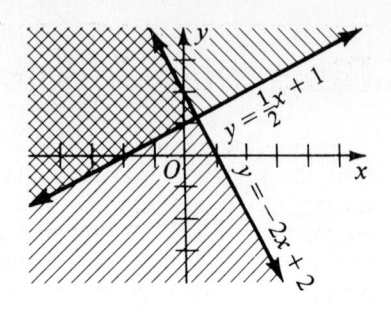

33. The figure shows the graph of which of the following systems of inequalities?

   a. $y \leq \frac{1}{2}x + 1$
      $y \leq -2x + 2$
   b. $y \leq \frac{1}{2}x + 1$
      $y \geq -2x + 2$
   c. $y \geq \frac{1}{2}x + 1$
      $y \leq -2x + 2$
   d. $y \geq \frac{1}{2}x + 1$
      $y \geq -2x + 2$

34. Which of the following pairs of triangles must be congruent?
   a.   b.   c.   d.

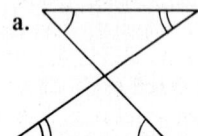

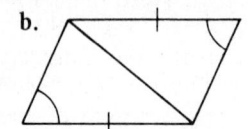

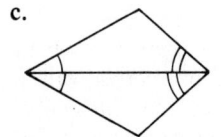

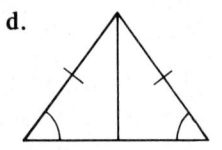

35. An urn contains a red ball, a white ball, and a green ball. A ball is drawn from the urn, then replaced, and a second ball is drawn. What is the probability that the same ball is drawn both times?
   a. $\frac{1}{9}$ b. $\frac{1}{3}$ c. $\frac{1}{2}$ d. $\frac{2}{3}$

36. If Keira had twice as much money as she has, she would have $1 more than three times as much as Eva has. Keira has $4 less than twice as much as Eva. How much money does Keira have?
   a. $9   b. $7   c. $22   d. $14

37. Solve for x: $a(x - c) = 2d$. Assume that $a \neq 0$ and $c \neq 0$.
   a. $\frac{a(x-c)}{2}$ b. $\frac{ac + 2d}{a}$ c. $\frac{ac + 2d}{c}$ d. $\frac{c + 2d}{a}$

38. Which of the following must be true if $\overleftrightarrow{AB} \parallel \overleftrightarrow{CD}$?
   a. $m\angle APQ = m\angle PRG$
   b. $m\angle APE + m\angle CQF = 180$
   c. $\overleftrightarrow{EF} \parallel \overleftrightarrow{GH}$
   d. $m\angle RPQ = m\angle RSQ$

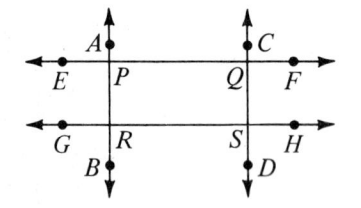

39. The fourth column of the truth table will be correct if the ? is replaced by:
   a. $\sim p \vee q$
   b. $\sim p \rightarrow q$
   c. $\sim p \wedge q$
   d. $q \rightarrow \sim p$

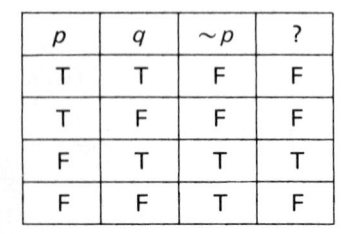

p	q	$\sim p$	?
T	T	F	F
T	F	F	F
F	T	T	T
F	F	T	F

40. Which ordered pair satisfies the given system of inequalities?
$$3x - y > 2$$
$$4x + 2y \geq 1$$
$$x \leq 3$$
   a. (2, 4)   b. (1, −1)   c. (4, 1)   d. (0, −3)

528   CHAPTER 14

Biographical Note   Emmy Noether

Emmy Noether (1882-1935) had a double gift: she could think clearly about complex theoretical algebra and she could teach its concepts so that students could understand them. Her influence on twentieth-century mathematics cannot be limited to the importance of her own research. It must also be measured by her many contributions to the development of others.

Noether was born in Erlangen, a town in southern Germany, where her father was a professor of mathematics at the University. During her student years, Noether occasionally lectured to her father's classes when he was ill.

After receiving her degree, Noether moved in 1916 to the University of Göttingen. There she came to maturity as a mathematician. Two well-known mathematicians, David Hilbert and Felix Klein, were at the university at that time. They were working on a general theory of relativity, and they welcomed Noether's help. As their associate, she became part of a very active group of mathematical thinkers, although positions on the university faculty were not open to women. She became interested in exploring the power of the axiomatic method as a basis of mathematical research. In 1920 she coauthored a paper on differential equations that brought increased recognition of her ability as a creative thinker.

Although much work was then being done on noncommutative algebra, class fields, abstract rings, and cyclic algebra, Noether's work was considered outstanding. As Albert Einstein said, "She discovered methods which have proved of enormous importance in the development of the present day younger generation of mathematicians."

When Hitler came into power in Germany, Noether lost the position to which she had finally been appointed at Göttingen. She came to the United States. Here she worked at Bryn Mawr College and also at the Institute for Advanced Studies in Princeton, New Jersey. Unfortunately, her teaching talents were not long available to American students, for Emmy Noether died suddenly when she was only fifty-three years old.

SYSTEMS OF LINEAR EQUATIONS

VOCABULARY REVIEW

Be sure that you understand the meaning of these terms:

system of linear equations, p. 501
linear programming, p. 518

MIXED REVIEW

1. Find the x-intercept and the y-intercept of the line with equation $6x - y = 4$.
2. When the radius of a circle decreases from 4 to 3, the circumference decreases by __?__ and the area decreases by __?__.
3. A right triangle has sides of lengths $x - 1$, x, and $x + 8$. Find the numerical length of the hypotenuse.
4. Find the area of the trapezoid shown in the diagram.
5. Factor: $6z^2 - 13z + 5$
6. Graph the solution set of $7 > 4 - x \geq -1$ if x is an integer.
7. How much money invested at $5\frac{1}{2}\%$ earns $19.25 interest?

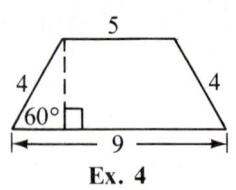

Ex. 4

Solve each system.

8. $7x - y = 12$
   $9x + 2y = -1$

9. $3x + 5y = 26$
   $4x - 3y = -4$

10. An urn contains 4 yellow and 6 red marbles. Two marbles are drawn from the urn. What is the probability that the marbles are different colors if the drawings are made with replacement? without replacement?

Solve.

11. $4x^2 + 25 = 20x$
12. $2y^2 + 4y = 1$
13. $2z^2 - 162 = 0$

14. If all the sides of quadrilateral $ABCD$ are congruent, must $ABCD$ be a parallelogram? a rhombus? a square?
15. When two parallel lines, r and s, are cut by transversal t, two alternate interior angles formed are supplementary. What can you conclude about the lines?
16. The perimeter of a rectangle is 38 cm. If the length is 5 cm more than the width, find the length and width.
17. Construct a truth table for $(p \wedge q) \rightarrow (p \vee q)$.
18. A sphere has radius 1.8. Find its surface area and its volume.

CHAPTER 14

The densities in kg/m³ of 39 gases and vapors are as shown in the table.

Density	Frequency
0-0.9	5
1-1.9	17
2-2.9	8
3-3.9	6
4-4.9	1
5-5.9	2

19. Construct a frequency polygon.
20. Construct a cumulative frequency histogram.
21. In which interval does the median value occur?
22. Estimate which density corresponds to the median.

23. Find an equation in slope-intercept form for the line passing through $(-5, -3)$ and $(7, -7)$.

Solve graphically.

24. $3x - 4y = 6$
    $5x + 2y = 10$

25. $y \leq 3x - 2$
    $y > 4$

26. How many different 5-digit numbers can be formed using the digits 1, 2, 3, 4, and 5 if digits cannot be repeated? How many are even?

27. In the diagram, $m \widehat{RS} = 40$ and $m \widehat{UR} = 70$. Find $m\angle UST$ and $m\angle UVT$.

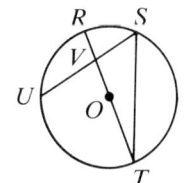

28. Simplify: $\dfrac{2}{1 + \sqrt{2}}$

29. What percent of 18 is 27?

30. Multiply: $(3 - 4x)(8 - x)$

31. Tracy has 98¢ in pennies and nickels. If she has twice as many pennies as nickels, how many nickels does she have?

32. If the measures of two complementary angles are in the ratio 8:7, find the measure of each.

33. An isosceles right triangle has a hypotenuse of length 8 cm. Find the area of the triangle.

34. Find the lateral area of a cone with radius 15 and height 8.

35. Use similar triangles to find the values of x and y in the figure at the right.

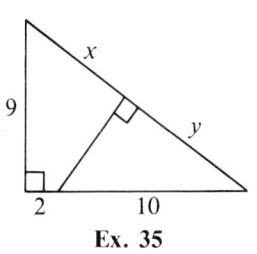

36. Find the GCF of $24a^2b^5c^3$ and $18a^3b^2c$.

37. Express $0.58\overline{3}$ as a fraction in lowest terms.

38. Solve $Ax + By = C$ for y. (Assume that $x \neq 0$ and $y \neq 0$.)

39. Write the converse of the statement "If $x = 2$, then $x^2 = 4$" and give the truth value of the converse.

Ex. 27

Ex. 35

SYSTEMS OF LINEAR EQUATIONS

Extra Practice

Chapter 1

A 1. Write out using the roster method: {years between 1979 and 1990}.
2. Draw a number line and graph the set $\{-3, 0, 2\}$.
3. Refer to the number line at the right.
   a. Name the graph of -1.
   b. Name the coordinate of point H.

$$\begin{array}{ccccc} C & H & A & I & R \\ \hline -2 & -1 & 0 & 1 & 2 \end{array}$$
Ex. 3

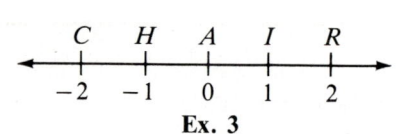

4. What is the opposite of -5?

Simplify.

5. $(45 + (-54)) + 10$
6. $-185 + 97$
7. $26 - 42$
8. $26 - (-42)$
9. $-132(41)$
10. $(-50)(-20)$
11. $\dfrac{-442}{-17}$
12. $1458 \div (-27)$
13. $\dfrac{8}{15} - \dfrac{11}{15}$
14. $\dfrac{3}{4} + \left(-\dfrac{5}{8}\right)$
15. $-0.21 - (-0.1)$
16. $0.07(-0.85)$
17. $-\dfrac{100}{27} \div \dfrac{25}{3}$
18. $1.71 \div 3$

B 19. Describe the set $\{5, 10, 15, 20, 25, \ldots\}$ by a rule.

Name the property illustrated by the statement.

20. $-1 + 6 = 6 + (-1)$
21. $-7(-1) = 7$

Simplify.

22. $-18 - (95 - 87)$
23. $(-10)(-8)(-6)$
24. $-59 + (-102) + 294 + (-3)$
25. $\dfrac{5}{6} - \dfrac{7}{12} + \dfrac{11}{18}$
26. $\left(-\dfrac{21}{24}\right)\left(\dfrac{36}{49}\right)\left(\dfrac{28}{3}\right)$
27. $18 \div (-3) \cdot 2 - 3^2$
28. $\dfrac{2(7 + 9)}{1 + 3 \cdot 3}$
29. $\dfrac{(0 - 6)(11 - 18)}{7^2 - 7}$
30. $((9 - 3)^2 - 1)(-1 + 0.6)$
31. $-14(2^3) + (-16)(-7)$

Chapter 2

A 1. If the domain of x is {positive integers}, find a value of x that makes the statement $x - 5 = 0$ true and a value that makes it false.

Write the negation. Find the truth values of the statement and its negation.

2. $0 > 4$
3. $6^2 = 3^2 \cdot 2^2$

4. State the hypothesis and the conclusion of the statement "$2k + 3$ is an odd integer if $2k + 1$ is an odd integer."

5. Construct a truth table for $\sim(p \wedge \sim q)$.

6. Write in symbols the converse, inverse, and contrapositive of $p \rightarrow \sim q$.

7. (Optional) Put the argument into symbolic form. Tell if it is valid. If it is, specify which argument pattern is used.

$$\text{If } x + 1 = 7, \text{ then } x = 6.$$
$$x \neq 6.$$
$$\text{Therefore, } x + 1 \neq 7.$$

For Exercises 8–15, take p, q, and r as follows:

$$p: \ 0 \cdot 0 = 0 \qquad q: \ 1 + 0 = 0 \qquad r: \ (-1)(-1) = 1$$

Write each statement in symbolic form and give its truth value.

8. $0 \cdot 0 \neq 0$ or $(-1)(-1) \neq 1$
9. $1 + 0 = 0$ only if $(-1)(-1) = 1$

Write each statement in words and give its truth value.

10. $p \wedge q$
11. $p \vee r$
12. $r \rightarrow \sim q$
13. $p \rightarrow q$

Give the truth value of each statement.

B 14. $(\sim p \vee q) \leftrightarrow \sim r$
15. $(p \wedge \sim q) \rightarrow r$

16. If the replacement set of x is $\{0, 1, 2, 3, 4, \ldots\}$, find the solution set of $4 - x = x - 4$.

17. If r, s, and t are true statements, find the truth value of $(r \wedge \sim s) \rightarrow \sim t$.

18. Construct a truth table for $(p \rightarrow q) \rightarrow (q \rightarrow p)$.

19. Write a statement that is logically equivalent to "If I learn computer programming, then I will get the job."

20. Is the statement $\sim(p \vee \sim p) \rightarrow q$ a tautology, a contradiction, or neither? Explain.

21. (Optional) Determine if the argument at the right is valid by expressing it as an implication and then using a truth table.

$$\sim(p \wedge q)$$
$$\underline{p}$$
$$\therefore \sim q$$

Chapter 3

A 1. Evaluate if $a = -2$, $b = 3$, and $c = -\frac{1}{3}$:   a. $2a - b$ b. $\frac{c}{-b}$

2. Solve if $\{-2, -1, 0, 1, 2\}$ is the domain of x.
      a. $-x = x + 4$ b. $x + 3 = 3$

Solve.

3. $g + 15 = 3$ 4. $d - \frac{2}{3} = -\frac{1}{3}$ 5. $\frac{n}{3} = -9$

6. $144 = 16t$ 7. $12m = -420$ 8. $-9 = 5v + 6$

Write as an algebraic expression.

9. The value in dollars of $4x$ five-dollar bills.

10. The sum of twice a number and 9.

Solve by writing and solving an equation.

11. The width of a rectangle is two-thirds the length. The perimeter is 80. Find the length and width.

12. A bank contains 18 coins, all dimes and nickels, worth $1.30. How many dimes are there?

Solve for the variable indicated.

13. $z = \frac{y - x}{2}$; y 14. $V = 2\pi rh$; r

B 15. Evaluate if $a = -2$, $b = 3$, and $c = -\frac{1}{3}$:   a. $\frac{a + b}{c}$ b. $c(6a - b)$

Solve.

16. $-8 = -8 + 2t$ 17. $z - \frac{5}{6} = \frac{2}{3}$ 18. $\frac{35}{27} = -\frac{7}{9}x$

19. $0.08y = 4.16$ 20. $0.3b = 13 - b$ 21. $3(r - 2) = 5r$

22. Show that $a + \frac{3}{4} = -\left(\frac{5}{4} - a\right)$ has no solution.

Solve.

23. Jerry is one year older than his sister, and in three years the sum of their ages will be 35. How old is each now?

24. The average cost of an item in 1978 was about one and a half times the cost in 1972. Find the cost in 1972 of a car costing $6000 in 1978.

Chapter 4

Express with an inequality symbol.

A 1. x is not equal to -1. 2. y is greater than -5.

Translate into a word sentence.

3. $y > 2$ 4. $-5 < x$ 5. $k \geq 100$

Write an inequality for the graph. Use x for the variable.

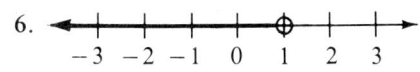

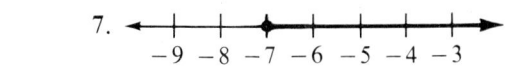

6. [number line with open circle at 1, from -3 to 3]

7. [number line with closed dot at -6, from -9 to -3]

Solve and graph the solution set.

8. $x < -2$ 9. $m - 4 \leq 1$ 10. $0 > y + 3$

11. $-14 \geq 6z$ 12. $-10r > 2$ 13. $6 - 3b \geq 18$

14. $\frac{2}{3}d - 7 \leq -1$ 15. $\frac{5}{4} + 8c \geq \frac{5}{4}$ 16. $7 > a - 1$

Solve and check.

17. Tina bikes at 6 km/h. If she bikes for at least 3.5 h, at least how far will she travel?

18. If one notebook costs $1.29, how many can you buy with a $10 bill?

19. Three more than half a number is not less than the number. Find the greatest possible value for the number.

Solve. Assume that the domain of the variable is {positive integers}.

B 20. $x - 7 < -2$ 21. $y + 3 > -1$ 22. $x - 6 < -6$

Solve.

23. $x + \frac{3}{2} > \frac{2}{5}$ 24. $-\frac{1}{3}n < -3$ 25. $\frac{5}{6}g < \frac{25}{3}$

26. $\frac{1}{2}(4 - 6a) \geq 0$ 27. $6 > -2x > -2$ 28. $2 < \frac{1}{2}y$

29. $5 - p < 6p - 9$ 30. $1 + z > 26 - 9z$ 31. $1 \leq 2k - 1 \leq 5$

32. The sum of a number and three times its opposite is at most -6. What can you conclude about the number?

33. The sum of an even integer and twice the next consecutive even integer is less than 46. Find the greatest possible values for the numbers.

EXTRA PRACTICE

Chapter 5

Exercises 1-3 refer to the diagram. Classify each statement as true or false.

A
1. $\overleftrightarrow{AB} \parallel \overleftrightarrow{CD}$
2. Points A, B, and D are contained by a single plane.
3. There is exactly one line through A, B, and C.

4. R and S have coordinates -3 and 7, respectively, on a number line.
   a. Find RS.   b. Find the coordinate of the midpoint of $\overline{RS}$.
5. Name each endpoint of $\overrightarrow{TV}$.

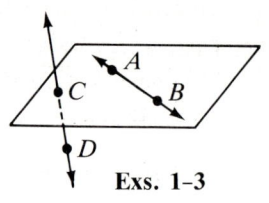

Exercises 6-9 refer to the diagram.
6. Name a right angle.
7. Name an obtuse angle and its measure.
8. If $\overrightarrow{OJ}$ bisects $\angle YOZ$, find the measure of $\angle JOZ$.
9. Find the measure of a complement of $\angle WOX$.

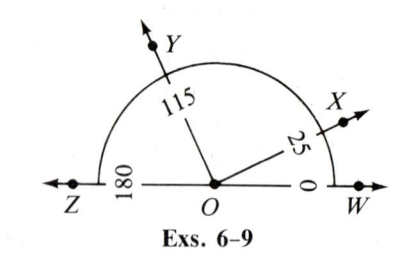

Exercises 10 and 11 refer to the diagram.
10. State the property that justifies the statement.
    a. If $a \parallel b$, then $m\angle 3 = m\angle 5$.
    b. If $m\angle 8 = m\angle 4$, then $a \parallel b$.
11. If $a \parallel b$ and $m\angle 1 = t$, find $m\angle 6$ in terms of t.

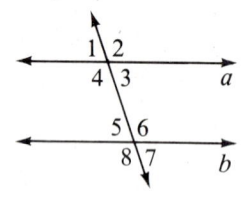

12. In $\triangle ABC$, $m\angle A = 2j + 20$ and $m\angle B = 50 - 2j$. Find the numerical measure of $\angle C$.
13. If a quadrilateral is equiangular must it also be equilateral?
14. Find the perimeter of a regular octagon whose sides are $1\frac{1}{2}$ units long.

B
15. If a line intersects one of two parallel planes, must it intersect the other?
16. The graph of $x \geq -3$ on a number line is best described as a __?__.
17. If two vertical angles have measures $17y - 37$ and $12y - 12$, then $y =$ __?__.
18. If the sum of the measures of a complement and a supplement of an angle is 192, find the measure of the angle.
19. Use the diagram for Exercises 10 and 11. If $m\angle 3 = 2x$ and $m\angle 6 = 3x$, what value of x implies that $a \parallel b$?
20. In $\triangle ABC$, $m\angle A = y - 13$, $m\angle B = 2y + 7$, and $m\angle C = 3y$. Find the numerical measure of each angle.
21. Sketch a regular pentagon and its diagonals.

Chapter 6

A 1. Draw line segments and an angle like those shown. Construct a triangle whose sides have the given lengths.

2. Draw an obtuse $\angle 1$. Construct a triangle with two sides congruent to $\overline{CD}$ and $\overline{EF}$ and the included angle congruent to $\angle 1$.

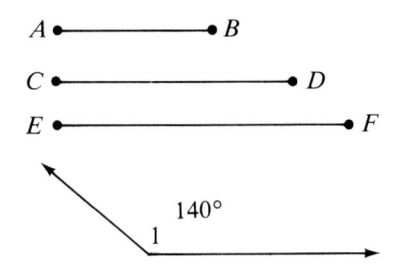

Can you tell from the given information that the two triangles are congruent? If so, give a reason (SSS, SAS, ASA).

3. 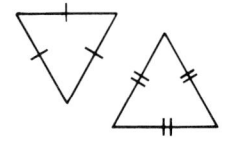 4. 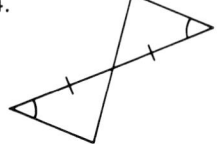 5.

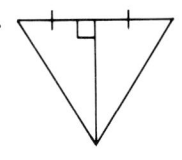

Classify each statement as true or false.

6. The diagonals of a rectangle always divide it into four congruent triangles.
7. One pair of sides of an isosceles trapezoid is congruent and parallel.
8. Find the perimeter and area of a rectangle 24 cm long and 19 cm wide.

Exercises 9–11 refer to the diagram.

9. Find the area of parallelogram $ABCD$.
10. Find the lengths of the bases of trapezoid $ADEB$.
11. Find the area of trapezoid $ADEB$.

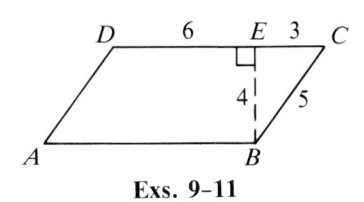

Exs. 9–11

12. A regular square pyramid has base edge 8 and slant height 10.
    a. Find the lateral area.   b. Find the total area.
13. Find the volume of a rectangular prism with $l = 9$, $w = 6.5$, and $h = 3$.

B 14. In the diagram, $\triangle RST \cong \triangle \underline{}$ by the $\underline{}$ method.

15. The measure of the vertex angle of an isosceles triangle is 15 less than three times the measure of a base angle. Find the measure of the vertex angle.
16. If a quadrilateral is equiangular, must it be a parallelogram? Explain.

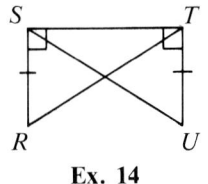

Ex. 14

17. A triangle has area 18 cm² and height 10 cm. Find the length of the base.
18. A right hexagonal prism has congruent base edges of length 10 and lateral area 180. Find the height of the prism.
19. Find the volume of a cube with total area 150 cm².

EXTRA PRACTICE   537

Chapter 7

A 1. Write $(-2)^5$ in simplest form.

2. Find the factors of 24.

Simplify and write in standard form. Give the degree of the polynomial and the coefficient of the term of greatest degree.

3. $-7 + x - 3x + 3x^2 - 1$

4. $-4y^4 + 7y + y^4 - y^3 - 8y + y^2$

Simplify.

5. $(5 - 3a^2 + a) - (a^3 - 7a^2 + 4a)$

6. $(-2cd^2)^2 \cdot c^2d$

7. $(7m - 4)(2m + 1)$

8. $\dfrac{16z^4 - 8z^2 + 2z}{-2z}$

9. Factor into the product of the greatest common monomial factor and another expression: $9j^3 - 15j^2k + 18j^4k^2$

Multiply.

10. $(7d - 2)^2$

11. $(5r - 3)(5r + 3)$

Factor.

12. $s^2 - 3s - 28$

13. $z^2 - 100$

14. $k^2 + 8k + 16$

15. Solve: $c^2 - 17c + 30 = 0$

16. The sum of a number and its square is 72. Find the number.

B Solve.

17. $(1 - 2b) + (5b - 7) = 0$

18. $4x^2 + 9 = 20x$

Simplify.

19. $(3y^2)^4 - (3y^4)^2$

20. $\dfrac{15z^7 - 5z^4 + 10z^6}{5z^3} - \dfrac{9z - 3z^3 + 9z^5}{3z}$

21. Multiply: $(2x - y)(3x^2 - 4xy + 5y^2)$

22. If the area of a triangle is $4x^2 + 8x$ and its height is $8x$, find its base.

23. Complete: $(m - \underline{\ ?\ })(m - 4) = m^2 - \underline{\ ?\ }m + 32$

Factor.

24. $121x^2 + 22x + 1$

25. $2y^2 - 16y - 40$

26. $28z^2 - 343$

27. The length of a rectangle is 3 less than twice the width. If the area is 135, find the length and width.

28. Take a certain negative number, multiply it by 2, subtract 1, and square the result to obtain 49. What is the number?

Chapter 8

A
1. Express 7.25 as a quotient of integers.
2. Write $\dfrac{9}{44}$ as a decimal.
3. Express 1.625 as a fraction in lowest terms.
4. Simplify $-\sqrt{\dfrac{9}{400}}$.
5. Use the table on **page** 281 to find an approximation for $\sqrt{33}$ to the nearest hundredth.

Simplify.

6. $\dfrac{\sqrt{72}}{\sqrt{8}}$
7. $(-3\sqrt{2})^2$
8. $\dfrac{2}{3}\sqrt{180}$
9. $\sqrt{\dfrac{1}{5}}$
10. $\dfrac{3}{4}\sqrt{7} - \dfrac{11}{4}\sqrt{7}$
11. $2\sqrt{50} - 7\sqrt{32}$
12. $(1+\sqrt{3})(2-\sqrt{3})$
13. $(4-2\sqrt{2})^2$
14. $\dfrac{4}{\sqrt{7}-3}$

Solve by using the quadratic formula.

15. $4x^2 - 8x + 3 = 0$
16. $6y^2 + 7y = 10$
17. $z^2 - 10z + 25 = 0$

Write as decimals.

B
18. $\dfrac{153}{160}$
19. $\dfrac{77}{108}$

20. Express $0.1\overline{53}$ as a fraction in lowest terms.
21. Find the square root of 0.0324.
22. Use the divide-and-average method to find an approximation of $\sqrt{307}$ to the nearest tenth.

Simplify.

23. $\sqrt{45} \cdot \sqrt{7} \cdot (-\sqrt{5})$
24. $\dfrac{-8\sqrt{3}}{\sqrt{15}}$
25. $\sqrt{3}(\sqrt{75}+\sqrt{20})$
26. $\dfrac{3}{\sqrt{7}} - \sqrt{28}$
27. $(5\sqrt{2}+\sqrt{10})^2$
28. $\dfrac{9}{3\sqrt{2}-2\sqrt{3}}$

Solve by using the quadratic formula.

29. $2x^2 - 3x - 1 = 0$
30. $5y^2 - 8y = -2$

EXTRA PRACTICE   539

Chapter 9

A
1. Seventy-two campers and counselors attend the Sunrise Day Camp. If the camper-counselor ratio is 5:1, how many campers attend?

2. Solve: $\dfrac{2y - 1}{30} = \dfrac{3y - 4}{40}$

3. If you can travel 252 km in 3.5 h, how far can you travel in 4.5 h at the same speed?

4. 3% of what number is 48?

Exercises 5–8 refer to the diagram.

5. Explain why the triangles must be similar.
6. Complete: $\triangle ABC \sim \triangle \underline{\ ?\ }$
7. Find CD and DE.
8. Find the ratio of the areas of the triangles.

Exs. 5–8

9. Two similar pyramids have heights of 12 cm and 15 cm. If the volume of the smaller pyramid is 192 cm³, find the volume of the larger pyramid.

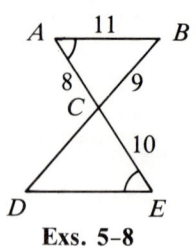

10. In right triangle XYZ, $XY = 12$ and $YZ = 9$. Find two possible values for XZ.

11. Is the triangle with sides of lengths $\sqrt{7}$, $\sqrt{2}$, and 3 a right triangle?

12. The hypotenuse of a 45°-45°-90° triangle has length $\sqrt{14}$. Find the length of each leg.

13. Find the length of the shorter leg of a 30°-60°-90° triangle when the length of longer leg is $4\sqrt{3}$.

B
14. Renee has $10 bills and $5 bills in the ratio of 2:5. If she has $135 in all, how many $10 bills does she have?

15. Solve: $\dfrac{4z - 7}{2z + 2} = \dfrac{2z + 1}{z + 4}$

16. The Nagles invested $3000, part at 6% and the rest at 9%. If the earnings from the 6% investment were $90 less than the earnings from the 9% investment, how much did they invest at each rate?

17. If $\triangle ABC \sim \triangle RST$, $\overline{AB} \cong \overline{BC}$, and $\overline{ST} \cong \overline{RT}$, find the measure of $\angle A$.

18. If $2x + 2$, $5x$, and $5x + 2$ are the lengths of the sides of a right triangle, find the value of x.

19. Find the length of the hypotenuse of an isosceles right triangle with area 9 cm².

Chapter 10

A 1. True or false? If $\overleftrightarrow{RS}$ intersects $\odot O$ and $\overline{OR}$ is a radius of $\odot O$, then R must be a point of tangency.

Find the values of x and y. O is the center of each circle.

2. 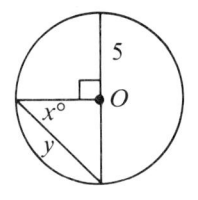
3. 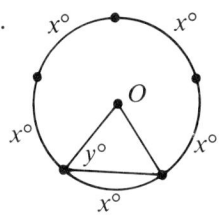
4.

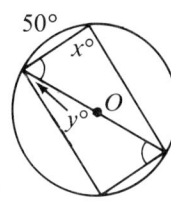

5. 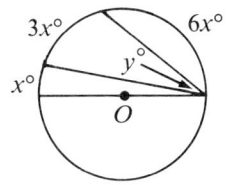
6. 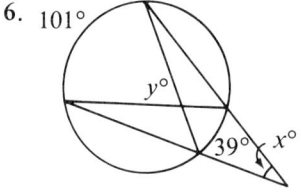
7.

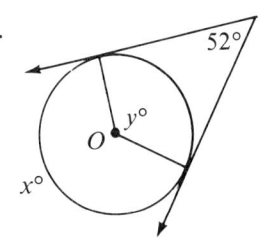

8. Find the exact circumference and the area of a circle with diameter $4\sqrt{3}$.

9. The circumferences of two circles are 4π cm and 8π cm. Find the ratio of their radii and the ratio of their areas.

10. A sphere has radius 5. Use $\pi \approx 3.14$ to find the surface area and the volume of the sphere to the nearest integer.

11. A cone has height 15 and radius 8. Find its total area and its volume.

B 12. Points M and N on a circle determine a major arc and a minor arc. If the measure of the major arc is 20 less than the square of the measure of the minor arc, find the measure of each arc.

13. Quadrilateral $ABCD$ is inscribed in a circle. If $m\angle A = 3x - 5$, $m\angle B = 2x + 9$, and $m\angle C = 5x - 23$, find the numerical measure of $\angle D$.

Find the value of x. Rays that appear to be tangent are tangent.

14. 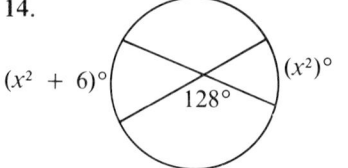
15.

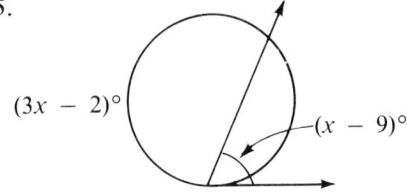

16. Square $ABCD$ is circumscribed about $\odot O$. If $\odot O$ has radius 4, find the exact area of the region inside the square and outside the circle.

EXTRA PRACTICE

Chapter 11

One card is drawn at random from a deck containing one 1, two 2's, three 3's, and four 4's.

A 1. Give the sample space for the experiment and the probability of each outcome.

2. Find the probability of getting an even number.

3. Find the probability of getting a factor of 6.

For the experiment of finding the nonnegative difference when two dice are rolled, find the probability of each event.

4. 0   5. 2   6. a number less than 4

7. In how many different ways can three children wash, rinse, and dry dishes?

8. How many 3-letter arrangements can be formed from the letters A, B, C, D, and E if repetition is allowed? if repetition is not allowed?

9. Evaluate $_6P_2$ and $_{50}P_4$.

Exercises 10 and 11 refer to the experiment in which the spinner shown is spun twice.

10. Draw a tree diagram and find the probability of each outcome.

11. Find the probability of each event.
    a. 4, 2   b. exactly one 4   c. even, even

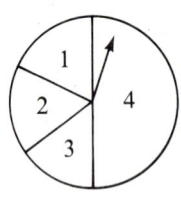

12. Find the probability of getting less than two tails when tossing three coins. Describe the complement of this event and find the probability of the complement.

13. Two cards are drawn at random from an ordinary deck. What is the probability of getting one black card and one red card if the drawing is made with replacement? without replacement?

B 14. Repeat Exercise 8 if the arrangement must begin with a vowel.

15. Refer to the experiment of Exercises 10 and 11. Find the probability that the sum of the two spins is less than 5.

16. Find the probability of getting exactly two heads when a coin is tossed four times.

17. An urn contains four yellow marbles and two red marbles. Three marbles are drawn with replacement. Find the probability of each outcome.

18. Repeat Exercise 17 if the drawings are made without replacement.

Chapter 12

Melissa kept the following record of her weekly earnings, in dollars, from a part-time job.

$$95, 70, 35, 60, 60, 100, 45, 30, 65, 75,$$
$$55, 40, 85, 35, 15, 85, 100, 80, 80, 65$$

A 1. Make a frequency table and a cumulative frequency table. Use intervals of 1-20, 21-40, 41-60, 61-80, and 81-100.

2. How many times did Melissa earn at least $61?

3. How many times did Melissa earn at most $40?

4. In which interval is Melissa's salary most likely to occur?

5. Construct a frequency histogram.

6. Construct a frequency polygon.

7. Construct a cumulative frequency polygon.

Find the mean, median, and each mode (if any).

8. 6, 4, 10, 3, 9, 4

9. 1.2, 1.1, 1.1, 1.0, 1.2

10. Find the first, second, and third quartile for the following data: 53, 54, 52, 50, 56, 50, 59, 56, 57, 50, 56, 51

11. Use the random number table on page 456 to choose a sample of 10 magazine subscribers from a subscription list of 4000.

An experiment consists of drawing a card from a deck composed of four 1's, three 2's, two 3's, and one 4.

12. State the possible outcomes and the probability of each one.

13. Find the expected value.

B 14. The average of the numbers 7, 10, 9, and x is 8. Find the value of x.

15. Repeat Exercise 14 if the median is 8.

16. Use the data in Exercises 1-7. What salary is at the 20th percentile?

17. Tell how you would use the random number table to simulate the experiment of tossing three coins. How would you determine the approximate probability of getting at least two heads?

An experiment consists of rolling two dice and finding the nonnegative difference of the numbers.

18. List all the possible outcomes and find the probability of each one.

19. Find the expected value.

EXTRA PRACTICE   543

Chapter 13

A 1. a. Graph the points $A(-4, 2)$, $B(0, 3)$, $C(1, -5)$, and $D(-2.5, 0)$.
   b. Name a point on the y-axis.
   c. Name a point in Quadrant IV.

Find the distance between each pair of points.

2. $(-3, -3), (-3, -8)$
3. $(-6, 2), (0, 2)$
4. $(4, -1), (-5, -1)$

Graph each equation.

5. $5x - 3y = -15$
6. $2y = 7$
7. $4x - 9y = 6$

Find the x-intercept and the y-intercept (if they exist).

8. $x + 3y = 9$
9. $y = -3x$
10. $x = -1$

11. Find the slope of the line through $(2, -3)$ and $(-7, 0)$.

12. Graph points $A(0, 1)$, $B(2, 4)$, $C(6, 2)$, and $D(6, -2)$. Use slopes to determine if quadrilateral $ABCD$ is a parallelogram, a trapezoid, or neither.

Find the slope, if it exists, of the line whose equation is given.

13. $2x - 3y = 1$
14. $2x = -8$
15. $y - 5 = 0$

16. Find the slope and y-intercept of the line with equation $2y = 6x - 3$.

17. Find an equation in slope-intercept form of the line with slope $-\frac{1}{4}$ that passes through $(8, 0)$.

18. Find an equation of the line with y-intercept 8 that is parallel to the line with equation $3x + y = 12$.

Graph each inequality.

19. $y \leqslant \frac{2}{3}x - 4$
20. $x + y < 5$
21. $6x - y > -3$

B 22. Sketch the triangle with vertices $X(-3, -2)$, $Y(-3, 5)$, and $Z(0, 1)$ and find its area.

23. Write $y + 2 = \frac{5}{3}(x - 1)$ in the form $Ax + By = C$.

24. If $(-3, -4)$ is on the line with equation $2x + By = -6$, find the value of B.

25. Use slopes to decide if $A(-8, 4)$, $B(-2, 0)$, and $C(0, -3)$ are collinear.

Find an equation in slope-intercept form of the line through the given points.

26. $(0, 4), (-2, 6)$
27. $(2, 5), (-4, -4)$
28. $(-1, 12), (1, -2)$

29. Graph the line with equation $y = 3x + 1$.

30. Graph: $5x - 2y \geqslant 15$

544   EXTRA PRACTICE

Chapter 14

Solve each system by graphing. If the system has exactly one solution, check your answer.

A
1. $4x + y = -2$
   $x - y = -3$
2. $2x - 3y = 6$
   $4x - 6y = 24$
3. $x + 3y = 6$
   $y = 4x + 2$

Use the substitution method to solve each system of equations.

4. $x - 2y = 0$
   $-3x + 6y = 0$
5. $x = 5 - y$
   $x - 2y = 9$
6. $2x - y = 2$
   $4x = 5(y - 1)$

Use the addition or subtraction method to solve each system of equations.

7. $3x + y = 5$
   $3x - y = 1$
8. $2x + 5y = 14$
   $6x - 7y = -2$
9. $11x + 12y = 16$
   $5x + 6y = 10$

10. Jane is four years more than six times as old as her son. If the sum of their ages is 39, how old is each?

Solve graphically.

11. $x < -y$
    $y \geqslant x - 2$
12. $y < 2x + 3$
    $x > -2$
13. $y < -\dfrac{1}{2}x$
    $x + y > 4$

14. (Optional) State the maximum value of $3x - y$ and the minimum value of $3x - y$ for the region shown in the diagram.

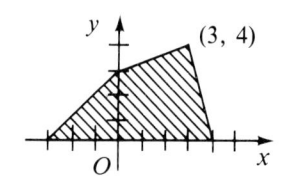

Solve each system.

B
15. $\dfrac{1}{3}x + \dfrac{3}{4}y = 3$
    $x + 2y = 7$
16. $7x - 5y = 18$
    $5x - 7y = 6$
17. $3x + 11y = -17$
    $5x - 9y = 40$

Solve graphically.

18. $3x - 2y < 0$
    $4x - 3y \geqslant 6$
19. $5x - 2y \leqslant 8$
    $7x + 5y > -21$

20. How many kilograms of pure salt and how many kilograms of a 20% salt solution must be combined to produce 4 kg of a 25% solution?

21. A scout troop paddles 10.5 km downstream in 1.5 h. The return trip upstream takes 2 h more. Find the rate of the current.

EXTRA PRACTICE

Algebra Review Exercises

Chapter 3

Solve.

A 1. $x + 3 = 12$ 2. $13 + c = 2$ 3. $2t = 5.2$

4. $-\dfrac{h}{6} = 1$ 5. $11t + 5 = 60$ 6. $\dfrac{3}{5}y - 4 = 2$

Solve for the variable indicated.

7. $A = rt + s;\ s$ 8. $P = A - 3t;\ A$ 9. $4(k - a) = RS;\ a$

Solve the problem by writing an equation and solving it. Check your answer.

10. Joan bought some rolls of tape and paid $7.36. How many rolls did she buy if the tape cost 32 cents per roll?

11. After Carla and Henry brought 16 tables into the gym, there were 126 tables. How many tables were there originally?

12. The sum of three consecutive even integers is 0. Find the three integers.

13. The sum of four consecutive even integers is -26. Find the integers.

B 14. A rope that is 12 m long is cut into two pieces. One piece is 3 m more than half the length of the other. Find the length of each piece.

15. Find four consecutive nonnegative integers with the property that the sum of the first two equals the difference of the last two.

16. There are 50 tables and 136 chairs in a school cafeteria. Some tables have 4 chairs and some have 2 chairs. How many have 4 chairs?

Evaluate the algebraic expression if $x = 4$, $y = -3$, and $z = \dfrac{1}{2}$.

17. $xz - \dfrac{y}{z}$ 18. $2x(y - z)$ 19. $-xy + \dfrac{z}{x}$

Chapter 4

Solve.

A 1. $r \geqslant 3$ 2. $c + 1 \leqslant 5$ 3. $\dfrac{3}{4}x - 4 \geqslant 8$

4. $3 + x < 2$ 5. $8 - s > 1$ 6. $2 - n < 4$

7. $3w > 12$ 8. $9f \leqslant 99$ 9. $-2b > 144$

10. $\dfrac{4}{5}h > -144$ 11. $5 + 3t < 5$ 12. $2 + \dfrac{1}{3}k \geqslant -1$

546   *ALGEBRA REVIEW EXERCISES*

13. $5 - 3z \leq 2$
14. $2n + 1 \leq n + 2$
15. $6w - 21 < w + 6$
16. $8 < g + 3 < 11$
17. $-7 < 14y < 14$
18. $-10 \leq 10x < 20$

Express with an inequality symbol or symbols. Then graph the solution set of the inequality.

19. a is at least 50.
20. p is greater than p subtracted from 3.
21. The sum of t and 1 is between 8 and 22.
22. -4 is less than z, and z is equal to or less than 2.
23. The difference 12 minus x is at most 23.

Solve, and graph the solution set.

B 24. $7m \leq -14$
25. $b + 5 > -2$
26. $5c + 4 > 3c$
27. $8 \leq 2x + 1 < 10$
28. $5 - t < 2 \leq 12 - t$
29. $0 < 5 - 2y \leq 10$
30. $-2 \leq \frac{3}{5}z - 7 \leq 1$
31. $-1 < 4(b - 3)$
32. $3(b + 1) \leq 2(4b - 1)$

Solve the problem by using an inequality.

33. The sum of two consecutive even integers and one-half the smaller is between 7 and 22. Find the largest possible values for the two integers.
34. The average of four integers is at least 95. Three of the integers are 50, 62, and 150. Find the smallest possible value for the fourth integer.
35. There are 15 coins consisting of nickels and dimes in a box. The value of the coins is at most $1.20. What is the greatest number of dimes the box can contain?

Chapter 7

Simplify.

A 1. $(4t + 5) + (-8t + 7)$
2. $(h^2 - 4h + 2) - (3h^2 - 2h - 7)$
3. $a^6 a^5$
4. $(-2x^5)(-5x^2)$
5. $(-2z)^2(-2z^2)$
6. $(c - 3)(3c - 1)$
7. $(3n + 2)(3n - 2)$
8. $(x + 3)(x + 3)$
9. $\dfrac{169b^3}{13b}$
10. $\dfrac{a^2x^2 + a^2x + ax^2}{ax}$

Factor.

11. $t^3 + 5t$
12. $2x^2 - x^3$
13. $x^2 - 3x + 2$
14. $2x^2 - 3x - 9$
15. $m^2 - 16$
16. $4m^2 + 4m + 1$

Solve.

17. $2x^2 + x = 0$
18. $m^2 - 8m + 16 = 0$
19. $t^2 - 5t = 24$
20. $36x^2 = 81$
21. $b^2 - 3b = 10$
22. $y^2 + 1 = 2y$

23. The square of a number added to twice the number equals 35. Find the number.

24. Rae is four times as old as her daughter. If the product of their ages is 256, how old is each?

Solve.

B 25. $36x^2 + 63x = 0$ 26. $3t^2 - 10t + 8 = 0$ 27. $t(3t + 10) = 32$
28. $4m^2 + 5m = 21$ 29. $9(x^2 + 1) + 30x = 0$ 30. $32 + 8x - 4x^2 = 0$
31. $2 + 3a - 2a^2 = 0$ 32. $(x + 1)^2 + (x + 2) = 91$ 33. $(2x + 1)(x + 2) = 5$

34. A flag has a striped design as shown in the diagram. The flag is 150 cm long and 80 cm wide. Find the value of x if the area of the design is half the area of the flag.

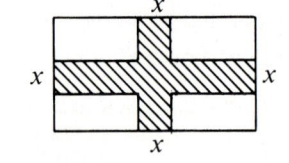

35. If $m^2 - n^2 = x$, $2mn = y$, and $m^2 + n^2 = z$, show that $x^2 + y^2 = z^2$.

36. The base of a triangle is 5 cm longer than its height, and the area of the triangle is 150 cm². Find the length of the base and the height of the triangle.

37. The lengths of the shorter base, the altitude, and the longer base of a trapezoid are three consecutive integers. Find the integers if the area of the trapezoid is 100 cm².

Chapter 8

Express as a fraction in lowest terms.

A 1. $3\frac{1}{8}$ 2. $\frac{111}{33}$ 3. 0.33
4. $0.\overline{3}$ 5. $0.1\overline{2}$ 6. $0.\overline{72}$
7. $\sqrt{1\frac{11}{25}}$ 8. $\sqrt{6\frac{1}{4}}$ 9. $\sqrt{0.\overline{4}}$

Simplify.

10. $\sqrt{54}$ 11. $\sqrt{108}$ 12. $\sqrt{\frac{2}{3}}\sqrt{\frac{3}{8}}$
13. $\sqrt{90}$ 14. $(2\sqrt{5})(5\sqrt{75})$ 15. $(1 + \sqrt{2})^2$
16. $(1 + \sqrt{2})(1 - \sqrt{2})$ 17. $\frac{\sqrt{14}}{\sqrt{2}}$ 18. $\frac{6}{\sqrt{2}}$
19. $\sqrt{27} + \sqrt{48} - \sqrt{75}$ 20. $2\sqrt{3}(2 + \sqrt{3})$ 21. $(3 - \sqrt{2})^2$

Solve.

22. $x^2 - 6x + 2 = 0$ 23. $x^2 + 3x - 1 = 0$ 24. $2x^2 + x - 2 = 0$
25. $x^2 - 1 - x = 0$ 26. $3x^2 - x - 1 = 0$ 27. $x^2 + 3 + 5x = 0$

B 28. Find the value of x^2 when $x = 1 + \sqrt{3}$.

29. Find the value of $x^2 + x$ when $x = \sqrt{\dfrac{1}{2}}$.

30. Simplify: $\sqrt{1\dfrac{1}{2}} \quad \sqrt{1\dfrac{1}{3}} \quad \sqrt{1\dfrac{1}{4}} \quad \sqrt{1\dfrac{1}{5}} \quad \sqrt{1\dfrac{1}{6}} \quad \sqrt{1\dfrac{1}{7}}$

Simplify.

31. $(\sqrt{3} + \sqrt{5})^2$

32. $\sqrt{3\dfrac{1}{5}} + \sqrt{7\dfrac{1}{5}} + \sqrt{16\dfrac{1}{5}}$

33. $(2 + \sqrt{5})(6 + \sqrt{5})$

34. $\dfrac{5}{\sqrt{7} - \sqrt{2}}$

35. $\dfrac{3}{1 - \sqrt{5}}$

36. $\dfrac{1 + \sqrt{2}}{2 + \sqrt{3}}$

Express as a fraction in lowest terms.

37. $0.1\overline{2}$

38. $0.7\overline{2}$

39. $0.\overline{2} + 0.\overline{7}$

Express in simplest form. Then use the table on page 281 to find an approximation for each square root to the nearest hundredth.

40. $\sqrt{171}$

41. $-\sqrt{368}$

42. $\sqrt{136}$

Chapter 9

Express each ratio in simplest form.

A 1. 180 cm to 3 m

2. 3 dozen eggs to 4 eggs

3. 12% of 8 to 8% of 12

4. 2% of x to x% of 3

5. $24\,ab^2c$ to $4a$

6. $(4x)^2$ to $4x^2$

Solve.

7. $\dfrac{n}{n+1} = \dfrac{8}{11}$

8. $\dfrac{t}{t-2} = \dfrac{4}{5}$

9. $\dfrac{k+1}{k+3} = \dfrac{2}{7}$

10. $\dfrac{2x+1}{x+2} = \dfrac{5}{7}$

11. $\dfrac{y-2}{y-1} = \dfrac{6}{y+3}$

12. $\dfrac{x+3}{x+2} = \dfrac{x+2}{2}$

Is the triangle whose sides have the given lengths a right triangle?

13. $\sqrt{2}, 1, \sqrt{3}$

14. $\sqrt{2} + 1, \sqrt{2} - 1, \sqrt{6}$

Solve.

15. Blood accounts for about 7% of the human body. How many kilograms of blood does a 75-kg person have?

16. An $875 computer is marked down to $805. Find the percent of discount.

B 17. $\dfrac{n}{n+3} = \dfrac{4}{n}$

18. $\dfrac{m}{m+4} = \dfrac{m-2}{4}$

19. $\dfrac{x+1}{x+2} = \dfrac{5}{2x-2}$

ALGEBRA REVIEW EXERCISES

20. A fraction equals $\frac{3}{7}$ and its numerator is 20 less than its denominator. Find the fraction.

21. A heart at rest pumps about 60 cm³ of blood per second. Find the percent of increase if it pumps about 300 cm³ per second during exercise.

22. Two positive numbers have a ratio of 3:4 and a sum of 84. Find the numbers.

23. The lengths of the legs of a right triangle are consecutive integers. The hypotenuse is 29 cm long. Find the lengths of the legs.

24. A stock was valued at $75 per share. Last week it lost 12% of its value. This week it gained 5%. What is its value now?

Chapter 13

Find the distance between each pair of points.

A
1. $(3, 9), (5, 25)$
2. $(-3, -9), (3, 9)$
3. $(4, -3), (6, 5)$

Find the slope of the line through each pair of points.

4. $(-1, 6), (6, 10)$
5. $(5, 55), (-5, 5)$
6. $(12, 8), (-21, 8)$

Find the x-intercept and the y-intercept of the line whose equation is given.

7. $6x + 5y = 30$
8. $x - 3y = 15$

9. Find the slope of the line with equation $x - 3y = 8$.

State whether or not the given ordered pair satisfies the given equality or inequality.

10. $3x + 2y = 1$; $(0, 2)$
11. $2x - 3y \geqslant 6$; $(0, -2)$

Find an equation for each line described.

12. Through the points $(0, 6)$ and $(-7, 0)$
13. Through the point $(4, 5)$ and with slope -3
14. Through the point $(5, -5)$ and parallel to the line with equation $x - 3y = 8$
15. With slope $\frac{4}{5}$ and y-intercept -2

16. Graph each inequality: a. $2x + y < 16$ b. $x \leqslant 10 + 3y$

A, B, and C are the points $A(3, 7)$, $B(7, -5)$, and $C(-8, 9)$.

B
17. Find the distance between A and C.
18. Find the slope of the line joining A and C.
19. Find an equation of the line joining B and C.

Find the slope of the line through each pair of points.

20. $(6, 3), (3, 6)$
21. $(5, 6), \left(\frac{1}{5}, \frac{1}{6}\right)$
22. $\left(2, \frac{2}{3}\right)\left(3, \frac{3}{2}\right)$

550   *ALGEBRA REVIEW EXERCISES*

Find the value of n.

23. $(n, 3)$ is on the line with equation $y = 3x + 2$.
24. The points $(6, 2), (8, -1)$, and $(2, n)$ are on the same line.
25. The line joining $(3, 6)$ and $(n, 7)$ has slope -2.

Find A or B if the graph of the given linear equation contains the given point.

26. $Ax + 3y = 3; (-3, 3)$
27. $-5x + By = 6; (4, -2)$

Chapter 14

For each pair of equations determine whether there are no solutions, one solution, or many solutions.

A
1. $y = 3x + 4$
   $y = x + 2$
2. $y = 2x + 1$
   $y = 2x + 7$
3. $y = 6x + 7$
   $2y = 12x + 14$

Solve by whatever method seems best.

4. $y = x - 5$
   $x + y = -1$
5. $y = 3x - 2$
   $y = 8 - 2x$
6. $2x + 3y = 5$
   $3x - y = -9$
7. $x - 6y = 1$
   $3x - 15y = 1$
8. $5x - 2y = 21$
   $3x + 4y = 10$
9. $3x + 4y = 8$
   $15x + 16y = 34$

Translate each sentence into an equation in two variables.

10. The perimeter of an isosceles triangle is 67 cm.
11. Some $20 bills and some $50 bills total $4410.
12. Solve each system graphically. a. $x + 3y \leqslant 5$ b. $y \leqslant 3x + 1$
    $2x + y \geqslant 6$ $x \leqslant y - 6$

Solve each problem using two equations in two variables.

13. Two angles are complementary. The measure of one is 18 more than three times the other. Find the measures of the two angles.

B
14. Find two numbers whose average is -9 and whose difference is 66.
15. The ratio of the probability that Nora will win the election to the probability that she will lose is $2:3$. Find the probability that she will win.
16. David invested some money at 7% and $6000 more at 5.5%. His interest income at the end of one year was $767.50. How much did he invest at each rate?
17. Ben and Molly drive in opposite directions. If Ben drives for 5 h and Molly drives for 3 h, they will be 510 km apart. If Molly drives for 5 h and Ben drives for 3 h, they will be 530 km apart. How fast does Ben drive?
18. Solve each system graphically: a. $3x - 2y \geqslant 2$ b. $5x + 2y \leqslant 7$
    $4x + 3y < 6$ $3x - 4y \geqslant 8$

ALGEBRA REVIEW EXERCISES

Programming in BASIC

1. Introduction to Computers

A computer contains a *central processing unit* (CPU), which does the work, and a *memory*, in which information is stored. It must also have *input* and *output* devices.

COMPUTER

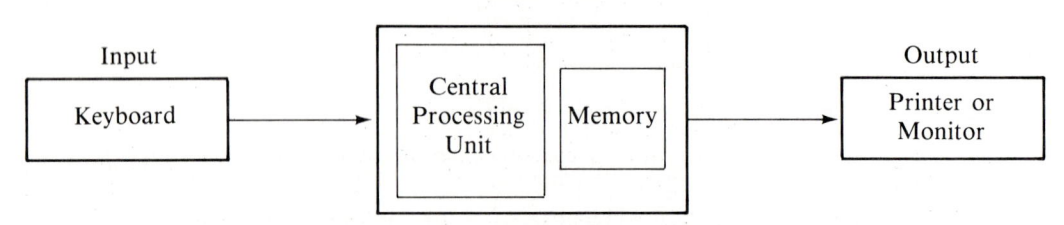

```
                    ┌─────────────────────┐
    Input           │ Central             │        Output
  ┌─────────┐       │ Processing  Memory  │     ┌──────────┐
  │Keyboard │─────▶ │ Unit                │────▶│Printer or│
  └─────────┘       │                     │     │ Monitor  │
                    └─────────────────────┘     └──────────┘
```

Computers are important because they can work with large amounts of data very fast. They work so rapidly that the directions for each project must be written out and stored ahead of time. Such a list of directions is called a *program*. Data and programs used with microcomputers (personal computers) are generally input by means of a *keyboard*. Everything that is typed in appears on the *monitor*, which is like a television screen. Sometimes a *printer* is also attached.

The actual instructions to the computer are given in *machine language*. Such a language may be written down by using 0's and 1's, where 0 and 1 may represent "off" and "on" switches, respectively. Numbers written with 0's and 1's are said to be written in the *binary system* in contrast to the decimal system, which we ordinarily use. In the decimal system, 10 means ten. In the binary system, 10 means two, 11 means three, 100 means four, and so on. Other systems that are sometimes used are the octal (base eight) and the hexademical (base sixteen).

Most programmers use one of the so-called "higher level" languages, such as **BASIC, COBOL, FORTRAN**, or Pascal. The computer contains a program, called a *compiler* or *interpreter*, that translates such a program into machine language. Since some form of **BASIC** is available on each of the microcomputers, as well as on larger computers, that language has been chosen for use in this book.

2. BASIC Symbols, PRINT, END, LET, TAB

BASIC uses many of the familiar signs and symbols of mathematics:

- `+` for addition
- `−` for subtraction
- `=` equals
- `( )` grouping symbols
- `>` is greater than
- `<` is less than

Other symbols are slightly different:

- `*` for multiplication
- `/` for division
- `X↑3` or `X^3` for x^3
- `>=` for $\geq$
- `<=` for $\leq$
- `<>` for $\neq$

552   *PROGRAMMING IN BASIC*

A computer program may be thought of as being made up of blocks like this:

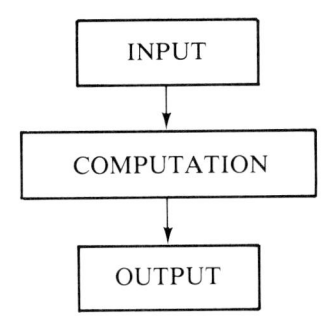

```
┌───────────────┐
│     INPUT     │
└───────┬───────┘
        ▼
┌───────────────┐
│  COMPUTATION  │
└───────┬───────┘
        ▼
┌───────────────┐
│    OUTPUT     │
└───────────────┘
```

However, sometimes the blocks may overlap and sometimes one block may contain several processes.

A BASIC program consists of a list of numbered lines, or *statements*. The simplest kind of program is a computation program like the one shown below (compare page 27). (Of course, from a practical standpoint, you would probably do a calculation like this on a calculator.)

```
10    PRINT ((3 * 4) - (32 / 4 ^ 2)) / 5
20    END
```

(Some versions of BASIC do not require an END statement.) Line 10 is a PRINT statement. It tells the computer to find the values of $((3*4) - (32/4^2))/5$, or 2, and print it. Thus, in this case, the input, computation, and output blocks are all represented in the one line.

It is customary to number the statements 10, 20, ..., as shown here, so that additional statements may be inserted later if needed.

BASIC handles *variables* much as you do in mathematics. There are several ways of giving a value to a variable. One is by using a LET statement, as shown in this example:

```
10    LET A = 25
20    LET B = 36
30    LET S = A + B
40    PRINT S
50    END
```

Line 10 means, "Give the value 25 to the variable A." The input block consists of lines 10-20. Line 30 contains the computation, and line 40 is the output. Lines 30 and 40 could have been combined as PRINT A + B. In either case, the result printed would be the sum of 25 and 36, or 61.

Take time out now to type this program into your computer. Then type the *command* RUN. A command has no line number because it is not part of a program. You can watch what happens on the monitor or on a printer.

3. INPUT, FOR-NEXT (STEP) Loop

Another way of giving a value to a variable is by using an INPUT statement. Change the preceding program to:

```
 5   PRINT "A = ";
10   INPUT A
15   PRINT "B = ";
20   INPUT B
```

Lines 5 and 15 will print exactly the letters and symbols enclosed in quotation marks. Each INPUT statement will print a question mark and wait for you to type in a response. The *semicolon* at the end of line 5 will make the question mark from line 10 print right after "A = ". Similarly, for lines 15 and 20. Type the command LIST to see what the revised program looks like, and then RUN it. (Compare this with Exercise 4 on page 28.)

The programs on page 100 show how an expression can be evaluated and how a formula can be used to solve a problem. Notice that an INPUT statement can be used to put in several values at a time. The computer would print only one question mark, but you would type in all the values asked for, separated by commas. Other examples are given on pages 215-216.

A computer is best used for complicated programs involving many repetitions. One way to take advantage of its great speed is to write a program that contains a *loop*. Then a short program can produce a great deal of output.

A simple type of loop is shown in Exercise 4 on page 28, in program lines 90-110. This is called a FOR-NEXT loop.

```
 90   FOR I = A TO B
100   PRINT I;" ";
110   NEXT I
```

If $A = 5$ and $B = 12$, then the value of the variable I will be printed as 5, 6, 7, 8, 9, 10, 11, 12 in succession.

The first block of this program contains the input; the second block contains the FOR-NEXT loop, which provides the output. When you RUN this program, notice that lines 30 and 80 "print" blank lines.

Exercises 6 and 10 on page 28 show how the STEP portion of the FOR-NEXT loop statements is handled. In general, we have the form:

```
FOR I = A TO B STEP C
```

IF STEP C is omitted, the STEP is taken to be 1. If $A < B$, so that I is increasing, then $C > 0$. If $A > B$, so that I is decreasing, then $C < 0$. See Exercise 10.

Now consider a slightly different FOR-NEXT loop:

```
10   FOR N = 1 TO 10
20   PRINT N,N * N,N ^ 3
30   NEXT N
40   END
```

Here the variable N is given the values 1, 2, . . . , 10. Thus, the input (line 10) and the computation and the output (line 20) are all contained in one block — the loop. The *commas* in line 20 will cause the values of N, N * N, and N^3 to be printed on the same line, but spaced apart. Type in this program and type the command RUN. The output will consist of ten lines and three columns.

Insert these lines at the beginning of the program:

```
5   PRINT "N","N^2","N^3"
7   PRINT
```

Now RUN the program again. The output should look something like the following.

N	N^2	N^3
1	1	1
2	4	8
3	9	27
4	16	64
5	25	125
6	36	216
7	49	343
8	64	512
9	81	729
10	100	1000

PRINT TAB(A) can be used to space out columns across a page when some distances other than the fixed spaces given by commas are desired.

TAB statements can also be used to draw graphs, as illustrated in the program on page 494. See also pages 522-523.

Microcomputers may have special graphics capabilities. Explore these in your user's manual.

4. IF-THEN (-ELSE), REM, GOTO

A very powerful tool of computer programming is the IF-THEN (-ELSE) construction. This is illustrated in the program on page 556. Lines 30-100 form a loop within which are two *branches* resulting from the test in line 40. If X < 6, then "*" is printed (line 90). Otherwise (else) "-" is printed (line 60). The "then" and "else" branches are emphasized in this program by inserting REM (remark) statements as in lines 50 and 80. REM statements appear in a listing of a program but do not affect its execution. The GOTO statement in line 70 is necessary to skip over the other branch.

The loop in lines 120-190 will print "0" when I has the value zero. Then the program ends.

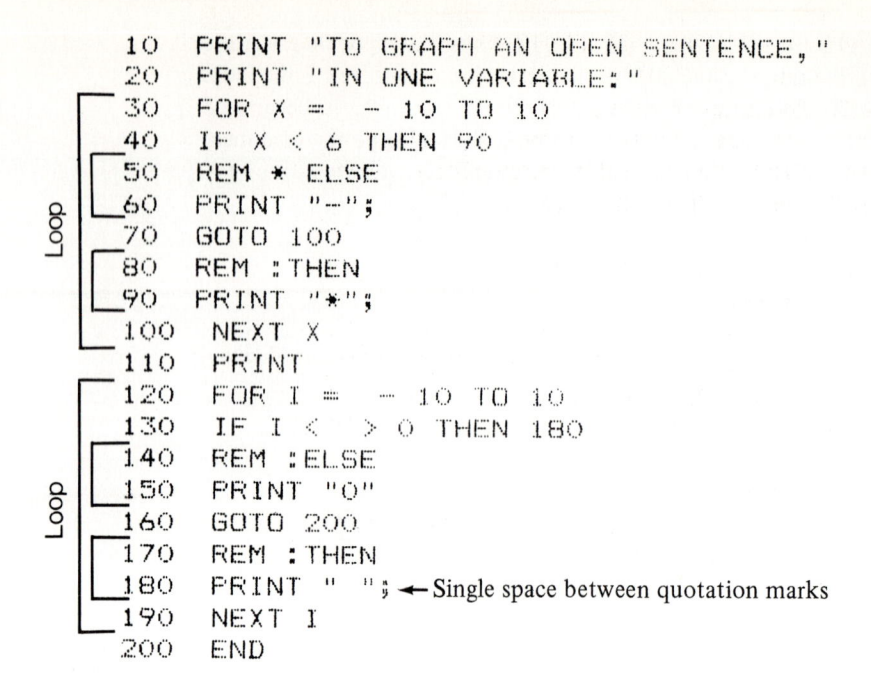

```
10   PRINT "TO GRAPH AN OPEN SENTENCE,"
20   PRINT "IN ONE VARIABLE:"
30   FOR X =  - 10 TO 10
40   IF X < 6 THEN 90
50   REM * ELSE
60   PRINT "-";
70   GOTO 100
80   REM : THEN
90   PRINT "*";
100  NEXT X
110  PRINT
120  FOR I =  - 10 TO 10
130  IF I < > 0 THEN 180
140  REM : ELSE
150  PRINT "O"
160  GOTO 200
170  REM : THEN
180  PRINT " ";   ← Single space between quotation marks
190  NEXT I
200  END
```

(Lines 30–100 form a Loop; lines 40–70 form an inner loop. Lines 120–190 form a Loop; lines 130–160 form an inner loop.)

RUN this program and compare the result with that on page 109. Notice that this program tests only *integers* from −10 to 10.

In this program, the inequality to be graphed appears in line 40. Therefore, this line must be changed for each exercise. Making this change:

$$40 \quad \text{IF } X + 4 < 10 \text{ THEN } 90$$

RUN the program again and notice that the result is the same. (Compare page 111.)

Now look back at the program on page 65. Make a copy of it, and mark off the loops for P (lines 30-200) and for Q (lines 80-190). Notice that the second loop is entirely within the first loop. Such loops are called *nested loops*.

Find the IF-THEN statements in this program, and insert REM: ELSE and REM: THEN statements where they are appropriate. (Another example of nested loops is shown in the program on page 338.)

5. String Variables, READ-DATA

The program on page 65 uses *string variables:* P$, Q$, C$. In this program, each of these string variables may have the *value* "T" or "F". A *string* in general is a set of characters enclosed in quotation marks.

The following fragment of a program shows how strings may be used in chatty kinds of programs.

```
20    PRINT "WHAT IS YOUR NAME";
30    INPUT N$
40    PRINT "HELLO ";N$;"!"
50    PRINT "I AM GLAD TO MEET YOU."
60    PRINT "DO YOU ENJOY WORKING WITH ";
70    PRINT "ME (YES/NO)";
80    INPUT A$
90    IF A$ = "YES" THEN 170
100     REM :ELSE
110     PRINT "I AM VERY SORRY, ";N$;"."
120     PRINT "PERHAPS IF YOUR TEACHER ";
130     PRINT "HELPED YOU..."
140     PRINT "PLEASE TRY ME AGAIN LATER."
150   GOTO 200
160     REM :THEN
170     PRINT "I AM VERY GLAD, ";N$;"."
180     PRINT "WE CAN LEARN LOTS ";
190     PRINT "OF THINGS TOGETHER."
200   END
```

Type in and RUN this program twice — once with the answer YES and again with the answer NO. (You do not need to type quotation marks around the answer here.) Notice that the program will go to the ELSE branch if you type in a string other than YES.

The program on page 165 uses the **READ-DATA** combination of statements to give values to a variable. Look at these lines:

```
10    READ D
170   DATA 28, 83, 45, 1, 30, 90, 66, 124, 56, 131
```

The first value of D to be read from line 170 is 28. With D = 28, the program RUNs through to line 140 or line 160 and then loops back to line 10. The next value of D is read from line 170, and so on. The program ends when the last value in the list of **DATA** has been read and used. Another way to end such a loop is to put a special number at the end of the **DATA** list, as is done in the program on page 337.

The **DATA** line may be put anywhere in the program. For instance, it may be put in as line 12 instead of line 170 if you wish to list all your input near the beginning of the program.

6. Functions: INT, ABS, SQR, RND

BASIC includes several *functions*. One of these is the *greatest integer function*, which is explained and used on page 261. (INT (3.9) = 3, and so on.)

To further clarify its use in factoring, we include here another program.

PROGRAMMING IN BASIC   557

```
10   PRINT "TO FIND FACTORS OF"
20   PRINT "A POSITIVE INTEGER W(>1):"
30   PRINT "WHAT IS YOUR POSITIVE INTEGER";
40   INPUT W
50   FOR F = 1 TO W / 2     ←There are no factors between W/2 and W.
60   LET Q = W / F
70   IF Q < F THEN 130
80   IF Q < > INT (Q) THEN 120
90   REM :ELSE
100  PRINT F;" AND ";Q;" ARE FACTORS OF ";W
110  REM :THEN
120  NEXT F
130  END
```

RUN this program for these values of W: 120, 567, 36, 37.

The greatest integer function is also used in rounding numbers. See page 293.

To find integral factors of an integer, we shall also use another function, the *absolute value function*. Thus, ABS(X) gives the absolute value of X. Make these changes in the program:

```
20   PRINT "AN INTEGER W: W<-1 OR W>1:"
30   PRINT "WHAT IS YOUR INTEGER";
50   FOR F = 1 TO ABS (W) / 2
70   IF ABS (Q) < F THEN 130
105  PRINT - F;" AND "; - Q;" ARE FACTORS OF ";W
```

RUN this revised program for the integers listed above and their negatives.

Because of line 70, this program stops computing factors when F becomes equal to Q or greater than Q. That is, we can stop the values of F at the *square root* (page 276) of W. BASIC has a special *square root function*. Thus, SQR(X) gives the non-negative square root of X. Make these changes in the program and RUN it again several times.

```
20   PRINT "AN INTEGER W (<> 0):"
50   FOR F = 1 TO SQR ( ABS (W))
70
```

Another function is the *random function*, RND, which is described on page 462.

In all ordinary computation, BASIC gives the answer correct to a fixed number of digits. For very large and very small numbers, it switches over to scientific notation as described on page 428. However, in some instances, a program can be written that will compute a result digit by digit as you would do if you were computing by hand. The following program will compute division digit by digit to express a proper (N < D) fraction as a decimal.

```
10   PRINT "TO FIND A DECIMAL FOR N/D:"
20   PRINT "INPUT N,D (0 < N < D)";
30   INPUT N,D
```

```
40   PRINT N;"/";D;" = 0.";
50   REM :   THE NUMERATOR IS THE FIRST
60   REM :   "REMAINDER" IN THE LOOP BELOW.
70   LET R = N
80   REM :   THIS LOOP COMPUTES AND PRINTS
90   REM :   ONE DIGIT, Q1, AT A TIME.
100  FOR I = 1 TO D + 3
110  LET A = R * 10
120  LET Q = A / D
130  LET Q1 = INT (Q)
140  PRINT Q1;
150  LET R = A - Q1 * D  ← New remainder
160  IF R = 0 THEN 200
170  NEXT I
180  PRINT " ... "
190  GOTO 210
200  PRINT "   TERMINATES"
210  END
```

RUN this program for 8/25, 2/3, and 13/22, and compare your results with those on page 272. Notice that line 100 will give D + 3 digits. RUN the program for 1/7, 2/7, 3/7, 4/7, 5/7, and 6/7, and write down the repetend in each case. What do you notice?

Make these changes in the program and RUN it.

```
20    LET N = 1
30    FOR D = 1 TO 20
210   NEXT D
220   END
```

7. Subscripted Variables, DIM, Sorting

Sometimes, especially in statistical work, it is necessary to work with a list of numbers. When working with a list of numbers, it is helpful to use a *subscripted variable*. Such a variable, X(I), is used in the programs on page 463 and is further described in Exercise 6 on page 464.

In the program in Exercise 5 on page 463, the values for the subscripted variable are READ from the DATA list as follows:

$X(1) = -2000$, $X(2) = -6$, $X(3) = -2$, $X(4) = 1$,
$X(5) = 7$, $X(6) = 26$, $X(7) = 41$, $X(8) = 117$,
$X(9) = .9999$

Since 9 values have been read, but only 8 values are to be used in the computation, line 70 reduces the value of I by 1.

This program can be made more flexible by using an INPUT statement. If more than 10 (or, in some cases, 11) values are to be used, a dimension (DIM) statement will be needed. Thus, the program might be changed as shown on page 560.

PROGRAMMING IN BASIC

```
4    PRINT "INPUT DATA (.9999 TO END):"
6    DIM X(20)
8    LET S = 0
10   LET I = 1
20   INPUT X(I)
130
```

In the program in Exercise 5, the DATA were arranged in order from the least to the greatest. It is possible to have the computer *sort* a list of values into this order if the values have been INPUT in any other order. To do this, insert these lines:

```
81   FOR J = 1 TO I - 1
82   FOR K = 1 TO I - J
83   IF X(K) < = X(K + 1) THEN 87
84   LET T = X(K)
85   LET X(K) = X(K + 1)
86   LET X(K + 1) = T
87   NEXT K
88   NEXT J
```

The sorting is done by making several passes through the list. Whenever

$$X(K) <= X(K+1),$$

the order is kept the same (line 83). But whenever

$$X(K) > X(K+1),$$

the order is reversed in lines 84–86. Notice that the value of X(K) must be held in a temporary variable T so that it can be given to X(K+1).

Add these lines to print the ordered list.

```
122   FOR J = 1 TO I - 1
124   PRINT X(J);",";
126   NEXT J
130   PRINT X(I)
```

LIST the revised program to see what it looks like. RUN this program when the DATA are INPUT in a different order: $-2000, 117, -6, 41, 7, -2, 1, 26, .9999$. The result should be the same.

Now RUN the program for the values listed in Exercises 7 and 11 on page 448. RUN the program for the list of rounded temperatures used in Exercises 1-5 at the bottom of page 444.

560   *PROGRAMMING IN BASIC*

Preparing for Regents Examinations

The Regents examination for Sequential Math–Course I is a comprehensive achievement test of the course objectives outlined in the New York State syllabus. It is prepared by a committee of teachers and State Education Department specialists. The test is scheduled for January, June, and August each year.

The test is divided into two parts. Part I consists of 35 short-answer and multiple-choice questions. Students must answer 30 questions, each worth 2 points. Part II consists of 7 analytical problems related to major strands, or topics, of the course. The student must answer 4 questions, each worth 10 points. A construction may also be asked on Part I.

In addition to its comprehensiveness, the test is designed to measure a student's ability to apply what he or she has learned to a variety of different problem settings. The course philosophy of topic integration and concept unification is clearly displayed in questions requiring insights and skills from several different strands. Most noteworthy is the integration of algebra and geometry, especially in the coordinate plane.

To prepare effectively for this examination, you should organize a list of the fundamental concepts, principles, and theorems. Then you should review and drill the various skills and procedures related to these topics. You should strive for both accuracy and the ability to apply these skills and strategies to solve problems.

To assist you in this preparation, go back to the Cumulative Reviews on pages 126–128, 264–266, 383–386, and 525–528. They provide ample practice for the short answer and multiple-choice sections of the test and help you sharpen test-taking skills such as eliminating choices that are clearly in error.

The following pages contain checklists for each topic as well as many sample questions. These questions are similar in form and style to those found on the Regents exam but are longer in order to provide comprehensive review of the topic being tested. A typical Regents question is shorter and more selective in its scope. The sample questions given here are designed to help you further organize course objectives in a manner consistent with Regents testing philosophy.

Logic – Checklist

Statements
  open sentences
  conjunctions ($p \wedge q$)
  disjunctions ($p \vee q$)
  conditions ($p \rightarrow q$)
  biconditionals ($p \leftrightarrow q$)
Negations

Truth tables
  equivalent statements
  tautologies
  contradictions
Related conditionals
  converse
  inverse
  contrapositive

Logic – Sample Problems

1. Let p be the statement "$x^2 = 25$" and q be the statement "$x = -5$" and let $S = \{-5, 0, 5\}$. For what values of S is the statement true?
   a. $p \wedge q$
   b. $p \vee q$
   c. $p \rightarrow q$
   d. $p \leftrightarrow q$
   e. $\sim p \rightarrow q$
   f. $p \rightarrow \sim q$
   g. the converse of $p \rightarrow q$
   h. the inverse of $p \rightarrow q$
   i. the contrapositive of $p \rightarrow q$

2. The conditional "If you pass this course, then you will not go to summer school" is given.
   a. Write a statement that is logically equivalent to the given statement.
   b. If you went to summer school, then what can you conclude about your grade in the course?
   c. If the statement is false, then what can you conclude about your grade in the course?
   d. If the statement is false, then did you go to summer school?
   e. Write the inverse of the statement.
   f. Write a statement that is logically equivalent to your answer from part (e) of this question.

3. Let p be the statement "It is raining," let q be the statement "Sheila is sad," and let r be the statement "The farmers are happy."
   a. Write in symbolic form.
      (1) It is not raining or the farmers are happy.
      (2) Sheila is sad and the farmers are happy.
      (3) If the farmers are not happy, then it is not raining.
      (4) Sheila is sad if and only if the farmers are happy.
   b. If p is true, q is true, and r is false, then what can you conclude about the truth value of the statement?
      (1) $\sim p \wedge q$ (2) $(p \vee q) \rightarrow r$ (3) $r \leftrightarrow \sim q$ (4) the converse of $r \rightarrow p$
   c. If the statement "The farmers are happy if it is raining" is true, then write another true conditional statement.

4. The statement $(p \rightarrow q) \leftrightarrow \sim(p \wedge \sim q)$ is given.
   a. Construct a truth table for the statement.
   b. Is the statement a tautology?
   c. If p represents the statement "Bob has a date" and if q represents the statement "He will go to the dance," write a statement that is equivalent to "If Bob has a date, then he will go to the dance."

Algebraic Skills – Checklist

Evaluating algebraic expressions
Solving linear equations
Solving linear inequalities
  relative to a given domain
  graphing solution set on a number line
  compound inequalities
    conjunctive
    disjunctive
Laws of exponents
Polynomials
  addition and subtraction
  multiplication
  division by a monomial
  factoring
    greatest common monomial factor
    trinomials into a product of binomials
Solving quadratic equations
Radical expressions
  approximating square roots
  simplifying (simple radical form)
  addition and subtraction
  multiplication
  division
  rationalizing the denominator

Ratios
Proportions
Percents
Systems of linear equations
  graphic solutions
  solving by substitution
  solving by addition or subtraction
Systems of linear inequalities-graphing
Problem solving
  by formula
  involving percents ($n\%$ of p is q)
  using equations
  using inequalities
  using quadratic equations
  using proportions
  using systems of equations
  using systems of inequalities

Algebraic Skills – Sample Problems

1. Evaluate if $x = 3$, $y = -2$, and $z = -\dfrac{1}{4}$.

   a. $2xy^2z$
   b. $(-2yz)^2$
   c. $xy - yz$
   d. $x^2 - 2xy + y^2$
   e. $\dfrac{x + y + z}{xyz}$
   f. $\dfrac{x}{z} - \dfrac{z}{y}$

   g. For what value of k is the expression $\dfrac{2yz}{k + x}$ undefined?

2. Solve.

   a. $3(5 + x) = 9$
   b. $4x + 7 = 2 - x$
   c. $\dfrac{x}{3} - \dfrac{x}{4} = 8$
   d. $0.3x - 2 = 4$
   e. $\dfrac{6}{x} = \dfrac{8}{12}$
   f. $x^2 - 7x = 0$
   g. $x^2 - 7x + 10 = 0$
   h. $3x^2 - 5x - 12 = 0$
   i. $x^2 = 36$
   j. $4x^2 - 9 = 0$
   k. $\dfrac{4}{x} = \dfrac{x}{16}$
   l. $\dfrac{x - 5}{x - 4} = \dfrac{x + 1}{x + 4}$
   m. $y = mx + b$
   n. $\begin{cases} 4x + y = 5 \\ 5x - y = 13 \end{cases}$
   o. $\begin{cases} 4x + 3y = 27 \\ y = 2x - 1 \end{cases}$

PREPARING FOR REGENTS EXAMINATIONS

3. Solve for x and represent the solution set graphically on a number line.
   a. $2x - 3 < 7$
   b. $5 - 4x \geq -3$
   c. $-4 \leq 3x + 2 \leq 8$
   d. $0 \geq -x - 4 > 3$

4. Perform the indicated operation. Express the answer in simplest form.
   a. $(6x^2 + 3x - 7) + (-4x^2 - 3x + 6)$
   b. $(6x^2 + 3x - 7) - (-4x^2 - 3x + 6)$
   c. $3x(5 - 2x)$
   d. $5 - (x + 4) + 6x$
   e. $(2x - 7)(x + 4)$
   f. $(3 - 2x)^2$
   g. $(3x - 2)(3x + 2)$
   h. $\dfrac{15x^6}{5x^2}$
   i. $\dfrac{-18x^3y^2z}{3x^2y^2z^2}$
   j. $(3x^4)(-2x^3)$
   k. $(-2x^2)^3$
   l. The mean (average) of $x + 2$ and $5x - 4$ is __?__.

5. Factor completely.
   a. $x^2 - 9x$
   b. $x^2 - 9$
   c. $x^3 - 9x$
   d. $x^2 - 9x + 20$
   e. $x^2 - 9x - 36$
   f. $5x^2 + 3x - 2$
   g. $3x^2 - 4x + 1$
   h. $9x^2 - 16$
   i. $7x^2 + 70x + 175$

6. Express in simplest radical form.
   a. $\sqrt{200}$
   b. $6\sqrt{54}$
   c. $\sqrt{\dfrac{4}{5}}$
   d. $(\sqrt{3})(2\sqrt{12})$
   e. $(-3\sqrt{5})^2$
   f. $\dfrac{\sqrt{75}}{\sqrt{3}}$
   g. $4\sqrt{2} - \sqrt{18}$
   h. $\sqrt{12} + \sqrt{27}$
   i. $\sqrt{3} + \sqrt{\dfrac{1}{3}}$
   j. The mean (average) of $\sqrt{50}$ and $\sqrt{72}$ is __?__.
   k. Find $\sqrt{22}$ to the nearest tenth.

7. Solve the system graphically and check.
   a. $\begin{cases} x + y = 3 \\ x - y = 5 \end{cases}$
   b. $\begin{cases} 2x + y = 7 \\ x - 2y = 1 \end{cases}$
   c. $\begin{cases} 2x + y > 4 \\ y \leq x \end{cases}$

8. Solve *algebraically*.
   a. Find three consecutive positive even integers such that twice the sum of the first two equals three times the third.
   b. The square of a certain positive number is four more than three times the number. Find the number.
   c. The sum of the squares of two consecutive positive even integers is 20. Find the two numbers.
   d. The sum of two numbers is 30 and their difference is 6. Find the two numbers.
   e. Bob invested $5000, some at 6% interest and the rest at 8%. If his total earnings were $352, how much did he invest at each rate?
   f. Sixty percent of a certain amount is $16 more than 40% of the same amount. Find the amount.
   g. Three out of every 5 students ride the bus to school. If there are 480 bus riders, then how many students are in the school?
   h. A test has 40 questions and Tom missed 12 of them. If all questions are weighted equally, what should Tom's score be on the test?

i. The sum of two numbers is 9 and the sum of their squares is 45. Find the two numbers.
j. If 7 is subtracted from half a certain number, the difference is 25. Find the number.

Geometry — Checklist

Points
Lines
  parallel
  perpendicular
  intersecting (transversals)
Segments
  midpoint
  measure
Angles
  types
  measure
  pairs of
    complementary
    supplementary
    vertical
    on parallel lines cut by transversal
  in a triangle
    interior
    exterior
    right triangle
    equilateral
    isosceles
Polygons
  triangles
  quadrilaterals
  regular
  others
Constructions
  copy segment
  copy angle
  triangle, given SSS
  triangle, given SAS
  triangle, given ASA
Congruence
  triangle
    methods of proving
    corresponding parts
  quadrilaterals
    properties of special quadrilaterals
    parallel or congruent sides
    congruent or supplementary angles
    diagonals

Measurement
  perimeter
  area
    triangles
    parallelograms
    trapezoids
  surface area
    cubes
    prisms
    pyramids
  volume
    cubes
    prisms
    pyramids
  similarity
    corresponding angles
    corresponding sides
    ratio of perimeters
    ratio of areas
    ratio of volumes
Pythagorean Theorem
Circles
  parts of
  angles of
    central
    inscribed
    tangent and chord
    two chords
    two secants
    secant and tangent
    two tangents
  measurement
    circumference
    area
    ratio of radii
    ratio of circumferences
    ratio of areas
Spheres
  volume and ratio of volumes
  surface area and ratio of areas
Cylinders–volume and surface area
Cones–volume and surface area

PREPARING FOR REGENTS EXAMINATIONS

Geometry – Sample Problems

Directions: In exercises involving radicals, leave answers in simplest radical form.

1. Given: Parallel lines l and m, with transversal n
   a. If $m\angle 4 = 130$, find $m\angle 5$.
   b. If the sum of the measures of $\angle 1$ and $\angle 5$ is 200, find $m\angle 2$.
   c. If the measures of angles 4 and 6 are in the ratio 5:4, find $m\angle 6$.
   d. If $m\angle 3 = x + 30$ and $m\angle 6 = 2x - 10$, find the value of x.

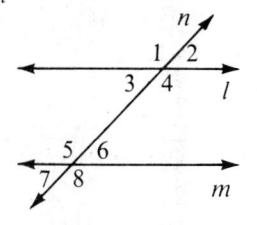

2. Given: $\overleftrightarrow{AC}$ and $\overleftrightarrow{BE}$ intersect at F;
   $\overrightarrow{FD} \perp \overleftrightarrow{BE}$; $m\angle CFD = 35$; $\overline{ED} \parallel \overleftrightarrow{AC}$
   Find the measure of each angle.
   a. $\angle BFD$ b. $\angle BFC$ c. $\angle AFB$
   d. $\angle AFE$ e. $\angle FDE$ f. $\angle FED$

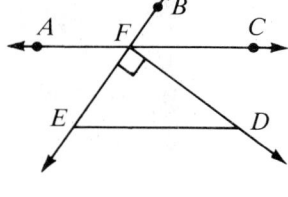

3. Given: $\overline{AB} \cong \overline{BC}$; $\overline{BE} \perp \overline{AC}$
   a. If $m\angle A = 40$, find $m\angle ABC$.
   b. If $m\angle BCD = 100$, find $m\angle ABC$.
   c. If $m\angle ABC = 32$, find $m\angle A$.
   d. If $m\angle A = 4x$ and $m\angle BCD = 5x$, find $m\angle ABC$.
   e. If $m\angle A = x$ and $m\angle BCD = 2x$, then $\triangle ABC$ is __?__.
   f. If $m\angle A = x$ and $m\angle ABC = 2x$, then $\triangle ABC$ is __?__.
   g. If AB is 4 cm more than twice AC and the perimeter of $\triangle ABC$ is 33 cm, find AC and AB.
   h. If BE is 5 cm more than AC and the area of $\triangle ABC$ is 12 cm^2, find AC, BE, AB, and the perimeter of $\triangle ABC$.

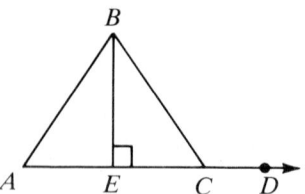

4. Given: Rectangle $ACDF$; square $ABEF$
   a. If $AF = 5$ cm and $ED = 3$ cm, find:
      (1) the perimeter of $ACDF$
      (2) the area of $\triangle AFE$
      (3) the area of trapezoid $ACDE$
      (4) AE
   b. If $BC:AC = 2:5$ and the perimeter of $ACDF$ is 48 cm, find:
      (1) AB and AE (2) the area of $ACDF$ (3) the area of $BCDE$
   c. If AC is 4 cm more than AF, and the area of $ACDF$ is 60 cm^2, find:
      (1) AF (2) AC (3) AE
      (4) the perimeter of $ACDE$
   d. If the perimeter of $ABEF$ is 28 cm, find the area of $ABEF$.
   e. If the area of $ABEF$ is 36 cm^2, find the perimeter of $ABEF$.
   f. If the perimeter of $ABEF$ is tripled, then:
      (1) each side of $ABEF$ is multiplied by __?__
      (2) the area of $ABEF$ is multiplied by __?__

PREPARING FOR REGENTS EXAMINATIONS

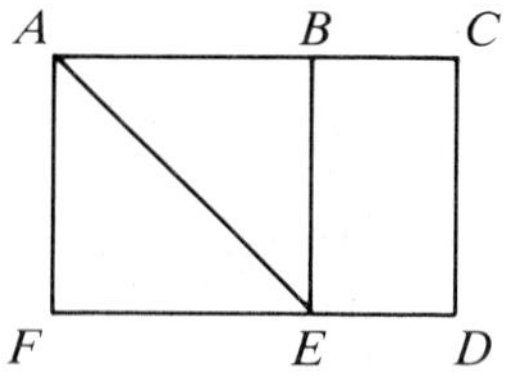

g. If $AB = 2x + 3$, find an expression for the perimeter of $ABEF$ in terms of x.
h. If $AC = 3x + 4$ and $BC = x - 2$, find an expression for the perimeter of $ACDF$ in terms of x.

5. Given: $\overline{BC} \parallel \overline{DE}$; $\overline{BC} \perp \overline{AD}$; $\overline{DE} \perp \overline{AD}$
   a. If $AB = 6$, $BD = 3$, and $BC = 8$, find:
      (1) DE (2) AC (3) AE
      (4) the ratio of the perimeters of $\triangle ABC$ and $\triangle ADE$
      (5) the ratio of the areas of $\triangle ABC$ and $\triangle ADE$
   b. If $AB = 5$, $AD = 15$, and $DE = 36$, find:
      (1) BC
      (2) the area of $\triangle ABC$
      (3) the area of trapezoid $BCED$
      (4) the perimeter of trapezoid $BCED$
   c. If $AB = x + 1$, $BC = x + 3$, $AC = 10$, and $AE = 25$, find:
      (1) the value of x (2) the area of $\triangle ABC$ (3) the area of $\triangle ADE$

6. A flagpole 12 m tall casts a shadow 9 m long. At the same time, a shadow 6 m long is cast by a nearby tree. How high is the tree?

7. Given: Rhombus $ABCE$ with altitude $\overline{CD}$
   a. If $m\angle ECD = 40$, find the measure of each angle.
      (1) $\angle CED$ (2) $\angle AEC$
      (3) $\angle BCE$ (4) $\angle B$
      (5) $\angle A$ (6) $\angle BCD$
   b. If $AE = 10$ cm and $ED = 6$ cm, find:
      (1) CD
      (2) the area of $\triangle CED$
      (3) the area of $ABCE$
      (4) the area of trapezoid $ABCD$

8. Given: Circle with center O; diameter $\overline{BC}$; $\overrightarrow{QP}$ tangent to $\odot O$ at A and $\overrightarrow{QR}$ tangent to $\odot O$ at C
   a. If $m\widehat{AC} = 80$, find the measure of each angle or arc.
      (1) $\angle AOC$ (2) $\angle ABC$
      (3) $\angle AOB$ (4) $\angle ACB$
      (5) $\angle OAC$ (6) $\angle BAO$
      (7) $\widehat{AB}$ (8) $\angle PQR$
      (9) $\angle QAO$ (10) $\angle QCO$
      (11) $\angle PAB$ (12) $\angle QAC$
   b. If $BC = 12$ cm, find:
      (1) AO
      (2) the circumference of $\odot O$ in terms of π
      (3) the area of $\odot O$ in terms of π

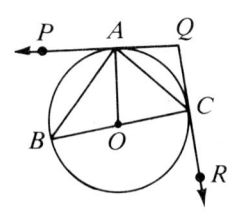

9. Given: Square *ABCD* inscribed within the larger circle and circumscribed about the smaller circle
   a. If $AB = 4$ cm, find:
      (1) the radius of the smaller circle
      (2) the area of the smaller circle
      (3) the circumference of the smaller circle
      (4) the diameter of the larger circle
      (5) the radius of the larger circle
      (6) the area of the larger circle
      (7) the circumference of the larger circle
      (8) the area of the shaded region
      (9) $m \stackrel{\frown}{AB}$
      (10) length of $\stackrel{\frown}{ABD}$
   b. Find:
      (1) the ratio of the radius of the smaller circle to the radius of the larger circle
      (2) the ratio of the area of the smaller circle to the area of the larger circle
   c. If the circumference of the larger circle is 20π, find the area of the larger circle.
   d. If the area of the larger circle is 36π, find the circumference of the larger circle.

10. Given: Cube with edges of length $3x$
    a. Express the volume in terms of x.
    b. Express the total surface area in terms of x.
    c. If the volume of the cube is 216 cubic units, find:
       (1) the value of x
       (2) the total surface area of the cube
    d. The edge of another cube has length $2x$. Determine the ratio of the volume of the larger cube to the volume of the smaller cube.

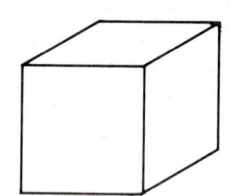

Probability – Checklist

Fundamental Counting Principle
Sample spaces
  table
  tree diagram
Probability
  fraction
  $0 \leqslant P(E) \leqslant 1$
  complementary events
Events: single stage
  coin flipping
  dice rolling
  card selections
  spinners
  urn

Events: multi-stage
  with replacement
  without replacement
Permutations
  $_nP_n = n!$

$_nP_r = \dfrac{n!}{(n-r)!}$

Probability — Sample Problems

1. Experiment: Coin flipping
   a. If one coin is flipped, find:
      (1) *P*(heads)   (2) *P*(heads or tails)   (3) *P*(heads and tails)
   b. If two coins are flipped, find:
      (1) *P*(two heads)   (2) *P*(no heads)   (3) *P*(at least one head)
   c. If three coins are flipped, find:
      (1) *P*(three tails)   (2) *P*(exactly two tails)   (3) *P*(at most two tails)

2. Experiment: Dice rolling
   a. If one die is rolled, find:
      (1) *P*(4)   (2) *P*(perfect square)   (3) *P*(factor of 6)
   b. If two dice are rolled and sums noted, find:
      (1) *P*(4)   (2) *P*(1)   (3) *P*(prime)
      (4) the number with the largest probability of occurrence

3. Experiment: Card selections from a standard deck of playing cards
   a. If one card is selected, find:
      (1) *P*(ace)   (2) *P*(diamond)   (3) *P*(black card)
      (4) *P*(face card)   (5) *P*(seven or spade)   (6) *P*(not a heart)
   b. If two cards are selected with replacement, find:
      (1) *P*(two tens)   (2) *P*(both red)   (3) *P*(both clubs)
      (4) *P*(neither card is a face card)
      (5) which has the greater probability of occurrence, both cards even or both cards odd
   c. If two cards are selected without replacement, find:
      (1) *P*(two tens)   (2) *P*(two red tens)   (3) *P*(both face cards)
      (4) *P*(neither card is a heart)

4. Experiment: Spinning the spinner shown
   a. If you spin this spinner once, find:
      (1) *P*(1)   (2) *P*(2)
      (3) *P*(4)   (4) *P*(even)
      (5) *P*(factor of 4)   (6) *P*(at most 3)
   b. If you spin this spinner twice, find:
      (1) *P*(first 1, then 2)   (2) *P*(both odd)
      (3) *P*(first even, then odd)   (4) *P*(neither is a 1)
      (5) *P*(first is larger than second)   (6) *P*(both same)

5. Experiment: Selecting from an urn containing five black marbles and two white marbles
   a. If one marble is chosen, find:
      (1) *P*(black)   (2) *P*(white)   (3) *P*(red)
   b. If two marbles are chosen with replacement, find:
      (1) *P*(exactly one white)   (2) *P*(both black)
      (3) *P*(neither black)   (4) *P*(at least one white)
      (5) *P*(both same color)   (6) *P*(at most one white)

(continued)

c. If two marbles are chosen without replacement, find:
   (1) *P*(exactly one black)   (2) *P*(both white)
   (3) *P*(neither white)   (4) *P*(at most one black)
   (5) *P*(different colors)   (6) *P*(at least one white)
d. If one marble is selected, and then one die is rolled, find:
   (1) the sample space   (2) *P*(white, odd)   (3) *P*(odd)
   (4) *P*(black)   (5) *P*(white or prime)   (6) *P*(black, 3 or 4)

6. Let $S = \{1, 2, 4, 6\}$.
   a. How many 4-digit numbers can be formed from the numbers in set S if no repetition is allowed?
   b. How many of these 4-digit numbers are even?
   c. How many of these 4-digit numbers are less than 2000?
   d. How many of these 4-digit numbers are greater than 4000?
   e. How many 4-digit numbers can be formed from the numbers in set S if repetition is allowed?
   f. How many 3-digit numbers can be formed from the numbers in S if no repetition is allowed?
   g. How many of these 3-digit numbers are odd?

7. How many 8-letter permutations can be formed from the letters of the word SUPERMAN?

8. Evaluate.
   a. 6!   b. $_4P_4$ c. $_5P_3$

9. There are five different sets of train tracks connecting Albany and New York City. In how many different ways can a train make a round trip between Albany and New York City?

Statistics — Checklist

Frequency
  tables
  histograms and polygons
Cumulative frequency
  tables
  histograms and polygons

Measures of central tendency
  mean
  mode
  median
Quartiles
Percentiles

Statistics — Sample Problems

1. Jamie bought a bag of marbles and sorted them by colors.
   a. How many marbles were in the bag?
   b. What is the mean number of marbles per color?
   c. If one marble was selected at random, find:
      (1) $P(red)$
      (2) $P(red, white, or blue)$
      (3) $P(orange)$
   d. If two marbles are selected at random, with replacement, find:
      (1) $P(two\ green)$
      (2) $P(at\ least\ one\ yellow)$
      (3) $P(red\ and\ blue)$
   e. If two marbles are selected at random without replacement, find:
      (1) $P(two\ yellow)$
      (2) $P(neither\ is\ green\ or\ blue)$
      (3) $P(first\ white,\ then\ red)$

2. Kim's test grades for this marking period are as follows: 78, 90, 87, 75, 88, 84, 90, 72, 86
   a. What is the current mean score?
   b. Identify the mode.
   c. Identify the median.
   d. What score must Kim get on the next test in order to achieve a mean of 85?

3. The table below shows the distribution of scores from thirty games of bowling.

Interval	Frequency	Cumulative Frequency
121-130	3	?
131-140	6	?
141-150	7	?
151-160	9	?
161-170	4	?
171-180	1	?

   a. Copy and complete the table.
   b. Draw a frequency histogram and a frequency polygon.
   c. Draw a cumulative frequency histogram and a cumulative frequency polygon.
   d. Which interval contains the median?
   e. Which interval contains the lower quartile?
   f. Which interval contains the 75th percentile?
   g. What percent of the scores were less than or equal to 140?
   h. What is the probability that a score selected at random will be greater than 160?

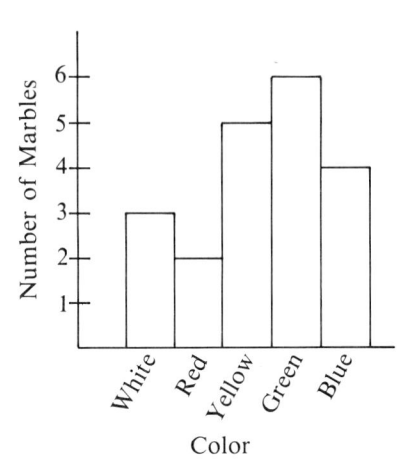

PREPARING FOR REGENTS EXAMINATIONS

Coordinate Geometry – Checklist

Coordinate axes
Plotting points
Quadrants
Distances
  horizontal and vertical
  diagonal
Graphing lines
  by table or list
  by x-intercept and y-intercept
  by slope-intercept method
Slope
  horizontal lines
  vertical lines
  diagonal lines
  parallel lines

Equations of lines
  given slope and y-intercept
  given slope and point on the line
  given two points on the line
  given one point on the line and the
    equation of a parallel line
Linear inequalities
  half-plane
    open
    closed
  boundary line
Systems
  linear equations
  linear inequalities

Coordinate Geometry – Sample Problems

1. Given: lines k, l, m, and n with the following equations:
    $k: x = 4$ $l: y = -3$ $m: y = 2x - 1$ $n: 2y + x = 8$
    a. Draw the graphs of k, l, m, and n on the same coordinate plane.
    b. Determine the slope of each line.
    c. Determine the y-intercept of each line. If none exists, write "none."
    d. Determine the x-intercept of each line. If none exists, write "none."
    e. Determine the coordinates of the point of intersection of lines m and n.
    f. Determine the solution set for the following system of linear equations:
    $$2y + x = 8$$
    $$x = 4$$
    g. If point $(3, z)$ lies on line m, then $z = \underline{\ ?\ }$.
    h. If point $(a, 1)$ lies on line n, then $a = \underline{\ ?\ }$.
    i. Determine the equation of a line parallel to line m that passes through point $(2, 5)$.
    j. Determine the x-intercept of the line determined in part (i).
    k. Find the distance between the y-intercepts of lines m and n.
    l. Find the distance between the x-intercepts of lines m and n.
    m. Find the distance between the x-intercept and the y-intercept of line n.
    n. Determine the solution set of the following system on your graph:
    $$y \geqslant 2x - 1$$
    $$y \leqslant -3$$
    o. Determine the coordinates of a point in the solution set found in part (n).
    p. Given the points $A(0, 4)$, $B(10, -1)$, and $C(0, -1)$, find:
      (1) the area of $\triangle ABC$ (2) the perimeter of $\triangle ABC$
    q. Given the points $D(1, 1)$, $E(4, 7)$, and $F(4, 1)$, find the area of $\triangle DEF$.

2. Given: points $A(0, 2)$, $B(6, 2)$, $C(4, -2)$, and $D(-2, -2)$
   a. Determine the slope of $\overleftrightarrow{AD}$.
   b. Find an equation of $\overleftrightarrow{AD}$.
   c. Determine the slope of $\overleftrightarrow{BC}$.
   d. Find an equation of $\overleftrightarrow{BC}$.
   e. Determine the slope of $\overleftrightarrow{AB}$.
   f. Find an equation of $\overleftrightarrow{AB}$.
   g. Determine the slope of $\overleftrightarrow{DC}$.
   h. Find an equation of $\overleftrightarrow{DC}$.
   i. Is quadrilateral $ABCD$ a parallelogram? Why?
   j. Find the area of $ABCD$.
   k. Find equations of $\overleftrightarrow{AC}$ and $\overleftrightarrow{BD}$.
   l. Use your answers from part (k) to determine the coordinates of the point E of intersection of diagonals $\overline{AC}$ and $\overline{BD}$.
   m. Use the Pythagorean Theorem and the results of part (l) to find:
      (1) EA (2) EB (3) EC (4) ED
   n. What conclusion can you draw concerning point E?
   o. Is $\triangle ADE \cong \triangle CBE$? Why?

ANSWERS TO SELF-TESTS

CHAPTER 1

SELF-TEST 1, page 7
1. {March, April}   2. {12, 14, 16, 18}   3. {the letters in the word MATH}
4. Answers may vary. For example, {the integers between 0 and 6}   5. B   6. F
7. 1   8. $^-5$

SELF-TEST 2, page 17
1. -163 2. -483 3. 27,181   4. 104   5. 223   6. 937   7. 2204
8. -1587 9. 3404   10. -5805 11. $-74,880$ 12. 1155   13. 23
14. -45 15. -98 16. 57

SELF-TEST 3, page 26
1. $\frac{25}{63}$ 2. $\frac{23}{30}$ 3. $\frac{7}{20}$ 4. $\frac{5}{24}$ 5. $\frac{5}{42}$ 6. $-\frac{1}{16}$ 7. $\frac{22}{21}$ 8. $-\frac{2}{15}$
9. 3   10. -2 11. 112   12. 4

CHAPTER 2

SELF-TEST 1, page 46
1. Answers will vary. For example, 12 and 13 make the statement true; 0 and 1 make it false.
2. {1, 2, 3, 4, 6, 8, 12, 24}   3. Wednesday is not the day in the middle of the week.
4. Elephants are small.   5. $5 > 7$ and $5 + 4 = 9$; False   6. $r \wedge s$; True   7. $5 > 7$ or $5 + 4 = 9$; True   8. $r \vee s$; False

SELF-TEST 2, page 60
1. If 12 is even, then 36 is odd; False   2. $r \to s$; True   3. true; true

4.
p	q	$p \wedge q$	$\sim p$	$(p \wedge q) \to \sim p$
T	T	T	F	F
T	F	F	F	T
F	T	F	T	T
F	F	F	T	T

5. Converse: If January is cold, then June is warm;   Inverse: If June is not warm, then January is not cold;   Contrapositive: If January is not cold, then June is not warm.   6. $w \to \sim r$
7. $p \leftrightarrow q$; True   8. They are logically equivalent.

CHAPTER 3

SELF-TEST 1, page 79
1. -9 2. 5   3. $5w - 10$ 4. no   5. yes; $\left\{\frac{5}{2}\right\}$ 6. 1

SELF-TEST 2, page 89
1. 15   2. 11   3. -4 4. -14 5. 8   6. -7 7. $-\frac{7}{4}$ 8. 8
9. -4 10. $\frac{7}{5}$ 11. -3 12. -20

574   *ANSWERS TO SELF-TESTS*

SELF-TEST 3, page 99
1. $3x - 6$ 2. $25(3n)$ 3. $a - 8$ 4. $2x + 8 = 16$ 5. $x + (x + 2) + (x + 4) = 54$; 16, 18, 20   6. $2(3w + 4) + 2w = 64$; length = 25, width = 7   7. $a + 6 = 2(a - 2)$; 10
8. $\frac{1}{3}n = -5$; $n = -15$; -15 9. $\frac{a + c}{2}$ 10. $2b - a$ 11. $\frac{2A}{b}$

CHAPTER 4

SELF-TEST 1, page 116
1. $n > 7$ 2. $y < 0$ 3. $x \neq 36$ 4. $t > -14$ 5. [number line]
6. [number line]   7. $\{t : t < 12\}$ 8. $\{n : n > -7\}$ 9. $\{x : x < 3\}$
10. $\{y : y > 10\}$ 11. $\{x : x < -3\}$ 12. $\{t : t < -21\}$

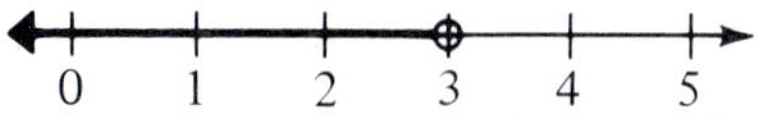

SELF-TEST 2, page 121
1. $x \leq -3$ 2. $3 < y < 8$ 3. $\{t : t \geq 6\}$ [number line]
4. $\{n : n \leq 10\}$ [number line]
5. $\{x : x \leq 1\}$ [number line]
6. $\{y : y \leq -20\}$ [number line]
7. $\{x : 6 \leq x < 8\}$ [number line]
8. $\{x : 0 \leq x < 2\}$ [number line]

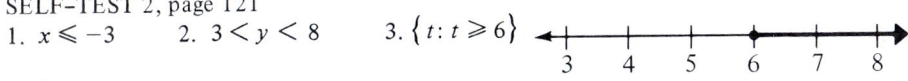

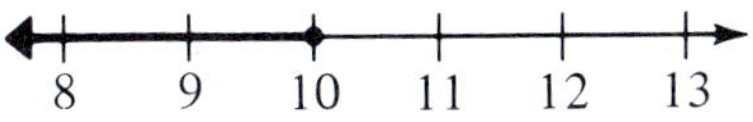

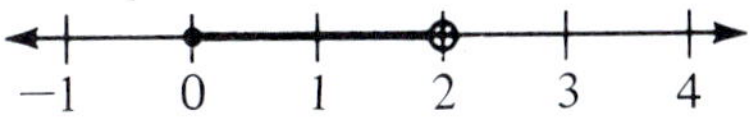

SELF-TEST 3, page 124
1. 4 and 9   2. 15 and 27   3. 22 and 49

CHAPTER 5

SELF-TEST 1, page 145
1. No; Yes   2. point C 3. $\overleftrightarrow{AB}$ 4. one   5. E
6. $\overline{FG} \cong \overline{GH}$ 7. 1   8. a. $\angle DOC$ or $\angle COB$ b. $\angle DOB$ or $\angle BOA$
c. $\angle DOA$ d. $\angle COA$ 9. 114   10. $\angle ZYV \cong \angle VYX$

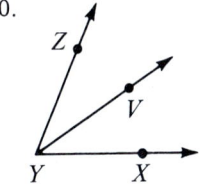

SELF-TEST 2, page 164
1. $\overline{TR} \perp \overline{QS}$ 2. $\angle RST$ 3. $\angle QRT$ and $\angle SRT$; $\angle 2$ and $\angle QTS$ 4. 23   5. 5
6. 102   7. $a; b$ 8. $s; t$ 9. 80   10. 58   11. 22 and 68   12. Yes; No
13. 135   14. 7.8 cm

ANSWERS TO SELF-TESTS   575

CHAPTER 6

SELF-TEST 1, page 185

1.

2.

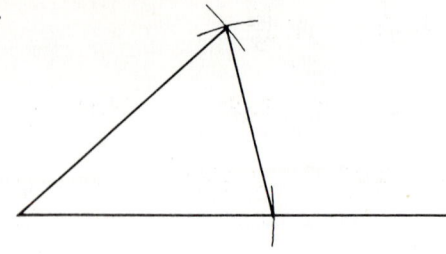

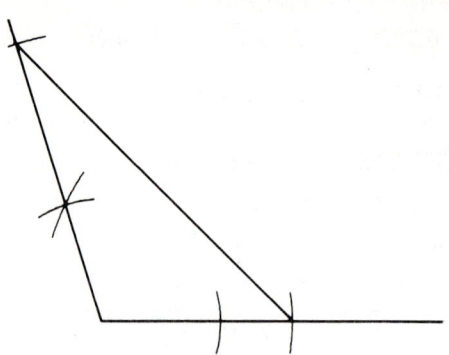

3. SAS   4. No   5. No   6. ASA   7. $\overline{AB} \cong \overline{CD}, \overline{BD} \cong \overline{DB}, \overline{AD} \cong \overline{CB}$, $\angle ABD \cong \angle CDB, \angle BDA \cong \angle DBC, \angle DAB \cong \angle BCD$

SELF-TEST 2, page 196
1. $m\angle 1 = m\angle 2 = 29$ 2. $m\angle 3 = 48, m\angle 4 = 84$ 3. vertex angle, 106; base angle, 37
4. a. YZW b. Z c. ZWY d. $\overline{ZW}$ 5. Both pairs of opposite sides are parallel; both pairs of opposite sides are congruent; two sides are both parallel and congruent.   6. 13
7. True   8. True   9. False   10. Answers may vary. For example:
11. a rectangle and a square

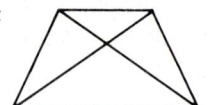

SELF-TEST 3, page 214
1. a. 25 cm   b. 28.5 cm^2 2. a. 46 m   b. 120 m^2 3. 8 cm   4. a. 10 m
b. 4.2 m^2 5. a. 78 cm   b. 270 cm^2 6. a. 100 m^2 b. 177 m^2 7. a. 300 cm^2
b. 525 cm^2 8. 80 m^2 9. 441 cm^3 10. 166.4 m^3

CHAPTER 7

SELF-TEST 1, page 237
1. 81   2. -32 3. $125x^2$ 4. $256y^3$ 5. $5t$ 6. $2z^2 + 10z$
7. $2s^3 + s^2 + 3s - 5$ 8. $-p^{10} - p^7 + p^5 + 2p^2$ 9. $7t - 3$ 10. $4z^2 + 2z - 5$
11. $4d - 8$ 12. $2q^3 - 4q^2 + 2q + 2$ 13. d^8 14. $-x^6 y^3 z^9$ 15. $28r^4 t^3$
16. $24x^4 y^4$

SELF-TEST 2, page 248
1. $10x^3 + 6x$ 2. $-2z^4 + 4z^3 - 10z$ 3. $9k^2 + 24k + 16$ 4. $2h^2 - 7h - 15$
5. t^5 6. $\dfrac{1}{27k^3}$ 7. $13xy^2$ 8. $3y^4 - 4y + y^2$ 9. 1, 2, 4, 8; 8
10. 1, 3, 5, 15; 15   11. $2x^2(2x + 1)$ 12. $ay(ay + y - a)$ 13. $z^2 - z - 30$
14. $s^2 - 4$ 15. $9q^2 + 6q + 1$ 16. $6t^2 + t - 15$

SELF-TEST 3, page 260
1. $(x + 3)(x + 1)$ 2. $(y + 5)(y - 2)$ 3. $(4h - 1)(2h + 1)$ 4. $(5s - 1)(2s + 3)$
5. $(t + 5)(t - 5)$ 6. $(k + 3)^2$ 7. $x^2 - 4$ 8. $9g^2 - 30g + 25$ 9. $\{-4, 3\}$
10. $\{4, 5\}$ 11. $\left\{-\dfrac{1}{3}, 5\right\}$ 12. $\left\{-\dfrac{5}{2}, \dfrac{7}{3}\right\}$ 13. -9 and -8 14. 7 and 9

576   *ANSWERS TO SELF-TESTS*

CHAPTER 8

SELF-TEST 1, page 276
1. 0.3125   2. 0.0$\overline{6}$ 3. 0.$\overline{06}$ 4. $\frac{13}{40}$ 5. $\frac{5}{11}$ 6. $\frac{13}{45}$

SELF-TEST 2, page 282
1. 30   2. −65   3. 28   4. −7.07   5. 9.90   6. −8.19   7. 14.14

SELF-TEST 3, page 288
1. 6$\sqrt{65}$ 2. −6   3. 12   4. 7$\sqrt{5}$ 5. $-\frac{3\sqrt{3}}{8}$ 6. $\frac{5\sqrt{11}}{11}$ 7. 20$\sqrt{7}$
8. $\sqrt{5}$ 9. $-3\sqrt{3}$

SELF-TEST 4, page 293
1. $9 - 4\sqrt{5}$ 2. $\sqrt{11} - 2$ 3. $2\sqrt{6} + \sqrt{2} - 2\sqrt{3} - 1$ 4. $\{-\frac{1}{2}, 2\}$
5. $\{-4 + 2\sqrt{3}, \ -4 - 2\sqrt{3}\}$

CHAPTER 9

SELF-TEST 1, page 313
1. $\frac{3}{10}$ 2. 24, 36, and 48   3. 32 and 20   4. 133 games   5. 28   6. 56
7. 520 km   8. 28 h   9. 234   10. 6%   11. 71.4%   12. 825 students

SELF-TEST 2, page 325
1. 15   2. 26   3. 27   4. 25:121   5. 9:49   6. 1:125   7. 202.5 cm^3

SELF-TEST 3, page 336
1. 9   2. $2\sqrt{6}$ 3. no   4. $8\sqrt{5}$ 5. $8\sqrt{2}$ 6. 7   7. $12\sqrt{2}$ 8. 12
9. $7\sqrt{3}$ 10. $12\sqrt{3}$

CHAPTER 10

SELF-TEST 1, page 352
1. $\overleftrightarrow{DC}$ 2. radius   3. chord   4. $x = 8$, $y = 108$, $z = 36$ 5. 302
6. 114   7. 27   8. 55

SELF-TEST 2, page 366
1. 43   2. 59   3. 114   4. 52   5. $m\angle 1 = 25$; $m\angle 2 = 102$ 6. 138
7. 22   8. 21

SELF-TEST 3, page 379
1. 16π m, 64π m^2 2. 10.9 cm   3. $\frac{3}{4}$, $\frac{9}{16}$ 4. 50 cm   5. 2304π cm^2
6. 4500π cm^2 7. 400π square units   8. 84π mm^2; 182π mm^2; 294π mm^3
9. 21π square units; 30π square units; $6\pi\sqrt{10}$ cubic units

CHAPTER 11

SELF-TEST 1, page 403
1. {Spring, Summer, Fall, Winter}; $\frac{1}{4}$ 2. {2, 3, 4, 5, 6, 7, 8, 9}; $\frac{1}{8}$
3. {111, 113}; $\frac{1}{2}$ 4. {pink, gray, olive}; $\frac{1}{3}$
5. $\frac{2}{3}$ 6. $\frac{1}{13}$ 7. $\frac{1}{2}$ 8. $\frac{2}{3}$ 9. $\frac{1}{12}$ 10. $\frac{1}{6}$ 11. $\frac{1}{8}$ 12. $\frac{7}{8}$

SELF-TEST 2, page 412

1. 6   2. 720   3. 120   4. $\frac{1}{6}$ 5. 42   6. 117,600   7. 210   8. 6720

SELF-TEST 3, page 420

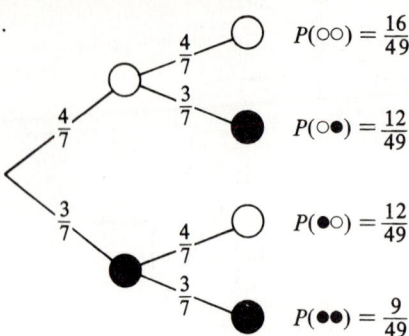

1.
$\frac{4}{7}$ ○ — $\frac{4}{7}$ ○ $P(○○) = \frac{16}{49}$
$\frac{3}{7}$ ● $P(○●) = \frac{12}{49}$
$\frac{3}{7}$ ● — $\frac{4}{7}$ ○ $P(●○) = \frac{12}{49}$
$\frac{3}{7}$ ● $P(●●) = \frac{9}{49}$

2. $\frac{24}{49}$ 3. $\frac{40}{49}$ 4. $\frac{33}{49}$ 5. Thursday, Saturday, $\frac{2}{7}$; all the other days, $\frac{5}{7}$ 6. Wednesday, Thursday, Saturday, $\frac{3}{7}$; all the other days, $\frac{4}{7}$ 7. HT, TH, HH, $\frac{3}{4}$; TT, $\frac{1}{4}$ 8. 5 + 1, 1 + 5, 2 + 4, 4 + 2, 3 + 3, 5 + 3, 3 + 5, 6 + 2, 2 + 6, 4 + 4, $\frac{5}{18}$; all other possible sums, $\frac{13}{18}$

SELF-TEST 4, page 428

1. $\frac{1}{169}$, $\frac{3}{676}$ 2. $\frac{1}{16}$, $\frac{3}{52}$ 3. $\frac{1}{676}$, $\frac{1}{663}$ 4. $\frac{144}{169}$, $\frac{141}{169}$ 5. $\frac{1}{2}$ 6. $\frac{1}{2}$ 7. $\frac{1}{2}$ 8. $\frac{1}{2}$

CHAPTER 12

SELF-TEST 1, page 445

1. 22 students   2. 51 students   3. 49 students   4. 73 students

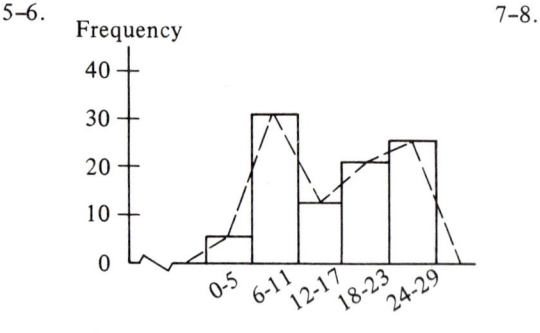

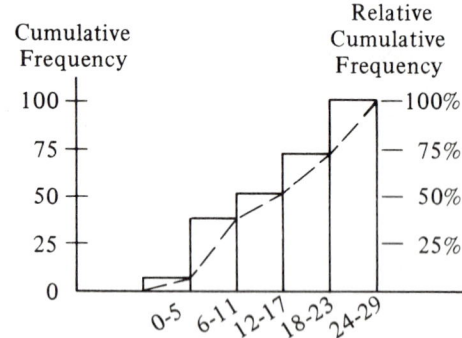

5–6. Frequency histogram (0-5, 6-11, 12-17, 18-23, 24-29)

7–8. Cumulative Frequency and Relative Cumulative Frequency graphs

SELF-TEST 2, page 454

1. 2; 2; 2   2. 38; 40; no mode   3. 18; 17; 20   4. $\frac{3}{5}$; $\frac{3}{5}$; no mode   5. 50   6. 35   7. 39   8. 38   9. 44   10. 49

SELF-TEST 3, page 461

1–2. Answers will vary. Examples are given.   1. 40, 63, 28, 09, 54, 26, 06, 73   2. 208, 031, 392, 298, 043, 217, 075, 350, 240, 152   3. Answers will vary. Actual probability is $\frac{1}{2}$.   4. Answers will vary. Actual probability is $\frac{3}{8}$.   5. 2   6. 0

578   *ANSWERS TO SELF-TESTS*

CHAPTER 13

SELF-TEST 1, page 479
1. $A(3, 1)$, $B(-1, 1)$, $C(-2, -3)$, $D(1, -3)$ 3. 6 4-7. Points chosen may vary.
2. [graph showing $E(-4, 1)$ and $F(2, 3)$]
4. [graph]
5. [graph]
6. [graph]
7. [graph]

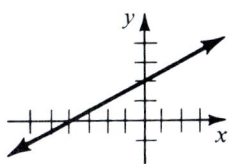

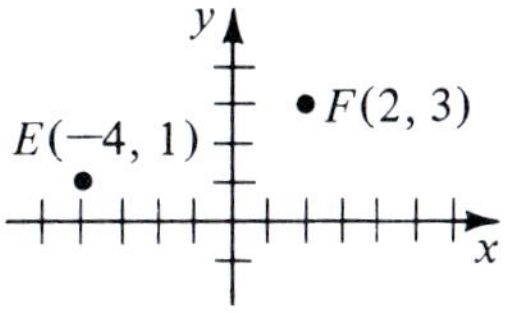

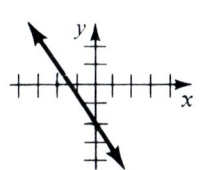

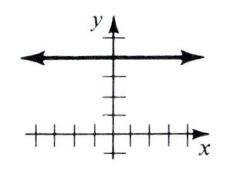

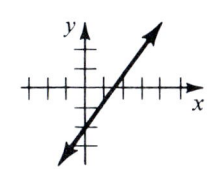

SELF-TEST 2, page 492
1. -1 2. slope of $\overleftrightarrow{AB}$ is $\frac{4}{9}$, slope of $\overleftrightarrow{BC}$ is $-\frac{4}{5}$, slope of $\overleftrightarrow{CD} = \frac{1}{2}$, slope of $\overleftrightarrow{AD} = -1$; not a parallelogram 3. $\frac{4}{7}$ 4. no slope 5. slope $= \frac{-5}{4}$, y-intercept $= 2$
6. $y = \frac{-2}{3}x + 3$ 7. $y = \frac{-3}{2}x + 2$ 8. $y = 2x + 1$
9. [graph] 10. [graph]
11. [graph] 12. [graph]

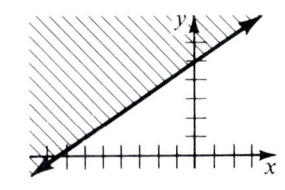

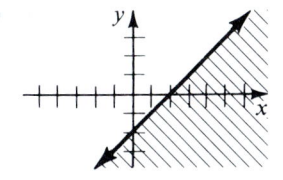

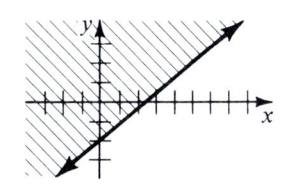

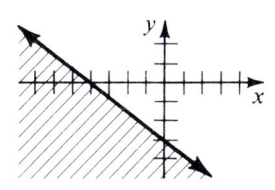

CHAPTER 14

SELF-TEST 1, page 511
1. $(3, -1)$ 2. $(2, 3)$ 3. $(-2, 2)$ 4. $(6, -2)$ 5. no solution 6. $(2, 5)$
7. $(1, 3)$ 8. $(3, 5)$ 9. $(4, -6)$

SELF-TEST 2, page 517
1. Margaret's age: 8; Janet's age: 16 2. 25 rectangular plots; 40 square plots
3. 76 km/h
4. [graph] 5. [graph] 6. [graph]

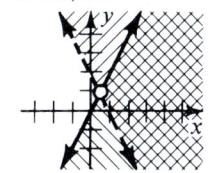

ANSWERS TO SELF-TESTS   579

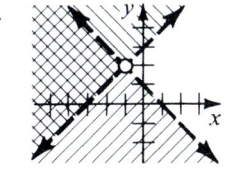

INDEX

AA, similarity by, 316
Abscissa, 472
Acute angle, 141, 157
Addition
  inverses for, 8
  on number line, 7-9
  of polynomials, 232
  properties of, 8-9, 24, 80, 110
  of radicals, 287
  of rational numbers, 17-19
  solving equations by, 80-81
  solving systems by, 507-509
  and subtraction, relationship between, 11
Addition or subtraction method, 507-509
Addition properties of inequalities, 110
Addition property of equivalent equations, 80
Additive inverse, 8
Adjacent angles, 146
Algebraic expression, 73-75, 89-90
Alternate interior angles, 151-152
Angle(s), 140
  acute, 141, 157
  adjacent, 146
  alternate interior, 151-152
  bisector of, 141
  in a circle, 349-362
  complementary, 146, 157
  congruent, 141
  construction of, 174-175
  corresponding, 151-152, 179, 314-316
  inscribed, 353-356
  measure of, 140
  obtuse, 141
  of a parallelogram, 189
  of a polygon, 160-161, 164, 314-316
  right, 141, 145-146
  straight, 141
  supplementary, 146
  of a triangle, 156-157, 186, 315-316, 331-333
  vertical, 146
  See also Problems
Applications, 103, 221, 341, 497
Approximation
  of probability, 424-426
  of square roots, 279-280
Arc, 174, 349-362
Archimedes, 367
Area, 197-208, 368-376
  of similar figures, 369, 320-322
  of similar solids, 273, 322-323
ASA, congruence by, 180
Associative properties, 9, 13
Average, arithmetic, 446-447
Axes, coordinate, 472

Base
  of a polygon, 198, 202-204
  of a solid, 207-208, 211-212, 376

Base of a power, 226
BASIC programming, 552-560
  See also Computer Exercises
Biconditional statement, 57-58
Binomials, 229
  factoring products of, 248-249, 251-252
  multiplying, 245-246, 251-252
  radicals in, 289
Biographical Notes, 69, 169, 267, 433, 529
Bisector
  of an angle, 141
  of a line segment, 194
Boundary of a half-plane, 489-491, 515-516

Careers, 31, 129, 297, 387, 467
Center
  of a circle, 174, 345
  of a sphere, 372
Central angle, 349
Chapter reviews. *See* Reviews
Chapter tests. *See* Tests
Chord, 345, 359
Circle, 174, 345-346
  angles in, 349-362
  arcs of, 174, 349-362
  area of, 368-369
  circumference of, 367-369
  circumscribed, 355-357
  inscribed, 362
Coefficient, 228
Coincident lines, 502, 506
Collinear points, 134, 475-477
Common denominator, 18
Common factor, 243
Commutative properties, 9, 13
Complementary angles, 146, 157
Complementary events, 418
Compound inequalities, 116-118, 515-516
Compound statement, 40
Computer, 552
Computer Exercises, 27-28, 65-66, 100, 124-125, 165-166, 215-216, 261, 293-294, 337-338, 380, 428-429, 462-463, 493-494, 522-523
Conclusion, 46, 60
Conditional statement, 46-48, 53-54
Cone, 376-377
Congruence
  of angles, 141
  of arcs, 350
  of circles, 346
  of line segments, 137
  of triangles, 173-186
Conjugates, 289
Conjunction, 40-41
Consecutive integers, 90
Constant term, 248
Constructions, 173-176
Contradiction, 55

Contrapositive, 53-54
Converse, 53-54
Convex polygon, 160
Coordinate(s), 5, 472
Coordinate plane, 472
Coordinate system, rectangular, 472
Coplanar points, 134
Corresponding angles
  of congruent triangles, 179
  of parallel lines, 151-152
  of similar polygons, 316-318
Corresponding lengths
  in congruent triangles, 179
  in similar polygons, 314-322
  in similar solids, 322-323
Counting numbers, 1
Counting principle, 404-406
Cube, geometric, 208, 322
Cube of a number, 226
Cube root, 279
Cumulative frequency, 439-443, 450-451
Cumulative reviews. *See* Reviews
Cylinder, 376

Data, 437-438
Decagon, 161
Decimals
  adding and subtracting, 19
  as approximations to square roots, 279-280
  irrational, 279
  multiplying and dividing, 21-22
  nonterminating, 272, 279
  in percents, 308-312
  repeating, 272-273, 275
  terminating, 272-275
Degree
  of a polynomial, 230, 248
  of a polynomial equation, 254
Denominator
  least common, 18
  rationalizing, 285-289
Diagonal
  of a parallelogram, 189, 193-194
  of a polygon, 160
  of a rectangle, 194, 327
  of a rhombus, 194
Diameter
  of a circle, 345, 367
  of a sphere, 372
Directed numbers, 5
Disjunction, 43-44
Distance
  between parallel lines, 192
  between points, 137, 471, 473
Distributive property of multiplication over addition, 24, 74-75
  applications of, 74-75, 81, 87, 237-238, 241, 287
Division, 15-16
  of polynomials, 240-241
  of radicals, 283-286
  of rational numbers, 22

Domain, 36, 48, 73, 77

Element of a set, 1
Empirical probability, 424-426
Empty set, 2, 36, 87
Endpoint, 137
Equation(s), 76-77
  degree of, 254
  equivalent, 77-78, 80, 82-83, 502
  of a line, 475-477, 485-487
  linear
    in one variable, 76-95
    systems of, 501-513
    in two variables, 475-487
  properties of, 80, 83
  quadratic, 254-255, 290-292
  root of, 254
  translating words into, 89-90
Equivalence
  of equations, 77-78, 80, 82-83, 502
  of inequalities, 110
  logical, 54-55, 58
Euclid, 133
Evaluating an expression, 73-75
Events, 395
  complementary, 418
  See also Probability
Expected value, 458-459
Experiment, random, 391-392
  See also Probability
Exponents, 226, 242
  rules of, 234-235, 240
Expression
  algebraic, 73-75, 89-90
  evaluating, 73-75
  linear, 518-519
  numerical, 23-24
Exterior angle, 151, 157
Extremes of a proportion, 305

Factor(s), 225-226, 243-244
  greatest common, 243
  prime, 243
  See also Factoring
Factorial, 408
Factoring
  difference of squares, 251-252
  integers, 225-226, 243
  monomials, 243
  polynomials, 243-244, 248-249, 251-252
  trinomial squares, 252
Formula, 97-98
  quadratic, 291-292
Fractions
  decimal form of, 272-273
  decimals expressed as, 274-275
  operations with, 17-18, 21-22
Frequency polygon, 443
Frequency table, 437-439

INDEX   **581**

Geometry
  basic concepts of, 133-161
  of the circle, 345-369
  of congruent triangles, 173-186
  of quadrilaterals, 188-194
  of right triangles, 326-333
  of similar polygons and solids, 314-323
  See also Area, Volume
Gradient, 471
Graph(s)
  in a coordinate plane, 471-473
  of an inequality, 109-111, 116-118, 489-491
  of a linear equation, 475-476
  on a number line, 4-5, 107-108, 137
  statistical, 441-443, 450-451
  of a system of equations, 501-502
  of a system of inequalities, 515-516
Greatest common factor, 243

Half-plane, 489, 515-516
Hemisphere, 372
Hexagon, 161
Histogram, 441-443
Hypotenuse, 326-327, 332-333
Hypothesis, 46

Identity properties, 8, 14
Implication, 46-48, 53-54
Inequalities, 107-108
  compound, 116-118
  equivalent, 110
  graphs of, 109-111, 116-118, 489-491, 515-516
  linear
    in one variable, 108-122
    systems of, 515-516
    in two variables, 489-491
  in linear programming, 518-519
  properties of, 110, 113
  in word problems, 121-122
Integers, 4
  consecutive, 90
  See also Problems
Intersection
  of geometric figures, 134
  of graphs, 501-502, 515-516
Inverse
  additive, 8
  of a conditional, 53-54
  multiplicative, 21
Inverse properties, 8, 21
Irrational number, 279

Lateral area, 207-208, 376
Least common denominator, 18, 234
Length(s)
  corresponding, 165, 314-323
  of a line segment, 137

Line(s), 133-135
  coincident, 502, 506
  equation of, 475-477, 485-487
  graph of, 475-477
  intersecting, 134, 501
  parallel, 135, 150-152, 480, 482, 502
  perpendicular, 145-146
  slope of, 480-482
  y-intercept of, 477, 485
Line segment(s), 187, 146
  bisector of, 194
  construction of, 174
Linear equation(s), 254
  in one variable, 76-95
  systems of, 501-513
  in two variables, 475-487
Linear expression, 518-519
Linear inequalities
  in one variable, 107-122
  systems of, 515-516
  in two variables, 489-491
Linear programming, 518-519
Linear term, 248
Logic
  arguments, valid, 60-63
  biconditional statement, 57-58
  compound statement, 40
  conclusion (consequent), 46, 60
  conditional statement (implication), 46-48, 53-54
  conjunction, 40-41
  contradiction, 55
  contrapositive, 53-54
  converse, 53-54
  disjunction, 43-44
  hypothesis (antecedent), 46
  inverse, 53-54
  logical equivalence, 54-55, 58
  negation, 38-39
  open sentence, 36, 48
  statement, 35-36, 38
  tautology, 55, 58
  truth table, 39, 50-52
  truth value, 35-36, 54-55
Logical equivalence, 54-55, 58
Lowest common multiple, 234
Lowest terms, 18

Major arc, 349
Mean (arithmetic average), 446-447
Means of a proportion, 305
Median, 446-451
Member of a set, 1
Metric units, list of, viii
Midpoint of a line segment, 137
Minor arc, 349
Mode, 446-447
Monomial(s), 228-230
  division by, 240-241
  factoring, 243
  multiplication of, 234-238

582   INDEX

Multiplication, 12-14
  of binomials, 245-246, 251-252, 289
  and division, relationship between, 15
  inverses for, 21
  of monomials, 234-235
  of polynomials, 237-238
  properties of, 13-14, 21, 24, 83, 113, 254
  of radicals, 283-284, 289
  of rational numbers, 21
  solving equations by, 82-84
Multiplication properties of inequalities, 113
Multiplication property of equivalent equations, 83
Multiplicative inverse, 21
Multiplicative property
  of −1, 14
  of zero, 14

Negation, 38-39
Negative numbers, 4-5
Number(s)
  comparing, 4, 107-108
  counting, 1
  directed, 5
  graphing, 4-5
  integers, 4, 90
  irrational, 279
  negative, 4-5
  opposite (additive inverse) of, 8
  positive, 4-5
  prime, 243
  rational, 17
  real, 4-5, 279
  whole, 1
Number line, 4-5, 107-108, 137
  addition on, 7-9
  inequalities on, 109-111, 116-118
Numerical expression, 23-24

Obtuse angle, 141
Octagon, 161
Open sentence, 36, 48
  See also Equation(s), Inequalities
Opposite of a number, 8
Order of operations, 23-24
Ordered pair, 472
Ordinate, 472
Origin, 4, 472
Outcome, 391-393
  See also Probability

Parallel lines, 135, 150-152, 480, 482, 502
Parallel planes, 135
Parallelogram, 188-194, 198-199
Pentagon, 161
Percent, 307-310
Percentile, 451
Perimeter(s), 93, 97, 161
  of similar polygons, 315

Permutations, 408-410
Perpendicular lines, 145-146
Pi (π), 279, 367
Plane(s), 133-135
  coordinate, 472
  half-, 489, 515-516
Point(s)
  coordinate(s) of, 5, 472
  distance between two, 137, 471, 473
  equation of line through two, 486-487
  geometric, 133-134
  of intersection, 501
  of tangency, 346
Polygon(s), 160-161
  circumscribed, 362
  cumulative frequency, 443, 450-451
  equiangular and equilateral, 161
  frequency, 443
  inscribed, 355-356
  regular, 161
  similar, 314-322
  See also under names of specific figures.
Polynomial(s), 225, 229-230
  adding and subtracting, 232
  degree of, 230, 248
  divided by monomial, 241
  factoring, 243-244, 248-252
  multiplying, 237-238, 245-246, 251-252
  quadratic, 248-249
  standard form of, 230
  terms of, 205-206, 224, 229-230, 248
Positive numbers, 4-5
Power(s), 22, 24, 226
  rules for, 234-235, 240
Prime number, 243
Principal square root, 276-277
Prism(s), 207-212
  similar, 322-323
Probability, 391-426
  counting principle, 404-406
  empirical, 424-426
  of an event, 395, 418
  of a many-stage experiment, 397-400, 412-414, 420-422
  of an outcome, 391-393
  permutations, 408-410
  rules about, 416-418
Problems
  age, 95-97, 123, 259, 511, 513, 515
  angle, 144, 149-150, 159, 187, 302, 402
  area, 201-202, 210-211, 259-260, 370-371
  circumference, 370-371
  distance, 512, 514-515
  integer, 94-97, 122-123, 259-260, 302, 513
  investment, 310, 513, 515
  mixture, 312, 512-514
  percent, 307-310, 312
  perimeter, 93-97, 123, 163, 513-514
  about right triangles, 326-327, 331-333
  about similar polygons, 314-316, 320-322
  volume, 213-214, 375, 378-379

INDEX 583

Problem solving
  equations used for, 93-95
  factoring used for, 257-259
  inequalities used for, 121-122
  linear programming used for, 518-519
  plan for, 94
  proportions used for, 304-309, 314-323
  quadratic equations used for, 257-258
  ratios used for, 301-302
  systems of equations used for, 511-513
Programming in BASIC, 552-560
Programming, linear, 518-519
Property(ies)
  addition, of equivalent equations, 80
  addition, of inequalities, 110
  associative, 9, 13
  commutative, 9, 13
  distributive, 24, 74-75
  identity, 8, 14
  inverse, 8, 21
  multiplication, of equivalent equations, 83
  multiplication, of inequalities, 113
  multiplicative, of -1, 14
  multiplicative, of zero, 14
  zero-product, 254
Proportions, 304-309, 314-323
Pyramid, 208, 212
Pythagorean Theorem, 326-330

Quadrant, 472
Quadratic equation, 254-255, 290-292
Quadratic formula, 291-292
Quadratic polynomial, 248-249
  factoring, 249, 251-252
Quadrilateral(s), 161
  inscribed, 356
  similar, 314-315, 320-321
  See also Parallelogram, Rectangle, Rhombus, Square, Trapezoid
Quartile, 449-451

Radical(s), 277
  approximating, 279-280
  binomials containing, 289
  simplifying, 283-287
Radius
  of a circle, 174, 345, 369
  of a solid, 372-374, 376
Random experiment, 391-392
Random sample, 455-457
Ratio, 301-302, 307-308, 314-323, 368-369, 373-374
  See also Percent, Proportion
Rational numbers, 17
  in decimal form, 271-273
  in fractional form, 274-275
  operations with, 17-22
Rationalizing denominators, 285-289
Ray, 137, 146
Real number, 4-5, 279
  See also under Properties

Reciprocal, 21
Rectangle, 97, 193-198
  diagonal of, 194, 327
Rectangular solid. *See* Prism(s)
Repeating decimal, 272-273, 275
Repetend, 272-273
Replacement set, 36, 48, 73
Reviews
  algebra, 546-551
  chapter, 29, 66-67, 100-101, 125, 166-167, 216-218, 262, 294-295, 338-339, 381-382, 429-430, 464-465, 495, 523-524
  cumulative, 126-128, 264-266, 383-386, 525-528
  mixed, 32, 70, 104, 130, 170, 222, 268, 298, 342, 388, 434, 468, 498, 530
  for Regents examinations, 561-573
  vocabulary, 32, 70, 104, 130, 170, 222, 268, 298, 342, 388, 434, 468, 498, 530
Rhombus, 193-194
Right angle, 141, 145-146
Right triangle, 157, 326-333
Root, 254
  See also Solution

Sample, 437, 455-457
  random, 455-457
Sample space, 391
SAS, congruence by, 180
Secant, 346, 360-361
Segment. *See* Line segment(s)
Self-Tests. *See* Tests
Semicircle, 349, 355
Set(s), 1-2
  element (member) of, 1
  empty, 2, 36, 87
  graphs of, 4-5
  of numbers, 1, 2, 4
  replacement, 36, 48, 73
  solution, 36, 77, 87, 108, 502, 506
Similar polygons, 314-322
Similar solids, 322-323, 373-374
Similar terms, 229, 232
Simplifying
  algebraic expressions, 74-75, 87
  numerical expressions, 23-24
  polynomials, 229-230
  radicals, 276-277, 283-289
Simulation, 455-457
Slope, 471, 480-482
Slope-intercept form, 485-487
Solids, similar, 322-323, 373-374
Solution
  of an equation, 77, 87, 254
  of an inequality, 108, 111, 117-118
  of an open sentence, 36
  of a system of equations, 501-509
  of a system of inequalities, 515-516
Solution set, 36, 77, 87, 108, 502, 506
Solving equations
  linear, in one variable, 76-95
  systems of, 501-513

proportions, 304-306
quadratic, 254-255, 290-292
Solving inequalities
  in one variable, 108-122
  systems of, 515-516
Sphere, 372-374
Square (geometric figure), 193, 320-321
Square(s)
  of a binomial, 252
  difference of, 251-252
  of a number, 226
  perfect, 277
Square root(s), 276-277
  approximating, 279-280
  binomials containing, 289
  principal, 276-277
  properties of, 283
  simplifying, 283-287
  table of, 281
SSS, congruence by, 179
Statement, 35-36, 38
  See also Logic
Statistics, 437
  charts and graphs, 441-443
  expectation, 458-459
  frequency tables, 437-441
  measures of central tendency, 446-447
  quartiles and percentiles, 449-451
  sampling and simulation, 455-457
Straight angle, 141
Substitution method, 504-506
Subtraction, 11
  of polynomials, 232
  of radicals, 287
  of rational numbers, 17-19
  solving systems by, 507-509
Supplementary angles, 146
Surface area, 208, 372, 376
Symbols, list of, viii
System
  coordinate, 471-472
  of equations, 501-513
  of inequalities, 515-516

Table
  of metric units, viii
  of random numbers, 456
  of square roots, 281
  of symbols, viii
Tangent, 346, 359, 361-362
Tautology, 55, 59
Terminating decimal, 272-275
Terms
  of a polynomial, 229-230, 248
  of a proportion, 305
  similar, 229, 232
Tests
  Chapter, 30, 67-68, 102, 126, 167-168, 218-220, 263, 295-296, 340, 382-383, 431-432, 466, 496, 524-525

Self-, 7, 17, 26, 46, 60, 79, 89, 99, 116, 121, 124, 145, 164-165, 185, 196-197, 214-215, 237, 248, 260, 276, 282, 288, 293, 313, 325, 336-337, 352-353, 366, 379, 403, 412, 420, 428, 445, 454-455, 461, 479, 492, 511, 517-518
  answers to, 574-579
  See also Reviews
Theorem, Pythagorean, 326-330
Transversals, 151
Trapezoid, 193, 203-204
  isosceles, 193
Triangle(s), 156, 161
  angles of, 156-157, 186
  area of, 202-203
  congruent, 173-186
  construction of, 175-176
  equiangular and equilateral, 186
  45°-45°-90°, 331-332
  isosceles, 186
  right, 157, 326-327, 330-333
  similar, 315-316, 319, 321
  30°-60°-90°, 331, 333
Trinomial(s), 229
  factoring, 248-252
Truth table, 39, 50-52
Truth value, 35-36, 54-55

Value
  expected, 458-459
  of an expression, 23-24, 73-74
  truth, 35-36, 54-55
  of a variable, 36, 48, 73
Variable, 36, 73
  domain (replacement set) of, 36, 48, 73, 77
  value of, 36, 48, 73
Vertex
  of an angle, 140
  of a polygon, 160
  of a solid, 208, 376
Vertical angles, 147
Volume, 211-212, 372-377
  of similar solids, 322-323, 374

Whole numbers, 1

x-axis, 472
x-coordinate, 472
x-intercept, 477

y-axis, 472
y-coordinate, 472
y-intercept, 477, 485-487

Zero, multiplicative property of, 14
Zero-product property, 254

SELECTED ANSWERS

Page 3 1. yes 3. no 5. no 7. {Tuesday, Thursday} 9. {Alabama, Alaska, Arizona, Arkansas} 11. {13, 14, 15, 16} 13. {0, 2, 4, 6, 8, 10, 12, 14} 15. {January, March, May, July, August, October, December} 17. Answers may vary. One example is given. {the days of the week beginning with the letter S} 19. {the counting numbers} 21. {the positive powers of 2} 23. {the prime numbers} 25. 3125; 15,625; 78,125 ($5^5, 5^6, 5^7$) 27. S, E, N (seven, eight, nine)

Pages 6–7 1. ⁻6 3. 3 5. C 7. G 9. 5 11. ⁻2 13. I 15. N 17–28. Graphs are given below.

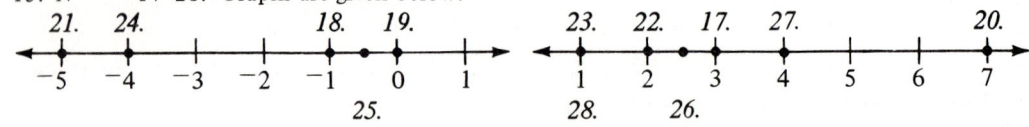

37. ⁻2 39. $\frac{1}{2}$ 41. 0 43. $1\frac{1}{2}$

Page 10 1. 32 3. 849 5. 14 7. −26 9. 33 11. −41 13. −50 15. 0 17. 138 19. −19 21. 0 23. 37 25. 0 27. 56 29. Inverse prop. for add. 31. Comm. prop. for add. 33. Assoc. prop. for add 35. Identity prop. for add. 37. Inverse prop. for add. 39. −(−3) = 3

Page 12 1. 12 3. −25 5. 13 7. −13 9. 89 11. −437 13. 558 15. −1062 17. −371 19. 594 21. −31 23. 59 25. −81 27. 63 29. −101 31. −61 33. −53 35. no

Pages 14–15 1. 156 3. 1665 5. −1161 7. −1206 9. 323 11. 1643 13. 188,175 15. 0 17. −35,109 19. 3792 21. −5681 23. 7920 25. 10,395 27. −2520 29. Assoc. prop. for mult. 31. Comm. prop. for mult. 33. Mult. prop of −1 35. Identity prop. for mult. 37. Mult. prop. of 0 39. Comm. prop. for mult. 41. Comm. prop. for add.

Page 16 1. 24 3. −5 5. −11 7. 19 9. −23 11. 15 13. −47 15. −17 17. 15 19. 65 21. −87 23. −113 25. 982 27. −963 29. (8 ÷ 4) ÷ 2 = 2 ÷ 2 = 1; 8 ÷ (4 ÷ 2) = 8 ÷ 2 = 4; there is no associative property for division.

Page 20 1. $-\frac{1}{4}$ 3. $-\frac{2}{5}$ 5. $-\frac{3}{8}$ 7. $\frac{2}{3}$ 9. $\frac{2}{5}$ 11. $\frac{2}{3}$ 13. $\frac{9}{10}$ 15. $\frac{3}{35}$ 17. $-\frac{7}{16}$ 19. $\frac{1}{4}$ 21. $-\frac{1}{24}$ 23. $\frac{1}{20}$ 25. $-\frac{7}{48}$ 27. $\frac{37}{45}$ 29. 0.60 31. −0.09 33. −0.27 35. 0.25 37. 1.96 39. 1.55 41. −2.28 43. −7.489 45. 0.267 47. $\frac{23}{24}$ 49. $\frac{1}{180}$ 51. −2 53. 0

586   *SELECTED ANSWERS*

Page 23   1. $-\dfrac{1}{3}$ 3. $-\dfrac{8}{3}$ 5. $\dfrac{5}{24}$ 7. 10   9. -0.0525 11. -1.408
13. $-\dfrac{6}{7}$ 15. $-\dfrac{8}{9}$ 17. -0.052 19. 50   21. -3 23. $\dfrac{2}{5}$ 25. -1.38
27. 0.0486   29. 4.1   31. -0.0038 33. 0.182   35. 0.485
37. $2\dfrac{5}{8}\left(2\dfrac{3}{4}\right) = \dfrac{21}{8}\left(\dfrac{11}{4}\right) = \dfrac{231}{32}$

Page 25   1. 61   3. 298   5. -4 7. 50   9. 135   11. 65   13. 3
15. 22   17. 18   19. 6   21. 5   23. 14   25. 324   27. 86   29. 50
31. Ident. prop. for mult.   33. Inv. prop. for add.   35. Assoc. prop. for add.
37. Comm. prop. for add.   39. Mult. prop. of 0   41. Comm. prop. for mult.
43. Identity prop. for mult.   45. 4225   47. 7225   49. 65,025

Page 29 · Chapter Review   1. $\{a, b, c, d, e\}$ 2. Answers may vary. Example: $\{$ the whole numbers between 4 and 8$\}$ 3. a, b, c
4. a. $^-2$ b. 0   5. a. H b. C 6. [number line with points at $-3, -2, 0, 1$]
7. -161 8. -643 9. 0   10. -8 11. 721   12. -226 13. -292
14. -150 15. -375 16. 4340   17. 0   18. -480 19. 23   20. -38
21. 35   22. -44 23. 0.09   24. 2.94   25. $\dfrac{1}{2}$ 26. $-\dfrac{1}{24}$ 27. $\dfrac{3}{2}$
28. $-\dfrac{6}{5}$ 29. $\dfrac{1}{6}$ 30. -10

Pages 32–33 · Mixed Review   1. 0.58   3. -21 5. -378 7. -120 9. 750
11. -57 13. $\{$the negative integers$\}$ 15. $-\dfrac{5}{14}$ 17. $\emptyset$ 19. -1 21. 18
23. -6 25. Inverse prop. for mult.   27. Assoc. prop. for add.   29. Distributive prop.
31. not well-defined   33. well-defined

Pages 37–38   1–9. Answers may vary. Examples are given.   1. x; 7, 8, 9; 1, 2, 3
3. y; Monday, Tuesday, Wednesday; January, Spring, March   5. x; 2, 4, 6; 3, 5, 7
7. it; football field, highway, hockey rink; ruler, car, sofa   9. it; 3, 5, 7; 4, 6, 9
11. always false   13. always false   15. sometimes true, sometimes false   17. sometimes true, sometimes false   19. $\{1, 2, 3, 4, 5, 6\}$ 21. $\{4, 5, 6, 7, 8\}$
23. $\{1, 3, 5, 7 \ldots\}$ 25. $\{8\}$ 27. $\{17\}$ 29. $\{0\}$ 31. F   33. F   35. F
37. Answers may vary. Example: $x + y = y + x$ 39. neither

Pages 39–40   1. Friday is not the last day of the week.   3. There is not a pot of gold at the end of the rainbow.   5. I am older than Bob.   7. $2 + 3 \neq 5$; T; F   9. 31 is not between 5 and 40; T; F   11. Monkeys live in trees; F; T   13. 5 is less than 2; T; F
15. It is not raining.   17. yes

19.

p	$\sim p$	$\sim(\sim p)$
T	F	T
F	T	F

p and $\sim(\sim p)$ always have the same truth value.

21. 6 is not less than 9; T; F   23. 4 is greater than 7; T; F   25. 16 is greater than 10; F; T

27. It is not true that all people have blond hair. Some people do not have blond hair.   29. It is not true that some students do not work hard. All students work hard.   31. It is not true that no other animal is taller than a giraffe. Some animal is taller than a giraffe.   33. "no p" or "all p"

SELECTED ANSWERS   587

Pages 42-43 1. April has 30 days and $3 \cdot 4 = 12$; T 3. $4 + 3 = 7$ and $3 \cdot 4 = 12$; T 5. April has 30 days and "mathematics" has 10 letters; F 7. April has 30 days and $3 \cdot 4 = 12$, and $4 + 3 = 7$; T 9. $4 + 3 = 7$, and $3 \cdot 4 = 12$ and April has 30 days; T 11. $r \wedge s$; F 13. $q \wedge r$; T 15. $q \wedge r \wedge p$; T 17. T 19. F 21. The moon is always full and the Yankees are not a baseball team; F 23. The moon is always full and $2 + 2 = 2 \cdot 2$, and motorcycles do not have four wheels; F 25. The Yankees are not a baseball team and the moon is not always full; F 27. F 29. F 31. F 33. T 35. p is true; q is false. ($\sim q$ is true.) 37. p and q are true.

Pages 44-45 1. $12 < 9$ or a triangle has three sides; T 3. 7 is between 2 and 5 or a triangle has three sides; T 5. $12 < 9$ or 32 is an odd integer; F 7. $12 < 9$ or a triangle has three sides, or 7 is between 2 and 5; T 9. 7 is between 2 and 5, or a triangle has three sides or $12 < 9$; T 11. $p \vee r$; F 13. $r \vee s$; F 15. $s \vee p$; F 17. T 19. T 21. $2 - 5 = 3$ or fires are not hot; F 23. $2 - 5 = 3$ or potholes are good for your car, or Earth is not the planet closest to the sun; T 25. Fires are not hot or $2 - 5 \neq 3$; T 27. T 29. F 31. T 33. T 35. p is false; q is true. ($\sim q$ is false.) 37. r is true; the truth values of p and q cannot be determined, but at least one of them is true. 39. $\sim p \wedge \sim q$ 41. $\sim p \vee \sim q$

Pages 48-50 1. hyp.: k is an even integer; conc.: $k + 1$ is an odd integer 3. hyp.: $z = -5$; conc.: $2z = -10$ 5. hyp.: Karen is Jack's niece; conc.: Jack is Karen's uncle. 7. If $3y - 2 = 7$, then $3y = 9$. 9. If the moon is made of green cheese, then mice will nibble. 11. If a rectangle has 4 sides, then $2 + 3 = 3 + 2$; T 13. If $2 + 3 = 3 + 2$, then 10 is not an integer; F 15. If 26 is between 20 and 24, then 10 is not an integer; T 17. If 10 is not an integer, then a rectangle has 4 sides; T 19. $s \rightarrow q$; T 21. $r \rightarrow q$; T 23. $r \rightarrow s$; T 25. $p \rightarrow q$ 27. $r \rightarrow s$ 29. If South America is a continent, then $3(6 - 1) \neq (3 \cdot 6) - (3 \cdot 1)$; F 31. If cows can't fly, then $3(6 - 1) \neq (3 \cdot 6) - (3 \cdot 1)$; F 33. If South America is a continent and $3(6 - 1) = (3 \cdot 6) - (3 \cdot 1)$, then cows can fly; F 35. If $3(6 - 1) = (3 \cdot 6) - (3 \cdot 1)$ or cows can fly, then South America is a continent; T 37. 0, 9 39. $-7, 0$ 41. 0, 5 43. If you are a student, then you like algebra. 45. If two numbers equal the same number, then they equal each other. 47. Either a rhombus is a rectangle or it is not a square. 49. Either these shoes are not too small or my feet will hurt. 51. $\sim p \vee q$

Page 56 1. $s \rightarrow r$; $\sim r \rightarrow \sim s$; $\sim s \rightarrow \sim r$ 3. $r \rightarrow \sim p$; $p \rightarrow \sim r$; $\sim r \rightarrow p$ 5. If $x + 2 = 3$, then $x = 1$; if $x \neq 1$, then $x + 2 \neq 3$; if $x + 2 \neq 3$, then $x \neq 1$. 7. If we win the game, then we can score a run; if we cannot score a run, then we will not win the game; if we do not win the game, then we can't score a run. 9. If $x > 4$, then $x > 2$; if $x \leq 2$, then $x \leq 4$; if $x \leq 4$, then $x \leq 2$. 11. If a figure is not a triangle, then it does not have 3 sides. 13. If Bill is not happy, then he did not pass his test. 15. tautology 17. tautology 19. neither 23 and 25. Answers may vary. One example is given. 23. If Lee is a man, then Lee is a person. 25. If $x = 2$, then $x^2 = 4$. 27. the contrapositive of the original statement

Pages 58-59 1. $p \leftrightarrow q$ 3. $r \leftrightarrow s$ 5. 8 is between 5 and 10 if and only if a square has 5 sides; F 7. 8 is between 5 and 10 if and only if a car has 6 wheels; F 9. A square has 5 sides if and only if a car has 6 wheels; T 11. $q \leftrightarrow r$; F 13. $r \leftrightarrow p$; T 15. $r \leftrightarrow q$; F 17. London is not the capital of England if and only if basketballs are not bigger than baseballs; T 19. $\frac{2}{3}$ is a rational number if and only if $3(6 + 2) \neq (3 \cdot 6) + 2$; T

21. Basketballs are not bigger than baseballs if and only if $\frac{2}{3}$ is not a rational number; T
23. logically equivalent   25. logically equivalent   27. not logically equivalent
29. Let p be "$4 > 3$"; q be "$x + 2 = 7$"; r be "today is Tuesday"; and s be "it's hot";
$(p \wedge q) \leftrightarrow (r \wedge s)$ 31. Let p be "it's raining"; q be "it's snowing"; r be "I don't go";
$(p \vee q) \leftrightarrow r$

Pages 63–64

1. p: $x + 1 = 10$
   q: $x = 9$
   P_1: $p \to q$
   P_2: p
   C: $\therefore q$ Valid; Law of Detachment

3. p: He's a glob.
   q: He's a nob.
   P_1: $p \to q$
   P_2: $\sim p$
   C: $\therefore \sim q$ Not valid.

5. p: The butler committed the crime.
   q: The maid didn't do it.
   P_1: $p \to q$
   P_2: $\sim p$
   C: $\therefore \sim q$ Not valid.

7. p: The butler did it.
   q: The maid didn't.
   P_1: $p \to q$
   P_2: p
   C: $\therefore q$ Valid; Law of Detachment.

9. p: It's a hob.
   q: It's a nob.
   r: It's a bob.
   P_1: $p \to q$
   P_2: $q \to r$
   C: $\therefore p \to r$ Valid; Law of the Syllogism.

11. $[(p \vee q) \wedge p] \to q$; not valid

13. $[(\sim p \vee q) \wedge p] \to q$; valid

15. $[(\sim p \to \sim q) \wedge q] \to p$; valid

17. p: x is a whole number.
    q: x is an integer.
    r: x is a real number.
    P_1: $p \to q$
    P_2: $q \to r$ (replaces $\sim r \to \sim q$)
    C_1: $\therefore p \to r$ (Law of the Syllogism)
    C_1: $p \to r$
    P_3: p
    C_2: $\therefore r$ (Law of Detachment)
    Conclusion: x is a real number.

19. p: You are a baby.
    q: You are small.
    r: You are tall.
    P_1: $p \to q$
    P_2: $q \to \sim r$
    C_1: $\therefore p \to \sim r$ (Law of the Syllogism)
    C_1: $p \to \sim r$
    P_3: r
    C_2: $\therefore \sim p$ (Law of Contrapositive Inference)
    Conclusion: You are not a baby.

Pages 66–67 · Chapter Review   1. statement; true   2. open sentence   3. $\{1, 3\}$
4. 1 is not greater than 2; false; true   5. Gold is not a mineral; true; false   6. a. Division by zero is defined; $\sim p$ b. false   7. false   8. false   9. true   10. $s \wedge r$; true
11. true   12. false   13. true   14. $r \vee p$; true   15. hypothesis: $2z = -8$; conclusion: $z = -4$ 16. If Trenton is the capital of New Jersey, then $\frac{1}{2} = \frac{2}{5}$; false
17. $p \to q$; false   18. false; true   19. false; false; true

20.

p	q	$p \vee q$	$p \rightarrow p \vee q$
T	T	T	T
T	F	T	T
F	T	T	T
F	F	F	T

21. contrapositive   22. converse   23. If $x \neq -1$, then $x + 5 \neq 4$.   24. yes; $q \rightarrow \sim p$ is the contrapositive of $p \rightarrow \sim q$.   25. $5 \cdot 5 = 5$ if and only if $5\left(\frac{4}{5} + \frac{1}{5}\right) = 5$; false

26.

p	q	$p \leftrightarrow q$
T	T	T
T	F	F
F	T	F
F	F	T

27. $r \leftrightarrow s$ is a tautology.
28. $p:$ I save money.
    $q:$ I conserve energy.
    $P_1:$ p
    $P_2:$ $q \rightarrow p$
    $C:$ $\therefore q$ Not valid

29. $p:$ $x = 2$
    $q:$ $x(x - 2) = 0$
    $P_1:$ $p \rightarrow q$
    $P_2:$ $\sim q$
    $C:$ $\therefore \sim p$ Valid;
    Law of Contrapositive Inference

Pages 70-71 · Mixed Review

1. a.

p	q	$p \vee q$	$(p \vee q) \rightarrow q$
T	T	T	T
T	F	T	F
F	T	T	T
F	F	F	T

b. neither

3. $\frac{9}{8}$ 5. -13 7. $\frac{9}{5}$ 9. {the consonants of the alphabet}   11. $x \not> 0$

13. [number line from 0 to 10, all points marked]   15. {1, 2, 3, 4}   17. 0.0104   19. $\frac{75}{76}$

21. -0.7 23. not necessarily   25. true, false   27. $-\frac{1}{2}$ 29. 9

31. sometimes true and sometimes false   33. false   35. true

Pages 75-76   1. 3   3. -28 5. 3   7. 0   9. -4 11. 18   13. 3
15. -1 17. -7 19. 9   21. 21   23. -14 25. -3 27. 35
29. 24   31. -21 33. $-\frac{7}{3}$ 35. $\frac{4}{6} = \frac{2}{3}$ 37. $3y + 18$ 39. $-2s - 8$
41. $-2c + 4$ 43. $5a + 5b$ 45. $xy - xz$ 47. $6a + 18$ 49. $11 + 2b$
51. $\frac{8}{xy}$ 53-57. Answers will vary. One example is given.   53. The number of days in w weeks   55. How old a person who is a years old now will be in 60 years   57. The cost of a phone call; 15¢ for the first minute and 13¢ for each of x additional minutes
59. 6

Pages 78-79   1. {1}   3. {3}   5. {2}   7. no   9. yes; {1}   11. no
13. -1 15. 3   17. 3   19. $\frac{3}{2}$ 21. 32   23. 4.5   25. 7   27. $\frac{2}{3}$
29. 1.2   31 and 33. Answers may vary. One example is given.   31. $2y + 1 = 7$
33. $5y - 4 = -4$ 35. {the real numbers}   37. {0, 10}   39. {3}   41. $\left\{\frac{5}{2}, 3\right\}$

Pages 81-82   1. 5   3. 5   5. −1   7. 9   9. −25   11. 0   13. −12
15. 33   17. −15   19. 2   21. $4\frac{3}{5}$ 23. $2\frac{1}{2}$ 25. 2.32   27. −3.55
29. 7.45   31. −21   33. −21   35. 12   37. 12   39. −4   41. $a = b$
42. $(a + c = b + c) \to a = b$ 43. converse   44. $(a = b) \leftrightarrow (a + c = b + c)$
45. Ex. 44 is another way of stating the addition property of equivalent equations.

Page 85   1. 2   3. $-\frac{7}{5}$ 5. −5   7. 9   9. −5   11. −9   13. 7
15. −15   17. 20   19. 35   21. 24   23. −21   25. 32   27. −9
29. −18   31. 16   33. 13.5   35. 65   37. −93   39. 1.7   41. 16
43. −9   45. $\frac{3}{2}$ 47. $\frac{20}{3}$ 49. $-\frac{12}{25}$ 51. $-\frac{2}{3}$ 53. $-\frac{3}{4}$ 55. $\frac{1}{10}$
57. $a = b$ 58. $(ca = cb) \wedge (c \neq 0); ((ca = cb) \wedge (c \neq 0)) \to (a = b)$ 59. None of these   60. No. It is not true for $c = 0$.   61. Yes

Page 88   1. 1   3. 10   5. −2   7. −1   9. $\frac{11}{2}$ 11. 4   13. $-\frac{1}{4}$
15. $\frac{16}{3}$ 17. 3   19. 2   21. 11   23. $-\frac{1}{3}$ 25. {the real numbers}
27. −5   29. no solution or $\emptyset$ 31. 1   33. 3   35. −18   37. $-\frac{7}{2}$
39. 18   41. −10   43. $-\frac{7}{4}$ 45. $-\frac{2}{3}$ 47. 34   49. $\frac{5}{2}$ 51. 1.5
53. 3   55. 13   57. −1   59. 95   61. 0.5   63. $\frac{3}{2}$ 65. {the real numbers}

Pages 91-93   1. $n + 6$ 3. $2n − 1$ 5. $3(n − 5)$ 7. $25q$ 9. $0.1d$
11. $0.01p$ 13. $10(3x)$ 15. $5.98r$ 17. $5200c$ 19. $f + 6$ 21. $x + 1$; $x + 2$ 23. $x + 2$; $x + 4$ 25. $2n + 5 = 25$ 27. $\frac{3n}{7} = 6$ 29. $5n − 8 = 17$
31. $2n + 3 = 3n − 16$ 33. $2l + 2\left(\frac{1}{2}l\right) = 32$ 35. $n + (n − 6) = 16$
37. $d + (d + 3) = 29$ 39. $j − 10 − \frac{1}{3}j$ 41. $n + (n + 1) + (n + 2) = −18$
43. $n = \frac{1}{2}(−n) + 6$ 45. $0.25q + 0.10(q + 5) = 2.95$

Pages 95-97   1. $x + 3x = 24; x = 6; 6$ and 18   3. $21d = 147; d = 7; 7$ days
5. $6b = 150; b = 25; 25$ km   7. $p + (p + 5) = 47; p = 21$; Punch is 21, Judy is 26.
9. $\frac{1}{6}p = 15; p = 90;$ $90   11. $n + 16 = 5n − 4; n = 5; 5$ 13. $2n + 27 = 49; n = 11; 11$
15. $0.5 + 0.05n = 1.30; n = 16; 16$ 17. $n + (n + 2) = 40; n = 19;$ 19 and 21
19. $n + (n + 1) + (n + 2) + (n + 3) = 110; n = 26; 26, 27, 28,$ and 29   21. $n + (n + 1) + (n + 2) = 2n; n = −3; −3, −2,$ and $−1$ 23. $(2n − 1) + n = 5; n = 2; 2n$ and $3n$
25. $4q = 2(60) + 38; q = 39.5; 39.5$ minutes   27. $\frac{n}{4} = n − 33; n = 44; 44$
29. $0.05(10) + 0.10n + 0.25(10 − n) = 1.95; n = 7; 7$ dimes, 3 quarters   31. $35r + 7(1.5r) = 227.5; r = 5;$ $5/h, $7.50/h   33. $−21, −20, −19$ 35. No solution   37. No solution
39. No solution

Page 99   1. $b = \frac{A}{h}; h \neq 0$ 3. $h = \frac{V}{lw}; l \neq 0, w \neq 0$ 5. $x = \frac{2a}{3}$ 7. $s = r + t$

SELECTED ANSWERS   591

9. $x = \dfrac{2y+1}{3}$ 11. $x = \dfrac{b+3}{2}$ 13. $f = \dfrac{2s}{n} - 1; n \neq 0$ 15. $b = \dfrac{2A}{h}; h \neq 0$
17. $x = 0$ 19. $P = \dfrac{A}{1+rt}; r \neq -\dfrac{1}{t}$ 21. $x = \dfrac{1}{a+1}; a \neq -1$ 23. $V = gt + k; g \neq 0$

Pages 100-101 · Chapter Review   1. 17   2. 2   3. −1   4. $7k - 14$ 5. $\{-1\}$
6. $\{2\}$ 7. $\{-2\}$ 8. no   9. yes; $\{-3\}$ 10. 31   11. −5.5   12. −7
13. −25   14. −76   15. 42   16. 4   17. −50   18. $-\dfrac{1}{4}$ 19. ∅
20. 2   21. {the real numbers}   22. $x + \dfrac{2}{3}x$ 23. $p + 5n$ 24. $2l + 2(l-4)$
25. $3(n+4) = n + 10$ 26. −2   27. 2 tens, 8 fives   28. 85, 73
29. $\dfrac{2A}{x-y}$ $(x \neq y)$ 30. $\dfrac{2A}{h} + y$ $(h \neq 0)$ 31. $x - \dfrac{2A}{h}$ $(h \neq 0)$

Pages 104-105 · Mixed Review   1. true   3. true   5. $\dfrac{5}{3}$ 7. $-\dfrac{11}{8}$ 9. $\dfrac{66}{65}$
11. $x = \dfrac{y-b}{m}$

13. a.

p	q	$\sim q$	$p \wedge \sim q$	$\sim p$	$(p \wedge \sim q) \to \sim p$
T	T	F	F	F	T
T	F	T	T	F	F
F	T	F	F	T	T
F	F	T	F	T	T

b. neither

15. $-\dfrac{1}{2}$ 17. $\left\{-\dfrac{4}{3}\right\}$ 19. always true   21. 7.28   23. ∅
25. {the real numbers}   27. $6(9-4) \neq (9-6)(6-4)$ or $2(3+5) \neq (2+3)(2+5)$; true

29.

p	q	$\sim p$	$\sim q$	$\sim p \wedge q$	$p \vee \sim q$	$\sim(p \vee \sim q)$
T	T	F	F	F	T	F
T	F	F	T	F	T	F
F	T	T	F	T	F	T
F	F	T	T	F	T	F

31. Comm. prop. for mult.   33. Inverse prop. for mult.   35. Answers may vary.
The statement is true for $x = -1, -2, -3$ and false for $x = 2, 3, 4$.   37. a. $5 + \dfrac{1}{2}x = x - 1$
b. 12

Pages 108-109   1. $y < 10$ 3. $n > 3$ 5. $x < -7$ 7. $n \neq -4(9)$ 9. x is less than 4.   11. n is greater than −3.   13. $\{1, 2, 3, \ldots, 19\}$ 15. {positive integers}
17. $\{5, 6, 7, \ldots\}$ 19. $\{1, 2, 3\}$ 21. $\{9, 10, 11, \ldots\}$ 23. ∅   25. $x > 18$
27. $x < 17$ 29. Answers may vary. Example: $2 < 3$ and $3 < 4$, so $2 < 4$.

Page 112   1. $x > -2$ 3. $x < 7$

5. [number line with open circle at 6, shaded left; marks 3, 4, 5, 6, 7]

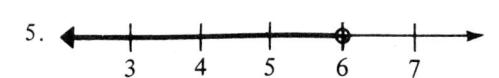

7. [number line with open circle at 0, shaded left; marks −2, −1, 0, 1, 2]

9. $\{x: x < 5\}$ 11. $\{x: x > 5\}$ 13. $\{y: y < -1\}$ 15. $\{y: y > 9\}$ 17. $\{x: x < 0\}$
19. $\{y: y < 2\frac{1}{2}\}$ 21. $\{x: x > 20\frac{3}{4}\}$ 23. $\{x: x < 4\frac{1}{2}\}$ 25. $\{y: y < 1\frac{1}{6}\}$
27. $\{x: x > a - b\}$ 29. $\{x: a - b < x\}$

Page 115

1. $\{x: x < 3\}$ [number line with open circle at 3, shaded left; marks 1, 2, 3, 4, 5]

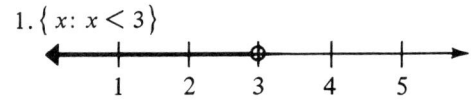

3. $\{y: y > -24\}$ [number line with open circle at −24, shaded right; marks −26, −24, −22]

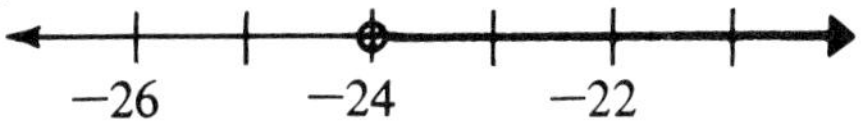

5. $\{x: x > -3\}$ [number line with open circle at −3, shaded right; marks −4, −3, −2, −1, 0]

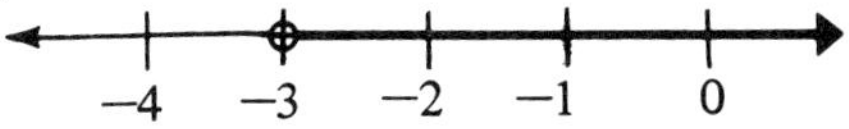

7. $\{x: x > 24\}$ [number line with open circle at 24, shaded right; marks 22, 24, 26]

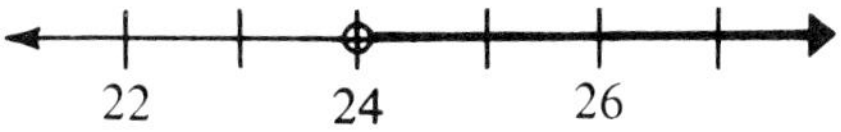

9. $\{t: t > -6\frac{2}{3}\}$ [number line with open circle between −7 and −6, shaded right; marks −8, −7, −6, −5, −4, −3, −2]

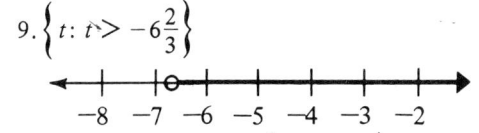

11. $\{x: x > \frac{2}{9}\}$ [number line with open circle between 0 and 1, shaded right; marks −2, −1, 0, 1, 2]

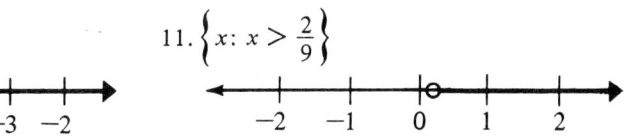

13. $\{x: x < 2\}$ 15. $\{t: \frac{7}{5} > t\}$ 17. $\{x: -4 > x\}$ 19. $\{x: x > 20\}$
21. $\{y: y < -12\}$ 23. $\{h: -4 > h\}$ 25. $\{x: x < \frac{7}{4}\}$ 27. $\{x: x < -\frac{1}{6}\}$
29. $\{x: \frac{34}{3} > x\}$ 31. Case 1 $(b > 0)$: $y > \frac{a}{b}$; Case 2 $(b < 0)$: $y < \frac{a}{b}$
33. Case 1 $(b > 0)$: $y < ab$; Case 2 $(b < 0)$: $y > \frac{a}{b}$

Pages 118–119   1. $d \leq 10$ 3. $3 < x < 9$ 5. $12 > t \geq 7$ 7. $0 > x \geq -11$

9. $\{y: y \leq 9\}$ [number line with closed dot at 9, shaded left; marks 2–10]

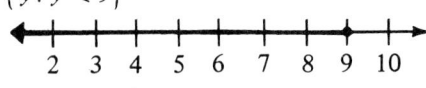

11. $\{x: 2 \leq x\}$ [number line with closed dot at 2, shaded right; marks −2 to 6]

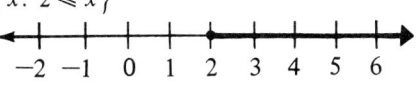

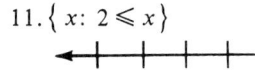

13. $\{n: n \geq -4\}$ [number line with closed dot at −4, shaded right; marks −6 to 0]

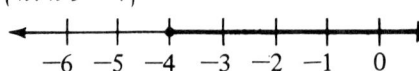

15. $\{y: 7 \geq y\}$ [number line with closed dot at 7, shaded left; marks 1 to 9]

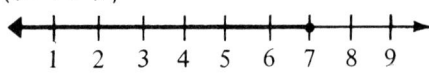

17. $\{x: x \geq 3\}$ [number line with closed dot at 3, shaded right; marks −2 to 6]

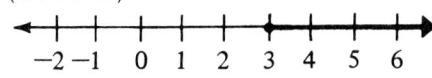

19. $\{t: t \geq -2\}$ [number line with closed dot at −2, shaded right; marks −4, −2, 0, 2, 4]

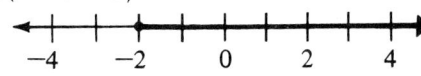

21. $\{x: 4 < x < 6\}$ [number line with open circles at 4 and 6, shaded between; marks −1 to 7]

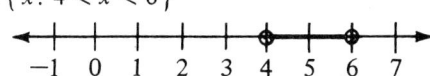

23. $\{y: 2 < y \leq 6\}$ [number line with open circle at 2, closed dot at 6, shaded between; marks −1 to 7]

SELECTED ANSWERS   593

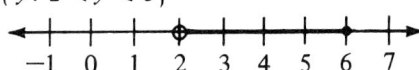

Page 120

1. $\{x: x < 2\}$

3. $\{n: n > 1\}$

5. $\{t: t \geq 2\}$

7. $\{y: y \leq -5\}$

9. $\{x: x < 1\}$ 11. $\{n: n > -10\}$ 13. $\{t: t < -4\}$ 15. $\{d: 1 \geq d\}$
17. $\{n: n < 2\}$ 19. $\{x: x \geq 2\}$ 21. $\{t: t < 20\}$ 23. $\{x: x < -1\}$
25. $\{s: s > -10\}$ 27. $\{t: 1 \geq t\}$ 29. $\{d: d \leq 1\}$ 31. $\{y: -2 < y \leq 3\}$
33. $\{x: \frac{50}{21} \leq x\}$ 35. $\{x: -3 \leq x \leq 15\}$ 37. $\{x: x \geq -1\}$

Pages 122-124 1. 5 and 25 3. 23 and 16 5. There are at least 570 students and at most 1292 students. 7. 28 L 9. One hour and fifty minutes 11. 21 and 23 13. length = 540 cm; width = 180 cm 15. 32 cm, 32 cm, 10 cm; 30 cm, 30 cm, 9 cm; 28 cm, 28 cm, 8 cm; 26 cm, 26 cm, 7 cm; 24 cm, 24 cm, 6 cm; 22 cm, 22 cm, 5 cm; 20 cm, 20 cm, 4 cm; 18 cm, 18 cm, 3 cm; 16 cm, 16 cm, 2 cm; 14 cm, 14 cm, 1 cm 17. 92

Page 125 · Chapter Review 1. $t > -8$ 2. r is less than 5. 3. $\{5, 6, 7, \ldots\}$
4. 1
5. (number line) 6. (number line)

7. $\{y: y > 4\}$ 8. $\{z: z < -1\}$ 9. $\{a: a < 6\}$ 10. $\{m: m > -3\}$
11. $\{x: x > -\frac{1}{8}\}$ 12. $\{b: 12 < b\}$ 13. 1 is greater than z and z is greater than -1.
14. $p \leq 14$
15. $\{c: c \geq 7\}$

16. $\{w: w \geq -2\}$

17. $\{n: n \leq -3\}$

18. $\{k: k < 0\}$

19. $\{w: w < -5\}$

20. $\{x: -1 < x \leq 0\}$

21. -3 and -12 22. 95 cm

Pages 126-128 · Cumulative Review: Chapters 1-4 1. c 2. b 3. d 4. d
5. a 6. a 7. a 8. a 9. a 10. d 11. c 12. a 13. c
14. c 15. b 16. d 17. d 18. c 19. b 20. d 21. b
22. c 23. a 24. a 25. b 26. a 27. d 28. b

Pages 130-131 · Mixed Review 1. New York City is not the capital of New York; true
3. (number line 0 to 5)

594 SELECTED ANSWERS

5.

p	q	$\sim p$	$\sim p \vee q$	$\sim(\sim p \vee q)$	$\sim q$	$p \wedge \sim q$
T	T	F	T	F	F	F
T	F	F	F	T	T	T
F	T	T	T	F	F	F
F	F	T	T	F	T	F

7. -24 9. never true 11. sometimes true 13. -0.002 15. 21.2
17. $\{s: s \geq 6\}$ 19. $-2\frac{7}{10}$ 21. $-\frac{1}{2}$ 23. $\{n: n > 9\}$ 25. Let x be an even integer. $x(x+2) = 168$ 27. $\{2, 3, 4, \ldots\}$ 29. 187.5 min 31. 15 years old
33. $-\frac{1}{2} > -\frac{2}{3}$ and $x \cdot (-1) = -x$; true 35. $-\frac{1}{2} > -\frac{2}{3}$ if and only if $x \cdot (-1) = -x$; true

37.

p	q	$p \wedge q$	$\sim(p \wedge q)$	$\sim p$	$\sim p \vee q$	$\sim(p \wedge q) \rightarrow \sim p \vee q$
T	T	T	F	F	T	T
T	F	F	T	F	F	F
F	T	F	T	T	T	T
F	F	F	T	T	T	T

neither

39. {Nebraska, Nevada, New Hampshire, New Jersey, New Mexico, New York, North Carolina, North Dakota} 41. false

Page 136 1. true 3. false 5. false 7. true 9. $\overleftrightarrow{BC}$ 11. one
13. infinitely many 15. infinitely many 17. yes; yes; no 19. one 21. one
23. $\underset{W\ X\ Y\ Z}{\bullet\!\!-\!\!\bullet\!\!-\!\!\bullet\!\!-\!\!\bullet}$ 25.   27.

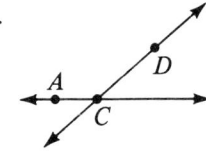

Pages 138-140 1. $\overleftrightarrow{KL}$ (or $\overleftrightarrow{LK}$) 3. $\overline{KL}$ (or $\overline{LK}$) 5. 5 7. 7 9. 2
11. 5 13. $\overline{YZ}$ and $\overline{WX}$: $\overline{VW}$ and $\overline{YX}$; $\overline{WY}$ and $\overline{ZX}$ (or $\overline{VX}$) 15. -1 17. no
19. 28 21. yes; C 23. no 25. yes; $\overline{BC}$ 27-33. Missing items in the table are given, reading down. 27. 12, 2 29. 8, 6 31. 10, 14 33. 4, $a+2$

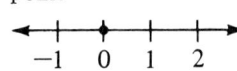

35. point 37. line segment 39. line

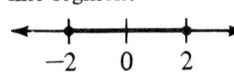

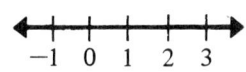

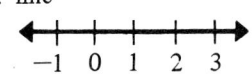

41. line 43. none of these

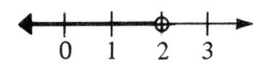

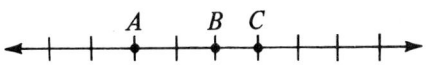

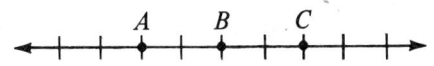

45. 47.

49. one 51. $\frac{a+b}{2}$

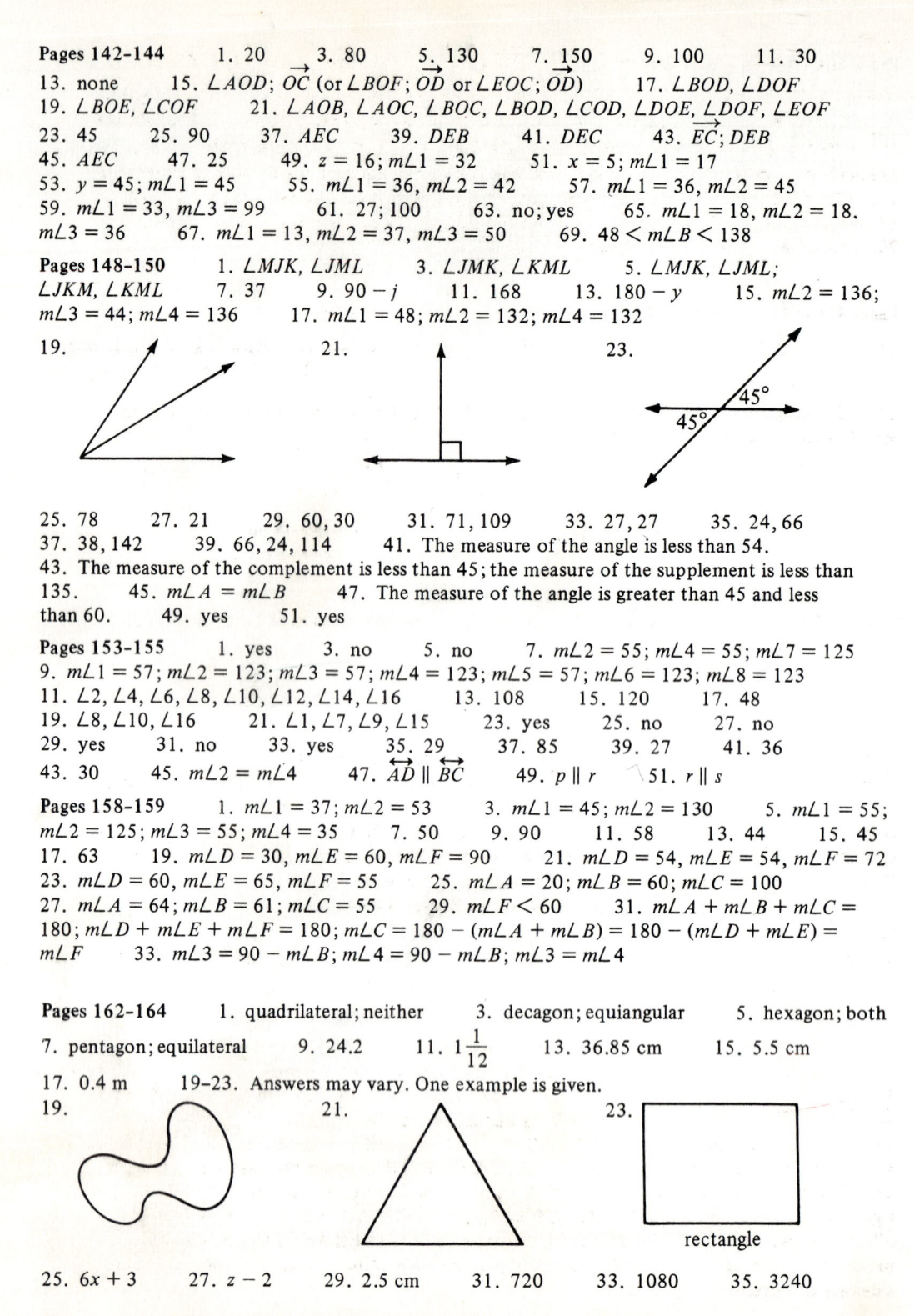

Pages 142–144 1. 20 3. 80 5. 130 7. 150 9. 100 11. 30
13. none 15. $\angle AOD$; $\overrightarrow{OC}$ (or $\angle BOF$; $\overrightarrow{OD}$ or $\angle EOC$; $\overrightarrow{OD}$) 17. $\angle BOD, \angle DOF$
19. $\angle BOE, \angle COF$ 21. $\angle AOB, \angle AOC, \angle BOC, \angle BOD, \angle COD, \angle DOE, \angle DOF, \angle EOF$
23. 45 25. 90 37. AEC 39. DEB 41. DEC 43. $\overrightarrow{EC}$; DEB
45. AEC 47. 25 49. $z = 16$; $m\angle 1 = 32$ 51. $x = 5$; $m\angle 1 = 17$
53. $y = 45$; $m\angle 1 = 45$ 55. $m\angle 1 = 36, m\angle 2 = 42$ 57. $m\angle 1 = 36, m\angle 2 = 45$
59. $m\angle 1 = 33, m\angle 3 = 99$ 61. 27; 100 63. no; yes 65. $m\angle 1 = 18, m\angle 2 = 18,$
$m\angle 3 = 36$ 67. $m\angle 1 = 13, m\angle 2 = 37, m\angle 3 = 50$ 69. $48 < m\angle B < 138$

Pages 148–150 1. $\angle MJK, \angle JML$ 3. $\angle JMK, \angle KML$ 5. $\angle MJK, \angle JML$;
$\angle JKM, \angle KML$ 7. 37 9. $90 - j$ 11. 168 13. $180 - y$ 15. $m\angle 2 = 136$;
$m\angle 3 = 44$; $m\angle 4 = 136$ 17. $m\angle 1 = 48$; $m\angle 2 = 132$; $m\angle 4 = 132$
19. 21. 23.

25. 78 27. 21 29. 60, 30 31. 71, 109 33. 27, 27 35. 24, 66
37. 38, 142 39. 66, 24, 114 41. The measure of the angle is less than 54.
43. The measure of the complement is less than 45; the measure of the supplement is less than 135. 45. $m\angle A = m\angle B$ 47. The measure of the angle is greater than 45 and less than 60. 49. yes 51. yes

Pages 153–155 1. yes 3. no 5. no 7. $m\angle 2 = 55$; $m\angle 4 = 55$; $m\angle 7 = 125$
9. $m\angle 1 = 57$; $m\angle 2 = 123$; $m\angle 3 = 57$; $m\angle 4 = 123$; $m\angle 5 = 57$; $m\angle 6 = 123$; $m\angle 8 = 123$
11. $\angle 2, \angle 4, \angle 6, \angle 8, \angle 10, \angle 12, \angle 14, \angle 16$ 13. 108 15. 120 17. 48
19. $\angle 8, \angle 10, \angle 16$ 21. $\angle 1, \angle 7, \angle 9, \angle 15$ 23. yes 25. no 27. no
29. yes 31. no 33. yes 35. 29 37. 85 39. 27 41. 36
43. 30 45. $m\angle 2 = m\angle 4$ 47. $\overleftrightarrow{AD} \parallel \overleftrightarrow{BC}$ 49. $p \parallel r$ 51. $r \parallel s$

Pages 158–159 1. $m\angle 1 = 37$; $m\angle 2 = 53$ 3. $m\angle 1 = 45$; $m\angle 2 = 130$ 5. $m\angle 1 = 55$;
$m\angle 2 = 125$; $m\angle 3 = 55$; $m\angle 4 = 35$ 7. 50 9. 90 11. 58 13. 44 15. 45
17. 63 19. $m\angle D = 30, m\angle E = 60, m\angle F = 90$ 21. $m\angle D = 54, m\angle E = 54, m\angle F = 72$
23. $m\angle D = 60, m\angle E = 65, m\angle F = 55$ 25. $m\angle A = 20$; $m\angle B = 60$; $m\angle C = 100$
27. $m\angle A = 64$; $m\angle B = 61$; $m\angle C = 55$ 29. $m\angle F < 60$ 31. $m\angle A + m\angle B + m\angle C = 180$; $m\angle D + m\angle E + m\angle F = 180$; $m\angle C = 180 - (m\angle A + m\angle B) = 180 - (m\angle D + m\angle E) = m\angle F$ 33. $m\angle 3 = 90 - m\angle B$; $m\angle 4 = 90 - m\angle B$; $m\angle 3 = m\angle 4$

Pages 162–164 1. quadrilateral; neither 3. decagon; equiangular 5. hexagon; both
7. pentagon; equilateral 9. 24.2 11. $1\frac{1}{12}$ 13. 36.85 cm 15. 5.5 cm
17. 0.4 m 19–23. Answers may vary. One example is given.
19. 21. 23.

rectangle

25. $6x + 3$ 27. $z - 2$ 29. 2.5 cm 31. 720 33. 1080 35. 3240

Pages 166-167 · Chapter Review 1. $\overleftrightarrow{BD}$ (or $\overleftrightarrow{DB}$) 2. one 3. $A, B, C; A, B, D, E$
(Answers may vary.) 4. $\overrightarrow{BE}$ (or $\overrightarrow{EB}$) 5. $\overline{AB}, \overline{BC}, \overline{BD}$ 6. 12 7. 90; right
8. 110; obtuse 9. 50; acute 10. $\angle UOV, \angle XOY; \angle VOX, \angle YOZ$ 11. $\overrightarrow{OU}, \overrightarrow{OX}$;
$\overrightarrow{OV}, \overrightarrow{OY}; \overrightarrow{OX}, \overrightarrow{OZ}$ 12. $\angle XOY, \angle UOV; \angle UOY$ 13. $90 < m\angle 2 < 180$ 14. Two
lines cut by a transversal are parallel if and only if alternate interior angles have equal measures.
15. $m\angle 1 = 138; m\angle 2 = 42; m\angle 4 = 42; m\angle 5 = 138; m\angle 6 = 42; m\angle 7 = 138; m\angle 8 = 42$
16. 45 17. $m\angle 1 = 58; m\angle 2 = 72; m\angle 3 = 122$ 18. 90; 66 19. 59, 60, 61
20. pentagon 21. equiangular 22. 4

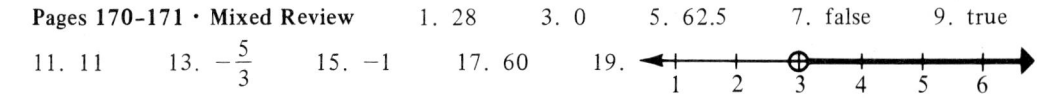

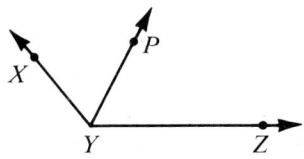

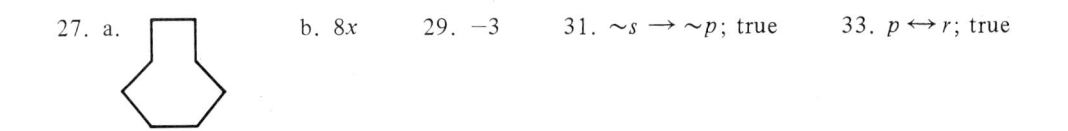

Pages 170-171 · Mixed Review 1. 28 3. 0 5. 62.5 7. false 9. true
11. 11 13. $-\frac{5}{3}$ 15. -1 17. 60 19. [number line with open circle at 3]
21. congruent, supplementary 23. $m\angle 9 = 85, m\angle 11 = 85, m\angle 12 = 95, m\angle 13 = 85,$
$m\angle 14 = 95, m\angle 15 = 85, m\angle 16 = 95$ 25.

27. a. [figure] b. $8x$ 29. -3 31. $\sim s \rightarrow \sim p$; true 33. $p \leftrightarrow r$; true

Pages 176-178 1. Use Construction 2. 3. Use Construction 3. 5. Use Construction 4. 7. Use Construction 5. 9. Use Construction 3. 11. Use Construction 5.
13. Answers may vary. Use Construction 3, Construction 4, or Construction 5. 15. no;
$QR + ST < UV$ 17. Check student's drawings. 19. Check student's drawings. Use
Construction 3. 21. Use Construction 1 to construct $\overline{DG}$ such
that $\overline{DG} \cong \overline{AB}$; use Construction 1 again to construct $\overline{GE}$ such that
$\overline{GE}$ is on $\overleftrightarrow{DG}$ and $\overline{GE} \cong \overline{AB}$. Repeat to construct line segments with
lengths $2BC$ and $2AC$; then use Construction 3 to construct the
desired triangle. 23. Use Construction 2 to construct
$\angle BAC \cong \angle 1$; use $\overrightarrow{AC}$ as the ray in Construction 2 to construct
$\angle CAD \cong \angle 2$; use a straightedge to draw $\overrightarrow{AE}$ on $\overleftrightarrow{AB}$; $\angle EAD$ is the desired angle.

Ex. 23

Pages 181-184 1. $\angle R, \angle J; \angle O, \angle E; \angle Y, \angle N; \overline{RO}, \overline{JE}; \overline{OY}, \overline{EN}; \overline{RY}, \overline{JN}$ 3. yes; ASA
5. no 7. yes; SSS 9. $\angle A, \angle D; \angle ABC, \angle DCB; \angle ACB, \angle DBC; \overline{AB}, \overline{DC}; \overline{AC}, \overline{DB}$;
$\overline{BC}, \overline{CB}$ 11. $\angle JML, \angle LKJ; \angle MLJ, \angle KJL; \angle LJM, \angle JLK; \overline{ML}, \overline{KJ}; \overline{MJ}, \overline{KL}; \overline{JL}, \overline{LJ}$
13. $\overline{ZW} \cong \overline{ZW}$ so $\triangle ZWX \cong \triangle ZWY$ by ASA. 15. $\angle JNK \cong \angle MNL$, since they are vertical
angles, so $\triangle JNK \cong \triangle MNL$ by ASA 17. $\angle DEC \cong \angle BEA$, since they are vertical angles,
so $\triangle DEC \cong \triangle BEA$ by SAS; $\angle DEA \cong \angle BEC$, since they are vertical angles, so $\triangle DEA \cong \triangle BEC$
by SAS. 19. $\angle S \cong \angle Y$ 21. $\overline{LN} \cong \overline{LQ}$ 23. $\overline{AD} \cong \overline{CB}$ 25. $\overline{MJ} \cong \overline{ML}$
27. a. $\angle 4$; when parallel lines are cut by a transversal, alternate interior angles have equal
measures. b. $\angle 5$ c. $\angle 6$; vertical angles are congruent. d. no; corresponding
sides are not congruent. e. no

Pages 187-188   1. $m\angle 1 = 67; m\angle 2 = 46$ 3. $m\angle 1 = 55; m\angle 2 = 55$
5. $m\angle 1 = 73; m\angle 2 = 107; m\angle 3 = 50$ 13. 42, 42, 96
15. a. If two angles of a triangle are congruent, then the sides opposite those angles are congruent.   b. $\triangle ADB$, $\triangle ADC$ c. $\angle B, \angle C; \overline{AD}, \overline{AD}; \angle BAD, \angle CAD$
d. $\angle CDA; m\angle BDA = 180 - (m\angle B + m\angle BAD) = 180 - (m\angle C + m\angle CAD) = m\angle CDA$ e. ASA (or AAS)   f. yes; corresponding parts of congruent triangles are congruent.   g. yes   17. $\overline{EB} \cong \overline{EC}$ so $\angle EBC \cong \angle ECB$; $\angle ABE$ and $\angle DCE$ are supplementary angles of congruent angles so $\angle ABE \cong \angle DCE$; $\overline{AB} \cong \overline{CD}$; $\triangle ABE \cong \triangle DCE$ by SAS.

	7.	9.	11.
Vertex angle	50	36	120
Base angle	65	72	30

Pages 191-192   1. no   3. yes   5. yes   7. $m\angle X = 65; m\angle Y = 115; m\angle Z = 65$
9. $x = 6$ 11. $x = 6$ 13. $x = 5; y = 7$
15. false   17. false

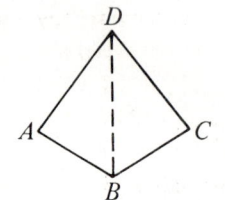

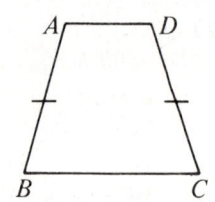

19. true; each side is a transversal between a pair of opposite sides; since same-side interior angles are supplementary, the opposite sides are parallel.   21. no; if $\overline{DC} \cong \overline{AB}$, then $y = 9$; if $\overline{DA} \cong \overline{CB}$, then $y = 5$.   23. (1) Definition of perpendicular lines (2) When two lines are cut by a transversal, the lines are parallel if and only if corresponding angles have equal measures. (3) Given (4) Both pairs of opposite sides are parallel. (5) Opposite sides of a parallelogram are congruent.

Pages 195-196   1. false   3. false   5. true   7. false   9. true   11. true
13. false
15. parallelogram   17. rhombus

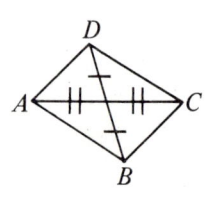

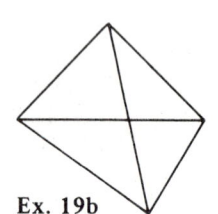

19. a. If the diagonals of a quadrilateral are congruent, then the quadrilateral is a rectangle.
21. (1) Definition of a rhombus. (4) SSS (5) Corresponding parts of congruent triangles are congruent. (7) Definition of perpendicular lines

Pages 200-202   1. 210 square units   3. 300   5. 1204   7. 3.4 cm^2 9. 286
11. 512   13. 197.2 cm^2 15. perimeter = 130 cm; area = 1000 cm^2 17. width = 16.4 m; perimeter = 79.8 m   19. perimeter = 46 cm; area = 117 cm^2 21. $AD = 6.1$ cm; $DX = 4.5$ cm   23. a. 36   b. 72   25. a. 30 cm b. 48 cm^2 27. 0.35 m^2 (or 3500 cm^2)   29. a. $75 \text{ cm}^2; 150 \text{ cm}^2$ b. Area is doubled.   c. Area is quadrupled.
31. 900 cm^2 33. area = 20; perimeter = 20   35. 80   37. a. width = 6 cm; length = 12 cm   b. 72 cm^2 39. (1) Definition of perpendicular lines (3) Opposite sides of a parallelogram are parallel. (5) Definition of congruent angles (6) Opposite sides of a parallelogram are congruent.

SELECTED ANSWERS

Pages 205-207   1. 64 cm^2 3. 38.85 m^2 5. $P = 80$ cm, $A = 120$ cm^2
7. 24 cm^2 9. 49 m^2 11. perimeter = 60 cm; area = 210 cm^2 13. 40 cm
15. 24 cm   17. (1) Use the formula $A = \frac{1}{2}h(b_1 + b_2)$. So, $A = \frac{1}{2} \cdot 4(4 + 7) = 22$.
(2) Add the areas of the triangle and the square. $A = \frac{1}{2}(3 \cdot 4) + (4 \cdot 4) = 22$.
19. perimeter = 84; area = 288   21. a. 6   b. $2\frac{2}{5}$

Pages 209-211   1. a. 60   b. 76   3. a. 160 cm^2 b. 358 cm^2 5. 60; 96
7. 240 cm^2; 384 cm^2 9. 336   11. 180 cm^2 13. 720   15. 88 cm^2
17. 150 cm^2 19. 960 cm^2 21. Let h = height, n = number of sides of the base, and $s_1, s_2, \ldots, s_n$ be the lengths of the sides of the base; then the perimeter, $p = s_1 + s_2 + \cdots + s_n$ and the lateral area $= hs_1 + hs_2 + \cdots + hs_n = h(s_1 + s_2 + \cdots + s_n) = hp$.
23. 150 cm^2 25. Area is quadrupled.   27. 32.4 m^2 29. approximately 289,440 cm^2

Pages 213-214   1. 36 cm^3 3. 200 cm^3 5. 99.5 cm^3 7. 150 cm^3
9. 3.8 m^3 11. 125 cm^3 13. Volume is multiplied by 8.   15. 21 cm
17. 13.75 m^3 19. 200

Pages 216-218 · Chapter Review   1. Construction 5   2. Construction 3   3. $\triangle IHG$, $\triangle IJK$; SAS   4. $\triangle TQR$, $\triangle TSR$; SSS   5. $\triangle EBL$, $\triangle LUE$; ASA   6. $m\angle 1 = 50$; $m\angle 2 = 65$ 7. $m\angle 3 = 52$; $m\angle 4 = 52$ 8. 48, 48, 84   9. 80   10. not possible
11. 100   12. yes; $\angle 1 \cong \angle 2$ so $\overline{ML} \parallel \overline{JK}$ and $\angle 3 \cong \angle 4$ so $\overline{MJ} \parallel \overline{LK}$. (Two lines cut by a transversal are parallel if and only if alternate interior angles have equal measures.) Then since both pairs of opposite sides of $JKLM$ are parallel, $JKLM$ is a parallelogram.   13. always
14. sometimes   15. sometimes   16. 195 cm^2 17. 2.88 m^2 18. 17.5
19. 96   20. a. 240 cm^2 b. 672 cm   21. a. 60   b. 96   22. 216 cm^3
23. 864 cm^3 24. 48

Pages 222-223 · Mixed Review   1. $\emptyset$ 3. yes; If three sides of one triangle are congruent to three sides of another triangle, then the triangles are congruent.   5. 53, 90
7. a. $V = \frac{s^2 h}{3}$ b. $h = \frac{3V}{s^2}$ 9. $m\angle 1 = 58, m\angle 2 = 72$ 11. -4
13. ←+−+−+−+−+−+−●−+−⊕−+−+−+→
    $\quad -9\ -8\ -7\ -6\ -5\ -4\ -3\ -2\ -1\ 0$ 15. 1   17. $-\frac{10}{13}$ 19. 7
21. $2x \leqslant 3^2$ 23. rectangle   25. 81   27. 144 cm^2 29. false
31. If $x + 0 \neq x$, then $x = -x$; true

Pages 227-228   1. $(-3)^3$ 3. x^2 5. $2z^2$ 7. $11b^2 c$ 9. 243   11. -512
13. 1,000,000   15. 72   17. $\frac{1}{9}$ 19. 0.04   21. $-9, -3, -1, 1, 3, 9$
23. $-16, -8, -4, -2, -1, 1, 2, 4, 8, 16$ 25. $-15, -5, -3, -1, 1, 3, 5, 15$ 27. $-17, -1,$ 1, 17   29. $16x^2$ 31. $3b^2$ 33. $6s^3$ 35. $24d^2$ 37. true   39. false

Page 231   1. monomial; 1   3. binomial; 2   5. monomial; 4   7. monomial; 0
9. monomial; 8   11. monomial; 4   13. trinomial; 3   15. trinomial; 2
17. trinomial; 9   19. $2m^2$ and $-3m^2$ 21. $8x$ and $9x$; $-10x^2$ and $5x^2$ 23. $7j^2 k$ and $-8j^2 k$ 25. $5x^2 + 2x + 1$ 27. $\frac{d^4}{2} - \frac{d^2}{3} + 9$ 29. $5m^3 - 4.5m - 4$

SELECTED ANSWERS   599

31. $-2r^5 + r^2 - 3$; 5; -2 33. $3x^2 + 2x$; 2; 3   35. $6b + 2$; 1; 6   37. $5z^5 + 5z^2$; 5; -5 39. $-a^2 + 4ab - b^2$; 2; -1 and 4   41. $-x^2y + 9x^2 - 7y^2$; 3; -1 43. $-8b^2c^2 + 5bc$; 4; -8 45, 47. Answers may vary. One example is given.   45. $x^3 + 2x + 1$; $8t^3 - 1$ 47. $9x^2 - 2x + 1$

Page 233   1. $-4x + 5$ 3. $-8y + 7$ 5. $-s^2 + 6s - 3$ 7. $-2ab + 7a - b$ 9. $-6r^2 - 12rs + 4s^2$ 11. $10c + 7$ 13. $-9z^2 - 6z + 13$ 15. $-y^4 + 12y^2 - 5y - 11$ 17. $4n - 6x - 12nx^2$ 19. $-13s^2t^2 - 10st - 13$ 21. $-7z^2 + 6z - 2$ 23. $-\frac{2}{7}s^4 - \frac{8}{7}s^2 + \frac{8}{7}$ 25. $0.7y^3 - 0.5y^2 + 0.5y - 1.5$ 27. 3   29. $\frac{1}{2}$ 31. 0   33. {the real numbers}   35. $\frac{7}{3}$ 37. 1   39. $4\frac{1}{2}$ 41. 21   43. $2\frac{2}{5}$ 45. 252

Pages 236-237   1. $9a^2$ 3. $4c^6$ 5. $100a^2b^2$ 7. $0.04j^2k^4$ 9. a^{13} 11. $-c^2$ 13. $-y^7$ 15. a^5b^{10} 17. $12a^5$ 19. $-42c^{12}$ 21. $24r^3s^4$ 23. $-10a^3b^3c^4$ 25. $80c^6d^9$ 27. $-36r^2st^{11}$ 29. a^{16} 31. $-24c^7d^6$ 33. $x^7y^5z^8$ 35. $r^{16}s^{19}$ 37. $9a^6b^6$ 39. $72x^6$ 41. 5   43. 4   45. 12   47. 18   49. 4

Pages 239-240   1. $143x^3 + 55x$ 3. $27z^4 - 21z^3 + 6z^2$ 5. $-2b^4 - 6b^3 + 10b^2$ 7. $8r^4s^2 - 12r^3s^3 + 24r^2s^4$ 9. $-12a^6b^5 + 8a^5b^4 - 36a^4b^3$ 11. $x^2 + 8x + 15$ 13. $2r^2 - 12r + 10$ 15. $s^2 - 36$ 17. $15x^2 + 2x - 8$ 19. $9t^2 + 12t + 4$ 21. $2x^3 - 3x^2 - 3x - 5$ 23. $8c^4 - 42c^2 - 135$ 25. $a^4 - 5a^3b + 5a^2b^2 + 2ab^3$ 27. $2a^3 + 5a^2b - ab^2 - 4b^3$ 29. $-10z^4 - 7z^3 + 13z^2 + 4z - 4$ 31. $12x + 8$ 33. $6y + 22$ 35. $\frac{1}{4}s^2 - 3s + 9$ 37. $4x^2 - 8x + 3$ 39. $(x + 7)(x + 2)$ 41. $k^3 - 19k - 30$ 43. $s^5 + 3s^4 - 25s - 75$ 45. $c^3 - 6c^2d + 12cd^2 - 8d^3$

Page 242   1. r^{10} 3. $\frac{1}{t^{10}}$ 5. 1   7. $-125d^3$ 9. $4x^5$ 11. $-22x^2$ 13. $\frac{5}{4}$ 15. $-\frac{xy}{14}$ 17. $2x^2 + x - 3$ 19. $-4y^5 + 2y^3 - y$ 21. $r^2k^2 - 5rk + 2$ 23. $2x^2y - y^2z^4 + 3x^4y$ 25. 10   27. $-2z^2 - 2z$ 29. $t^2 - 8t + 6$ 31. $2^0 = 1$ (Then $2^5 \cdot 2^0 = 2^5 \cdot 1 = 2^5$ and $2^{5+0} = 2^5$; also $\frac{2^5}{2^5} = 1$ and $2^{5-5} = 2^0 = 1$.)   33. $2^{-3} = \frac{1}{2^3}$

Pages 244-245   1. $2 \cdot 2 \cdot 2$ 3. $5 \cdot 7$ 5. $2 \cdot 2 \cdot 2 \cdot 2 \cdot 3 \cdot 3$ 7. 16   9. $12x^2$ 11. $2a^2b^2$ 13. $7xyz^2$ 15. $8x^4$ 17. $22a^2b$ 19. $2x + 1$ 21. $x^2 + 3x + 2$ 23. $4x^2(2 + x)$ 25. $4ab(3a - b)$ 27. $2x(2x^2 + x - 3)$ 29. $a^2(12a^3 - 3 + 4a^2)$ 31. $ax(ax + x - a)$ 33. $cz(c^2z^2 + cz + 1)$ 35. $a^2x^2y^2(xy^2 + ay - x)$ 37. $4x^3(2 + 3x^2)$ 39. $5a^2$ 41. $x + 3$ 43. $5x$

Page 247   1. $x^2 + 3x + 2$ 3. $z^2 + z - 2$ 5. $s^2 - s - 30$ 7. $p^2 - 6p + 8$ 9. $b^2 + 6b + 9$ 11. $d^2 - 1$ 13. $x^2 + 12x + 27$ 15. $z^2 + 9z - 36$ 17. $s^2 - 14s + 48$ 19. $k^2 - 16k + 63$ 21. $h^2 + 14h + 49$ 23. $x^2 - 100$ 25. $8x^2 + 8x + 2$ 27. $5z^2 + 31z + 30$ 29. $6b^2 + 14b + 4$ 31. $35q^2 + 29q + 6$ 33. $9h^2 + 6h + 1$ 35. $10d^2 + 29d - 21$ 37. $14n^2 - 38n + 20$ 39. $64k^2 - 4$ 41. $16z^2 - 9$ 43. 6   45. 1; 3   47. 5; 3   49. 6; 17   51. 5; 0   53. 3 and 7   55. 25 and 10   57. 5 and -5 59. 7; 8   61. 9; 8   63. 5; 5   65. 11; 11

Pages 250-251 1. $(x+6)(x+1)$ 3. $(z+7)(z-1)$ 5. $(s+4)(s+2)$
7. $(h+9)(h-2)$ 9. $(p+2)(p+2)$ 11. $(u-6)(u+5)$ 13. $(a-5)(a-4)$
15. $(x+5)(x+5)$ 17. $(z-3)(z+3)$ 19. $(m+13)(m-6)$ 21. 3; 1
23. 3; 3 25. 7; 5 27. 3 29. 3 31. 2; 7 33. $2(x+1)(2x+1)$
35. $(3z+4)(5z+2)$ 37. $(3k+1)(2k-1)$ 39. $2(n+1)(n-3)$
41. $(5a-1)(2a-1)$ 43. $(4d-3)(2d-5)$ 45. $(3p-1)(3p+1)$
47. $(5u-2)(5u+2)$ 49. $(3x+1)^2$ 51. $(2z+7)^2$ 53. $(x^2+2)(x^2+1)$
55. $(4t^2+5)(2t^2+3)$ 57. $(7s^4+3)(3s^4-2)$ 59. $(3u^2-5)(3u^2+5)$
61. $(y-1)(y+1)(y^2-5)$

Pages 253-254 1. $(x+8)(x-8)$ 3. $(z+9)(z-9)$ 5. $(4q+1)(4q-1)$
7. $(12h+11)(12h-11)$ 9. $(z+10)^2$ 11. $(p-9)^2$ 13. $(k-11)^2$
15. x^2-4 17. $4z^2-1$ 19. $25t^2-16$ 21. $x^2+8x+16$
23. z^2+6z+9 25. $25g^2+20g+4$ 27. $25h^2+110h+121$ 29. $3(x+5)(x-5)$
31. $6(y+3)(y-3)$ 33. $5(2n+3)(2n-3)$ 35. $3(s+1)^2$ 37. $5(n+4)^2$
39. $9(s+3)^2$ 41. $(2y+1)^2$ 43. $4(6y-5)^2$

Pages 256-257 1. $\{1\}$ 3. $\{-1, 1\}$ 5. $\{2, 1\}$ 7. $\{-4, 3\}$ 9. $\{-2, -7\}$
11. $\{3, -2\}$ 13. $\{-5, 2\}$ 15. $\{2, 3\}$ 17. $\left\{-\frac{2}{3}, -\frac{3}{2}\right\}$ 19. $\left\{-\frac{3}{5}, -\frac{1}{3}\right\}$
21. $\left\{\frac{2}{3}, -\frac{3}{2}\right\}$ 23. $\left\{\frac{3}{4}, \frac{5}{2}\right\}$ 25. $\{-12, 14\}$ 27. $\{15, 19\}$ 29. $\left\{\frac{3}{5}, -\frac{1}{5}\right\}$
31. $\{40, 50\}$ 33. $x^2-3x+2=0$ 35. $x^2-4x-5=0$ 37. $x^2+x-42=0$
39. $x^2+13x+36=0$ 41. $\{0, -3, 2\}$ 43. $\{0, -6, 4\}$ 45. $\{-1, 1, -2, 2\}$
47. $\left\{-\frac{1}{2}, \frac{1}{2}\right\}$

Pages 259-260 1. 6 and 7 3. −8 and −7 5. 6 and 8 7. 5 and 7
9. width = 5 m; length = 7 m 11. width = 8 m; length = 11 m 13. 10
15. 6 and 7 17. 12 and 14 19. 4 cm 21. width = 4 m; length = 5 m
23. 9, 10, and 11

Page 262 · Chapter Review 1. $2y^4$ 2. $-\frac{8}{27}$ 3. 1, 2, 3, 5, 6, 10, 15, 30
4. 3; 2 5. $-5t^5+3t^3+t+6$ 6. $-m^2+4m$ 7. a. $15r^2+6r-2$
b. $-r^2+6r-6$ 8. a. $a^2b^4-4ab^2+11$ b. $7a^2b^4-4ab^2-9$ 9. g^{10}
10. $9x^4y^6$ 11. $-6b^2c^4d^3$ 12. $-6n^3+4n^2$ 13. $4y^2-4y-24$
14. $4a^2-28a+49$ 15. $\frac{1}{64g^2h^2}$ 16. $-\frac{r^4s^2}{7}$ 17. $-2w^3+w-\frac{3}{w}$
18. $3^3 \cdot 5$ 19. $4a^3b^2$ 20. $8xy^2(5x^4+2y)$ 21. $6b^2(6b+2b^4-3)$
22. $z^2+12z+35$ 23. $16h^2-24h+9$ 24. $9d^2-9d-10$ 25. $(s-15)(s+3)$
26. $(4k-3)(2k-1)$ 27. $(10m-7)(10m+7)$ 28. $(q-13)^2$ 29. $\{-9\}$
30. $\left\{-\frac{2}{5}, \frac{1}{2}\right\}$ 31. −10 and −8 32. width = 6 m; length = 18 m

Pages 264-266 · Cumulative Review: Chapters 1-7 1. c 2. d 3. b
4. d 5. a 6. b 7. a 8. b 9. d 10. a 11. c 12. b
13. b 14. a 15. c 16. d 17. c 18. d 19. a 20. a
21. b 22. d 23. a 24. d 25. a 26. d 27. d 28. c
29. b 30. d 31. d 32. a 33. b 34. b 35. c 36. b
37. c 38. d

Pages 268–269 · Mixed Review 1. $-27r^5s^{10}$ 3. -16 5. 128 dimes   7. $\frac{2}{3}, -\frac{2}{3}$
9. $\{z: z \geq 50\}$ 11. 2   13. $\angle A \cong \angle R, \angle B \cong \angle S, \angle C \cong \angle T, \overline{AB} \cong \overline{RS}, \overline{AC} \cong \overline{RT}, \overline{BC} \cong \overline{ST}$
15. 5   17. 60°   19. $-8, -10$ 21. yes; no; yes   23. $\{-2, -1, 0, 1\}$
25. $(y-8)(y-1)$ 27. Answers will vary.   29. Their intersection is a line.   31. 14
33. $x + 20 = 3x$ 35. $-7 < y < 14$ 37. [number line from -2 to 4 with open circle at -1 and closed circle at 3]

Pages 273–274   1–5. Answers may vary. One example is given.   1. $\frac{64}{2}$ 3. $-\frac{5}{2}$
5. $\frac{4}{10}$ 7. 0.25   9. -0.7 11. -1.625 13. 3.2   15. $-0.\overline{2}$ 17. $-0.\overline{63}$
19. $0.\overline{296}$ 21. $0.\overline{42}$ 23. $0.2\overline{27}$ 25. 0.015625   27. $0.4\overline{950}$ 29. $0.\overline{076923}$
31. $\frac{78}{100} = 0.78; 0.78, 0.\overline{78}, 0.\overline{78}$ 33. $0.\overline{93}, 0.9\overline{39}, 9.\overline{39}$ 35. $\frac{3}{4} = 0.75; 0.\overline{34}, 0.75, 3.4$

Page 276   1. $\frac{4}{5}$ 3. $\frac{3}{25}$ 5. $\frac{107}{500}$ 7. $\frac{1}{9}$ 9. $\frac{13}{99}$ 11. $\frac{104}{333}$ 13. $\frac{26}{45}$
15. $\frac{1}{10}$ 17. $\frac{401}{3300}$ 19. $\frac{17}{33}$ 21. $<$ 23. $>$ 25. $=$

Pages 278–279   1. 15   3. -14 5. -35 7. 12   9. -55 11. 70
13. -27 15. 36   17. 19   19. $-\frac{4}{9}$ 21. $\frac{16}{15}$ 23. $-\frac{7}{30}$ 25. 12
27. -31 29. -61 31. 85   33. 0.1   35. -0.07 37. a. $x = 5; y = 3$
b. 30 cm²   c. 64 cm²   d. 8 cm   39. $-2xy$ 41. $-15a$ 43. $7x^2$
45. $\frac{4a}{5b}$ 47. -2 49. 4   51. 10 cm

Page 282   1. 4.58   3. 9.06   5. -7.35 7. -8.60 9. 4.90   11. 5.39
13. 6.24   15. -15.10 17. -24.98 19. 4.44   21. 7.33, $\sqrt{54}$, 7.35
23. 13.34, 13.36, $\sqrt{179}$ 25. $\sqrt{450}$, 21.22, 21.23   27. yes; 324   29. no
31. no   33. no

Pages 284–285   1. $\sqrt{21}$ 3. $24\sqrt{6}$ 5. $-60\sqrt{35}$ 7. $\sqrt{21}$ 9. $7\sqrt{2}$
11. $9\sqrt{11}$ 13. 120   15. 21   17. 4   19. 15   21. 45   23. 15
25. 0.6   27. 0.8   29. 3   31. $\sqrt{30}$ 33. 20   35. 25
37. a. $\sqrt{42}$ cm (≈ 6.48 cm)   b. $\sqrt{84}$ cm (≈ 9.17 cm)   c. $(2\sqrt{42})^2 = 168$ cm²
d. The area of Square C is twice that of Square B e. four times as great   39. a^2b^2
41. a^5 43. a^2b^3

Pages 286–287   1. $2\sqrt{5}$ 3. $-10\sqrt{11}$ 5. $18\sqrt{10}$ 7. $6\sqrt{11}$ 9. $5\sqrt{3}$
11. $5\sqrt{2}$ 13. $2\sqrt{2}$ 15. $\frac{3\sqrt{14}}{2}$ 17. $-\frac{\sqrt{7}}{5}$ 19. $\frac{4\sqrt{2}}{9}$ 21. $\frac{5\sqrt{3}}{3}$
23. $\frac{\sqrt{14}}{7}$ 25. $-\frac{\sqrt{3}}{3}$ 27. $\frac{\sqrt{3}}{9}$ 29. $5\sqrt{2}$ 31. $36\sqrt{5}$ 33. $2\sqrt{53}$; 14.56
35. $5\sqrt{5}$; 11.18   37. $-2\sqrt{39}$; -12.49 39. $7x\sqrt{x}$ 41. $2x^4\sqrt{2}$
43. $3bc\sqrt{2c}$

Page 288   1. $14\sqrt{7}$ 3. $-21\sqrt{5}$ 5. $-5\sqrt{3}$ 7. $9\sqrt{15}$ 9. $2\sqrt{3}$
11. $3\sqrt{13}$ 13. $19\sqrt{7}$ 15. 0   17. $9\sqrt{3}$ 19. $\frac{12\sqrt{5}}{5}$ 21. $\frac{5\sqrt{6}}{6}$
23. $45\sqrt{21}$ 25. 6   27. 50   29. $10x\sqrt{2}$ 31. $x = \frac{1}{4}$ 33. $x = \frac{2}{5}$

602   SELECTED ANSWERS

Page 290   1. $3; 54 - 14\sqrt{5}$ 3. $4; 44$ 5. $1; \sqrt{15} + 3\sqrt{2} + \sqrt{10} + 2\sqrt{3}$
7. $2; 10 + 4\sqrt{6}$ 9. $4; x^2 - 3$ 11. $\dfrac{\sqrt{13} - 2}{3}$ 13. $\dfrac{6 + \sqrt{3}}{3}$ 15. $\dfrac{5\sqrt{10} + 10}{6}$
17. $74 + 40\sqrt{3}$ 19. $84 + 24\sqrt{6}$ 21. -157 23. $\dfrac{35\sqrt{6} + 21\sqrt{3}}{41}$
25. $(x + \sqrt{3})(x - \sqrt{3})$ 27. $\dfrac{3 - \sqrt{2}}{7}$

Page 292   1. $\{4, 3\}$ 3. $\left\{3, -\dfrac{1}{6}\right\}$ 5. $\left\{\dfrac{1}{5}, -\dfrac{2}{3}\right\}$ 7. $x^2 + 3x - 4 = 0$, $\{1, -4\}$
9. $2y^2 + 3y - 5 = 0$; $\left\{1, -\dfrac{5}{2}\right\}$ 11. $7w^2 + 47w - 14 = 0$; $\left\{\dfrac{2}{7}, -7\right\}$
13. $\left\{\dfrac{1 + \sqrt{41}}{2}, \dfrac{1 - \sqrt{41}}{2}\right\}$ 15. $\left\{\dfrac{-3 + \sqrt{29}}{10}, \dfrac{-3 - \sqrt{29}}{10}\right\}$ 17. $\left\{\dfrac{2 + \sqrt{11}}{7}, \dfrac{2 - \sqrt{11}}{7}\right\}$
19. $\{4 + \sqrt{15}, 4 - \sqrt{15}\}$; $\{7.87, 0.13\}$ 21. $\left\{\dfrac{1 + \sqrt{17}}{8}, \dfrac{1 - \sqrt{17}}{8}\right\}$; $\{0.64, -0.39\}$

Page 294 · Chapter Review   1. $0.58\overline{3}$ 2. 0.104 3. $-1.0\overline{15}$ 4. $\dfrac{8}{25}$ 5. $\dfrac{48}{37}$
6. $\dfrac{311}{900}$ 7. -90 8. $-\dfrac{7}{18}$ 9. 13 10. 4.80 11. -9.43 12. 11.27
13. -10 14. $3\sqrt{13}$ 15. 48 16. $-6\sqrt{7}$ 17. $\dfrac{3\sqrt{3}}{10}$ 18. $2\sqrt{3}$
19. $-4\sqrt{11}$ 20. $\dfrac{7\sqrt{7}}{2}$ 21. 0 22. 4 23. $\dfrac{2\sqrt{15} + 6}{3}$ 24. $132 - 10\sqrt{35}$
25. $\left\{\dfrac{3}{4}, -\dfrac{1}{2}\right\}$ 26. $\left\{\dfrac{-2 + \sqrt{5}}{2}, \dfrac{-2 - \sqrt{5}}{2}\right\}$

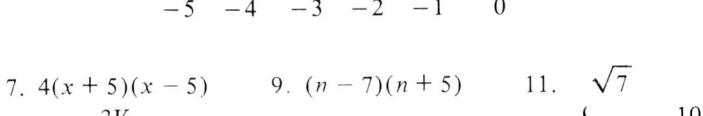

Pages 298–299 · Mixed Review   1. [number line from -5 to 0]   3. true
5. B (point diagram with A, C, D)   7. $4(x + 5)(x - 5)$ 9. $(n - 7)(n + 5)$ 11. $\sqrt{7}$
13. $h = \dfrac{3V}{B}$ 15. $m\angle T = 42$, $m\angle S = 96$ 17. $\left\{n: n \geq \dfrac{10}{3}\right\}$
19. $\left\{-\dfrac{1}{3}\right\}$ 21. $\{-2 + \sqrt{2}, -2 - \sqrt{2}\}$ 23. yes, ASA

25.

p	q	$\sim p$	$\sim p \vee q$
T	T	F	T
T	F	F	F
F	T	T	T
F	F	T	T

27. length = 10 cm, width = 6 cm   29. $x^2 - 7$ 31. $6v^9 - 27v^3 + 33$
33. $3x^3 - 9x^2 + 8x - 8$; 3 35. 11.9 37. Two lines cut by a transversal are parallel if and only if alternate interior angles have equal measures.   39. $39, 60, 81$

SELECTED ANSWERS   **603**

Pages 303-304 1. 1:3   3. 2:1   5. 3:5   7. 2:5:6   9. 2:7   11. 6:11
13. Rita: 816 votes; Chris: 612 votes   15. 54°, 36°   17. 750 tickets   19. 20 cm
21. −24, −54, −12   23. 1:4   25. 18 nickels, 27 dimes   27. 12 cm by 32 cm
29. 90°   31. 11:7   33. 123:2   35. $m\angle A = 48$, $m\angle B = 96$, $m\angle C = 36$

Pages 306-307 1. 7   3. 15   5. 24.5   7. 5   9. 10   11. 5
13. $350   15. 880 g   17. a. 420 cm   b. 24 cm   19. 24 min   21. 8
23. $6\sqrt{3}$ 25. −2   27. 7 L

Pages 310-313 1. 154   3. 55%   5. 120   7. 22.32   9. 350   11. 35%
13. 25%   15. 74%   17. $28,000   19. $16,000   21. $237.25   23. 78%
25. $450   27. a. $4500   b. $4275   c. $4250   29. $800 at 9%; $200 at 7%
31. $6,000   33. 18.27   35. 20 L   37. 348 g

Pages 317-319 1. 2:5   3. 3.6   5. RST   7. △RST, △RPQ   9. $PQ = 6$
11. 21   13. 12   15. $w = 104$, $x = 3.2$, $y = 5.6$, $z = 4.4$ 17. 10.5   19. 6
21. 6.4   23. 15.6   25. 50 m   27. 7.5   29. 8.4   31. (2) Corresponding angles of similar polygons are congruent   (3) Since an altitude is perpendicular to the base, the angles formed are right angles and all right angles are congruent.   (4) AA   (5) Corresponding sides of similar polygons are in proportion.   33. 1:1   35. (1) $\overline{CD}$ is the altitude to $\overline{AB}$ (Given)   (2) $CD \perp AB$ (Definition of altitude)   (3) $\angle BDC$ and $\angle BCA$ are right angles. (Definition of perpendicular lines)   (4) $\angle BDC \cong \angle BCA$ (All right angles are congruent.)   (5) $\angle CBD \cong \angle ABC$ (Identity)   (6) $\triangle CBD \sim \triangle ABC$ (AA)   37. Since $\triangle CBD \sim \triangle ABC$, we have $\dfrac{BC}{BD} = \dfrac{AB}{BC}$. Simplifying, we get $(BC)^2 = AB \cdot BD$.

Pages 323-325 1. 4:9   3. 7:12   5. 16:9   7. 8 cm   9. 27:1
11. 192 cm³   13. 98   15. 93.75   17. $\sqrt{2}$ times as long   19. a. 300
b. 192   21. $(12 + 12\sqrt{2})$ cm

Pages 328-331 1. 13   3. 12   5. 3   7. 6   9. no   11. yes
13. yes   15. yes   17. 5.4 m   19. a. 6, 8   b. 9, 12   c. 30, 40
d. 1.2, 1.6   21. $\sqrt{7}$ 23. 6   25. 4   27. 10   29. In a triangle, the square of the length of one side is equal to the sum of the squares of the lengths of the other two sides if and only if the triangle is a right triangle.   31. a. 3   b. 5   c. 3   d. $\sqrt{34}$
e. 48   33. 6, 8, 10   35. 12, 35, 37   37. 0.75 m   39. 16 mm
41. $AB \cdot AD + AB \cdot BD$, $AD + BD$, AB 43. $2\sqrt{3}$ 45. 41 cm

Pages 334-336 1. $2\sqrt{2}$ 3. 8   5. 2   7. $2\sqrt{2}$ 9. $5\sqrt{3}$, 10
11. 13, $13\sqrt{3}$ 13. 4, 8   15. $3\sqrt{2}$, $2\sqrt{6}$ 17. $15\sqrt{2}$ 19. $72\sqrt{3}$
21. 9   23. $3\sqrt{2}$ 25. 90   27. 10   29. $22\sqrt{3}$ 31. a. $\dfrac{\sqrt{3}}{2}$ b. $\dfrac{\sqrt{3}}{4}$
c. $\dfrac{3\sqrt{3}}{2}$ d. $\dfrac{9\sqrt{3}}{4}$ e. 1:9   33. $10\sqrt{3} - 10$ 35. $10\sqrt{3} + 10\sqrt{2} + 10$
37. $\sqrt{2}$ 39. 54   41. $\dfrac{\sqrt{2}}{2}$

Page 338 · Chapter Review   1. 51:200   2. 6:4:1   3. −35, −14   4. 14
5. 5   6. 6.75 kg   7. 864   8. 18%   9. $3600   10. a. $\triangle ABC \sim \triangle DBA$
b. $\dfrac{AB}{DB} = \dfrac{BC}{BA} = \dfrac{AC}{DA}$ 11. 27.6   12. 9:25   13. 15 cm   14. 640 cubic units
15. yes   16. yes   17. 72 cm   18. $x = 9$, $y = 9\sqrt{3}$ 19. $x = 45$, $y = 5\sqrt{2}$
20. $x = 6\sqrt{3}$, $y = 6$

Pages 342-343 • Mixed Review   1. 9

3.
p	q	p ∨ q	~(p ∨ q)	~p	~(p ∨ q) → ~p
T	T	T	F	F	T
T	F	T	F	F	T
F	T	T	F	T	T
F	F	F	T	T	T

5. ASA   7. $(y-5)(y+3)$ 9. $\frac{7\sqrt{3}}{6}$ 11. $\frac{\sqrt{3}}{2}$
13. ⟵——•——+——+——+——•——⟶
       -8 -7 -6 -5 -4 -3
15. $127.50   17. 1, 3, 11, 33
19. The measure of ∠1 is greater than 50.   21. $b = \frac{4-a^2}{a}, a \neq 0$ 23. 0.92
25. $49 - 12\sqrt{5}$ 27. 9 cm   29. $\frac{3}{5}, 1$ 31. $\frac{2}{3}, \frac{2}{3}$ 33. $\{y: y > 1\frac{1}{2}\}$
35. no; yes   37. $4\sqrt{3}$ 39. 7

Pages 346-348   1. a. diameter   b. chord   c. secant   d. tangent   e. radius   f. secant   g. chord   h. radius   3. true   5. false   7. true   9. $x = 6$; $y = 2$; $z = 2\sqrt{13}$ 11. $x = 35$; $z = 110$; $y = 70$ 13. $x = 100$; $y = 100$; $z = 40$ 15. $z = 90$; $x = y = 45$ 17. (2) Perpendicular lines intersect to form right angles   (3) A right angle has measure 90; two angles are congruent if they have equal measures.   (4) All radii of a circle are congruent.   (5) If two sides of a triangle are congruent, the angles opposite those sides are congruent.   (6) MBO; MCO   (7) Corresponding parts of congruent triangles are congruent.

Pages 351-352   1. $m\angle 1 = 126$; $m\angle 2 = m\angle 3 = 27$; $m\angle 4 = 54$; $m\angle 6 = m\angle 5 = 63$ 3. $x = 35$; $y = 145$; $z = 145$ 5. $x = 45$; $y = 135$; $z = 180$ 7. $w = 130$; $x = 25$; $y = 130$; $z = 50$ 9. $x = 45$ 11. $x = 50$ 13. $x = 80$; $y = 140$; $z = 20$ 15. 7 cm   17. 57   19. c. $m\angle QOR = 2m\angle QSR$

Pages 357-359   1. $m\angle 1 = 75$; $m\angle 2 = 70$; $m\angle 3 = 35$ 3. $m\angle 1 = 70$; $m\angle 2 = 35$ 5. $m\angle 1 = 15$; $m\angle 2 = 20$ 7. $x = 120$; $y = 65$; $z = 110$ 9. $w = 44$; $x = 90$; $y = 65$; $z = 136$ 11. 75   13. 95   15. (2), (3) The measure of an inscribed angle is $\frac{1}{2}$ the measure of its intercepted arc (Case 1)   (4) Substitution   (5) Distributive Property   17. Let $m\stackrel{\frown}{BCD} = x$; $m\angle A = \frac{1}{2}x$; $m\angle C = \frac{1}{2}(360 - x)$; $m\angle A + m\angle C = \frac{1}{2}x + 180 - \frac{1}{2}x = 180$ 19. $m\angle BAC = 16$ and $m\angle ACB = 114$ or $m\angle BAC = 64$ and $m\angle ACB = 66$

Pages 363-365   1. $m\angle 1 = 10$; $m\angle 2 = 10$; $m\angle 3 = 25$; $m\angle 4 = 25$; $m\angle 5 = 35$; $m\angle 6 = 15$; $m\angle 7 = 70$; $m\angle 8 = 85$; $m\angle 9 = 85$; $m\angle 10 = 110$ 3. 148   5. 110   7. 156   9. 38   11. 254   13. a. $m\angle BCA = \frac{1}{2}m\stackrel{\frown}{CA} = m\angle BAC$ b. $\overline{BA} \cong \overline{BC}$; if two angles of a triangle are congruent, the sides opposite those angles are congruent.   c. equilateral   15. 50   17. 22   19. 11   21. $m\angle A = 54$; $m\angle C = 88$; $m\angle PRQ = 71$; $m\angle RQP = 63$

Pages 370-371   1. $6\pi\sqrt{2}$ cm; 18π cm²   3. $24\pi k$ mm; $144\pi k^2$ mm²   5. 31.4; 78.5   7. 17.6; 24.64   9. 339.1 m   11. The radius of the larger circle is six times

SELECTED ANSWERS

that of the smaller.   13. $4\sqrt{3}$ 15. 1:9   17. a. 25π b. 4π c. 21π
19. $8\pi - 16$ 21. $8 - 2\pi$ 23. a. $4\sqrt{3}$ b. $\dfrac{4\sqrt{3}}{3}$ c. $\dfrac{16\pi}{3} - 4\sqrt{3}$

Pages 374-375   1. $d = 6$ cm; S.A. $= 36\pi$ cm^2; $V = 36\pi$ cm^3 3. $d = 2.4$ m; S.A. $= 5.76\pi$ m^2; $V = 2.304\pi$ m^3 5. $r = 6$ m; $d = 12$ m; $V = 228\pi$ m^3 7. $d = 2\sqrt{2}$; S.A. $= 8\pi$; $V = \dfrac{8\sqrt{2}}{3}\pi$ 9. $r = 1$; $d = 2$; $V = \dfrac{4\pi}{3}$ 11. multiplied by 8
13. $r = \sqrt{3.6}$; S.A. $= 14.4\pi$ 15. 1:9   17. $\dfrac{148}{3}\pi$ cubic units   19. $\dfrac{256\pi}{3}$; 288π

Pages 378-379   1. L.A. $= 6\pi\sqrt{2}$; T.A. $= 6\pi\sqrt{2} + 4\pi$; $V = 6\pi$ 3. $r = 3$; T.A. $= 42\pi$; $V = 36\pi$ 5. $r = 8$; L.A. $= 48\pi$; T.A. $= 176\pi$ 7. $4\pi\sqrt{2}$; $4\pi\sqrt{2} + 4\pi$; $\dfrac{8\pi}{3}$
9. 4   11. Brand X   13. Volume is doubled.   15. 3.75 cm
17. $(250\sqrt{3} - 160\pi)$ cm^3 19. $r = 5$; $h = 2$ 21. $3h$

Page 381 · Chapter Review   1. e   2. a   3. b   4. h   5. g   6. $x = 100$, $y = 40$, $z = 80$ 7. $x = 90$, $y = 45$, $z = 2\sqrt{2}$ 8. $x = 50$, $y = 110$, $z = 55$
9. $x = 29$, $y = 29$, $z = 90$ 10. $x = 168$, $y = 114$, $z = 36$ 11. $x = 38$, $y = 54$, $z = 22$
12. 62.8 cm, 314 cm^2 13. 22 m   14. 7:1   15. 12π, $4\pi\sqrt{3}$ 16. The volume is one-eighth as large.   17. L.A. $= 609\pi$, T.A. $= 1050\pi$, $V = 2940\pi$ 18. L.A. $= 126\pi$, T.A. $= 224\pi$, $V = 441\pi$ 19. 8 cm

Pages 383-386 · Cumulative Review: Chapters 1-10   1. b   2. a   3. c   4. b
5. b   6. b   7. d   8. a   9. d   10. c   11. d   12. c   13. a
14. a   15. a   16. a   17. d   18. c   19. d   20. a   21. b   22. d
23. b   24. d   25. b   26. c   27. d   28. d   29. b   30. a   31. c
32. b   33. d   34. c   35. a   36. c   37. c   38. c   39. c   40. d
41. c   42. b

Pages 388-389 · Mixed Review   1. no; $3^2 + (2\sqrt{5})^2 \neq 5^2$ 3. 8 and 15; -8 and -15
5. no; yes   7. 18.75%   9. $2\sqrt{6} + 4$ 11. $-\dfrac{21}{4}$ 13. $\dfrac{5}{2}, -\dfrac{5}{2}$ 15. $\dfrac{2}{3}, \dfrac{5}{2}$
17. 68   19. $\dfrac{7\sqrt{2}}{2}$ 21. 75 km   23. $\sqrt{34}$ 25. -16.8 27. 36, 63, 81.
29. sphere   31. $V = \dfrac{1}{3}\pi r^2 h$; $h = \dfrac{3V}{\pi r^2}$ 33. $\angle A \cong \angle C$ and $\angle B \cong \angle D$ (If 2 ∥ lines are cut by a trans., alt. int. $\angle$s are $\cong$.); $\triangle ABX \sim \triangle CDX$ (AA)   35. 0.075

Page 394   1. $\dfrac{1}{26}$ 3. $\dfrac{1}{15}$ 5. $\dfrac{1}{2000}$ 7. 1   9. $\dfrac{1}{2}$ 11. {purple, yellow, orange}; purple: $\dfrac{1}{2}$; yellow: $\dfrac{1}{4}$; orange: $\dfrac{1}{4}$ 13. {brown, gray, yellow}; brown: $\dfrac{1}{4}$; gray: $\dfrac{1}{4}$; yellow: $\dfrac{1}{2}$
15. {1¢, 5¢, 10¢, 25¢}; 1¢: $\dfrac{1}{8}$; 5¢: $\dfrac{1}{8}$; 10¢: $\dfrac{1}{4}$; 25¢: $\dfrac{1}{2}$ 17. {A, 2, 3, ..., J, Q, K of spades, clubs, diamonds, and hearts}; $\dfrac{1}{52}$ 19. {A, 2, 3, ..., J, Q, K of spades, clubs, diamonds, and hearts}; probability of picking a jack: $\dfrac{1}{13}$ 21. {A, 2, 3, ..., J, Q, K of spades, clubs, diamonds, and hearts}; probability of picking a diamond: $\dfrac{1}{4}$

Pages 396-397   1. $\frac{1}{2}$ 3. $\frac{1}{2}$ 5. 1   7. $\{1,2,3,4,5,6\}$; $\{2,3,4,5,6\}$; $\frac{5}{6}$ 9. $\{1,2,3,4,5,6\}$; $\{2,3,5\}$; $\frac{1}{2}$ 11. $\{$Jan., Feb., Mar., Apr., May, June, July, Aug., Sept., Oct., Nov., Dec.$\}$; $\{$Mar., Apr., May, Aug.$\}$; $\frac{1}{3}$ 13. $\{$A, 2, 3, ..., J, Q, K of spades, clubs, diamonds, and hearts$\}$; $\{$A, 2, 3, ..., J, Q, K of clubs$\}$; $\frac{13}{52}$ or $\frac{1}{4}$ 15. $\{$A, 2, 3, ..., J, Q, K of spades, clubs, hearts, and diamonds$\}$; $\{$J, Q, K of spades, clubs, hearts, and diamonds$\}$; $\frac{3}{13}$ 17. $\frac{1}{2}$ 19. $\frac{3}{4}$ 21. $\frac{1}{8}$ 23. $\frac{1}{4}$ 25. $\frac{5}{8}$ 27. $\frac{5}{8}$ 29. $\frac{8}{13}$ 31. $\frac{4}{13}$ 33. $\frac{1}{2}$

Pages 401-403   1. $\frac{1}{4}$ 3. $\frac{1}{4}$ 5. $\frac{1}{4}$ 7. $\{$RH, RT, BH, BT, YH, YT, GH, GT$\}$; $\frac{1}{8}$ 9. $\frac{1}{2}$ 11. $\frac{3}{4}$ 13. $\frac{1}{2}$ 15. $\frac{5}{12}$ 17. $\frac{5}{12}$ 19. $\frac{1}{3}$ 21. $\{$HHH, HHT, HTH, HTT, THH, THT, TTH, TTT$\}$; $\frac{1}{8}$ 23. $\{$DFL, DFT, DFD, DSL, DST, DSD, RFL, RFT, RFD, RSL, RST, RSD, MFL, MFT, MFD, MSL, MST, MSD$\}$; $\frac{1}{18}$ 25. $\frac{3}{8}$ 27. $\frac{1}{2}$ 29. $\frac{7}{8}$ 31. $\frac{1}{8}$ 33. $\frac{7}{16}$ 35. $\frac{2}{3}$ 37. $\frac{7}{9}$

Pages 407-408   1. 24   3. 720   5. 20   7. 60   9. 48   11. 24   13. 12   15. $\frac{1}{5}$ 17. $\frac{3}{5}$ 19. $\frac{4}{5}$ 21. $\frac{2}{5}$ 23. 15,120   25. 3024   27. 3024   29. $\frac{1}{2}$ 31. $\frac{1}{5}$

Pages 410-411   1. 6   3. 720   5. 720   7. 720   9. 3024   11. 9900   13. 210   15. 151,200   17. 4896   19. 336   21. 2176   23. $\frac{5}{34}$ 25. $\frac{5}{9}$ 27. $\frac{29}{34}$ 29. 56   31. 7   33. 7

Pages 414-415   1. BB: $\frac{1}{36}$; BW: $\frac{5}{36}$; WB: $\frac{5}{36}$; WW: $\frac{25}{36}$ 3. $\frac{35}{36}$ 5. $\frac{11}{36}$

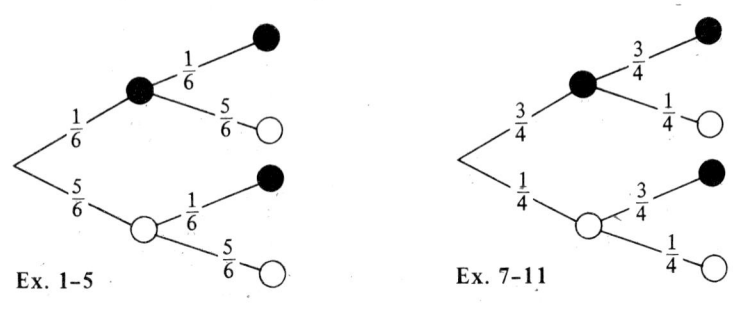

Ex. 1–5

Ex. 7–11

7. BB: $\frac{9}{16}$; BW: $\frac{3}{16}$; WB: $\frac{3}{16}$; WW: $\frac{1}{16}$ 9. $\frac{7}{16}$ 11. $\frac{15}{16}$ 13. BB: $\frac{1}{16}$; BR: $\frac{1}{16}$;

BG: $\frac{1}{8}$; RB: $\frac{1}{16}$; RR: $\frac{1}{16}$; RG: $\frac{1}{8}$; GB: $\frac{1}{8}$; GR: $\frac{1}{8}$; GG: $\frac{1}{4}$ 15. $\frac{1}{8}$ 17. $\frac{1}{2}$
19. BBB: $\frac{1}{27}$; BBW: $\frac{2}{27}$; BWB: $\frac{2}{27}$; BWW: $\frac{4}{27}$; WBB: $\frac{2}{27}$; WBW: $\frac{4}{27}$; WWB: $\frac{4}{27}$; WWW: $\frac{8}{27}$
21. $\frac{1}{2704}$ 23. $\frac{1}{4}$ 25. $\frac{51}{52}$ 27. $\frac{1}{52}$ 29. $\frac{1}{2}$ 31. 0   33. 0   35. $\frac{1}{16}$
37. $\frac{9}{32}$ 39. $\frac{1}{4}$ 41. 0   43. $\frac{1}{16}$

Pages 419–420   1. {HT, TH, TT}; {HH}   3. {H2, H4, H6}; {H1, H3, H5, T1, T2, T3, T4, T5, T6}   5. Rolling two 1's or two 6's; rolling anything else   7. $\frac{3}{4}$; $\frac{1}{4}$ 9. $\frac{1}{4}$; $\frac{3}{4}$
11. $\frac{1}{18}$; $\frac{17}{18}$ 13. $\frac{51}{52}$ 15. $\frac{12}{13}$ 17. $\frac{23}{26}$ 19. $\frac{1}{25}$; $\frac{24}{25}$ 21. $\frac{4}{25}$; $\frac{21}{25}$
23. $\frac{27}{64}$; $\frac{37}{64}$ 25. $\frac{9}{64}$; $\frac{55}{64}$

Page 423   1. $\frac{1}{169}$; $\frac{1}{221}$ 3. $\frac{1}{16}$; $\frac{1}{17}$ 5. $\frac{1}{676}$; $\frac{1}{1326}$ 7. $\frac{1}{169}$; $\frac{4}{663}$
9. $\frac{144}{169}$; $\frac{188}{221}$ 11. $\frac{1}{4}$; $\frac{25}{102}$ 13. $\frac{16}{49}$; $\frac{2}{7}$ 15. $\frac{12}{49}$; $\frac{2}{7}$ 17. $\frac{24}{49}$; $\frac{4}{7}$ 19. $\frac{1}{16}$; $\frac{1}{22}$
21. $\frac{25}{144}$; $\frac{5}{33}$ 23. $\frac{5}{48}$; $\frac{5}{44}$ 25. $\frac{5}{36}$; $\frac{5}{33}$ 27. $\frac{5}{36}$; $\frac{5}{33}$ 29. $\frac{4}{9}$; $\frac{16}{33}$

Page 427   1–5. Answers will vary. For comparison purposes the *theoretical* probability is given.   1. $\frac{1}{2}$ 3. Odd: $\frac{1}{2}$; even: $\frac{1}{2}$ 5. 0 heads: $\frac{1}{8}$; 1 head: $\frac{3}{8}$; 2 heads: $\frac{3}{8}$; 3 heads: $\frac{1}{8}$ 7. Example: choose several pages at random from the telephone directory and count the occurrence of each digit.

Page 429 · Chapter Review   1. {Claire, Li-Hui, Beth, Ivan}; $\frac{1}{4}$ each   2. {−1, 0, 1, 2, 3, 4, 5, 6, 7, 8, 9}; $\frac{1}{2}$ 3. {G, W, Y} P(G) = $\frac{1}{3}$ P(W) = $\frac{1}{4}$ P(Y) = $\frac{5}{12}$ 4. $\frac{1}{2}$ 5. $\frac{2}{5}$
6. $\frac{2}{13}$ 7. $\frac{11}{12}$ 8. $\frac{1}{4}$ 9. $\frac{4}{9}$ 10. 60   11. 125   12. 24   13. 11,880
14. 720   15. 15,120   16. $P(\bullet\bullet) = \frac{4}{49}$
17. $\frac{29}{49}$ 18. $\frac{24}{49}$ 19. P(HHT, HTH, THH) = $\frac{3}{8}$ P(HHH, HTT, THT, TTH, TTT) = $\frac{5}{8}$ $P(\bullet\circ) = \frac{10}{49}$
20. P(HHH, HHT, HTH, HTT, THH, THT, TTH, TTT) = 1   P(∅) = 0   21. P(∅) = 0 $P(\circ\bullet) = \frac{10}{49}$
P(Mon., Tues., Wed., Thurs., Fri., Sat., Sun) = 1
22. $\frac{4}{49}$, $\frac{1}{21}$ 23. $\frac{20}{49}$, $\frac{10}{21}$ 24. $\frac{10}{49}$, $\frac{5}{21}$ $P(\circ\circ) = \frac{25}{49}$
25. $\frac{1}{3}$ 26. $\frac{1}{4}$

27. Answers may vary. Example: The engineer could choose a certain number of batteries, say 1000, and test them to determine the number of defective batteries. From this information, the engineer could approximate the required probability.

Pages 434-435 · Mixed Review   1. T.A. = 90π, $V = 100\pi$

3.
p	q	$\sim q$	$p \vee \sim q$	$\sim(p \vee \sim q)$	$\sim p$	$\sim p \wedge q$
T	T	F	T	F	F	F
T	F	T	T	F	F	F
F	T	F	F	T	T	T
F	F	T	T	F	T	F

5. $\emptyset$ 7. with replacement: $\dfrac{3}{4}$, without replacement: $\dfrac{4}{5}$ 9. $m\angle Z = 60$; $XZ = 4$

11. a. $(x-12)(x+4)$ b. $(2y+1)(y+12)$ 13. 100π 15. 76   17. $\dfrac{4}{13}$

19. $124 - 70\sqrt{3}$ 21. $-1 - \sqrt{3}$ 23. 120 cm^2 25. 6720   27. height = $5\sqrt{3}$ cm, area = $25\sqrt{3}$ cm^2 29. yes; yes   31. $-1(-1) \neq -1$; true   33. $\dfrac{8}{3}$ 35. $\dfrac{1}{9}$

Pages 440-441

1.
Interval	Frequency	Cumulative Frequency
70-75	1	1
76-81	0	1
82-87	4	5
88-93	7	12
94-99	6	18
100-105	2	20

3. 15   5. 88-93   7. 3   9. 6

Pages 444-445

2-3.

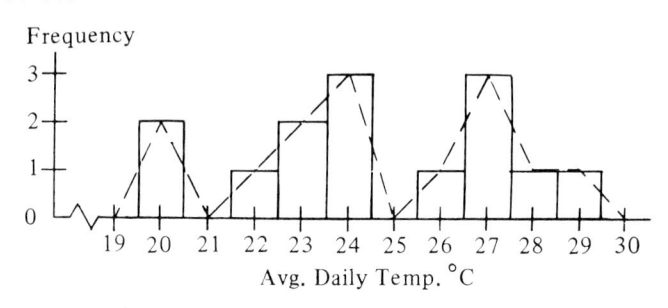

SELECTED ANSWERS   609

4-5.

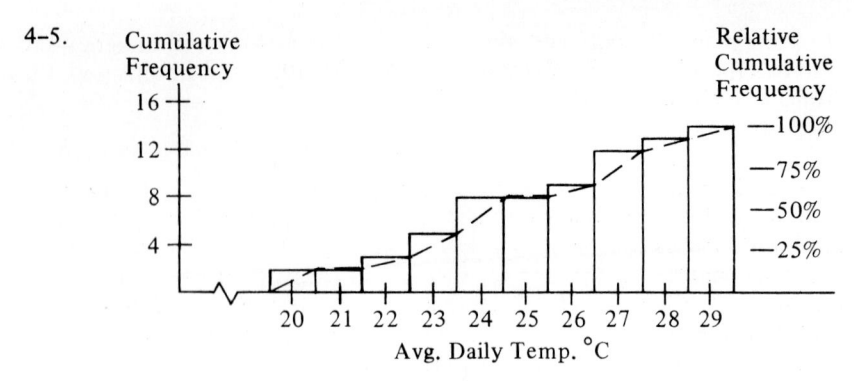

[Cumulative Frequency / Relative Cumulative Frequency histogram with ogive for Avg. Daily Temp. °C, 20–29]

6-7.

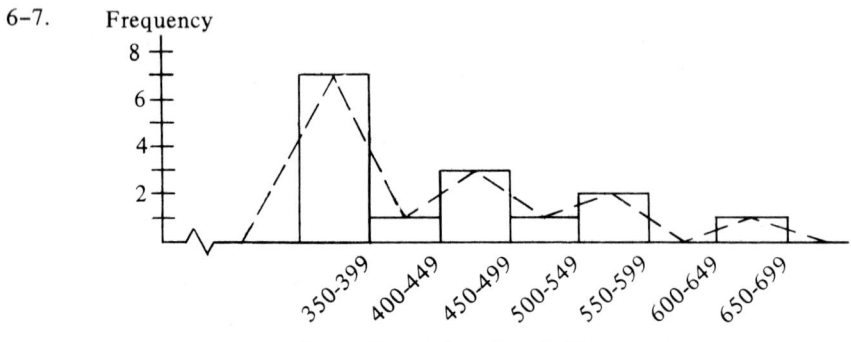

[Frequency histogram with frequency polygon for Take-offs and Landings in Thousands, 350–399 through 650–699]

8-9.

[Cumulative Frequency / Relative Cumulative Frequency histogram with ogive for Take-offs and Landings in Thousands, 350–399 through 650–699]

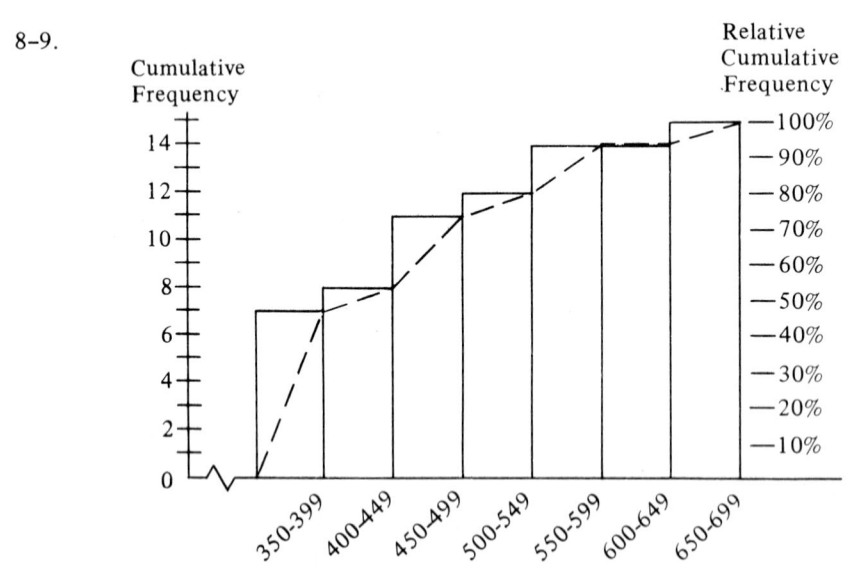

Page 448   1. 3; 2; 1   3. 65; 67; none   5. 14.2; 14; 14, 15   7. 82.4; 84; 85
9. $\frac{1}{3}$; $\frac{1}{3}$; none   11. 14.36; 11.65; none   13. 0.098; 0.12; none   15. 46
17. mean larger than the median   19. about the same   21. mean slightly larger than the median   23–29. Answers may vary.   23. 6, 8, 10, 12, 14   25. 40, 55, 63, 63, 72
27. 15, 20, 20, 25, 45   29. 10, 40, 50, 100, 200

Pages 452-454   1. 71; 73; 78   3-27. Answers may vary.   3. 8; 12; 20   5. 83   7. 88   9. 73   11. 62 km/h   13. 40 km/h   15. 60 km/h   17. 85%   19. 35%   21. 110 h   23. 29   25. 46   27. 38

Pages 457-458   5-9. Answers may vary. An example simulation is given.   5. Let even numbers represent heads and odd numbers represent tails. Choose a random sample of 2-digit numbers and count the number of integers containing exactly one even digit.   7. Choose a random sample of digits and count the number of fives. Ignore 0 and digits greater than 6.   9. Choose a random sample of 2-digit numbers. Let the first digit represent the toss of the coin with heads represented by an even digit and tails by an odd digit. Ignore numbers whose units digit is zero or greater than 6. Count the numbers whose digits are both even.

Pages 460-461   1. 3   3. $-\frac{1}{12}$ 5. a. 0: $\frac{1}{4}$; 1: $\frac{1}{2}$; 2: $\frac{1}{4}$ b. 1   7. win $20   9. win $687.50   11. b. 12.25   13. 50¢; yes

Page 464 · Chapter Review

1.

km/L	Freq.	Cum. Freq.
10	2	2
11	2	4
12	1	5
13	2	7
14	3	10

2. 5 time   3. twice   4-5. Frequency histogram

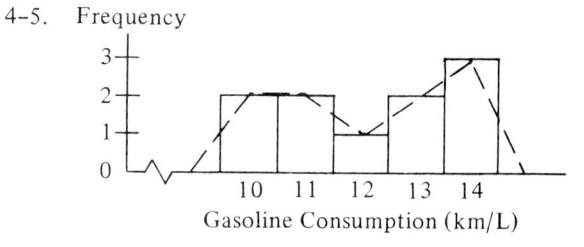

6-7. Cumulative Frequency graph

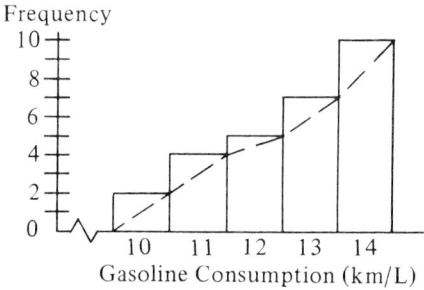

8. 12.2; 12.5; 14   9. 6; 6.5; 2 and 9   10. 0.58; 0.53; none   11. 20   12. 25   13. 26   14. 19   15-16. Answers may vary.   15. 1950, 1468, 0767, 1260, 1256   16. 5313, 7587, 1950, 3187, 3450, 6397, 4894, 4615   17. Choose a random sample of digits and count the twos.   18. 2   19. 1

Pages 468-469 · Mixed Review   1. 60   3. 10 cm   5. $11\frac{1}{3}$; $11\frac{1}{2}$; 12   7. $x = 51$, $y = 78$, $z = 51$ 9. number line with open circles at 1 and 2   11. $\frac{5}{18}$ 13. $\frac{9}{16}$; $\frac{7}{16}$ 15. Frequency polygon of Quiz Scores

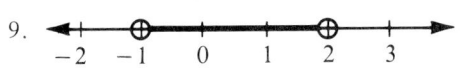

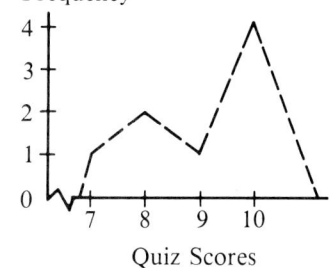

17. 8, 9.5, 10   19. 0; 1   21. 90   23. $\left\{\dfrac{-1+\sqrt{3}}{2}, \dfrac{-1-\sqrt{3}}{2}\right\}$ 25. 64

27.

p	q	$p \wedge q$	$q \rightarrow (p \wedge q)$
T	T	T	T
T	F	F	T
F	T	F	F
F	F	F	T

29. 8   31. Let even numbers represent heads and odd numbers represent tails. Choose a random sample of 3-digit numbers, and divide the number of integers containing 3 even digits by the number of 3-digit numbers in your sample.

Pages 474-475   1. $A(-1, 2)$; $B(-2, -3)$; $C\left(1\dfrac{1}{2}, 0\right)$; $D\left(0, -2\dfrac{1}{2}\right)$; $E\left(-2\dfrac{1}{2}, \dfrac{1}{2}\right)$; $F\left(2\dfrac{1}{2}, -\dfrac{1}{2}\right)$ 2-13.   15. 3   17. 10   19. 8   21. isosceles trapezoid   23. 27 square units   25. 24 square units   27. 18 square units   29. 48 square units   31. $(-4, -6)$ 33. $(8, -1)$ 35. $\left(-2\dfrac{1}{2}, 20\right)$ or $\left(-2\dfrac{1}{2}, -16\right)$ 37. (6, 2) or $(-2, 8)$; $AB = 10$

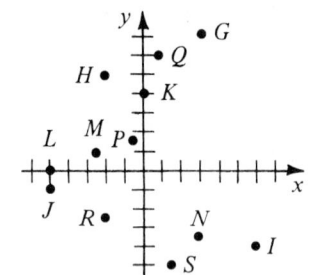

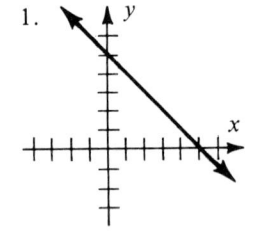

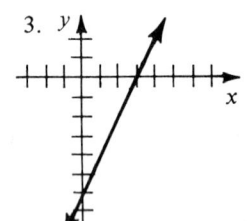

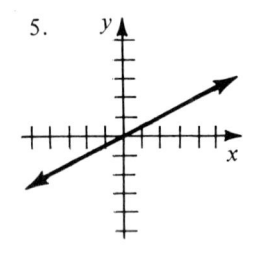

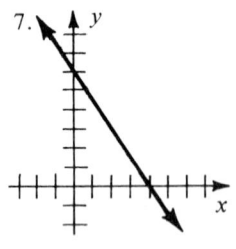

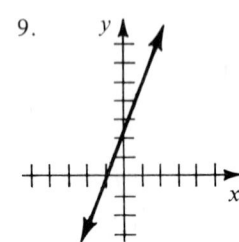

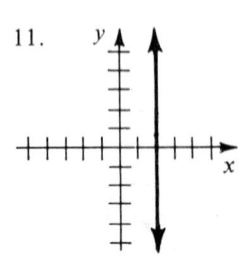

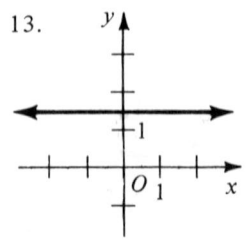

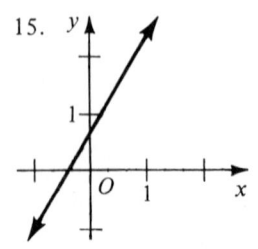

Pages 478-479
1.   3.   5.   7.
9.   11.   13.   15.
17. 4; 4   19. $\dfrac{5}{4}$; -5 21. 0; 0   23. none; 4   25. $2x - y = 2$

27. $5x + y = 4$ 29. $A = 1$ 31. $B = 7$ 33. $A = \dfrac{5}{2}$ 35. $B = -\dfrac{5}{2}$

37. $(0, -4)$ 39. $(-1, 1)$ 41. x-intercept $= \dfrac{C}{A}$; y-intercept $= \dfrac{C}{B}$

Pages 484-485   1. $\dfrac{1}{2}$ 3. -3 5. $\dfrac{3}{4}$ 7. slope not defined   9. no; slope of $\overleftrightarrow{AD} = \dfrac{1}{5}$; slope of $\overleftrightarrow{BC} = \dfrac{1}{4}$ 11. yes; slope of $\overleftrightarrow{AB} = \dfrac{1}{2} =$ slope of $\overleftrightarrow{CD}$; slope of $\overleftrightarrow{AD} = 2 =$ slope of $\overleftrightarrow{BC}$ 13. $-\dfrac{3}{4}$ 15. $\dfrac{4}{3}$ 17. slope undefined

19. $\dfrac{3}{2}$ 21. -2 23. 1 25. 0 27. -2 29. $\dfrac{2}{3}$ 31. $r > 2$

33. $\overleftrightarrow{AB} = -\dfrac{3}{2}$; $\overleftrightarrow{BC} = -\dfrac{3}{2}$; yes   35. $\overleftrightarrow{AB} = -\dfrac{5}{3}$; $\overleftrightarrow{BC} = -2$; no

37.

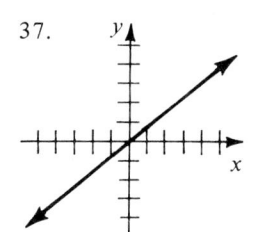

39.

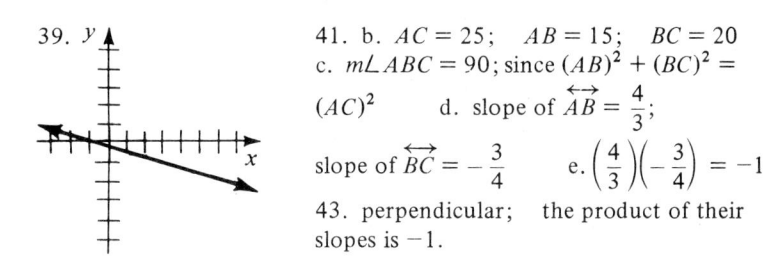

41. b. $AC = 25$; $AB = 15$; $BC = 20$
c. $m\angle ABC = 90$; since $(AB)^2 + (BC)^2 = (AC)^2$ d. slope of $\overleftrightarrow{AB} = \dfrac{4}{3}$; slope of $\overleftrightarrow{BC} = -\dfrac{3}{4}$ e. $\left(\dfrac{4}{3}\right)\left(-\dfrac{3}{4}\right) = -1$

43. perpendicular; the product of their slopes is -1.

Pages 487-488   1. -3; 2 3. 2; $-\dfrac{5}{4}$ 5. 2; -5 7. $2x - y = 1$

9. $0x + y = -2$ 11. $3x + y = 5$ 13. $y = 3x - 4$ 15. $y = -\dfrac{1}{2}x + 3$

17. $y = 6$ 19. $y = 2x + 5$ 21. $y = \dfrac{1}{2}x - 5$ 23. $y = -\dfrac{3}{2}x + 10$

25. $y = -x - 2$ 27. $y = -2x - 1$ 29. $y = \dfrac{2}{3}x - \dfrac{7}{3}$ 31. $y = \dfrac{3}{2}x + 1$

33. $y = \dfrac{1}{2}x - 3$ 35. $y = 6$ 37. $y = \dfrac{1}{2}x + 4$ 39. $y = \dfrac{2}{3}x - 4$

41. $y = \dfrac{3}{2}x + \dfrac{7}{2}$ 43. $y = -\dfrac{4}{3}x$

45.   47. 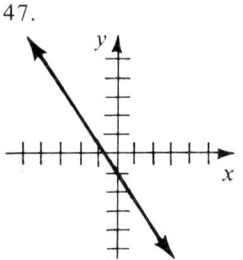   49. 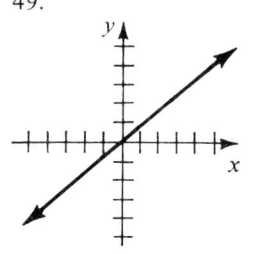   51.

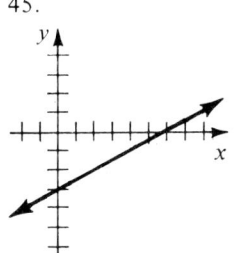

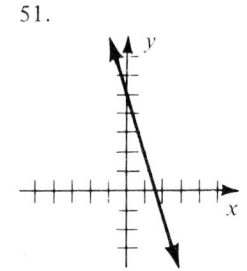

SELECTED ANSWERS   613

Pages 491–492 1. no 3. yes 5. no

7.

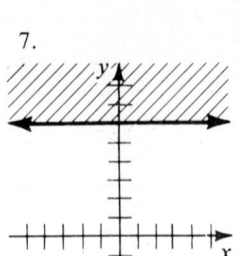

9.

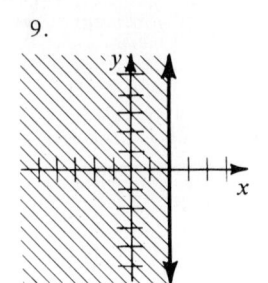

11.

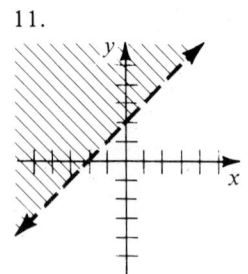

13.

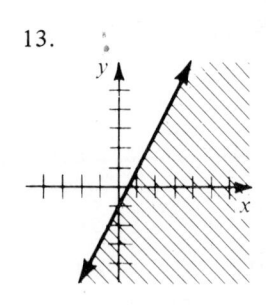

15.

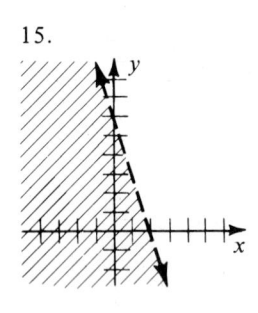

17.

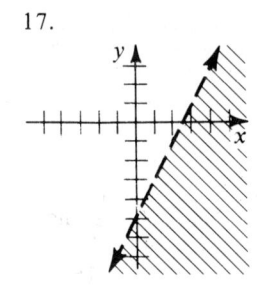

19.

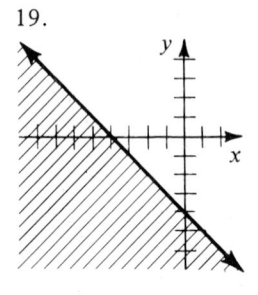

21.

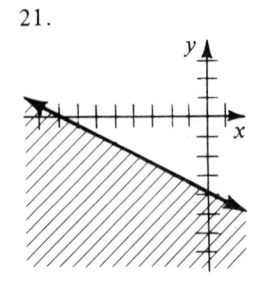

23.

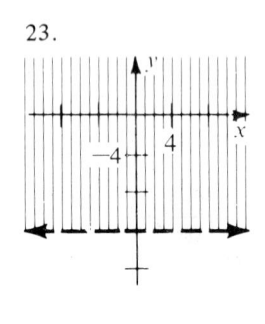

25.

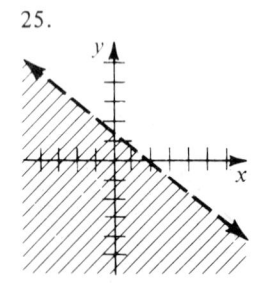

27.

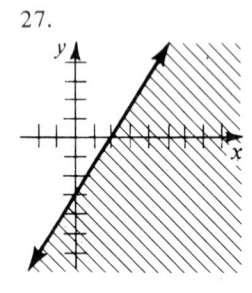

29.

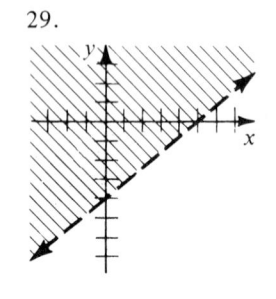

31.

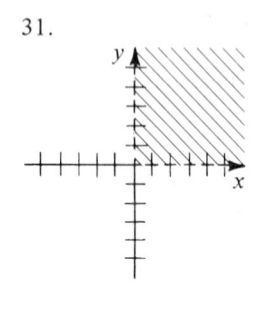

33.

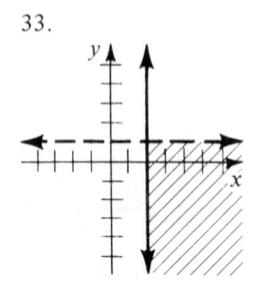

35.

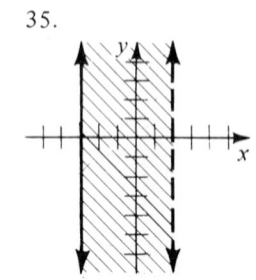

37. $y > 0$; $x < 0$

614 SELECTED ANSWERS

Page 495 · Chapter Review

1. $A(-3, 1)$; $B(3, -1)$; $C(4, 3)$; $D(-5, -2)$
2. 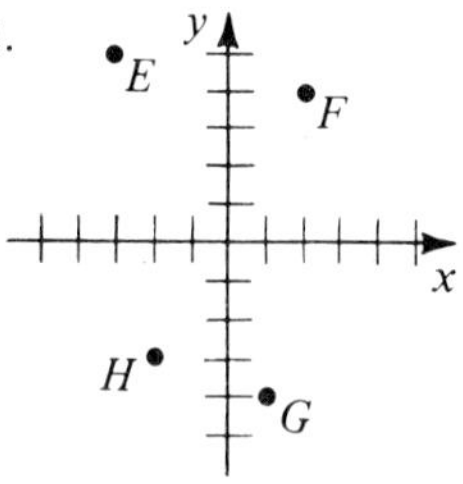 [graph showing points E, F, H, G]
3. 10   4. 4   5. -2; 5   6. $\frac{1}{3}$; $\frac{1}{4}$

7. 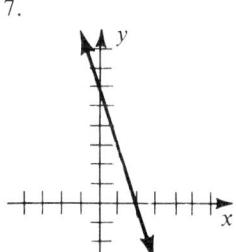 [graph] 8. 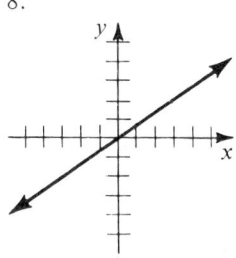 [graph] 9. 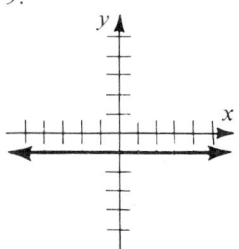 [graph] 10. 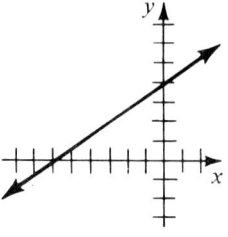 [graph]

11. $-\frac{7}{4}$ 12. $\frac{1}{3}$ 13. $\frac{1}{4}$ 14. yes; slope of $\overline{RS} = -\frac{1}{10}$ and slope of $\overline{TU} = -\frac{1}{10}$, slope of $\overline{ST} = -\frac{8}{5}$ and slope of $\overline{UR} = -\frac{8}{5}$ 15. $\frac{2}{3}$; -4 16. $y = \frac{2}{5}x - 9$
17. $y = \frac{1}{2}x + 4$ 18. $y = \frac{3}{4}x - 2$

19. 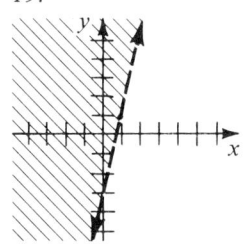 [graph] 20. 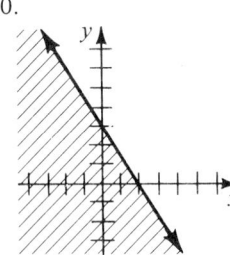 [graph] 21. 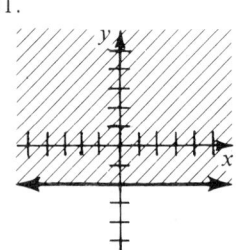 [graph] 22. 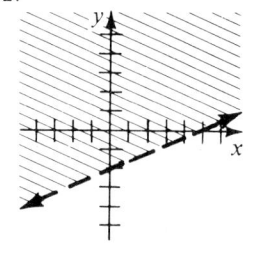 [graph]

Pages 498-499 · Mixed Review   1. $y = 4$ 3. 80   5. a. $\frac{11}{13}$ b. $\frac{1}{26}$ 7. $3\sqrt{2}$ cm
9. $\frac{1}{16}$; $\frac{1}{17}$ 11. 3¢   13. $84   15. $\frac{169\pi - 240}{8}$ 17. Answers may vary. 70, 71, 72, 73, 74   19. a. [graph] b. [graph]

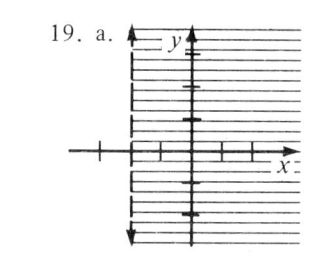

SELECTED ANSWERS   615

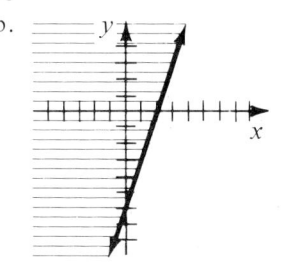

21. 16   23.

Time (min)	Frequency
17	1
18	3
19	4
20	2

Frequency graph (dashed line) and Cumulative Frequency histogram for Running Times in Minutes (17–20).

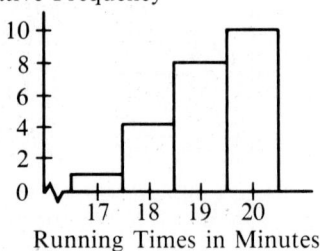

25. 840   27. $0.3\overline{6}$ 29. $\{y: y \leq 20\}$

31.

p	q	$p \wedge q$	$\sim p$	$(p \wedge q) \rightarrow \sim p$
T	T	T	F	F
T	F	F	F	T
F	T	F	T	T
F	F	F	T	T

; neither

33. perimeter = 56; area = 192

Pages 503–504   1. (1, 1)   3. (4, 1)   5. (−1, 3)   7. no solution
9. $\{(x, y): x + 3y = -2\}$ 11. (−3, −4)   13. $\left(2\frac{1}{2}, 4\frac{1}{2}\right)$ 15. $\left(2\frac{1}{2}, 4\right)$
17. $\left(-3\frac{1}{2}, 2\frac{1}{2}\right)$ 19. 9 square units

Page 507   1. (2, 5)   3. (7, −1)   5. no solution   7. $\{(x, y): x = y + 4\}$
9. (−2, −2)   11. (3, 1)   13. (−10, 12)   15. (−7, −8)   17. (−4, −9)
19. (2, −1)   21. $\left(\frac{8}{7}, -\frac{23}{14}\right)$ 23. $\left(2, -\frac{1}{2}\right)$ 25. $(b, 0)$ 27. $\left(\frac{b+d}{a-c}, \frac{ad+bc}{a-c}\right)$

Page 510   1. (2, 2)   3. (1, 0)   5. (3, −2)   7. (−5, 10)   9. (5, −11)
11. (−11, 3)   13. (4, −3)   15. (1, 2)   17. (3, 1)   19. (7, −3)
21. (−5, −6)   23. no solution   25. equivalent   27. (7, 20)   29. (−9, −8)
31. $\left(\frac{5b}{3a}, -\frac{4}{3}\right)$ 33. $\left(\frac{-6a}{a^2+b^2}, \frac{6b}{a^2+b^2}\right)$

Pages 513–515   1. 13 and 27   3. Adam: 33 years old; Eva: 23 years old
5. length: 14 cm; width: 7 cm   7. 49, 131   9. 16 $10-bills, 21 $5-bills

11. 87.5 km/h   13. 676 $5 tickets, 284 $8 tickets   15. 8 L of 10% solution, 2 L of 20%
17. 8.5 kg of peanuts, 3 kg of almonds   19. 42 dimes, 5 quarters   21. 26 km/h
23. $6500 in the 90-day notice account; $1500 in the regular savings account   25. 9 km/h

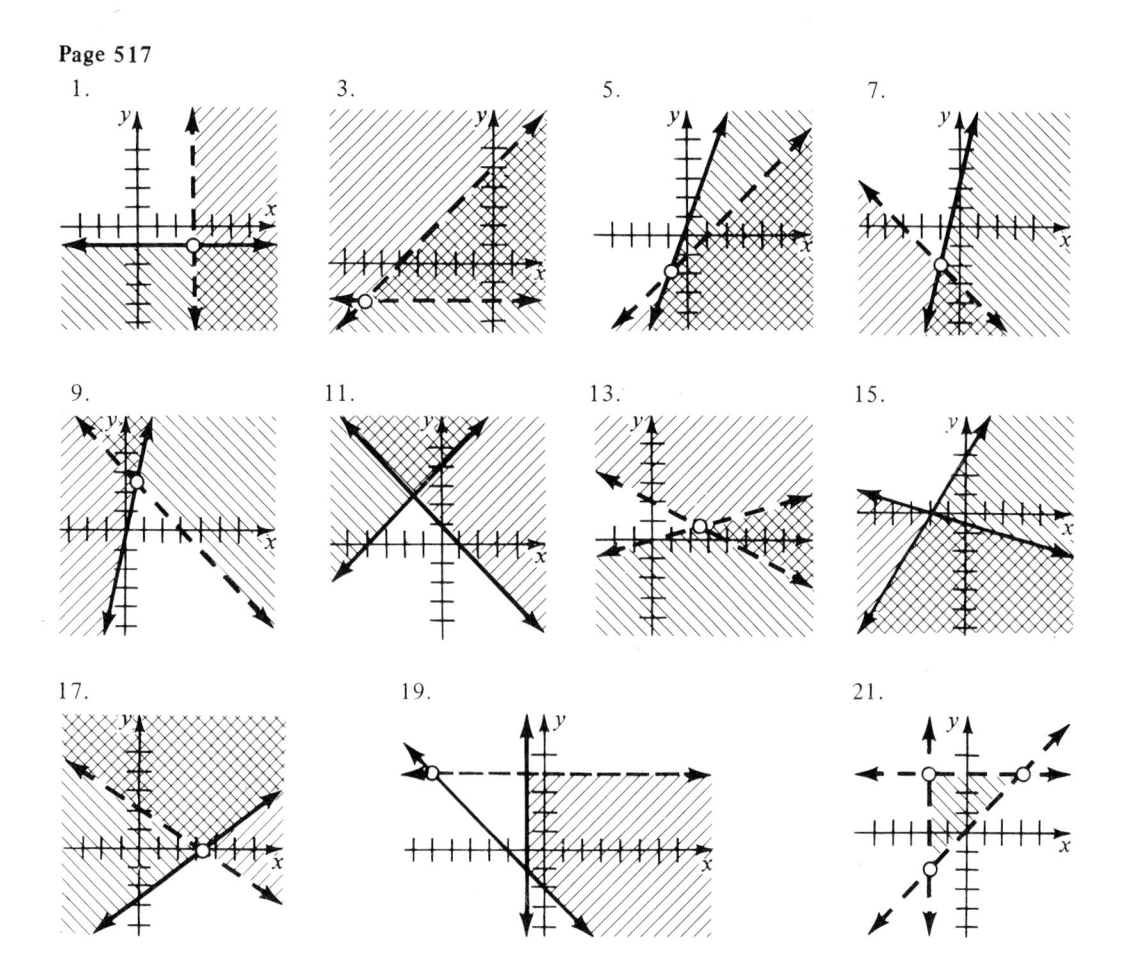

Page 517

1.

3.

5.

7.

9.

11.

13.

15.

17.

19.

21.

Pages 520–521   1. a. $(4, 5)$ b. 23   c. $(0, 0)$ d. 0   3. a. $(6, 0)$
b. 36   c. $(-3, 0)$ d. -18 5. a. $(5, 2)$ b. 41   c. $(0, 4)$ d. -8
7. a. $(0, 6)$ b. 24   c. $(7, 1)$ d. -10 9. b. $(0, 0), (0, 2), (4, 0)$
c. $2(0) + 7(2) = 14$ 11. b. $(0, 0), (0, 5), (3, 2)$ c. $0 + 2(5) = 10$ 13. b. $(0, 3)$,
$(4, 1)$ c. $3(0) + 2(3) = 6$ 15. Minimum cost is $2370; 12 orders from Bluebell and
22 orders from Pleasant View   17. A maximum profit of $76,000; 12,000 standard
models, 8000 deluxe models

Page 523 · Chapter Review   1. $(2, 4)$ 2. no solution   3. $(-3, -4)$ 4. $(-1, 1)$
5. $(6, -2)$ 6. equivalent   7. $(-3, -2)$ 8. $(2, 2)$ 9. $(0, -4)$ 10. 25 and
11   11. 12 L of 30% acid solution;   8 L of 25% acid solution

SELECTED ANSWERS   617

12.

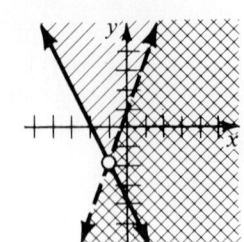

13.

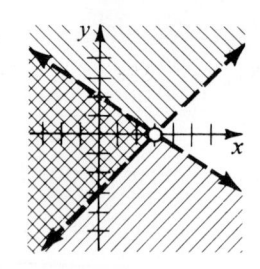

14.

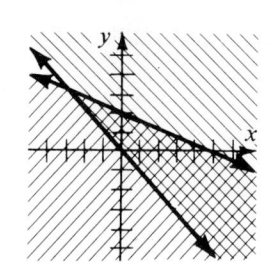

15. maximum at (4, 0), 8; minimum at (2, 6), −14
16. maximum at (8, 2), 10; minimum at (0, 3) and (3, 5), −9

Pages 525-528 · Cumulative Review: Chapters 1-14   1. a   2. c   3. b   4. c
5. d   6. a   7. d   8. c   9. b   10. b   11. c   12. d   13. a
14. d   15. a   16. a   17. d   18. a   19. d   20. c   21. c   22. a
23. b   24. d   25. b   26. c   27. a   28. c   29. d   30. a   31. c
32. a   33. c   34. c   35. b   36. d   37. b   38. b   39. c   40. b

Pages 530-531 · Mixed Review   1. x-intercept $= \frac{2}{3}$, y-intercept $= -4$ 3. 29

5. $(3z - 5)(2z - 1)$ 7. $350   9. (2, 4)   11. $\frac{5}{2}$ 13. 9, −9   15. $t \perp r$ and $t \perp s$

17.

p	q	$p \wedge q$	$p \vee q$	$(p \wedge q) \rightarrow (p \vee q)$
T	T	T	T	T
T	F	F	T	T
F	T	F	T	T
F	F	F	F	T

19. 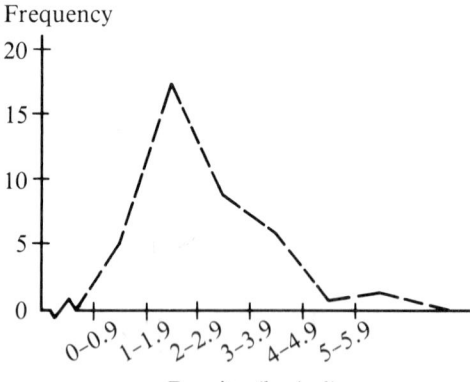   Frequency

Density (kg/m³)

21. 1–1.9   23. $y = -\frac{1}{3}x - \frac{14}{3}$

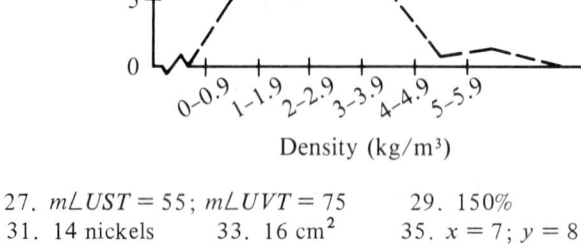

25.

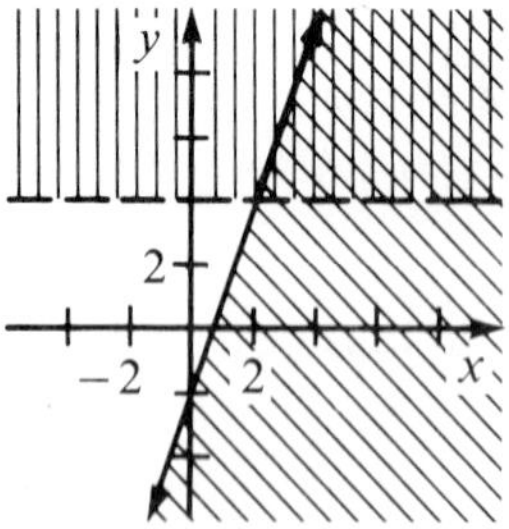

27. $m\angle UST = 55$; $m\angle UVT = 75$ 29. 150%
31. 14 nickels   33. 16 cm²   35. $x = 7$; $y = 8$
37. $\frac{7}{12}$ 39. If $x^2 = 4$, then $x = 2$; false

EXTRA PRACTICE

Chapter 1, page 532 1. {1980, 1981, 1982, 1983, 1984, 1985, 1986, 1987, 1988, 1989}
3. a. H b. -1 5. 1 7. -16 9. -5412 11. 26 13. $-\dfrac{1}{5}$
15. -0.11 17. $-\dfrac{4}{9}$ 19. {the positive multiples of 5} 21. Mult. prop. of -1
23. -480 25. $\dfrac{31}{36}$ 27. -21 29. 1 31. 0

Chapter 2, page 533 1. Answers may vary for a value that makes the statement false. 5; 7
3. $6^2 \neq 3^2 \cdot 2^2$; true; false

5.
p	q	$\sim q$	$p \wedge \sim q$	$\sim(p \wedge \sim q)$
T	T	F	F	T
T	F	T	T	F
F	T	F	F	T
F	F	T	F	T

7. $p: x + 1 = 7$
$q: x = 6$
$p_1: p \to q$
$p_2: \sim q$
$\overline{C: \therefore \sim p}$ Valid;
Law of Contrapositive Inference

9. $q \to r$; true 11. $0 \cdot 0 = 0$ or $(-1)(-1) = 1$; true 13. If $0 \cdot 0 = 0$, then $1 + 0 = 0$; false 15. true 17. true 19. If I will not get the job, then I will not learn computer programming.
21. $[\sim(p \wedge q) \wedge p] \to \sim q$;

p	q	$p \wedge q$	$\sim(p \wedge q)$	$\sim(p \wedge q) \wedge p$	$\sim q$	$[\sim(p \wedge q) \wedge p] \to \sim q$
T	T	T	F	F	F	T
T	F	F	T	T	T	T
F	T	F	T	F	F	T
F	F	F	T	F	T	T

valid

Chapter 3, page 534 1. a. -7 b. $\dfrac{1}{9}$ 3. -12 5. -27 7. -35 9. $20x$
11. $2l + 2 \cdot \dfrac{2}{3}l = 80$; length = 24; width = 16 13. $y = 2z + x$ 15. a. -3 b. 5
17. $\dfrac{3}{2}$ 19. 52 21. -3 23. Jerry is 15 years old; his sister is 14 years old.

Chapter 4, page 535 1. $x \neq -1$ 3. y is greater than 2 5. k is greater than or equal to 100 7. $x \geq -7$
9.  [number line from 2 to 7, dots at 3 and 6]
11. 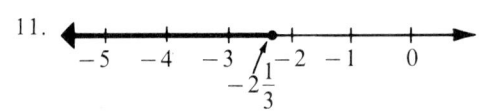 [number line from -5 to 0, point at $-2\tfrac{1}{3}$]
13. 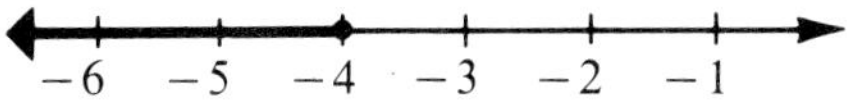 [number line from -6 to -1]
15. 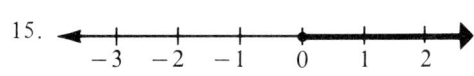 [number line from -3 to 2]
17. 21 km 19. 6 21. {positive integers} 23. $\left\{x: x > -1\dfrac{1}{10}\right\}$ 25. $\{g: g < 10\}$
27. $\{x: -3 < x < 1\}$ 29. $\{p: p > 2\}$ 31. $\{k: 1 \leq k \leq 3\}$ 33. 12 and 14

SELECTED ANSWERS 619

Chapter 5, page 536 1. false   3. false   5. T   7. $\angle WOY$; $m\angle WOY = 115$ or $\angle XOZ$; $m\angle XOZ = 155$ 9. 65   11. $180 - t$ 13. no   15. yes   17. 5   19. 36   21.

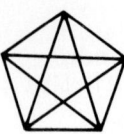

Chapter 6, page 537   1. Use Construction 3.   3. no   5. yes; SAS   7. false   9. 36   11. 30   13. 175.5   15. 102   17. 3.6 cm   19. 125 cm^3

Chapter 7, page 538   1. -32 3. $3x^2 - 2x - 8$; 2; 3   5. $-a^3 + 4a^2 - 3a + 5$ 7. $14m^2 - m - 4$ 9. $3j^2(3j - 5k + 6j^2k^2)$ 11. $25r^2 - 9$ 13. $(z - 10)(z + 10)$ 15. 15, 2   17. 2   19. $72y^8$ 21. $6x^3 - 11x^2y + 14xy^2 - 5y^3$ 23. 8; 12   25. $2(y - 10)(y + 2)$ 27. length = 15, width = 9

Chapter 8, page 539   1. Answers may vary. $\frac{725}{100}$ 3. $\frac{13}{8}$ 5. 5.75   7. 18   9. $\frac{\sqrt{5}}{5}$ 11. $-18\sqrt{2}$ 13. $24 - 16\sqrt{2}$ 15. $\left\{\frac{3}{2}, \frac{1}{2}\right\}$ 17. $\{5\}$ 19. $0.71\overline{296}$ 21. 0.18   23. $-15\sqrt{7}$ 25. $15 + 2\sqrt{15}$ 27. $60 + 20\sqrt{5}$ 29. $\left\{\frac{3 + \sqrt{17}}{4}, \frac{3 - \sqrt{17}}{4}\right\}$

Chapter 9, page 540   1. 60 campers   3. 324 km   5. $\angle A \cong \angle E$ (Given); $\angle ACB \cong \angle ECD$ (Vertical angles are congruent.); $\triangle ACB \sim \triangle ECD$ (AA)   7. $CD = \frac{45}{4}$, $DE = \frac{55}{4}$ 9. 375 cm^3 11. yes   13. 4   15. 10   17. 60   19. 6 cm

Chapter 10, page 541   1. false   3. $x = 72$, $y = 54$ 5. $x = 18$, $y = 27$ 7. $x = 232$, $y = 128$ 9. $\frac{1}{2}$; $\frac{1}{4}$ 11. T.A. = 200π; $V = 320\pi$ 13. 119   15. 76

Chapter 11, page 542   1. $\{1, 2, 2, 3, 3, 3, 4, 4, 4, 4\}$; $P(1) = \frac{1}{10}$, $P(2) = \frac{1}{5}$, $P(3) = \frac{3}{10}$, $P(4) = \frac{2}{5}$ 3. $\frac{3}{5}$ 5. $\frac{2}{9}$ 7. 6   9. 30; 5,527,200   11. a. $\frac{1}{12}$ b. $\frac{1}{2}$ c. $\frac{4}{9}$ 13. $\frac{1}{2}$; $\frac{26}{51}$ 15. $\frac{1}{6}$ 17. $P(3\ Y) = \frac{8}{27}$, $P(2\ Y, 1\ R) = \frac{4}{9}$, $P(1\ Y, 2\ R) = \frac{2}{9}$, $P(3\ R) = \frac{1}{27}$

Chapter 12, page 543   1.

Interval	Frequency	Cumulative Frequency
1–20	1	1
21–40	4	5
41–60	4	9
61–80	6	15
81–100	5	20

3. 5

620   SELECTED ANSWERS

5. 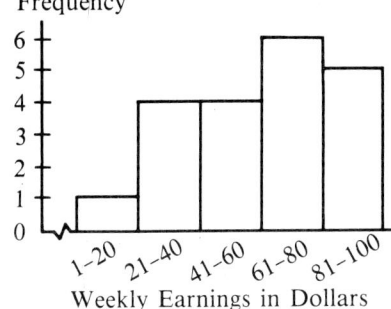 Frequency

[Histogram: Weekly Earnings in Dollars]

7. 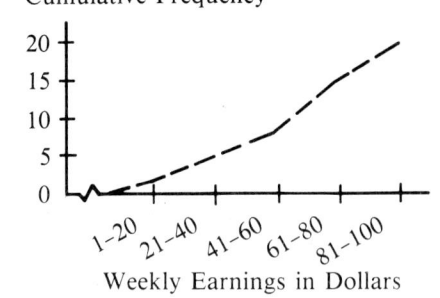 Cumulative Frequency

[Graph: Weekly Earnings in Dollars]

9. 1.12; 1.1; 1.1, 1.2   11. Answers may vary.   13. 2   15. Answers may vary. 5
17. Answers may vary.   19. $\frac{35}{18}$

Chapter 13, page 544   1. a. 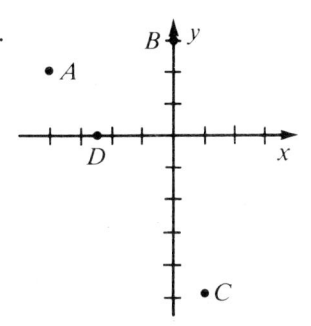 [graph showing points A, B, C, D on coordinate plane]   b. B   c. C   3. 6

5. 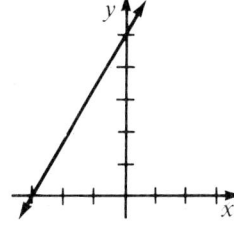 [graph]   7. 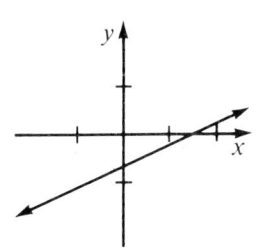 [graph]   9. 0; 0   11. $-\frac{1}{3}$ 13. $\frac{2}{3}$

15. 0   17. $y = -\frac{1}{4}x + 2$ 19. 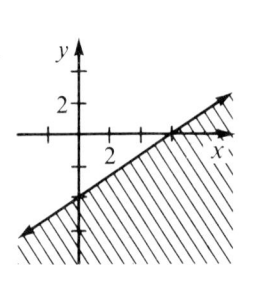 [graph]   21. 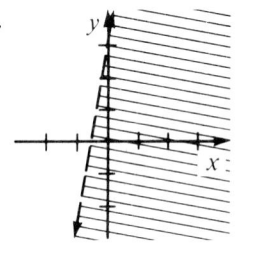 [graph]

23. $5x - 3y = 11$ 25. slope of $\overleftrightarrow{AB} = -\frac{2}{3}$, slope of $\overleftrightarrow{BC} = -\frac{3}{2}$, slope of $\overleftrightarrow{AC} = -\frac{7}{8}$; ∴ the points are not collinear.   27. $y = \frac{3}{2}x + 2$

29. [graph]

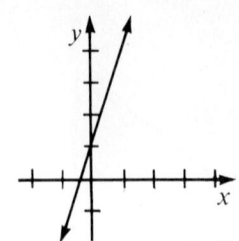

Chapter 14, page 545   1. $(-1, 2)$ 3. $(0, 2)$ 5. $\left(\dfrac{19}{3}, -\dfrac{4}{3}\right)$ 7. $(1, 2)$
9. $(-4, 5)$ 11. [graph]   13. [graph]   15. $(-9, 8)$

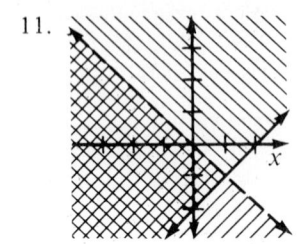

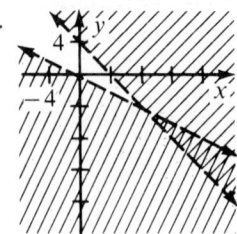

17. $\left(3\dfrac{1}{2}, -2\dfrac{1}{2}\right)$ 19. [graph]   21. 2 km/h

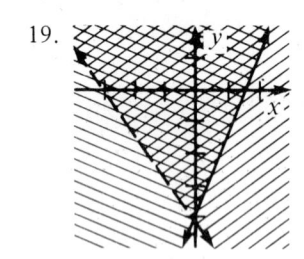

ALGEBRA REVIEW EXERCISES

Chapter 3, page 546   1. 9   3. 2.6   5. 5   7. $s = A - rt$ 9. $a = \dfrac{4K - RS}{4}$
11. 110 tables   13. no solution   15. 0, 1, 2, 3   17. 8   19. $12\dfrac{1}{8}$

Chapter 4, pages 546-547   1. $\{r: r \geq 3\}$ 3. $\{x: x \geq 16\}$ 5. $\{s: s < 7\}$
7. $\{w: w > 4\}$ 9. $\{b: b < -72\}$ 11. $\{t: t < 0\}$ 13. $\{z: z \geq 1\}$
15. $w: w < 5\dfrac{2}{5}$ 17. $\left\{y: -\dfrac{1}{2} < y < 1\right\}$
19. $a \geq 50$ [number line]

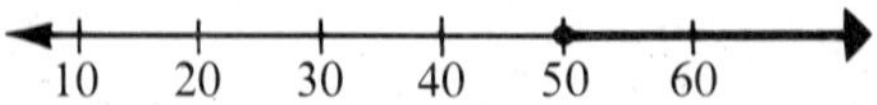

21. $8 < t + 1 < 22$ [number line]

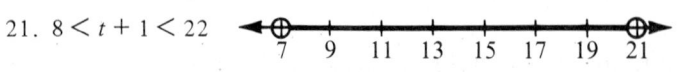

23. $12 - x \leq 23$ [number line]

622   SELECTED ANSWERS

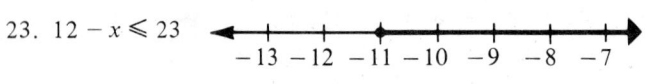

25. $\{b: b > -7\}$

27. $\{x: 3\frac{1}{2} \leq x < 4\frac{1}{2}\}$

29. $\{y: -2\frac{1}{2} \leq y < 2\frac{1}{2}\}$

31. $\{b: b > 2\frac{3}{4}\}$

33. 6 and 8   35. 9 dimes

Chapter 7, pages 547-548   1. $-4t + 12$ 3. a'' 5. $-8z^4$ 7. $9n^2 - 4$
9. $13b^2$ 11. $t(t^2 + 5)$ 13. $(x - 2)(x - 1)$ 15. $(m - 4)(m + 4)$
17. $\{0, -\frac{1}{2}\}$ 19. $\{8, -3\}$ 21. $\{5, -2\}$ 23. $-7, 5$ 25. $\{0, -\frac{7}{4}\}$
27. $\{-\frac{16}{3}, 2\}$ 29. $\{-\frac{1}{3}, -3\}$ 31. $\{-\frac{1}{2}, 2\}$ 33. $\{\frac{1}{2}, -3\}$
35. $(m^2 - n^2)^2 + (2mn)^2 = m^4 - 2m^2n^2 + n^4 + 4m^2n^2 = m^4 + 2m^2n^2 + n^4 = (m^2 + n^2)^2$;
∴ $x^2 + y^2 = z^2$ 37. 9, 10, 11

Chapter 8, pages 548-549   1. $\frac{25}{8}$ 3. $\frac{33}{100}$ 5. $\frac{4}{33}$ 7. $\frac{6}{5}$ 9. $\frac{2}{3}$
11. $6\sqrt{3}$ 13. $3\sqrt{10}$ 15. $3 + 2\sqrt{2}$ 17. $\sqrt{7}$ 19. $2\sqrt{3}$ 21. $11 - 6\sqrt{2}$
23. $\{\frac{-3 + \sqrt{13}}{2}, \frac{-3 - \sqrt{13}}{2}\}$ 25. $\{\frac{1 + \sqrt{5}}{2}, \frac{1 - \sqrt{5}}{2}\}$ 27. $\{\frac{-5 + \sqrt{13}}{2}, \frac{-5 - \sqrt{13}}{2}\}$
29. $\frac{1 + \sqrt{2}}{2}$ 31. $8 + 2\sqrt{15}$ 33. $17 + 8\sqrt{5}$ 35. $\frac{3 + 3\sqrt{5}}{-4}$ 37. $\frac{11}{90}$
39. 1   41. $-4\sqrt{23}$; -19.18

Chapter 9, pages 549-550   1. $\frac{3}{5}$ 3. $\frac{1}{1}$ 5. $\frac{6b^2c}{1}$ 7. $\{\frac{8}{3}\}$ 9. $\{-\frac{1}{5}\}$
11. $\{0, 5\}$ 13. yes   15. 5.25 kg   17. $\{6, -2\}$ 19. $-\frac{3}{2}, 4$ 21. 400%
23. 20 cm, 21 cm

Chapter 13, pages 550-551   1. $2\sqrt{65}$ 3. $2\sqrt{17}$ 5. 5   7. 5; 6   9. $\frac{1}{3}$
11. yes   13. $y = -3x + 17$ 15. $y = \frac{4}{5}x - 2$ 17. $5\sqrt{5}$ 19. $y = -\frac{14}{15}x + \frac{23}{15}$
21. $\frac{175}{144}$ 23. $\frac{1}{3}$ 25. $\frac{5}{2}$ 27. -13

Chapter 14, page 551   1. one solution   3. many solutions   5. (2, 4)
7. $(-3, -\frac{2}{3})$ 9. $(\frac{2}{3}, \frac{3}{2})$ 11. $20x + 50y = 4410$ 13. 18, 72   15. $\frac{2}{5}$
17. 60 km/h

SELECTED ANSWERS   623

PREPARING FOR REGENTS EXAMINATIONS

Logic
1. a. -5 b. $-5, 5$ c. $-5, 0$ d. $-5, 0$ e. $-5, 5$ f. $0, 5$ g. $-5, 0, 5$ h. $-5, 0, 5$
   i. $-5, 0$
3. a. (1) $\sim p \vee r$ (2) $q \wedge r$ (3) $\sim r \rightarrow \sim p$ (4) $q \leftrightarrow r$
   b. (1) F   (2) F   (3) T   (4) F
   c. If the farmers are not happy, then it is not raining.

Algebra Skills
1. a. -6 b. 1 c. $-6\frac{1}{2}$ d. 25 e. $\frac{1}{2}$ f. $-12\frac{1}{8}$ g. -3
3. a.

   b.

   c.

   d. $\emptyset$
5. a. $x(x - 9)$ b. $(x - 3)(x + 3)$ c. $x(x - 3)(x + 3)$ d. $(x - 5)(x - 4)$
   e. $(x - 12)(x + 3)$ f. $(5x - 2)(x + 1)$ g. $(3x - 1)(x - 1)$ h. $(3x - 4)(3x + 4)$
   i. $7(x + 5)^2$
7. a. b. c.

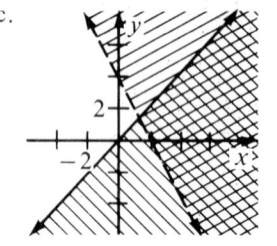

Geometry
1. a. 130   b. 80   c. 80   d. 40
3. a. 100   b. 20   c. 74   d. 20   e. equilateral   f. a right triangle   g. $AC = 5, AB = 14$
   h. $AC = 3$ cm, $BE = 8$ cm, $AB = \dfrac{\sqrt{265}}{2}$ cm, perimeter $= 3 + \sqrt{265}$ cm
5. a. (1) 12   (2) 10   (3) 15   (4) 2:3   (5) 4:9
   b. (1) 12   (2) 30   (3) 240   (4) 84
   c. (1) 5   (2) 24   (3) 150
7. a. (1) 50   (2) 130   (3) 50   (4) 130   (5) 50   (6) 90
   b. (1) 8 cm   (2) 24 cm^2 (3) 80 cm^2 (4) 104 cm^2
9. a. (1) 2 cm   (2) 4π cm^2 (3) 4π cm   (4) $4\sqrt{2}$ cm   (5) $2\sqrt{2}$ cm   (6) 8π cm^2
   (7) $4\sqrt{2}\pi$ cm   (8) $(16 - 4\pi)$ cm^2 (9) 90   (10) $3\sqrt{2}\pi$ cm
   b. (1) $\sqrt{2}:2$ (2) 1:2
   c. 100π
   d. 12π

624   SELECTED ANSWERS

Probability

1. a. (1) $\frac{1}{2}$ (2) 1 (3) 0
   b. (1) $\frac{1}{4}$ (2) $\frac{1}{4}$ (3) $\frac{3}{4}$
   c. (1) $\frac{1}{8}$ (2) $\frac{3}{8}$ (3) $\frac{7}{8}$

3. a. (1) $\frac{1}{13}$ (2) $\frac{1}{4}$ (3) $\frac{1}{2}$ (4) $\frac{3}{13}$ (5) $\frac{4}{13}$ (6) $\frac{3}{4}$
   b. (1) $\frac{1}{169}$ (2) $\frac{1}{4}$ (3) $\frac{1}{16}$ (4) $\frac{100}{169}$ (5) both cards even
   c. (1) $\frac{1}{221}$ (2) $\frac{1}{1326}$ (3) $\frac{11}{221}$ (4) $\frac{19}{34}$

5. a. (1) $\frac{5}{7}$ (2) $\frac{2}{7}$ (3) 0
   b. (1) $\frac{20}{49}$ (2) $\frac{25}{49}$ (3) $\frac{4}{49}$ (4) $\frac{24}{49}$ (5) $\frac{29}{49}$ (6) $\frac{45}{49}$
   c. (1) $\frac{10}{21}$ (2) $\frac{1}{21}$ (3) $\frac{10}{21}$ (4) $\frac{11}{21}$ (5) $\frac{10}{21}$ (6) $\frac{11}{21}$
   d. (1) {B1, B2, B3, B4, B5, B6, W1, W2, W3, W4, W5, W6} (2) $\frac{1}{7}$ (3) $\frac{1}{2}$ (4) $\frac{5}{7}$
   (5) $\frac{9}{14}$ (6) $\frac{5}{21}$

7. 40,320
9. 25 ways

Statistics

1. a. 20 marbles b. 4 marbles c. (1) $\frac{1}{10}$ (2) $\frac{9}{20}$ (3) 0
   d. (1) $\frac{9}{100}$ (2) $\frac{7}{16}$ (3) $\frac{1}{25}$
   e. (1) $\frac{1}{19}$ (2) $\frac{9}{38}$ (3) $\frac{3}{190}$

3. a.

Interval	Frequency	Cumulative Frequency
121–130	3	3
131–140	6	9
141–150	7	16
151–160	9	25
161–170	4	29
171–180	1	30

3. b. Frequency

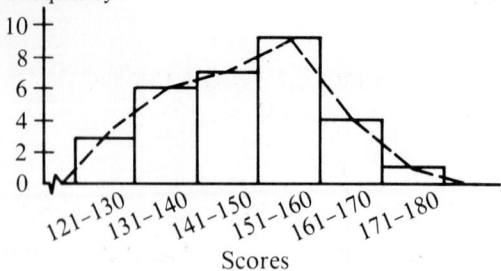

c. Cumulative Frequency

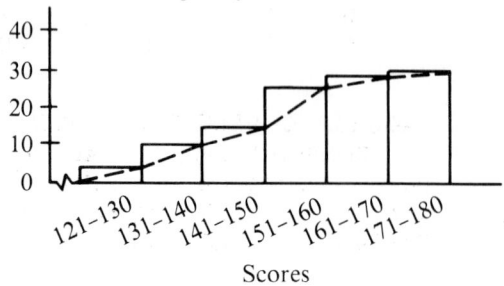

d. 141–150   e. 131–140   f. 151–160   g. 30%   h. $\frac{1}{6}$

Coordinate Geometry

1. a.

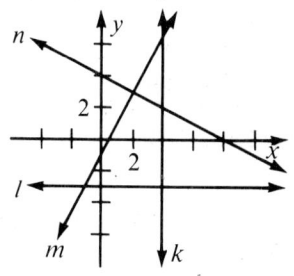

b. k: no slope; l: 0; m: 2; n: $-\frac{1}{2}$ c. k: none; l: -3; m: -1; n: 4

d. k: 4; l: none; m: $\frac{1}{2}$; n: 8   e. (2, 3)   f. (4, 2)   g. 5   h. 6   i. $y = 2x + 1$

j. $-\frac{1}{2}$ k. 5   l. $7\frac{1}{2}$ m. $4\sqrt{5}$ n.

o. Answers may vary. One possible answer is $(-3, -5)$.

p. (1) 25   (2) $15 + 5\sqrt{5}$ q. 9

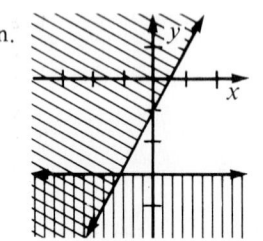

ANSWERS TO COMPUTER EXERCISES

PAGES 27-28 · COMPUTER EXERCISES
1. 2   2. a. $(15 - 3) * (15 + 3)$
 b. $15 - 3 * 15 + 3$ c. $3 * 4 \uparrow 2 - 18$
 d. $(3 * 4) \uparrow 2 - 18$
 e. $(3 * 8 - 4 * 5 \uparrow 2)/(7 + 3 * 4)$
 f. $(18 \uparrow 6 + (39 - 3) * 2)/4$
3. a. 10 PRINT $(15 - 3) * (15 + 3)$ 216
 b. 10 PRINT $15 - 3 * 15 + 3$ -27
 c. 10 PRINT $3 * 4 \uparrow 2 - 18$ 30
 d. 10 PRINT $(3 * 4) \uparrow 2 - 18$ 126
 e. 10 PRINT $(3 * 8 - 4 * 5 \uparrow 2)$
    $/(7 + 3 * 4)$ -4
 f. 10 PRINT $(18 \uparrow 6 + (39 - 3) * 2)/4$
    $8.50307E + 06$
   (The last digit may vary depending upon the computer used.)
4. Results will vary.   5. Results will vary.
6. When B is even, the even numbers from A to B are printed out. When B is odd, the even numbers from A to B $-$ 1 are printed out.
7. When B is even, the odd numbers from A to B $-$ 1 are printed out. When B is odd, the odd numbers from A to B are printed out.
8. When A is odd, even numbers beginning with A + 1 are printed. When A is even, odd numbers beginning with A + 1 are printed.
9. For A = -10 and B = 10, the printout is $-9, -8, -7, \ldots, 7, 8, 9$.   10. The integers between A and B are printed in descending order.

PAGES 65-66 · COMPUTER EXERCISES
1. a. | P | Q | P?2 |
       |---|---|-----|
       | T | T | T |
       | T | F | F |
       | F | T | F |
       | F | F | F |
 b. and
2. a. | P | Q | P?2 |
       |---|---|-----|
       | T | T | T |
       | T | F | T |
       | F | T | T |
       | F | F | F |
 b. or

3. | 130 | IF P = Q THEN 170 |
P	Q	P?2
T	T	T
T	F	F
F	T	F
F	F	T

4. | P | Q | P AND Q | P OR Q |
    |---|---|---------|--------|
    | 1 | 1 | 1 | 1 |
    | 1 | 0 | 0 | 1 |
    | 0 | 1 | 0 | 1 |
    | 0 | 0 | 0 | 0 |

PAGE 100 · COMPUTER EXERCISES
1. a. Results will vary.   b. Answers will vary.   2. a. $X = 1$ b. $X = 4$
c. $X = -1$ 3. If $A = C$, then $A - C = 0$. Division by zero is not allowed.

PAGES 124-125 · COMPUTER EXERCISES
1. If A is nonnegative, the inequality sign remains unchanged. If A is negative, the inequality sign must be reversed.
2. a. $3X + 8 < 2; 3X < -6; X < -2$
   b. $-8X + 17 < -7; -8X < -24; X > 3$
   c. $2X - 8 < 13; 2X < 21; X < 10.5$
3. Results will vary.
4. 20 "   AX + B $>$ C, A $<>$ 0"
   70 PRINT A; "X + "; B; "$>$"; C
   80 PRINT A; "X $>$"; C $-$ B
   90 IF A $>$ 0 THEN 120

PAGES 165-166 · COMPUTER EXERCISES
1. a, b. Results will vary.   2. a. 180
b. 360   c. 540   d. 1440
e. 3240   f. 17640

PAGES 215-216 · COMPUTER EXERCISES
1. a. PERIMETER = 14, AREA = 10
   b. PERIMETER = 108, AREA = 473
   c. PERIMETER = 150, AREA = 1134
   d. PERIMETER = 31.6, AREA = 61.92
2. a. 10 PRINT "INPUT BASE, HEIGHT";
      20 INPUT B, H
      30 PRINT "AREA ="; $0.5 * B * H$
      40 END

SELECTED ANSWERS

b. 10 PRINT "INPUT HEIGHT, BASE 1, BASE 2";
   20 INPUT H, A, B
   30 PRINT "AREA ="; 0.5 * H * (A + B)
   40 END
3. a. 60 LET A = 2 * (L * W + W * H + L * H)
   70 LET V = L * W * H
   b. Results will vary.
4. a. THERE IS A TRIANGLE.
   b. THERE IS NO TRIANGLE.
   c. THERE IS A TRIANGLE.
   d. THERE IS NO TRIANGLE.

PAGE 261 · COMPUTER EXERCISES
1. a. and b. Results will vary.
   c. 10 PRINT "PROGRAM FINDS IF A IS A FACTOR OF B"
      20 PRINT "ENTER A AND B";
      30 INPUT A, B
      40 IF INT(B/A) = B/A THEN 70
      50 PRINT A; "IS NOT A FACTOR OF"; B
      60 STOP
      70 PRINT A; "IS A FACTOR OF"; B
      80 END
2. a. 1, 5, 6   b. 1, 14, 49
   c. 2, 17, 30   d. 3, 29, 40
   e. 15, −26, 8 f. 21, −39, −6

PAGES 293–294 · COMPUTER EXERCISES
1. Results will vary. 2. The square roots to the nearest 100th are 6, 9.59, 8.66, 13, 26.15, 52.23, 16, .04.
3. a. 10 READ N
      20 IF N = .99999 THEN 60
      30 PRINT N, SQR(N)
      40 GOTO 10
      50 DATA 36, 92, 75, 169, 2728, 256, .00143, .99999
      60 END
   Change line 30 as follows:
      PRINT N, INT (1000 * SQR(N) + .5)/1000

338 · COMPUTER EXERCISES
1.5; 9, 12, 3, 10.5, 12,
   b. The sum is 64.5, the larger polygon (see line 90).
.75, 6.25, 5, 1.25,
triples are: (3, 4, 5),
(8, 6, 10), (5, 12, 13), (15, 8, 17),
(12, 16, 20), (7, 24, 25), (24, 10, 26),
(21, 20, 29), (16, 30, 34), (9, 40, 41),
(35, 12, 37), (32, 24, 40), (27, 36, 45),
(20, 48, 52), (11, 60, 61), (48, 14, 50),
(45, 28, 53), (40, 42, 58), (33, 56, 65),
(24, 70, 74), (13, 84, 85), (63, 16, 65),
(60, 32, 68), (55, 48, 73), (48, 64, 80),
(39, 80, 89), (28, 96, 100), (15, 112, 113)
b. (3, 4, 5), (5, 12, 13), (8, 15, 17),
(7, 24, 25), (20, 21, 29), (9, 40, 41),
(12, 35, 37), (11, 60, 61), (28, 45, 53),
(33, 56, 65), (13, 84, 85), (16, 63, 65),
(48, 55, 73), (39, 80, 89), (15, 112, 113)
c. Results will vary.
3. a. 10 PRINT "INPUT THE LENGTH OF THE HYPOTENUSE";
   20 INPUT H
   30 PRINT "THE LENGTH OF THE SHORTER LEG IS"; H/2
   40 PRINT "THE LENGTH OF THE LONGER LEG IS"; (H/2) * SQR(3)
   50 END
3. b. 10 PRINT "INPUT THE LENGTH OF THE LONGER LEG";
   20 INPUT L
   30 PRINT "THE LENGTH OF THE SHORTER LEG IS"; L/SQR(3)
   40 PRINT "THE LENGTH OF THE HYPOTENUSE IS"; 2 * L/SQR(3)
   50 END

PAGE 380 · COMPUTER EXERCISES
1. Results will vary. For N = 16, $\pi \approx 3.13699$; for N = 512, $\pi \approx 3.14157$.
2. a. 10 PRINT "INPUT THE RADIUS OF THE SPHERE";
   20 INPUT R
   30 LET P = 3.14159
   40 PRINT "THE VOLUME IS"; (4/3) * P * R3
   50 PRINT "THE SURFACE AREA IS"; 4 * P * R2
   60 END
   b. 10 PRINT "INPUT THE CIRCUMFERENCE";
   20 INPUT C
   30 LET P = 3.14159
   40 R = C/(2 * P)
   50 PRINT "THE AREA IS"; P * R2
   60 END

PAGES 428-429 · COMPUTER EXERCISES
1. 53461.9, 534619., 5.34619E+06, 5.34619E+07, 5.34619E+08
2. a. 4638.63 b. 51.091 c. 938842. d. 146970000.
3. a. Results will vary. b. 9

PAGES 462-464 · COMPUTER EXERCISES
1. Results will vary. 2. Results will vary.
3. a. Results will vary. b. Results will vary.
   c. 10 FOR I = 1 TO 20
      20 PRINT INT(8 * RND(0)) + 1
      30 NEXT I
      40 END
   d. 10 FOR I = 1 TO 20
      20 PRINT INT(8 * RND(0)) + 2
      30 NEXT I
      40 END
4. Results will vary. 5. a. The mean is −227. The median is 4. b. Results will vary. 6. a. 5, 7, 12 b. 9, 6, −2, −7, 4

PAGES 493-494 · COMPUTER EXERCISES
1. a. If $x_1 = x_2$, $x_2 - x_1 = 0$ and line 40 will produce an ERROR message. Testing for $x_1 = x_2$ in line 30 must precede the computation of m in line 40 so that the ERROR message caused by dividing by zero can be avoided. b. $y = 1x - 1$ c. $x = -2$ d. $y = 0x - 5$
2. a. 10 PRINT "INPUT A, B, C";
     20 INPUT A, B, C
     30 PRINT "y = "; −A/B; "x + "; C/B
     40 END
   The output will be $y = -1.5x + 3$.
   b. To eliminate the possibility of dividing by 0 if B = 0. Insert in the program in part (a):
     25 IF B = 0 THEN 37
     33 GO TO 40
     37 PRINT "THIS EQUATION CANNOT BE WRITTEN IN SLOPE-INTERCEPT FORM."

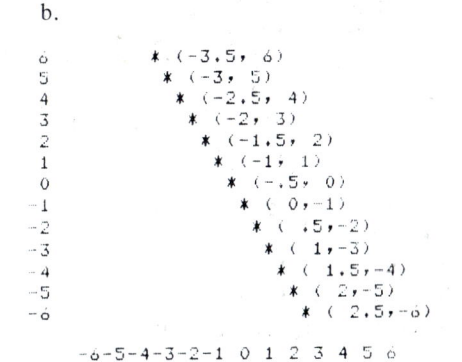

3. a.

```
 6                         * ( 5, 6)
 5                         * ( 4, 5)
 4                         * ( 3, 4)
 3                         * ( 2, 3)
 2                         * ( 1, 2)
 1                         * ( 0, 1)
 0                         * (-1, 0)
-1                         * (-2,-1)
-2                         * (-3,-2)
-3                         * (-4,-3)
-4                         * (-5,-4)
-5                         * (-6,-5)
-6
   -6-5-4-3-2-1 0 1 2 3 4 5 6
```

b.

```
 6        * (-3.5, 6)
 5        * (-3, 5)
 4        * (-2.5, 4)
 3        * (-2, 3)
 2        * (-1.5, 2)
 1        * (-1, 1)
 0        * (-.5, 0)
-1        * ( 0,-1)
-2        * ( .5,-2)
-3        * ( 1,-3)
-4        * ( 1.5,-4)
-5        * ( 2,-5)
-6        * ( 2.5,-6)
   -6-5-4-3-2-1 0 1 2 3 4 5 6
```

c.

```
 6              * (-2, 6)
 5              * (-2, 5)
 4              * (-2, 4)
 3              * (-2, 3)
 2              * (-2, 2)
 1              * (-2, 1)
 0              * (-2, 0)
-1              * (-2,-1)
-2              * (-2,-2)
-3              * (-2,-3)
-4              * (-2,-4)
-5              * (-2,-5)
-6              * (-2,-6)
   -6-5-4-3-2-1 0 1 2 3 4 5 6
```

Selected Answers

d.

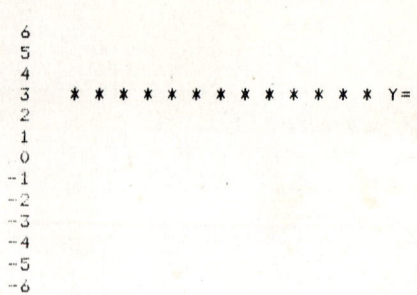

```
 6
 5
 4
 3   * * * * * * * * * * * *   Y= 3
 2
 1
 0
-1
-2
-3
-4
-5
-6
   -6-5-4-3-2-1 0 1 2 3 4 5 6
```

b.

(−.5, 2)

c.

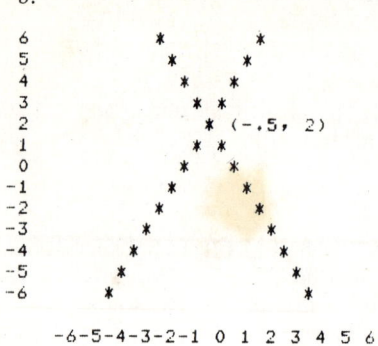

PAGES 522–523 · COMPUTER EXERCISES

1. a. (14, −17)   b. (2, −1)   c. No common solution. The lines are parallel.
d. Infinitely many common solutions. The lines coincide.

2. a. (1, 1)

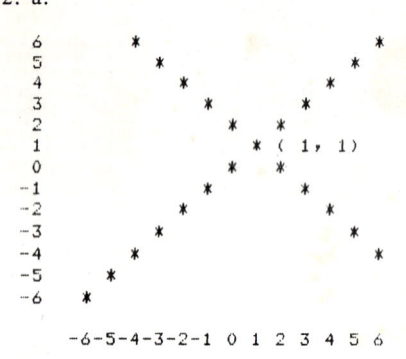

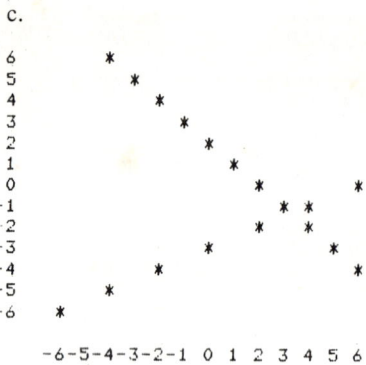

d. $\left(\dfrac{10}{3}, -\dfrac{4}{3}\right)$

630   SELECTED ANSWERS